Statistical Advances in the
Biomedical Sciences

BICENTENNIAL
1807
WILEY
2007
BICENTENNIAL

THE WILEY BICENTENNIAL—KNOWLEDGE FOR GENERATIONS

*E*ach generation has its unique needs and aspirations. When Charles Wiley first opened his small printing shop in lower Manhattan in 1807, it was a generation of boundless potential searching for an identity. And we were there, helping to define a new American literary tradition. Over half a century later, in the midst of the Second Industrial Revolution, it was a generation focused on building the future. Once again, we were there, supplying the critical scientific, technical, and engineering knowledge that helped frame the world. Throughout the 20th Century, and into the new millennium, nations began to reach out beyond their own borders and a new international community was born. Wiley was there, expanding its operations around the world to enable a global exchange of ideas, opinions, and know-how.

For 200 years, Wiley has been an integral part of each generation's journey, enabling the flow of information and understanding necessary to meet their needs and fulfill their aspirations. Today, bold new technologies are changing the way we live and learn. Wiley will be there, providing you the must-have knowledge you need to imagine new worlds, new possibilities, and new opportunities.

Generations come and go, but you can always count on Wiley to provide you the knowledge you need, when and where you need it!

WILLIAM J. PESCE
PRESIDENT AND CHIEF EXECUTIVE OFFICER

PETER BOOTH WILEY
CHAIRMAN OF THE BOARD

Statistical Advances in the Biomedical Sciences

Clinical Trials, Epidemiology, Survival Analysis, and Bioinformatics

Edited by

ATANU BISWAS

Applied Statistics Unit, Indian Statistical Institute, Calcutta, India

SUJAY DATTA

Department of Statistics, Texas A&M University, College Station, Texas, USA

JASON P. FINE

Departments of Statistics and Biostatistics and Medical Informatics, University of Wisconsin—Madison, Madison, Wisconsin, USA

MARK R. SEGAL

Department of Epidemiology and Biostatistics, University of California, San Francisco, California, USA

WILEY-INTERSCIENCE
A JOHN WILEY & SONS, INC., PUBLICATION

Wiley Bicentennial Logo: Richard J. Pacifico

Library of Congress Cataloging-in-publication Data:

Statistical advances in the biomedical sciences : clinical trials, epidemiology, survival analysis, and
bioinformatics/[edited by] Atanu Biswas . . . [et al.].

p. ; cm.— (Wiley series in probability and statistics)
Includes bibliographical references and indexes.
ISBN 978-0-471-94753-0 (cloth : alk. paper)

1. Medicine—Research—Statistical methods. 2. Biology—Research—Statistical methods. 3. Clinical
trials—Statistical methods. 4. Epidemiology—Statistical methods. 5. Survival analysis (Biometry).
6. Bioinformatics. I. Biswas, Atanu, 1970- II. Series. [DNLM: 1. Models, Statistical. 2. Biomedical
Research. 3. Clinical Trials. 4. Computational Biology—methods. 5. Epidemiologic Methods.
6. Survival Analysis. WA 950 S7961 2008]
 R853.S7S73 2008
 610.72'7—dc22

Printed in the United States of America
10 9 8 7 6 5 4 3 2 1

To my teacher and mentor, Professor Uttam Bandyopadhyay
A. Biswas

To my dearest cousins Rupa, Rina, and Arunangshu
S. Datta

To Amy, Sierra, and Jacob
M. R. Segal

Contents

11. An Overview of the Semi–Competing Risks Problem   177

Limin Peng, Hongyu Jiang, Rick J. Chappell, and Jason P. Fine

12. Tests for Time-Varying Covariate Effects within Aalen's Additive Hazards Model   193

Torben Martinussen and Thomas H. Scheike

13. Analysis of Outcomes Subject to Induced Dependent Censoring: A Marked Point Process Perspective   209

Yijian Huang

14. Analysis of Dependence in Multivariate Failure-Time Data   221

Li Hsu and Zoe Moodie

15. Robust Estimation for Analyzing Recurrent-Event Data in the Presence of Terminal Events   245

Rajeshwari Sundaram

30. Models for Carcinogenesis   547

Anup Dewanji

Index   569

Preface

It is widely believed that the twenty-first century will be the century of biotechnology, which, in turn, will lead to unprecedented breakthroughs in the medical sciences revolutionizing every aspect of medicine from drug discovery to healthcare delivery. The rapid advancement in high-performance computing that took place in the last quarter of the twentieth century has indeed been a driving force in this revolution, enabling us to generate, store, query, and transfer huge amounts of medical data. As is well known, this is where statisticians come into the picture—lending their expertise in extracting information from data and converting that information to knowledge.

The key role that statisticians have been playing in this *information revolution* in the medical sciences has created new challenges and posed difficult problems before our own discipline, whose solutions have often necessitated new statistical techniques, new approaches, or even new modes of thinking. These have been the motivating force behind an astonishing flurry of research activities in biostatistics since the mid-1990s, which have been well documented in the contemporary journals and books. Since the involvement of statistics in the medical sciences is almost always interdisciplinary in nature, there has been a surge of activities on another front, namely, helping experts in the biomedical sciences (as well as practitioners in related fields) learn the basic concepts of statistics quickly and familiarizing statisticians with the medical parlance at the same time.

So a closer look at the books and monographs that have come out in these areas in the last 20 years will reveal four broad categories:

- Expository introductions to basic statistical methodology with examples and datasets from the biomedical sciences (e.g., O. J. Dunn, *Basic Statistics: A Primer for the Biomedical Sciences*, Wiley; B. Brown and M. Hollander, *Statistics: A Biomedical Introduction*, Wiley; R. P. Runyon, *Fundamentals of Statistics in the Biological, Medical and Health Sciences*, Duxbury; B. S. Everitt, *Modern Medical Statistics: A Practical Guide*, Arnold Publishing Company, or R. F. Woolson and W. R. Clarke, *Statistical Methods for the Analysis of Biomedical Data*, Wiley)

- Advanced monographs and textbooks on some special topics in statistics that are relevant to special types of biomedical data (e.g., J. K. Lindsey, *Nonlinear Models for Medical Statistics*, Oxford Statistical Science Series; W. J. Ewens and G. R. Grant, *Statistical Methods in Bioinformatics*, Springer-Verlag; R. G. Knapp and M. C. Miller III, *Clinical Epidemiology and Biostatistics*, Harwal Publishing Company; J. F. Lawless, *Statistical Models and Methods for Lifetime Data*, Wiley; or E. Marubini and

M. G. Valsecchi, *Analyzing Survival Data from Clinical Trials and Observational Studies*, Wiley)

- Encyclopedic collections or handbooks of concepts and methodology (e.g., B. S. Everitt, *Medical Statistics from A to Z: A Guide for Clinicians and Medical Students*, Cambridge Univ. Press; D. J. Balding, M. Bishop, and C. Cannings, eds., *Handbook of Statistical Genetics*, Wiley; C. R. Rao and R. Chakraborty, eds., *Handbook of Statistics 8: Statistical Methods in Biological and Medical Sciences*, Elsevier)

- Historical accounts (e.g., O. B. Sheynin, *On the History of Medical Statistics*, Springer-Verlag).

In addition to these, there are a few examples of a fifth kind. These are *edited volumes* of peer-reviewed articles encompassing several aspects of statistical applications in many areas of the biomedical sciences (e.g., Y. Lu and J.-Q. Fang, eds., *Advanced Medical Statistics*, World Scientific Publishers; B. S. Everitt and G. Dunn, eds., *Statistical Analysis of Medical Data: New Developments*, Arnold Publishing Company; B. G. Greenberg, ed., *Biostatistics: Statistics in Biomedical, Public Health and Environmental Sciences*, Elsevier). These edited volumes, which are a "snapshot" of the contemporary developments in statistical methodology for dealing with complex problems in the biomedical sciences, are neither introductory nor encyclopedic in nature. However, they have some distinct advantages. Unlike the advanced textbooks or monographs in certain specialized areas of biostatistics, they are not narrow in their coverage—usually offering a "wide angle" view of the contemporary methodological developments and technical innovations. Additionally, since they are not expected to cover every single concept and every single innovation in the field (which encyclopedias and handbooks try to do in a few hundred pages), they have a better opportunity of going in-depth and showing real-life data analysis or case study examples. Finally, the articles they contain are often firsthand accounts of research reported by the researchers themselves, as opposed to secondhand accounts provided by the author(s) of a textbook or monograph.

So when we first contemplated bringing out a book that summarizes some of the major developments in statistics in the biomedical context, this is the format we chose. A number of eminent researchers in the four major areas of the modern-day biomedical sciences where statistics has made its mark, namely, *Clinical Trials, Epidemiology, Survival Analysis*, and *Bioinformatics*, have contributed 30 carefully prepared and peer-reviewed articles. In addition, there are a few more that do not exactly fit into those areas but are strongly relevant to the overall theme. Each of the three edited volumes mentioned in the preceding paragraph is a valuable resource to students, practitioners, and researchers. But each of them has its own limitations. In proposing this volume, our motivation was to overcome many of the shortcomings of its predecessors and to combine their best features. The contributors have been carefully chosen so as to cover as much ground as possible in each broad area. Although the chapters are independent of one another, the chapter sections have been organized so as to make the thematic transition between them as smooth as possible. A structural uniformity has been maintained across all the chapters, each starting with an introduction that discusses the general concepts and describes the biomedical problem under focus. The subsequent sections provide more specific details on concepts, methods, and algorithms, with the primary emphasis on applications. Theoretical derivations and proofs are, for the most part, relegated to the appendix in chapters that contain such items. Each chapter ends with a concluding section that summarizes the main ideas in the chapter or points to future research directions.

From the beginning, our intention has been to target this book to a broad readership. Not only is it intended to be a useful reference or supplementary study material for doctoral or advanced master's-level students in different areas of biostatistics and the medical sciences, but also an informative resource guide to researchers in both academia and industry. Practitioners, such as professional consultants or research support staff, will also get a lot out of this volume. People with different backgrounds will probably benefit from different aspects of it. Medical researchers will get to see the arsenal of statistical "weapons" available to them and be more aware of how to collect information in the right way for these statistical techniques to be applicable. They will receive valuable guidelines regarding what to expect and what not to expect from statistics, that is, about proper interpretation and drawbacks of statistical methodology. Statisticians, on the other hand, will get an opportunity to put their current research in the broader perspective of the biomedical sciences and at the same time, pick up some useful suggestions for their future applied research. Statistical consultants who work with clients from the medical field or the pharmaceutical industry will gain a better understanding of the "party at the other end of the table." Even the hardcore mathematical statisticians may pick up some new directions for their pathbreaking theoretical work. Any university or research institute with a medical or public health program or a graduate-level statistics/biostatistics curriculum offers advanced courses in one or more of the four major areas covered in our volume. We believe that this volume will nicely supplement the primary textbooks used in those courses.

We admit that the 30 chapters included here are not all written at the same level of technicality or clarity, but personally, we view this disparity positively. This allows a reader to appreciate the great diversity in the training and expertise of people who are currently doing research in biomedical statistics. Perhaps the reader will realize that routinely used phrases such as "applied statistics," "interdisciplinary research," or "real-life datasets" are also subject to interpretation.

In summary, when we first contemplated this project, our primary goals were to

- Come up with a well-organized and multifaceted presentation of cutting-edge research in biomedical applications of statistics under one umbrella.
- Provide new directions of research or open problems in each area for future researchers and a detailed list of references in order to facilitate self-study.
- Do all these in a way accessible to people outside academia as well.

How far we have succeeded in achieving these goals is for the reader to judge. But we thank the contributing authors for trying their best to bring us closer to these goals by adhering to the guidelines we provided and the deadlines we set. And we do believe that our volume will be popular and will stand out among comparable pieces of work because:

- There have not been too many previous attempts to summarize contemporary research in several aspects of a large interdisciplinary area such as this in a compact volume.
- In spite of the technical nature of the subject matter, the style of presentation in our book maintains a certain degree of lucidity, aided by an informative introduction that explains the problem in simple terms, a substantial literature survey that puts each topic in perspective, and an adequate number of real-life examples and/or case studies, which will make this volume suitable for a wider audience.
- A great deal of care has been taken by the editors to avoid the usual "incoherent cut-and-paste nature" of edited volumes and ensure smooth thematic transitions between chapters within each of the five major parts.

A project like this would never be successful without the unselfish and silent contributions made by a number of distinguished colleagues who kindly accepted our invitations to referee one or more chapters or shared their opinions and expertise with us at various stages of the project—from proposal evaluation to proofreading. We deeply appreciate their contributions and offer them our sincerest gratitude.

It has been a pleasure to work with the editorial and production staff at Wiley—from the initial planning stage to the completion of the project. Susanne Steitz-Filler was patient and encouraging right from the beginning, and so was Steve Quigley. Without Susanne's editorial experience and the technical prowess of the Wiley production team, it would be a much more daunting task for us, and we are truly grateful to them. Also, one of the editors (Datta) would like to thankfully acknowledge the support received in the form of a faculty grant and reduced academic responsibilities from Northern Michigan University. At the same time, he is thankful to the Texas A&M University for supporting project through an R25 grant from the National Institutes of Health.

Both collectively and individually, we express indebtedness to our colleagues, students, and staff at our home institutions. Last but not least, we would like to lovingly acknowledge the never-ending support and encouragement from our family members that kept us going all the way to the finish line.

A. Biswas
S. Datta
J. Fine
M. Segal

Acknowledgments

To those colleagues and well-wishers who kindly donated their time and efforts to the successful completion of this project, we express our sincerest gratitude and deepest appreciation. Without their enthusiastic and unselfish support, it would be much tougher for us—to put it mildly. A number of colleagues shared the burden of editing with us by graciously accepting our request to referee one or more articles. We consider it an honor and a privilege to mention all of them by name and acknowledge their contributions.

AB, SD, JPF, and MRS

Moulinath Banerjee	Sudipto Banerjee	Tathagata Banerjee
Gopal Basak	Rebecca Betensky	Rahul Bhattacharya
Nilanjan Chatterjee	Ying Qing Chen	Sourish Das
Anup Dewanji	Kim-Anh Do	Nancy Flournoy
Debashis Ghosh	Saurabh Ghosh	David Glidden
Sarah Holte	Jenny Huang	Jian Huang
Alan Hubbard	Edward Ionides	Sunduz Keles
Uriel Kitron	Charles Kooperberg	Debasis Kundu
Lynn Kuo	Gordon Lan	Hongzhe Li
Geoff McLachlan	Giovanni Parmigiani	Limin Peng
John Pinney	Xing Qiu	Glen Satten
Rob Scharpf	Douglas Schaubel	Pranab Sen
Saunak Sen	Kerby Shedden	Debajyoti Sinha
Nigel Stallard	John Staudenmeyer	Anastasios Tsiatis
Yuanyuan Xiao	Lance Waller	Hulin Wu
Donglin Zeng	Lanju Zhang	Hongwei Zhao

Contributors

Jean-François Angers, Departement de Mathematiques et de Statistique, Universite de Montreal, CP 6128 succ Centre-Ville, Montreal, QC H3C 3J7, Canada, email: angers@DMS.umontreal.CA

Raji Balasubramanian, BG Medicine, Inc. 610 Lincoln Street (N), Waltham, MA 02451, email: rbalasub@bg-medicine.com

Uttam Bandyopadhyay, Dept. of Statistics, University of Calcutta, 35 Bally-gunge Circular Road, Kolkata 700019, West Bengal, India, email: ubandyopadhyay@yahoo.com

Tathagata Banerjee, Indian Institute of Management Ahmedabad, Vastrapur, Ahmedabad 380015, Gujarat, India, email: tathagata@iimahd.ernet.in

Ayanendranath Basu, Applied Stat. Unit, Indian Statistical Institute, 203 B.T. Road, Kolkata 700108, West Bengal, India, email: ayanbasu@isical.ac.in

Richard W. Bean, ARC Centre in Bioinformatics (Institute for Molecular Bioscience), Univ. of Queensland, St. Lucia, Brisbane, Queensland 4072, Australia, email: rbean@maths.uq.edu.au

Rahul Bhattacharya, Ashutosh College, Kolkata, West Bengal, India

Atanu Biswas, Applied Statistics Unit, Indian Statistical Institute, 203 B.T. Road, Kolkata 700108, West Bengal, India, email: atanu@isical.ac.in

Carles Bretó, Department of Ecology and Evolutionary Biology, Kraus Natural Science Building, 830 North University Avenue, University of Michican, Ann Arbor, MI 48109-1048, email: cbreto@umich.edu

Bradley M. Broom, Dept. of Biostatistics and Applied Math., M.D. Anderson Cancer Center, 1515 Holcombe Avenue, Box 237, Houston, TX 77030, USA, email: bmbroom@mdanderson.org

Hrishikesh Chakraborty, Statistics and Epidemiology Division, RTI International, Research Triangle Park 3040 Cornwallis, NC 27709, USA, email: hchakraborty@rti.org

Richard Chappell, Dept. of Biostat. & Medical Informatics, Univ. of Wisconsin, Room K6-430, Clinical Sciences Center, 600 Highland Ave., Madison, WI 53706, USA, email: chappell@biostat.wisc.edu

Nilanjan Chatterjee, Biostatistics Branch, Division of Cancer Epidemiology and Genetics, National Cancer Institute (NIH, DHHS), 6120 Executive Blvd., Rockville, MD 20852, email: chattern@mail.nih.gov

Hyungwon Choi, Department of Biostatistics, School of Public Health, 109 Observatory Rd., University of Michigan, Ann Arbor, MI 48109-2029, USA, email: hwchoi@umich.edu

Anup Dewanji, Applied Stat. Unit, Indian Statistical Institute, 203 B.T. Road, Kolkata 700108, West Bengal, India, email: dewanjia@isical.ac.in

Kim-Anh Do, Dept. of Biostatistics, M.D. Anderson Cancer Center, 1515 Holcombe Avenue, Houston, TX 77030, USA, email: kim@mdanderson.org

Jason Fine, Dept. of Biostat. & Medical Informatics, Univ. of Wisconsin, Room K6-420, Clinical Sciences Center, 600 Highland Ave., Madison, WI 53792-4675, email: fine@biostat.wisc.edu

Nancy Flournoy, Dept. of Statistics, University of Missouri 222 Mathematical Sciences Building Columbia, MO 65211, USA, email: flournoyn@missouri.edu

Robert Gentleman, Program in Computational Biology, Division of Public Health Sciences, Fred Hutchinson Cancer Research Center, 1100 Fairview Ave. (N), M2-B876, P.O. Box 19024, Seattle, WA 98109, USA, email: rgentlem@fhcrc.org

Debashis Ghosh, Dept. of Biostatistics, School of Public Health, 109 Observatory Rd., University of Michigan, Ann Arbor, MI 4810-2029, USA, email: ghoshd@umich.edu

Saurabh Ghosh, Human Genetics Unit, Indian Statistical Institute, 203 B.T. Road, Kolkata 700108, West Bengal, India.

Li Hsu, Biostatistics and Biomathematics Program, Public Health Sciences Division, Fred Hutchinson Cancer Research Center, 1100 Fairview Ave. (N), P.O. Box 19024, M2-B500, Seattle, WA 98109, USA, email: lih@fhcrc.org

Yangxin Huang, Department of Epidemiology and Biostatistics, University of South Florida, MDC56, 13201 Bruce B. Downs Blvd., Tampa, FL 33612, USA, email: yhuang@hsc.usf.edu

Yijian Huang, Department of Biostatistics, Emory University, Atlanta, GA 30322, USA, email: yhuang5@emory.edu

Edward L. Ionides, Dept. of Statistics, University of Michigan, 1085 South University Avenue, Ann Arbor, MI 48109, email: ionides@umich.edu

Anastasia Ivanova, Dept. of Biostatistics, School of Public Health, CB# 7420, 3103-C McGavran-Greenberg Hall, University of North Carolina, Chapel Hill, NC 27599-7420, USA, email: aivanova@bios.unc.edu

Hongyu Jiang, Dept. of Biostat., Center for Biostatistics in AIDS Research, Harvard Univ., Room 413, Building 1,677 Huntington Ave., Boston, MA 02115, USA, email: hjiang@hsph.harvard.edu

Timothy D. Johnson, Dept. of Biostat, Univ. of Michigan, SPH II, M4218, 1420 Washington Heights, Ann Arbor, MI 48 109-2029, USA, email: tditdj@umich.edu

Aaron A. King, Department of Mathematics, East Hall, 530 Church Street, University of Michigan, Ann Arbor, MI 48109, USA, email: aaron.king@umich.edu

Michael R. Kosorok, Dept. of Biostat. & Medical Informatics, Univ. of Wisconsin, K6-428 Clinical Sciences Center, 600 Highland Ave., Madison, WI 53792-4675, USA, email: kosorok@biostat.wisc.edu

Lynn Kuo, Dept. of Statistics, Univ. of Connecticut, 215 Glenbrook Rd., Unit 4120, CLAS Building, Storrs, CT06269, USA, email: lynn@stat.uconn.edu

Hongzhe Li, Dept. of Biostat. and Epidemiology, Univ. of Pennsylvania School of Medicine, 920 Blockley Hall, 423 Guardian Dr., Philadelphia, PA 19104-6021, USA, email: hli@cceb.upenn.edu

Brenda MacGibbon, Departement de Mathematiques, Universite du Quebec a Montreal, Case Postale 8888, succursale Centre-Ville, Montreal, Quebec, H3C 3P8, Canada, email: brenda@math.uqam.ca OR macgibbon.brenda@uqam.ca

Partha P. Majumder, Human Genetics Unit, Indian Statistical Institute, 203 B.T. Road, Kolkata, 700108, India, email: ppm@isical.ac.in

Torben Martinussen, Dept. of Natural Sciences, Royal Veterinary and Agricultural Univ., Thorvaldsensvej 40, DK-1871, Frederiksberg C, Denmark, email: torbenm@dina.kvl.dk

Geoffrey J. McLachlan, Department of Mathematics, University of Queensland, St. Lucia, Brisbane, Queensland 4072, Australia, email: gjm@maths.uq.edu.ac

Zoe Moodie, Biostatistics and Biomathematics Program, Public Health Sciences Division, Fred Hutchinson Cencer Research Center, 1100 Fairview Ave. (N), P.O. Box 19024, Seattle, WA, USA 98109, email: zmoodie@scharp.org

Shu-Kay (Angus) Ng, Dept. of Math., Univ. of Queensland, St. Lucia, Brisbane, Queensland 4072, Australia, email: skn@maths.uq.edu.au

Anne-Michelle Noone, Dept. of Biostat, School of Public Health, University of Michigan, 1420 Washington Heights, Ann Arbor, MI 48109, USA, email: amnoone@umich.edu

Limin Peng, Emory University, Atlanta, GA 30322, USA, Rollins School of Public Health, email: lpeng@sph.emory.edu

Surupa Roy, St. Xavier's College, Kolkata, West Bengal, India

Thomas H. Scheike, Department of Biostatistics, Øster Farimagsgade 5B, Univ. of Copenhagen, P.O.B. 2099, DK-1014, Copenhagen K, Denmark, email: ts@biostat.ku.dk *OR* ts@pubhealth.ku.dk

Denise Scholtens, Dept. of Preventive Medicine, Northwestern Univ. Medical School, 680 North Lake Shore Dr., Suite 1102, Chicago, IL 60611, USA, email: dscholtens@northwestern.edu

Mark R. Segal, Department of Epidemiology and Biostatistics, University of California, San Francisco, CA, USA, email: mark@biostat.ucsf.edu

Kerby Shedden, Dept. of Statistics, Univ. of Michigan 461 West Hall, Ann Arbor, MI 48109-1107, USA, email: kshedden@umich.edu

Nigel Stallard, Warwick Medical School, The University of Warwick, Coventry, CV4 7AL, U.K., email: N.Stallard@warwick.ac.uk

Rajeshwari Sundaram, Biometry and Mathematical Statistics Branch; Division of Epidemiology, Statistics and Prevention Research; National Institute of Child Health and Human Development, National Institutes of Health, Rockville, MD 20852, USA, email: sundaramr2@mail.nih.gov

Lance Waller, Dept. of Biostat,, Rollins School of Public Health, Emory University 1518 Clifton Rd. (NE), Atlanta, GA 30340, USA, email: lwallerfg@sph.emory.edu

Yuedong Wang, Dept. of Sta. and Applied Probability College of Letters and Science, Univ. of California, 5509 South Hall, Santa Barbara, CA 93106-3110, USA, email: yuedong@pstat.ucsb.edu

Hulin Wu, Department of Biostatistics and Computational Biology, University of Rochester, 601 Elmwood Ave., Box 630, Rochester, NY 14642, USA, email: hwu@bst.rochester.edu

Fang Yu, Dept. of Statistics, Univ. of Connecticut, 215 Glenbrook Rd., U-4120, CLAS Building, Storrs, CT 06269, USA, email: fangyu@stat.uconn.edu *OR* fangyu.fang@gmail.com

Yifang Zhao, Dept. of Statistics, Univ. of Connecticut, 215 Glenbrook Rd., U-4120, CLAS Building, Storrs, CT 06269, USA, email: yifang@stat.uconn.edu

PART I

Clinical Trials

CHAPTER 1

Phase I Clinical Trials

Anastasia Ivanova

Department of Biostatistics, University of North Carolina, Chapel Hill, North Carolina

Nancy Flournoy

Department of Statistics, University of Missouri, Columbia, Missouri

1.1 INTRODUCTION

Phase I trials are conducted to find a dose to use in subsequent trials. They provide data on the rate of adverse events at different dose levels and provide data for studying the pharmacokinetics and pharmacology of the drug. Dose-finding studies that involve therapies with little or no toxicity often enroll healthy volunteers and usually have a control group. Trials in oncology and other life-threatening diseases such as HIV enroll patients because treatments are usually highly toxic and to enroll healthy volunteers would not be ethical. The primary outcome for phase I trials in oncology and HIV is typically dose-limiting toxicity. Such studies require different design strategies.

In Section 1.2, we review dose-finding procedures used in healthy volunteers. In Section 1.3 we describe dose-finding procedures for trials with toxic outcomes enrolling patients. In Section 1.4, we list some other design problems in dose finding.

1.2 PHASE I TRIALS IN HEALTHY VOLUNTEERS

Buoen et al. [7] reviewed designs that are used for studying first-time-in-human drugs by looking at 105 studies published in five major pharmacology journals since 1995. In this section we briefly summarize their findings. Bouen et al. found that first-time-in-human studies usually enroll healthy volunteers; most are placebo-controlled and more than half are double-blind. The placebo group is included to reduce observer bias and sometimes to

Statistical Advances in the Biomedical Sciences, edited by Atanu Biswas, Sujay Datta,
Jason P. Fine, and Mark R. Segal

3

enable comparison of the active drug with placebo. Usually three to eight dose levels are investigated. Doses are selected using linear, logarithmic, Fibonacci, modified Fibonacci dose escalation patterns, or some combinations of these. The popular modified Fibonacci procedure escalates doses in relative increments of 100%, 65%, 50%, 40%, and 30% thereafter.

The simplest pattern of dose administration being used in first-time-in-human studies is the *parallel single-dose design* in which a single dose is administered once. Multiple administrations of the same dose are referred to as *parallel multiple-dose design*. Parallel dose administration was found to be the most frequently used procedure in first-time-in-human studies. In a typical trial with parallel dose administration, subjects are assigned in cohorts consisting of eight subjects, with six assigned to the active treatment and two assigned to a control. All treated subjects in a cohort receive the same dose. Doses are increased by one level for each subsequent cohort. The trial is stopped when an unacceptable number of adverse events is observed, the highest dose level is reached, or for other reasons. The "target dose," the dose recommended for future trials, is usually determined on the basis of the rates of adverse events at dose levels studied and/or on pharmacokinetic parameters.

More complex dose administration patterns were found to involve the administration of several different dose levels to each patient. In such trials, the healthy subjects are given some rest time between administrations to minimize the carryover effect. One such pattern is referred to as an *alternating crossover design*. An example of an alternating crossover design for a study with six doses is as follows:

Cohort 1:	Dose 1	REST	Dose 4
Cohort 2:	Dose 2	REST	Dose 5
Cohort 3:	Dose 3	REST	Dose 6

Another dose administration pattern is the grouped crossover escalation. An example of this pattern for a trial with four dose levels is as follows:

Cohort 1

Subject 1	Placebo	Dose 1	Dose 2
Subject 2	Dose 1	Placebo	Dose 2
Subject 3	Dose 1	Dose 2	Placebo

Cohort 2

Subject 1	Placebo	Dose 3	Dose 4
Subject 2	Dose 3	Placebo	Dose 4
Subject 3	Dose 3	Dose 4	Placebo

Sheiner et al. [41] reviewed parallel and crossover designs and methods for analyzing the data obtained in such studies. They point out ethical problems and a lack of representativeness in these designs. Sheiner et al. [41] advocated using a dose administration pattern that they call the *dose escalation design*:

> According to the dose-escalation design all subjects are given a placebo dose first. If after some predefined time period the response fails to satisfy a certain clinical endpoint and no unacceptable toxicity is seen, the dose is increased by one level. This process is repeated at

each dose level until either the clinical endpoint is reached or the highest dose is attained. If the response is adequate at any dose, the dose is maintained at that level for the duration of the study.

The main obstacle to using this design is the lack of formal statistical methods for data analysis.

Girard et al. [17] studied the effects of several confounding factors on trials that use parallel dose, crossover and dose escalation designs by simulations. They concluded that the presence of nonresponders biases the estimate of the dose producing 50% of the maximum effect, in all three designs. However, other confounders such as carryover effects only bias the results of trials in which the dose escalation design is used.

Buoen et al. [7] conclude that, although "the development of study designs and evaluation methods for cancer trials is extensive, ... formal statistically based methods ... are unusual in phase I dose-escalation trials in healthy volunteers." This lack and the recognition of need present both challenges and opportunities to the statistical research community.

1.3 PHASE I TRIALS WITH TOXIC OUTCOMES ENROLLING PATIENTS

In many phase I trials in which the subjects are patients, rather than healthy volunteers, the goal is to find the dose that has a prespecified toxicity rate. This is particularly true in oncology. In these trials, the primary outcome is typically binary: dose-limiting toxicity? Yes or no. For example, the dose-limiting toxicity (DLT) in radiotherapy and chemotherapy studies is usually defined as treatment-related nonhematological toxicity of grade 3 or higher or treatment-related hematological toxicity of grade 4 or higher. The *maximally tolerated dose* (MTD) is statistically defined as the dose at which the probability of DLT is equal to the some prespecified rate Γ. The typical underlying model assumption is that the probability of toxicity is a nondecreasing function of dose, even though decreasing toxicity rates at high doses have been observed [43].

Preclinical studies in animals often attempt to determine the dose with approximately 10% mortality (e.g., the murine LD_{10}). In first-in-human toxicity studies, one-tenth or two-tenths of the dose considered to be equivalent to the murine equivalent, expressed in milligrams per meter squared (mg/m^2), is generally used as a starting dose in escalation procedures. The starting dose is anticipated to be $5-10$-fold below the dose that would demonstrate activity in humans. In trials with oral drugs, only certain doses can be used; therefore, the set of possible doses is fixed in advance. The set of possible doses is often chosen according to the modified Fibonacci sequence.

In dose-finding trials in oncology, patients may receive a single dose of a drug or multiple administrations of the same dose. To address ethical concerns similar to those of Sheiner et al. [41] and to shorten trial duration, Simon et al. [42] introduced acceleration titration designs. Such designs allow intrapatient dose escalation if no toxicity is observed in a patient at the current dose. A patient goes off study or the patient's dose is reduced if toxicity is observed. Although appealing from an ethical perspective, this approach is not widely used for the same reason as in the hesitation to use Sheiner's dose escalation design. In the rest of this chapter, we review methods with parallel dose administration.

One cannot begin to detail all designs that have been used with parallel administration for dose finding in patients with dose-limiting toxicity. Some popular procedures are ad hoc, as are the designs used in healthy volunteers. Others were developed with various desirable characteristics. We discuss the most popular procedures, but our choice is admittedly biased by our own interests.

1.3.1 Parametric versus Nonparametric Designs

Designs for dose finding can be classified as parametric or nonparametric. Non-parametric designs are attractive because they are easy to understand and implement; the decision rules are intuitive and their implementation does not involve complicated calculations. By *nonparametric*, we mean that no parametric representation of the dose–response relationship is used in the design's treatment allocation rule. In this chapter, we discuss several Markovian and Markovian-motivated non-parametric up-and-down designs and the $A + B$ designs of Lin and Shih [31]. We also discuss non-parametric designs in which the treatment allocation rule is based on isotonic estimates of the dose–response function. These are called *isotonic designs*.

Then we describe some parametric designs that assume one- or two-parameter models for the dose–toxicity relationship. Popular parametric designs include the continual reassessment method [33] and escalation with overdose control [2].

With the Markovian and Markovian-motivated designs, treatment assignments typically cluster unimodally around a specific dose, and the key to their effectiveness is to select design parameters so as to center the treatment distribution judiciously [11]. For example, for toxicity studies with increasing dose–response functions, these designs can be constructed to cluster treatments around the unknown dose with prespecified "target" toxicity rate Γ.

In other designs that allow multiple escalations and deescalations of dose, treatment assignments first fluctuate around the MTD and then converge assignments to the MTD. Such designs include, for example, the continual reassessment method [33] and isotonic designs [29].

1.3.2 Markovian-Motivated Up-and-Down Designs

In up-and-down designs, the next dose assigned is never more than one level distant from the dose given to the current cohort of patients. Such designs are appealing in dose-limiting toxicity studies because of the potentially devastating consequences of abruptly making major changes in dosing. Many ad hoc up-and-down procedures exist, including the most widely cited design in oncology, that is, the $3 + 3$ design [44,28]. The $3 + 3$ design is a special case of the $A + B$ designs [31]. It is important in trials with patients who are critically ill not to assign too many patients to low, ineffective doses. The $A + B$ designs address this concern by assigning A patients to the lower doses and assigning $A + B$ patients to doses closer to the target.

Before describing the $A + B$ designs, we review a fundamental theorem that is useful for characterizing the Markovian up-and-down design. Let p_k, q_k, and r_k denote the probability of increasing, decreasing, and repeating dose d_k, respectively. Assume that these probabilities depend only on d_k, $k = 1, \ldots, K$. Furthermore, assume that p_k decreases with dose, whereas q_k increases with dose. Let d_κ denote the largest dose such that $p_{\kappa-1} \geq q_\kappa$. The stationary distribution for Markov chain designs with transition probabilities p_k, q_k, r_k exists uniquely if the Markov chain is recurrent, irreducible, and aperiodic. Under these conditions, Durham and Flournoy [11] proved that the stationary distribution of the dose assignments is unimodal and the mode occurs at d_κ. Additionally, if $p_{\kappa-1} = q_\kappa$, then the mode spans $d_{\kappa-1}$ as well as d_κ.

Convergence of the dose assignments to their stationary distribution is reached exponentially rapidly, so asymptotic results apply well with a relatively small number of treatment assignments, regardless of the initial dose. Because of the discreteness of the dose space, as a practical approximation, we say that a Markovian up-and-down design "targets" d_κ if $p_\kappa = q_\kappa$; treatments will cluster unimodally around this dose. Alternatively, we say that the design targets the toxicity rate Γ, for which $P\{\text{toxicity} \mid d_\kappa\} = \Gamma$. Markovian up-and-down designs can be characterized using this and other asymptotic and finite sample theory for Markov chains. Techniques are given in Durham and Flournoy [12], Durham et al. [13], Giovagnoli and Pintacuda [16], and Bortot and Giovagnoli [5].

A corollary of the Durham–Flournoy theorem is that treatments from the traditional up-and-down design of Dixon and Mood [10] are distributed unimodally around $d_\kappa = LD_{50}$, regardless of the underlying (increasing) dose–response model. In this procedure, the dose is decreased if a toxicity is observed and increased otherwise. So $p_k = P\{\text{toxicity} \mid d_k\}$ and $q_k = 1 - p_k = P\{\text{toxicity} \mid d_k\}$ (except at $k = 1$ or K). Solving $p_k = q_k$ yields $p_\kappa = 0.50$.

Durham and Flournoy [11,12] generalized the Dixon–Mood decision rule by using a biased coin, together with the Durham–Flournoy theorem, to provide a procedure that targets any given toxicity rate Γ. This procedure was not well received in oncology trials because clinicians were averse to using randomization in phase I treatment allocation rules.

Using cohorts at each dose, the Durham–Flournoy theorem was employed by Gezmu and Flournoy [15] to devise treatment allocation rules without randomization that still target a given toxicity rate Γ. However, the set of possible targets is limited by the group size. Some examples they give of Γ that are possible with groups of size 2 are 0.29, 0.50, and 0.71; with groups of size 3, they are 0.21, 0.35, 0.50, 0.65, and 0.79; and with groups of size 4, they are 0.16, 0.27, 0.38, 0.39, 0.50, 0.61, 0.62, 0.73, and 0.84. Procedures for values of Γ greater than 0.5 are useful for efficacy studies, but not toxicity studies. Gezmu and Flournoy [15] show that each of these target values can be found as a direct application of the Durham–Flournoy theorem; details justifying this application are given by Ivanova et al. [25]. Antognini et al. [1] generalize the Gezmu–Flournoy group procedure to target any $\Gamma \in (0,1)$ by introducing a randomization procedure. This is clever, but will probably not be any more attractive to oncologists than was the biased coin design of Durham and Flournoy [11].

Ivanova et al. [25] take a different approach to adapting the group up-and-down design so that it will target any given $\Gamma \in (0,1)$. They call their procedure the *cumulative cohort design*, which is as follows.

Cumulative Cohort Design Suppose that the most recent assignment was to dose d_j. Let $\hat{q}_j$ be the cumulative proportion of toxicities at d_j, and let $\Delta > 0$ denote a design parameter. Then

1. If $\hat{q}_j \leq \Gamma - \Delta$, the next group of subjects is assigned to dose d_{j+1}.
2. If $\hat{q}_j \geq \Gamma + \Delta$, the next group of subjects is assigned to dose d_{j-1}.
3. If $\Gamma - \Delta < \hat{q}_j < \Gamma + \Delta$, the next group of subjects is assigned to dose d_j.

Appropriate adjustments are made at the lowest and highest doses.

An intuitive choice of the parameter $\Delta > 0$ in the cumulative cohort design is close to 0. For example, with $\Delta = 0.01$ and moderate sample sizes, the dose will be repeated if the estimated toxicity rate is exactly equal to Γ, and changed otherwise. Ivanova et al. [25] suggested choosing Δ to maximize the total number of subjects assigned to the MTD over a set of dose–toxicity scenarios. For example, for moderate sample sizes they recommended using $\Delta = 0.09$ if $\Gamma = 0.10$, 0.15, 0.20, or 0.25; $\Delta = 0.10$ if $\Gamma = 0.30$ or 0.35; $\Delta = 0.12$ if $\Gamma = 0.40$; and $\Delta = 0.13$ if $\Gamma = 0.45$ or 0.50. Ivanova et al. [25] demonstrated via simulations that $\Delta = 0.01$ and choosing their recommended values of Δ yield similar frequency of correctly selecting the MTD. However, the cumulative cohort design with their recommended D values assigns significantly more patients to the MTD.

The $A + B$ designs as given by Lin and Shih [31] begin like the first run of a Markovian group up-and-down design, but the design is switched when the dose would otherwise be repeated and stopped (for designs without deescalation) when the dose would otherwise be decreased.

$A + B$ *Design*: Let A and B be positive integers. Let c_L, c_U, and C_U be integers such that $0 \leq c_L < c_U \leq A$, $c_U - c_L \geq 2$, and $c_L \leq C_U < A + B$. Let $X_A(d_j)$ be the number of toxicities in a cohort of size A assigned to dose d_j, and let $X_{A+B}(d_j)$ be the number of toxicities in a cohort of size $A + B$. Subjects are treated in cohorts of size A starting with the lowest dose. Suppose that the most recent cohort was a cohort of A subjects that has been treated at dose $d_j, j = 1, \ldots, K - 1$. Then

1. If $X_A(d_j) \leq c_L$, the next cohort of A subjects is assigned to dose d_{j+1}.
2. If $c_L < X_A(d_j) < c_U$, the cohort of B subjects is assigned to dose d_j; then, if in the combined cohort assigned to d_j, $X_{A+B}(d_j) \leq C_U$, the next cohort of size A receives dose d_{j+1}; otherwise the trial is stopped.
3. If $X_A(d_j) \geq c_U$, the trial is stopped.

The dose that is one level below the dose where unacceptable numbers of toxicities are observed ($\geq c_U$ toxicities in a cohort of size A or $> C_U$ toxicities in a cohort of size $A + B$) is the estimated MTD.

In an $A + B$ design, the frequency of stopping dose escalation at a certain level depends on toxicity rate at this dose as well as on toxicity rate at all lower dose levels. Ivanova [21] used the Durham–Flournoy theorem to derive recommendations for constructing escalation designs and explains how to compute the toxicity rate Γ that will be targeted by any given $A + B$ design. The algorithm for selecting parameters A, B, c_L, c_U, and C_U for a given target quantile Γ is as follows (where Bin = binomial distribution):

1. Find A, c_L, and c_U, $0 \leq c_L < c_U \leq A$, $c_U - c_L \geq 2$, so that Γ_A, the solution to the equation $\Pr\{\text{Bin}(A, \Gamma_A) \leq c_L\} = \Pr\{\text{Bin}(A, \Gamma_A) \geq c_U\}$, is equal to or slightly exceeds Γ.
2. Set B (the choice $A \leq B$ yields more efficient designs), and given that Γ_{A+B} is the solution to the equation $\Pr\{\text{Bin}(A + B, \Gamma_{A+B}) \leq C_U\} = 0.5$, find C_U such that $C_U/(A + B) < \Gamma < \Gamma_{A+B}$.

The 3 + 3 design is a special case of the $A + B$ design with $A = B = 3$, $c_L = 0$, $c_U = 2$, and $C_U = 1$ that target quantiles around $\Gamma = 0.2$. Applying the algorithm above, we obtain

1. $\Gamma_A = 0.35$ is the solution of the equation $\Pr\{\text{Bin}(3, \Gamma_A) \leq 0\} = \Pr\{\text{Bin}(3, \Gamma_A) \geq 2\}$; $\Gamma_A = 0.35$ is slightly higher than $\Gamma = 0.2$.
2. $\Gamma_{A+B} = 0.26$ is the solution of the equation $\Pr\{\text{Bin}(3 + 3, \Gamma_{A+B}) \leq 2\} = 0.5$, and $C_U(A + B) = 0.17$. Hence, approximate bounds for Γ targeted by the 3 + 3 design are $0.17 < \Gamma < 0.26$.

Exact probability calculations and simulation studies for several dose–response scenarios by Reiner et al. [39], Lin and Shih [31], Kang and Ahn [26,27] and He et al. [20] are consistent with the theoretical calculation above establishing that the 3 + 3 design selects a dose with toxicity rate near 0.2. He et al. [20] also showed that if dose levels are selected close to each other, the mean toxicity rate at the dose selected by the $3 + 3$ design is slightly lower than the dose selected by trials with a sparser set of possible dose levels.

1.3.3 Isotonic Designs

Isotonic designs assume that the dose–toxicity relationship is isotonic and use isotonic estimates of the toxicity rates in the treatment allocation rule. We first review isotonic estimation of the toxicity rates, which are maximum-likelihood estimates for the isotonic model of the data. Let N (d_j, n) be the number of patients assigned to dose d_j, and let $X(d_j, n)$ be the number of toxicities at d_j after n patients have been treated. Define $\hat{q}_j = X(d_j, n)/N_j(n)$ for all $j \in \{1, \ldots, K\}$ such that $N(d_j, n) > 0$, and let $(\hat{q}_1, \ldots, \hat{q}_k)$ be the vector of these proportions. The vector of isotonic estimates $(\tilde{q}_1, \ldots, \tilde{q}_K)$ can be obtained from $(\hat{q}_1, \ldots, \hat{q}_K)$ by using the pool adjacent violators algorithm (see, e.g., Ref. 3). At the end of the trial the dose with the value $\tilde{q}_i$ closest to Γ is the estimated MTD. If there are two or more such doses, the highest dose with the estimated value below Γ is chosen. If all the estimated values at these doses are higher than Γ, the lowest of these doses is chosen. The cumulative cohort decision rule [25] described in Section 1.3.1 when used with isotonic estimates of toxicity rates is an isotonic design. A few other isotonic designs have been proposed, including the isotonic design of Leung and Wang [29]. Ivanova and Flournoy [24] compared several isotonic designs with the cumulative cohort design via simulations for a number of target quantiles and dose–toxicity models and concluded that the cumulative cohort design performs better than others.

1.3.4 Bayesian Designs

Parametric methods require assumptions about the model for the dose–toxicity relationship. In addition, Bayesian methods require priors on the model parameters. The continual reassessment method (CRM) is a Bayesian design proposed in 1990 [36]. The CRM starts with a working model for the dose–toxicity relationship. Let $y_i = 1$ if the ith patient experiences toxicity and let $y_i = 0$ otherwise, $i = 1, \ldots, n$. For example

$$F(d, \theta) := P\{y_i = 1 \mid d\} = [(\tanh d + 1)/2]^\theta. \tag{1.1}$$

The CRM uses Bayes' theorem to update a prior distribution $g(\Sigma)$ of Σ, for example, $g(\theta) = \exp(-\theta)$. After each patient's response is observed, the mean posterior density of the parameter is computed. Let $x_i \in D$ be the dose received by the ith patient. So after the nth patient's response, $\Omega_n = \{(x_1, y_1), \ldots, (x_n, y_n)\}$ are the accumulated data and

$$\hat{\theta}^{(n)} = E(\theta \mid \Omega_n) = \int_0^\infty \theta f(\theta \mid \Omega_n) d\theta \tag{1.2}$$

is the posterior mean of θ. Here $f(\theta \mid \Omega_n) = L_{\Omega n}(\theta) g(\theta) / \int_0^\infty L_{\Omega n}(u) g(u) du$ and $L_{\Omega n}(\theta)$ is the likelihood function.

In the CRM, no prespecified set of doses is required and subjects are assigned one at a time. However, doses can be restricted to a prespecified ordered set $D = \{d_1, \ldots, d_K\}$ [34]. In this case, the model above can also be written as $F(d_i, \theta) = b_i^\theta$, where $(b_1, \ldots, b_k)$ is a set of constants, $b_i = (\tanh d_i + 1)/2$.

The first patient receives the first dose level, $x_1 = d_1$. Assume that n patients have been assigned so far. The dose to be administered to the next patient is the dose x_{n+1} such that the absolute difference between $\Pr\{y = 1 \mid x_{n+1}, \hat{\theta}^{(n)}\}$ and Γ is minimized. If a prespecified set D is chosen, this quantity is minimized over D. Dose x_{n+1} can be used as an estimate of the MTD after n patients have been assigned. Other estimators were explored by O'Quigley [35]. Necessary conditions for the CRM to converge to the target dose were given in Shen and O'Quigley [40], and more relaxed conditions were given by Cheung and Chappell [9]. Also, subjects can be assigned in groups [14,28,18] to shorten the total duration of the trial.

The CRM is a special case of a Bayesian decision procedure with the next dose x_{n+1} selected to maximize the gain function [47]:

$$G(\hat{\theta}^{(n)}, d) = (F(d, \hat{\theta}^{(n)}) - \Gamma)^{-2}. \tag{1.3}$$

Another Bayesian design for dose-finding studies is the escalation with overdose control [2]. This design is from a class of Bayesian feasible designs. It uses a loss function to minimize the predicted amount by which any given patient is overdosed. Bayesian decision procedures for dose-finding studies were described in McLeish and Tosh [32], Whitehead and Brunier [47], and Whitehead and Williamson [48]. Leung and Wang [30] point out that the CRM is a myopic strategy and might not be globally optimal. A globally optimal strategy requires comparison of all possible sets of actions that could be taken, and this remains computationally formidable for designs having more than three dose levels [19].

1.3.5   Time-to-Event Design Modifications

If a follow-up time is required for each patient as, for example, in many radiation therapy trials, the dose-finding trial can be impractically long. Cheung and Chappell [8] suggested a modification of the CRM that allows treatment assignments to be staggered so as to shorten the trial duration.

In the original CRM [33], the calculation of the posterior mean of θ at the time when the $(n + 1)$th patient enters the trial is based on the likelihood

$$L_n(\theta) = \prod_{i=1}^{n} F(x_i, \theta)^{y_i} \{1 - F(x_i, \theta)\}^{1-y_i}, \tag{1.4}$$

where $F(x_i, \theta)$ is a working model as before. Cheung and Chappell [8] introduced the so-called TITE-CRM for trials with long follow-up. They redefined the toxicity rate at dose d_i to be the probability of observing toxicity at d_i during a time period of length T after initiation of therapy. Data for the ith patient, $i = 1, \ldots n$, are $\{x_i, y_{i,n}, u_{i,n}\}$ when the $(n + 1)$st patient enters the trial, where x_i is the dose, $y_{i,n}$ is the toxicity indicator, and $u_{i,n}$ is the time that has elapsed from the moment when the ith patient entered the trial to the time $(n + 1)$th patient enters the trial.

Cheung and Chappell [8] suggested using a weighted likelihood for TITE-CRM:

$$\tilde{L}_n(\theta) = \prod_{i=1}^{n} \{w_{i,n} F(x_i, \theta)\}^{y_i} \{1 - w_{i,n} F(x_i, \theta)\}^{1-y_i}, \tag{1.5}$$

where $w_{i,n}$ is the weight assigned to the ith observation prior to entry of the $(n + 1)$th patient. For example, a weight of $w_{i,n} = \min(\mu_{i,n}/T, 1)$ reflects an assumption that the density of time to toxicity is flat in $(0, T)$. Other choices for weights can be considered [8].

Similar modifications can be applied to any treatment allocation rules that are based on the likelihood function. In particular, the isotonic designs can be extended using this idea for trials with long follow-up. Such extension of the cumulative cohort design is described in Ivanova et al. [25].

1.4 OTHER DESIGN PROBLEMS IN DOSE FINDING

Below we list various other design problems that arise in the dose-finding context. We have not included designs for bivariate outcomes, but note that dose-finding designs whose goals combine toxicity with efficacy form a growing area of research. Otherwise, we apologize in advance if we have overlooked one of your favorites.

Ordered Groups Sometimes patients are stratified into two subpopulations, for example, heavily pretreated and not, where the first subpopulation is more likely to experience toxicity. The goal is to find two MTDs, one for each subpopulation. One of the subpopulations is often very small, rendering the running of two separate trials, one for each subpopulation, unfeasible. O'Quigley and Paoletti [37] proposed a parametric design for this problem. Their method is an extension of the CRM. Ivanova and Wang [22] proposed an isotonic approach where bivariate isotonic regression is used to estimate toxicity rates in both populations simultaneously.

Multitreatment Trials Multi-treatment trials are very common. The goal is usually to find the maximum tolerated dose combination. Often only the dose of one agent is varied, with doses of all the other agents held fixed. Thall et al. [45] propose a Bayesian design for trials with two agents in which the doses of both agents are changed simultaneously.

Ivanova and Wang [22] and Wang and Ivanova [46] considered a two-agent trial where two doses of one of the agents, say, the second agent, have already been selected. The problem is to find two maximum tolerated doses of the first agent, one MTD for each dose of the second agent. Ivanova and Wang [22] described an isotonic design, and Wang and Ivanova [46] described a Bayesian design for the problem.

Ordinal Outcomes Toxicity in oncology, and many other settings, is measured as an ordinal variable. Bekele and Thall [4] gave an example of a dose-finding trial where different grades of toxicity are combined to obtain a toxicity score for each patient. The goal was to find the dose with a certain weighted sum of probabilities of toxicity grades corresponding to different toxicity types. They [4] suggested a Bayesian design for this problem. Ivanova [21] described a trial where three grades of toxicity (none, mild, and dose-limiting) are combined in a single score. A design in the spirit of the $A + B$ designs to target the dose with the score of 0.5 was used in that trial [21].

Paul et al. [38] considered a different problem in which, target toxicity rates are specified for each grade of toxicity. The goal is to find the vector of doses that have the prespecified rates of toxicity. A multistage random-walk rule with a multidimensional isotonic estimator is proposed.

Finding a Maximum Tolerated Schedule In chemotherapy trials treatment is usually administered in cycles. The goal is to find a maximum tolerated schedule for an agent used in chemotherapy administration. Braun, et al. [6] presented a parametric design for this problem.

1.5 CONCLUDING REMARKS

We have given an overview of dose-finding designs. There has been much progress in the area of dose-finding designs; new dose-finding problem are being formulated and new methods developed. Statistical methods for dose-finding designs are most advanced for trials in oncology and other life-threatening diseases. Ad hoc designs, such as the $3 + 3$ or $A + B$ designs, are often criticized for being inflexible with regard to their objectives. It is true that $A + B$ designs do not converge to a certain quantile because they invoke the stopping rule and use small sample size. Increasing the size of the second cohort or using an $A + B + C$ design will lead to better performance of these types of design. The major limitation of and $A + B$ design is that no modifications of the design exist to use the design in trials with delayed toxicity outcome. The CRM had been shown to converge to the MTD under certain conditions. It performs very well for small to moderate sample sizes. The CRM can be used for trials with delayed outcomes [8]. However, attempts to design a stopping rule (for use with the CRM) that performs very well have been unsuccessful. Therefore, one needs to specify the total sample size in advance when the CRM is used. A number of publications on isotonic designs or "model free" designs have appeared in the literature and are discussed by Ivanova and Flournoy [24]. These designs do not use any assumption other then toxicity monotonicity. As extension of nonparametric designs, isotonic designs allow using all the data available to obtain toxicity estimates. From the parametric design perspective, model-free designs bring flexibility to modeling when needed.

REFERENCES

1. Antognini, A. B., Bortot, P., and Giovagnoli, A., Randomized group up and down experiments, *Ann. Ins. Statist. Math.* **58** (in press).

2. Babb, J., Rogatko, A., and Zacks, S., Cancer phase I clinical trials: Efficient dose escalation with overdose control, *Statist. Med.* **17**, 1103–1120 (1998).

3. Barlow, R. E., Bartholomew, D. J., Bremner, J. M., and Brunk, H. D., *Statistical Inference under Order Restrictions*, Wiley, London, New York, 1972.

4. Bekele B. N. and Thall F. T., Dose-finding based on multiple toxicities in a soft tissue sarcoma trial, *J. Am. Statist. Assoc.* **99**, 26–35 (2004).

5. Bortot, P. and Giovagnoli, A., Up-and-down designs of the first and second order, *J. Statist. Plan. Infer.* **134**, 236–253 (2005).

6. Braun, T. M., Thall, P. F., and Yuan, Z., Determining a maximum tolerated schedule of a cytotoxic agent, *Biometrics* **61**, 335–343 (2005).

7. Buoen, C., Bjerrum, O. J., and Thomsen, M. S., How first-time-in-human studies are being performed: A survey of phase I dose-escalation trials in healthy volunteers published between 1995 and 2004, *J. Clin. Pharmacol.* **45**,1123–1136 (2005).

8. Cheung, Y. K. and Chappell, R., Sequential designs for phase I clinical trials with late-onset toxicities, *Biometrics* **56**, 1177–1182 (2000).

9. Cheung, Y. K. and Chappell, R., A simple technique to evaluate model sensitivity in the continual reassessment method, *Biometrics* **58**, 671–674 (2002).

10. Dixon, W. J. and Mood, A. M., A method for obtaining and analyzing sensitivity data, *J. Am. Statist. Assoc.* **43**, 109–126 (1954).

11. Durham, S. D. and Flournoy, N., Random walks for quantile estimation, in *Statistical Decision Theory and Related Topics V*, Berger, J. and Gupta, S., eds., Springer-Verlag, New York, 1994, pp. 467–476.

12. Durham, S. D. and Flournoy, N., Up-and-down designs I: Stationary treatment distributions, in *Adaptive Designs. Lecture-Notes Monograph Series*, Flournoy, N. and Rosenberger, W. F. eds., Institute of Mathematical Statistics, Hayward, CA, Vol. 25, 1995, pp. 139–157.

13. Durham, S. D., Flournoy, N., and Haghighi, A. A., Up-and-down designs II: Exact treatment moments, in *Adaptive Designs. Lecture-Notes Monograph Series*, Flournoy, N. and Rosenberger, W. F., eds., Institute of Mathematical Statistics, Hayward, CA, Vol. 25, 1995, pp. 158–178.

14. Faries, D., Practical modifications of the continual reassessment method for phase I cancer clinical trials, *J. Biopharm. Statist.* **4**, 147–164 (1994).

15. Gezmu, M. and Flournoy, N., Group up-and-down designs for dose finding, *J. Statist. Plan. Infer.* **136**, 1749–1764 (2006).

16. Giovagnoli, A. and Pintacuda, N., Properties of frequency distributions induced by general up-and-down methods for estimating quantiles, *J. Statist. Plan. Infer.* **74**, 51–63 (1998).

17. Girard, P., Laporte-Simitsidis, S., Mismetti, P., Decousus, H., and Boissel, J. P., Influence of confounding factors on designs for dose-effect relationship estimates, *Statist. Med.* **14**, 987–1005 (1995).

18. Goodman, S. N., Zahurak, M. L., and Piantadosi, S., Some practical improvements in the continual reassessment method for phase I studies, *Statist. Med.* **14**, 1149–1161 (1995).

19. Hardwick, J. P., Oehmke, R., and Stout, Q. F., A program for sequential allocation of three Bernoulli populations, *Comput. Statist. Data Anal.* **31**, 397–416 (1999).

20. He, W., Liu, J., Binkowitz, B., and Quan, H., A model-based approach in the estimation of the maximum tolerated dose in phase I cancer clinical trials, *Statist. Med.* **25**, 2027–2042 (2006).

21. Ivanova, A., Escalation, up-and-down and A + B designs for dose-finding trials, *Statist. Med.* **25**, 3668–3678 (2006).

22. Ivanova, A. and Wang, K., A nonparametric approach to the design and analysis of two-dimensional dose-finding trials, *Statist. Med.* **23**, 1861–1870 (2004).

23. Ivanova, A. and Wang, K., Bivariate isotonic design for dose-finding with ordered groups, *Statist. Med.* **25**, 2018–2026 (2006).

24. Ivanova, A. and Flournoy, N., Comparison of isotonic designs for dose-finding, Submitted (in press).

25. Ivanova, A., Flournoy, N., and Chung, Y., Cumulative cohort design for dose-finding, *J. Statist. Plan. Infer.* (in press).

26. Kang, S. H. and Ahn, C., The expected toxicity rate at the maximum tolerated dose in the standard phase I cancer clinical trial design, *Drug Inform. J.* **35**(4), 1189–1200 (2001).

27. Kang, S. H. and Ahn, C., An investigation of the traditional algorithm-based designs for phase I cancer clinical trials, *Drug Inform. J.* **36**, 865–873 (2002).

28. Korn, E. L., Midthune, D., Chen, T. T., Rubinstein, L. V., Christian, M. C., and Simon, R. M., A comparison of two phase I trial designs, *Statist. Med.* **13**, 1799–1806 (1994).

29. Leung, D. H. and Wang, Y. G. Isotonic designs for phase I trials, *Control. Clin. Trials* **22**, 126–138 (2001).

30. Leung, D. H. and Wang, Y. G., An extension of the continual reassessment method using decision theory, *Statist. Med.* **21**, 51–63 (2002).

31. Lin, Y. and Shih, W. J., Statistical properties of the traditional algorithm-based designs for phase I cancer clinical trials, *Biostatistics* **2**, 203–215 (2001).

32. McLeish, D. L. and Tosh, D. Sequential designs in bioassay, *Biometrics* **46**, 103–116 (1990).

33. O'Quigley, J., Pepe, M., and Fisher, L., Continual reassessment method: A practical design for phase I clinical trials in cancer, *Biometrics* **46**, 33–48 (1990).

34. O'Quigley, J. and Chevret, S., Methods for dose-finding studies in cancer clinical trials: A review and results of a Monte Carlo study, *Statist. Med.* **10**, 1647–1664 (1991).

35. O'Quigley, J., Estimating the probability of a toxicity at recommended dose following a phase I clinical trial in cancer, *Biometrics* **48**, 853–862 (1992).

36. O'Quigley, J., Pepe, M., and Fisher, L. Continual reassessment method: A practical design for phase I clinical trials in cancer, *Biometrics* **46**, 33–48 (1990).

37. O'Quigley, J. and Paoletti, X., Continual reassessment method for ordered groups, *Biometrics* **59**, 430–440 (2003).

38. Paul, R. K., Rosenberger, W. F., and Flournoy, N., Quantile estimation following non-parametric phase I clinical trials with ordinal response, *Statist. Med.* **23**, 2483–2495 (2004).

39. Reiner, E., Paoletti, X., and O'Quigley, J., Operating characteristics of the standard phase I clinical trial design, *Comput. Statist. Data Anal.* **30**, 303–315 (1999).

40. Shen, L. Z. and O'Quigley, J., Consistency of continual reassessment method under model misspecification, *Biometrika* **83**, 395–405 (1996).

41. Sheiner, L. B., Beal, S. L., and Sambol, N. C., Study designs for dose-ranging, *Clin. Pharmacol. Ther.* **46**, 63–77 (1989).

42. Simon, R., Friedlin, B., Rubinstein, L., Arbuck, S. G., Collins, J., and Christian, M. C., Accelerated titration designs for phase I clinical trials in oncology, *J. Nat. Cancer Inst.* **89**(15), 1138–1147 (1997).

43. Simpson, D. G. and Margolin, B. H., Recursive nonparametric testing for dose-response relationships subject to downturns at high doses, *Biometrika* **73**, 589–596 (1986).

44. Storer, B. E., Design and analysis of phase I clinical trials, *Biometrics* **45**, 925–937 (1989).

45. Thall, P., Millikan, R., Mueller, P., and Lee, S., Dose finding with two agents in phase I oncology trials, *Biometrics* **59**, 487–496 (2003).

46. Wang, K. and Ivanova, A., Two-dimensional dose finding in discrete dose space, *Biometrics* **61**, 217–222 (2005).

47. Whitehead, J. and Brunier, H., Bayesian decision procedures for dose determining experiments, *Statist. Med.* **14**, 885–893 (disc: p897–899) (1995).

48. Whitehead, J. and Williamson, D. Bayesian decision procedures based on logistic regression models for dose-finding studies, *J. Biopharm. Statist.* **8**, 445–467 (1998).

CHAPTER 2

Phase II Clinical Trials

Nigel Stallard
Warwick Medical School, The University of Warwick, UK

2.1 INTRODUCTION

2.1.1 Background

This chapter is concerned with biostatistical aspects of the design and analysis of phase II clinical trials. Although the nature of phase II clinical trials varies considerably between different therapeutic areas and research institutions, such trials are usually small-scale studies intended to help us decide whether to continue clinical evaluation of the experimental therapy in further larger-scale trials. The small sample size and decision-making functions are in contrast to the phase III clinical trials considered in the previous chapter, and it is these features that mean that special statistical approaches are required. The development of such approaches has been the focus of much statistical work since the mid-1970s. This work will be described briefly below along with remaining challenges.

The main focus of the chapter will be phase II clinical trials of new drug therapies. In particular, many of the methods that are described below have been developed for the evaluation of anticancer treatments. A review of earlier work in this area is given by Mariani and Marubini [32]. Many of the issues and methods discussed, however, have wider relevance. Problems in the conduct of phase II drug trials are similar to those found in other small clinical trials and proof-of-concept studies, in which decisions are to be made on the basis of evidence from a small number of experimental subjects [18]. The approaches described in these chapters are thus also applicable in these areas.

The chapter starts with a discussion of the place of the phase II trial in the clinical development program and a description of some of the types of phase II trial that are commonly conducted.

Statistical Advances in the Biomedical Sciences, edited by Atanu Biswas, Sujay Datta,
Jason P. Fine, and Mark R. Segal
Copyright © 2008 John Wiley & Sons, Inc.

2.1.2   The Role of Phase II Clinical Trials in Clinical Evaluation of a Novel Therapeutic Agent

Following preclinical development of a new therapeutic agent the first clinical trials, that is, trials involving human subjects, are the phase I trials. As the experimental therapy under investigation has not previously been used in humans, the primary focus of these trials is the assessment of tolerability and safety of the therapy. Phase I clinical trials are usually conducted using healthy volunteers. The subjects are treated using a dose escalation design, so that the first subjects are exposed to lower doses, and all subjects are closely monitored for adverse reactions. Typically 10 or 20 subjects are enrolled in the study. In oncology and other areas in which treatments may be associated with severe side effects, healthy volunteer subjects are not used. In this case the subjects are typically patients for whom other treatments have failed. These patients may have advanced disease, so in many cases successful treatment with the experimental therapy is not anticipated, and the main objective of the trial is again the investigation of tolerability. Whether based on healthy volunteers or patients, the aim of the phase I study is to determine (1) whether the therapy can safely be used and (2) the dose or range of doses that can be tolerated without leading to an unacceptably high level of side effects, and so are suitable for use in further clinical studies.

At the other end of the clinical assessment program from phase I clinical trials are the phase III clinical trials. These trials aim to provide definitive evidence of treatment efficacy and are primarily intended to support a licence submission to regulatory authorities. A large sample size is usually required, with some phase III clinical trials collecting data from several thousands of patients. The trial will include a control treatment, which in some cases may be a placebo control, and patients will be randomized to receive either the experimental or control treatment. In order to minimize bias, randomization will usually be *double-blind*, so that neither the patient nor the administering clinician or other clinical staff know which treatment is being given to which patients. The setting of the trial is chosen to match actual clinical practice as closely as possible, so that a typical patient population will be chosen, often with the trial conducted in a number of clinical centers. Usually two phase III clinical trials will be conducted, often with one conducted in North America and one in Europe. The focus of the phase III trial is the assessment of efficacy, with the aim of providing conclusive evidence via a hypothesis test of treatment effectiveness. Safety data and often pharmacoeconomic data are also collected, however, to allow a full picture of the treatment to emerge.

The phase I study is followed by one or more phase II studies. These studies may include the first testing in patients, and are thus the first studies in which evaluation of treatment efficacy can be made. The range of studies are described in more detail below, and in some cases a number of phase II studies of the same therapy may be conducted in different patient populations and with different designs and objectives.

The large sample size needed to provide evidence of efficacy and safety in a phase III clinical trial means that such trials are time-consuming and expensive. The phase II studies are typically much smaller-scale, with 50–100 subjects, and so can be conducted more quickly and cheaply. The purpose of these studies, therefore, is to allow a relatively inexpensive evaluation of treatment efficacy. Although this does not provide the definitive evidence obtained from the larger phase III clinical trial, it does provide an indication of whether the allocation of resources for further phase III evaluation is likely to be justified. The aim of the phase II clinical trial can thus be seen as enabling a decision to be made regarding future evaluation of the therapy. This is in contrast to the phase III trial, in which the main focus of statistical analysis is a hypothesis test or estimation.

In order to provide a decision regarding further evaluation to be made in a timely fashion, the primary endpoint in the phase II trial will be a rapidly observable response. This may be a surrogate for long-term endpoints that will be assessed in subsequent phase III trials. The endpoint used will very often be binary with a yes/no response, and it for this sort of endpoint that the majority of statistical work on phase II trials has been conducted. The rapidly observable endpoint in phase II enables the use of *sequential methods* in which the accumulating data on treatment efficacy are monitored through the course of the trial. This monitoring will be in addition to the monitoring of safety data, which remains important given the small number of patients who have previously been treated with the experimental therapy.

2.1.3 Phase II Clinical Trial Designs

There is a considerable range in the types of phase II clinical trials that are conducted. In the simplest trial design, all patients receive the same treatment, with no control group and no variation in dose, formulation, or treatment regimen. This type of design is common in phase II clinical trials in oncology, where, for ethical reasons, a placebo control cannot be used, and patients may already have received unsuccessful treatments with standard therapies. In the remainder of this chapter, we will refer to this type of study as a *single-treatment pilot study*. It is this type of study that has received the most attention from statistical researchers. This is probably for two reasons: (1) the simplicity of this approach means that development and evaluation of new methods is most straightforward, and (2) much of the statistical work has been conducted in cancer centers where such approaches are most common. While the demand for a concurrent control group in phase III can easily be justified in order to reduce bias in treatment comparisons, the need for control groups in phase II clinical trials is less clear. Given the small sample size, it may be more appropriate to collect as much data as possible on the experimental treatment and comparing these to historical control data that are often available. Even in cases where use of a control group is not proscribed on ethical grounds, therefore, a single-arm phase II trial might be considered preferable.

An increase in the complexity of the phase II clinical trial can come from the addition of a concurrent control group in a randomized trial. Such designs are common in proof-of-concept studies, in which an initial assessment of efficacy relative to a control treatment is sought, and are also used in small clinical trials in areas other than drug development. Such trials resemble a small-scale phase III clinical trial, and many of the statistical methods proposed for this type of design are similar to those used in a phase III trial setting. Studies of this type will be termed *comparative studies* below.

In both single-treatment pilot studies and comparative studies, as there is a single experimental treatment included in the trial, the decision to be made at the end of the trial is whether to continue with further development of the therapy in phase III. In a sequential trial, at each interim analysis a decision must also be made as to whether the trial should continue.

A further increase in complexity arises when more than one dose, formulation, or experimental treatment is included in the trial, either with or without a control treatment. In this case, in addition to deciding whether evaluation should proceed with a phase III clinical trial, a decision of which dose or treatment should be used must also be made. If interim analyses are conducted, ineffective doses or treatments may also be dropped through the course of the trial.

In addition to the variation in phase II clinical designs, there are a number of different statistical approaches that have been used in the development of approaches to the design and analysis of phase II trials. In particular, researchers have developed methods based on the use of frequentist, Bayesian, and decision-theoretic paradigms. Approaches based on these three viewpoints are described in the next three sections of this chapter. In each case, the section starts with a short review of the statistical methodology used. As stated above, the outcome of the phase II clinical trial is a decision of whether to proceed to phase III testing. This means that the phase II trial designs may be seen as providing a decision rule defining which observed datasets will lead to continuing to phase III and which to abandoning development of the experimental therapy. For the Bayesian and decision-theoretic designs the decision rule may be explicitly stated. For the frequentist designs the decision will be assumed to be based on the result of the hypothesis test. This decisionmaking role means that this chapter focuses on phase II trial *designs*, that is, on the decision rule, rather than dealing specifically with the analysis of the data obtained in the trial.

2.2 FREQUENTIST METHODS IN PHASE II CLINICAL TRIALS

2.2.1 Review of Frequentist Methods and Their Applications in Phase II Clinical Trials

The *frequentist*, or *classical*, statistical approach, as described in considerably more detail elsewhere (see, e.g., Ref. 29), focuses on hypothesis testing and the control of error rates. Inference centers on some parameter, which in the phase II setting is chosen to summarize the efficacy of the experimental treatment. We will denote this parameter by θ, with larger values of θ assumed to correspond to improved efficacy of the experimental treatment, and suppose that we wish to test some null hypothesis H_0 that θ is equal to some specified null value θ_0. In a comparative study, θ is usually taken to correspond to the difference in success rates on the experimental and control treatments and θ_0 is taken to be zero so that H_0 corresponds to the success rates being equal. In a single-treatment study, θ_0 is chosen so that H_0 corresponds to the success rate being equal to some specified value judged to be barely acceptable. Rejection of the null hypothesis in the direction of improved efficacy for the experimental treatment will lead to the conclusion that the treatment is superior to the control or to the specified value, and so indicate that further evaluation in phase III is warranted. If the parameter θ is truly equal to θ_0, the rejection of H_0 is called a *type I error*. Also of concern is the *type II error*. This corresponds to failure to reject H_0 when in fact θ is equal to some *alternative* value θ_1. The value θ_1 is chosen to correspond to some clinically relevant improvement in efficacy over the control treatment or target value. The randomness of the data from the phase II trial means that it is impossible to avoid both type I and type II errors. Well-known statistical tests have been constructed so as to control the probability of type I errors to some specified level, generally denoted by α. The probability of type II errors then depends on the sample size, and is usually controlled to be at most some level denoted by β, with $1 - \beta$ called the *power* of the test.

Frequentist methods have received near-universal use in the analysis of phase III clinical trials, where the need for definitive proof of efficacy is well-matched with the form of the hypothesis test, and the objectivity of the frequentist method in contrast to the Bayesian methods described below are considered appealing. In the phase III setting, it is common to set the type I error rate to be 0.05. A power value of 0.9 is common, and values such as 0.8 or 0.95 are also used. As discussed above, the sample size required for phase III clinical trials is often large. This is required to attain these error rates for values of θ_1 considered clinically relevant. In a phase II setting, the need for a smaller sample size than usually required for phase III clinical trials means that some compromise must be made on either α, β, or both. In a phase III setting if the calculated sample size is prohibitive it is most common to increase the type II error rate β, that is, to reduce the power. The justification for this is that a type I error corresponds to indication that a new treatment is superior to the control when it is actually no better. As drug registration relies on the results of this test, a type I error thus represents a risk to future patients. The risk of such errors must thus be limited to an acceptable level. The type II error, in contrast, can be viewed as a risk mainly to the trial sponsor, so that an increase in the rate of this type of error is more acceptable. In the phase II setting the position is rather different. Errors of type I now correspond to an ineffective treatment continuing to phase III. While this may lead to a waste of resources, the fact that the phase III testing is likely to correct the type I error means that it does not represent a consumer risk. The risk associated with a type II error might be considered greater, since this corresponds to erroneously abandoning development of an experimental therapy that is actually effective. In this case there will be no further testing to reverse the incorrect conclusion of the phase II trial. This contrast to phase III clinical trials was pointed out by Schoenfeld [34]. He concluded that in the phase II setting the control of power was more important than the control of type I error rate, and suggested that, in order to reduce the sample size to an acceptable level for a phase II clinical trial, a value of α as large as 0.25 could be considered.

It is described above how the use of rapidly observable endpoints in phase II clinical trials means that sequential monitoring is possible, and how the small number of previously treated patients makes sequential monitoring ethically desirable. In a frequentist setting, the inclusion of a number of interim analyses in a clinical trial impacts on the error rates. Suppose that at each of a series of interim analyses, a hypothesis test of H_0 is conducted with the type I error rate for that individual test controlled to be α. If, at any one of these analyses, H_0 is rejected, the trial will stop and be judged to have yielded definitive evidence of a treatment effect. The chance to reject H_0 at any one of the interim analyses means that the overall type I is increased above the planned level α. To maintain the overall type I error rate at a level of α requires adjustment of the individual hypothesis tests. Methods for this adjustment have been developed following the seminal work on sequential analysis by Wald [57] and Barnard [2]. Overviews are given by Whitehead [61] and Jennison and Turnbull [23]. On the whole, the methods rely on assumptions of normality, and so are most suitable for use with large samples. In the setting of phase II clinical trials, in which large sample approximations may hold poorly, the sequential analysis methods must be modified.

2.2.2 Frequentist Methods for Single-Treatment Pilot Studies

As described above, in the single-treatment pilot study, all patients receive the same treatment. Assuming a binary (success/fail) response, the data can thus be summarized by the number of successes, which will be denoted by S with observed value s, and the number of patients n. Inference will focus on the unknown success rate, which will be denoted by p. If responses from individual patients are considered to be independent and identically distributed, S follows a binomial distribution, $S \sim \text{Bin}(n, p)$. As the sample size in phase II is relatively small, it is generally feasible to work with this binomial distribution directly in development of statistical methods rather than using large-sample approximations as might be more common in a phase III clinical trial setting.

If the phase II clinical trial is conducted in a single stage, that is, without any interim analyses, the decision rule at the end of the trial will be to continue clinical development in phase III if S is sufficiently large, that is, if $S \geq u$ for some u. In the frequentist setting, the decision will be based on a test of the null hypothesis $H_0 : p = p_0$, where p_0 is chosen to be some value of p that is of minimal clinical importance, such that if $p = p_0$, it would be desirable to abandon development of the new therapy. The alternative value, p_1, is chosen to be a value of p of clinical significance, so that if $p = p_1$, it would be desirable to continue with development of the therapy. The type I error rate and power of the hypothesis test are then given respectively by

$$\Pr(S \geq u; p_0) = \sum_{i=u}^{n} \binom{n}{i} p_0^i (1 - p_0)^{n-i},$$

and

$$\Pr(S \geq u; p_1) = \sum_{i=u}^{n} \binom{n}{i} p_1^i (1 - p_1)^{n-i}.$$

Direct computation of these binomial probabilities allows values of n and u to be chosen to control the error rates at chosen levels α and β, where, as discussed above, α may take a value rather larger than conventionally used in a phase III clinical trial setting. The discrete nature of the binomial distribution, particularly when the sample size is small, means that the error rates may in some cases be smaller than α and β, giving a conservative test. The calculations required can be performed using exact single-sample settings on sample size software such as nQuery Advisor (ADALTA, www.CreActive.net).

Sequential, or multistage frequentist methods for single-arm studies extend the single-stage exact binomial method just described. A very well known approach is the two-stage design due to Simon [36]. Initially, the number of successes S_1 from some n_1 patients are observed. If S_1 is too small, say, $S_1 \leq l_1$, the trial will stop at this point and development of the therapy be abandoned. Otherwise, the trial will continue, with treatment of a further $n_2 - n_1$ patients, giving a total of n_2 in all. If, after continuation to include n_2 patients, the total number of successes S_2 is less than or equal to some l_2, development of the therapy will be abandoned. Otherwise, it will continue into phase III. The probability of continuing to phase III for a true success rate of p is given by

$$\Pr(\text{phase III}; p) = 1 - \Pr(\text{abandon at first stage}; p)$$
$$- \Pr(\text{continue at first stage and abandon at second}; p).$$

The probability of abandoning at the first stage is equal to

$$\sum_{i=0}^{l_1} \binom{n_1}{i} p^i (1 - p)^{n_1 - i},$$

while the probability of continuing at the first stage and abandoning at the second is equal to

$$\sum_{i=l_1+1}^{n_1} \Pr(S_1 = i \text{ and } S_2 - S_1 \leq l_2 - i; p),$$

which is equal to zero if $i > l_2$ as $S_2 - S_1$ cannot then be less than or equal to $l_2 - i$, since this is less than zero. Since $S_2 - S_1 \sim \text{Bin}(n_2 - n_2, p)$, with $S_2 - S_1$ independent of S_1, the probability of continuing at the first stage and abandoning at the second is equal to

$$\sum_{i=l_1+1}^{\min\{n_1, l_2\}} \sum_{j=0}^{l_2-i} \binom{n_1}{i} p^i (1 - p)^{n_1 - i} \binom{n_2 - n_1}{j} p^j (1 - p)^{n_2 - n_1 - j}.$$

This expression allows calculation of the probabilities under $p = p_0$ and $p = p_1$ of proceeding to phase III, that is, the type I error rate and the power of the sequential procedure considered as an hypothesis test. Calculation of the probability of stopping at the end of the first stage also enables the expected sample size to be found for p equal to p_0 or p_1. Simon proposes a numerical search be conducted to find all designs, that is, sets of values of n_1, n_2, l_1, and l_2, that have type I and type II error rates no greater than α and β, respectively, and among these to find the design that minimizes the expected sample size when $p = p_0$. The increased choice of sample sizes and critical values for two stages relative to a single stage means that the planned error rates are typically much more closely attained for a two-stage design. This, together with the chance to stop the trial early for poorly performing therapies, leads to an expected sample size that is smaller than the sample size for the eqivalent single-stage design.

The rationale for minimizing the expected sample size for $p = p_0$ is that it is undesirable to expose more patients than necessary to an ineffective treatment, but that if the treatment is effective, for example, if $p = p_1$, a larger sample size is acceptable. An alternative view is that it is desirable to reach a conclusion in phase III as quickly as possible for a range of true values of p, since if the therapy is effective, it is desirable to proceed to phase III in as timely a fashion as possible. This has led to suggestions to minimize the expected sample size under $(p_0 + p_1)/2$, to minimize the average of the expected sample sizes under p_0 and p_1, or to minimize the maximum expected sample size [35], in each case subject to constraint of the error rates. A similar argument would suggest that an upper critical value u_1 could be set for the end of the first stage, with progression to phase III without continuation to the second stage of the phase II clinical trial if $S_1 \geq u_1$.

The approach of the Simon design has been modified by Lin and Shih [28] to allow the sample size for the second stage to depend on the results from the first. The approach has also been extended to include three stages by Chang et al. [11], Ensign et al. [17], and Chen [12], who suggest that the tables given by Ensign et al. are inaccurate.

If more than three stages are to be included in the phase II clinical trial, the search for sample sizes and critical values to satisfy the error rate constraints and minimize the sample size in the way proposed by Simon can become computationally prohibitive. An alternative approach more suitable for a larger number of interim analyses can be based on the sequential analysis methods more common in phase III clinical trials, such as the spending function method proposed by Lan and DeMets [27] and described by Jennison and Turnbull [23]. An exact single-sample binomial spending function method was proposed by Stallard and Todd [43], which, although developed for a large-sample test for small values of p_0 and p_1, could be used in a phase II clinical trial setting. Alternatively, single-sample sequential methods based on large-sample approximations have been proposed for use in phase II clinical trials by Fleming [20] and by Bellisant et al. [5].

2.2.3 Frequentist Methods for Comparative Studies

As discussed above, a comparative phase II clinical trial, in which a single experimental therapy is compared with a control treatment, is similar in design to a small-scale phase III clinical trial. This means that much of the frequentist statistical methodology for trials of this type reflects the "standard" methods that have been developed in that setting.

A common test for comparison of binary responses from two groups of patients is the χ^2 test. Although based on asymptotic theory, the χ^2 test has been shown to be accurate for relatively small sample sizes, so that this test can be used in the analysis of comparative phase II clinical trials as in phase III. For very small sample sizes, or for extreme observed success rates, an exact alternative such as Fisher's exact test can be used as an alternative. Details of the χ^2 test and Fisher's exact test are given elsewhere (see, e.g., Ref. 6), and both tests are widely implemented in statistical computer packages.

Sample size determination for a trial comparing two groups in terms of a binary response is generally based on normal approximations to the binomial distribution. Formulas are given by, for example, Machin et al. [31] and are available in most commercially available sample size software.

2.2.4   Frequentist Methods for Screening Studies

In a screening study several experimental treatments, $T_1, T_2, \ldots, T_k$, are compared, and possibly also compared with a concurrent control treatment, T_0. In a sequential trial, treatments may be dropped from the study at interim analyses. At the end of the trial, either all experimental treatments will be abandoned or one will be selected to continue into phase III. The definition of frequentist error rates is less straightforward than in the single-arm or comparative settings discussed above. If all treatments are identical to the control, or in an uncontrolled trial have success rate equal to some p_0 chosen to represent a value for which further testing is unacceptable, it is desirable to discontinue any treatment. Denoting by p_i the success rate for treatment T_i, we thus wish to control the probability Pr(select any treatment to continue to phase III $| p_1 = \cdots = p_k = p_0$), and can view this as being analogous to a type I error rate. Specification of a probability analogous to the power in a comparative trial is rather more difficult. A common formulation is to require a high probability of selecting some treatment, say, T_1, to proceed to phase III if the success rate for this treatment is superior to that for the control by some specified margin δ_0 and the success rate for all the other treatments is not superior to that for the control by more than some specified margin δ_1 less than δ_0. Controlling this probability is equivalent to controlling the probability Pr(select treatment T_1 to continue to phase III $| p_1 = p_0 + \delta_0, p_2 = \cdots = p_k = p_0 + \delta_1$). A single-stage procedure to control these error rates was proposed by Dunnett [16], while Thall et al. [51,52] proposed two-stage procedures in which all except the treatment seen to be most effective at the interim analysis are dropped from the trial.

2.3   BAYESIAN METHODS IN PHASE II CLINICAL TRIALS

2.3.1   Review of Bayesian Methods and Their Application in Phase II Clinical Trials

In the frequentist approach, the value of the parameter of interest, which was denoted above by θ, was fixed, for example, to be θ_0, and the resulting distribution of the random data was considered. Inference thus focuses on a comparison of the observed data with the distribution that would be expected if the null hypothesis were actually true. In contrast, under the *Bayesian* paradigm, the parameter θ is itself considered to be a random variable, and inference focuses on what can be said about the distribution of θ. Thus we might obtain the expected value of θ, or the probability that it exceeds some specified value such as θ_0.

Since θ is a random variable, it must have some distribution even before any data are observed, and this *prior distribution* must be specified in advance. The prior distribution of θ, the density function of which will be denoted by $h_0(\theta)$, may be updated by observation of data. If data x are observed, conditional on these data the distribution of θ is given by *Bayes' theorem* to have density equal to

$$h(\theta \mid x) = \frac{f(x;\theta)h_0(\theta)}{\int f(x;\theta)h_0(\theta)d\theta'}$$

where $f(x; \theta)$ is the likelihood (or probability for discrete data) of x given the value of θ, so that the the the numerator is the joint density of x and θ, and the integral in the denominator (which is replaced by a sum for discrete data) runs over the range of θ, giving the marginal density of x. This distribution is called the *posterior distribution* of θ given the observed data x.

Specification of the prior distribution for θ is an essential part of the Bayesian approach, and the choice of prior distribution influences the posterior distribution obtained and hence any inference that is based on the posterior distribution. The method can thus be considered as combining prior opinion with the observed data to update that opinion. This can be seen as either a disadvantage or an advantage of the Bayesian method depending on the viewpoint taken. Many researchers consider that the lack of objectivity makes the method inappropriate for use in the analysis of phase III clinical trials, when a definitive result is sought. In phase I and phase II clinical trials, or in phase III trial design, as opposed to analysis, however, incorporation of all possible information, including prior opinion, in the decisionmaking process arising from the analysis of the trial data may be considered more appropriate.

The calculations required to obtain the posterior distribution can be burdensome, and much of the more recent progress in Bayesian methods in general has been due to advances in this computation. In the phase II clinical trial setting, it is common to choose a form of prior distribution for which the computations can be completed using analytical methods. Assuming that we have binary data, with the number of successes observed in a single group following a binomial Bin(n, p) distribution, if the parameter p has a prior beta distribution, Beta(a, b), that is, with prior density

$$h_0(p) = \frac{p^{\alpha-1}(1-p)^{b-1}}{B(a, b)}$$

for some choice of $a, b > 0$, where $B(a, b)$ is the beta function, $\int_0^1 p^{a-1}(1-p)^{b-1}\, dp$, the posterior distribution after observation of s successes is proportional to $p^{a+s-1}(1-p)^{n-s+b-1}$, and so is also of the beta form. The use of a beta prior is thus mathematically convenient. The beta family includes a wide range of unimodal prior distributions on [0, 1], including flat, J-shaped and U-shaped densities depending on the values of a and b. A prior distribution, such as the beta prior for binomial data, that leads to a posterior distribution of the same form is called a *conjugate prior*.

In a single-arm or comparative phase II clinical trial setting, the decision of whether to proceed to phase III evaluation is based on comparison of the parameter p with either a target value p_0 in a single-arm trial or with the corresponding parameter for a control treatment, which we will also denote by p_0. From a Bayesian viewpoint, the distinction between comparison with a target p_0 or the success rate for a control treatment is a fairly minor one. In either case, p_0 may be given a prior distribution. The only difference is whether this is updated by data to give a posterior distribution. In either a single-arm or a comparative study, then, we may focus on the difference between p and p_0, and in particular on whether this difference exceeds some target value, δ. The posterior probability that $p - p_0$ exceeds δ is given by

$$\int_{p_0=0}^{1-\delta} \int_{p=p_0+\delta}^{1} h(p \mid s, n)h(p_0 \mid s_0, n_0)dp\, dp_0, \tag{2.1}$$

where s_0 denotes the number of successes out of n_0 observations on the control arm in a comparative study, or $h(p_0 \mid s_0, n_0)$ is taken to be $h_0(p_0)$, the prior density for p_0, in a single-arm study.

In Bayesian methods, inference is based on the posterior distribution, which depends on the data through the likelihood function. Since the likelihood is unaffected by any stopping rule, interim analyses can be conducted without any adjustment to the posterior distribution obtained, or to any inference drawn from the data. Data monitoring in a phase II clinical trial setting can thus be carried out by monitoring the posterior probability given by (2.1) at a number of interim analyses as the data accumulate through the course of the trial. A number of Bayesian methods that have been proposed for phase II studies are discussed in the remainder of this section. These are all based on monitoring the probabilities of the form given by (2.1).

Sequential monitoring of Bayesian posterior probabilities of a form similar to (2.1) for normally distributed data has also been proposed in the phase III clinical trial setting, for example, by Spiegelhalter et al. [39]. In fact, the underlying principle is considerably older, discussed, for example, by Novick and Grizzle [33] in 1965.

2.3.2   Bayesian Methods for Single-Treatment Pilot Studies, Comparative Studies, and Selection Screens

Thall and Simon [50] proposed a Bayesian method for monitoring single-arm phase II clinical trials. They assume a binary response, with the number of successes taken to follow a binomial distribution, and propose that posterior probabilities of the form given by (2.1) be calculated at a number of interim analyses through the course of the trial. They illustrate the method with an example in which the posterior probabilities are calculated after the response from every patient starting at the tenth patient and ending at the sixty-fifth, when the trial will be stopped. The first analysis is made after the tenth patient to prevent stopping on the basis of data from a very small number of patients. After each patient's response is observed, the probability given by (2.1) is calculated with $\delta = 0$ and $\delta = \delta_0$, a value chosen to correspond to some desirable advantage of the experimental therapy relative to the probability p_0 corresponding to some notional control treatment. If the probability for $\delta = 0$ is sufficiently large, say, larger than some critical value λ_U, it is judged that there is sufficient evidence to conclude that the experimental therapy is superior to the control treatment. In this case the trial is stopped and development of the therapy will continue in phase III. If the probability for $\delta = \delta_0$ is too small, say, less than or equal to some λ_L, it is judged that there is sufficient evidence to conclude that the experimental therapy is not superior to the control by the required margin δ_0. In this case the trial will be stopped and development of the experimental therapy will be abandoned. If neither of these criteria is met, the trial will continues and the next patient's response is observed.

The properties of the Thall–Simon design depend on the choice of λ_U and λ_L as well as on the prior distributions for p and p_0 (the latter is also the posterior distribution since no patients are randomized to receive the control treatment in this single-arm trial). They propose that the prior distribution for p should be a beta distribution, since, as discussed above, this is the conjugate form and so enables analytical calculation of the posterior distribution, and suggest that the sum $a + b$ should be small, say, with $2 \leq a + b \leq 10$ since this means that the prior distribution is relatively noninformative. A more informative beta prior distribution, that is, with larger $a + b$, is proposed for the control treatment success probability p_0, reflecting the relative abundance of information on the efficacy of a standard treatment. The values λ_U and λ_L determine how strong the evidence of efficacy or lack of efficacy for the experimental treatment must be before the phase II trial is stopped. Thall and Simon suggest that the values of λ_U and λ_L should be chosen by considering the frequentist properties of the decision-making procedure for a range of values of p considered to be fixed.

Since the probability (2.1) can be calculated in the single-arm setting using the prior distribution for p_0 and in the comparative setting using the posterior distribution updated with the observed data from the control arm, the method proposed by Thall and Simon [50] can be used in both single-arm and comparative phase II clinical trials. In comparative trials, the posterior probabilities that the success rate for the experimental treatment either exceeds that for the control treatment or exceeds it by a margin δ_0 are monitored in exactly the same way as in the single-arm approach.

Thall and Estey [49] have proposed an approach similar to that of Thall and Simon [50] for monitoring phase II selection screens. In this approach, patients are randomized between a number of experimental treatments, $T_1, T_2, \ldots, T_k$ without a concurrent control treatment. After every response the posterior probability, given all the observed data, that the success rate for each treatment exceeds some target p_0, which in this case is considered fixed, is calculated. Since the data from each treatment are independent, if the success rates for individual treatments have independent priors, the posteriors are independent. For treatment T_i, with success rate denoted by p_i, it is thus sufficient to consider $\Pr(p_i > p_0 \mid s_{im}, m)$ where s_{im} denotes the number of successes from the first m patients receiving treatment i. It can be seen that this probability is of the same form as (2.1) except that p_0 is now considered fixed rather than given some prior distribution. Treatment T_i will be dropped from the trial unless $\Pr(p_i > p_0 \mid s_{im}, m)$ exceeds some specified critical value. At the end of the trial, after some predetermined number of patients have been treated, the best remaining treatment will proceed to phase III provided this treatment has $\Pr(p_i > p_0 \mid s_{im}, m)$ sufficiently large.

2.4 DECISION-THEORETIC METHODS IN PHASE II CLINICAL TRIALS

A number of biostatistical researchers, following from initial work in the 1960s [1,13], have suggested the use of Bayesian decision-theoretic methods in the design and analysis of clinical trials. The aim of such an approach is to model the decision-making process, leading to a decision that is optimal in that it maximizes the value of some specified *utility function* that expresses the preferences of the decisionmaker. The utility function is a function of the unknown parameter of interest, in this case the success rate for the experimental therapy p, and so has unknown value. Following observation of data, a posterior distribution for the parameter can be obtained, and this can be used to calculate a posterior expected utility associated with each possible action that can be taken. A rational decision-maker whose preferences are accurately modeled by the utility function would then choose the action with the largest posterior expected utility.

Early work on the integration of decision-theoretic methods into clinical trials [1,13] generally focused implicitly on phase III clinical trials, assuming that following a successful trial, the new therapy would be immediately available for general use. Later work focused on phase II, where, as discussed above, the outcome of the trial is a decision of whether to conduct further trials with the new therapy. Since this is an action that is generally within the control of the trial sponsor, the approach seems more appropriate in this setting. Decision-theoretic methods for single-arm phase II clinical trials have been proposed by Sylvester and Staquet [46], Sylvester [45] (see also the correction to this paper [21]), and Brunier and Whitehead [8]. These authors based their utility function on the number of extra successes associated with development of the experimental treatment relative to continuation with treatment with the standard. A similar approach has been taken by Carlin et al. [10] for comparative studies.

Whitehead [59] proposed a method for the design of selection screens based on optimiz-ation of the expected success rate for the selected treatment. A similar method [60] considers the phase II trial specifically in the context of a drug development program and aims to design the phase II selection screen so as to maximise the probability of final success in subsequent phase III evaluation.

An alternative view is to attempt to construct a utility function reflecting the financial gains and losses associated with the drug development process. This has been attempted by Stallard [40,41] and Leung and Wang [30] in the single-arm trial setting.

2.5 ANALYSIS OF MULTIPLE ENDPOINTS IN PHASE II CLINICAL TRIALS

As described in the introduction to this chapter, although the primary endpoint in a phase II trial is some measure of treatment efficacy, the fact that the therapy under investigation may have been previously administered to a relatively small number of patients means that analysis of safety data is also important. The monitoring of safety data in the trial is particularly import-ant in oncology, where new treatments are often associated with severe, at times fatal, toxicity. In this indication, a reduction in the rate of dose-limiting toxicity may be considered almost as important as an improvement in efficacy.

Recognizing the importance of sequential monitoring of safety data in the phase II trial, some biostatistical researchers have developed methods that formally combine the safety and efficacy endpoints in a single analysis. This means that at interim analyses development of a new treatment may be terminated because of toxicity problems, absence of evidence, or a com-bination of the two.

Frequentist methods that allow monitoring of toxicity and efficacy were proposed by Bryant and Day [9] and Conaway and Petroni [14]. They consider monitoring the number of patients who *respond*, that is for whom the treatment is effective, and the number of patients experien-cing toxicity. Their decision rules are of the form that lead to development of the therapy being abandoned at interim analyses or at the final analysis if either the number of successes is too low or the number of toxicities is too high. Extending the frequentist approach for a single end-point described above, they assume that the number of responses and the number of patients demonstrating toxicity both follow binomial distributions with probability parameters that are denoted by p_r and p_t, respectively. Their aim is to construct a two-stage test in a similar fashion to Simon [36]. The critical values for numbers of responses and toxicities are chosen so as to control the type I error rate corresponding to the probability of proceeding to phase III when p_r and p_t take some "unacceptable" values p_{r0} and p_{t0}, and the type II error rate corresponding to the probability of not proceeding to phase III when p_r and p_t take some "promising" values p_{r1} and p_{t1}. A difficulty arises since the numbers of successes and toxicities are not necessarily independent. If p_{00}, p_{01}, p_{10}, and p_{11} denote respectively probabil-ities of neither response nor toxicity, toxicity without response, response without toxicity, and both response and toxicity, the lack of independence can be specified by the odds ratio $p_{00}p_{11}/p_{01}p_{10}$. As the probability of proceeding to phase III depends on this odds ratio as well as the marginal probabilities p_r and p_t, it must be considered when determining critical values for the decision rule. Conaway and Petroni [14] propose specifying the value of the odds ratio in advance. Bryant and Day [9] suggest ensuring that the error rates are controlled for all values of the odds ratio, and also consider control of the probability of proceeding to phase III over all sets of (p_r, p_t) values with $p_r \leq p_{r0}$ and $p_t \geq p_{t0}$ rather than just at (p_{r0}, p_{t0}).

A criticism of the methods of Conaway and Petroni [14] and Bryant and Day [9] is that in their division of the set of (p_r, p_t) values into those that are "acceptable" and those that are not, they consider only p_r and p_t individually. In reality a small deterioration in safety might be acceptable if it is accompanied by a large improvement in efficacy, and vice versa. This has led to the development of frequentist methods that base control of error rates on more complex null regions [15,47,48].

Bayesian methods for monitoring multiple endpoints such as safety and efficacy have also been proposed [53,54]. In these methods a decision of whether to abandon development is based on the posterior probabilities $\Pr(p_r > p_{r0} + \delta \,|\, x)$ and $\Pr(p_t > p_{t0} + \delta \,|\, x)$ where now p_r, p_t, p_{r0}, and p_{t0} are taken to be random variables with specified prior distributions. An alternative approach is taken by Thall et al. [55], who use a utility function combining the probability of success and toxicity into a single measure of treatment performance, in this case in a screening trial. Other authors have also considered combining endpoints in a single utility function, and obtained designs to optimize this using a decision-theoretic method in a single-arm trial [42] or screening trial [19].

2.6 OUTSTANDING ISSUES IN PHASE II CLINICAL TRIALS

The ultimate aim of the phase II clinical trial is to provide reliable information on the safety and effectiveness of the new therapy in as safe, timely, and cost-effective a manner as possible. This information can then be utilized to aid in the decision of whether further clinical development of the new therapy is justified. It is this aim that provides the motivation for the biostatistical work on the design of phase II trials that has been described above. The continuing desire to make phase II trials more informative, safer, or quicker through improved design means that statistical methods in the area are not static. In this final section of the chapter a number of areas of focus for current research are briefly described.

As described above, a large part of the work on phase II clinical trials has been directed toward trials in cancer. This has led to methods in which a single treatment group is evaluated in terms of a rapidly observable binary outcome such as response to treatment. Traditional *cytotoxic* anticancer drugs aim to kill the tumor cells. Unfortuately they may also be highly toxic to other healthy cells, leading to severe side effects. This means that monitoring of safety data is very important. More recent advances in the treatment of cancer have led to the increased development of *cytostatic* drugs. In contrast to cytotoxic treatments, these are agents that do not directly kill cancer cells but act to limit the growth of tumors, for example, by restricting their blood supply. These drugs are typically much less toxic than conventional anticancer treatments. This means that different phase II trial designs may be more appropriate [26]. The primary measure of efficacy in a phase II trial may no longer be a binary success/fail outcome but some continuous measure such as tumor burden, and the formal monitoring of toxicity may be less important. Phase II trials for cytotoxic drugs might thus more closely resemble those for new therapies in other indications. As less attention has been focused outside oncology, this is an area where there is a need for further biostatistical work.

An area of considerable recent interest has been the combination of phases II and III into a single clinical trial. Such an approach would lead to an acceleration of the development program of a new therapy both by essentially including the phase II patients in the definitive phase III evaluation and by removing the hiatus necessary for administrative reasons between the end of a phase II clinical trial and the start of a phase III program. Methods have been proposed [3,44] to control the overall frequentist type I error rate of a clinical trial that acts as both a selection screen and a comparison between the selected treatment and a control. Much recent work in this area has been based on the *adaptive design* approach proposed by Bauer and Köhne [4], since this enables control of the overall type I error rate while allowing great flexibility in terms of design modification. A description of the application of this methodology to combining phases II and III and a review of more recent work in this area is given by Bretz et al. [7].

Adjusting for treatment selection is one of the problems that must be overcome to combine phases II and III. Another is the common change in primary endpoint between the two clinical development phases. Bayesian [22] and frequentist [56] methods to address this change of endpoint have also been proposed. In spite of the considerable recent work, there remain many statistical and practical challenges in this area, including the incorporation of multiple endpoint data and estimation of treatment effects at the end of the trial.

This chapter has followed the majority of the biostatistical literature in the area of phase II clinical trial design in dividing the designs according to whether they are constructed using frequentist, Bayesian, or decision-theoretic viewpoints. In practice, any phase II design provides a decision rule indicating whether development of the therapy under investigation should be abandoned at an interim or final analysis or continue into phase III. As discussed in detail by Wang et al. [58], the division into Bayesian or frequentist designs is artificial. It is possible to calculate frequentist error rates for all designs. Similarly, given prior distributions, it is possible to calculate posterior probabilities of the form (2.1) for a dataset at an interim or final analysis in any phase II trial. In some cases [40] it is also possible to determine a utility function with respect to which a given design is optimal. This suggests that however designs are obtained, a broader view of their properties should be taken than is generally currently the case. Such an approach would enable comparison of designs obtained using different statistical paradigms and ensure that the properties of any designs used are fully understood.

The need for distinct biostatistical methodology for phase II clinical trials arises from their unique position in the clinical testing process for an experimental therapy. As discussed above, this means that the data generated by the phase II trial act to inform a decision that is generally within the power of the trial sponsor. Appropriate designs for phase II clinical trials should thus reflect a full knowledge of other parts of the clinical testing process. This is the aim of decision-theoretic methods such as that proposed by Whitehead [60]. Extending this view further suggests that rather than focusing on the phase II trial alone, optimal design approaches should be used across the whole clinical development program for a product or even across the whole portfolio of drugs under development by a large pharmaceutical company. While such an approach has been discussed by a number of researchers (see, e.g., Refs. 37, 38 and 24), the enormous complexity of this problem means that a great deal of work remains to be done.

REFERENCES

1. Anscombe, F. J., Sequential medical trials, *J. Am. Statist. Assoc.* **58**, 365–383 (1963).

2. Barnard, G. A., Sequential tests in industrial statistics, *J. Roy. Statist. Soc.* **8**(Suppl.) 1–26 (1946).

3. Bauer, P. and Kieser, M., Combining different phases in the development of medical treatments within a single trial, *Statist. Med.* **18**, 1833–1848 (1999).

4. Bauer, P. and Köhne, K., Evaluation of experiments with adaptive interim analyses, *Biometrics* **50**, 1029–1041 (1994).

5. Bellisant, E., Benichou, J., and Chastang, C., Application of the triangular test to phase II cancer clinical trials, *Statist. Med.* **9**, 907–917 (1990).

6. Bland, M., *An Introduction to Medical Statistics*. Oxford University Press, Oxford, 2000.

7. Bretz, F., Schmidli, H., König, F., Racine, A., and Maurer, W., Confirmatory seamless phase II/III clinical trials with hypothesis selection at interim: General concepts, *Biometr. J.* (in press).

8. Brunier, H. C. and Whitehead, J., Sample sizes for phase II clinical trials derived from Bayesian decision theory, *Statist. Med.* **13**, 2493–2502 (1994).

9. Bryant, J. and Day, R., Incorporating toxicity considerations into the design of two-stage phase II clinical trials, *Biometrics* **51**, 1372–1383 (1995).

10. Carlin, B. P., Kadane, J. B., and Gelfand, A. E., Approaches for optimal sequential decision analysis in clinical trials, *Biometrics* **54**, 964–975 (1998).

11. Chang, M. N., Therneau, T. M., Weiand, H. S., and Cha, S. S., Designs for group sequential phase II clinical trials, *Biometrics* **51**, 1372–1383 (1987).

12. Chen, T. T., Optimal three-stage designs for phase II cancer clinical trials, *Statist. Med.* **16**, 2701–2711 (1997).

13. Colton, T., A model for selecting one of two medical treatments, *J. Am. Statist. Assoc.* **58**, 388–400 (1963).

14. Conaway, M. R. and Petroni, G. R., Bivariate sequential designs for phase II trials, *Biometrics* **51**, 656–664 (1995).

15. Conaway, M. R. and Petroni, G. R., Designs for phase II trials allowing for a trade-off between response and toxicity, *Biometrics* **52**, 1375–1386 (1996).

16. Dunnett, C. W., Selection of the best treatment in comparison to a control with an application to a medical trial, in *Design of Experiments: Ranking and Selection*, Santer, T. J. and Tamhane, A. C., eds., Marcel Dekker, New York, 1984.

17. Ensign, L. G., Gehan, E. A., Kamen, D. S., and Thall, P. F., An optimal three-stage design for phase II clinical trials, *Statist. Med.* **13**, 1727–1736 (1994).

18. Evans, C. H. and Ildstad, S. T., eds., *Small Clinical Trials: Issues and Challenges*, National Academy Press, Washington, DC, 2001.

19. Fan, S. H. and Wang, Y.-G., Decision-theoretic designs for dose-finding clinical trials with multiple outcomes, *Statist. Med.* (in press).

20. Fleming, T. R., One-sample multiple testing procedure for phase II clinical trials, *Biometrics* **38**, 143–151 (1982).

21. Hilden, J., Reader reaction: Corrected loss calculation for phase II trials, *Biometrics* **46**, 535–538 (1990).

22. Inoue, L. Y. T., Thall, P. F., and Berry, D. A., Seamlessly expanding a randomised phase II trial to phase III, *Biometrics* **58**, 823–831 (2002).

23. Jennison, C. and Turnbull, B. W., *Group Sequential Methods with Applications to Clinical Trials*, Chapman & Hall/CRC, Boca Raton, FL, 2000.

24. Julious, S. A. and Swank, D. J., Moving statistics beyond the individual clinical trial: Applying decision science to optimize a clinical development plan, *Pharm. Statist.* **4**, 37–46 (2005).

25. Kelly, P. J., Stallard, N., and Todd, S., An adaptive group sequential design for phase II/III clinical trials that select a single treatment from several, *J. Biopharm. Statist.* **15**, 641–658 (2005).

26. Korn, E. L., Arbuck, S. G., Pluda, J. M., Simon, R., Kaplan, R. S., and Christian, M. C., Clinical trial designs for cytostatic agents: Are new approaches needed?, *J. Clin. Oncol.* **19**, 265–272 (2001).

27. Lan, K. K. G. and DeMets, D. L., Discrete sequential boundaries for clinical trials, *Biometrika* **70**, 659–663 (1983).

28. Lin, Y. and Shih, W. J., Adaptive two-stage designs for single-arm phase IIA cancer clinical trial, *Biometrics* **60**, 482–490 (2004).

29. Lindgren, B. W., *Statistical Theory*, Macmillan, New York, 1968.

30. Leung, D. H.-Y. and Wang, Y.-G., Reader reaction: A Bayesian decision approach for sample size determination in phase II trials, *Biometrics* **57**, 309–312 (2006).

31. Machin, D., Campbell, M. J., Fayers, P. M., and Pinol, A. P. Y., *Sample Size Tables for Clinical Studies*, Blackwell Science Ltd., Oxford, 1997.

32. Mariani, L. and Marubini, E., Design and analysis of phase II cancer trials: A review of statistical methods and guidelines for medical researchers, *Int. Statist. Rev.* **64**, 61–88 (1996).

33. Novick, M. R. and Grizzle, J. E., A Bayesian approach to the analysis of data from a clinical trial, *J. Am. Statist. Assoc.* **60**, 81–96 (1965).

34. Schoenfeld, D., Statistical considerations for pilot studies, *Int. J. Radiat. Oncol. Bio. Phys.* **6**, 371–374 (1980).

35. Shuster, J., Optimal two-stage designs for single arm phase II cancer trials, *J. Biopharm. Statist.* **12**, 39–51 (2002).

36. Simon, R., Optimal two-stage designs for phase II clinical trials, *Control. Clin. Trials* **10**, 1–10 (1989).

37. Senn, S., Some statistical issues in project prioritization in the pharmaceutical industry, *Statist. Med.* **15**, 2689–2702 (1996).

38. Senn, S., Further statistical issues in project prioritization in the pharmaceutical industry, *Drug Inform. J.* **32**, 253–259 (1998).

39. Spiegelhalter, D., Freedman, L. S., and Parmar, M. K. B., Bayesian approaches to randomised trials, *J. Roy. Statist. Soc. Ser. A* **157**, 357–416 (1994).

40. Stallard, N., Sample size determination for phase II clinical trials based on Bayesian decision theory, *Biometrics* **54**, 279–294 (1998).

41. Stallard, N., Decision-theoretic designs for phase II clinical trials allowing for competing studies, *Biometrics* **59**, 402–409 (2003).

42. Stallard, N., Thall, P. F., and Whitehead, J., Decision theoretic designs for phase II clinical trials with multiple outcomes, *Biometrics* **55**, 971–977 (1999).

43. Stallard, N. and Todd, S., Exact sequential tests for single samples of discrete responses using spending functions, *Statist. Med.* **19**, 3051–3064 (2000).

44. Stallard, N. and Todd, S., Sequential designs for phase III clinical trials incorporating treatment selection, *Statist. Med.* **22**, 689–703 (2003).

45. Sylvester, R. J., A Bayesian approach to the design of phase II clinical trials, *Biometrics* **44**, 823–836 (1988).

46. Sylvester, R. J. and Staquet, M. L., Design of phase II clinical trials in cancer using decision theory, *Cancer Treat. Rep.* **64**, 519–524 (1980).

47. Thall, P. F. and Cheng, S.-C., Treatment comparisons based on two-dimensional safety and efficacy alternatives in oncology trials, *Biometrics* **55**, 746–753 (1999).

48. Thall, P. F. and Cheng, S.-C., Optimal two-stage designs for clinical trials based on safety and efficacy, *Statist. Med.* **20**, 1023–1032 (2001).

49. Thall, P. F. and Estey, E. H., A Bayesian strategy for screening cancer treatments prior to phase II clinical evaluation, *Statist. Med.* **12**, 1197–1211 (1993).

50. Thall, P. F., Simon, R., Practical Bayesian guidelines for phase IIB clinical trials, *Biometrics* **50**, 337–349 (1994).

51. Thall, P. F., Simon, R., and Ellenberg, S. S., Two-stage selection and testing designs for comparative clinical trials, *Biometrika* **75**, 303–310 (1988).

52. Thall, P. F., Simon, R., and Ellenberg, S. S., A two-stage design for choosing among several experimental treatments and a control in clinical trials, *Biometrics* **45**, 537–547 (1989).

53. Thall, P. F., Simon, R., and Estey, E. H., Bayesian sequential monitoring designs for single-arm clinical trials with multiple outcomes, *Statist. Med.* **14**, 357–379 (1995).

54. Thall, P. F. and Sung, H.-G., Some extensions and applications of a Bayesian strategy for multiple outcomes on clinical trials, *Statist. Med.* **17**, 1563–1580 (1998).

55. Thall, P. F., Sung, H.-G., and Estey, E. H., Selecting therapeutic strategies based on efficacy and death in multicourse clinical trials, *J. Am. Statist. Assoc.* **97**, 29–39 (2002).

56. Todd, S. and Stallard, N., A new clinical trial design combining phases II and III: sequential designs with treatment selection and a change of endpoint, *Drug Inform. J.* **39**, 109–118 (2005).

57. Wald, A., *Sequential Analysis*, Wiley, New York, 1947.

58. Wang, Y.-G., Leung, D. H.-Y., Li, M., and Tan, S.-B., Bayesian designs with frequentist and Bayesian error rate considerations, *Statist. Meth. Med. Res.* **14**, 445–456 (2005).

59. Whitehead, J., Designing phase II studies in the context of a programme of clinical research, *Biometrics* **41**, 373–383 (1985).

60. Whitehead, J., Sample sizes for phase II and phase III clinical trials, an integrated approach, *Statist. Med.* **5**, 459–464 (1986).

61. Whitehead, J., *The Design and Analysis of Sequential Clinical Trials*, Wiley, Chichester, 1997.

CHAPTER 3

Response-Adaptive Designs in Phase III Clinical Trials

Atanu Biswas
Indian Statistical Institute, Kolkata, India

Uttam Bandyopadhyay
University of Calcutta, Kolkata, India

Rahul Bhattacharya
Ashutosh College, Kolkata, India

3.1 INTRODUCTION

Patients arrive sequentially, for a phase III clinical trial, and are to be randomized to one of the existing treatments. The standard practice is to assign equal numbers of subjects to each treatment arm. But this type of assignment ignores the effectiveness of the treatments reflected through the available responses from both the treatments and hence results in subjecting more patients to inferior treatments than what is ethically desired. These drawbacks can be best illustrated by the results of a relatively recent clinical trial. For illustration, we consider the zidovudine trial reported by Connor et al. [30]. The trial aimed to evaluate the hypothesis that the antiviral zidovudine therapy (referred to by the trade name AZT) reduces the risk of maternal-to-infant HIV transmission. A standard randomization scheme was used to ensure equal numbers of patients in both AZT and placebo groups, resulting in 239 pregnant women receiving AZT and 238 receiving placebo. Here the endpoint was whether the newborn infant was HIV-infected. It was observed that 60 newborns were HIV-negative in the placebo group and 20 newborns were HIV-positive in the AZT group. These statistics revealed the harsh reality that 3 times as many infants on placebo were sentenced to death by the transmission of HIV while in the womb. It seems, therefore, logical to think that

Statistical Advances in the Biomedical Sciences, edited by Atanu Biswas, Sujay Datta,
Jason P. Fine, and Mark R. Segal
Copyright © 2008 John Wiley & Sons, Inc.

33

more allocation to AZT could save more newborn lives. This compels the experimenter to utilize the accrued data at each stage to set the assignment decision for the next stage.

Randomized clinical trials play a crucial role in experiments to determine which of the treatments shows superiority. Since any medical experiment involves human beings, there is an ethical imperative to provide the best possible medical care for the individual patient. This compels to develop an allocation procedure where ethical and logistical considerations must always drive the equation and the resulting mathematics. So whenever the accrued data reveal the superiority of a treatment arm, the randomization procedure should be biased in favor of this arm to ensure more allocation to this treatment. Allocation designs attempting to achieve this goal are called *response-adaptive designs* or simply *adaptive designs*. These designs were first formulated to identify the better treatment in the context of a two-treatment clinical trial. The preliminary ideas can be found in Thompson [65] and Robbins [53]. These works were followed by a flurry of activity, starting with studies by Anscombe [2] and Colton [29]. More history, including later developments, can be found in Rosenberger and Lachin [56], Rosenberger [55], and a book-length treatment by Rosenberger and Lachin [57].

The organization of this chapter is as follows. Section 3.2 describes several available adaptive designs for binary treatment responses. Note that most of the adaptive design literature is given in this direction. Section 3.3 describes designs for binary responses in the presence of covariates. Adaptive designs for categorical responses and continuous responses are discussed in Sections 3.4 and 3.5, respectively. Optimal adaptive designs are provided in Section 3.6, and delayed response in adaptive designs are described in Section 3.7. Biased coin designs are discussed in Section 3.8. The real adaptive clinical trials are outlined in Section 3.9 Section 3.10 illustrates and compares different designs using both real datasets and simulation. Section 3.11 ends the chapter with some concluding discussions.

3.2 ADAPTIVE DESIGNS FOR BINARY TREATMENT RESPONSES

3.2.1 Play-the-Winner Design

Adaptive designs perhaps started with the excellent work of Robbins [53] in the context of designing a sequential trial. But it is Marvin Zelen who made the first significant contribution in this direction by the pioneering concept of play-the-winner (PW) for binary response trial. To be specific, suppose that we have two treatments and patients enter the clinical trial sequentially, to be assigned to either treatment. The trial outcome is either a success or a failure, and the response depends solely on the treatment given. Then PW rule assigns the opposite treatment to the next patient if the previous patients' response was a failure and the same treatment if the previous patient was a treatment success. This rule is deterministic and hence carries with it the selection bias. A practical drawback of this rule is that no clear-cut explanation is given on how to proceed when patient's responses are not immediate. However, one can use the finally obtained response to determine the next patient's allocation.

3.2.2 Randomized Play-the-Winner Design

A natural question following PW is whether it is ethically justified to repeat the successful treatment blindly. The answer is in the negative, mainly because (1) the unpredictability of treatment assignment, a fundamental requirement in any clinical trial [57] is not ensured; and (2) the last successful treatment may have a lower success rate, so this response should not be given much importance to determine the next assignment. Wei and Durham [70] modified Zelen's [73] rule by using all the past information of allocation and responses in an appropriate manner. They named the modified alocation procedure as randomized-play-the-winner (RPW) rule. The allocation procedure starts with an urn with a fixed equal number of balls (say, α), for each of the two treatments. To randomize an incoming subject, a ball is drawn, the corresponding treatment assigned, and the ball replaced. An additional number (say, β) of balls of the same type are added to the urn if the patient's response is a success and the same number of balls of the opposite kind are added if the patients' response is a failure. This rule is referred to as RPW(α, β). The intuitive idea behind the procedure is to favor the treatment doing better and provide some ethical gain through it. Moreover, the allocation probability for each incoming patient depends on all the previous response-and-allocation history and thus the ethical drawback of PW is expected to be avoided. For RPW(0, 1), the limiting allocation proportion of patients treated by one of the treatments, say, the kth, can be found to be $(1/q_k)/(1/q_1 + 1/q_2)$, $k = 1, 2$, with qk as the failure rate of the kth treatment. Therefore, the limiting proportion is seen to be inversely proportional to the failure rate, indicating a lower number of allocations to the treatment with higher failure rate. The same expression can be found for the PW rule.

This procedure can be effectively applied for more than two treatments. Wei [69] provided a multitreatment version of the RPW rule (called MRPW), with explanation facilitated by an urn model. Here the urn starts with K types of balls, α balls of each type. An entering subject is treated by drawing a ball from the urn with replacement. If success occurs, an additional $(K - 1)\beta$ balls of the same type are added to the urn, whereas for a failure, β balls of each of the remaining types are added to the urn. Bandyopadhyay and Biswas [13] obtained the limiting allocation proportion to kth treatment for MRPW as

$$\frac{1}{K} + \frac{\sum_{j=1}^{K} p_j/K - \left(\sum_{j=1}^{K} p_j/q_j\right)/\sum_{j=1}^{K} 1/q_j}{Kq_k}$$

Ivanova et al. [47] also investigated the theoretical properties of the same rule through a simulation study.

3.2.3 Generalized Pólya's Urn (GPU)

Urn models have long been recognized as a valuable mathematical tool for assigning subjects in a clinical trial. Among various urn models, the Pólya urn (also known as the *Pólya–Eggenberger urn*) model is the most popular one. It was derived to tackle the problem of contagious diseases [35]. Athreya and Karlin [3] successfully embed this urn scheme into a continuous-time branching process to provide important limiting results. Wei [69] generalized the abovementioned urn model to develop a response adaptive randomization procedure known as the *generalized Pólya urn* (GPU) or *generalized Friedmans urn* (GFU) in the literature. This rule provides a nonparametric treatment assignment procedure for comparing $K \geq 2$ treatments in a clinical trial. A general description of the GPU model is as follows. An urn contains particles of K types representing K treatments. Patients arrive sequentially and are to be randomized to the treatments. At the outset, the urn contains a vector $\mathbf{Y}_0 = (Y_{01}, Y_{02}, \ldots, Y_{0K})$ of balls of type $1, 2, \ldots, K$. When an eligible subject arrives, a ball is selected from the urn and replaced. If it was of type i, the ith treatment is assigned and the response is observed. Depending on the response, a random number d_{ij} of additional balls are added to the urn of type $j = 1, 2, \ldots, K$. This procedure is repeated sequentially up to n stages. Let $N_k(n)$ be the number of times a type k ball being selected in the first n stages [i.e., in the context of a clinical trial $N_k(n)$] denotes the number of subjects assigned to treatment k among the first n subjects. Then first-order asymptotics of GFU are determined by the generating matrix $\mathbf{H} = (E(d_{ij}))$, i, $j = 1, \ldots, K$. Under certain regularity conditions, it is proved that with probability 1, $\lim_{n \to \infty} N_k(n)/n = v_j, j = 1,2, \ldots, K$, where $v = (v_1, v_2, \ldots, v_K)$ is the normalized left eigenvector of $\mathbf{H}$ with respect to its largest eigenvalue [3,6,7]. The random mechanism for adding balls at each draw is attractive. Allowing the number of balls to be added to depend on past history of responses and allocations, a variety of response adaptive procedures are developed from GPU. It is interesting to note that RPW(α, β) is a particular case of GPU for two treatments with generating matrix

$$H = \begin{pmatrix} \beta p_1 + \alpha q_1 & \alpha p_1 + \beta q_1 \\ \alpha p_2 + \beta q_2 & \beta p_2 + \alpha q_2 \end{pmatrix},$$

where $p_k (= 1 - q_k)$ is the success rate of the kth treatment, $k = 1, 2$.

Since the early 1990s, several generalizations to GPU (or GFU) have been made. Smythe [63] defined an extended Pólya urn, where the expected number of balls added at each stage is held fixed, that is, $E(d_{ij}) = d > 0$ and $d_{ij} \geq 0$ for $j \neq i$. He suggested not replacing the type i ball drawn and allowed removal of additional type i balls from the urn, of course satisfying the restriction that one cannot remove more balls of a certain type that are present in the urn. Relaxing these conditions, Durham and Yu [32] propose a rule (called "modified play-the-winner") that adds balls to the urn only if there is a success, but the urn remains unchanged if there is a failure.

The next major generalization of GFU is the introduction of a nonhomogeneous generating matrix, where the expected number of balls added to the urn changes across the draws. Bai and Hu [6] showed that under certain assumptions, the usual limiting results hold. The next generalization allows the number of balls added at a draw to depend on previous draws. Andersen et al. [1] introduced the idea for a K-treatment clinical trial where a success on treatment i results in the addition of a type i ball and a failure causes the addition of "fractional balls" of remaining types, proportionate to the urn composition at the previous stage. They did not investigate the theoretical properties although. Then, Bai et al. [9] considered a similar nonhomogeneous urn model, and explored these theoretical properties. According to their formulation, a success on treatment i results in the addition of a type i ball, whereas for a failure on the ith treatment results in adding balls of the remaining types, proportionally on the basis of their previous success rates.

3.2.4 Randomized Pólya Urn Design

Suppose that the performance of one of the $K(\geq 2)$ treatments is relatively poor. Therefore it seems unethical to add balls corresponding to this least favorable treatment as a result of another treatment's failure, as in RPW. Consequently Durham et al. [31] modified the RPW rule by introducing what they called a *randomized Pólya urn*. Naturally the allocation design is referred to as a *randomized Pólya urn* (RPU) design. This is a success-driven design; that is, it allows the urn composition to change depending on the success on different treatment arms. The allocation procedure starts with an urn containing balls of K types, representing K possibly unrelated treatments. When a subject arrives, a ball is drawn with replacement, its type is noted, and the subject is assigned to the represented treatment arm. If the response is a success, a ball of the same type is added to the urn, but for a failure the urn remains unchanged. Thus a success on a particular treatment rewards the treatment by adding balls of its corresponding color, while a failure on this treatment leaves the urn unaltered; thus, other treatments are not rewarded on the basis of a particular one's failure. This design can be embedded in the family of continuous-time–pure birth process with linear birth rate, and this embedding enables one to obtain limiting behavior of the urn with much ease. Durham et al. [31] proved that if $p^* = \max_{1 \leq k \leq K} p_k$ is unique, then, with probability 1

$$\lim_{n \to \infty} \frac{N_k(n)}{n} = 1 \quad \text{if} \quad p^* = p_k$$

$$= 0 \text{ otherwise.}$$

Therefore, the allocation procedure assigns the far-future patients to the treatment with highest success probability.

3.2.5 Birth-and-Death Urn Design

As a logical extension of RPU, Ivanova et al. [47] developed a birth-and-death urn design (BDU), which is the same as RPU except that whenever a failure occurs on treatment k, a type k ball is removed from the urn. The term "birth" therefore refers to the addition of a ball to the urn, and removal of a ball from the urn signifies "death." BDU is an improvement over RPW because, in case of a failure, it accounts not only for the number of balls corresponding to the opposite treatments but also for the number of balls corresponding to the treatment on which a failure just occurred. Detailed investigation of the distributional (both exact and asymptotic) properties can be found in Ivanova et al. [47].

3.2.6 Birth-and-Death Urn with Immigration Design

A problem with the BDU is that when a particular treatment is harmful, the type of balls corresponding to the treatment will eventually become extinct. This immediately led to the generalization of the BDU with immigration (BDUI after Ivanova et al. [47]), where a random mechanism is considered that adds balls to the urn at a constant rate. The rule can be described as follows. The urn starts with balls of K types representing K treatments and aK, $(a \geq 0)$, immigration balls. Assignment of an entering subject is made by drawing a ball with replacement from the urn. If it is an immigration ball, it is replaced and two additional balls, one ball of each type, are added to the urn and the next ball is drawn. The procedure is repeated until a ball other than the immigration ball is obtained. If a ball of a treatment type is obtained, the subject is given that treatment and an outcome is observed. If a success is observed, a ball of the selected type is returned to the urn, and for a failure, a ball of that type is removed. The procedure continues sequentially with the entrance of the next subject. Ivanova et al. [47] discussed the convergence properties of a BDUI.

3.2.7 Drop-the-Loser Urn Design

The latest addition in the family of BDUs with immigration is the drop-the-loser (DL) rule developed by Ivanova and Durham [44] and Ivanova [42]. The urn initially contains balls of $K + 1$ types, balls of types $1, 2, \ldots, K$ represent treatments, and balls of type 0 are called "immigration balls." As in BDUI, when a subject arrives, a ball is removed from the urn. If it is an immigration ball, it is replaced and K additional balls, one of each treatment type is added to the urn. The procedure is repeated until a ball representing a treatment is obtained. If a success is observed, then the ball is returned to the urn; if a failure is observed, then one ball of that type is withdrawn.

The RPU, the BDUI, and the DL rules can all be regarded as special cases of an adaptive urn scheme with immigration for ternary responses [45]. Given treatment i, let λ_i be the probability of adding a type i ball to the urn, let η_i be the probability that the urn composition remains unchanged, and let μ_i be the probability of removing a type i ball from the urn, where $\lambda_i + \eta_i + \mu_i = 1$ for $i = 1, 2, \ldots, K$. Whenever an immigration ball is selected, it is replaced and one ball of each type is added to the urn. The number of immigration balls is a. For all these designs, a type i ball is added to the urn following a success on treatment i and a type i ball is removed from the urn following a failure on treatment i. We also admit the possibility of no response. If $\lambda_i = P\{\text{success}|i\} = p_i$, $\eta_i = P\{\text{failure}|i\} = q_i$, $\mu_i = 0$, and $a = 0$, we have the RPU rule considered by Durham et al. [31]. For $\lambda_i = p_i$, $\eta_i = 0$, $\mu_i = q_i$, and $a > 0$, we have the BDUI rule with a common immigration process. When $\lambda_i = 0$, $\eta_i = p_i$, $\mu_i = q_i$, and $a > 0$, we have the DL rule developed by Ivanova and Durham [44]. Embedding these designs in a family of continuous-time birth-and-death processes with common immigration, Ivanova and Flournoy [45] studied various urn characteristics, both exact and asymptotic. Through a simulation study, these rules are compared in Ivanova and Rosenberger [46]. They have noted that BDUI has the least proportion of failures and the GPU has this proportion largest among all. However, these rules are always an improvement over the equal allocation.

Limiting proportion of subjects assigned to a particular treatment is of fundamental interest in any clinical trial. We provide the available proportions in Table 3.1 (where $p^* = \max_{1 \le j \le K} p_j$):

It is easily observed that these ratios are greater than $\frac{1}{K}$ if the kth treatment is the best. Moreover, the ratios vary from $\frac{1}{K}$ according to the degree of superiority of the kth treatment (except for RPU). For two treatments, the limiting allocation proportions for PW, RPW, and DL rules are the same. Precise rates of divergence of allocation proportion for BDU/BDUI when $p^* \ge 0.5$ can be found in Ivanova et al. [47].

3.2.8 Sequential Estimation-Adjusted Urn Design

Zhang et al. [74] proposed a multitreatment allocation scheme targeting a prespecified allocation proportion within the framework of an urn model [called the *sequential estimation-adjusted urn* (SEU) model]. Let Θ be the matrix of treatment parameters for the K treatments. Also let $\rho_j(\Theta)$ be the target allocation proportion for treatment j. Then they suggested adding $\rho_j(\hat{\Theta}_{n-1})$ particles of type j to the urn at stage n, $j = 1, 2, \ldots, K$ where

Table 3.1   Limiting Allocation Proportion to kth Treatment for Different Designs

Design	Limiting Proportion of Allocation
MRPW	$\frac{1}{K} + \dfrac{\sum_{j=1}^{K}(p_j/K) - \sum_{j=1}^{K}(p_j/q_j)/\sum_{j=1}^{K}(1/q_j)}{Kq_k}$
RPU	1 or 0 according to whether p^* is unique or not
BDU/BDUI	$\dfrac{1/p_k - q_k}{\sum_{i=1}^{K} 1/p_i - q_i}$ if $p^* \le 0.5$
DL	$\dfrac{1/q_k}{\sum_{i=1}^{K} 1/q_i}$

$\hat{\Theta}_n$ is the sample estimate of Θ after n stages. Then, under certain conditions, it is shown that almost surely

$$\frac{N_k(n)}{n} \to \frac{\rho_k(\Theta)}{\sum_{j=1}^{K} \rho_j(\Theta)}.$$

The importance of this model is that it (1) can be used to target any specified allocation proportion and (2) enjoys certain desired asymptotic properties under some widely satisfied conditions. For example, suppose that we want to achieve the same allocation proportion as in a two-treatment RPW rule. Then, the urn design will be as follows. At the $(n + 1)$st stage, regardless of what the response of the nth patient is, we add $\hat{q}_{n2}/(\hat{q}_{n1} + \hat{q}_{n2})$ particles of type 1 and $\hat{q}_{n1}/(\hat{q}_{n1} + \hat{q}_{n2})$ particles of the opposite kind to the urn, where $\hat{q}_{nk}$ is the estimate of q_k after n responses, $k = 1, 2$. Then, it is shown [74] that, almost surely, $N_1(n)/n \to q_1/(q_1 + q_2)$, and as $n \to \infty$, we obtain

$$\sqrt{n}\left(\tfrac{N_{1n}}{n} - \tfrac{q_1}{q_1+q_2}\right) \to \mathcal{N}(0, \sigma_s^2)$$

in distribution, where $\sigma_s^2 = q_1 q_2 (12 - 5q_1 - 5q_2)/(q1 + q2)^3$. We observe that σ_s^2 can be evaluated for any $0 \le q_1, q_2 \le 1$ but the corresponding expression is not straightforward for calculation of the RPW rule when $q_1 + q_2 \ge 0.5$ [49]. Moreover, further investigation revealed that σ_s^2 is much smaller than that provided by RPW rule when $q_1 + q_2 \le 0.5$.

3.2.9 Doubly Adaptive Biased Coin Design

All the rules discussed so far, except that of Zhang et al. [74], were developed with an aim to allocate more patients to the better treatment, and hence cannot target any prespecified allocation proportion. Eisele [36] and Eisele and Woodroofe [37] propose a more complicated allocation design for two treatments [called *doubly adaptive biased coin design* or DBCD] to target any desired allocation proportion ρ to treatment 1. They defined a function $g(x, y)$ from $[0, 1]^2$ to $[0,1]$ that bridges the current allocation proportion to the target allocation satisfying the following regularity conditions: (1) g is jointly continuous, (2) $g(x, x) = x$, (3) $g(x, y)$ is strictly decreasing in x and strictly increasing in y on $(0, 1)^2$, and (4) g has bounded derivatives in both arguments. The procedure then allocates patient $j + 1$ to treatment 1 with probability $g(N_{1j}/j, \hat{\rho}_j)$, where $\hat{\rho}_j$ is the estimated target allocation after the jth stage. However, the properties of the DBCD depend heavily on the choice of an appropriate allocation function g. Eisele and Woodroofe [37] gave a set of conditions that the allocation function g should satisfy. These conditions are very restrictive and are usually difficult to check. In fact, Melfi et al. [50] pointed out that the suggested choice of g by Eisele and Woodroofe [37] violated their own regularity conditions. Hu and Zhang [40] define the following family of allocation functions having nice interpretative properties

$$g^{(\alpha)}(0, \rho) = 1, \quad g^{(\alpha)}(1, \rho) = 0,$$

$$g^{(\alpha)}(x, \rho) = \frac{\rho(\rho/x)^\alpha}{\rho(\rho/x)^\alpha + (1 - \rho)(1 - \rho/1 - x)^\alpha},$$

where $\alpha \ge 0$. The parameter α controls the randomness of the procedure. Different choices of α produce different allocation procedures. For $\alpha = 0$, we have $g^{(\alpha)}(x, \rho) = \rho$, which leads to the sequential maximum-likelihood procedure (SMLE) [58], where at each stage ρ is estimated, preferably by the method of maximum likelihood, and the next incoming subject is assigned to treatment 1 with this probability. Properties of the SMLE procedure targeting two-treatment Neyman allocation are explored in Melfi et al. [50]. On the contrary, a large value of α provides an allocation design with smaller variance. Therefore, α should be chosen to reflect the tradeoff between the degree of randomization and the variation. Hu and Zhang [40] have shown that, under some favorable conditions, $\lim_{n\to\infty} N_1(n)/n = \rho$, with probability 1 where ρ depends on the success rates of the two treatments. A generalization of Eisele's procedure together with some related asymptotic results for $K \geq 2$ treatments can also be found in Hu and Zhang [40].

To indicate the importance of this family of allocation rules, we provide an example. For two treatments with success rates p_1 and p_2, the RPW rule maintains a limiting allocation proportion $(1 - p_2)/(2 - p_1 - p_2)$ to treatment 1. Now, one can use DBCD to target the same allocation proportion. Then $\rho(p_1, p_2) = (1 - p_2)/(2 - p_1 - p_2)$, and the design is as follows. At the first stage, n_0 patients are assigned to each treatment. After $m \geq (2n_0)$ patients are assigned, we let $\hat{p}_{mk}$ be the sample estimator of p_k, $k = 1, 2$. At the $(m + 1)$st stage, the $(m + 1)$st patient is given treatment 1 with probability $g(N_{1m}/m, \hat{p}_m)$ and to treatment 2 with the remaining probability, where $\hat{p}_m$ is the estimated value of ρ after m stages.

3.3   ADAPTIVE DESIGNS FOR BINARY TREATMENT RESPONSES INCORPORATING COVARIATES

3.3.1   Covariate-Adaptive Randomized Play-the-Winner Design

Response-adaptive procedures are considered to be valuable statistical tools in clinical trials. Even though the literature is vast in adaptive designs, the effort to incorporate covariate information still lacks maturity. The treatment allocation problem in the presence of covariate can be found in Begg and Iglewicz [18], where the optimum design theory is used to provide a deterministic allocation rule. Quite naturally, when patients are heterogeneous, their responses to treatment are influenced by the respective covariate information. For example, consider a single covariate, suitably categorized with $(G + 1)$ ordered grades $0, 1, \ldots, G$. Grade 0 is for the most favorable condition and grade G for the least favorable condition of a patient. Then it seems reasonable to favor the treatment with a success in the least favorable condition, and assign less subjects to the treatment with failure in the most favorable condition. Keeping all these aspects in mind, Bandyopadhyay and Biswas [12] developed an RPW-type urn design with covariate, called *adaptive RPW* (ARPW). They have considered a single nonstochastic covariate with $(G + 1)$ ordered grades $0, 1, \ldots, G$, ordered as earlier. The allocation procedure starts with an urn containing two types (say, 1 and 2) of treatment balls, α balls of each type. An entering subject with grade u, is treated by drawing a ball from the urn with replacement. If success occurs, an additional $(u + t)\beta$ balls of the same type and $(G - u)\beta$ balls of the opposite kind are added to the urn. On the other hand, when a failure occurs, an additional $u\beta$ balls of the same kind and $(t + G - u)\beta$ balls of the opposite kind are added to the urn. Clearly, the treatment with a success at $u = G$, the least favorable condition, is rewarded by addition of a higher number of balls of the same type, whereas a failure at $u = 0$, the most favorable condition, adds the smallest number of balls of the same type to the urn. This was the basic motivation of the design. Starting from a particular response model involving covariates, they [12] have set decision rules to identify the better treatment and established some related asymptotic results. A guideline is also provided to accommodate more than one covariate.

3.3.2 Treatment Effect Mappings

The next set of contributors in this field is Rosenberger et al. [60], who, using the approach of *treatment effect mappings*, developed a covariate adjusted adaptive allocation rule for binary responses. In their procedure, when the nth patient arrives to be randomized to the treatments, the current treatment effect difference (effect of treatment 1 minus effect of treatment 2), computed from patients who have responded thus far, is mapped to $P_n \in [0, 1]$. A random number $U_n \in [0, 1]$ is generated and the nth patient is assigned to treatment 1 or 2 according to whether $U_n \leq P_n$ or $U_n > P_n$. They used logistic regression model and argued in favor of allocating a patient to a treatment with probability proportional to estimated covariate adjusted odds ratio. Through a simulation study they observed that, for greater treatment effects, the procedure will have similar power to that of equal allocation with reduced rate of treatment failures.

3.3.3 Drop-the-Loser Design with Covariate

More recently Bandyopadhyay et al. [16] modified the DL rule of Ivanova [23] to consider the heterogeneity of the subjects. A categorized ordinal covariate with levels 0,1, is introduced with the abovementioned ordering. The urn setup is similar to that in DL rule except that the covariate information of any entering subject is reasonably used to determine the next patient's allocation. If a success occurs for patient j with covariate value 1(0), the ball is returned to the urn with probability 1(p). However, if a failure occurs with covariate value 0(1), the ball is replaced with probability 0($1 - p$). The same procedure is carried out for the next entering patient. This is referred to as *drop-the-loser with covariate* (DLC) design. Thus the allocation of an entering patient is skewed in toward the treatment with a success at the least favorable condition in the last assignment. Assuming the covariate to be random, generating functions of various urn characteristics and related asymptotic results are developed by embedding this urn scheme into a continuous-time Markov process.

Most of the urn designs discussed so far are birth processes, and accordingly the variability is too high. In fact, the standard deviations of the proportion of allocation for these designs are so high that an allocation that is less than one or two standard deviation(s) from the expectation leads less than 50% of patients treated by the better treatment, in case of a two-treatment experiment with binary outcomes. The more recently introduced DL rule is a death process, and consequently the variation is quite low as it is known that death processes have less variability than do birth processes. Hu and Rosenberger [39] observed that the DL rule has the least variability among the available adaptive designs for binary responses. Starting from a covariate-involved response model, it is shown that satisfying the ethical requirements, DLC is less variable than the original DL rule.

3.4 ADAPTIVE DESIGNS FOR CATEGORICAL RESPONSES

In several biomedical studies the responses include pain and postoperative conditions, which are often measured in an ordinal categorical scale such as *nil, mild, moderate*, or *severe*. In real situations, categories are clubbed together to apply the available allocation procedures. Yet, the adaptive designs using the categorical responses are more sensible than the designs with transformed binary responses in any case, as the former use complete categorical responses. In 2001 Bandyopadhyay and Biswas [13] provided an upgraded version of the RPW rule to incorporate the categorical responses. This is an urn design where possible responses are denoted by $0, 1, 2, \ldots, l$ and the urn starts with α balls of both types 1 and 2.

For a response $j(= 0, 1, \ldots, l)$ from treatment 1(2), an additional $j\beta$ balls of type 1(2) along with $(l - j)\beta$ balls of kind 2(1) are added to the urn. They have investigated the properties of the allocation design both numerically and theoretically. However, this is a birth process and hence suffers from high variability. To this end, Biswas et al. [26] proposed a categorical version of the DL rule. The treatment-assigning procedure is same as the original rule except that at the ith stage, the ball drawn is replaced with a probability $\pi_k(Z_i)$, $k = 1, 2$, where the subjects' response is Z_i when assigned to treatment k. Because of the ordinal nature of the responses, it requires $\pi_k(j)$ to be nondecreasing in $j = j/l$, the authors have explored the properties of such a design and observed the lower rate of variability than existing designs.

3.5   ADAPTIVE DESIGNS FOR CONTINUOUS RESPONSES

3.5.1   Nonparametric-Score-Based Allocation Designs

What we have discussed so far relies on the binary responses of the subjects. However, in many clinical trials, the primary outcome is the length of time from treatment to an event of interest, such as death, relapse, or remission. In most of the available works in the literature, there have been suggestions as to how to make a continuous response dichotomous by setting some appropriate threshold [11,66]. For outcomes with a continuous nature, Rosenberger [54] introduced the idea of a treatment effect mapping, in which allocation probabilities are some functions of the current treatment effect. Let g be a continuous function from R to $[0,1]$, such that $g(0) = 0.5$, $g(x) > 0.5$ if $x > 0$, and $g(x) < 0.5$ otherwise. Let Δ be some measure of the true treatment effect, and let $\hat{\Delta}_j$ be the observed value of Δ after j responses, where $\hat{\Delta}_j > (<0)$ if treatment 1 is performing better (worse) than treatment 2, and $\hat{\Delta}_j = 0$ if the two treatments are performing equally well after j responses. Then, Rosenberger [54] suggested assigning the jth subject to treatment 1 with probability $g(\hat{\Delta}_{j-1})$. It is presumed (but not formally proved) that for such an allocation procedure the limiting allocation procedure to treatment 1 would be $g(\Delta)$, for any function g. Rosenberger [54] formulated the idea of *treatment effect mapping* in the context of a linear rank test, where Δ is the normalized linear rank test and $g(x) = (1 + x)/2$. Rosenbeger and Seshaiyer [59] used the mapping $g(x) = (1 + x)/2$, with Δ as the centered and scaled log-rank statistic to derive an adaptive allocation rule for survival outcomes, but it was not studied explicitly. Another application of treatment effect mapping can be found in Yao and Wei [72], with

$$g(x) = \frac{1}{2} + xr \quad \text{if} \quad |xr| \le 0.4$$

$$= 0.1 \quad \text{if} \quad xr < -0.4$$

$$= 0.9 \quad \text{if} \quad xr > 0.4,$$

where r is a constant reflecting the degree to which one wishes to adapt the trial and Δ is the standardized Gehan–Wilcoxon test statistic. The rule of Rosenberger et al. [58], considered earlier, is also an example of treatment effect mapping for binary responses with $g(x) = 1/(1 + x)$. The intuitive appeal of "treatment effect mapping" is that the patients are allocated according to the currently available magnitudes of the treatment effect. Bandyopadhyay and Biswas [15] developed a two-treatment allocation-cum-testing procedure using a nonparametric methodology. They have used an urn mechanism where after each response the urn is updated according to the value of a statistic based on an Wilcoxon–Mann–Whitney type [and subsequently called the *Wilcoxon–Mann–Whitney-type adaptive design* (WAD)]. They have studied the design theoretically and obtained some asymptotic results together with an exactly distribution free solution for generalized Behrens–Fisher problem.

3.5.2 Link-Function-Based Allocation Designs

The work of Bandyopadhyay and Biswas [13] is perhaps the first attempt where a response adaptive randomization procedure has been developed for continuous responses in the presence of prognostic factors. They considered a simple linear model with known error variance for the responses. This two-treatment allocation rule can be viewed as a treatment effect mapping with $g(x)$ as the distribution function of a $N(0,T^2)$ random variable and Δ as the usual treatment difference. Here T is referred to as a *tuning constant*. This rule assigns a larger proportion of patients to the better treatment consistently. When a sufficiently large number of patients are treated, Bandyopadhyay and Biswas [13] show that the limiting allocation proportion to treatment 1 is $\Phi(\Delta/T)$. However, it is pointed out that use of this design amounts to some loss in efficiency in estimation of the treatment difference. Nevertheless, this kind of loss is general to any allocation design, where the allocation is skewed in favor of a particular (obviously the better) treatment.

This rule considers only univariate responses but the reality is that the responses may be multivariate in many situations (see Ref. 48, Ch. 15). Other limitations of this design are the assumption of known error variance and lack of any treatment–covariate interaction. Moreover, the design is not covariate-adjusted and is not straightforward to extend to multi-treatment situations. In a more recent work, Biswas and Coad [21] generalized the earlier design to develop a multitreatment, covariate-adjusted adaptive rule for multivariate continuous responses. They have used an weighted treatment effect mapping of possible treatment differences. An extensive simulation study revealed that the proposed procedure successfully assigned more patients to the better treatment without much sacrifice in power when testing the equivalence of treatment effect vectors.

3.5.3 Continuous Drop-the-Loser Design

As indicated earlier, the error rate of allocation proportions is an important consideration in evaluating the performance of a response-adaptive allocation design. Ivanova [42] introduced the DL rule to reduce variability. However, it was based on the binary treatment responses. Ivanova et al. [43] later developed a drop-the-loser-type design for continuous responses, and subsequently called it a *continuous drop-the-loser* (CDL) design. The allocation is carried out in the same way as in the two treatment binary response trial, except that the continuous response is categorized by means of some suitable cutoff value. A variation of this rule is also provided that suggests replacing the ball drawn with probability depending on the outcome observed. This maintains a lower rate of variability than the available competitors. Simulations also show that the performance of the procedure is worth mentioning for unequal treatment variances.

3.6 OPTIMAL ADAPTIVE DESIGNS

There are two competing goals in a clinical trial with an adaptive design. One is to optimize some criteria given certain precision of estimation or certain power of the test; the other, to skew the allocation toward the treatment doing better. Hardwick and Stout [38] review several criteria that one may wish to optimize. The list included the expected number of treatment failures, the total expected sample size, the expected number of allocations to the inferior treatment, or the total expected cost. The idea is to find an optimal allocation ratio R^* according to the selected criterion by fixing the variance of the test statistic and then framing the randomization procedure that targets R^*. Therefore, any optimal rule assigns treatment 1 to the $(j + 1)$st patient with probability $R^*(\hat{\theta}_j)/(1 + R^*(\hat{\theta}_j))$, where $\hat{\theta}_j$ is any estimator of the unknown parameter θ after j patients have responded. This development is consistent with the framework of Jennison and Turnbull [48, Ch. 17].

For binary treatment responses, Rosenberger et al. [58] derived the *optimal allocation rule* [the Rosenberg–Stallard–Ivanova–Harper–Ricks (RSIHR) rule] for minimizing the expected number of treatment failures under the fixed power. This rule targets the ratio $R^*(p_1, p_2) = \sqrt{p_1/p_2}$, where p_k is the success rate of the kth ($k = 1, 2$) treatment. Again, the allocation minimizing the sample size for fixed variance is the well-known Neyman allocation. Melfi et al. [50] studied the randomized design that targets this proportion, namely, $R^*(\sigma_1, \sigma_2) = \sigma_1/\sigma_2$, where σ_k is the variability of the kth treatment. Under certain regular assumptions, it is proved that the optimal allocation ratio $R^*(\theta)/(1 + R^*(\theta))$, to treatment 1 is attained in the limit.

Hu and Rosenberger [39] conducted a simulation study to compare some of these optimal rules. It is observed that the Neyman allocation assigns fewer patients to the better treatment when the treatments are highly successful. Computing the overall failure proportions, it is indicated that features of Neyman allocation are undesirable for highly successful treatments, and RSIHR is the most effective allocation in terms of preserving power and protecting patients.

Zhang and Rosenberger [75] developed an optimal allocation design for normal responses that minimizes the total expected response maintaining a fixed variance of the estimated treatment comparison. This is regarded as a DBCD procedure targeting a specified allocation procedure. This rule is compared with the DBCD procedure targeting the Neyman allocation, the Bandyopadhyay–Biswas rule [13], and the failure-saving rule due to Biswas and Mandal [25], and superiority of the procedure is claimed through a simulation study.

3.7 DELAYED RESPONSES IN ADAPTIVE DESIGNS

In much of the work concerning adaptively designed clinical trials, the authors have assumed instantaneous patient responses. Typically, however, clinical trials do not result in immediate responses and the usual urn models are simply inappropriate for today's long-term survival trials, where outcomes may not be ascertainable for many years. However, in many clinical trials a particular patient's response may not be obtained before the entry of the next subject and we may experience a delayed response. Consequently, the adaptation is carried out when outcomes become available, and this does not involve any additional logistical complexities. Wei [71] suggested such an adaptation in the context of RPW(α, β) by introducing an indicator ε_{ji}, $j < i$, which takes the value 1 or 0 according to whether the response of patient j occurs before patient i is randomized or not. But he did not explore the theoretical properties. Later Bandyopadhyay and Biswas [11] explored the theoretical properties assuming $P(\varepsilon_{ji} = 1) = \pi_{i-j}$, a constant depending on the lag $i - j$ only. In real practice, however, the pattern of delay varies for different treatments. Moreover, a failure may be obtained more quickly than a success. Therefore the simple model for delay described above is no longer applicable in practice. Possible generalizations of the simple model can be found in Biswas [19]. Interestingly, it is observed that these limiting proportions are not affected by the presence of delay, that is, that the limiting composition of the urn remains the same as that of a simple immediate-response model. Bai et al. [8] also considered the possibility of delay for multinomial responses within the framework of an urn model. It is established that the limiting distribution is not affected if the patients have independent arrival times and that time to response has a distribution that depends on both the treatment assigned and the patient's response. The effect of delayed responses in the context of binary outcome multitreatment clinical trials has also been investigated [46] through a simulation study.

For continuous responses Bandyopadhyay and Biswas [13] briefly mentioned the possibility of delayed responses and suggested performing the adaptation procedure with the available data. Biswas and Coad [21] gave a full mathematical treatment of this problem assuming an exponential rate of entrance of patients in the context of a general multitreatment adaptive design. Delay-adjusted procedures for two-treatment continuous clinical trials are also available in Zhang and Rosenberger [75], where delays are assumed to be exponentially distributed. But they relied on a simulation study to explore the effects of possible delays. In all the designs discussed above, it is observed that presence of delay has little effect on the performances of the clinical trials.

3.8 BIASED COIN DESIGNS

We now consider a class of sequential designs that are not response-adaptive, as the responses are completely ignored while assigning the next patient. The purpose of such design is to prevent potential biases as well as ensure that the trial will be approximately balanced whenever it is stopped. The origin of these designs can be found in the work of Efron [34], where it is referred as the "biased coin design" (BCD). Atkinson [4] extended and modified BCD to achieve balance over prognostic factors and subsequently studied by Smith [62] and Burman [27]. All these rules were derived with the idea of reducing the variance of the estimated treatment comparison using optimum design theory. Some randomization was also introduced, but in an ad hoc manner. To include both these aspects, Ball et al. [10] suggested a biased-coin-type design within a Bayesian framework that combines both the variability and randomness. Their proposal was to maximize the utility $U = U_V - \gamma U_R$ to determine different allocation probabilities. The contribution of U_V is to provide estimates with low variance, whereas U_R provides randomness. Here γ is the tradeoff coefficient between the two aspects. To obtain meaningful assignment proportions, Ball et al. [10] suggested taking U_V as some function of the posterior precision matrix and U_R as the well-known entropy function. It is shown that this design asymptotically provides equal allocation of all treatments.

Atkinson and Biswas [5] extended this approach to provide a skewed Bayesian design that, in the long run, allocates a specified proportion of patients to a particular treatment. In a numerical study, it is revealed that the extension to skewed allocations does not greatly increase the loss due to imbalance.

3.9 REAL ADAPTIVE CLINICAL TRIALS

Some real-life applications of adaptive allocation designs are also cited in the literature, although the number of real adaptive trials is very small to date. In phase I clinical trials, an ad hoc adaptive design has been widely used for many years [67], even though the poor operating characteristics of this design have been well documented [52]. Iglewicz [41] reports one use of data-dependent allocation in an unpublished application by Professor M. Zelen to a lung cancer trial. The randomized play-the-winner (PW) rule of Wei and Durham [70] has been used in at least three clinical trials: the Michigan ECMO trial [17] and two trials of fluoxetine in depression to treat outpatients sponsored by Eli Lilly [66]. The Michigan ECMO was a trial of extracorporeal membrane oxygenation (ECMO) to treat newborns of respiratory failure, where 12 patients were treated, and out of the 12 patients, only one infant was treated by the conventional medical therapy (CMT) and the infant died. All the 11 infants treated by ECMO survived. But this trial created lot of controversy due to only one allocation to CMT, and it might have pushed the application of adaptive design toward the back, to some extent. Some description of the fluoxetine trial is given in the next section. Ware [68] described another clinical trial based on ECMO using an outcome-dependent allocation conducted with his medical colleagues at Boston's Children's Hospital Medical Center and Brigham and Women's Hospital. This was a two-stage clinical trial. Rout et al. [61] applied the PW rule, and Muller and Schafer [51], also reported some adaptive clinical trials. Biswas and Dewanji [22–24] reported an adaptive trial involving patients of rheumatoid arthritis in which the observations were longitudinal and an extension of the randomized PW rule was implemented. Although some other adaptive clinical trials have been carried out, the number of real clinical trials in which adaptive allocation procedures have been used remains small.

3.10   DATA STUDY FOR DIFFERENT ADAPTIVE SCHEMES

3.10.1   Fluoxetine Trial

Despite the attractive property of assigning a larger number of subjects to the better treatment on an average, only a few real adaptive trials have been reported. For the illustration of the proposed procedure, we consider the data from the fluoxetine trial of Tamura et al. [66], which attempted to reveal the effect of an antidepressant drug on patients suffering from depressive disorder. In this trial, the patients were classified according to their shortened rapid-eye-movement latency (REML), which is presumed to be a marker for endogenous depression. A primary measure of clinical depression was taken as the total of the first 17 items of the *Hamiltonian depression scale* (HAMD$_{17}$), where a higher value of HAMD$_{17}$ indicates a severe depression. Patients receiving therapy for at least 3 weeks who exhibited 50% or greater reduction in HAMD$_{17}$ were defined to have a positive response (i.e., a success). However, patients' responses were not readily available, and the adaptation was based on surrogate outcomes using RPW rules.

As we consider homogeneity among the subjects, we apply the allocation methodology on the patients correctly assigned to the shortened REML stratum. This will not result in loss of generality, as separate adaptive allocation schemes were performed in each stratum. Then we have 45 patients, where observing the final response (either a success or a failure) is our study endpoint. We consider 39 patients and ignore patient numbers 56, 73 (misclassified), 57, 63, 79, and 88 (final response not yet available). In the study, we therefore have data from 39 patients, 19 of whom are treated by fluoxetine and 20 by placebo. We find the empirical distributions of treatment responses from the data and treat them as the true ones. Then, we obtain $\hat{p}_A = (11/19)$ and $\hat{p}_B = (7/20)$, where $p_A(p_B)$ is the success probability of fluoxetine (placebo).

Table 3.2   Data Study for Different Adaptive Schemes

Design	Allocation Proportion of Fluoxetine (SD)	Overall Failure Proportion (SD)
PW	0.603 (0.072)	0.512 (0.083)
RSIHR	0.609 (0.130)	0.513 (0.088)
DL	0.582 (0.065)	0.519 (0.082)
RPW	0.591 (0.110)	0.515 (0.083)

The results of Table 3.2 came from a simulated clinical trial with 5000 repetitions considering various allocation designs. Thus we observe that the performance of all the adaptive schemes is more or less similar, in terms of both allocation proportion and failure rate.

3.10.2   Pregabalin Trial

To illustrate the need of the adaptive procedures, we consider the real clinical trial conducted by Dworkin et al. [33]. It was a randomized, placebo-controlled trial with an objective of evaluating the efficacy and safety of pregabalin in the treatment of *postherpetic neuralgia* (PHN). There were $n = 173$ patients, 84 of whom received the standard therapy placebo and 89 were randomized to pregabalin. The primary efficacy measure was the mean of the last seven daily pain ratings, as maintained by patients in a daily diary using the 11-point numerical pain rating scale (0 = no pain, 10 = worst possible pain); therefore, a lower score (response) indicates a favorable situation. After the 8-week duration of the trial, it was observed that pregabalin-treated patients experienced a higher decrease in pain score than did patients treated with placebo. We use the final mean scores, specifically, 3.60 (with SD = 2.25) for pregabalin and 5.29 (with SD = 2.20) for placebo as the true ones for our purpose with an appropriate assumption regarding the distribution for pain scores.

The results listed in Table 3.3 were obtained as follows. Simulations with 5000 repetitions of a response-adaptive trial were performed for $n = 173$ patients with a $N(3.60, 2.25^2)$ distribution for pregabalin and a $N(5.29, 2.20^2)$ distribution for placebo. Allocation probabilities are updated according to the rule considered. Since a lower response is desirable, any response greater than the estimated simple combined mean of responses can be regarded as a failure.

Table 3.3   Data Study for Different Adaptive Schemes under Normal Assumption

Procedure	Allocation Proportion to Pregabalin (SD)	Overall Failure Proportion (SD)
BB ($T = 2$)	0.703 (0.068)	0.441 (0.042)
CDL	0.581 (0.037)	0.478 (0.038)
Optimal	0.509 (0.10)	0.499 (0.04)
Rosenberger	0.554 (0.07)	0.486 (0.06)
Equal	0.500 (0.04)	0.500 (0.04)

The CDL has the least variability, but the BB design is most ethical in terms of allocating a larger proportion (almost 20% more than that of the competitors) of patients in favor of the better treatment.

3.10.3 Simulated Trial

Consider a hypothetical clinical trial where patients enter sequentially and are to be randomized to one of the two treatments 1 and 2. We assume that the response is binary and instanteneous with p_k as the success rate for treatment k, $k = 1, 2$. Then, for different values of (p_1, p_2), we simulate the trial using different allocation procedures with trial size $n = 80$. Results are given in Table 3.4. We have chosen three pairs of values of (p_1, p_2), reflecting different rates of success, from highly successful to least successful, of the treatments. Treatment 1 is taken to be the better; that is, we always considered combinations of (p_1, p_2) with $p_1 > p_2$.

Except for optimal and equal allocation rules, the allocation proportion to treatment 1 is always greater than $1/2$, reflecting the benefit of a response adaptive procedure. It is observed that the allocation proportion to the better treatment in optimal rule of Melfi et al. [50] is less than $1/2$ whenever the better treatment possesses lower variability. Therefore the rule is not ethically attractive. RPW and DL rules generally maintained a lower rate of failure, even lower than those of RSIHR! This is surprising because RSIHR is an *optimal* rule developed with an aim to minimize overall failure proportion. Thus optimality may not ensure the absolute fulfillment of the objective (e.g., minimization of treatment failures in this case). We also observe that performance levels of RPW and DL are very similar except that the latter possesses

Table 3.4 Comparison of Different Binary Response-Adaptive Procedures

Procedure	(p_1, p_2)	Allocation Proportion to Treatment 1 (SD)	Overall Failure Proportion (SD)
RPW	(0.9, 0.7)	0.658 (0.16)	0.167 (0.049)
	(0.7, 0.5)	0.607 (0.103)	0.378 (0.059)
	(0.3, 0.1)	0.561 (0.049)	0.788 (0.047)
PW	(0.9, 0.7)	0.743 (0.097)	0.151 (0.045)
	(0.7, 0.5)	0.624 (0.067)	0.375 (0.057)
	(0.3, 0.1)	0.561 (0.028)	0.789 (0.047)
DL	(0.9, 0.7)	0.631 (0.067)	0.174 (0.041)
	(0.7, 0.5)	0.602 (0.058)	0.378 (0.056)
	(0.3, 0.1)	0.558 (0.028)	0.789 (0.046)
Optimal	(0.9, 0.7)	0.413 (0.077)	0.218 (0.046)
	(0.7, 0.5)	0.478 (0.062)	0.404 (0.052)
	(0.3, 0.1)	0.589 (0.078)	0.782 (0.047)
RSIHR	(0.9, 0.7)	0.532 (0.059)	0.194 (0.043)
	(0.7, 0.5)	0.541 (0.063)	0.390 (0.054)
	(0.3, 0.1)	0.615 (0.083)	0.777 (0.049)
Equal	(0.9, 0.7)	0.500 (0.054)	0.198 (0.044)
	(0.7, 0.5)	0.500 (0.055)	0.397 (0.054)
	(0.3, 0.1)	0.500 (0.054)	0.800 (0.044)

the lower variability of allocation proportions. Thus we conclude that among the adaptive rules for binary treatment responses, RPW seems to be the best ethical alternative.

3.11 CONCLUDING REMARKS

In this present chapter, we covered different directions of work in phase III response-adaptive designs. We discussed the theoretical models, their properties, the applications, and tried to provide a comparative discussion of various designs under different situations through simulations and real data examples. We tried to provide a wide range of references that might provide some guidance for the researchers and practitioners in this area. Note that we did not discuss some topics in this context, due to space constraints. For example, there are some more recent works demonstrating applications of adaptive designs in the longitudinal response scenario. We did not discuss this. Any interested reader can go through the papers by Sutradhar et al. [64] and Biswas and Dewanji [22–24] to get an overview of this. Again, we restricted our discussions to allocations. Adaptive designs are now used for many real situations, such as for interim monitoring to decide on the trial or to decide on the ultimate sample size. We did not discuss these issues here.

Although most of the work in the literature has been carried out from a frequentist viewpoint, the essence of adaptive designs are basically Bayesian. Here, the data dictate the allocation pattern at any stage, based on some prior weight on equivalence. This is what the Bayesian philosophy says. In a Bayesian paradigm, the posteriors are obtained following some rules, but in the adaptive designs the allocation probabilities are set ad hoc, based on the data. However, as yet no attempt has been made to frame the adaptive design in a proper Bayesian way. Only Biswas and Angers [20] considered a continuous-response two-treatment setup in the presence of covariates. The setup is similar to that of Bandyopadhyay and Biswas [13]. They have suggested computing the predictive density of a suitable link function that bridges the past history. In a simulation study, they have indicated that the proposed design reaches the level of desirability.

Finally we note that although there has been a considerable amount of interest in adaptive trials more recently, its use is still not adequate. We feel that some bridges should be built between the statisticians and the experimenters to fill this gap. Moreover, the designs should be developed in a simple way, but taking into account the realities faced by the practitioners. This might result in more and more applications of adaptive designs in the near future.

REFERENCES

1. Andersen, J., Faries, D., and Tamura, R. N., A randomized play-the-winner design for multi-arm clinical trials, *Commun. Statist. Theory Meth.* **23**, 309–323 (1994).

2. Anscombe, F. J., Sequential medical trials, *J. Am. Statist. Assoc.* **58**, 365–384 (1963).

3. Athreya, K. B. and Karlin, S., Embedding of urn schemes into continuous time Markov branching processes and related limit theorems, *Ann. Math. Statist.* **39**, 1801–1817 (1968).

4. Atkinson, A. C., Optimum biased coin designs for sequential clinical trials with prognostic factors, *Biometrika* **69**, 61–67 (1982).

5. Atkinson, A. C. and Biswas, A., Bayesian adaptive biased coin design for sequential clinical trials, *Biometrics* **61**, 118–125 (2005).

6. Bai, Z. D. and Hu, F., Asymptotic theorems for urn models with non-homogeneous generating matrices, *Stochast. Process. Appl.* **80**, 87–101 (1999).

7. Bai, Z. D. and Hu, F., Asymptotics in randomized urn models, *Ann. Appl. Probab.* **15**, (1B), 914–940 (2005).

8. Bai, Z. D., Hu, F. and Rosenberger, W. F., Asymptotic properties of adaptive designs for clinical trials with delayed response, *Ann. Statist.* **30**, 122–129 (2002).

9. Bai, Z. D., Hu, F. and Shen, L., An adaptive design for multi-arm clinical trials, *J. Multivar. Anal.* **81**, 1–18 (2002).

10. Ball, F. G., Smith, A. F. M. and Verdinelli, I., Biased coin designs with a Bayesian bias, *J. Statist. Plan. and Infer.* **34**, 403–421 (1993).

11. Bandyopadhyay, U. and Biswas, A., Delayed response in randomized play-the-winner rule: A decision theoretic outlook, *Calcutta Statist. Assoc. Bull.* **46**, 69–88 (1996).

12. Bandyopadhyay, U. and Biswas, A., Allocation by randomized play-the-winner rule in the presence of prognostic factors, *Sankhya B* **61**, 397–412 (1999).

13. Bandyopadhyay, U. and Biswas, A., Adaptive designs for normal responses with prognostic factors, *Biometrika* **88**, 409–419 (2001).

14. Bandyopadhyay, U. and Biswas, A., Selection procedures in multi-treatment clinical trials, *Metron* **60**, 143–157 (2002).

15. Bandyopadhyay, U. and Biswas, A., An adaptive allocation for continuous response using Wilcoxon-Mann-Whitney score, *J. Statist. Plan. Infer.* **123**, 207–224 (2004).

16. Bandyopadhyay, U., Biswas, A. and Bhattacharya, R., A covariate adjusted two-stage allocation design for binary responses in randomized clinical trials. *Stat. Med.* **26**, (in press).

17. Bartlett, R. H., Roloff, D. W., Cornell, R. G., Andrews, A. F., Dillon, P. W. and Zwischenberger, J. B., Extracorporeal circulation in neonatal respiratory failure: A prospective randomized trial, *Pediatrics* **76**, 479–487 (1985).

18. Begg, C. B. and Iglewicz, B., A treatment allocation procedure for sequential clinical trials, *Biometrics* **36**, 81–90 (1980).

19. Biswas, A., Generalized delayed response in randomized play-the-winner rule, *Communi. Statist. Simul. Comput.* **32**, 259–274 (2003).

20. Biswas, A. and Angers, J. F., A Bayesian adaptive design in clinical trials for continuous responses. *Statist. Neerland.* **56**, 400–414 (2002).

21. Biswas, A. and Coad, D. S., A general multi-treatment adaptive design for multivariate responses, *Sequential Anal.* **24**, 139–158 (2005).

22. Biswas, A. and Dewanji, A., A randomized longitudinal play-the-winner design for repeated binary data, *Austral. New Z. J. Statist.* **46**, 675–684 (2004).

23. Biswas, A. and Dewanji, A. Inference for a RPW-type clinical trial with repeated monitoring for the treatment of rheumatoid arthritis, *Biometr. J.* **46**, 769–779 (2004).

24. Biswas, A. and Dewanji, A., Sequential adaptive designs for clinical trials with longitudinal response, in *Applied Sequential Methodologies*, Mukhopadhyay, N., Chattopadhyay, S. and Datta, S., eds., Marcel Dekker, New York, 2004, pp. 69–84.

25. Biswas, A. and Mandal, S., Optimal adaptive designs in phase III clinical trials for continuous responses with covariate, in *moDa 7—Advances in Model-Oriented Design and Analysis*, Bucchianico, A., Di., Lauter, H. and Wynn, H.P., eds., Physica-Verlag, Heidelberg, 2004, pp. 51–58.

26. Biswas, A., Huang, W. T. and Bhattacharya, R., An adaptive design for categorical responses in phase III clinical trials (preprint, 2006).

27. Burman, C. F., *On Sequential Treatment Allocations in Clinical Trials*, Dep. Mathematics, Univ. Goteberg, 1996.

28. Coad, D. S. and Ivanova, A., Bias calculations for adaptive urn designs, *Sequent. Anal.* **20**, 91–116 (2001).

29. Colton, T., A model for selecting one of two medical treatments, *J. Am. Statist. Assoc.* **58**, 388–400 (1963).

30. Connor, E. M., Sperling, R. S., Gelber, R., Kiselev, P., Scott, G., O'Sullivan, M. J., Vandyake, R., Bey, M., Shearer, W., Jacobson, R. L., Jimminez, E., O'Neill, E., Bazin, B., Delfraissy, J., Culname, M., Coombs, R., Elkins, M., Moye, J., Stratton, P. and Balsey, J., Reduction of maternal-infant transmission of human immunodeficiency virus of type 1 with zidovudine treatment, *New Engl. J. Med.* **331**, 1173–1180 (1994) (report written for the Pediatrics AIDS Clinical Trial Group Protocol 076 Study Group).

31. Durham, S. D., Flournoy, N. and Li, W., Sequential designs for maximizing the probability of a favorable response, *Can. J. Statist.* **3**, 479–495 (1998).

32. Durham, S. D. and Yu, C. F., Randomized play-the-leader rules for sequential sampling from two populations, *Probabil. Eng. Inform. Sci.* **4**, 355–367 (1990).

33. Dworkin, R. H., Corbin, A. E., Young, J. P., Sharma, U., LaMoreaux, L., Bockbrader, H., Garofalo, E. A. and Poole, R. M., Pregabalin for the treatment of posther-petic neuralgia. A randomized, placebo-controlled trial, *Neurology* **60**, 1274–1283 (2003).

34. Efron, B., Forcing a sequential experiment to be balanced, *Biometrika* **58**, 403–417 (1971).

35. Eggenberger, F. and Pólya, G., Uber die Statistik Verketteter Vorgange, *Z. Angew. Math. Mech.* **3**, 279–289 (1923).

36. Eisele, J., The doubly adaptive biased coin design for sequential clinical trials, *J. Statist. Plan. Infer.* **38**, 249–262 (1994).

37. Eisele, J. and Woodroofe, M., Central limit theorems for doubly adaptive biased coin designs, *Ann. Statist.* **23**, 234–254 (1995).

38. Hardwick, J. and Stout, Q. F., Flexible algorithms for creating and analyzing adaptive sampling procedures, in *New Developments and Applications in Experimental Design*, Flournoy, N., Rosenberger, W. F. and Wong, W. K., eds., Hayward, Institute of Mathematical Statistics, CA, 1998, pp. 91–105.

39. Hu, F. and Rosenberger, W. F., Optimality, variability, power: Evaluating response adaptive randomisation procedures for treatment comparisons, *J. Am. Statist. Assoc.* **98**, 671–678 (2003).

40. Hu, F. and Zhang, L. X., Asymptotic properties of doubly adaptive biased coin design for multi-treatment clinical trials, *Ann. Statist.* **32**, 268–301 (2004).

41. Iglewicz, B., Alternative designs: Sequential, multi-stage, decision theory and adaptive designs, in *Cancer Clinical Trials: Methods and Practice*, Buyse, M. E., Staquet, J. and Sylvester, R. J., eds., Oxford Univ. Press, 1983, pp. 312–334.

42. Ivanova, A., A play-the-winner type urn model with reduced variability, *Metrika* **58**, 1–13 (2003).

43. Ivanova, A., Biswas, A. and Lurie, H., Response adaptive designs for continuous outcomes, *J. Statist. Plan. Infer.* **136**, 1845–1852 (2006).

44. Ivanova, A. and Durham, S. D., *Drop The Loser Rule*, Technical Report TR-00-01, Univ. North Carolina, Chapel Hill, 2000.

45. Ivanova, A. and Flournoy, N., A birth and death urn for ternary outcomes: Stochastic processes applied to urn models, in *Probability and Statistical Models with Applications*, Charalambides, C. A., Koutras, M. V. and Balakrishnan, N. eds., Chapman & Hall, Boca Raton, FL, 2001, pp. 583–600.

46. Ivanova, A. and Rosenberger, W. F., A comparison of urn designs for randomized clinical trials of $K > 2$ treatments, *J. Biopharm. Statist.* **10**, 93–107 (2000).

47. Ivanova, A., Rosenberger, W. F., Durham, S. D. and Flournoy, N., A birth and death urn for randomized clinical trials: Asymptotic methods, *Sankhya B* **62**, 104–118 (2000).

48. Jennison, C. and Turnbull, B. W., *Group Sequential Methods with Applications to Clinical Trials*. Chapman & Hall/CRC, Boca Raton, FL, 2000.

49. Matthews, P. C. and Rosenberger, W. F., Variance in randomized play-the-winner clinical trials, *Statist. Probabil. Lett.* **35**, 233–240 (1997).

50. Melfi, V. F., Page, C. and Geraldes, M., An adaptive randomized design with application to estimation, *Can. J. Statist.* **29**, 107–116 (2001).

51. Muller, H.-H. and Schafer, H., Adaptive group sequential designs for clinical trials: Combining the advantages of adaptive and of classical group sequential approaches, *Biometrics* **57**, 886–891 (2001).

52. O'Quigley, J., Pepe, M. and Fisher, L., Continual reassessment method: A practical design for phase I clinical trials in cancer, *Biometrics* **46**, 33–48 (1990).

53. Robbins, H., Some aspects of the sequential design of experiments, *Bull. Am. Math. Soc.* **58**, 527–535 (1952).

54. Rosenberger, W. F., Asymptotic inference with response-adaptive treatment allocation designs, *Ann. Statist.* **21**, 2098–2107 (1993).

55. Rosenberger, W. F., New directions in adaptive designs, *Statist. Sci.* **11**, 137–149 (1996).

56. Rosenberger, W. F. and Lachin, J. M., The use of response-adaptive designs in clinical trials, *Control. Clin. Trials* **14**, 471–484 (1993).

57. Rosenberger, W. F. and Lachin, J. L., *Randomisation in Clinical Trials: Theory and Practice*, Wiley, New York, 2002.

58. Rosenberger, W. F., Stallard, N., Ivanova, A., Harper, C. and Ricks, M., Optimal adaptive designs for binary response trials, *Biometrics* **57**, 173–177 (2001).

59. Rosenberger, W. F. and Seshaiyer, P., Adaptive survival trials, *J. Biopharm. Statist.* **7**, 617–624 (1997).

60. Rosenberger, W. F., Vidyashankar, A. N. and Agarwal, D. K., Covariate adjusted response adaptive designs for binary responses, *J. Biopharm. Statist.* **11**, 227–236 (2001).

61. Rout, C. C., Rocke, D. A., Levin, J., Gouw's, E. and Reddy, D., A reevaluation of the role of crystalloid preload in the prevention of hypotension associated with spinal anesthesia for elective cesarean section, *Anesthesiology* **79**, 262–269 (1993).

62. Smith, R. L., Sequential treatment allocation using biased coin designs, *J. Roy. Statist. Soc. Ser. B* **46**, 519–543 (1984).

63. Smythe, R. T., Central limit theorems for urn models, *Stochast. Process. Appl.* **65**, 115–137 (1996).

64. Sutradhar, B. C., Biswas, A. and Bari, W., Marginal regression for binary longitudinal data in adaptive clinical trials, *Scand. J. Statist.* **32**, 93–114 (2005).

65. Thompson, W. R., On the likelihood that one unknown probability exceeds another in the view of the evidence of the two samples, *Biometrika* **25**, 275–294 (1933).

66. Tamura, R. N., Faries, D. E., Andersen, J. S. and Heiligenstein, J. H., A case study of an adaptive clinical trial in the treatment of out-patients with depressive disorder, *J. Am. Statist. Assoc.* **89**, 768–776 (1994).

67. Vann Hoff, D. D., Kuhn, J., and Clark, G. M., Designs and conduct of phase I trials, in *Cancer Clinical Trials*, Buyse, M. E., Staquet, M. J. and Sylvester, R. J., eds., Oxford Univ. Press, New York, 1984, pp. 210–220.

68. Ware, J. H., Investigating therapies of potentially great benefit: ECMO. (with discussions), *Statist. Sci.* **4**, 298–340 (1989).

69. Wei, L. J., The generalized Polya's urn design for sequential medical trials, *Ann. Statist.* **7**, 291–296 (1979).

70. Wei, L. J. and Durham, S., The randomized play-the-winner rule in medical trials, *J. Am. Statist. Assoc.* **73**, 840–843 (1978).

71. Wei, L. J., Exact two-sample permutation tests based on the randomized play-the-winner rule in medical trials, *Biometrika* **75**, 603–606 (1988).

72. Yao, Q. and Wei, L. J., Play the winner for phase II/III clinical trials, *Statist. Med.* **15**, 2413–2423 (1996).

73. Zelen, M., Play the winner rule and the controlled clinical trial, *J. Am. Statist. Assoc.* **64**, 131–146 (1969).

74. Zhang, L. X., Hu, F. and Cheung, S. H., Asymptotic theorems of sequential estimation-adjusted urn models, *Ann. Appl. Probabil.* **16**, 340–369 (2006).

75. Zhang, L. X. and Rosenberger, W. F., Response-adaptive randomization for clinical trials with continuous outcomes, *Biometrics* **62**, 562–569 (2006).

CHAPTER 4

Inverse Sampling for Clinical Trials: A Brief Review of Theory and Practice

Atanu Biswas

Indian Statistical Institute, Kolkata, India

Uttam Bandyopadhyay

University of Calcutta, Kolkata, India

4.1 INTRODUCTION

Determination of sample size in a clinical trial is always an important issue as the samples are quite costly. Particularly when there is significant treatment difference, an early stopping, possibly keeping a fixed power of the test, might result in substantial reduction in sample sizes, and hence save precise administrative and ethical costs.

A fixed sample size trial sets a prefixed number of samples, say n_0, and randomizes the entering patients among the competing treatments by a sampling design. In contrast, an inverse sampling prefixes a certain number of events from the trial, and stops sampling as soon as the number of that events reaches that prefixed number. Such an "event" may be "success", "failure", "responses below/above a certain threshold", and so on. Thus the sample size of the trial will be random, but of course we can look at the expectation and variance of the sample size. Consider the following simple example. Suppose that we have a single sample case, where the subjects are treated by a single treatment. The response X has Bernoulli (p) distribution, $0 < p < 1$, and the successive responses are independent. Suppose that we want to test

$$H_0 : p = p_0 \quad \text{against} \quad H_1 : p > p_0,$$

Statistical Advances in the Biomedical Sciences, edited by Atanu Biswas, Sujay Datta,
Jason P. Fine, and Mark R. Segal

The fixed sample size test can be carried out by fixing the sample size n_0, and using the test statistic

$$T_{n_0} = \sum_{i=1}^{n_0} Z_i,$$

where the Z_i terms are independent copies of X, which takes the value 1 for a success and the value 0 for a failure. Thus T_{n_0} denotes the number of successes from the n_0 subjects, and T_{n0} follows binomial (n_0, p) distribution. As $E_{H1}(T_{n_0}) > E_{H0}(T_{n_0})$, a right-tailed test can be carried out by using T_{n_0}. Such a test is also a *uniformly most powerful* (UMP) test.

4.1.1   Inverse Binomial Sampling

Alternatively, for the Bernoulli situations described above, we can frame a test for H_0 against H_1 using inverse binomial sampling [20,35] in the following way. Let r be a prefixed positive integer. Observe X sequentially and stop sampling as soon as the number of successes reaches r. Let N_r be the number of X terms observed. Naturally N_r has a negative binomial (r,p) distribution, and hence

$$E_{H_1}(N_r) < E_{H_0}(N_r).$$

This suggests that a left-tailed test based on N_r is appropriate. It can also be seen that such a test is UMP. Moreover, for every given positive integer c, there exists a random variable T_c having the binomial (c,p) distribution with

$$[N_r \leq c] <=> [T_c \geq r],$$

and hence, for any $p \in (0,1)$, we obtain

$$\beta(p) = P_p(N_r \leq c) = P_p(T_c \geq r).$$

This shows that, given size and power, we can always frame an N_r test and a fixed sample size test assuming that T_c has the same size and the power. Moreover, the N_r test keeps smaller expected sample size under H_1. In practice we can get an N_r test in the following way. Given r and c [or the level of significance $\alpha \in (0,1)$], we stop sampling at that n for which $\sum_{i=1}^n Z_i \leq c$ or $= r$, whichever is earlier. Then H_0 is accepted or rejected according to whether $\sum_{i=1}^n \leq c$ or $\sum_{i=1}^n Z_i > c$. This gives, for any p, the expected sample size for the N_r test as

$$S(p) = E_p(N_r | N_r \leq c)P_p(N_r \leq c) + (c+1)P_p(N_r \geq c+1),$$

where

$$P_p(N_r) = \binom{n-1}{r-1} p^r (1-p)^{n-r}, \quad n = r, r+1, \ldots.$$

Thus, whatever $p \in [p_0, 1]$ may be, we always have $S(p) \leq c$. In a sequential testing procedure a term such as $S(p)$ is also called the *average sample number* (ASN) function.

In some applications, sometimes it would be appropriate to consider the problem of testing

$$H_0 : p = p_0 \quad \text{against} \quad H_1' : p \neq p_0.$$

If we use binomial sampling, then a fixed size test based on n observations would be to reject H_0 if and only if T_n is too large or too small. Such a test, as in Lehmann [25], is UMP unbiased. This test, as earlier, does not correspond to the inverse binomial two-sided test. Here we have to modify the inverse binomial sampling in the following way. Let r and s be two prefixed positive integers. Observe Xs sequentially, and stop sampling as soon as r successes or s failures, whichever is earlier, are obtained. Let N' be the number of trials to meet such an objective. Then, for any positive integer $c'(\geq s)$, we have

$$[N' \leq c'] <=> [T_{c'} \geq r] \cup [T_{c'} \leq c' - s],$$

and hence

$$\beta'(p) = P_p(N' \leq c') = P_p(T_{c'} \geq r) + P_p(T_{c'} \leq c' - s).$$

Now, if we write

$$N' = \min(N_r, \bar{N}_s),$$

where N_r is as before and $\bar{N}_s$ is the number of trials required to obtain the sth failure, we have, for any $p \in (0,1)$,

$$E_{H'}(N') < E_{H_0}(N'),$$

and hence, as before, a left-tailed test based on N' is appropriate. If c' is the level $\alpha \in (0,1)$ cutoff point for this test, then c' can be obtained from the relation

$$P_{H_0}(N' \leq c') \leq \alpha.$$

The probability mass function (pmf) of N' is

$$P_p(N'') = \binom{n'-1}{r-1} p^r (1-p)^{n'-r} + \binom{n'-1}{s-1} (1-p)^s p^{n'-s},$$

$$n' = \min(r, s), \dots, (r+s-1).$$

Stopping rules play a central role in the theory of sequential analysis [30]. Here the variables N_r and N' can be interpreted as the stopping variables connected with renewal theory. Thus N_r

and N' can be written as

$$N_r = \min\left\{n : n \geq 1, \sum_{i=1}^{n} Z_i = r\right\}$$

and

$$N' = \min\left\{n : n \geq 1, \sum_{i=1}^{n} Z_i = r \quad \text{or} \quad \sum_{i=1}^{n} (1 - Z_i) = s\right\}.$$

Hence all the asymptotic results related to these inverse binomial trials can follow from that of renewal theory. Moreover, using N_r, the maximum-likelihood estimator (MLE) of p is r/N_r. This is biased as

$$p < E\left(\frac{r}{N_r}\right) < \left(\frac{r}{r-1}\right)p.$$

In fact, $(r-1)/(N_r - 1)$ is the minimum-variance unbiased estimator of p. In the subsequent sections, we have provided various stopping rules as an extension or generalization of inverse binomial sampling. Unlike renewal theory, the variables associated with those rules are, in general, dependent, and hence the properties of renewal theory are not applicable here in a straight forward manner.

4.1.2   Partial Sequential Sampling

Two-treatment comparisons are the most common in clinical trials, and hence we focus at the two sample problems. There are many practical situations in which the observations from one of these treatments are easy and relatively inexpensive to collect, while the sample observations corresponding to the other population are costly and difficult to obtain. For example, in a clinical trial, observations X on a standard treatment may be easily available, and the same Y for a new treatment may be difficult to obtain. In such a situation, one would like to gather data (may be large) on X and collect only enough observations necessary to reach a decision regarding the problem under consideration. To achieve this goal, we consider collecting Y observations in a sequential manner, with sampling terminated following some stopping rule.

We consider the following setup, where F_1 and F_2 may be two continuous univariate distribution functions. We want to test the null hypothesis

$$H_0 : F_1 = F_2 \tag{4.1}$$

against a class of one-sided alternatives H_1. If we restrict F_2 as (1) $F_2(x) = F_1(x - \mu)$, $-\infty < x$, $\mu < \infty$, or (2) $F_2(x) = F_1(x \exp(-\mu))$, $x > 0$, $-\infty < \mu < \infty$, then the hypotheses can be reduced to

$$H_0 : \mu = 0 \quad \text{against} \quad \mu > 0.$$

In connection with the problem of testing the abovementioned null hypothesis against H_1, several partial sequential designs were proposed and studied [41,27,28,16]. For details, one can also see the book by Randles and Wolfe [29].

4.2 TWO-SAMPLE RANDOMIZED INVERSE SAMPLING FOR CLINICAL TRIALS

We consider a simple setup where we have two competing treatments A and B, having binary responses, with success probabilities p_1 and p_2, respectively; that is, if the responses to A and B are represented by X and Y, then X follows Bernoully (p_1) and Y follows Bernoulli (p_2). Suppose that each patient is randomly (50:50) allocated to either A or B, by tossing a fair coin. Here we want to test the null hypothesis $H_0 : p_1 = p_2$ against the one-sided alternative $H_1 : p_1 > p_2$. Here, for the ith patient, we define a pair of indicator variables (δ_i, Z_i), where δ_i is the indicator of assignment, which takes the value 1 if the ith patient is treated by A, and 0 if the ith patient is treated by B, and where Z_i is as in binomial sampling. A fixed-sample-size test can be based on $(\sum_{i=1}^{n_0} \delta_i Z_i, \sum_{i=1}^{n_0} (1 - \delta_i) Z_i)$, which are the number of successes by A and B, respectively, for a prefixed total sample size n_0. Clearly, for such random sampling, δ_i follows Bernoulli (0.5), independently of each other, which gives $\sum_{i=1}^{n_0} \delta_i$ follows binomial (n_0, 0.5). Hence, given $\sum_{i=1}^{n_0} \delta_i = m$, we have $\sum_{i=1}^{n_0} \delta_i Z_i \sim$ Bin (m, p_1), and consequently, $E \left(\sum_{i=1}^{n_0} \delta_i Z_i \right) = n_0 p_1 / 2$. In a similar way, $E \left(\sum_{i=1}^{n_0} (1 - \delta_i) Z_i \right) = n_0 p_2 / 2$. A suitable test statistic may be based on

$$ T_{n_0} = \frac{\sum_{i=1}^{n_0} \delta_i Z_i}{\sum_{i=1}^{n_0} \delta_i} - \frac{\sum_{i=1}^{n_0} (1 - \delta_i) Z_i}{\sum_{i=1}^{n_0} (1 - \delta_i)}, $$

and a right-tailed test can be suggested.

For an inverse sampling, the stopping rule can be set as in using N_r as a test statistic. Here $E_{H_0}(Z_i) = p_1$, regardless of whether the patient is treated by A or B. But $E_{H_1}(Z_i \mid \delta_i) = \delta_i p_1 + (1 - \delta_i) p_2$, and $E_{H_1}(Z_i) = (p_1 + p_2)/2$. Thus $E_{H_1}(Z_i) \; E_{H_0}(Z_i)$. Consequently, N_r is expected to be larger under H_0 than under H_1. A left-tailed test as earlier can be suitably suggested.

In the group-sequential framework, inverse sampling was carried out by Bandyopadhyay and Biswas [4,5].

4.2.1 Use of Mann–Whitney Statistics

The test can be carried out by inverse sampling using statistics different from the negative binomial type. Here, under pairwise sampling, we consider a situation in a much simpler scenario where the patients are taken in pairs and are randomly treated by the two treatments. Suppose that we want to test the hypotheses of Section 4.1.2.

We restrict ourselves in pairwise sampling for the purpose of comparison, where the ith pair is (X_i, Y_i) with $X \sim F_1$ and $Y \sim F_2$.

For the ith pair, we define $Z_i = 1$ if $Y_i > X_i$, and $Z_i = 0$ otherwise. Obviously, Z_i are independently and identically distributed as Bernoulli (p) with

$$ p = \int F_1(x) dF_2(x), $$

which is equal to $\frac{1}{2}$ under H_0, and is greater than $\frac{1}{2}$ under H_1. There is substantial gain in sample size over the fixed sample size (maintaining close power). The ASN of the inverse sampling procedure is much less than the ASN for the fixed-sample-size procedure with the same power.

For inverse sampling, some improvement in power can be achieved by considering other complicated score functions than simple Z_i for the ith pair. For example, we can consider the Mann–Whitney scores. Here, after drawing the ith pair, the differences $Y_j - X_k$, ($1 \leq j$, $k \leq i$), are observed instead of observing the differences $Y_i - X_i$. The indicator Z_{jk} takes the value 1 if $Y_j - X_k > 0$ and 0 otherwise. This leads to the following sequence of Mann–Whitney statistics:

$$U_n = \sum_{j=1}^{n} \sum_{k=1}^{n} Z_{ij}, \quad n \geq 1,$$

and the stopping variable N_q can be defined as

$$N_q = \min\{n : U_n \geq q(r)\},$$

where $q(r)$ is an integer-valued quadratic function of r. Clearly, N_q is stochastically smaller under H_1 than under H_0, resulting in a gain in sample size. The power of a test will depend on the statistic that we choose to carry out the test. However, for some suitably chosen test statistic, the power will increase. Bandyopadhyay and Biswas [8] studied this in detail.

4.2.2   Fixed-Width Confidence Interval Estimation

Consider the following sequence of estimators

$$\hat{\theta}_n = \frac{\sum_{i=1}^{n} \delta_i Z_i}{\sum_{i=1}^{n} \delta_i} - \frac{\sum_{i=1}^{n} (1 - \delta_i) Z_i}{\sum_{i=1}^{n} (1 - \delta_i)}$$

of some parameter θ measuring the treatment difference (e.g., the difference between two population means). Thus $\hat{\theta}_n$ is only the difference in proportion of successes ($p_1 - p_2$) if the observations are binary. Otherwise, it is the difference in average responses. Suppose that $\hat{\theta}_n$ is consistent for θ, and, for some $V^2 > 0$, as $n \to \infty$, we have

$$\sqrt{n}(\hat{\theta}_n - \theta) \xrightarrow{d} N(0, V^2).$$

Then, for some γ, we can choose $n \geq \gamma$, such that

$$\lim_{d \downarrow 0} P(|\hat{\theta}_n - \theta| < d) \geq 1 - \alpha,$$

where γ is the smallest integer exceeding $(uV/d)^2$, and $1 - \Phi(u) = \alpha/2$ with $\Phi(\cdot)$ as the cumulative distribution function of an $N(0, 1)$ random variable. In practice, V^2 is not known. Let $\hat{V}_{n^2}$ be a sequence of strongly consistent estimators of V^2 based on $\{(\delta_i, Z_i), i = 1, \ldots, n\}$. We then introduce the stopping variable

$$\tau = \min\left\{n : n \geq m, n \geq \frac{u^2 \hat{V}_n^2}{d^2}\right\}$$

with m as the initial sample size, which means that we always take a sample size greater than it. Then we have (see Ref. 19, Ch. 10)

$$\lim_{d \downarrow 0} P(|\hat{\theta}_\tau - \theta| < d) \geq 1 - \alpha$$

provided, as $d \to 0$, that

(i) $\tau/v \xrightarrow{P} 1$,

(ii) $\hat{V}_\tau^2 \xrightarrow{P} V^2$,

(iii) $\sqrt{\tau}(\hat{\theta}_\tau - \hat{\theta}_v) \xrightarrow{P} 0$.

This can be shown in many simple randomized clinical trial situations as well as for some adaptive designs.

4.2.3 Fixed-Width Confidence Interval for Partial Sequential Sampling

Let $X_m = (X_1, \ldots, X_m)'$ be a random sample of fixed size $m(\geq 1)$ on X, and let $\{Y_n, n \geq 1\}$, be a sequence of observations on Y. Let $\hat{\theta}_n$ be a sequence of estimators based on X_m and $(Y_1, \ldots, Y_n)$, $n \geq 1$. Suppose for each m that there is a positive integer $r = r(m)$ such that, as $m \to \infty$, we have $r \to \infty$. We also assume that as $m \to \infty$

$$\sup_{n \geq r} |\hat{\theta}_n - \theta| \longrightarrow 0$$

in probability. Then, for given $d(> 0)$, the random variable

$$N(d) = \sup\{n \geq 1 : |\hat{\theta}_n - \theta| \geq d\}$$

is related to a partially sequential fixed-width confidence interval of θ in the sense that there exists a positive integer $v = v(m)$ such that asymptotically, for given $\alpha \in (0, 1)$,

$$P(|\hat{\theta}_n - \theta| < d \quad \text{for all} \quad n \geq v) = 1 - \alpha. \tag{4.2}$$

The random variable of the type $N(d)$ is studied by Hjort and Fenstad [21] in connection with sequential fixed-width confidence interval estimation. A terminal version of (4.2) is that there exists a positive integer $v^* = v^*(m)$ (different from v) such that asymptotically

$$P(|\hat{\theta}_n - \theta| < d) \geq 1 - \alpha \quad \text{for all} \quad n \geq v^*. \tag{4.3}$$

Under the sequential setup, (4.3) has been studied by many researchers. For details, one can consult the book by Ghosh et al. [19]. Bandyopadhyay et al. [13] studied (4.3) under partial sequential setup.

4.3   AN EXAMPLE OF INVERSE SAMPLING: BOSTON ECMO

This is a real clinical trial that employed inverse sampling effectively. Details on the trial and data description with discussions were presented by Ware [36] followed by discussions by quite a few experts in clinical trial statistics. The purpose of the trial was to evaluate extracorporeal membrane oxygenation (ECMO) for treatment of persistent pulmonary hypertension of the newborn (PPHN). The mortality rate among the infants with severe PPHN treated with conventional medical therapy (CMT) was 80% or higher for many years. ECMO treatment of PPHN was introduced in 1977, and by the end of 1985, several centers reported survival rates of 80% or more in infants treated with ECMO. Bartlett et al. [14] reported a randomized trial for ECMO versus CMT, which was an adaptive trial based on the randomized play-the-winner (PW) rule [37,38]. In that trial, out of 12 patients, only one infant was treated by the placebo, and later died. The other 11 infants, treated with ECMO, all survived. This trial received lot of cristicism as, due to the adaptive sampling, only one patient was treated by the CMT, and very little information on CMT was gathered.

To balance the ethical and scientific concerns, Dr. Ware and his colleagues designed a two-stage trial. They considered a family of study designs where a maximum of prefixed r deaths are allowed in either treatment group. The treatments were selected by a randomized permuted block design with blocks of size 4. When r deaths occur in one of the treatment groups, randomization ceases and all subsequent patients are assigned to the other treatment until r deaths occur in that arm or until the number of survivors is sufficient to establish the superiority of that treatment arm, using a test procedure based on the conditional distribution of the number of survivors in one treatment given the total number of survivors.

In the trial, in the late 1980s, patients were randomized in blocks of 4, and treatments were assigned randomly to the first 19 patients. Of these 19 patients, 10 received CMT, including patient 19, and 4 died. Here r was taken as 4. In the second stage, all the remaining 9 patients received ECMO and all survived.

For details of the trial and the analysis, we refer to Ware [36]. But at this point we want to emphasize that inverse sampling can be carried out in real trials in some way following this fashion, possibly by modifying the design in an appropriate way, which might help in stopping the trial earlier and saving the lives of some patients who are exposed to the inferior treatment during the trial.

4.4   INVERSE SAMPLING IN ADAPTIVE DESIGNS

Adaptive designs or response adaptive designs are used in clinical trials in order to allocate a larger number of patients for the better treatment. This is done by choosing the allocation probability of a patient to either treatment depending on the response and allocation history of the patients allocated so far. Consider the setup described in Section 4.3. Here δ_i takes the values 1 or 0 with probabilities π_i and $1 - \pi_i$, where $\pi_i = \pi_i(\delta_1, \ldots, \delta_{i-1}; Z_1, \ldots, Z_{i-1})$. The functional form of π_i depends on the particular adaptive design under consideration. Some such designs are the randomized PW rule [37,38], the success-driven design [17], the drop-the-loser (DL) rule [23] for binary responses, and the link-function-based design [10] optimal designs of Atkinson and Biswas [1,2] for continuous responses.

For such designs, the δ_i terms are of dependent sequence, and quite naturally the Z_i terms are also dependent as they depend on δ_i values. Thus one can use the same stopping rule as in N_r, and a similar test as well. But the properties, the exact and the asymptotic distributions, will be quite different in this dependent setup. Bandyopadhyay and Biswas [6,7] considered such inverse-sampling-based analysis in the setup of the randomized PW rule.

For binary treatment responses, Sobel and Weiss [31] suggested the combination of the PW rule [42], an adaptive design, and an inverse sampling scheme. The suggestion was to stop sampling when a prefixed number of successes were observed from either of the two treatments. Later Hoel [22] modified the sequential procedure of Sobel and Weiss [31], and introduced another stopping variable, which takes both the number of successes and failures into account. Specifically, they suggested stopping the experiment when the number of successes of a treatment *plus* the number of failures by the other treatment exceeds a prefixed threshold. Subsequently, Fushimi [18], Berry and Sobel [15], Kiefer and Weiss [24], and Nordbrock [26] considered the PW allocation with more complicated stopping rules. Wei and Durham [38] extended the PW rule of Zelen [42] to obtain the randomized PW rule. They studied the properties of this allocation rule for fixed sample size and also for the stopping rule proposed by Hoel [22]. Bandyopadhyay and Biswas [6,7,9,11,12] considered the randomized PW allocation and the stopping rule of Sobel and Weiss [31]. Baldi Antognini and Giovagnoli [3] considered the estimation of treatment effects in sequential experiments for comparing several treatments when the responses belong to the exponential family, suggesting that the experiment will be stopped when the absolute value of the sum of responses to each treatment reaches a given value. For two treatments, that rule becomes

$$N = \inf_{n \in \mathcal{N}} \left\{ \left| \sum_{i=1}^{n} \delta_i X_i \right| \geq r_1 \quad \text{and} \quad \left| \sum_{i=1}^{n} (1 - \delta_i) Y_i \right| \geq r_2 \right\}. \tag{4.4}$$

For binary responses, (4.4) becomes a lower threshold for the number of observed successes by each treatment, and for one treatment only it reduces to the classical inverse binomial sampling. For normal responses, (4.4) reduces to the sampling scheme of Tweedie [35]. Combining the sequential ML design with this stopping rule (4.4), Baldi Antognini and Giovagnoli [3] showed that strong consistency and asymptotic normality of the MLEs still hold approximately.

In adaptive design, the amount of research is still inadequate. Stallard and Rosenberger [32] observed that "most of the theoretical development and practical implementation of adaptive designs has assumed a fixed sample size."

4.5 CONCLUDING REMARKS

Inverse sampling is designed specifically for the purpose of estimation, usually with the objective of attaining a confidence interval with fixed width. In the early–mid-1940s, inverse binomial sampling was discussed by Haldane [20] and Tweedie [34] and inverse normal sampling, by Stein [33]. Both these methods are discussed in the book of Wetherill and Glazebrook [39, Ch. 8]; see also Whitehead [40, Ch. 5] for a note.

Inverse sampling for a multitreatment clinical trial can be similarly designed. It is true that the number of real applications of inverse sampling is still not adequate. This can be successfully done if the gap between the statisticians working in this area and the practitioners can be bridged. This can effectively reduce the sample sizes, especially in trials where there is prior belief of substantial treatment difference, as in the Boston ECMO trial.

REFERENCES

1. Atkinson, A. C. and Biswas, A., Bayesian adaptive biased coin design for sequential clinical trials, *Biometrics* **61**, 118–125 (2005).

2. Atkinson, A. C. and Biswas, A., Adaptive biased-coin designs for skewing the allocation proportion in clinical trials with normal responses, *Statist. Med.* **24**, 2477–2492 (2005).

3. Baldi Antognini, A. and Giovagnoli, A., On the large sample optimality of sequential designs for comparing two or more treatments, *Sequent. Anal.* **24**, 205–217 (2005).

4. Bandyopadhyay, U. and Biswas, A., Some nonparametric group sequential tests for two population problems, *Calcutta Statist. Assoc. Bull.* **45**, 73–91 (1995).

5. Bandyopadhyay, U. and Biswas, A., Sequential-type nonparametric tests for ordered bivariate alternatives, *Sequent. Anal.* **15**, 123–144 (1996).

6. Bandyopadhyay, U. and Biswas, A., Some sequential tests in clinical trials based on randomized play-the-winner rule, *Calcutta Statist. Assoc. Bull.* **47**, 67–89 (1997).

7. Bandyopadhyay, U. and Biswas, A., Sequential comparison of two treatments in clinical trials: A decision theoretic approach based on randomized play-the-winner rule, *Sequent. Anal.* **16**, 65–92 (1997).

8. Bandyopadhyay, U. and Biswas, A., Sequential-type nonparametric test using Mann-Whitney statistics, *J. Appl. Statist.* **26**, 301–308 (1999).

9. Bandyopadhyay, U. and Biswas, A., Some sequential-type conditional tests in clinical trials based on generalized randomized play-the-winner rule, *Metron* **LVIII**, 187–200 (2000).

10. Bandyopadhyay, U. and Biswas, A., Adaptive designs for normal responses with prognostic factors, *Biometrika* **88**, 409–419 (2001).

11. Bandyopadhyay, U. and Biswas, A., Test of Bernoulli success probability in inverse sampling for nearer alternatives using adaptive allocation, *Statist. Neerland.* **56**, 387–399 (2002).

12. Bandyopadhyay, U. and Biswas, A., Nonparametric group sequential designs in randomized clinical trials, *Austral. New Z. J. Statist.* **45**, 367–376 (2003).

13. Bandyopadhyay, U. Das, R., and Biswas, A., Fixed width confidence interval of $P(X < Y)$ in partial sequential sampling scheme, *Sequent. Anal.* **22**, 75–93 (2003).

14. Bartlett, R. H., Roloff, D. W., Cornell, R. G., Andrews, A. F., Dillon, P. W., and Zwischenberger, J. B., Extracorporeal circulation in neonatal respiratory failure: A prospective randomized trial, *Pediatrics* **76**, 479–487 (1985).

15. Berry, D. A. and Sobel, M., An improved procedure for selecting the better of two binomial populations, *J. Am. Statist. Assoc.* **68**, 979–984 (1973).

16. Chatterjee, S. K. and Bandyopadhyay, U., Inverse sampling based on general scores for nonparametric two-sample problems, *Calcutta Statist. Assoc. Bull.* **33**, 35–58 (1998).

17. Durham, S. D., Flournoy, N., and Li, W., Sequential designs for maximizing the probability of a favorable response, *Can. J. Statist.* **3**, 479–495 (1998).

18. Fushimi, M., An improved version of a Sobel-Weiss play-the-winner procedure for selecting the better of two binomial populations, *Biometrika* **60**, 517–523 (1973).

19. Ghosh, M., Mukhopadhyay, N., and Sen, P. K., *Sequential Estimation*, Wiley, New York, 1997.

20. Haldane, J. B. S., A labour-saving method of sampling, *Nature (Lond.)* **155**, 49–50 (1945).

21. Hjort, N. L. and Fenstad, G., On the last time and the number of times an estimator is more than ε from its target value, *Ann. Statist.* **20**, 469–489 (1992).

22. Hoel, D. G., An inverse stopping rule for play-the-winner sampling, *J. Am. Statist. Assoc.* **67**, 148–151 (1972).

23. Ivanova, A., A play-the-winner type urn model with reduced variability, *Metrika* **58**, 1–13 (2003).

24. Kiefer, J. E. and Weiss, G. H., Truncated version of a play-the-winner rule for choosing the better of two binomial populations, *J. Am. Statist. Assoc.* **69**, 807–809 (1974).

25. Lehmann, E. L., *Theory of Point Estimation*, 2nd ed., Springer-Verlag, New York, 2001.

26. Nordbrock, E., An improved play-the-winner sampling procedure for selecting the better of two binomial populations, *J. Am. Statist. Assoc.* **71**, 137–139 (1976).

27. Orban, J. and Wolfe, D. A., Distribution-free partially sequential placement procedures, *Commun. Statist. Ser. A* **9**, 883–904 (1980).

28. Orban, J. and Wolfe, D. A., A class of distribution free two sample tests based on placements, *J. Am. Statist. Assoc.* **77**, 666–671 (1982).

29. Randles, R. H. and Wolfe, D. A., *Introduction to the Theory of Nonparametric Statistics*, Wiley, New York, 1979.

30. Siegmund, D., *Sequential Analysis: Tests and Confidence Intervals*, Springer, Heidelberg, 1985.

31. Sobel, M. and Weiss, G. H., Play-the-winner rule and inverse sampling in selecting the better of two binomial populations, *J. Am. Statist. Assoc.* **66**, 545–551 (1971).

32. Stallard, N. and Rosenberger, W. F., Exact group-sequential designs for clinical trials with randomized play-the-winner allocation, *Statist. Med.* **21**, 467–480 (2002).

33. Stein, C., A two-sample test for a linear hypothesis whose power is independent of the variance, *Ann. Math. Statist.* **16**, 243–258 (1945).

34. Tweedie, M. C. K., Inverse satistical variates, *Nature (Lond.)* **155**, 453 (1945).

35. Tweedie, M. C. K., Statistical properties of inverse Gaussian distributions, *Ann. Math. Statist.* **28**, 362–377 (1957).

36. Ware, J. H., Investigating therapies on potentially great benefit: ECMO (with discussions), *Statist. Sci.* **4**, 298–340 (1989).

37. Wei, L. J., The generalized Polya's urn for sequential medical trials, *Ann. Statist.* **7**, 291–296 (1979).

38. Wei, L. J. and Durham, S., The randomized play-the-winner rule in medical trials, *J. Am. Statist. Assoc.* **73**, 838–843 (1978).

39. Wetherill, G. B. and Glazebrook, K. D., *Sequential Methods in Statistics*, 3rd ed., Chapman & Hall, London, 1986.

40. Whitehead, J., *The Design and Analysis of Sequential Clinical Trials*, 2nd ed., Wiley, New York, 1997.

41. Wolfe, D. A., On the class of partially sequential two-sample test procedures, *J. Am. Statist. Assoc.* **72**, 202–205 (1977).

42. Zelen, M., Play the winner rule and the controlled clinical trial, *J. Am. Statist. Assoc.* **64**, 131–146 (1969).

CHAPTER 5

The Design and Analysis Aspects of Cluster Randomized Trials

Hrishikesh Chakraborty

Statistics and Epidemiology, RTI International, Research Triangle Park, North Carolina

5.1 INTRODUCTION: CLUSTER RANDOMIZED TRIALS

In a simple randomized trial, an individual is the unit of randomization, but in a cluster randomized trial (CRT), a group is the unit of randomization. For example, in a simple drug trial or in a vaccine trial, individual subjects are randomized to a drug/vaccine or placebo group or to different competing drug arms. Also known as *group randomized trials*, CRTs randomize groups such as hospitals, clinicians, medical practices, schools, households, villages, communities, or administrative boundaries. Cluster randomized trials, where clusters can be formed on the basis of natural grouping or geographic boundaries, are accepted as the gold standard for the evaluation of new health interventions [41] such as neonatal mortality rate, episiotomy rate, and postpartum hemorrhage rate.

Hayes and Bennett [41] identified several reasons for adopting cluster randomized trials:

1. Some intervention trials, such as hospital intervention trials and educational intervention trials, have to be implemented at the cluster level to avoid the resentment or contamination that could occur if certain interventions were provided for some individuals but not others in a cluster.

2. CRTs are preferred to capture the mass effect on disease of applying an intervention to a large proportion of community members, such as reduction of early neonatal mortality by providing advanced training to birth attendants in a rural setting.

3. Cluster randomized trials are useful after efficacy has been established at the individual level and there is a desire to measure intervention effectiveness at the community level. Many of the difficulties encountered in the design and analysis of cluster randomized trials arise from their dual nature, focusing, on both the individual and the cluster.

Statistical Advances in the Biomedical Sciences, edited by Atanu Biswas, Sujay Datta, Jason P. Fine, and Mark R. Segal

There are several advantages to using a cluster randomized design, Clusters are physically separated from each other and interventions are administered to the whole group; all group members receiving the same treatment can help minimize contamination by stopping or reducing the spread of an intervention to the control group [11,47,68]. Cluster randomized designs can be used for convenience, to aid in cost reduction, to eliminate potential ethical problems, to increase administrative efficiency, and to provide less intrusive randomization [20,26,68]. Cluster randomized trials can also be used for implementing interventions that can be conducted only at the cluster level, such as the hospital or community level [41].

The main disadvantage of cluster randomized trials is that participants within a given cluster often tend to respond in a similar manner, and thus their data can no longer be assumed to be independent of one another. Therefore, there are two sources of correlation: between and within clusters. Between-cluster (intercluster) correlation measures the variation in outcomes across clusters (intracluster). Within-cluster correlation occurs when subjects in a cluster are influenced by common factors, such as age, ethnicity, gender, geographic, socioeconomic, and political factors [11,47].

Some studies have incorrectly analyzed trial data as though the unit of allocation had been the individual participant. This incorrect analysis is often referred to as "unit of analysis error" [87] because the unit of analysis is different from the unit of allocation. If the clustering is ignored and CRTs are analyzed as though individuals had been randomized, resulting P values will be artificially small, resulting in false-positive conclusions that the intervention had an effect. Because individuals within clusters tend to be more alike, the independent information contributed in a cluster randomized sample is usually less than that of an individually randomized trial; thus, the power of the study is reduced [41,46].

Numerous studies have shown problems with the reporting of CRTs. Divine et al. [14] reviewed 54 published papers on physicians' behavior from a broad selection of journals and found that 70% used the wrong unit of analysis. After reanalyzing the data, they found only four statistically significant measures in the original analysis, whereas eight of nine studies had reported statistically significant findings when they used the wrong unit of analysis. A similar study of 21 public health published papers showed that only 57% accounted for clustering in their analyses [74]. Several other reviewers have found similar results [18,32]. Furthermore, intraclass correlation (ICC) values are not reported in published literature. For example, MacLennan et al. [54] found that ICCs were reported in only 6 out of 149 trials, Eldridge et al. [30], in 13 out of 152 trials; and Isaakidis and Ioannidis [44] in only 1 out of 51 trials.

Two approaches are used in cluster randomized trials: one drawing cluster-level inferences and the other drawing, individual-level inference. To draw cluster-level inferences for cluster randomized trials, we need to assess outcomes only at the level of the cluster, keeping the unit of analysis the same as the unit of randomization. We might measure a dichotomous outcome (e.g., whether the practices hospital, or community, was a "success" or a "failure") or a continuous outcome (e.g., the percentage of individuals in the cluster who benefited). In both scenarios, we obtain one outcome measurement from each randomized unit, and we perform the analysis as if the groups were individuals by using the standard methods. This approach has two major limitations:

1. we may end up with fewer data points than a simple trial involving substantially fewer participants because cluster randomized trials are likely to randomize fewer clusters than most simple trials and hence have lower statistical power. For example, a trial might randomize 12 communities with a total of 12,000 inhabitants. Analyzing by community, we end up with only 12 observations.

2. Not all groups are the same size, and we would give the same weight to a village of 1000 inhabitants as to a village of 50 inhabitants.

The Guideline trial [35] is a more recently published CRT in which the unit of randomization and the unit of analysis are the same. As an example, the Guideline trial is a two-arm cluster randomized controlled trial using hospitals as units of randomization. Nineteen hospitals in three urban districts of Argentina and Uruguay were randomized to either (1) a multifaceted behavioral intervention to develop and implement guidelines about episiotomy use and management of the third stage of labor, or (2) a control group that continued with the usual in-service training activities. The main outcomes to be assessed were the use of episiotomies and of oxytocin during the third stage of labor.

On the other hand, when we randomize by cluster and draw inferences similar to those from the individually randomized trial, complications and statistical challenges arise in the design and analysis of the trial. In this situation, we need to account for within- and between-cluster correlation in the design and analysis of the cluster clinical trials. The FIRST BREATH trial [9] is one such trial where randomization was done by community and the inference will be drawn at the individual level. The FIRST BREATH trial is a cluster randomized controlled trial to assess the effect of training and implementation of a neonatal resuscitation education program for all birth attendants in intervention clusters on 7-day neonatal mortality in communities across six different countries. The primary hypothesis is that, among infants with birth weights of 1500 g or more born in the selected communities, an educational intervention based on the Neonatal Resuscitation Program (NRP) will decrease early neonatal mortality (7 days) by at least 20% (from 25 to 20 deaths per 1000 live births) compared to control communities. The clusters for this study are communities, defined as geographic areas characterized by physical or administrative boundaries with an average of 500 births per year. Each community within a research unit is randomized into either the control or the intervention group. The birth attendants in all communities are trained in the revised *essential newborn care* (ENC) training, and the birth attendants in the intervention communities also receive additional training in the American Academy of Pediatrics Neonatal Resuscitation Program.

The remainder of this chapter is organized as follows. Section 5.2 reviews the methods for calculating the intracluster correlation coefficient for categorical and continuous variables and the methods used to calculate the confidence interval for ICC. Section 5.3 discusses the sample size calculation for cluster randomized trials. Section 5.4 discusses the analysis methods related to cluster randomized trial data, and Section 5.5 discusses the major issues related to the cluster randomized trial and future directions.

5.2 INTRACLUSTER CORRELATION COEFFICIENT AND CONFIDENCE INTERVAL

When we randomize by cluster and draw inference, the individual level, we need to account for the within- and between-cluster correlations. The intracluster correlation coefficient (ICC) measures the degree of similarity between responses from subjects within the same cluster. Because cluster members are similar, the variance within a cluster is less than that expected from randomly assigned individuals. The degree to which the within-cluster variance is less than would be otherwise expected can be expressed as ICC. For example, when we randomize communities in the FIRST BREATH trial, residents in one community may share the same resources, nutrition, education, and healtcare, causing their outcomes to more likely be similar to each other than to the outcomes of residents of a different community. The ICC value describes the extent to which two members of a community are more similar than two people from different communities. There are several different methods of calculating the ICC. The most popular is the analysis of variance (ANOVA) method with a formula derived by Fleiss [36] that uses mean square values from a one-way ANOVA to estimate the ICC. The ICC, denoted by ρ (rho), is calculated by dividing the between-cluster variance by the total variance. If $\rho = 0$, then individuals within the same cluster are no more correlated with each other than with individuals in different clusters. If $\rho = 1$, then there is no variability within a cluster, and individuals within the same cluster respond identically [11]. The ANOVA method was originally proposed for continuous variables, but various authors have subsequently shown that the method is valid for both categorical and continuous variables [31,61]. For the continuous response, let $Y_{ij}(i = 1, \ldots, k; j = 1, \ldots, m_i)$ be the response for k clusters with the ith cluster containing m_i individuals and $N = \sum_{i=1}^{k} m_i$. The mean response for the ith group is $\bar{Y}_i = \sum_{j=1}^{m_i} Y_{ij}/m_i$, and the grand mean of all observations is $\bar{Y} = \sum_{i=1}^{k} \sum_{j=1}^{m_i} Y_{ij}/N$. The ANOVA estimator of ICC is defined as

$$\hat{\rho} = \frac{\text{MSB} - \text{MSW}}{\text{MSB} + (m - 1)\text{MSW}},$$

where m is the cluster size, MSB is the mean square between clusters defined as $\text{MSB} = \frac{1}{k-1}\sum m_i(\bar{Y}_{i.} - \bar{Y}..)^2$, and MSW is the mean square within clusters defined as $\text{MSW} = \frac{1}{N-k}\sum \sum (\bar{Y}_{ij} - \bar{Y}_{i.})^2$ When cluster size varies, we can replace the cluster size m with the average cluster size m_0, where $m_0 = \bar{m} + \{\sum_i (m_i - \bar{m})^2/N(N - 1)\bar{m}\}$.

For the binary response case, let us introduce the responses $X_{ij}(i = 1, \ldots, k; j = 1, \ldots, m_i)$ for k clusters with the ith cluster containing m_i individuals. Then the total number of successes in the ith cluster is $Y_i = \sum_{j=1}^{m_i} X_{ij}$. For binary data, MSB, MSW, and m_0 can be defined as follows:

$$\text{MSB} = \frac{1}{k-1}\left\{\sum_{i=1}^{k}\frac{Y_i^2}{m_i} - \frac{1}{N}\left(\sum_{i=1}^{k}Y_i\right)^2\right\}, \quad \text{MSW} = \frac{1}{N-k}\left\{\sum_{i=1}^{k}Y_i - \sum_{i=1}^{k}\frac{Y_i^2}{m_i}\right\},$$

$$m_0 = \frac{1}{k-1}\left\{N - \frac{1}{N}\sum_{i=1}^{k}m_i^2\right\},$$

where K is the number of clusters, N is the total number of subjects in the sample, and m_i is the number of subjects in the ith cluster [21,68].

The average cluster size approximation tends to slightly underestimate the required sample size, but the effect will be negligible if the resulting total sample size requirement in each group is moderately large (≥ 100). A conservative approach would be to replace average cluster size with the largest expected cluster size in the sample [19].

For continuous variables, the intralcluster correlation ρ must satisfy the inequality $\rho \geq -1/(m_{max} - 1)$, where m_{max} is the size of me largest group [69]. For binary variables, a much more stringent constraint [65] can be shown as

$$\rho \geq \frac{-1}{(m_{max} - 1)} + \frac{w(1 - w)}{m_{max}(m_{max} - 1)\pi(1 - \pi)},$$

where $\Pr(X_{ij} = 1) = \pi$, $w = \pi m_{max} - \text{int}(\pi m_{max})$, and $\text{int}(\pi m_{max})$ denotes the integer part.

In addition to the ANOVA method, several methods have been proposed and refined by several authors to estimate the ICC for binary data. These include moment estimators [48,80,86,90], estimators with direct probabilistic interpretation [37,55], estimators based on direct calculation of correlation within each group [45,17,72,53], and extended quasilikelihood and pseudolikelihood estimators [63,8,57]. Ridout et al. [69] performed an extensive simulation to compare several of these methods. Their simulation study shows that the ANOVA estimators, a few of the moment estimators, and an estimator with a direct probabilistic interpretation all performed well with low bias and smaller standard deviations. There are also additional common correlation models published by different authors who assumed beta binomial distribution [49,65], correlated binomial distribution [49,1] and correlated probit distribution [64]. ICC also has been applied to conduct a sensitivity analysis where ICC is used to measure the effectiveness of an experimental treatment [3].

Although in theory the ICC could be negative, in practice this almost never occurs. If it becomes negative, the researcher usually assumes it to be zero and analyzes the data using the methods for simple randomized trials. In most human studies, ICC values are between 0 and 1 [47,2]. Chakraborty et al. [9,10] presented a simulation technique for calculating an ICC estimate and its 95% confidence interval for various cluster size and number combinations for binary responses when the ICC was unknown.

There are several ways to calculate confidence intervals for the ICC. These include the following methods based on approximation to the F distribution: a procedure based on modifying the solution for the balanced case and the Thomas–Hultquist procedure [81,15], a procedure based on Fisher's transformation [25], a procedure based on the standard error of the ICC estimate including the confidence limit based on Smith's formula [77] and the formula derived by Swiger et al. [78], and a confidence limit based on maximum-likelihood theory [27,75,76]. Donner and Koval [28] showed that the procedure based on Fisher's transformation is a highly accurate approximation of the true variance of ICC estimates in a broad range of circumstances in moderately large sample sizes. Dormer and Wells [29] conducted a Monte Carlo simulation study under the one-way random-effect model to compare six different confidence interval methods to obtain the two-sided confidence intervals; they conclude that the method based on the large-sample standard error of the sample ICC derived by Smith provides consistently good coverage for all ICC values and recommend its use over the other methods.

5.3 SAMPLE SIZE CALCULATION FOR CLUSTER RANDOMIZED TRIALS

Designs commonly used in cluster randomized trials include completely randomized, stratified, and matched pair. In a completely randomized design, interventions are allocated randomly to clusters. This design is suitable when randomizing a large number of clusters. For a small number of clusters, completely randomized designs are likely to produce unbalanced treatment groups with respect to baseline characteristics. In a stratified design, clusters are grouped in homogenous strata and are then randomly allocated to interventions. Stratification by cluster size is often regarded as advantageous, not only to achieve overall balance in the number of individuals assigned to each arm of the trial but also because cluster size may be a surrogate for within-cluster dynamics that are predictive of outcomes. Other common stratification factors include geographic area and socioeconomic status. A matched-pair design is an extreme form of stratification, where each stratum consists of only two clusters randomly assigned to different arms. The main advantage of this design is its very tight and explicit balancing of important baseline risk factors [16].

In a cluster randomized trial, groups of individuals are randomized together rather than individually. The sample size required for a CRT depends on the magnitude of the ICC. To obtain an accurate sample size estimate in any cluster design setting, one must account for the variation within and between clusters. The ICC is the amount of variation in the data that can be explained by the variation between clusters [6,7]. Since the variance is affected by the cluster design, the sample size required for a certain power and effect size is also affected. Because the sample size is directly proportional to the variance, we can simply use the standard methods and multiply the results by the appropriate variance inflation factor (VIF), also called the *design effect* $[1 + (m - 1)\rho]$, where m is the average cluster size and ρ is the estimate of the ICC calculated on the basis of a presumed value for the ICC and estimated cluster size. If the cluster size varies, then m can be replaced by the average cluster size m_0 for a slight underestimate or the maximum cluster size for a conservative estimate [25]. The design effect accounts for the similarities among clustered subjects, because there is a net loss of independent data. The design effect is the ratio of the total number of subjects required using cluster randomization to the number required using individual randomization [47]. The design effect will always be greater than one, although it may take values close to one. The larger the ICC, the larger the design effect and the more subjects are needed [46].

For example, in a two-arm simple randomized trial, let P_1 and P_2 indicate the population proportions of interest for the intervention and control group, respectively. Then the required number of subjects per group for a two proportion difference in a cluster trial is

$$n = \frac{(Z_{\alpha/2} + Z_\beta)^2 [P_1(1 - P_1) + P_2(1 - P_2)][1 + (m - 1)\rho]}{(P_1 - P_2)^2},$$

where Z_α and Z_β correspond to the critical values for a normal distribution for error rates α and β. However, if it is a cluster randomized trial, then the required sample size must be multiplied by the quantity $[1 + (m - 1)\rho]$ to account for the cluster trial.

During the design phase, an ICC value is often based on the most relevant estimate from earlier studies or from pilot study data. Since those estimates are often imprecise, researchers may use a more conservative upper 95% confidence limit for ICC, an extremely conservative approach that usually requires a larger sample size. It has been shown using a Bayesian simulation approach that allowances can be made for ICC imprecision when designing cluster randomized trials [83]. In planning trials, the advantages of a design that randomizes clusters of individuals must be weighed against the disadvantages in terms of statistical power, cross-contamination, and cost.

Two major difficulties arise in a sample size calculation for clustered randomized studies: (1) the number of units in each cluster, called *cluster size*, tends to vary with a certain distribution; and (2) observations within each cluster are correlated. Dormer and Klar [25] present a sample size formula for clustered binary data when cluster sizes are constant. Their test statistic is based on the binary proportion estimator obtained by assigning equal weights to all units Lee and Dubin [51] develop a sample size formula for clustered binary data with variable cluster sizes. They propose estimating the binary proportion by assigning equal weights to clusters regardless of their sizes to simplify the derivation of their sample size formula.

It is well known that the application of standard sample size approaches to cluster randomization designs may lead to seriously underpowered studies, and that the application of standard statistical methods could lead to spurious statistical significance. The severity of this problem increases not only with the magnitude of the intracluster correlation but also with the average cluster size. Increasing the number of clusters is a more efficient.method of gaining statistical power than is increasing total sample size. Sometimes it is easier to add observations to existing clusters than to add more clusters; however, increasing the average cluster size can increase the power only to a certain point [16].

Chakraborty et al. [9] explain how to use a simulation technique at the design phase of a trial to estimate the required sample size by simulating the ICC estimate and its 95% confidence interval for various cluster sizes and number of cluster combinations for binary responses. A common design effect is usually assumed across intervention groups during the analysis of CRT data. But it is not true in most cases after the end of the intervention period. To combat this problem, Chakraborty et al. [10] used a simulation technique to show that the ICC value depends on the effect size distribution in addition to the cluster size and number of clusters. They also showed how to adjust for the ICC value at the design phase of the trial according to the prediction that the overall effect size will change at the end of the trial.

Hayes and Bennett [41] published a different set of formulas for sample size calculation for rates per person-year, proportions, and means for both unmatched and pair-matched trials where they expressed the formulas in terms of the coefficient of variation (SD/mean) of cluster rate, proportions, or means. The main limitations of this method are that they assumed the observed cluster rates or means or proportions to be approximately normally distributed, all clusters to be of equal size, and the between-cluster coefficient of variation to be equal in both treatment groups. If all of these assumptions hold for a given study, this may be a simple alternative method to implement.

5.4 ANALYSIS OF CLUSTER RANDOMIZED TRIAL DATA

Statistical methods for the analysis of cluster randomized trials are not well established compared to those for individually randomized trials. Fisher's classical theory of experimental design assumes that the experimental unit that is randomized is the unit of analysis [58]. The statistical challenges of cluster randomization trials arise because inferences are frequently applied at the level of individual subjects while randomization took place at the cluster level. The discrepancy between the unit of randomization and the analytic unit means that the standard statistical methods for analysis are not applicable [16]. Cornfield [12] brought the analytical implications of cluster randomization to widespread attention in the health research community. Donnar and Klar [22,23] have provided a review of the extensive development that has occurred since.

While analyses of CRT data are more problematic than analyses of data from a straightforward individual randomization trial, in some situations cluster randomization is the only practical option for addressing certain questions. When a cluster design is used, it is essential that the analysis address the clustering approach appropriately [5]. If clustering is ignored during data analysis, the within-cluster variance and between-cluster variance will be mixed, leading to an underestimate of the overall variance and providing inaccurately small p values and narrow confidence intervals [25,66,73]. This problem of erroneous statistical significance increases with the magnitude of the ICC and average cluster size [25]. As we know, the individuals within a cluster are not independen, and applying traditional statistical methods to cluster randomized observations is not feasible without adjusting for the correlation.

There are statistical techniques for appropriate analyses of cluster randomized trials in cluster-level and individual-level analyses. In cluster-level analysis, also called *analysis by allocation unit*, we summarize individual observations within a cluster to a single summary measure, such as the cluster mean or proportion, and then use standard statistical methods to analyze these summary measures as if they were primary observations. This approach reduces the sample size to the number of clusters, reducing statistical power and degrees of freedom of the test. There is no ICC adjustment issue in this type of analysis because the randomization unit and the analysis unit are both the same. Most of the standard statistical analysis techniques can be used to draw cluster-level inference including a simple *t* test, weighted and unweighted linear regression, and random-effects meta-analysis. For example, compare two groups in a cluster randomized trial using the *t* test, applied to cluster-specific outcome measures, and weighted by the number of patients in each cluster [22,23,46]. Continuous outcome variables from a paired cluster randomized trial can be analyzed using the paired *t* test at the cluster level. This analysis is fully efficient when there is no variability in cluster size in a balanced design scenario [16]. Some researchers [39,56] prefer to use permutation tests rather than the paired *t* test to avoid the normality and homogeneity assumptions. But other researchers [24,42] found that in fairly small samples the *t* test is a remarkably robust to departure from the underlying homogeneity of variance and normality assumptions. Rosner and Hennekens [71] showed that a *t* test can be used to adjust for unaccounted baseline differences between treatment groups for matching case–control and cohort studies.

CRTs have many parallels in meta-analysis since meta-analysis also involves combining information from different units (trials) of varying sizes. Random effect meta-analysis pools the summary statistics across clusters rather than across studies and uses a maximum-likelihood estimation method [13,82]. The parameter estimates from different analyses are expected to differ substantially only if there are large differences in cluster size and/or cluster-specific outcome proportions.

When the randomization unit and the inference unit are different—when we randomize by cluster and draw conclusions about individual characteristics—we must be very careful to choose the correct analysis method. Any analysis method that accounts for clustering in some way would be appropriate for analyzing cluster clinical trial data. Parameter estimates, and in particular their standard errors, differ markedly depending on the choice of analysis method, even when the analysis methods are based on a common underlying principle. The simple analysis strategy is to ignore the clustering and apply a standard statistical approach, treating individual data as independent observations but using the variance inflation factor $[1 + (m - 1)\rho]$, where m is the average cluster size and ρ is the estimate of the ICC, to adjust the variance before hypothesis testing. If the cluster size varies, m can be replaced by the average cluster size m_0 to correct the variance used in calculating the test statistics [60]. The degrees of freedom for the revised test statistics are based on the number of clusters, not the total sample size.

Several different statistical methods allow analysis at the level of the individual while also accounting for the clustering in the data. The ideal information to extract from a cluster randomized trial is a direct estimate of the required effect measure (e.g., odds ratio with its confidence interval) from an analysis that properly accounts for the cluster design:

1. Binary variable analysis might be based on a standard logistic regression with robust standard errors, where the standard logistic regression model uses but adjusts the standard errors to allow for clustering, and the robust standard errors are calculated using

the "sandwich" variance estimator [43,88] and modified by Rogers [70] to allow for clustering. The regression coefficients (e.g., log odds ratios) estimators are identical to those for the standard logistic regression model because those are unaffected by this procedure.

2. Mixed-effects linear models are often used to analyze the continuous outcome data from completely or stratified CRT using the generalized leas-square method [79,85,89].

3. Generalized estimating equations (GEEs) extend the standard logistic regression model to allow for clustering. This is achieved by specifying a correlation matrix that describes the association between different individuals in the same cluster in terms of additional parameters [52].

Different correlation matrix types can be assumed, and if the sample size is large enough, both the regression coefficients and their standard errors are correct in the sense that they are consistently estimated whether robust standard errors are specified. The parameter estimates from GEE should not be interpreted as corresponding to the parameter estimates from random-effects models. Parameter estimates from GEE are described as "population-averaged" interpretations [86,91] because they are averaged across the values of the cluster-level random effect in the context of longitudinal data analysis. Other types of analysis such as multilevel modeling [40] or hierarchical linear modeling [4,67], and "variance components analysis" can also be used to analyze cluster randomized trial data. In addition to specifying the primary data analysis in advance, different sensitivity analysis methods can be considered for cluster clinical trials including presenting results using different analysis methods.

5.5 CONCLUDING REMARKS

We all recognize that clusters are made up of individuals, that there may be more individuals in one cluster than in another, and that the intralcluster correlation co-efficient plays an important role in design and analysis of CRTs. Intracluster correlation may appear small compared with other types of correlations, but small values can have a substantial impact on design and analysis of CRTs. Ignoring the small correlation may lead to standard errors for intervention effects that are too small, confidence intervals that are too narrow, and P values that are too small. Calculating and reporting the ICC is another important aspect of the CRT because different methods are available to calculate the ICC, and different software packages can provide different ICC results [84]. Adjustments for covariates also play a very important role in calculating ICC value; adjusting for covariates usually leads to smaller ICCs because some of the between-cluster variation can be explained by cluster-level factors [34]. There is a need for more publications presenting the ICC values from different studies; different kinds of variables can be stored in a central database, and an advanced computer interactive simulation program could be required for an ICC estimate to design studies.

Despite the advanced and well-established principles of the design and analysis of CRT, there remains considerable uncertainty about the relative merits of the different methods, and further illustrations of the alternatives and their performance in different settings are required. There are several ways to conduct CRT, and the choices regarding both the method of analysis and the variables included in the model can make important differences to the conclusions. In the context of estimating variance and covariance, parameters by different procedures appear to perform better in certain situations and with certain types of outcome variables [33]. Methodological studies fail to provide clear guidance as to the best approaches or methods to implement in different trial scenarios for planning and analyzing CRT for individual-level inference. For example, we need methods to deal with covariate adjustments for varying lengths of follow-up of subjects, to analyze ordinal, multinomial and time-to-event data, and to implement analysis with missing values at both the individual and the cluster levels. In the absence of proper methodological direction, we must check the sensitivity of the conclusions and model assumptions very carefully before interpreting CRT results. For Bayesian analysis, one needs to check the impact of changing the assumed priors. Murray [61] reviewed the recent methodological developments regarding the design and analysts of a CRT and concluded that the methods required for a CRT are not as simple as those for randomized clinical trials but there are several readily available methods for the design and analysis of a CRT.

REFERENCES

1. Altham, P. M. E., Two generalizations of the binomial distribution, *Appl. Statist.* **27**, 162–167 (1978).

2. Baskerville, B., Hogg, W., and Lemelin, J., The effect of cluster randomization on sample size in prevention research, *J. Fam. Practice* **50**, W241–W246 (2001).

3. Bradley, R. A. and Schumann, D. E. W., The comparison of sensitivities of similar experiments: Applications, *Biometrics*, **13**, 496–510 (1957).

4. Bryk, A. S. and Raudenbush, S. W., *Hierarchical Linear Models: Applications and Data Analysis Methods*, Sage Publications, Newbury Park, CA, 1992.

5. Campbell, M. K. and Grimshaw, J. M., Cluster randomized trials: Time for improvement, *Br. Med. J.* **317**, 1171–1172 (1998).

6. Campbell, M., Grimshaw, J., and Elbourne, D., Intracluster correlation coefficients in cluster randomized trials: Empirical insights into how should they be reported, *BioMed Central Med. Res. Methodol.*, **4**(9) (2004).

7. Campbell, M., Thomas, S., Ramsay, C., MacLennan, G., and Grimshaw, J., Sample size calculator for cluster randomized trials, *Comput. Biol. Med.* **34**, 113–125 (2004).

8. Carroll, R. J. and Ruppert, D., *Transformation and Weighting in Regression*, Chapman & Hall, London, 1988.

9. Chakraborty, H., Bartz, J., Carlo, A. W., Hartwell, T. D., and Write, L. L., A simulation-based technique to estimate intra-cluster correlation for categorical variables, *Proc. 26th Annual Meeting of the Society for Clinical Trials Conf.*, Portland, OR, 2005.

10. Chakraborty, H., Bartz, J., Carlo, A. W., Hartwell, T. D., and Write, L. L., Sample size adjustment to maintain power in cluster randomized trials, *Clin. Trials* **3**(Suppl.) 208 (2006).

11. Chuang, J., Hripcsak, G. and Heitjan, D., Design and analysis of controlled trials in naturally clustered environments, *J. Am. Med. Inform. Assoc.* **9**, 230–238 (2002).

12. Cornfield, J., Randomization by group: A formal analysis, *Am. J. Epidemiol.* **108**, 100–102 (1978).

13. DerSimonian, R. and Laird, N., Meta-analysis in clinical trials, *Control. Clin. Trials* **7**, 177–188 (1986).

14. Divine, G. W., Brown, J. T., and Frazier, L. M., Unit of analysis error in studies about physicians̆ patient care behavior: Studies about physicians' patient care behavior, *J. Gen. Intern. Med.* **7**, 623–629 (1992).

15. Donner, A., The use of correlation and regression in the analysis of family resemblance, *Am. J. Epidemiol.* **110**, 335–342 (1979).

16. Donner, A., Some aspects of the design and analysis of cluster randomization trials, *Appl. Statis.* **47**(1) 95–113 (1998).

17. Donner, A., A review of inference procedures for the intraclass correlation coefficient in the one-way random effect model, *Int. Statist. Rev.* **54**, 67–82 (1986).

18. Donner, A., Brown, K. S., and Brasher, P., A methodological review of nontherapeutic intervention trials employing cluster randomization, 1979–1989, *Int. J. Epidemiol.* **19**, 795–800 (1990).

19. Donner, A., Birkett, N., and Buck, C., Randomization by cluster: Sample size requirements and analysis, *Am. J. Epidemiol.* **114** (6), 906–914 (1981).

20. Donner, A. and Donald, A., Analysis of data arising from a stratified design with the cluster as unit of randomization, *Statist. Med.* **6**, 43–52 (1987).

21. Donner, A. and Eliasziw, M., Methodology for inferences concerning familial correlations: A review, *J. Clin. Epidemiol.* **44**, 449–455 (1991).

22. Donner, A. and Klar, N., Methods for comparing event rates in intervention studies when the unit of allocation is a cluster, *Am. J. Epidemiol.* **140**, 279–289 (1994).

23. Donner, A. and Klar, N., Cluster randomisation trials in epidemiology: theory and application, *J. Statist. Plan. Infer.* **42**, 37–56, (1994).

24. Donner, A. and Klar, N., Statistical considerations in the design and analysis of community intervention trials, *J. Clin. Epidemiol.* **49**, 435–439 (1996).

25. Donner, A. and Klar, N. S., *Design and Analysis of Cluster Randomisation Trials in Health Research*, Hodder Arnold, London, 2000.

26. Donner, A. and Klar, N., Pitfalls of and controversies in cluster randomized trials, *Am. J. Public Health* **94**, 416–422 (2004).

27. Donner, A. and Koval, J. J., The estimation of intraclass correlation in the analysis of family data, *Biometrics* **36**, 19–25 (1980).

28. Donner, A. and Koval, J. J., A note on the accuracy of Fishers approximation to the large-sample variance of an intraclass correlation, *Commun. Statist. Simul. Comput.* **12**, 443–449 (1983).

29. Donner, A. and Wells, G., A comparison of confidence interval methods for the intraclass correlation coefficient, *Biometrics* **42**, 401–412 (1986).

30. Eldridge, S. M., Ashby, D., Feder, G. S., Rudnicka, A. R., and Ukoumunne, O. C., Lessons for cluster randomized trials in the twenty-first century: A systematic review of trials in primary care, *Clin. Trials* **1**, 80–90 (2004).

31. Elston, R. C., Response to query: Estimating "heritability" of a dichotomous trait, *Biometrics* **33**, 232–233 (1977).

32. Ennett, S. T., Tobler, N. S., Rignwalt, C. L., and Flewelling, R. L., How effective is drug abuse resistance education? A meta-analysis of Project DARE outcome evaluations, *Am. J. Public Health* **84**, 1394–1401 (1994).

33. Evans, B. A., Feng, Z., and Peterson, A. V., A comparison of generalized linear mixed model procedures with estimating equations for variance and covariance parameter estimation in longitudinal studies and group randomized trials, *Statist. Med.* **20**(22), 3353–3373 (2001).

34. Feng, Z., Diehr, P., Yasui, Y., Evans, B., Beresford, S., and Koepsell, T. D., Explaining community-level variance in group randomized trials, *Statist. Med.* **18**, 539–556 (1999).

35. Fernando, A., Buekens, P., Bergel, E., Belizan, J. M., Kropp, N., and Wright, L., A cluster randomized controlled trial of a behavioral intervention to facilitate the development and implementation of clinical practice guidelines in Latin American maternity hospitals: The Guidelines Trial: Study protocol, *BioMed Central Women's Health* **5**(4), (2005).

36. Fleiss, J. L., Reliability of measurement, In *The Design and Analysis of Clinical Experiments*, Wiley, New York, 1986, pp. 1–32.

37. Fleiss, J. L. and Cuzick, J., The reliability of dichotomous judgments: Unequal numbers of judges per subject, *Appl. Psychol. Meas.* **3**, 537–542 (1979).

38. Fisher, R. A., *Statistical Methods for Research Workers*, Oliver and Boyd, Edinburgh, 1925.

39. Gail, M. H., Byar, D. P., Pechacek, T. F., and Corle, D. K., Aspect of statistical design for the community intervention trial for smoking cessation (COMMIT), *Control. Clin. Trials* **13**, 6–21 (1992).

40. Goldstein, H., *Multilevel Statistical Models*, Edward Arnold, London and Halstead Press, New York, 1995.

41. Hayes, R. J. and Bennett, S., Simple sample size calculation for cluster-randomized trials, *Int. Epidemiol. Assoc.* **28**, 319–326 (1999).

42. Heeren, T. and d'Agostino, R., Robustness of the two-independent-samples *t*-test when applied to ordinal scale data, *Statist. Med.* **6**, 79–90 (1987).

43. Huber, P. J., *Robust Statistics*, Wiley, New York, 1981.

44. Isaakidis, P. and Ioannidis, J. P. A., Evaluation of cluster randomized controlled trials in sub-saharan Africa, *Am. J. Epidemiol.* **158**, 921–926 (2003).

45. Karlin, S., Cameron, P. E., and Williams, P., Sibling and parent-offspring correction with variable family age, *Proc. Nat. Acad. Sci. USA* **78**, 2664–2668 (1981).

46. Kerry, S. and Bland, J., Trials which randomize practices I: How should they be analysed?, *Fam. Practice* **15**, 80–83 (1998).

47. Killip, S., Mahfoud, Z., and Pearce, K., What is intracluster correlation coefficient? Crucial concepts for primary care researchers, *Ann. Fam. Med.* **2**, 204–208 (2004).

48. Kleinman, J. C., Proportions with extraneous variance: Single and independent samples, *J. Am. Statist. Assoc.* **68**, 46–54 (1973).

49. Kupper, L. L. and Haseman, J. K., The use of a correlated binomial model for the analysis of certain toxicological experiments, *Biometrics* **34**, 69–76 (1978).

50. Laird, N., Lange, N., and Stram, D., Maximum likelihood computations with repeated measures: Application of the EM algorithm, *J. Am. Statist. Assoc.* **82**, 97–105 (1987).

51. Lee, E. and Dubin, N., Estimation and sample size considerations for clustered binary responses, *Statist. Med.* **13**, 1241–252 (1994).

52. Liang, K. Y. and Zeger, S. L., Longitudinal data analysis using generalized linear models, *Biometrikcu* **72**, 13–22 (1986).

53. Lipsitz, S. R., Laird, N. M., and Brennan, T. A., Simple moment estimates of the k-coefficient and its varianve, *Appl. Statist.* **43**, 309–323 (1994).

54. MacLennan, G. S., Ramsay, C. R., Mollison, J., Campbell, M. K., Grimshaw J.M., and Thomas, R. E., Room for improvement in the reporting of cluster randomised trials in behaviour change research, *Control. Clin. Trials* **24**, 69S–70S (2003).

55. Mak, T. K., Analysing interclass correlation for dichotomous variables, *Appl. Statist.* **37**, 344–352 (1988).

56. Maritz, J. S. and Jarrett, R. G., The use of statistics to examine the association between fluoride in drinking water and cancer death rates, *Appl. Statist.* **32**(2), 97–101 (1983).

57. McCullagh, P. and Nelder, J. A., *Generalized Linear Models*, 2nd ed., Chapman & Hall, London, 1989.

58. McKinlay, S. M., Stone, E. J., and Zucker, D. M., Research design and analysis issues, *Health Educ. Q.* **16**, 307–313 (1989).

59. Mickey, R. M. and Goodwin, G. D., The magnitude and variability of design effects for community intervention studies, *Am. J. Epidemiol* **137**, 9–18 (1993).

60. Murray, D. M., *Design and Analysis of Group-Randomized Trials*, Oxford Univ. Press, Oxford, UK, 1998.

61. Murray, D. M., Design and analysis of group-randomized trials: A review of recent methodological developments, *Public Health Matters* **94**(3), 423–432 (2004).

62. Murray, D. M., Brenda, L. R., Peter, J. H., Arthur, V. P., Dennis, V. A., Anthony, B., Gilbert, J. B., Richard, I. E., Brian, R. F., Robert, F., Greg, J. G., Pat, M. M., Mario, O., MaryAnn, P., Cheryl, L. P., and Steven, P. S. Intraclass correlation among common measures of adolescent smoking: Estimates, correlates and applications smoking prevention studies, *Am. J. Epidemiol.* **140**(11), 1038–1050 (1994).

63. Nelder, J. A. and Pregibon, D., An extended quasi-likelihood function, *Biometrikas* **74**, 221–232 (1987).

64. Ochi, Y. and Prentice, L. R., Likelihood inference in a correlated probit regression model, *Biometrika* **71**, 531–543 (1984).

65. Prentice, R. L., Binary regression using an extended beta-binomial distribution, with discussion of correlation induced by covariate measurement errors, *J. Am. Statist. Assoc.* **81**, 321–327 (1986).

66. Puffer, S., Torgerson, D., and Watson, J., Evidence for risk of bias in cluster randomised trials: Review of recent trials published in three general medical journals, *Br. Med. J.* **327**, 785–788 (2003).

67. Raudenbush, S. W., Statistical analysis and optimal design for cluster randomized trials, *Psychol. Meth.* **2**(2) 173–185 (1997).

68. Reading, R., Harvey, I., and Mclean, M., Cluster randomised trials in maternal and child health: Implications for power and sample size, *Arch. Dis. Child* **82**, 79–83 (2000).

69. Ridout, M., Demetrio, C., and Firth, D., Estimating intraclass correlation for binary data, *Biometrics* **55**, 137–148 (1999).

70. Rogers, W. H., sgl7: Regression standard errors in clustered sampler, *Stata Tech. Bull.* **13**, 19–23 (1993).

71. Rosner, B. and Hennekens, C. H., Analytic methods in matched pair epidemiological studies, *Int. J. Epidemiol.* **7**, 367–372 (1978).

72. Schouten, H. J. A., Nominal scale agreement among observers, *Psychometrica* **51**, 453–466 (1986).

73. Schulz, K. F., Subverting randomisation in controlled trials, *J. Am. Med. Assoc.* 456–458 (1995).

74. Simpson, J. M., Klar, N., and Donner, A., Accounting for cluster randomisation—a review of primary prevention trials, 1990 through 1993, *Am. J. Public Health* **85**, 1378–1383 (1995).

75. Smith, C. A. B., Estimating genetic correlations, *Ann. Hum. Genet.* **44**, 265–284 (1980).

76. Smith, C. A. B., Further remarks on estimating genetic correlations, *Ann. Hum. Genet.* **44**, 95–105 (1980).

77. Smith, C. A. B., On the estimation of intraclass correlation, *Ann. Hum. Genet.* **21**, 363–373 (1956).

78. Swiger, L. A., Harvey, L. R., Everson, D. O., and Gregory, K. E., The variance of intraclass correlation involving groups with one observation, *Biometrics* **20**, 81–826 (1964).

79. Stiratelli, R., Laird, N. M., and Ware, J. H., Random effects models for serial observations with binary response, *Biometrics* **40**, 961–971 (1984).

80. Tamura, R. N. and Young, S. S., A stabilized moment estimator for the beta-binomial distribution, *Biometrics* **43**, 813–824 (1987) [correction: **50**, 321 (1994)].

81. Thomas, J. D. and Hultquist, R. A., Interval estimation for the unbalanced case of the one-way random effects model, *Ann. Statist.* **6**, 582–587 (1978).

82. Thompson, S. G., Pyke, S. D., and Hardy, R. J., Design and analysis of paired cluster randomized trails: An application of meta-analysis techniques, *Statist. Med.* **16**, 2063–2079 (1997).

83. Turner, R. M., Prevost, A. T., and Thompson, S. G., Allowing for imprecision of the intracluster correlation coefficient in the design of cluster randomized trials, *Statist. Med.* **23**, 1195–1214 (2004).

84. Ukoumunne, O. C., Gulliford, M. C., Chinn, S., Sterne, J. A., and Burney, P. G., Methods for evaluating area-wide and organization-based interventions in health and health care: A systematic review, *Health Technol. Assess.* **3**(5), iii–92 (1999).

85. Ware, J. H., Linear models for the analysis of longitudinal studies, *Am. Statist.* **39**(2), 95–101 (1985).

86. Williams, D. A., Extra-binomial variation in logistic linear models, *Appl. Statist.* **31**, 144–148 (1982).

87. Whiting-O'Keefe, Q. E., Henke, C., and Simborg, D.W., Choosing the correct unit of analysis in medical care experiments, *Med. Care.* **22**, 1101–1114 (1984).

88. White, H., A heteroskedasticity-consistent covariance matrix estimator and a direct test for heteroskedasticity, Econometrica **48**, 817–838 (1980).

89. Wolfinger, R. D. and O'Connell, M., Generalized linear mixed models: A pseudo-likelihood approach, *J. Statist. Comput. Simul.* **48**, 233–243 (1993).

90. Yamamoto, E. and Yanagimoto, T., Moment estimators for the binomial distribution, *J. Appl. Statist.* **19**, 273–283 (1992).

91. Zeger, S. L. and Liang, K.-Y., An overview of methods for the analysis of longitudinal data, *Statist. Med.* **11**, 1825–1839 (1992).

PART II

Epidemiology

CHAPTER 6

HIV Dynamics Modeling and Prediction of Clinical Outcomes in AIDS Clinical Research

Yangxin Huang

Department of Epidemiology and Biostatistics, University of South Florida, Tampa, Florida

Hulin Wu

Department of Biostatistics and Computational Biology, University of Rochester, New York

6.1 INTRODUCTION

A virological marker, the number of human immunodeficiency virus type 1 (HIV-1) RNA copies in plasma (viral load), is currently used to evaluate anti-HIV therapies in AIDS clinical trials. Antiretroviral treatment of HIV-1-infected patients with highly active antiretroviral therapies (HAART), consisting of reverse transcriptase inhibitor (RTI) drugs and protease inhibitor (PI) drugs, results in several orders of magnitude reduction of viral load. The rapid decay in viral load can be observed in a relatively short term [24,25,34], and it either can be sustained or may be followed by a resurgence of virus within months [19]. The resurgence of virus may be caused by drug resistance, noncompliance, pharmacokinetics problems, and other factors during therapy. Mathematical models, describing the dynamics of HIV and its host cells, have been of essential importance in understanding the biological mechanisms of HIV infection, the pathogenesis of AIDS progression, and the role of clinical factors in antiviral activities.

Many HIV dynamic models have been proposed by AIDS researchers [6,12,14,23–25,32,34,29] since the mid-1990s to provide theoretical principles in guiding the development of treatment strategies for HIV-infected patients, and have been used to quantify short-term dynamics. Unfortunately, these models are of limited utility in interpreting long-term HIV

Statistical Advances in the Biomedical Sciences, edited by Atanu Biswas, Sujay Datta, Jason P. Fine, and Mark R. Segal
Copyright © 2008 John Wiley & Sons, Inc.

dynamic data from clinical trials. The main reason is that few parameters of these models can be estimated uniquely from viral load data, because simplified and linearized models have often been used to characterize the viral dynamics based on observed viral load data [12,23,24,32,34]. Although these models are useful and convenient for quantifying short-term viral dynamics, they cannot be used to characterize more complex long-term virological response. Huang et al. [15–17] developed a set of relatively simplified models, a system of differential equations with time-varying parameters, to characterize long-term viral dynamics. In the models, they considered several factors related to the resurgence of viral load, such as the pharmacokinetics, and compliance with treatment and drug susceptibility, and thus these models are flexible enough to quantify long-term HIV dynamics.

Bayesian statistics has made great progress in recent years. For various models, parameter estimation and statistical inference are carried out via the Markov chain Monte Carlo (MCMC) procedures [10,11,17,26,31]. The Bayesian method for HIV dynamic models was investigated by Han et al. [11], Huang et al. [17], and Putter et al. [26]. Han et al. and Putter et al. considered a dynamic model with the assumption that the number of uninfected target cells remained constant during a treatment, and they used short-term viral load data only to estimate parameters. In addition, they did not consider the fact of variability in drug susceptibility (resistance) and adherence in the presence of antiretroviral therapy. Huang et al. [16,17] extended the model to characterize long-term viral dynamics described by a system of nonlinear differential equations with time-varying drug efficacy.

Although prediction methods for deterministic models have been proposed under the Bayesian framework in other research fields such as prediction of whale populations [27], those models are essentially different from HIV dynamic models. In this chapter, we consider a model designed to characterize long-term viral dynamics developed by Huang et al. [16,17] and combine the Bayesian analytic methods and mixed-effect modeling to investigate individual/population predictions of clinical outcomes based on the proposed model. Predictions of clinical outcomes are very important for clinicians in developing individualized treatments, making clinical decisions, and optimizing a treatment strategy.

The technical details on the Bayesian analysis of hierarchical nonlinear mixed-effect models can be found in the literature [11,16,17,31]. We employ the model and estimation approach proposed by Huang et al. [16,17] to address the predictions of clinical outcome in this chapter the remainder of which is organized as follows. In Section 6.2, we briefly describe the viral dynamic model and treatment effect models. The Bayesian modeling approach for hierarchical nonlinear mixed-effect models for predictions of virological responses is discussed in Section 6.3. A simulation study is presented to illustrate our methodology in Section 6.4. We apply the proposed methodology to a clinical dataset and present the results in Section 6.5. Finally, the chapter concludes with some discussions in Section 6.6.

6.2 HIV DYNAMIC MODEL AND TREATMENT EFFECT MODELS

Details of the HIV dynamic models and treatment effect models are described in Huang et al. [16,17]. For completeness, a brief summary of these models is given as follows.

6.2.1 HIV Dynamic Model

Mathematical models for HIV dynamics have been developed since the mid-1980s. The detailed surveys can be found in Perelson and Nelson [25], Nowak and May [21], and

Tan and Wu [29]. We consider a simplified HIV dynamic model with antiviral treatment as follows [15–17]

$$\frac{d}{dt}T = \lambda - d_T T - [1 - \gamma(t)]k\,TV,$$

$$\frac{d}{dt}T^* = [1 - \gamma(t)]k\,TV - \delta\,T^*, \tag{6.1}$$

$$\frac{d}{dt}V = N\delta T^* - cV,$$

where the three differential equations represent three compartments: target uninfected cells (T), infected cells (T^*), and free virions (V). The parameter λ (day^{-1} mm^{-3}) represents the rate at which new T cells are created from sources within the body, such as the thymus, d_T (day^{-1}) is the death rate of T cells, k (day^{-1} mm^{-3}) is the infection rate without treatment, δ (day^{-1}) is the death rate of infected cells, N is the number of new virions produced from each infected cell during its lifetime, and c (day^{-1}) is the clearance rate of free virions. The time-varying parameter $\gamma(t)$ is the antiviral drug efficacy at treatment time t, as defined in Section 6.2.2. In this model, the difference between the RTI and PI drug actions is not considered, but is expected to have only a small effect on long-term HIV dynamics and model predictions. If we assume that the system of Equations (6.1) is in a steady state before initiating antiretroviral treatment, then it is easy to show that the initial conditions for the system are

$$T_0 = \frac{c}{kN}, \quad T_0^* = \frac{cV_0}{\delta N}, \quad V_0 = \frac{\lambda N}{c} - \frac{d_T}{k}. \tag{6.2}$$

If the regimen is not 100% effective (does not provide perfect inhibition), the system of ordinary differential equations cannot be solved analytically. The solutions to (6.1) then have to be evaluated numerically. Let $\beta = (\phi, c, \delta, \lambda, \rho, N, k)^T$ denote a vector of parameters, where ϕ is a parameter in the treatment effect model presented below. In the estimation procedure, we only need to evaluate the logarithmic difference between observed data and numerical solutions of $V(t)$, so there is no need for an explicit solution of Equation (6.1).

6.2.2 Treatment Effect Models

Within the population of HIV virions in a human host, there is likely to be genetic diversity and corresponding diversity in sensitivity to the various antiretroviral (ARV) agents. In clinical practice, genotypic or phenotypic tests can be performed to determine the sensitivity of HIV-1 to ARV agents before a treatment regimen is selected. Here we use the phenotypic marker, the median inhibitory concentration (IC$_{50}$) [20] to quantify agent-specific drug susceptibility. To model within-host changes over time in IC$_{50}$ due to the emergence of new drug-resistant mutations, we use the following function [16]

$$\text{IC}_{50}(t) = \begin{cases} I_0 + \dfrac{I_r - I_0}{t_r}t & \text{for} \quad 0 < t < t_r \\[2ex] I_r & \text{for} \quad t \geq t_r, \end{cases} \tag{6.3}$$

where I_0 and I_r are respective values of IC$_{50}(t)$ at baseline and timepoint t_r at which the resistant mutations dominate. If $I_r = I_0$, no new drug-resistant mutation is developed during

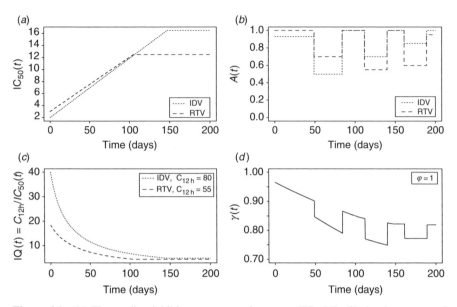

Figure 6.1 (*a*) The median inhibitory concentration curve [$IC_{50}(t)$]; (*b*) the timecourse of adherence [$A(t)$]; (*c*) the timecourse of inhibitory quotient [$IQ(t)$]; (*d*) the timecourse of drug efficacy [$\gamma(t)$].

treatment. Although more complicated models for median inhibitory concentration have been proposed according to the frequencies of resistant mutations and cross-resistance patterns [3,30], in clinical studies or clinical practice it is common to collect IC_{50} values only at baseline and failure time as designed in A5055. Thus, this function may serve as a good approximation. As examples, such functions for the ritonavir (RTV) and indinavir (IDV) drugs are plotted in Figure 6.6.1a.

Poor adherence to a treatment regimen is one of the major causes of treatment failure [2,18]. Patients may occasionally miss doses, may misunderstand prescription instructions, or may miss multiple consecutive doses for various reasons. These deviations from prescribed dosing affect drug exposure in predictable ways. We use the following model to represent adherence:

$$A_d(t) = \begin{cases} 1 & \text{for} \quad T_k < t \leq T_{k+1}, \text{ if all doses are taken in } [T_k, T_{k+1}] \\ R_d & \text{for} \quad T_k < t \leq T_{k+1}, \text{ if } 100R_d\% \text{ doses are taken in } [T_k, T_{k+1}], \end{cases} \quad (6.4)$$

where $0 \leq R_d < 1$ ($d = 1,2$), with R_d indicating the adherence rate for drug d (in our study, we focus on the two PI drugs of the prescribed regimen). Time T_k denotes the adherence evaluation time at the kth clinical visit. As an example, Figure 6.6.1b shows the effect of adherence over time for RTV and IDV drugs.

The HAART, containing two or more nucleoside/nonnucleoside reverse transcriptase inhibitors (RTIs) and protease inhibitors (PI), has proved to be effective in reducing the amount of virus in the blood and tissues of HIV-infected patients. In most viral dynamic studies [5,7,25,35], investigators assumed that the drug efficacy was constant over treatment time. Drug efficacy may actually vary, however, because the concentrations of ARV drugs and other factors (e.g., emergence of drug-resistant mutations) vary during treatment [5,8,25], and thus the drugs may not be perfectly effective. Also, patients' viral load may rebound as a result of drug resistance, nonadherence, and other factors [9]. To model the relationship of drug exposure and resistance with antiviral efficacy, we employ the following modified E_{max} model [28] to represent the time-varying drug efficacy for two ARV agents within a class (e.g., the two PI drugs IDV and RTV)

$$\gamma(t) = \frac{IQ_1(t)A_1(t) + IQ_2(t)A_2(t)}{\phi + IQ_1(t)A_1(t) + IQ_2(t)A_2(t)}, \tag{6.5}$$

where $IQ_d(t) = C_{12h}^d/IC_{50}^d(t)$ ($d = 1, 2$) denotes the inhibitory quotient (IQ) [13]; C_{12h}^d and IC_{50}^d ($d = 1, 2$) are the trough levels of drug concentration in plasma (measured 12 h after the doses had been taken) and the median inhibitory concentrations for the two agents, respectively. Note that C_{12h} could be replaced by other pharmacokinetic parameters such as the area under the plasma concentration-time curve (AUC). Although $IC_{50}(t)$ can be measured by phenotype assays *in vitro*, it may not be equivalent to the $IC_{50}(t)$ *in vivo*. The parameter ϕ is used to quantify the conversion between *in vitro* and *in vivo* IC_{50} that can be estimated from clinical data. The value of $\gamma(t)$ ranges from 0 to 1. If $\gamma(t) = 1$, the drug is 100% effective, whereas if $\gamma(t) = 0$, the drug has no effect. Note that if C_{12h}^d, $A_d(t)$, and $IC_{50}^d(t)$ are measured from a clinical study and ϕ can be estimated from clinical data, then the time-varying drug efficacy $\gamma(t)$ can be estimated for the whole period of antiviral treatment. Similarly, we can model the combined drug efficacy of an HAART regimen with both PI and RTI agents. Lack of adherence reduces the drug exposure, which can be quantified by Equation (6.4), and thus, on the basis of formula (6.5), reduces the drug efficacy, which, in turn, can affect virological response. The examples of the timecourses of the inhibitory quotients and the drug efficacy $\gamma(t)$ with $\phi = 1$, $C_{12h}^1 = 80$ and $C_{12h}^2 = 50$ for two PI drugs are shown in Figures 6.1c and 6.1d, respectively.

6.3 STATISTICAL METHODS FOR PREDICTIONS OF CLINICAL OUTCOMES

6.3.1 Bayesian Nonlinear Mixed-Effects Model

A number of studies investigated various statistical methods, including Bayesian approaches, to fit viral dynamic models using short-term viral load data [23,11,34,35,33]. Huang et al. [16,17] extended the existing methods to model long-term HIV dynamics. In this chapter, we focus on the predictions of virological response under the setting of a hierarchical Bayesian nonlinear model.

We denote the number of subjects by n and the number of measurements on the ith subject by m_i. For notational convenience, let $\boldsymbol{\mu} = (\ln \phi, \ln c, \ln \delta, \ln \lambda, \ln d_T, \ln N, \ln k)^T$, $\boldsymbol{\theta}_i = (\ln \phi_i,$ $\ln c_i, \ln \delta_i, \ln \lambda_i, \ln d_{Ti}, \ln N_i, \ln k_i)^T$, $\Theta = \{\boldsymbol{\theta}_i = 1, \ldots, n\}$ $\Theta_{\{i\}} = \{\boldsymbol{\theta}_l, l \neq i\}$ and $\mathbf{Y} = \{y_{ij}, i =$ $1, \ldots, n; j = 1, \ldots, m_i\}$. Let $f_{ij}(\boldsymbol{\theta}_i, t_j) = \log_{10}(V_i(\boldsymbol{\theta}_i, t_j))$, where $V_i(\boldsymbol{\theta}_i, t_j)$ denotes the numerical solution of the differential equations (6.1) for the ith subject at time t_j. Let $y_{ij}(t)$ and $e_i(t_j)$ denote the repeated measurements of common logarithmic viral load and a measurement error with mean zero, respectively. The Bayesian nonlinear mixed-effects model can be written in the following three stages [4,15–17]:

Stage 1: *Within-Subject Variation*:

$$y_i = f_i(\theta_i) + e_i, \quad e_i|\sigma^2,\theta_i \sim \mathcal{N}(0,\sigma^2 I_{m_i}) \tag{6.6}$$

Here, $y_i = (y_{i1}(t_1), \ldots, y_{im_i}(t_{m_i}))^T$, $f_i(\theta_i) = (f_{i1}(\theta_i, t_1), \ldots, f_{im_i}(\theta_i, t_{m_i}))^T$, $e_i = (e_i(t_1), \ldots, e_i(t_{m_i}))^T$

Stage 2: *Between-Subject Variation*:

$$\theta_i = \mu + b_i, \quad [b_i|\Sigma] \sim \mathcal{N}(0,\Sigma) \tag{6.7}$$

Stage 3: *Hyperprior Distributions*:

$$\sigma^{-2} \sim \text{Ga}(a,b), \quad \mu \sim \mathcal{N}(\eta,\Lambda), \quad \Sigma^{-1} \sim \text{Wi}(\Omega,v) \tag{6.8}$$

where the mutually independent Gamma (Ga), normal ($\mathcal{N}$), and Wishart (Wi) prior distributions are chosen to facilitate computations [4]. The values of hyper-parameters were determined from previous studies and literature [11,12,21–25,32].

Following the studies by Davidian and Giltinan [4] and Gelfand et al. [10], we have shown [16] from (6.6)–(6.8) that the full conditional distributions for the parameters σ^{-2}, μ, and Σ^{-1} may be written explicitly as

$$[\sigma^{-2}|\mu,\Sigma^{-1},\Theta,Y] \sim \text{Ga}\left(a + \frac{\sum_{i=1}^n m_i}{2}, \left\{\frac{1}{b} + \frac{1}{2}\sum_{i=1}^n \sum_{j=1}^{m_i} [y_{ij} - f_{ij}(\theta_i,t_j)]^2\right\}^{-1}\right) \tag{6.9}$$

$$[\mu|\sigma^{-2},\Sigma^{-1},\Theta,Y] \sim \mathcal{N}\left((n\Sigma^{-1} + \Lambda^{-1})^{-1}\left(\Sigma^{-1}\sum_{i=1}^n \theta_i + \Lambda^{-1}\eta\right), (n\Sigma^{-1} + \Lambda^{-1})^{-1}\right) \tag{6.10}$$

$$[\Sigma^{-1}|\sigma^{-2},\mu,\Theta,Y] \sim \text{Wi}\left(\left[\Omega^{-1} + \sum_{i=1}^n (\theta_i - \mu)(\theta_i - \mu)^T\right]^{-1}, n+v\right) \tag{6.11}$$

Here, however, the full conditional distribution of each θ_i, given the remaining parameters and the data, cannot be calculated explicitly. The distribution of $[\theta_i|\sigma^{-2}, \mu, \Sigma^{-1}, \Theta_{\{i\}}, Y]$ has a density function that is proportional to

$$\exp\left\{\frac{-\sigma^{-2}}{2}\sum_{j=1}^{m_i}[y_{ij} - f_{ij}(\theta_i,t_j)]^2 - \frac{1}{2}(\theta_i - \mu)^T\Sigma^{-1}(\theta_i - \mu)\right\} \tag{6.12}$$

The Markov chain Monte Carlo (MCMC) scheme for drawing samples from the posterior distributions of all parameters in the three-stage model presented above is obtained by iterating between the following two steps: (1) sampling from one of the conditional distributions (6.9)–(6.11) and (2) sampling from the expression (6.12). To implement an MCMC algorithm, the Gibbs sampler is used here to update σ^{-2}, μ, and Σ^{-1}, while we update θ_i ($i = 1, 2, \ldots, n$) using the Metropolis–Hastings algorithm. See Huang et al. [15–17] for detailed discussions of the Bayesian modeling approach, including the choice of the hyperparameters and implementation of the MCMC procedures [31].

6.3.2 Predictions Using the Bayesian Mixed-Effects Modeling Approach

In this section, we propose the methods for predictions of virological responses. We apply the proposed deterministic antiviral response model to characterize long-term viral dynamics and use the Bayesian modeling approach for predictions. We investigate two prediction problems: (1) predicting the virological response for a new subject, and (2) predicting future virological responses for one of the individuals who has some data available. A method for doing this is to calculate the posterior predictive distribution of responses based on the model specified by (6.6)–(6.8) and the clinical data.

Let Y denote the data from all patients for which posterior distributions of all population and individual parameters $\Psi = (\sigma^{-2}, \mu, \Sigma^{-1})$ and θ_i ($i = 1, 2, \ldots, n$) are available. Denote by y^* the virological response (viral load in $\log_{10}$ scale) for an individual. Then the posterior predictive distribution of interest is $p(y^*|Y)$, which can be expressed as

$$p(y^*|Y) = \int p(y^*, \theta^*, \Psi|Y) d\theta^* \, d\Psi = \int p(y^*|\theta^*, \Psi) p(\theta^*|\Psi) p(\Psi|Y) d\theta^* \, d\Psi \quad (6.13)$$

where θ^* denotes the parameter vector for the patient of interest. If the patient is new and no information is available, the population dynamic parameter μ can be used. We denote the G usable iterations from the MCMC sampler by

$$\{\theta_i^{(g)}, \Psi^{(g)}\} = \{\theta_i^{(g)}, (\sigma^{-2})^{(g)}, \mu^{(g)}, (\Sigma^{-1})^{(g)}\}, \quad g = 1, \ldots, G. \quad (6.14)$$

For the prediction problem of a new patient, denote the conditional predictive distribution by $p(y^*|\Psi, Y)$. We can now obtain a Monte Carlo estimator of $p(y^*|Y)$ by using the iterations of MCMC sampler in (6.14):

$$\hat{p}(y^*|Y) = \frac{1}{G} \sum_{g=1}^{G} p(y^*|\Psi^{(g)}, Y) \quad (6.15)$$

Let us now consider the second prediction problem, predicting future virological responses for one of the n patients, for example, patient k. Denote the conditional predictive distribution of y^* by $p(y^*|\theta_k, \Psi, Y)$. We can again obtain a Monte Carlo estimator of $p(y^*|Y)$ by using the iterations of MCMC sampler in 6.14:

$$\hat{p}(y^*|Y) = \frac{1}{G} \sum_{g=1}^{G} p(y^*|\theta_k^{(g)}, \Psi^{(g)}, Y). \quad (6.16)$$

In (6.15) and (6.16), we used the MCMC iterations to compute the predictive distributions using their known functional forms. We can also incorporate this step directly into the MCMC sampler by adding either $p(y^*|\Psi, Y)$ or $p(y^*|\theta_k, \Psi, Y)$ to the set of conditional distributions (6.9)–(6.11), from which we sample. We then obtain the MCMC iterations $(y^*)^{(g)}$, $g = 1, 2, \ldots, G$ as part of the simulation output, and we can readily analyze them.

6.4   SIMULATION STUDY

In this section, we present a simulation example to illustrate the proposed Bayesian prediction approach. The scenario we consider is as follows. We simulate a clinical trial with 20 HIV-1-infected patients receiving long-term antiviral treatment. For each patient, we assume that measurements of viral load are taken at 15–30 timepoints ranging from day 0 to day 200 of follow-up. We consider the Bayesian nonlinear mixed-effects model (6.6)–(6.8), but for illustration purposes and for computational convenience, we propose to estimate only the two parameters $\log c$ and $\log \delta$ that are identifiable in our model [17], and assume that the other five parameters ($\log \phi$, $\log \lambda$, $\log \rho$, $\log N$, $\log k$) are fixed to be ($\log \phi$, $\log \lambda$, $\log \rho$, $\log N$, $\log k$) = (2.5, 4.6, -2.0, 6.9, -9.6). These values were chosen from previous studies in the literature [7,21,25]. From the discussion in Section 6.3.1, the prior distribution for $\mu = (\log c, \log \delta)^T$ was assumed to be $\mathcal{N}(\eta, \Lambda)$, where Λ is a diagonal matrix. The details of the prior construction for unknown parameters are discussed in Huang et al. [16]. Thus, the values of hyperparameters are chosen as follow: $a = 4.5$, $b = 9.0$, $\nu = 5.0$, $\eta = (1.1, -1.0)^T$, $\Lambda = \text{diag}(10^3, 10^3)$, $\Omega = \text{diag}(2.5, 2.5)$.

Note that the noninformative priors are chosen for both $\log c$ and $\log \delta$. The values of the hyperparameters were determined based on several studies in the literature [11,23]. In addition, the data for the pharmacokinetic factor (C_{12h}), phenotype marker (baseline and failure IC$_{50}$ s), adherence and the baseline viral load (V_0) were taken from an AIDS clinical trial study (Section 6.5). The true individual dynamic parameters, $\log c_i$ and $\log \delta_i$, are generated by $\log c_i = \log c + b_{1i}$ and $\log \delta_i = \log \delta + b_{2i}$, where $\log c = 1.1$ and $\log \delta = -1.0$ are the true values of population parameters, and both b_{1i} and b_{2i} are random effects following a normal distribution with mean 0 and standard deviation 0.2.

On the basis of the generated true individual parameters and five known constant parameters, as well as clinical factor data $[C_{12h}, IC_{50}, \text{and } A(t)]$, the observations y_{ij} (the common logarithm of total viral load) are generated by perturbing the solution to the differential equations (6.1) with a within-subject measurement error, $y_{ij} = \log_{10}(V_{ij}) + e_i$, where V_{ij} is the numerical solution for viral load obtained from the differential equations (6.1) for the ith subject at time t_j. It is assumed that the within-subject measurement error e_i is normally distributed with $\mathcal{N}(0, 0.1^2)$. We apply the proposed Bayesian prediction approach to estimate the dynamic parameters via the MCMC procedure.

We consider the following two prediction scenarios: (1) completely removing the data of a simulated patient when estimating dynamic parameters, with the objective of predicting the viral load responses of this patient; and (2) removing only some of the late measurements of viral load, and trying to use the remaining data to predict the future viral load responses. As an example, Figure 6.2 displays the predicted curves beginning at a point denoted by the circle with generated viral load data in $\log_{10}$ scale (solid circles) for two subjects: subject 14 (21 viral load measurements generated) with viral rebound; subject 20 (30 viral load measurements generated) with a rapid decay of viral load in the short term, followed by a rebound of the virus. We show the predicted curve in the case of completely removing the data for the

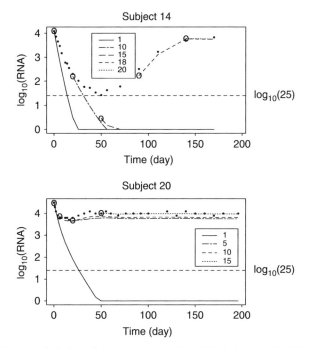

Figure 6.2 Generated viral load data in $\log_{10}$ scale (solid circles) and individual prediction curves beginning at the point denoted by the circle for the two representative subjects. The values in the legend denote the number of viral load measurements used for predicting future virological responses. The HIV-1 RNA measurements below a limit of detection of 25 copies/mL are imputed by 25 copies/mL (dashed horizontal line).

prediction subject in Figure 6.2 (corresponding to legend boxes (1)). In this case, we used the estimated population parameters and the baseline viral load from this subject to predict future virological responses. We can see that the prediction power is very poor in this case since no subject-specific information is available and only population parameter estimates are used.

However, if some data from this subject (i.e., some observed viral load measurements) are available after initiation of treatment, the subject-specific information can be combined with the information from other subjects together to fit the model and to predict future virological responses of this subject. The results indicate that the predictions have been greatly improved. It is seen from Figure 6.6.2 that the more information from this subject is provided, the better predictions are achieved. Note that the numbers in the legend denote the numbers of viral load measurements available for predicting future virological responses.

6.5 CLINICAL DATA ANALYSIS

We apply the proposed methodology to the data from an AIDS clinical study. This study was a phase I/II, randomized, open-label, 24-week comparative study of the pharmacokinetic, tolerability, and antiretroviral effects of two regimens of indinavir (IDV) and ritonavir (RTV), plus two nucleoside analog reverse transcriptase inhibitors on HIV-1-infected subjects failing protease inhibitor-containing antiretroviral therapies [1]. The 44 subjects were randomly assigned to the two treatment arm A [IDV 800 mg twice daily (q12h) + RTV 200 mg q12h] and arm B (IDV 400 mg q12h + RTV 400 mg q12h). Out of the 44 subjects, 42 subjects are included in the analysis; of the remaining 2 subjects, 1 was excluded from the analysis since the pharmacokinetic parameters were not obtained and the other was excluded since PhenoSense HIV could not be completed on this subject because of an atypical genetic sequence that causes the viral genome to be cut by an enzyme used in the assay. Plasma HIV-1 RNA (viral load) measurements were taken at days 0, 7, 14, 28, 56, 84, 112, 140, and 168 of follow-up. The data for pharmacokinetic parameters (C_{12h}), phenotype marker (baseline and failure IC_{50}s) and adherence from this study were also used in our modeling. The adherence data were determined from the pillcount data. More detailed description of this study can be found in the publication by Acosta et al. [1].

Similar to the simulation study discussed in Section 6.4, the prior distribution for $\mu = ($log ϕ, log c, log δ, log λ, log ρ, log N, log $k)^T$ is assumed to be $\mathcal{N}(\eta, \Lambda)$, where Λ is a diagonal matrix. We chose the values of the hyperparameters [7,11,21,23] as follows:

$$a = 4.5, \quad b = 9.0, \quad \nu = 8.0, \quad \eta = (2.5, 1.1, -1.0, 4.6, -2.3, 6.9, -9.0)^T,$$

$$\Lambda = \text{diag}(1000.0, 0.0025, 0.0025, 0.0025, 0.0025, 0.0025, 0.001),$$

$$\Omega = \text{diag}(1.25, 2.5, 2.5, 2.0, 2.0, 2.0, 2.0).$$

The MCMC techniques introduced in Section 6.3 were used to obtain the prediction results, which are summarized below. Figure 6.3 presents the observed viral load data in $\log_{10}$ scale (solid circles) and the predicted curves (solid) beginning at the point denoted by the circle as well as the corresponding 95% prediction credible intervals (dotted curves) for two subjects: one subject with viral rebound and one subject with virological success. We find the prediction results to have patterns similar to what we observed in the simulation study: (1) the prediction power for a new subject is very poor (see Fig. 6.3, legend boxes (1)) (2) we should notice in Figure 6.3 that since the viral load measurements below a limit of detection of 25 copies/mL are imputed by 25 copies/mL, while our method can predict the exact viral load values [i.e., those below the dashed horizontal line of $\log_{10}(25)$]. In this sense, when we predict future virological responses for one of the 42 subjects based on partial information about this subject, the more the amount of information from this individual, the better the predictions. We find, not surprisingly, that the predicted values are closer to the observed values when more information from this subject is used. The 95% prediction credible interval associated with each predicted value generally covers the observed value in almost all cases where enough information is available. This suggests that the proposed Bayesian prediction approach under the framework of the nonlinear mixed-effect model performs reasonably well.

6.6 CONCLUDING REMARKS

This chapter uses the MCMC techniques to estimate dynamic parameters in a hierarchical nonlinear mixed-effects model and to make predictions of antiviral response. We have presented a simulation example and an actual AIDS clinical trial study to illustrate how the proposed Bayesian procedure can be applied to HIV dynamic studies for predictions of antiviral responses.

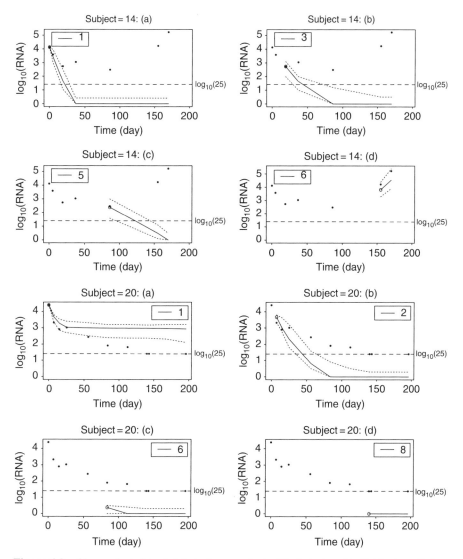

Figure 6.3 Observed viral load data in $\log_{10}$ scale (solid circles) and individual prediction curves (solid) started at the point denoted by the circle as well as the corresponding 95% prediction credible intervals (dotted curves) for the two representative subjects. The values in the legend denote the number of viral load measurements used for predicting future virological responses. The HIV-1 RNA measurements below a limit of detection of 25 copies/mL are imputed by 25 copies/mL (dashed horizontal line).

We investigated two prediction problems for an individual patient. If a patient's baseline characteristics are available, the antiviral response model established from a similar patient population can be used to predict the outcomes under the assumption of the same treatment. However, as expected, our results reveal that the prediction power is low in this case since the subject-specific information is limited. If the antiviral response data from the patient are partially available in addition to the baseline information, the predictions for this particular patient can be greatly improved since both the subject-specific information and the information borrowed from other patients are used for predictions. In addition, the proposed Bayesian approach can be employed to incorporate the prior information from existing clinical studies to increase the prediction power. We can dynamically update the prediction results, as soon as new additional data from this patient are available. Our results from both the simulation example and real data analysis confirm these arguments. The dynamic predictions of antiviral responses for individual patients will be useful for clinicians to develop individualized treatments and to make clinical decisions.

We notice that there exist some limitations and difficulties in predicting antiviral response for individual patients. In fact, patients' behavior during treatment is difficult to predict, and some unexpected events may occur during the treatment period. Thus, it is not easy to accurately predict the antiviral response for a particular individual. For example, in many cases we may not be able to predict viral load rebound until we have observed the rebound data. This is one limitation of our current model in which not all the parameters are identifiable on the basis of the measurements of viral load only. Although we have used the (informative) prior information for some of the population parameters, it can only solve the unidentifiability problem for the population parameters. For the parameters of individual patients, the identifiability problem may still exist, which may result in poor prediction for individual response, although it may still produce a good fit to the observed data. We believe that if there is no identifiability problem and there are enough data in the early stage of the treatment before viral rebound, our method should be able to predict the viral load rebound (virologic response failure). We are actively investigating this problem now. We hope to report more successful prediction results in the near future.

In clinical practice or clinical studies, we may not have frequent measurements of viral load and other data, which will make the prediction difficult. This is not only because we do not have enough data and information in this case but also because many unexpected and unpredictable events such as emergence of drug resistance, drug holidays, or other noncompliance to the therapy may occur between two clinical visits or measurements. Even in this case, our results suggest that more frequent clinical visits and monitoring are necessary in order to prevent treatment failure.

In summary, we have proposed combining a mechanism-based mathematical model and the Bayesian inference approach for antiviral response predictions for AIDS patients. Although the basic models and methodologies are not new, the application of these methods in the new model settings for this particular biomedical problem is innovative. However, further studies are warranted in order to make the proposed approaches for practical use. We expect that similar ideas and the developed Bayesian prediction methods can also be applied to other biomedical fields.

ACKNOWLEDGMENT

We thank Drs. John G. Gerber, Edward P. Acosta and other A5055 and DACS 210 study investigators for allowing us to use the clinical data from their study. This work was supported in part by NIH research grants RO1 AI052765, RO1 AI055290, and AI27658.

REFERENCES

1. Acosta, E. P., Wu, H., Hammer, S. M. et al. Comparison of two indinavir/ritonavir regimens in the treatment of HIV-infected individuals. *J. AIDS* **37**, 1358–1366 (2004).

2. Besch, C. L., Compliance in clinical trials, *AIDS* **9**, 1–10 (1995).

3. Bonhoeffer, S., Lipsitch, M., and Levin, B. R., Evaluating treatment protocols to prevent antibiotic resistance, *Proc. Natl. Acad. Sci. USA* **94**, 12106–12111 (1997).

4. Davidian, M. and Giltinan, D. M., *Nonlinear Models for Repeated Measurement Data*, Chapman & Hall, London, 1995.

5. Ding, A. A., and Wu, H., Relationships between antiviral treatment effects and biphasic viral decay rates in modeling HIV dynamics. *Math. Biosci.* **160**, 63–82 (1999).

6. Ding, A. A. and Wu, H., A comparison study of models and fitting procedures for biphasic viral decay rates in viral dynamic models, *Biometrics*, **56**, 16–23 (2000).

7. Ding, A. A. and Wu, H., Assessing antiviral potency of anti-HIV therapies *in vivo* by comparing viral decay rates in viral dynamic models, *Biostatistics* **2**, 13–29 (2001).

8. Dixit, N. M. and Perelson, A. S., Complex patterns of viral load decay during antiretroviral therapy: Influence of pharmacokinetics and intracellular delay, *J. Theor. Bio.* **226**, 95–109 (2004).

9. Fitzgerald, A. P., DeGruttola, V. G., and Vaida, F., Modelling HIV viral rebound using non-linear mixed effects models, *Statist. Med.* **21**, 2093–2108 (2002).

10. Gelfand, A. E., Hills, S. E., Racine-Poon, A., and Smith, A. F. M., Illustration of Bayesian inference in normal data models using Gibbs sampling, *J. Am. Statist. Assoc.* **85**, 972–985 (1990).

11. Han, C., Chaloner, K., and Perelson, A. S., Bayesian analysis of a population HIV dynamic model, in *Case Studies in Bayesian Statistics*, Gatsoiquiry, C., Kass, R. E., et al. eds. Springer-Verlag, New York, 2002, Vol. 6, pp. 223–237.

12. Ho, D. D., Neumann, A. U., Perelson, A. S., Chen, W., Leonard, J. M., and Markowitz, M., Rapid turnover of plasma virions and CD4 lymphocytes in HIV-1 infection, *Nature* **373**, 123–126 (1995).

13. Hsu, A., Isaacson, J., Kempf, D. J., et al. Trough concentrations-EC_{50} relationship as a predictor of viral response for ABT-378/ritonavir in treatment-experienced patients, *Proc. 40th Interscience Conf. Antimicrobial Agents and Chemotherapy.* San Francisco, Poster session **171**, 2000.

14. Huang, Y. Rosenkranz, S. L., and Wu, H., Modeling HIV dynamics and antiviral responses with consideration of time-varying drug exposures, sensitivities and adherence, *Math. Biosci.* **184**, 165–186 (2003).

15. Huang, Y. and Wu, H., Bayesian estimation of individual parameters in an HIV dynamic model using long-term viral load data, in *Deterministic and Stochastic Models of AIDS Epidemics and HIV Infections with Intervention*, Tan, W. Y., Wu, H., eds., World Scientific, Singapore, 2005, Ch. 15, pp. 361–383.

16. Huang, Y. and Wu, H., A Bayesian approach for estimating antiviral efficacy in HIV dynamic model, *J. Appl. Statist.* **33**, 155–174 (2006).

17. Huang, Y., Liu, D., and Wu, H., Hierarchical Bayesian methods for estimation of parameters in a longitudinal HIV dynamic system, *Biometrics* **62**(2), 413–423 (2006).

18. Ickovics, J. R. and Meisler, A. W., Adherence in AIDS clinical trial: A framework for clinical research and clinical care, *J. Clin. Epidemiol* **50**, 385–391 (1997).

19. Kaufmann, G. R., et al. Patterns of viral dynamics during primary human immunodeficiency virus type 1 infection, *J. Infect. Dis.* **178**, 1812–1815 (1998).

20. Molla, A. et al. Ordered accumulation of mutations in HIV protease confers resistance to ritonavir, *Nature Med.* **2**, 760–766 (1996).

21. Nowak, M. A. and May, R. M., *Virus Dynamics: Mathematical Principles of Immunology and Virology*, Oxford Univ. Press, Oxford, UK, 2000.

22. Perelson, A. S., Kirschener, D. E., and Boer, R. D., Dynamics of HIV infection of $CD4^+T$ cells, *Math. Biosci.* **114**, 81–125 (1993).

23. Perelson, A. S., Neumann, A. U., Markowitz M., Leonard, J. M., and Ho, D. D., HIV-1 dynamics *in vivo*: Virion clearance rate, infected cell life-span, viral generation time, *Science* **271**, 1582–1586 (1996).

24. Perelson, A. S. Essunger, P., Cao, Y., Vesanen, M., Hurley, A., Saksela, K., Markowitz, M., and Ho, D. D., Decay characteristics of HIV-1-infected compartments during combination therapy, *Nature* **387**, 188–191 (1997).

25. Perelson, A. S. and Nelson, P. W., Mathematical analysis of HIV-1 dynamics *in vivo*, *SIAM Rev.* **41**(1), 3–44 (1999).

26. Putter, H., Heisterkamp, S. H., Lange, J. M. A., and De Wolf, F., A Bayesian approach to parameter estimation in HIV dynamical models, *Statist. Med.* **21**, 2199–2214 (2002).

27. Raftery, A. E., Givens, G. H., and Zeh, J. E., Inference from a deterministic population dynamics model for bowhead whales, *J. Am. Statist. Asso.* **90**, 403–430 (1995).

28. Sheiner, L. B., Modeling pharmacodynamics: Parametric and nonparametric approaches. in *Variability in Drug Therapy: Description, Estimation, and Control*, Rowland, M., et al., eds., Raven Press, New York, 1985, pp. 139–152.

29. Tan, W. Y. and Wu, H., *Deterministic and Stochastic Models of AIDS Epidemics and HIV Infections with Intervention*, World Scientific, Singapore, 2005.

30. Wainberg, M. A., et al. Effectiveness of 3TC in HIV clinical trials may be due in part to the M184V substation in 3TC-resistant HIV-1 reverse transcriptase, *AIDS* **10** (suppl.) S3–S10 (1996).

31. Wakefield, J. C., The Bayesian analysis to population pharmacokinetic models, *J. Am. Statist. Assoc.* **91**, 62–75 (1996).

32. Wei, X., Ghosh, S. K., et al. Viral dynamics in human immunodeficiency virus type 1 infection, *Nature* **373**, 117–122 (1995).

33. Wu, H., Statistical methods for HIV dynamic studies in AIDS clinical trials, *Statist. Meth. Med. Res.* **14**, 171–192 (2005).

34. Wu, H., Ding, A. A., and de Gruttola, V., Estimation of HIV dynamic parameters, *Statist. Med.* **17**, 2463–2485 (1998).

35. Wu, H. and Ding, A. A., Population HIV-1 dynamics *in vivo*: Applicable models and inferential tools for virological data from AIDS clinical trials, *Biometrics* **55**, 410–418 (1999).

CHAPTER 7

Spatial Epidemiology

Lance A. Waller

Emory University, Atlanta, Georgia

7.1 SPACE AND DISEASE

The spatial distribution of cases of disease often captures the imagination of health researchers and the general public, based primarily on the notion that the observed pattern of incidence or prevalence can provide insight into the underlying mechanisms driving disease incidence, its progression, and the design and implementation of effective public health responses. Historical examples include John Snow's famous maps of cholera incidence in London neighborhoods and early maps of yellow fever incidence in relation to features of cities and docks [57,27]. More recent examples include reports of clusters of cancer cases near hazardous-waste sites. However, quantifying such hypotheses through statistical inference is a difficult task due to often subtle signals within multiple layers of noisy, nonindependent, observational data from multiple agencies collected for multiple purposes, typically other than the spatial epidemiology issue at hand. As a result, few applications have the luxury of a research design optimized for the questions of interest, or an experimental setting within which to conduct inference controlling for potential confounding factors. For these reasons, spatial epidemiologic studies encounter many complications in addition to those in traditional studies, including some unique to the geographic setting.

The field of spatial statistics involves the statistical analysis of observations with associated locations in space. These observations rarely follow a Gaussian distribution and are not independent, two mainstays in the development of statistical methods. In addition, asymptotic results take on a different flavor depending on whether we consider an increasing number of observations within a fixed study area (infill asymptotics) or an increasing number of observations in an increasing study area (increasing domain asymptotics). In response to these issues, a wide variety of statistical techniques for spatial epidemiologic inference have emerged more recently, coalescing into a collection of approaches addressing specific questions.

Statistical Advances in the Biomedical Sciences, edited by Atanu Biswas, Sujay Datta,
Jason P. Fine, and Mark R. Segal
Copyright © 2008 John Wiley & Sons, Inc.

The field of spatial epidemiology is the subject of several lengthy texts [22,23,34,54] and we present only an overview of a few particular issues here, accompanied by brief examples. We encourage the interested reader to follow up in the referenced material for more detailed development and applications.

In the sections below, we provide an overview of statistical methods commonly applied to gain insight on epidemiologic questions based on spatially referenced data. We begin with an overview of relevant spatial questions typically addressed in spatial epidemiology and of the typical data structures available, and then review and illustrate methods based on general topics.

7.2   BASIC SPATIAL QUESTIONS AND RELATED DATA

Often, our main goals in the analysis of spatial data mirror those from nonspatial data; namely, we seek to describe data patterns, measure associations, and assess variability. In the specific case of epidemiology, this often falls to assessment of associations between exposure and disease. For spatial epidemiology, we typically don't have precisely the information we want and we seek to use spatial information to fill in for data we cannot easily measure.

More specifically, consider the following basic inferential epidemiologic questions, each with a spatial dimension:

Q1. Can we quantify spatial trends and/or patterns in the location of cases? Does the risk of disease appear to vary over space?

Q2. Can we quantify spatial trends and/or patterns in regional counts of incidence or prevalence (e.g., the number of cases reported within each of a set of census regions)? Again, does the risk of disease appear to vary over space?

Q3. Can we predict ambient exposure levels at locations where no measurement is taken, based on measurements from several point locations?

Q4. Can we measure associations between disease risk at particular locations, accounting for residual spatial correlation in model error terms?

Next, consider the types of data typically accompanying each question:

D1. Residential locations of cases and controls.

D2. Reported counts of incident or prevalent cases and population sizes from census regions.

D3. Continuous observations of exposure levels at each of a number of monitoring locations.

D4. Local measures of disease incidence, prevalence, or risk and associated exposure measures.

Texts on spatial statistics often categorize methods by the data type available, but Waller and Gotway [54] note a close correspondence between the data structures D1–D4 and their associated underlying inferential questions of interest Q1–Q4. In this chapter we consider more recent developments addressing questions Q1 and Q3 (spatial point patterns and spatial prediction, respectively), and comment on relationships to questions Q2 and Q4.

7.3 QUANTIFYING PATTERN IN POINT DATA

We begin with question Q1, how we quantify the spatial pattern of observed events in space. If we consider the locations of events as random variables distributed in space, we may use a spatial point process to describe the pattern of events within our study area, denoted D. Basic questions include whether events are clustered or regularly distributed in space, and a spatial Poisson process offers a convenient reference model of spatial randomness. Diggle [16, p. 50] and Stoyan et al. [48, p. 33] define a homogeneous spatial Poisson process according to the following criteria:

(i) The number of events occurring within a finite region $A \subseteq D$ is a Poisson random variable with mean $\lambda|A|$ for some positive constant λ and $|A|$ denoting the area of A.

(ii) Given the total number of events n occurring within the area A, the locations of the events represent an independent random sample of n locations within A, where each point is equally likely to be chosen as an event location.

Cressie [13, p. 634] lists the following equivalent definition of a homogeneous spatial Poisson process:

(a) The numbers of events in nonoverlapping regions are statistically independent,

(b) For any region A within the study area

$$\lim_{|A| \to 0} \frac{\text{Pr[exactly one event in } A]}{|A|} = \lambda > 0,$$

where $|A| = $ the area of the region A, and

(c)

$$\lim_{|A| \to 0} \frac{\text{Pr[two or more events in } A]}{|A|} = 0.$$

These definitions rely on a distinction between a *point* (any location within the study area where an event could occur) and an *event* (a location within the study are where an event did occur within a particular realization of the process). Properties (i) and (a) provide motivation for extensions for the models of regional counts for question Q2, leading to development of spatial Poisson regression models with residual spatial correlation [54, Ch. 9; 4, Ch. 5].

A few features of spatial Poisson processes merit mention: (1) the constant λ represents both the *intensity* of the process and the expected number of events occurring per unit area; (2) a Poisson process assumes that event locations are independent of one another, that is, that the occurrence of an event at one location does not influence the probability of events occurring at any other locations; (3) properties (i) and (ii) provide a recipe for simulating realizations from a spatial Poisson process, enabling Monte Carlo assessments of deviations of observed patterns from a null distribution defined by the Poisson process. For instance, if we want to assess evidence of clustering of observations, we may define a test statistic summarizing some pattern aspect, calculate its value in the observed data, and then compare this value to a histogram of values obtained under repeated simulations under a Poisson process [8,54]. As a result, Monte Carlo methods are widely used in the analysis of spatial point processes, as illustrated in the example below.

If we wish to allow the intensity to vary over space, we may define a *heterogeneous* spatial Poisson process with intensity $\lambda(s)$, a function that varies (typically smoothly) across locations within the study area $s \in D$. Properties (i) and (ii) are revised to

(**i'**) The number of events occurring within a finite region A follows a Poisson distribution with mean $\int_A \lambda(s)\,ds$.

(**ii'**) Given the total number of events n occurring within an area A, the events are distributed according to a (spatial) probability distribution proportional to $\lambda(s)$.

Spatial Poisson processes provide a convenient framework for modeling spatial point patterns, but are not the only set of models for doing so. The Poisson process assumes that all events are independent of one another and models all spatial pattern via the heterogeneous intensity function; that is, we model all pattern through the *first order* (mean) properties of the process. Models assuming interdependence between event locations (e.g., contagion processes) also allow pattern due to *second-order* properties of the process. It is mathematically impossible to distinguish first-order patterns from second-order patterns based on a single realization of a point process as one could describe an aggregation equally well through a locally increased mean number (intensity) of independent observations, through a constant mean number (intensity) of dependent observations, or some combination of the two [5; 54, p. 137].

In practice, one often assumes independence between events associated with chronic outcomes (e.g., cancers) and seeks to identify spatial variation in the risk of disease by estimating the underlying intensity function associated with cases. However, the intensity of cases alone can be misleading without due consideration of the spatial distribution of the population at risk, since more cases will be observed where more persons are at risk under the null model of a constant risk of disease. As a result, most modern studies of spatial point patterns of disease incorporate a set of "controls" or nondiseased individuals sampled from the population at risk. The analyst then compares the spatial pattern of cases with that of the controls and quantifies differences to identify case patterns of interest.

To illustrate this point, consider the following dataset originally presented in Cuzick and Edwards [14]. Figure 7.1 represents the residential locations of 62 cases of childhood leukemia diagnosed between 1974 and 1982 in the North Humberside region of the United Kingdom. Also shown are the residential locations of 143 controls sampled at random from the birth register for the same years. Note the concentration of cases and controls in the southern region representing the higher population density found in the city of Hull. Without the controls, the concentration of cases might seem suspicious, illustrating the importance of considering a heterogeneous process with intensity adjusted for spatial patterns in the population at risk. Additional analyses of these data appear in Lawson [34] and Diggle [18].

Since we wish to compare patterns between cases and controls, we begin by estimating the intensity functions for each, denoted $\lambda_1(s)$ for cases and $\lambda_0(s)$ for controls. Here $s = (u,v)$ denotes any location within our study area D illustrated by the polygon containing observed case and control events in Figure 7.1.

Since an intensity function is simply a nonnormalized spatial density function, it is natural to use kernel-based smoothing methods to provide nonparametric intensity estimates [56]. The use of kernel-based intensity function estimates to identify regions of different patterns between cases and controls have been proposed for some time [10,36] and developed in depth by Kelsall and Diggle [27–29]. The approach provides both visual and inferential output addressing the questions of interest, but applications in the literature are somewhat limited by software

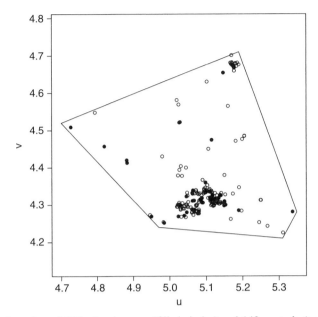

Figure 7.1 Location of 62 leukemia cases (filled circles) and 143 controls (open circles) in the North Humberside region of the United Kingdom (data originally from Cuzick and Edwards [14]. Cases were diagnosed between 1974 and 1982 and controls sampled from the birth register for the same years.

availability. The analyses below were implemented in the R language [43] using the libraries splancs and KernSmooth.

To start, suppose we estimate $\lambda_1(s)$ by

$$\tilde{\lambda}_1(s) = \frac{1}{|D|b} \sum_{i=1}^{n_1} \mathrm{Kern}\left(\frac{\|s - s_i\|}{b}\right), \qquad (7.1)$$

where s is a location within D; s_i, $i = 1, \ldots, n_1$ represent the locations of n_1 cases; $\|s - s_i\|$ denotes the distance between a point location s and an observed event location s_i; Kern($\cdot$) is a kernel function satisfying $\int_D \mathrm{Kern}(s)ds = 1$; $|D|$ denotes the area of D; and b denotes a smoothing parameter (bandwidth). We define $\tilde{\lambda}_0(s)$ similarly.

A note on the scaling factor $1/|D|$ is in order. Scaling by $1/|D|$ results in a kernel estimate that integrates to $N/|D|$, the average number of events per unit area, and omitting the $1/|D|$ term generates a kernel estimate integrating to N. Wand and Jones [56] suggest omitting the scaling factor $1/|D|$, and Diggle [17] suggests scaling by $1/|D|$ to provide an estimate expressed as average event counts rather than a probability. Perhaps stricter notation would define the estimate in Equation (7.1) as "proportional to" rather than "equal to" the estimated intensity function, but, in a sense, scaling by $1/N$ (density), 1 (intensity), or $1/|D|$ (expected events per area) is somewhat irrelevant for visualization of the local peaks and valleys for a particular process.

Kelsall and Diggle [27,28] note that, under an assumption of independent heterogeneous Poisson processes for cases and controls, when we condition on the observed numbers of cases and controls, the data are equivalent to two independent random samples from (spatial) density function

$$f(s) = \lambda_1(s) \Big/ \int_D \lambda_1(s^*)ds^*$$

for cases and

$$g(s) = \lambda_0(s) \Big/ \int_D \lambda_0(s^*)ds^*$$

for controls, where s^* represents any location within D. Conditional on the observed case and control totals (n_1 and n_0), Kelsall and Diggle [27,28] build inference based on the natural logarithm of the ratio of the two spatial densities

$$r(s) = log\{f(s)/g(s)\},$$

a quantity related to the logarithm of the *relative risk* of observing a case rather than a control at location s in D.

To illustrate the approach, Figure 7.2 gives the kernel density estimates for cases and controls in the North Humberside data for a common bandwidth of 0.05 distance units. We note an overall similarity between the general patterns in Figure 7.2. Taking the ratio of the two elements, we obtain the (log) relative risk surface shown in Figure 7.3.

We see some spatial variation in the log relative risk surface, most notably a generally decreasing west-to-east trend with a marked increase on the westernmost edge. For inference, we need to assess the variability of the estimated log relative risk surface under a null hypothesis of constant risk everywhere [i.e., a flat (log) relative risk surface]. Kelsall and Diggle [28]

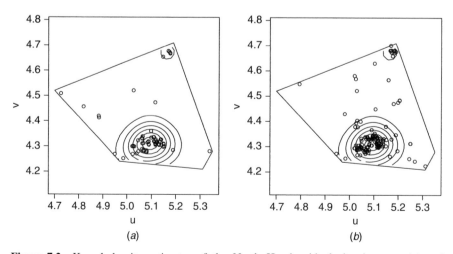

Figure 7.2   Kernel density estimates of the North Humberside leukemia cases (*a*) and controls (*b*) in North Humberside with bandwith set to 0.05.

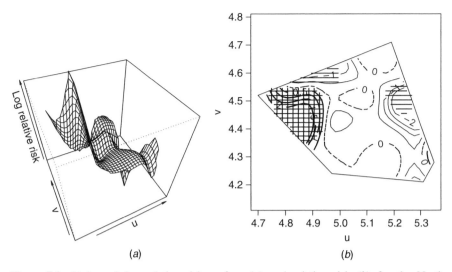

Figure 7.3 Estimated log relative risk surface (*a*) and relative risk (*b*) for the North Humberside leukemia data with bandwidth set to 0.05 for both cases and controls.

offer a convenient way to operationalize this null hypothesis, conditional on the observed case and control locations, based on a "random labeling" of n_1 of the $n_1 + n_0$ observed locations as cases. For a grid of points, we calculate the observed log relative risk surface illustrated in Figure 7.3*a*. For each of a large number of random samples of n_1 cases from the set of $n_1 + n_0$ observed locations (500 such relabelings in our examples below), we recalculate the log relative risk surface at the same grid of points, then compare the log relative risk estimate at each location to those simulated under the random labelings.

We note that the random labeling inferences represent pointwise rather than simultaneous confidence bounds across all grid locations, due to the large number of grid points where we make comparisons.

For the North Humberside data, we mark grid locations in Figure 7.3*b* with the observed log relative risk value falling above the 97.5th percentile of random labeling values by a " + " and those falling below the 2.5th percentile by a " − ." We note that the peak on the western side of the study area and the troughs on the northern and eastern edges each fall outside the random labeling tolerance intervals. Examination of the data in Figure 7.1 reveals that all three areas represent low-density regions within the study area. In particular, the peak covers an area with very few cases and no controls.

To investigate the robustness of the significant departures in Figure 7.3, we consider broader bandwidths to incorporate more information in the sparsely represented regions of the study area. Kelsall and Diggle [28] stress the theoretical and practical importance of a shared bandwidth value to maintain comparability between the two density estimates and offer a cross-validation algorithm for identifying the bandwidth minimizing the mean integrated squared error between the estimate and the true (unobserved) density surface. While identifying a single, optimal bandwidth has merit, Silverman [46] also notes the value of exploring the stability of observed structures across bandwidth values, an approach that we take in our illustrative example here. To illustrate the point, consider the density estimates and associated log relative risk surfaces in Figures 7.4 and 7.5. As one would expect, we see the peaks

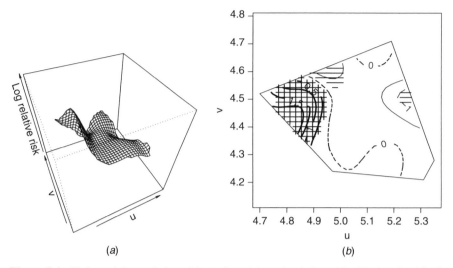

Figure 7.4 Estimated log relative risk surface (*a*) and relative risk (*b*) for the North Humberside leukemia data with bandwidth set to 0.07 for both cases and controls.

and valleys attenuated, but note that the western peak remains outside the range of values obtained under 500 random labeling simulations, suggesting a "significant" increase in the local relative risk.

The preceding example reveals several items meriting additional comments. First, we note the impact of bandwidth selection on the general appearance and smoothness of the estimated log relative risk surface. In addition, we note the impact of sparsely populated (or at least

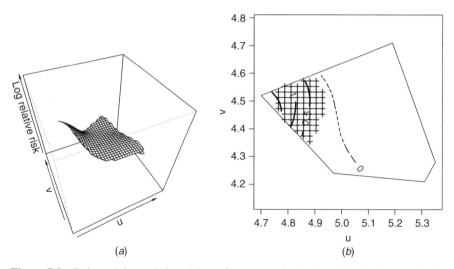

Figure 7.5 Estimated log relative risk surface (*a*) and relative risk (*b*) for the North Humberside leukemia data with bandwidth set to 0.1 for both cases and controls.

sparsely sampled) areas on the estimated surface, particularly when such areas are near the edge of the study area. Kernel estimates are particularly variable near these edges since we do not include information from any cases or controls falling outside of the study region boundary. This issue of "edge effects" has seen some discussion in the spatial analysis literature [34], but adjustments remain largely ad hoc and rarely (if ever) see different adjustments for different sorts of edges (e.g., a coastline represents a true edge beyond which no cases can occur, while a county or country boundary is purely a political distinction with little causal association with local disease risk). Finally, the illustration also reminds us of the epidemiologic distinction between relative and attributable risks since the observed elevated estimate of log relative risk corresponds to only a small number of actual cases.

The log relative risk surface represents one way of exploring spatially referenced point data for local "clusters" of increased risk. Waller and Gotway [54, Chs. 6, 7] review a wide variety of additional approaches for cluster detection, noting that each approach uses its own mathematical definition of a "cluster" and, as a result, different methods may detect differing evidence of clusters (and, in fact, different clusters) within the same dataset.

To illustrate this point, we briefly describe the popular spatial scan statistic developed by Kulldorff [31] and available in the freeware package SaTScan [32], and then apply it to the North Humberside leukemia data. Scan statistics consist of a moving window where one calculates the statistic of interest (here the local relative risk) inside and outside the window and seeks to identify the window (or windows) providing the most extreme values. In the particular instance of the spatial scan statistic developed by Kulldorff [31], we consider circular windows of varying sizes (with windows ranging from those containing a single case or control to those containing half of the sampled population), and for each window we calculate a likelihood ratio statistic for the hypothesis of equal risk inside and outside of the window. More specifically, let $n_{1,\text{in}}$ and $n_{\text{in}} = (n_{1,\text{in}} + n_{0,\text{in}})$ denote the number of case locations and persons at risk (number of case *and* control locations) inside a particular window, respectively, and similarly define $n_{1,\text{out}}$ and $n_{\text{out}} = (n_{1,\text{out}} + n_{0,\text{out}})$ for outside the window. The statistic of interest is the maximum of the local likelihood ratio statistics, for the Poisson case

$$\max_{\text{All windows}} \left(\frac{n_{1,\text{in}}}{n_{\text{in}}}\right)^{n_{1,\text{in}}} \left(\frac{n_{1,\text{out}}}{n_{\text{out}}}\right)^{n_{1,\text{out}}} I\left(\frac{n_{1,\text{in}}}{n_{\text{in}}} > \frac{n_{1,\text{out}}}{n_{\text{out}}}\right), \tag{7.2}$$

where $I(\cdot)$ denotes the indicator function, so we only maximize over windows where the observed rate inside the window exceeds that outside the window.

We obtain a likelihood ratio via Equation (7.2) within each window and identify the window(s) yielding the highest value. Rather than using distributional results for each likelihood ratio statistic (which would result in multiple testing problems), Kulldorff [31] instead proposes a Monte Carlo test addressing the following question: "How unusual is the highest observed local likelihood ratio statistic?" For each of a large number of simulated assignments of case locations, we again find the window (among those considered) and the largest local likelihood ratio statistic. The test statistic obtained from the original observed data is ranked against those values obtained from the simulated data, thereby providing a Monte Carlo p value associated with the "most likely cluster."

It is important to note two features:

1. We note that the observed likelihood ratio statistic for the most likely cluster is ranked against the maximum statistic from each simulation, regardless of location. This avoids the multiple comparison problem in a clever way, but requires careful explanation

to understand what precisely is meant by significance statements regarding the most likely cluster.

2. The qualifying phrase "among those considered" added above in reference to the spatial window also defines an aspect of the proper context for interpretation.

Specifically, the set of windows considered defines a family of potential clusters and our analysis assesses the most unusual case aggregation among these. As an example, consider a long, linear cluster of increased risk. In order to "capture" the cluster within our circular windows, we will either have a small circle containing part of the cluster or a large circle containing most of the cluster but also a large area experiencing the null, background risk. The first example loses statistical power due to a smaller local sample size; the second, due to the diluted relative risk averaged over the larger window. More recent implementations of SaTScan allow elliptical clusters to generalize the set of potential clusters. Other more recent developments include the upper level set approach of Patil and Taillie [42] and an approach based on minimum spanning trees [2], which broaden the class of potential clusters at the expense of increases in computation time.

We apply the spatial scan statistic (using SaTScan, version 3.0) [32] to the North Humberside leukemia data presented in Figure 7.1. Figure 7.6 indicates the most likely clusters by arrows (both having the same local relative risk value). These clusters are quite small, both contain four cases out of four individuals at risk (with 1.22 cases expected under a null hypothesis of constant risk), and the Monte Carlo p value associated with this value is 0.648, based on 999 simulations.

Comparing the results of the spatial scan statistic to those of the log relative risk surface brings home several conceptual issues. First, note that the basic question of interest is the

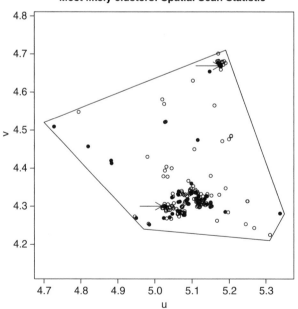

Figure 7.6 Most likely clusters identified by the spatial scan statistic for the North Humberside leukemia data. Neither is statistically significant, based on 999 Monte Carlo simulations.

same in both cases; namely, we wish to find areas ("clusters") inconsistent with a null hypothesis of equal risk. In this example, the data are the same, but the different methods operationalize our question slightly differently. The most likely clusters for the scan statistic are much smaller than the suspicious areas raised by the log relative risk surface. Both of the most likely clusters are found in areas of higher population density. Because of their small size and the concentration of nearby case and control event locations, these aggregations are unlikely to be detected by kernel smoothing methods, due to the use of a single, fixed bandwidth across the entire study area. Regardless of the resulting statistical significance, the choice of a bandwidth small enough to detect very local excesses is unlikely to be effective in summarizing patterns in the more sparsely populated sections of the study area. In addition, the area most suspicious for the log relative risk surface is unlikely to be detected by a circular scan statistic due to its elongated shape along the edge of the study area.

To wrap up our discussion of point patterns in spatial epidemiologic literature and our North Humberside example, we note that neither approach presented above (nor many other proposed methods found in the literature) is necessarily more "correct" than any other. Rather, each test examines a class of potential deviations from the null setting, and draws conclusions from that examination. Our example presents two specific approaches, but the same general principle applies to all methods for detecting clusters different methods define (explicitly or implicitly) the sorts of deviations under consideration, and different methods may provide different results based the types of clusters present; hence no single method will provide a comprehensive assessment of the presence or absence of clusters. Rather, applying different methods to the same dataset may provide insight into the type of clusters potentially found within the data [53].

We conclude this section by noting that the field of spatial point process modeling is far broader than the two methods presented here and direct interested readers to general surveys in Lawson [34], Waller and Gotway [54, Chs. 5–8], and especially more recent texts by Møller and Waagepetersen [40] and Baddeley et al. [3]. In addition, the subfield of cluster detection includes many additional approaches. In particular, the text by Lawson and Denison [35]; papers by Anselin [1], Getis and Ord [24], Ord and Getis [41], and Tango [49]; and the comprehensive review by Kulldorff in 2006 [33] provide entrance to additional families of analytic cluster detection techniques. These publications and the references therein provide a wider examination of different classes of methods and inferential questions and techniques than those presented here.

7.4   PREDICTING SPATIAL OBSERVATIONS

We next consider another important component of spatial epidemiology, namely, the prediction of local exposures across the study area D , based on a finite number of observations taken at point locations within D . Many methods central to spatial prediction have their roots in the geology and mining literature, and the field of *geostatistics* is focused on the mathematics and associated inferential methods of spatial prediction.

The basic elements of spatial prediction follow a very intuitive structure:

1. We assume that observations are spatially correlated with observations taken close together to be more closely related than those taken far apart (also known in the geography literature as "Tobler's first law of geography," after the eminent geographer Waldo Tobler [50].

2. An accurate estimate of spatial correlation as a function of distance should allow us to combine information across sites and predict the outcome at any location relative to the observed measurements and their locations relative to the prediction location.

Taken together, spatial prediction is a two-component process; we first estimate the correlation (covariance) as a function of space and then use this covariance function to combine observations and create a set of predictions across the study area. In the case of a linear prediction of Gaussian data (a weighted average of observed measurements with weights dependent on the covariance function), the process is often referred to as *kriging* in honor of D. G. Krige, a South African mining engineer (Krige 1951).

The geostatistical literature is large and varied. Cressie [13] and Chilès and Delfiner [12] offer comprehensive coverage of classical statistical inference in geostatistics, while Stein [47] expands its theoretical basis, and Wackernagel [51] focuses on multivariate setting. Webster and Oliver [58] give an applied introduction to geostatistics, and Waller and Gotway [54, Ch. 8] explore results in the setting of predicting exposure values for public health.

More recent methodologic advances of particular interest to spatial epidemiology involve the use of hierarchical models for spatial prediction, often in a Bayesian setting using Markov chain Monte Carlo (MCMC) algorithms for inference. We focus attention on this formulation here, drawing primarily from Diggle, Tawn and Moyeed et al. [19] and the text by Banerjee et al. [4].

We illustrate the model development and application on a dataset involving soil samples and dioxin contamination originally published in Zirschky and Harris [59] and used as a case study in Waller and Gotway [54]. In 1971, a truck carrying dioxin-contaminated waste dumped part of its load along a road in rural Missouri. In 1983, the United States Environmental Protection Agency (USEPA) collected soil samples in a systematic manner along the road at varying distances, and at a higher frequency in the immediate area of the spill. Figure 7.7 shows the sampling locations with circles of area proportional to the measured log concentration (with concentration measured in $\mu g/kg$ of soil) of dioxin taken at that location. As we might expect, concentrations are highest in the vicinity of the spill [near coordinates (15,30)] and along the roadway (the line $Y = 30$).

Several basic assumptions underly spatial prediction. Most of these may be generalized, but we present a straightforward example here to illustrate the approach:

1. The first assumption is that of *stationarity*, specifically the assumption that the spatial correlation structure is the same across the entire study area. This assumption provides a sense of replication for estimation of the spatial covariance function, since we often observe only one dataset in any particular application.

2. One often assumes *isotropy*, that is, that the spatial covariance declines with distance in the same manner in all directions.

3. Spatial prediction often assumes Gaussian measurements taken at each location. Under the Gaussian assumption, classical kriging methods provide the *best linear unbiased prediction* (BLUP) for each location in the study area.

4. Finally, for simplicity, we will assume a constant mean for all observations in our development, that is, that all spatial patterns are due to spatial covariance among the observations.

In our dioxin example, we maintain these assumptions for ease of exposition. However, for the assumptions to best apply, we transform the data, first dividing the original X coordinate by

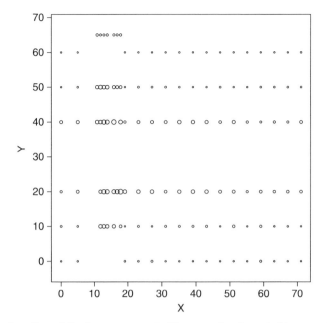

Figure 7.7 Location of dioxin measurements. The area of each symbol is proportional to the natural logarithm of the concentration as measured in $\mu g/kg$ of soil.

50 to better follow isotropy then considering the log concentrations to better meet the Gaussian assumption. The latter transformation results in the BLUP of the transformed values and is no longer a BLUP of the dioxin concentration itself. Related bias adjustments are provided by Cressie [13], but we focus here on the basic approach and compare the classical and more recent hierarchical formulations of spatial prediction using the transformed dataset for illustration.

Let $Z(s)$ denote the outcome of interest (dioxin in our example), measured at location s within study area D. Let $Z(s_i)$ denote our n measured values taken at locations s_i, $i = 1, \ldots, n$. In classical geostatistics we seek a BLUP; hence our goal is to obtain an unbiased prediction $\tilde{Z}(s_0)$ for any location s_0 in D, defined by a linear combination of observed values

$$\tilde{Z}(s_0) = \sum_{i=1}^{n} \eta_i Z(s_i),$$

with weights η_i to minimize the prediction error $[Z(s_0) - \tilde{Z}(s_0)]$, typically summarized in classical geostatistics by the *mean square prediction error* (MSPE):

$$\text{MSPE} = E\left[\left(Z(s_0) - \tilde{Z}(s_0)\right)^2\right].$$

In short, we need to find the set of weights $\{\eta_i, i = 1, \ldots, n\}$ minimizing the MSPE under the given unbiasedness constraint.

Since we assume a constant mean, it should come as no surprise that the optimal weights will depend on the spatial covariance between the observed locations and any prediction location of interest. The standard classical approach typically considers the *semivariogram*, denoted $\gamma(h)$, defined via

$$2\gamma(\|s_i - s_j\|) = \text{Var}(Z(s_i) - Z(s_j)), \tag{7.3}$$

for locations s_i and s_j within D. In other words, the semivariogram is defined as one-half of the variance of the contrast between observations taken at distance $h = \|s_i - s_j\|$ apart. If we assume that this function is the same for distance h regardless of the locations and relative orientation of s_i and s_j, then the semivariogram is stationary and isotropic, respectively. The semivariogram is related to the spatial covariance function, $C(h)$, specifically

$$\gamma(h) = C(0) - C(h),$$

and, as such, the semivariogram must meet conditions necessary to ensure a positive–definite variance–covariance matrix for all measurements (observed or not) within the study area. On the basis of this relationship, and since we generally assume positive spatial correlation declining with distance, the semivariogram will typically be an increasing function of distance, often rising to a "sill" value representing the variance between observations taken far enough apart to be effectively independent. The semivariogram has been slightly preferred over the covariance function in classical geostatistics, due primarily to ease and accuracy of estimation over those for the covariance function, among other, more technical reasons.

Given the semivariogram $\gamma(h)$, we obtain the optimal prediction weights $\eta_1, \ldots, \eta_n$ as solutions to the *kriging equations*

$$\eta = \Gamma^{-1}\gamma, \tag{7.4}$$

where

$$\eta = (\eta_1, \ldots, \eta_n, m)',$$
$$\gamma = (\gamma(s_0 - s_1), \ldots, \gamma(s_0 - s_n), 1)',$$

where the elements of Γ are

$$\Gamma_{ij} = \begin{cases} \gamma(s_i - s_j) & i = 1, \ldots, n; \\ & j = 1, \ldots, n; \\ 1 & i = n+1; \quad j = 1, \ldots, n; \\ & j = n+1; \quad i = 1, \ldots, n; \\ 0 & i = j = n+1. \end{cases}$$

So (7.4) becomes

$$\begin{bmatrix} \eta_1 \\ \eta_2 \\ \vdots \\ \eta_n \\ m \end{bmatrix} = \begin{bmatrix} \gamma(s_1 - s_1) & \cdots & \gamma(s_1 - s_n) & 1 \\ \gamma(s_2 - s_1) & \cdots & \gamma(s_2 - s_n) & 1 \\ \vdots & \ddots & \vdots & \vdots \\ \gamma(s_n - s_1) & \cdots & \gamma(s_n - s_n) & 1 \\ 1 & \cdots & 1 & 0 \end{bmatrix}^{-1} \begin{bmatrix} \gamma(s_0 - s_1) \\ \gamma(s_0 - s_2) \\ \vdots \\ \gamma(s_0 - s_n) \\ 1 \end{bmatrix}.$$

The equations above derive from minimization of the MSPE with an additional Lagrangian multiplier m included to guarantee an unbiased predictor.

Computationally, note that we must calculate η for each prediction location, s_0. However, only the vector γ changes with the prediction location. Since Γ depends only on the data but not the prediction locations, we need invert Γ only once and then multiply by the associated γ vector to obtain a prediction for any s_0 in D.

The minimized MSPE, also known as the *kriging variance*, derives from the same elements and is given by

$$\sigma_k^2(s_0) = \eta'\gamma$$

$$= \sum_{i=1}^{n} \lambda_i \gamma(s_0 - s_i) + m \tag{7.5}$$

$$= 2\sum_{i=1}^{n} \lambda_i \gamma(s_0 - s_i) - \sum_{i=1}^{n}\sum_{j=1}^{n} \eta_i \eta_j \gamma(s_i - s_j),$$

From Equations (7.4) and (7.5) we see that if we have the semivariogram function, we have all we need to provide BLUP predictions at any location. As a result, there is a considerable literature on *variography*, or the estimation of the semivariogram from observed data. One typically estimates the semivariogram from the observed data contrasts, often averaging over pairs of observations taken the same (or nearly the same) distance apart. Such averages often provide an *empirical semivariogram*, to which one fits a *theoretical semivariogram* defined as a parametric function of distance. Commonly used parametric families are cataloged throughout the spatial statistical literature, for example, in Cressie [13] and Waller and Gotway [54, Ch. 8]. Parametric semivariogram families are often defined in terms of the semivariogram's limiting value as distance approaches zero (the "nugget"), the semivariogram's limiting value as distance increases (the "sill"), and the distance beyond which observations are effectively independent (the "range").

Figure 7.8 illustrates the empirical semivariogram for the dioxin data, shown with dots representing one-half of the average variation between contrasts observed at given distances apart. The lines represent the best-fitting theoretical semivariograms from the exponential semivariogram model defined by

$$\gamma(h; c_0, c_e, a_e) = \begin{cases} 0 & h = 0 \\ c_0 + c_e\{1 - \exp(-h/a_e)\} & h > 0, \end{cases} \tag{7.6}$$

where $c_0 \geq 0$ denotes the nugget effect, $c_e \geq 0$ denotes the partial sill, and $a_e > 0$, where $3a_e$ denotes the *effective range* (traditionally defined as the distance at which the autocorrelation is 0.05). We fit the theoretical semivariograms using both least squares (ordinary and weighted) and likelihood-based (maximum likelihood and restricted maximum likelihood) method in the geoR library for R [45]. The observed difference between the two types of estimator suggest a skewed likelihood across the parameter space.

At this point, one would select a "best" semivariogram from the estimates, and then condition on this estimated function to define the vector γ in the kriging equations to obtain both the point predictions at a set of prediction locations as well as the associated kriging variances. For Gaussian data, these will allow construction of pointwise prediction intervals at each location.

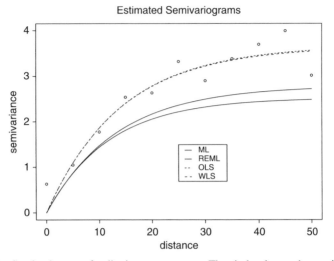

Figure 7.8 Semivariograms for dioxin measurements. The circles denote the empirical semi-variogram and the lines denote the best-fitting exponential semivariogram under maximum likelihood (ML), restricted maximum likelihood (REML), ordinary least squares (OLS), and weighted least squares (WLS).

As noted above, the classical kriging approach is typically a two-stage process: estimating the semivariogram and then solving the kriging equations. Common criticisms of the classical kriging approach include concern that the uncertainty associated with the estimation of the semivariogram is not adequately reflected in the prediction errors expressed in the kriging variances. In addition, the typical semivariogram estimation procedure outlined above is itself a two-step procedure involving construction of the point estimates defining the empirical semivariogram, and then estimation of the parameters in the selected theoretical semivariogram family. Finally, the asymptotic properties of semivariogram estimation are nontrivial, depending on whether one uses infill [47] or increasing domain [13, pp. 100–101] asymptotics.

More recently attention has turned toward an effort to express spatial prediction in a more cohesive manner addressing the uncertainty in all components in such a way as to accurately reflect the entire process. The preceding description of classical spatial prediction followed a rather utilitarian approach, highlighting each step of the typical analytical process. The basic theoretical construction underlying both classical and Bayesian kriging, on the other hand, is rather elegant when expressed in a hierarchical fashion, but until relatively recently could not be readily implemented as such. However, with the advent of Markov chain Monte Carlo (MCMC) techniques, a general computational framework for addressing hierarchical models now exists and more recent statistical publications move toward bringing these ideas and algorithms to the field of spatial prediction. This is not to say that MCMC solves all problems, as its implementation is often slow and much more computationally demanding than the classical approach outlined above.

Statistical prediction can be regarded in a Bayesian framework where the optimal predictor is defined by the conditional expectation of $Z(s_0)$ given the observed data. In a more formal Bayesian statement, the point prediction defined by the conditional expectation

$E[Z(s_0)|Z(s_1), \dots, Z(s_n)]$ represents the posterior predictive expectation of $Z(s_0)$ marginalizing over the posterior distribution of any model parameters. For Gaussian data, this conditional expectation is a linear function of the data (motivating the BLUP formulation presented above), but for non-Gaussian data this conditional expectation need not be linear in the data, so the conditional expectation provides a more general framework for prediction, provided the expectation (and appropriate uncertainty measures) can be calculated.

More specifically, one can briefly contrast classical and hierarchically specified spatial prediction following exposition found in the literature [19,21,4,55]. Let Z denote the vector of data, S the unobserved spatial random field of all observations in D, and θ a vector of model parameters, for our purposes the set of covariance parameters. In effect, S is the set of values $Z(S_0)$ for all s_0 in D. It is helpful to distinguish Z and S in model development, since we model the stochastic structure of S and seek inference regarding model parameters based on observations in Z. Our prediction goal is to obtain the conditional distribution of S given Z and θ, expressed in general notation as $[S|Z, \theta]$. Diggle and Ribeiro [21] note that Bayes' theorem provides

$$[S|Z,\theta] = \frac{[Z,S|\theta]}{[Z|\theta]} = \frac{[Z|S,\theta][S|\theta]}{\int [Z|S,\theta][S|\theta]dS}. \tag{7.7}$$

This development conditions on a known value of the set of covariance parameters θ, and summarizes the classical spatial prediction setting in a general and elegant hierarchical form. As illustrated in the development above, in practice, one typically estimates θ from the data, and then calculates the conditional expectation above (conditional on both the data Z and the parameter estimates) via the kriging equations.

The hierarchical structure in Equation (7.7) allows both frequentist and Bayesian implementation. A frequentist approach builds the likelihood from the hierarchical components and then requires calculation of the associated predictive distribution for inference, allowing one to incorporate the variability associated with the covariance function (or, equivalently, the semivariogram) into the likelihood at the cost of more complicated computation. In contrast, a Bayesian view of Equation (7.7) assumes a prior distribution $[\theta]$ for the unknown parameters, and then marginalizes over the posterior distribution of the parameters given the data $[\theta|Z]$, yielding the posterior predictive distribution

$$[S|Z] = \int [S|\theta,Z][\theta|Z]d\,\theta. \tag{7.8}$$

Bayesian kriging draws point and interval predictions from this posterior predictive distribution. While simple in theory, application of the full Bayesian approach typically encounters complicated or intractable integrals, and also relies on advanced computing through MCMC sampling from the desired posterior distributions for inference. Both the frequentist and Bayesian implementations require either advanced computing or simplifying assumptions (e.g., treating the estimated semivariogram as fixed, as in the classical approach). We focus here on the Bayesian implementation of the hierarchical structure because of its increasing application in the literature and the advent of more generally available MCMC code for such models, and we apply both Bayesian and classical kriging to the dioxin data to compare results.

For comparability with the classical development above, suppose that our data are Gaussian with constant mean, β_0; that is, suppose

$$Z \sim \text{MVN}(\beta_0, \Sigma_Z), \tag{7.9}$$

where β_0 represents an n vector of the constant mean β_0 and Σ_Z the $n \times n$ variance–covariance matrix of the Z terms. (We use the intercept notation β_0 to highlight where one might add additional parameters in a regression-type model of the mean.) Next, suppose that we have a parametric covariance function $C(h; \theta)$, defined up to the unknown model parameters θ. For our example, we will use the isotropic, stationary exponential covariance function family, corresponding to the exponential semivariogram defined above. The covariance function defines the elements of Σ_Z based on the distances between pairs of observation locations:

$$\Sigma_{Z_{ij}} = C(||s_i - s_j||, \theta).$$

The latent random field S serves as a data generator for Z, in the sense that the covariance function $C(h, \theta)$ is defined for *any* location s in D. For any set of n locations, we obtain a vector of observations Z with the multivariate normal distribution defined in Equation (7.9), with $C(h, \theta)$ and the relative locations of observations defining Σ_Z. For simplicity, we assume no measurement error and set $Z(s_i) = S(s_i)$.

The next step in the model definition is to define prior parameters for model parameters β_0 and $\theta = (c_e, c_0, a_e)$. In most applications, the likelihood structure for the mean parameter β_0 is quite strong, allowing very vague prior specifications. Prior specification for covariance parameters in θ is somewhat more complicated, reviewed by Waller [55] and summarized here. One could ignore θ by specifying a conjugate inverse Wishart prior for Σ_Z [11, pp. 459–470; 15], but note that the inverse Wishart does not limit attention to specifically spatial covariance structures and, similar to discussions of clustering approaches in the preceding section, may not focus attention on the set of models that we are particularly interested in exploring. A more common practice is to define individual conjugate prior distributions for each parameter within θ. While practical, this approach still requires care as noted by Berger et al. [6,7], who consider reference and Jeffreys' priors for variance–covariance parameters in a Gaussian random field. This setting provides one of the few examples where applying Jeffreys' prior independently to each element of θ yields an improper posterior distribution, suggesting a need for further work and especially for care in transferring seemingly sensible priors from the nonspatial to the spatial setting.

For the dioxin example, we assign a flat prior to β_0, assume a nugget effect of $c_0 = 0$, and a prior proportional to the reciprocal of the sill c_e (variance of independent observations). Rather than consider the effective range parameter a_e, directly, we instead model the rate of exponential decay in the covariance function, denoted ϕ. We assign a discrete prior based on 51 equally spaced values between zero and twice the maximum observed distance between sampling locations. We fit the model using the R library geoR [45]. Samples from the posterior distributions of all three parameters appear in Figure 7.9. We find a clear posterior signal for β_0 (compared to the assumed flat prior). For ϕ, Figure 7.9 includes a thin line illustrating the assigned prior distribution and a thick line representing a kernel estimate of the posterior distribution. We see both the posterior moving away from the prior and tightening around its (posterior) mean value.

Figure 7.10 shows the empirical semivariogram values from Figure 7.8 and the semivariograms corresponding to the exponential covariance function evaluated at the estimated posterior mean, median, and mode. The similarity between the two figures is reassuring and

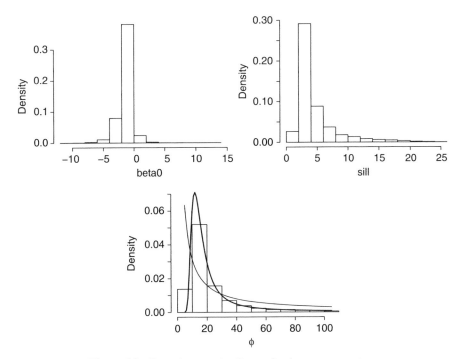

Figure 7.9 Posterior samples for semivariogram parameters.

suggests accurate implementation of the Bayesian approach. We note that the skewed posterior distributions of the sill and ϕ result in the difference between the curves based on posterior mean and medians and those based on the posterior mode (which is, not surprisingly, quite similar to the maximum-likelihood estimates shown in Fig. 7.8).

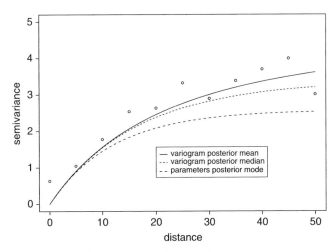

Figure 7.10 Bayesian semivariograms for dioxin measurements.

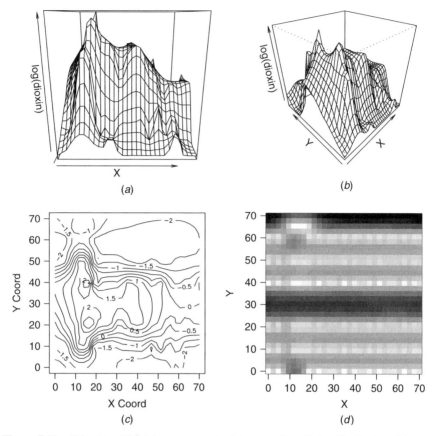

Figure 7.11   Kriged log(dioxin) measurements: (*a*) mean of predictive distribution; (*b*) mean of predictive distribution; (*c*) mean of predictive distribution; (*d*) local variance of predictive distribution.

Figure 7.11 shows the posterior mean predicted levels of log(dioxin) (the posterior mean surface *S*) as perspective plots (from two different orientations) in the top row and as a contour plot in the lower left. The image plot in the lower right shows the spatial pattern in the variance of the posterior predictive distribution. Predictive variance is higheset in areas with few observations, in particular, note the band of high predictive variances along the road ($Y = 30$), representing the band of no observations.

Figure 7.12 illustrates the posterior predictive mean and associated 2.5th and 97.5th percentiles drawn on the basis of 5000 samples from the posterior predictive distribution taken along transects for four different values of *X*. Note the widening of the posterior predictive distribution in the area near the road ($Y = 30$) corresponding to the increased predictive variance associated with the lack of data taken in this area. Also, note the "tightening" of the prediction errors near the sampling rows near $Y = 10,20,40,50,60$.

An interesting feature appears in Figure 7.12 for $X = 15$, the predictions taken through the area of highest observed dioxin concentration values. Note that the mean posterior prediction dips down between the peak values taken on either side of the road. This is due to our basing prediction on a distance–decay correlation function. If we move beyond the effective range

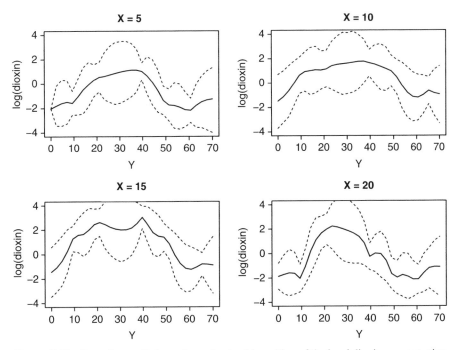

Figure 7.12   Posterior predictive values of natural logarithm of the local dioxin concentration for selected transects perpendicular to the road. The thick line denotes the posterior predictive mean value and the dashed lines represent the 2.5th and 97.5th percentiles of 5000 samples from the posterior predictive distribution.

estimated by our covariance (or semivariogram) function, observations would be (nearly) independent of one another and each would receive nearly the same weight, resulting in the sample mean in the BLUP case or a posterior estimate of the overall (assumed fixed) mean β_0. In short, spatial prediction adjusts toward neighboring values according to how closely correlated we expect those values to be to our desired prediction. If no observations are close, the method resorts to an estimate of the background mean.

Results based on classical results (solving the kriging equations in terms of the variogram estimates shown in Fig. 7.8) are quite similar to those shown here for Bayesian kriging. As a result, one might wonder what we gain from the extra model specification (setting prior distributions) and the extra computational effort required for MCMC implementation over the basic matrix calculations required for classical kriging. For Gaussian data with a covariance structure easily modeled by standard parametric covariance functions and satisfying the basic assumptions of stationarity and isotropy, there may be little gain other than a sense of completeness in the definition of the full probability model incorporating both covariance estimation and outcome prediction.

However, the Bayesian hierarchical framework also sets the stage for broader extensions than does the classical framework. To see this, consider the descriptions above. Both the classical setting and the Bayesian setting derive from a basic definition of conditional probability in Equation (7.7). At this point the two descriptions (above and in most of the spatial statistical literature) diverge in a manner mirroring a contrast seen in the frequentist and Bayesian literature. At the risk of oversimplification, the classical (frequentist) description often builds on the basic framework in the following manner. Given the data are Gaussian with a fixed mean, we know (or can derive) that the predictor will be a linear function of the observed data. We would like the weights in this linear predictor to minimize the mean square prediction error, so we define the MSPE as a function of the weights and optimize subject to an unbiasedness constraint, obtaining the MSPE as part of the optimization.

We find the optimal weights are a function of the semivariogram, and next seek to estimate the semivariogram in a accurate and consistent manner in order to provide the best set of prediction weights. The prediction goal has been achieved by a series of theoretical derivations and accompanying calculations, effectively achieving the larger task by a series of focused smaller tasks, each built on appropriate theoretical results. Again, at the risk of oversimplification, in the Bayesian description the initial goal of prediction is again cast in the setting of a conditional probability. This conditional probability is reexpressed as a marginalization of a hierarchical probability model with parameters defining the overall mean and covariance function (or semivariogram if preferred). The inclusion of prior probabilities places prediction in the setting of a posterior distribution, the primary inferential tool for Bayesian statistics. The details of interest in development relate to the structure of the full probability model, such as defining the likelihood via the multivariate normal distribution given the covariance structure in Equation (7.9), next specifying this covariance structure given its parameters, and finally specifying the prior distributions for these parameters. The computational implementation falls to MCMC, perhaps a complicated MCMC requiring care in setup and implementation, but an MCMC algorithm nonetheless.

Here we note a subtle difference in value between the settings, specifically, a difference in what is regarded as the "cool" part of the derivation. In the classical development, value is placed on a deconstruction of the problem at hand into a series of steps motivated and validated by statistical theory, each step with an accurate and efficient mode of calculation. Even in a frequentist evaluation of a hierarchical likelihood, similar steps occur [38]. In the Bayesian development however, value is placed on an accurate formulation of the problem in terms of a joint posterior distribution defined through an interconnected set of hierarchical components, each justified in its own right and fitting together in a manner that guarantees a proper posterior distribution incorporating all sources of potential information.

With this distinction in mind, an advantage of the Bayesian approach appears when we consider changes to the components of the probabilistic structure. Suppose that we no longer have Gaussian data, but rather observations following some other distribution. In the classical setting, this impacts one of the first steps in the process, namely, the equivalence between the desired conditional expectation and a linear combination of observations. One approach is to transform the data to achieve a distribution closer to a Gaussian distribution (as in the dioxin example). The problem is not insurmountable; in fact, log Gaussian (or more generally trans-Gaussian kriging) is widely used, but it requires reconstruction of one of the key components of the classical approach. For the Bayesian formulation, moving from a Gaussian distribution requires reformulation of the likelihood function and can introduce complicated identifiability issues as the mean and covariance parameters may no longer be orthogonal, yet the basic method is still in place. This is not to say that the Bayesian approach is necessarily easier to adapt, but one can argue that the required adjustments impact less of the basic structure than they do in classical kriging.

7.5 CONCLUDING REMARKS

In the discussion above, we focus attention on developing statistical issues for two of the four general questions of interest in spatial epidemiology. The resulting examples illustrate important issues underlying the analysis of spatially referenced data ranging from the interpretation of spatial cluster detection to contrasts between classical and Bayesian spatial prediction. While the approaches and issues may seem very different at first glance, the two areas highlight current converging directions of development in spatial epidemiology:

1. Note the similarity between the spatial intensity function $\lambda(s)$ in Section 7.3 and the latent random field S in Section 7.4. The hierarchical structure outlined regarding spatial prediction suggests extension to the point process setting, as discussed in Møller and Waagepetersen [40] and Diggle et al. [20]. In particular, Diggle et al. [20] note that parametric (often hierarchical) approaches to the analysis of spatial point processes often provide increased accuracy, while nonparametric (e.g., the kernel-based approaches described above) are often more robust to violations of the assumed parametric models.

2. While the definition of a spatial Poisson process motivates the construction of spatial Poisson regression models popular in disease mapping [34,54,4,52], the hierarchical framework also offers a way to build inference for aggregated counts of underlying (latent) point patterns, even for data collected at differing levels of aggregation. Best et al. [9] provide an example of such an approach.

The field of spatial epidemiology is much larger than the two detailed areas considered here, and further classes of analytic methods appear in the literature addressing additional epidemiologic questions for additional forms of available data. A prime example is the use of remote sensing data in epidemiologic investigations. Robinson [44] provides a thorough review of statistical techniques for remote sensing data in public health, and Goovaerts et al. [25] define methods linking both of the analytic areas considered above (geostatistical prediction and cluster detection) within the setting of remote sensing data.

In conclusion, spatial epidemiology offers much opportunity for continued methodologic development in order to provide accurate, reliable inference on important public health issues.

ACKNOWLEDGMENT

This work is supported in part by grant NIEHS R01 ES007750. The opinions expressed herein are solely those of the author and may not reflect those of the National Institutes of Health or the National Institute of Environmental Health Sciences.

REFERENCES

1. Anselin, L., Local indicators of spatial association: LISA, *Geogr. Anal.* **27**, 93–116 (1995).

2. Assunção, R., Costa, M., Tavares, A., and Ferreira, S., Fast detection of arbitrarily shaped disease clusters, *Statist. Med.* **25**, 723–742 (2006).

3. Baddeley, A., Gregori, P., Mateau, J., Stoica, R., and Stoyan D., *Case Studies in Spatial Point Process Modeling*, Springer, New York, 2006.

4. Banerjee, S., Carlin, B. P., and Gelfand, A. E., *Hierarchical Modeling and Analysis of Spatial Data*, Chapman & Hall/CRC, Boca Raton, FL, 2004.

5. Bartlett, M. S., The spectral analysis of two-dimensional point processes, *Biometrika* **51**, 299–311 (1964).

6. Berger, J. O. and De Oliveira, S. B., Objective Bayesian analysis of spatially correlated data, *J. Am. Statist. Assoc.* **96**, 1361–1374 (2001).

7. Berger, J. O., De Oliveira, S. B., Ren, C., and Wittig, T. A. Correction, *J. Am. Statist. Assoc.* **98**, 779 (2003).

8. Besag, J. and Diggle, P. J., Simple Monte Carlo tests for spatial pattern, *Appl. Statist.* **26**, 327–333 (1977).

9. Best, N. G., Ickstadt, K., and Wolpert, R. L., Spatial Poisson regression for health and exposure data measured at disparate resolutions, *J. Am. Statist. Assoc.* **95**, 1076–1088 (2000).

10. Bithell, J. An application of density estimation to geographical epidemiology, *Statist. Med.* **9**, 691–701 (1990).

11. Box, G. E. P. and Tiao, G. C., *Bayesian Inference in Statistical Analysis*, Wiley, New York, 1973.

12. Chilès, J.-P. and Delfiner, P., *Geostatistics: Modeling Spatial Uncertainty*, Wiley, New York, 1999.

13. Cressie, N. A. C., *Statistics for Spatial Data*, rev. ed. Wiley, New York, 1993.

14. Cuzick, J. and Edwards, R., Spatial clustering for inhomogeneous populations (with discussion), *J. Roy. Statist. Soc. Ser. B* **52**, 73–104 (1990).

15. Daniels, M. and Kass, R., Nonconjugate Bayesian estimation of covariance matrices in hierarchical models, *J. Am. Statist. Assoc.* **94**, 1254–1263 (1999).

16. Diggle, P. J., *Statistical Analysis of Spatial Point Patterns*, Academic Press, London, 1983.

17. Diggle, P. J., Overview of statistical methods for disease mapping and its relationship to cluster detection, in *Spatial Epidemiology: Methods and Applications*, Elliott, P., Wakefield, J. C., Best, N. G., and Briggs, D. J., eds., Oxford Univ. Press, Oxford, UK, 2000, pp. 87–103.

18. Diggle, P. J., *Statistical Analysis of Spatial Point Patterns*, 2 ed., Oxford Univ. Press, Oxford, UK, 2003.

19. Diggle, P., Tawn, J., and Moyeed, R., Model-based geostatistics (with discussion), *Appl. Statist.* **47**, 299–350 (1998).

20. Diggle, P. J., Mateau, J., and Clough, H. E., A comparison between parametric and nonparametric approaches to the analysis of replicated spatial point patterns, *Adv. Appl. Probabil.*, **32**, 331–343 (2000).

21. Diggle, P. J. and Robeiro Jr., P. J., Bayesian inference in Gaussian model-based geostatistics, *Geogr. Environ. Model.* **6**, 129–146 (2002).

22. Elliott, P., Cuzick, J., English, D., and Stern, R., *Geographical and Environmental Epidemiology: Methods for Small-Area Studies*, Oxford Univ. Press, Oxford, UK, 1992.

23. Elliott, P., Wakefield, J. C., Best, N. G., and Briggs, D. J., *Spatial Epidemiology: Methods and Applications*, Oxford Univ. Press, Oxford, UK, 2000.

24. Getis, A. and Ord, K., The analysis of spatial association by distance statistics, *Geogr. Anal.* **24**, 189–207 (1992).

25. Goovaerts, P., Jacquez, G. M., and Marcus, A., Geostatistical and local cluster analysis of high resolution hyperspectral imagery for detection of anomalies, *Remote Sens. Environ.* **95**, 351–367 (2005).

26. Koch, T., *Cartographies of Disease: Maps, Mapping, and Medicine*, ESRI Press, Redlands, CA (2005).

27. Kelsall, J. and Diggle, P. J., Kernel estimation of relative risk, *Bernoulli* **1**, 3–16 (1995).

28. Kelsall, J. and Diggle, P. J., Non-parametric estimation of spatial variation in relative risk, *Statist. Med.* **14**, 2335–2342 (1995).

29. Kelsall, J. and Diggle, P. J., Spatial variation in risk: A nonparametric binary regression approach, *Statist. Med.* **17**, 559–573 (1998).

30. Krige, D. G., A statistical approach to some basic mine evaluation problems on the Witwaterstrand, *J. Chem. Metallurg. Mining Soc. S. Afr.* **52**, 119–139 (1951).

31. Kulldorff, M., A spatial scan statistic, *Commun. Statist. Theor. Meth.* **26**, 1487–1496 (1997).

32. Kulldorff, M. and Information Management Services, Inc., *SaTScan v. 3.0: Software for the Spatial and Space-Time Scan Statistics*, National Cancer Institute, Bethesda, MD, 2002.

33. Kulldorff, M., Tests of spatial randomness adjusted for inhomogeneity: A general framework, *J. Am. Statist. Assoc.* **101**, 1289–1305 (2006).

34. Lawson, A. B., *Statistical Methods in Spatial Epidemiology,* Wiley, Chichester, UK, 2001.

35. Lawson, A. B. and Denison, D. G. T., *Spatial Cluster Modelling,* Chapman & Hall/CRC, Boca Raton, FL, 2002.

36. Lawson, A. B. and Williams, F. L. R., Applications of extraction mapping in environmental epidemiology, *Statist. Med.* **12**, 1249–1258 (1993).

37. Lawson, A. B. and Williams, F. L. R., *An Introductory Guide to Disease Mapping,* Wiley, Chichester, UK, 2001.

38. Lee, Y. and Nelder, J. A., Hierarchical generalized linear models, *J. Roy. Statist. Soc. Ser. B* **58**, 619–656 (1996).

39. Lee, Y. and Nelder, J. A., Double hierarchical generalized linear models, *J. Roy. Statist. Soc. Ser. C* **55**, 139–167 (2006).

40. Møller, J. and Waagepetersen, R., *Statistical Inference and Simulation for Spatial Point Patterns*, Chapman & Hall/CRC, Boca Raton, FL, 2004.

41. Ord, K. and Getis, A., Local spatial autocorrelation statistics: Distributional issues and an application, *Geogr. Anal.* **27**, 286–306 (1995).

42. Patil, G. P. and Taillie, C., Upper level set scan statistic for detecting arbitrarily shaped hotspots, *Environ. Ecol. Statist.* **11**, 183–197 (2004).

43. R Development Core Team, *R: A Language and Environment for Statistical Computing*, R Foundation for Statistical Computing, Vienna, Austria, 2005 (ISBN 3-900051-07-0, URL http://www.R-project.org).

44. Robinson, T. P., Spatial statistics and geographical information systems in epidemiology and public health, *Adv. Parasitol.* **47**, 81–128 (2000).

45. Ribeiro Jr., P. J. and Diggle, P. J., geoR: A package for geostatistical analysis, *R-NEWS* **1**(2), 15–18 (2001) (ISSN 1609-3631).

46. Silverman, B. W., *Density Estimation for Statistics and Data Analysis*, Chapman & Hall/CRC, Boca Raton, FL, 1986.

47. Stein, M. L., *Interpolation of Spatial Data: Some Theory for Kriging*, Springer, New York, 1999.

48. Stoyan, D., Kendall, W. S. and Mecke, J., *Stochastic Geometry and Its Applications*, 2 ed., Wiley, New York, 1995.

49. Tango, T., A class of tests for detecting "general" and "focused" clustering of rare diseases, *Statist. Med.* **14**, 2323–2334 (1995).

50. Tobler, W., A computer movie simulating urban growth in the Detroit region, *Econ. Geogr.* **46**, 234–240 (1970).

51. Wackernagel, H., *Multivariate Geostatistics*. Springer, Berlin, 1995.

52. Wakefield, J., Best, N. G., and Waller, L.A., Bayesian approaches to disease mapping, in *Spatial Epidemiology: Methods and Applications*, Elliott, P., Wakefield, J. C., Best N. G., and Briggs D. J., eds., Oxford Univ. Press, Oxford, UK, 2000, 106–127.

53. Waller, L. A. and Jacquez, G. M., Disease models implicit in statistical tests of disease clustering, *Epidemiology* **6**, 584–590 (1995).

54. Waller, L. A. and Gotway, C. A., *Applied Spatial Analysis of Public Health Data*, Wiley, Hoboken, NJ, 2004.

55. Waller, L. A., Bayesian thinking in spatial statistics, in *Handbook of Statistics*, Vol. 25: *Bayesian Thinking: Modeling and Computation*, Dey, D. K. and Rao, C. R., eds., Elsevier–North Holland, Amsterdam, 2005.

56. Wand, M. P. and Jones, M. C., *Kernel Smoothing*, Chapman & Hall/CRC, Boca Raton, FL, 1995.

57. Walter, S. D., Disease mapping: A historical perspective, in *Spatial Epidemiology: Methods and Applications*, Elliott, P., Wakefield, J. C., Best, N. G., and Briggs D. J., eds., Oxford Univ. Press, Oxford, UK, 2000, pp. 223–239.

58. Webster, R. and Oliver, M. A., *Geostatistics for Environmental Scientists*, Wiley, Chichester, UK, 2001.

59. Zirschky, J. H. and Harris, D. J., Geostatistical analysis of hazardous waste site data, *J. Environ. Eng. ASCE* **112**, 770–784 (1986).

CHAPTER 8

Modeling Disease Dynamics: Cholera as a Case Study

Edward L. Ionides

Department of Statistics, University of Michigan, Ann Arbor, Michigan

Carles Bretó

Department of Ecology and Evolutionary Biology, University of Michigan, Ann Arbor, Michigan

Aaron A. King

Department of Mathematics, University of Michigan, Ann Arbor, Michigan

8.1 INTRODUCTION

Disease dynamics are modeled at a population level in order to create a conceptual framework to study the spread and prevention of disease, to make forecasts and policy decisions, and to ask and answer scientific questions concerning disease mechanisms such as discovering relevant covariates. Population models draw on scientific understanding of component processes, such as immunity, duration of infection, and mechanisms of transmission, and investigate how this understanding relates to population-level phenomena. There are several compelling reasons to consider disease processes at this population scale:

1. Anthropogenic change, in land use, climate, and biodiversity has many potentially large public health impacts [1]. Predicting the future effects of changes to a complex system is difficult. Retrospective studies of the relationship between climate and disease prevalence over space [27] and over time [58] can facilitate predictions and inform policy decisions [48]. A major challenge in retrospective studies is to disentangle the

Statistical Advances in the Biomedical Sciences, edited by Atanu Biswas, Sujay Datta, Jason P. Fine, and Mark R. Segal
Copyright © 2008 John Wiley & Sons, Inc.

extrinsic effects of climate or other environmental drivers from the intrinsic disease dynamics [46].

2. The effectiveness of medical treatment and vaccination strategies for certain infectious diseases, such as malaria and cholera, is limited by drug resistance, genetic shift, and poor medical infrastructure in affected regions. This leads to an emphasis of controlling the disease by behavioral and environmental interventions. An ability to model the disease dynamics can be used to forecast the danger of a major epidemic [65], a step toward implementing effective interventions.

3. Emerging infectious diseases pose a significant public health threat. Many important emerging infectious diseases are zoonotic, that is, endemic animal diseases that cross over to humans. Examples include HIV/AIDS from chimpanzee and sooty mangabey [28], severe acute respiratory syndrome (SARS) from bats [49], and avian flu [52]. Epidemics are best prevented by early containment of outbreaks. Containment strategies may be evaluated using population models [52]. Alternatively, one can attempt to monitor and control the disease in the animal population to reduce contact between humans and infected animals. This can be facilitated by employing population models to gain an understanding of the dynamics of the disease in the animal population.

Since the pioneering work of Ross [59] and Kermack and McKendrick [41], mathematical modeling has been a mainstay of epidemiologic theory. It has also long been recognized that disease models arising in epidemiology are closely related to population models arising in ecology [7]; the population dynamics of an infectious disease arise from the interaction of host and pathogen species in the context of their environment. This chapter explores some new developments in statistical inference for nonlinear dynamical systems from time-series data, using cholera in Bangladesh as a case study.

8.2   DATA ANALYSIS VIA POPULATION MODELS

A mainstay of population modeling is the compartment model, where the population is divided into groups that can be considered homogeneous. The classical Susceptible-Infected-Removed (SIR) compartment model [41,7] groups N_t individuals as susceptible (S_t), infected (I_t), and recovered or removed (R_t). Exposed classes, age-structured classes, and geographically structured classes are just some of many possible extensions. Population models may use continuous or discrete time, take continuous or discrete values, and be stochastic or deterministic. Real-world processes are continuous-time, discrete-valued, and stochastic. Stochasticity arises from demographic noise (variability due to uncertainty of individual outcomes, such as the number of contacts made with an infected individual) and from environmental noise (such as variability due to weather, or economic events affecting the whole population). To a first approximation, demographic stochasticity has variance linear in population size, and environmental stochasticity has variance quadratic in population size, although more subtle distinctions can be made [22]. Models must also choose to be mechanistic or phenomenological, really a continuous scale tradeoff between incorporating scientific understanding and aiming for a simple description of relationships observed in data [20]. Developing techniques that draw on understanding of population dynamics, while also permitting statistical inference about unknown model parameters and exploration of relevant covariates, is a topic of current research interest [9].

Data are often aggregated over time and space, such as weekly or monthly counts per region. This has led to the use of discrete-time models for data analysis. Finkenstädt and

Grenfell [24] and Koelle and Pascual [46] represent the state of the art for data analysis via discrete-time mechanistic modeling, using a Taylor series to generate a log-linear model with unobserved variables reconstructed via backfitting. There are several reasons to prefer continuous-time models:

1. For discrete-time models, the sampling frequency affects the models available and the interpretation of the resulting parameters. The underlying continuous-time processes are most naturally modeled in continuous time.
2. Continuous-time modeling facilitates the inclusion of covariates measured at various frequencies.
3. Continuous-time disease models have been studied much more extensively from the mathematical perspective than for their discrete-time counterparts [5,3,30,17]. This focus represents both that continuous-time models more accurately reflect the real properties of the systems and that such models are relatively easy to analyze. Most data analysis, on the other hand, has made use of discrete-time formulations, which can be fitted to discretely sampled data in a relatively straightforward fashion. However, the dynamics of discrete-time nonlinear systems are frequently at odds with those of their continuous-time analogs [53,25], a fact that can complicate the interpretation of the parameters of discrete-time models.

Strategies appropriate for fitting continuous-time models to discretely observed data include atlas methods [66], gradient matching [21], and approaches based on nonlinear forecasting [39]. Likelihood-based analysis (frequentist or Bayesian) has largely been overlooked because finding the likelihood involves the difficult task of integrating out unobserved variables. Maximum-likelihood estimates (MLEs) have some considerable advantages:

1. *Statistical Efficiency*—the MLE is typically efficient (makes good use of limited data).
2. *Transformation Invariance*—for example, estimates do not depend on whether the model is written using a log or natural scale.
3. *Asymptotic Results*—the second derivative of the log likelihood at its maximum can be used to give approximate standard errors. This means that simulations to understand the variability in estimates are seldom necessary.
4. *Model Selection*—likelihoods are comparable between different models for the same data. In particular, a χ^2 approximation is often appropriate: if p parameters are added to a model and the increase in the log likelihood is large compared to a $(\frac{1}{2})\chi_p^2$ random variable, then the fit is a statistically significant improvement.

Bayesian analysis is also attractive, since previous research may be available to provide an informed prior. Bayesian methods have been used for population models [64,14]. For this chapter we consider MLE methods, but the computational issue of integrating out unobserved variables arises in a similar way with Bayesian methods.

Evaluation of the likelihood and determination of the conditional distribution of unobserved variables given data are computationally approachable in a broad class of time-series models known as *state space models* (SSMs). SSMs have been proposed as a unifying framework for ecological modeling [64]. Likelihood based inference has been shown to outperform other more ad hoc statistical model-fitting criteria for population models incorporating process noise and observation error [16]. The linear, Gaussian SSM [38] became fundamental to engineering, for signal processing and control theory [2], and found applications in economics [29]. Early attempts to handle nonlinear SSMs were plagued by the lack of computational ability to evaluate the likelihood, so inference resorted to ad hoc methods [2]. Brillinger et al. [11] provides an early ecological application of nonlinear SSMs.

The development of Monte Carlo methods for nonlinear SSMs, combined with increases in computational capability, has made likelihood-based inference feasible for increasingly general nonlinear SSMs. This gives the modeler considerable freedom to write down an appropriate model without undue concern for inferential feasibility. There are two main approaches to Monte Carlo inference for SSMs: sequential Monte Carlo [26,18,4] and Markov chain Monte Carlo (MCMC) [61]. This chapter focuses on sequential Monte Carlo (SMC), which is more widely used for SSMs and simpler to implement. A careful comparison between SMC and MCMC is still, to the authors' knowledge, an unresolved problem.

8.3 SEQUENTIAL MONTE CARLO

An SSM is a partially observed Markov process. The unobserved Markov process, x_t, called the *state process*, takes values in a *state space* $\mathcal{X}$. The *observation process* y_t takes values in an *observation space* $\mathcal{Y}$, and y_t is assumed to be conditionally independent of the past given x_t. Here, we take $\mathcal{X}$ to be $\mathcal{R}^{d_x}$ and $\mathcal{Y}$ to be $\mathcal{R}^{d_y}$. There is also a vector of unknown parameters $\theta \in \mathcal{R}^{d_\theta}$. We suppose that observations take place at discrete times, $t = 1, \dots, T$. We further suppose that all required densities exist, and we adopt a convention that $f(\cdot \mid \cdot)$ is a generic density that is then specified by its arguments. We write concatenated observations as $y_{1:t} =_1 , \dots , y_t)$. For the case $t = 0, y_{1:0}$ is defined to be an empty vector. The properties of a state space model are

$$f_\theta(x_t \mid x_{1:t-1}, y_{1:t-1}) = f_\theta(x_t \mid x_{t-1}), \tag{8.1}$$

$$f_\theta(y_t \mid x_{1:t}, y_{1:t-1}) = f_\theta(y_t \mid x_t). \tag{8.2}$$

The dependence on θ will be written explicitly only when necessary for clarity. In principle, the assumed Markov structure in (8.1) and (8.2) allows the likelihood, $f_\theta(y_{1:T})$, to be found recursively via the identities

$$f(x_t \mid y_{1:t-1}) = \int f(x_{t-1} \mid y_{1:t-1}) f(x_t \mid x_{t-1}) \, dx_{t-1}, \tag{8.3}$$

$$f(x_t \mid y_{1:t}) = \frac{f(x_t \mid y_{1:t-1}) f(y_t \mid x_t)}{\int f(x_t \mid y_{1:t-1}) f(y_t \mid x_t) \, dx_t}, \tag{8.4}$$

$$f(y_t \mid y_{1:t-1}) = \int f(y_t \mid x_t) f(x_t \mid y_{1:t-1}) \, dx_t, \tag{8.5}$$

$$f(y_{1:T}) = \prod_{t=1}^{T} f(y_t \mid y_{1:t-1}). \tag{8.6}$$

In practice, this requires solving potentially challenging integrals. Following Kitagawa [43], de Valpine and Hastings [16] showed how these integrals could be solved numerically for relatively simple population models. For more complex models, one may employ an SMC method such as Algorithm 8.1.

Algorithm 8.1: Sequential Monte Carlo (SMC)

Initialize: Let $\{X_{0,j}^F, j = 1, \ldots, J\}$ be a sample draw from $f(x_0)$. These J realizations are commonly termed "particles." Each particle will give rise to a trajectory through the state space with distribution $f(x_t \mid y_{1:t})$.

FOR $t = 1$ to T

- **Move particles according to unconditional state process:** Make $X_{t,j}^P$ a draw from $f(x_t \mid x_{t-1} = X_{t-1,j}^F)$. Then $\{X_{t,j}^P\}$ has approximate marginal distribution $f(x_t \mid y_{1:t-1})$. $\{X_{t,j}^P\}$ is said to solve the prediction problem at time t.

- **Calculate conditional likelihood of new observation:** Estimate $f(y_t \mid y_{1:t-1})$ by $(1/J) \sum_{j=1}^{J} f(y_t \mid x_t = X_{t,j}^P)$.

- **Prune particles according to likelihood given data:** Generate $X_{t,j}^F$ by resampling from $\{X_{t,j}^P\}$ with probability proportional to $w_j = f(y_t \mid x_t = X_{t,j}^P)$ using Algorithm 8.2 (below). Then $\{X_{t,j}^F\}$ has approximate marginal distribution $f(x_t \mid y_{1:t})$. Then $\{X_{t,j}^F\}$ is said to solve the filtering problem at time t.

END FOR

Calculate Log Likelihood: $\log f(y_{1:T}) = \sum_{t=1}^{T} \log f(y_t \mid y_{1:t-1})$.

Algorithm 8.2: Systematic Resampling

Input: J particles $\{X_{t,j}^P, j = 1, \ldots, J\}$ with weights $\{w_j = f(y_t \mid x_t = X_{t,j}^P)\}$

Calculate Cumulative Sum of Normalized Weights: FOR $j = 1$ to J set $c_j = (\sum_{k=1}^{j} w_k)/(\sum_{k=1}^{J} w_k)$

Resample Cumulative Sum at Intervals of $1/J$:
Set $i = 1$ and $u \sim U[0,1]$
FOR $j = 1$ to J

- WHILE $(j - u)/J > c_i$ set $i = i + 1$
- Set $X_{t,j}^F = X_{t,i}^P$. This resampling generates a tree structure, where $X_{t,j}^F$ is said to descend from $X_{t-1,i}^F$.

END FOR

Output: J particles $\{X_{t,j}^F, j = 1, \ldots, J\}$

The reader is referred to the literature [4,18,51] for extensive discussions of Algorithms 8.1 and 8.2, with many possible variations. Algorithm 8.1 can be fine-tuned to be computationally more efficient in many ways. A more critical issue, in the authors' opinion, is how to use the output of Algorithm 8.1 for effective inference. Although Algorithm 8.1 is widely applicable for calculating the likelihood at a fixed value of θ, complications arise for both Bayesian and MLE methods, which must compare likelihoods for different values of θ.

Bayesian inference might appear straightforward; simply add θ to the state space. The initial particles are then drawn from $f(x_0, \theta)$, and the particle filter will then produce a sample from $f(\theta \mid y_{1:T})$. Each particle at time t has exactly the same value of θ as does its ancestor at time $t - 1$, and the prior distribution on θ is updated via the SMC algorithm, giving particles with successful values of θ more descendants. The catch is that the SMC algorithm degenerates when there is no variability in the θ component of the state process after $t = 0$. Heuristically, the particles in SMC evolve by natural selection according to their plausibility given the data. Particles whose θ component are fixed over time are analogous to natural selection without mutation, which produces only limited scope for evolution. One solution to this is to allow the parameter to vary slowly with time by adding noise [44]. In cases where this modification to the model is considered unacceptable, Liu and West [50] showed how to add noise to the parameters but balance this by simultaneously contracting the parameter distribution toward its mean. The method of Liu and West [50] has been applied to ecological models by Thomas et al. [64] and Newman and Lindley [54].

The difficulty for finding the MLE is that the likelihood is calculated with Monte Carlo error. One useful tool for optimizing functions calculated via Monte Carlo is the method of common random numbers [63, Sec. 14.4], which involves fixing the seed of the random-number generator. This method requires synchronization of the Monte Carlo randomness, which is not directly applicable to SMC techniques. General stochastic optimization methods of the Robbins–Monro type [57,42,63] are not applicable for problems where there are many unknown parameters and each function evaluation is a considerable computational expense. The elegant method of Hürzeler and Künsch [33] for calculating local likelihood surfaces is also not readily applicable to relatively difficult problems—it is more computationally intensive than standard SMC methods such as Algorithm 8.1. Ionides et al. [35] showed how to find the MLE by taking a limit where the noise, added in a way similar to that in Kitagawa [44], shrinks to zero. This novel method is described in Algorithm 8.3 and is applied to a cholera population model in Section 8.4.

Algorithm 8.3 is appropriate when information about parameters arrives steadily throughout a time series. Heuristically, it gains computational efficiency because the parameter estimate is being constantly updated throughout each iteration. Each iteration would correspond to one evaluation of the likelihood for a general-purpose optimization algorithm. In Section 8.4, $N = 20$ or $N = 30$ iterations are sufficient to optimize a stochastic function of 13 variables, without availability of analytic derivatives. This computational efficiency is critical when each iteration takes around 30 min to compute.

In certain situations, such as estimating the initial value vector x_0, information about a parameter does not arrive steadily throughout a time series. In this case, Algorithm 8.3 is not effective. If $\{x_t\}$ is stationary, then x_0 can be drawn from the stationary distribution. If $\{x_t\}$ is not stationary, one can either pick some more arbitrary distribution for x_0 or treat x_0 as an unknown parameter (in the frequentist sense). We choose to do the latter, and estimate x_0 by maximum likelihood simultaneously with θ by applying Algorithm 8.4, the theoretical justification of which is similar to that for Algorithm 8.3 [35]. The value of T_0 in Algorithm 8.4 should be as small as possible such that $y_{T_0+1:T}$ contains negligible additional information about x_0, beyond that contained in $y_{1:T_0}$. This compromise is known as *fixed-lag smoothing* [2].

Algorithm 8.3: MLE via Iterated Filtering

Initialize: Select $\theta^{\text{hi}} > \theta^{(1)} > \theta^{\text{lo}}$ to be vectors giving a plausible initial value and range for the parameters. Select scalars $0 < \alpha < 1$, C, and N.

FOR $n = 1$ to N

• **Apply SMC** (Algorithm 8.1) with θ included in the state space as a time-varying parameter, evolving as

$$\theta_0 \sim N_{d_\theta}(\theta^{(n)}, C\Sigma_n),$$
$$\theta_t \mid \theta_{t-1} \sim N_{d_\theta}(\theta_{t-1}, \Sigma_n) \quad for \quad t = 2, \ldots, T,$$

where the covariance matrix Σ_n is defined by $[\Sigma_n]_{ii}^{1/2} = [(\theta^{\text{hi}} - \theta^{\text{lo}})/2\sqrt{T}]_i \alpha^{n-1}$ and $[\Sigma_n]_{ij} = 0$ for $i \neq j$. Each particle is now a pair, e.g., $(X_{t,j}^F, \theta_{t,j}^F)$.

• **Calculate Updated Estimate:**

$$\hat{\theta}_t = (1/J) \sum_{j=1}^{J} \theta_{t,j}^F \quad \text{for } 1 \leq t \leq T,$$

$$V_1 = (C + 1)\Sigma_n,$$

$$V_{t+1} = \frac{\sum_{j=1}^{J} (\theta_{t,j}^F - \hat{\theta}_t)(\theta_{t,j}^F - \hat{\theta}_t)'}{J - 1} + \Sigma_n \quad \text{for } 1 \leq t \leq T - 1,$$

$$\theta^{(n+1)} = V_1 \left(\sum_{t=1}^{T-1} (V_t^{-1} - V_{t+1}^{-1}) \hat{\theta}_t + V_T^{-1} \hat{\theta}_T \right).$$

END FOR

The MLE is estimated as $\hat{\theta} = \theta^{(N+1)}$.

Algorithm 8.4: MLE via Iterated Filtering, for Initial Values

Initialize: Select $x_0^{\text{hi}} > x_0^{(1)} > x_0^{\text{lo}}$ to be vectors giving a plausible initial value and range for the initial values. Select scalars $0 < \alpha < 1$, T_0 and N.

FOR $n = 1$ to N

• **Apply SMC** (Algorithm 8.1) with $X_{0,j}^F \sim N_{d_x}(x_0^{(n)}, \Phi_n)$ where $[\Phi_n]_{ii}^{1/2} = [(x_0^{\text{hi}} - x_0^{\text{lo}})/2]_i \alpha^{n-1}$ and $[\Phi_n]_{ik} = 0$ for $i \neq k$. For each particle $X_{t,j}^F$, track label of the corresponding initial value, denoted $a(t, j)$. In the terminology of Algorithm 8.2, $X_{t,j}^F$ descends from $X_{0,a(t,j)}^F$.

• **Calculate Updated Estimate:** $x_0^{(n+1)} = (1/J) \sum_{j=1}^{J} X_{0,a(T_0,j)}^F$

END FOR

The MLE is estimated as $\hat{x}_0 = x_0^{(N+1)}$.

Algorithms 8.3 and 8.4 are different variations on the same theme of using limiting Bayesian posterior distributions to find MLEs. Both algorithms can be combined, so that one filtering iteration updates estimates of all estimated parameters, including initial value parameters.

8.4 MODELING CHOLERA

Cholera is a diarrheal disease endemic to the Ganges delta region [60]. Global pandemics have occurred throughout recent history. The current (seventh) pandemic started in 1960 and has seen the O1 serogroup become established in various locations throughout South Asia, Africa, and South America. Cholera is caused by virulent strains of *Vibrio cholerae*, a bacterium that can live and grow in brackish, warm water. Human-to-human transmission can be direct, through contact with stool from infected individuals; or indirect, via the environment. There is not a clear distinction between these two paths; we separate them by supposing that the increase in force of infection depending on the number of infected individuals is due to human-to-human transmission. The environmental reservoir is taken to be responsible for the background force of infection (extrapolating to a situation with no infected humans). A compartment model describing the basic features of disease transmission is diagrammed in Figure 8.1. Formally, the diagram in Figure 8.1 corresponds to a set of equations:

$$dS_t = dN_t^{BS} - dN_t^{SI} - dN_t^{SD} + dN_t^{R^kS},$$

$$dI_t = dN_t^{SI} - dN_t^{IR^1} - dN_t^{ID},$$

$$dR_t^1 = dN_t^{IR^1} - dN_t^{R^1R^2} - dN_t^{R^1D}.$$

$$\vdots$$

$$dR_t^k = dN_t^{R^{k-1}R^k} - dN_t^{R^kS} - dN_t^{R^kD}$$

Here, time is measured in months; S_t is the number of individuals in class S (susceptible) and the infinitesimal dS_t is defined such that $S_t = S_0 + \int_0^t dS_u$. For example, N_t^{SI} corresponds to the total number of individuals who have passed from S to I by time t. The k recovered classes allow for flexibility in modeling the time from infection to loss of immunity, at which point an individual becomes newly susceptible. This temporary immunity, with a duration

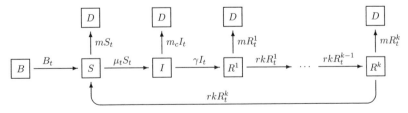

Figure 8.1 Compartment model for cholera. Each individual is in S (susceptible), I (infected), or one of the classes R^j (recovered). Transitions to B and D model birth and death, respectively. The arrows show possible transitions, with superscripts showing transition rates.

of 3–10 years, is believed to be a key feature of the population dynamics of cholera. We use a model developed by Ionides et al. [35]:

$$dN_t^{SI} = \mu_t S_t dt + \psi_t S_t \, dW_t,$$
$$\mu_t = \beta_t I_t / P_t + w,$$
$$\psi_t = e I_t / P_t. \tag{8.7}$$

Here, population size P_t is interpolated from available census data, and is presumed to be accurately known; seasonal transmissibility is modeled as $\log(\beta_t) = \sum_{j=0}^{5} b_j s_j(t)$, where $\{s_j(t), \, j = 0, \ldots, 5\}$ is a periodic cubic B-spline basis; e is an *environmental stochasticity* parameter, modeling noise on the environmental scale (with infinitesimal variance propotional to S_t); w corresponds to a nonhuman *reservoir* of disease; $\beta_t I_t / P_t$ is *human-to-human* infection; $1/\gamma$ gives the mean time to recovery; $1/r$ is the mean time to loss of immunity following recovery, with k giving the shape of this distribution; and m and m_c are the death rates among uninfected and infected individuals, respectively. The remaining transition equations were modeled deterministically:

$$dN_t^{IR^1} = \gamma I_t \, dt; \quad dN_t^{R^{j-1} R^j} = rk R_t^{j-1} \, dt;$$
$$dN_t^{R^k S} = rk R_t^k \, dt; \quad dN_t^{SD} = m S_t \, dt;$$
$$dN_t^{ID} = m_c I_t \, dt; \quad dN_t^{R^j D} = m R_t^j \, dt;$$
$$dN_t^{BS} = dP_t + dN_t^{SD} + dN_t^{ID} + \sum_{j=1}^{k} dN_t^{R^j D}. \tag{8.8}$$

Defining $C_t = N_t^{ID} - N_{t-1}^{ID} = \int_{t-1}^{t} dN_t^{ID}$, the number of cholera mortalities between monthly observation times, the data on observations data on observed mortality were modeled conditional on C_t as $y_t \sim \mathcal{N}[\rho C_t, \rho(1-\rho)C_t + \tau^2 \rho^2 C_t]$ with reporting rate ρ. The variance component $\rho(1-\rho)C_t$ models demographic stochasticity via binomial sampling variation. Environmental stochasticity is modeled via $\tau^2 \rho^2 C_t^2$, which dominates demographic variability for large C_t and is found to be appropriate when fitting (8.10) and (8.11) to data. The dominance of environmental stochasticity has been assumed implicitly in previous analyses of similar data, by modeling additive noise of variance τ^2 in $\log(\rho C_t)$ [24,46]. Demographic variability is nonnegligible when C_t is small, and can be included in our framework without adding any additional parameters.

Continuous-state population models, such as the model given by (8.7) and (8.8), are more convenient for data analysis than discrete-state population models. Theoretical results and simulation studies of population models often resort to demographic (Poisson) variability, using the rates in Figure 8.1 to define a continuous-time Markov chain. Apart from the inherent appropriateness of discrete populations, the Markov chain approach has the advantage that no extra parameters, beyond the rates, are needed to describe the stochasticity. However, demographic stochasticity alone is not always sufficient to describe observed variations in data; for cholera, demographic stochasticity is entirely inadequate. If extra variability has to be introduced, stochastic differential equations (SDEs) provide a simple way to do this. SDEs are a natural extension to the ordinary differential equation (ODE) systems already used for describing population dynamics. Other examples of the use of SDEs to provide a framework for modeling and data analysis include those by Kendall [40], Brillinger and Stewart [13],

Brillinger et al. [12], and Ionides et al. [36]. There are several misconceptions about SDEs that explain why they are not currently more widely used for modeling. These are listed below, with refutations:

1. The theory of SDEs is inaccessible and obscure. However, numerical solutions to SDEs are now well established [45,31]. This allows development and exploration of models that would be difficult to investigate analytically. In particular, application of the inference methodology in Algorithms 8.3 and 8.4 for the models in (8.7) and (8.8) requires only numerical solution of the system of SDEs.

2. There may be little reason to think that Gaussian white noise is a plausible stochastic driver for the system under investigation. Supplying random coefficients to an ODE or Markov chain adds lower-frequency "colored noise". However, most practical time-series models, such as the autoregressive moving average (ARMA) framework [62], use white noise as the basic building block. This noise is often modeled as Gaussian, for convenience, and the data may sometimes be transformed to increase the plausibility of this assumption. Solutions to SDEs driven by Gaussian white noise include almost all non-Gaussian continuous-time, continuous-sample-path Markov processes [55]. Smooth, low frequency noise can be modeled by adding white noise to a derivative of the process of interest.

3. There has been much discussion in theoretical modeling literature concerning different possible interpretations of an SDE. The two most popular interpretations are the Itô and Stratonovich solutions [55]. The distinction, involving the exact way the SDE is solved as a limit of finite sums, should have little scientific relevance. Meaningful scientific conclusions should not depend on the choice of interpretation of SDE [36]. Numerical solution is most straightforward for the Itô solution, so that is the one adopted here.

8.4.1 Fitting Structural Models to Cholera Data

Maximizing a nonconvex function of more than a few variables is seldom routine, especially when the function is evaluated by Monte Carlo methods. Algorithm 8.3 provides a way to leverage the special structure of an SSM for optimization, but diagnostic checks are necessary before one has confidence in the results. Beyond the standard approach of trying various initial values $[\theta^{(1)}, \theta^{lo}$ and $\theta^{hi}]$, one should assess the choice of the two variables α and C for Algorithm 8.3. If α is too small, the rapid decrease in step size in Algorithm 8.3 may leave the algorithm stranded, unable to reach the maximum. This is analogous to excessively rapid cooling in simulated annealing [63]. If α is too large, insufficient cooling will occur within a reasonable computation time. These issues can be diagnosed by plotting $\theta^{(n)}$ against n for several values of α and $\theta^{(1)}$, looking for consistent convergence. The term C is a dimensionless constant controlling the initial dispersion of the parameter values, relative to their random perturbations through time. If C is too small, the algorithm converges slowly. If C is too large, the algorithm is less stable and converges erratically. This can be assessed by the same type of convergence plot as used for α, or by the observation that a good choice of C is one which makes V_t fairly stable as a function of t.

The likelihood surface near the convergence point $\hat{\theta}$ can be further examined by "sliced likelihood" plots. Setting $\lambda(\theta) = \log f_\theta(y_{1:T})$, the sliced likelihood for $\hat{\theta}_i$ plots $\lambda(\hat{\theta} + c\delta_i)$ against $\hat{\theta}_i + c$, where δ_i is a vector of zeros with a one in the ith position. If $\hat{\theta}$ is at (or near) the maximum of each sliced-likelihood plot, then $\hat{\theta}$ is (approximately) a local maximum of $\lambda(\theta)$. Computing sliced likelihoods requires moderate computational effort, linear in the dimension of θ. A smoothed fit (as suggested by Ionides [34]) is made to the sliced log likelihood, because $\lambda(\hat{\theta} + c\delta i)$ is calculated with a Monte Carlo error. Figure 8.2 shows a convergence and sliced-likelihood plot for a simulation study, presented in Ionides et al. [35], using

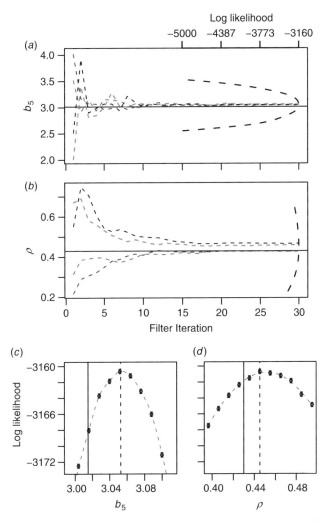

Figure 8.2 Examples of convergence plots for a simulation from (8.7) and (8.8) with four different starting points, validating the convergence of Algorithms 8.3 with $\alpha = 0.9$ and $C = 20$. The dotted parabolic line corresponds to a sliced likelihood through $\hat{\theta}$. (c,d) Corresponding closeups of the sliced likelihood. The dashed vertical line is at $\hat{\theta}$ and the solid vertical line is at the true value of θ. The simulation was carried out with $\rho = 0.43$, $e = 0.289$, $b_0 = -1.48$, $b_1 = 2.42$, $b_2 = 0.02$, $b_3 = -0.98$, $b_4 = 0.02$, $b_5 = 3.02$, $\tau^2 = 0.02$, $w = 2.5 \times 10^{-6}$, $m_c = 1.19$, $\gamma = 1$, $k = 4$, $1/r = 120$, and $1/m = 600$. The last four of these parameters were treated as known, and the remaining parameters were estimated, using Algorithm 8.3 with $J = 9000$.

the cholera model in (8.7) and (8.8). The deviation between the MLE and the true parameter value is due to the finite length (50 years) of the simulated dataset. In some generality, the MLE for state space models is consistent and asymptotically normally distributed [37].

Sliced likelihoods can be used to generate standard errors, since calculating $\lambda(\hat{\theta} + c\delta_i)$ involves finding $\log f_{\hat{\theta}+c\delta_i}(y_t \mid y_{1:t-1})$. Regressing $\log f_{\hat{\theta}+c\delta_i}(y_t \mid y_{1:t-1})$ on c gives an estimate of $(\partial/\partial\theta_i) \log f_{\hat{\theta}}(y_t \mid y_{1:t-1})$, giving rise to an estimate of the observed Fisher information

$$[\hat{\mathcal{I}}_F]_{ij} = \sum_{t=1}^{T} \frac{\partial}{\partial\theta_i} \log f_{\theta}(y_t \mid y_{1:t-1}) \frac{\partial}{\partial\theta_j} \log f_{\theta}(y_t \mid y_{1:t-1}) \tag{8.9}$$

where the derivatives are evaluated at $\theta = \hat{\theta}$. This leads to a corresponding estimate $\hat{\mathcal{I}}_F^{-1}$ for the covariance matrix of $\hat{\theta}$.

A superior way to find confidence intervals is via a profile likelihood [6]. If θ is partitioned into two components ζ and η, then the profile log likelihood of η is defined [6] by $\lambda_{(p)}(\eta) = \sup_\zeta \lambda(\zeta, \eta)$. The optimization required for the profile likelihood can be carried out using Algorithm 8.3. Calculating the profile likelihood for each parameter therefore requires approximately N times the computational effort of the sliced likelihood (typically, N is between 20 and 30). The optimization also introduces additional Monte Carlo variability over a simple likelihood evaluation. Figure 8.3 shows the profile likelihood for a parameter of the model in (8.7) and (8.8). This parameter was selected because the profile likelihood confidence interval constructed in Figure 8.3, of width 0.27, was considerably different from the approximation using (8.9), of width 0.10. This rather large discrepancy arose because the quadratic approximation in (8.9) is overly optimistic when some nonlinear combination of the parameters is poorly estimable. The extra computation required to calculate a profile likelihood is

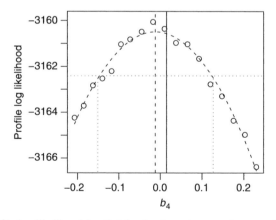

Figure 8.3 Profile log likelihood $\lambda_{(p)}(b_4)$ for the August seasonal parameter. The log likelihood was maximized over all parameters excluding b_4 (circles) and was then smoothed (dashed line) using nonparametric regression [34,15]. The dotted lines show the construction of an approximate 95% confidence interval, given by $\{b_4 : 2[\lambda_{(p)}(\hat{b}_4) - \lambda_{(p)}(b_4)] < \chi^2_{0.95}(1)\}$, where $\chi^2_{0.95}$ is the 0.95 quantile of a χ^2 random variable with one degree of freedom and $\hat{b}_4 = \arg\max \lambda_{(p)}(b_4)$.

evidently worthwhile for a parameter of particular interest. The quadratic approximation can be calculated more routinely, to get a general idea of the scale of uncertainty.

The model in (8.7) and (8.8) was fitted to historical data for Dhaka, Bangladesh, [10,58,46], shown in Figure 8.4*a*. Our resulting estimate of the seasonal transmissibility β_t is shown in Figure 8.4*b*. Observed mortality is seen to have two seasonal peaks that appear later than the peaks in transmissibility. The winter dip in mortality has been ascribed to reduced environmental viability of *V. cholerae* in colder temperatures. The early January local minimum in transmissibility is consistent with the early January minimum in mean temperature in Dhaka. The summer dip in mortality has been ascribed to dilution of *V. cholerae* due to monsoon rainfall. The monsoon season in Dhaka is from May to September, with greatest average rainfall in July. Fitting (8.7) and (8.8), the transmissibility is seen to decrease too soon to be explained fully by rainfall. Snowmelt from the Himalayas is one candidate to explain this discrepancy.

Investigating residuals is a routine diagnostic check in time-series and regression analyses. The most basic residuals to consider for SSMs are the standardized prediction residuals

$$u_t(\hat{\theta}) = [\text{Var}_{\hat{\theta}}(y_t \mid y_{1:t-1})]^{-1/2}(y_t - E_{\hat{\theta}}(y_t \mid y_{1:t-1})),$$

although there are other possibilities [35,19]. Checking whether the residuals are approximately uncorrelated is a way to test the goodness of fit of the model. Residuals also have an important role in the search for covariates. Inasmuch as the model successfully captures the intrinsic dynamics of the disease, the residuals are left with the system noise plus signal from the extrinsic variables, such as climate. From this point of view, features that the intrinsic model cannot capture are as important as those that it can! A more flexible model might fit the data better, but only by explaining variation that in fact has some

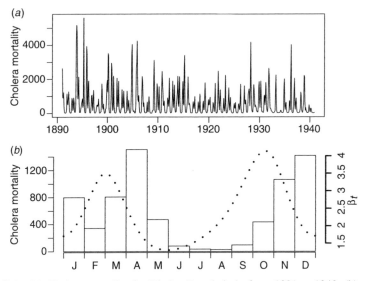

Figure 8.4 (*a*) Cholera mortality for Dhaka, Bangladesh, from 1891 to 1940; (b) monthly averages of Dhaka cholera mortality (boxes) and the seasonal transmissibility β_t (dotted line) from fitting (8.7) and (8.8).

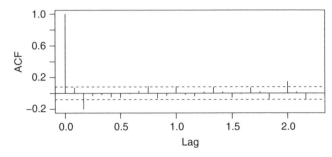

Figure 8.5 Sample autocorrelation function for the standardized residuals when fitting (8.7) and (8.8) to the data in Figure 8.4.

extrinsic origin. The next step after identifying covariates is to include them in the model. This is not necessarily an easy task—even explaining seasonality can be a challenge [56]. For example, both rainfall and drought can initiate cholera epidemics. The low-frequency component of residuals from a time-series model fit to cholera data has been found to match various plausible environmental drivers, such as rainfall, river discharge, and El Niño indices [47]. Fitting the model of (8.7) and (8.8) results in less than perfectly white residuals (see Fig. 8.5). The residuals nevertheless give evidence of increased cholera infection in Dhaka after the monsoon during El Niño conditions [35], and this association is not evident from the original time series. How best to include environmental covariates in a mechanistic model is a topic for future investigation. However, the methodology in Section 8.3 provides both a tool to identify covariates and a flexible framework for including them in a mechanistic way.

8.5 CONCLUDING REMARKS

Six key areas requiring further development for time-series analysis of population data were identified by Bjørnstad and Grenfell [9]. They may be summarized as follows: (1) including measurement error in mechanistic models, (2) mechanistic modeling of environmental forcing, (3) employing ecologically realistic continuous-time models, (4) reconstructing unobserved variables, (5) identifying interactions, and (6) spatiotemporal modeling. The cholera modeling example in Section 8.4 demonstrates that the SSM approach in Section 8.3 can be used to address issues 1–4. In addition, likelihood-based model comparison then provides an approach to issue 5. In principle, one can write down a spatiotemporal SSM to address issue 6. In practice, the dimension of the state space typically scales linearly with the number of spatial locations considered, and high-dimensional state spaces increase the numerical burden on the SMC method. For large spatiotemporal problems, such as data assimilation in atmospheric and oceanographic science, SMC is not feasible. Related techniques have been developed for data assimilation [23,32], employing an ensemble of numerical solutions of a spatiotemporal model to approximate the conditional distribution given data. Alternatively, spatiotemporal variability can be incorporated through random-effect models [67,68,8]. More progress is necessary before SMC techniques can be routinely applied to spatiotemporal data. However, SMC provides an effective and flexible tool for partially observed stochastic nonlinear dynamical systems of moderate dimension, allowing freedom to develop models based on scientific principles rather than on methodological constraints.

ACKNOWLEDGMENTS

The authors acknowledge many helpful discussions with Mercedes Pascual and Menno Bouma; the latter provided the cholera data for Figure 8.4. The authors were supported by National Science Foundation grant 0430120.

REFERENCES

1. Aguirre, A. A., Ostfeld, R. S., Tabor, G. M., House, C., and Pearl, M. C., eds., *Conservation Medicine*, Oxford Univ. Press, New York, 2002.
2. Anderson, B. D. and Moore, J. B., *Optimal Filtering*, Prentice-Hall, Englewood Cliffs, NJ, 1979.
3. Anderson, R. M. and May, R. M., *Infectious Diseases of Humans*, Oxford Univ. Press, Oxford, UK, 1991.
4. Arulampalam, M. S., Maskell, S., Gordon, N., and Clapp, T., A tutorial on particle filters for online nonlinear, non-Gaussian Bayesian tracking, *IEEE Trans. Signal Process.* **50**, 174–188 (2002).
5. Bailey, N. T. J., *The Mathematical Theory of Infectious Diseases and Its Application*, 2nd ed., Charles Griffin, London, 1975.
6. Barndorff-Nielsen, O. E. and Cox, D. R., *Inference and Asymptotics*, Chapman & Hall, London, 1994.
7. Bartlett, M. S., *Stochastic Population Models in Ecology and Epidemiology*, Wiley, New York, 1994.
8. Berliner, L. M., Wikle, C. K., and Cressie, N., Long-lead prediction of Pacific SSTs via Bayesian dynamic modeling, *J. Climate* **13**, 3953–3968 (2000).
9. Bjørnstad, O. N. and Grenfell, B. T., Noisy clockwork: Time series analysis of population fluctuations in animals, *Science* **293**, 638–643 (2001).
10. Bouma, M. J. and Pascual, M., Seasonal and interannual cycles of endemic cholera in Bengal 1891–1940 in relation to climate and geography, *Hydrobiologia* **460**, 147–156 (2001).
11. Brillinger, D. R., Guckenheimer, J., Guttorp, P., and Oster, G., Empirical modelling of population time series: The case of age and density dependent rates, in *Some Questions in Mathematical Biology*, Oster, G., ed., American Mathematical Society, Providence, RI, 1980, pp. 65–90.
12. Brillinger, D. R., Preisler, H. K., Ager, A. A., Kie, J. G., and Stewart, B. S., Employing stochastic differential equations to model wildlife motion, *Bull. Brazil. Math. Soc.* **33**, 385–408 (2002).
13. Brillinger, D. R. and Stewart, B. S., Elephant-seal movements: Modelling migration, *Can. J. Statist.* **26**, 431–443 (1998).

14. Clark, J. S. and Bjørnstad, O. N., Population time series: Process variability, observation errors, missing values, lags, and hidden states, *Ecology* **85**, 3140–3150 (2004).

15. Cleveland, W. S., Grosse, E., and Shyu, W. M., Local regression models, in *Statistical Models in S*, Chambers, J. M. and Hastie, T. J., eds., Chapman & Hall, London, 1993.

16. de Valpine, P. and Hastings, A., Fitting population models incorporating process noise and observation error, *Ecol. Monogr.* **72**, 57–76 (2002).

17. Diekmann, O. and Heesterbeek, J. A. P., *Mathematical Epidemiology of Infectious Diseases: Model Building, Analysis and Interpretation*, Wiley, Chichester, UK, 2000.

18. Doucet, A., de Freitas, N., and Gordon, N. J., eds., *Sequential Monte Carlo Methods in Practice*, Springer, New York, 2001.

19. Durbin, J. and Koopman, S. J., *Time Series Analysis by State Space Methods*, Oxford Univ. Press, Oxford, UK, 2001.

20. Ellner, S. P., Bailey, B. A., Bobashev, G. V., Gallant, A. R., Grenfell, B. T., and Nychka, D. W., Noise and nonlinearity in measles epidemics: Combining mechanistic and statistical approaches to population modeling, *Am. Naturalist* **151**, 425–440 (1998).

21. Ellner, S. P., Seifu, Y., and Smith, R. H., Fitting population dynamic models to time-series data by gradient matching, *Ecology* **83**, 2256–2270 (2002).

22. Engen, S., Bakke, O., and Islam, A., Demographic and environmental stochasticity: Concepts and definitions, *Biometrics* **54**, 840–846 (1998).

23. Evensen, G. and van Leeuwen, P. J., Assimilation of geostat altimeter data for the Agulhas Current using the ensemble Kalman filter with a quasigeostrophic model, *Monthly Weather Rev.* **124**, 58–96 (1996).

24. Finkenstädt, B. F. and Grenfell, B. T., Time series modelling of childhood diseases: A dynamical systems approach, *Appl. Statist.* **49**, 187–205 (2000).

25. Glass, K., Xia, Y., and Grenfell, B. T., Interpreting time-series analyses for continuous-time biological models—measles as a case study, *J. Theor. Biol.* **223**, 19–25 (2003).

26. Gordon, N., Salmond, D. J., and Smith, A. F. M., Novel approach to nonlinear/ non-Gaussian Bayesian state estimation, *IEE Proc. F* **140**(2), 107–113 (1993).

27. Greene, S. K., Ionides, E. L., and Wilson, M. L. (2006). Patterns of influenza-associated mortality among US elderly by geographic region and virus subtype, 1968–1998, *Am. J. Epidemiol.* (prepublished online) (2006).

28. Hahn, B. H., Shaw, G. M., de Cock, K. M., and Sharp, P. M., AIDS as a zoonosis: Scientific and public health implications, *Science* **287**, 607–614 (2000).

29. Harvey, A. C., *Forecasting, Structural Time Series Models and the Kalman Filter*, Cambridge Univ. Press, 1989.

30. Hethcote, H. W. The mathematics of infectious diseases, *SIAM Rev.* **42**, 599–653 (2000).

31. Higham, D. J. An algorithmic introduction to numerical simulation of stochastic differential equations, *SIAM Rev.* **43**, 525–546 (2001).

32. Houtekamer, P. L. and Mitchell, H. L., Data assimilation using an ensemble Kalman filter technique, *Monthly Weather Rev.* **129**, 123–137 (2001).

33. Hürzeler, M. and Künsch, H. R., Approximating and maximising the likelihood for a general state-space model, in *Sequential Monte Carlo Methods in Practice*, Doucet, A., de Freitas, N., and Gordon, N. J., eds., Springer, New York, 2001, pp. 159–175.

34. Ionides, E. L., Maximum smoothed likelihood estimation, *Statistica Sinica* **15**, 1003–1014 (2005).

35. Ionides, E. L., Bretó, C., and King, A. A., Inference for nonlinear dynamical systems, *Proc. Natl. Acad. Sci. USA* **103**, 18438–18443 (2006).

36. Ionides, E. L., Fang, K. S., Isseroff, R. R., and Oster, G. F., Stochastic models for cell motion and taxis, *J. Math. Biol.* **48**, 23–37 (2004).

37. Jensen, J. L. and Petersen, N. V., Asymptotic normality of the maximum likelihood estimator in state space models, *Ann. Statist.* **27**, 514–535 (1999).

38. Kalman, R. E., A new approach to linear filtering and prediction problems, *J. Basic Eng.* **82**, 35–45 (1960).

39. Kendall, B. E., Ellner, S. P., McCauley, E., Wood, S. N., Briggs, C. J., Murdoch, W. M., and Turchin, P., Population cycles in the pine looper moth: Dynamical tests of mechanistic hypotheses, *Ecol. Monogr.* **75**(2), 259–276 (2005).

40. Kendall, D. G., Pole-seeking Brownian motion and bird navigation, *J. Roy. Statist. Soc. Ser. B* **36**, 365–417 (1974).

41. Kermack, W. O. and McKendrick, A. G., A contribution to the mathematical theory of epidemics, *Proc. Roy. Soc. Lond. A* **115**, 700–721 (1927).

42. Kiefer, J. and Wolfowitz, J., Stochastic estimation of the maximum of a regression function, *Ann. Math. Statist.* **23**, 462–466 (1952).

43. Kitagawa, G., Non-Gaussian state-space modelling of non-stationary time series, *J. Am. Statist. Assoc.* **82**, 1032–1063 (1987).

44. Kitagawa, G., A self-organising state-space model, *J. Am. Statist. Assoc.* **93**, 1203–1215 (1998).

45. Kloeden, P. E. and Platen, E., *Numerical Soluion of Stochastic Differential Equations*, 3rd ed., Springer, New York, 1999.

46. Koelle, K. and Pascual, M., Disentangling extrinsic from intrinsic factors in disease dynamics: A nonlinear time series approach with an application to cholera, *Am. Naturalist* **163**, 901–913 (2004).

47. Koelle, K., Rodó, X., Pascual, M., Yunus, M., and Mostafa, G., Refractory periods and climate forcing in cholera dynamics, *Nature* **436**, 696–700 (2005).

48. Kovats, R. S. and Bouma, M., Retrospective studies: Analogue approaches to describing climate variability and health, in *Environmental Change, Climate and Health*, Martens, P. and McMichael, A. J., eds., Cambridge Univ. Press, Cambridge, UK, 2002, pp. 144–171.

49. Li, W., Shi, Z., Yu, M., Ren, W. et al., Bats are natural reservoirs of SARS-like coronaviruses, *Science* **310**, 676–679 (2005).

50. Liu, J. and West, M., Combining parameter and state estimation in simulation-based filtering, in *Sequential Monte Carlo Methods in Practice*, Doucet, A., de Freitas, N., and Gordon, N. J., eds., Springer, New York, 2001, pp. 197–224.

51. Liu, J. S., *Monte Carlo Strategies in Scientific Computing*, Springer, New York, 2001.

52. Longini, I. M., Nizam, A., Xu, S., Hanshaoworakul, K. W., Cummings, D. A. T., and Halloran, M. E., Containing pandemic influenza at the source, *Science* **309**, 1083–1087 (2005).

53. May, R. M., Simple mathematical models with very complicated dynamics, *Nature* **261**, 459–467 (1976).

54. Newman, K. B. and Lindley, S. T., Accounting for demographic and environmental stochasticity, observation error and parameter uncertainty in fish population dynamic models, *N. Am. J. Fish. Manage.* (in press).

55. Øksendal, B., *Stochastic Differential Equations*, 5th ed., Springer, New York, 1998.

56. Pascual, M. and Dobson, A., Seasonal patterns of infectious diseases, *PLoS Med.* **2**, 18–20 (2005).

57. Robbins, H. and Monro, S., A stochastic approximation method, *Ann. Math. Statist.* **22**, 400–407 (1951).

58. Rodó, X., Pascual, M., Fuchs, G., and Faruque, A. S. G., ENSO and cholera: A nonstationary link related to climate change?, *Proc. Natl. Acad. Sci. USA* **99**, 12901–12906 (2002).

59. Ross, R., An application of the theory of probabilities to the study of a priori pathometry, Part I, *Proc. Roy. Soc. Lond. Ser. A* **92**(638), 204–230 (1916).

60. Sack, D. A., Sack, R. B., Nair, G. B., and Siddique, A. K., Cholera, *Lancet* **363**, 223–233 (2004).

61. Shephard, N. and Pitt, M. K., Likelihood analysis of non-Gaussian measurement time series, *Biometrika* **84**, 653–667 (1997).

62. Shumway, R. H. and Stoffer, D. S., *Time Series Analysis and Its Applications*, Springer, New York, 2000.

63. Spall, J. C., *Introduction to Stochastic Search and Optimization*, Wiley, Hoboken, NJ, 2003.

64. Thomas, L., Buckland, S. T., Newman, K. B., and Harwood, J., A unified framework for modelling wildlife population dynamics, *Austral. New Z. J. Statist.* **47**, 19–34 (2005).

65. Thomson, M. C., Doblas-Reyes, F. J., Mason, S. J., Hagedorn, S. J., Phindela, T., Morse, A. P., and Palmer, T. N., Malaria early warnings based on seasonal climate forecasts from multi-model ensembles, *Nature* **439**, 576–579 (2006).

66. Turchin, P., *Complex Population Dynamics—Theoretical/Empirical Synthesis*, Princeton Univ. Press, 2003.

67. Wikle, C. K., Hierarchical models in environmental science, *Int. Statist. Rev.* **71**, 181–199 (2003).

68. Wikle, C. K., Berliner, L. M., and Cressie, N., Hierarchical Bayesian space-time models, *Environ. Ecol. Statist.* **5**, 117–154 (1998).

CHAPTER 9

Misclassification and Measurement Error Models in Epidemiologic Studies

Surupa Roy
St. Xavier's College, Kolkata, India

Tathagata Banerjee
Indian Institute of Management, Ahmedabad, India

9.1 INTRODUCTION

In standard regression analysis we assume that the predictors x are directly observable without any errors. This is seldom true. Very often in epidemiologic studies, for some reason, the predictors are not directly observable. Instead, measurements on its surrogates z are available. The true predictor x is a perturbation of z. In the measurement error literature this is known as the *Berkson error model*. In such cases it is usually assumed that x is a linear function of z plus an error. The classical measurement error model, on the other hand, assumes that x is directly observable and z is a perturbation of x, that is, that measurements of x are subject to errors. Thus z is x plus the measurement error. The substitution of z for x complicates the analysis of the observed data when the purpose of analysis is inference about a model defined in terms of x. Finding statistical models and methods for analyzing data that arise in either of these ways is known as a *measurement error problem*. In epidemiologic studies, most often we encounter the Berkson error model. In this chapter we would mostly be concerned with this model.

Until the late 1970s, measurement error models were developed mostly for continuous responses. Excellent introductions to linear measurement errors involving continuous responses are provided by Madansky [33], Kendall and Stuart [29, Ch. 29], and Fuller [20]. Historically, early research in measurement error problems was driven by applications in

Statistical Advances in the Biomedical Sciences, edited by Atanu Biswas, Sujay Datta,
Jason P. Fine, and Mark R. Segal

physical sciences, especially in astronomy, and soon thereafter by applications in econometrics, whereas today much of the current research in this area is driven by applications in health sciences. The seminal article by Carroll et al. [11] on measurement error in binary regression marks the shift in emphasis from linear to nonlinear modeling. The article is noteworthy for breaking ground in the applications of measurement errors in health sciences in particular and in the study of generalized linear models with measurement error in general. This article literally opened the floodgate and was followed by a spate of contributions in this area by a host of researchers.

A classical measurement error model is specified in terms of three submodels: (1) an outcome model connecting the response y to the true predictor x, (2) a measurement error model specifying the distribution of z (the perturbed value of x) given the true predictor x, and (3) the assumptions that are made regarding x itself. Two types of assumption are usually made about x; in a *functional model*, x is assumed to be fixed but unknown and in a *structural model*, it is assumed to have a probability distribution. However, in the Berkson model we have only two submodels: an outcome model and a model giving the conditional distribution of x denoted by the surrogate z. We refer the reader to Fuller [20] for an extensive discussion of linear measurement error models; for nonlinear models, to Carroll et al. [12,13]. It is found that ignoring measurement errors in both linear and nonlinear models in general results in attenuation of the estimates of regression parameters [33,20,49,50,5,45,12,47,53]. Consequently, different methods under different model assumptions have been proposed to remove the bias in the estimates. Likelihood-based methods [58,56,57] methods based on instrumental variables [1–3,52], the estimating-equation-based method [7,9,10,26], the Nakamura method [35,51,55], and the simulation extrapolation (SIMEX) method [58,59,31,60] are the major ones. Compared to the classical framework, relatively few works have been done from the Bayesian point of view [15,21,32,39,40,23,25,6]. Also there are a few papers that consider either the effect of measurement errors in covariates on nonparametric regression [18,61,16] or a nonparametric method to adjust for mismeasured covariate data [37].

Besides the measurement errors in the predictors, in epidemioiogic studies, very often the binary responses y are subject to classification errors and are not observable. Instead, the manifest response $\tilde{y}$ is observable. However, the model is defined in terms of y. Thus, replacing y with $\tilde{y}$ makes the analysis of data more difficult. We refer to Gustafson [24] for a detailed discussion of the impact of these errors on the parameter estimates and its Bayesian adjustments.

In this chapter, instead of discussing the existing methods, we choose to discuss the likelihood-based adjustments in some interesting models arising in epidemiologic studies, most of which are yet to be published. The primary reason is that the three books mentioned above have extensive discussions on most of the existing problems and the methods for circumventing them. Also, in a restricted space, we find it difficult to do justice to our task as reviewers if we at all attempt to discuss the existing methods in some detail.

We begin with a few examples of epidemiologic studies just to illustrate the situations where models that we discuss in Sections 9.3–9.5 may be applied. In Section 9.3, we consider a regression model that incorporates substitution of (y, x) by ($\tilde{y}$, z). This model is then extended to bivariate binary responses in Section 9.4. In practice, in certain situations we may get mixed binary and continuous outcomes. In Section 9.5 we extend our discussion to such models. The methodology developed in Section 9.3 is illustrated with atom bomb survivor data in Section 9.6. Finally, in Section 9.7 we give the concluding remarks.

9.2 A FEW EXAMPLES

9.2.1 Atom Bomb Survivors Data

In a lifespan study among 86,520 survivors of atom bomb explosion in Hiroshima and Nagasaki, data are collected on the number of deaths due to cancer among the survivors corresponding to different dose groups of radiation exposure. Clearly the true radiation dose is unobservable; instead an estimate of this dose is obtained using DS86, a dosimetry system that acts as a surrogate. A close inspection of the data shows an unexpected pattern in the proportion of cancer deaths, which suddenly decreases in the last two dose categories, contrary to what is expected [46]. This is due to the misclassification of cancer deaths as noncancer deaths on death certificates. Thus the data are contaminated not only with measurement errors in predictors but also with classification errors in binary responses. This example is considered in more detail in Section 9.6.

9.2.2 Coalminers Data

A cohort of 18,282 coalminers aged 20–64 years are examined for the presence of wheeze and breathlessness. The data are collected from a short questionnaire where each respondent was classified as suffering or not suffering from breathlessness and wheeze. In this instance each response factor has two levels and all four combinations are possible. Data are provided corresponding to different age groups, which are at intervals of 4 years. Here the covariate is the actual age of the miner. However, as a surrogate, we take the midpoint of the age interval to which the miner happens to belong. Clearly, in addition to the x values being affected by *grouping* errors—a specific kind of measurement error [30], they may also be affected by *recording* errors, which does not sound unrealistic considering the poor awareness among the educationally disadvantaged people about their exact ages. Also, in this study, misclassification of the bivariate binary response is likely. Ekholm and Palmgren [17] pointed out that the coalminer data were contaminated with classification errors. Thus, the data are contaminated by both measurement errors in covariates and classification errors in the bivariate response.

9.2.3 Effect of Maternal Dietary Habits on Low Birth Weights in Babies

The study of the effect of dietary habits of mothers on low birth weight (LBW) of babies in a given population is an important health research problem. Suppose that, for a newborn baby, the mixed binary continuous responses are the household income (y_2) and whether the newborn baby is LBW or not LBW (y_1). The covariates affecting the binary response may include the dietary habits of the mother, the mother's age at childbirth, age at marriage, smoking status of mother, and whether the mother suffered from any major disease during pregnancy. In particular, some of the covariates related to dietary habits (e.g., daily protein intake) may not be observable, but their estimates may be obtained by personal interview. These estimates would work as surrogates. In this case we would be interested in estimating the regression coefficients of dietary habits on the occurrence of LBW (y_1) and also its (y_1) correlation with income (y_2).

9.3 BINARY REGRESSION MODELS WITH TWO TYPES OF ERROR

In this section, a binary regression model is developed using a general link function that incorporates two important and independent sources of error, namely, measurement errors in covariates and classification errors in the binary response. Suppose that y denotes the latent or true binary response, $\tilde{y}$ the manifest binary response, $X_{p \times 1}$ the true predictor, and $Z_{p \times 1}$ its surrogate. Let ε_0 and ε_1 denote the probabilities of misclassification, which are assumed to be independent of the covariate values:

$$P(\tilde{y} = 1 \mid y = 0) = \varepsilon_0; \; P(\tilde{y} = 0 \mid y = 1) = \varepsilon_1. \tag{9.1}$$

For a fixed $X = x$, it is assumed that

$$P(y = 1 \mid x) = g^{-1}(\beta_0 + x^T \beta), \tag{9.2}$$

where $g(\cdot)$ is an appropriate link function and (β_0, β^T) are the regression parameters. Now a simple probability calculation yields

$$P(\tilde{y} = 1 \mid x) = P(\tilde{y} = 1 \mid y = 0)P(y = 0 \mid x) + P(\tilde{y} = 1 \mid y = 1)P(y = 1 \mid x)$$

$$= \varepsilon_0 + (1 - \varepsilon_0 - \varepsilon_1)g^{-1}(\beta_0 + x^T \beta). \tag{9.3}$$

Note that if $\varepsilon_0 + \varepsilon_1 = 1$, Equation (9.3) becomes independent of (β_0, β^T) and thus the manifest response does not contain any information about the regression parameters. Thus it would be unreasonable to use this model if the probability of either kind were greater than 0.5. Neuhaus [36] has shown that for scalar β, ignoring errors on responses produces biased covariate effect estimates. If $\tilde{\beta}$ denotes the naive estimate of β and $\hat{\beta}$ the estimate obtained under the correct model, then it can be shown that $\tilde{\beta} \cong \hat{\beta}H'(0)$, where $H'(0)$ is given by

$$\frac{(1 - \varepsilon_0 - \varepsilon_1)\phi(\beta_0)}{\phi[\Phi^{-1}\{\varepsilon_0 + (1 - \varepsilon_0 - \varepsilon_1)\Phi(\beta_0)\}]},$$

$$\frac{(1 - \varepsilon_0 - \varepsilon_1)\exp(\beta_0)}{\{\varepsilon_0 + (1 - \varepsilon_1)\exp(\beta_0)\}\{1 - \varepsilon_0 + \varepsilon_1 \exp(\beta_0)\}}$$

for probit and logit link functions, respectively. Here Φ and ϕ denote the cumulative distribution function (cdf) and the probability distribution function (pdf) of a standard normal distribution, respectively. Clearly, $0 \le H'(0) \le 1$. Besides these link functions, the result also holds for any link function based on the inverse of a cdf. Thus for such links, ignoring errors in response leads to attenuated estimates of the regression coefficients and the attenuation factor is given by $H'(0)$.

Notice that the correct model incorporating the classification errors is given by (9.3). Now we need to incorporate measurement errors. Since we have assumed a Berkson model, the measurement error process in our case is represented by $f(x \mid z)$. Assuming nondifferential measurement error, we obtain

$$P(y = 1 \mid z) = \int P(y = 1 \mid x) f(x \mid z) dx = \int g^{-1}(\beta_0 + x^T \beta) f(x \mid z) dx. \qquad (9.4)$$

The multiple integral in (9.4) is not tractable analytically in all situations. In the case of a probit link function and under the assumption that $x \mid z \sim N_p(z, \Sigma(z))$, it can be shown that

$$P(y = 1 \mid z) = \Phi(\gamma_0 + \gamma^T z), \qquad (9.5)$$

where

$$\gamma_0 = \frac{\beta_0^T}{(1 + \beta^T \Sigma(z) \beta)^{1/2}} \quad \text{and} \quad \gamma^T = \frac{\beta^T}{(1 + \beta^T \Sigma(z) \beta)^{1/2}}. \qquad (9.6)$$

In the case of a logistic regression model, the integral has no closed-form solution. However, a simple technique often works just as well. The technique is to approximate the logistic by the probit. For $c \cong 1.70$, it is well known that $H(v) = \Phi(v/c)$ [28,63,34]. For estimating the regression parameters the maximum-likelihood estimate (MLE) of γ_0, γ^T is first derived and then the MLE of (β_0, β^T) is obtained from it by inverting the relationships given by (9.6). Also, it is to be noted that the MLE of (β_0, β^T) exists provided that $\gamma^T \Sigma(z) \gamma < 1$. It is also evident from expression (9.6) that ignoring measurement error attenuates the estimate of the regression coefficients. Now, combining models (9.3) and (9.5), the conditional probability of the manifest response given the surrogates is given by

$$P(\tilde{y} = 1 \mid z) = \varepsilon_0 + (1 - \varepsilon_0 - \varepsilon_1) \int g^{-1}(\beta_0 + x^T \beta) f(x \mid z) dx \qquad (9.7)$$

For the parameters in Equation (9.7) to be identifiable, validation, replication, or instrumental data will typically be required to estimate [12]. It should also be mentioned here that if all the observations lie in the central part of the probit or logit function, then simultaneous estimation of $\varepsilon_0, \varepsilon_1$ and (γ_0, γ^T), and hence of (β_0, β^T), clearly falls through since, in that case, $g^{-1}(\gamma_0 + \gamma^T z)$ can be well approximated by a linear function. In such situations, estimation of the regression coefficients is possible only when independent estimates of ε_0 and ε_1 are available from external validation studies.

An extensive simulation study was carried out with x_i terms generated from univariate $N(z_i, \sigma^2)$, $i = 1, 2, \ldots, n$, where σ^2 is a prefixed value of the measurement error variance. The results are reported for a number of such prefixed choices of σ^2 such as 0.01, 0.5, and 1.0 and for varying $(\varepsilon_0, \varepsilon_1)$. For the purpose of simulation study, β_0 is taken to be 0 and $\beta_1 = 1$. Samples of size $n = 10,000$ were selected and the simulation was repeated 500 times. The average of the estimated values of the parameters was obtained along with the standard errors (given in parentheses), calculated from the inverse of the Fisher information matrix. In Table 9.1 we list one such result corresponding to $(\sigma^2, \varepsilon_0, \varepsilon_1) = (0.5, 0.05, 0.05)$ for the naive model (M_N), the model incorporating measurement errors only (M_M), the model incorporating classification errors only (M_C) and the model incorporating both the errors (M_{MC}).

The results show that the classification error dominates in terms of its effect on the estimates. One can also observe that the joint effect of die errors measured by the attenuation of

Table 9.1 $(\sigma^2, \varepsilon_0, \varepsilon_1) = (0.5, 0.05, 0.05)$

Estimate	Model M_{N}	Model M_{C}	Model M_{M}	Model M_{MC}
$\hat{\beta}_0$	0.0003 (0.016)	0.002 (0.029)	0.000 (0.024)	0.001 (0.036)
$\hat{\beta}_1$	0.558 (0.009)	0.820 (0.029)	0.608 (0.012)	1.000 (0.055)
$\hat{\varepsilon}_0$	— (—)	0.050 (0.005)	— (—)	0.050 (0.006)
$\hat{\varepsilon}_1$	— (—)	0.050 (0.005)	— (—)	0.050 (0.006)

the estimates of the regression coefficients is less than the sum effect of the errors. Finally, the study reveals that the model incorporating both the classification errors and measurement errors works best in all the situations. For details and further results of the simulation study, we refer the reader to Roy et al. [44].

A bivariate extension of this model is discussed in the next section. The interesting new feature of this model is that it involves too many error probabilities as unknown parameters. To make the model parsimonious, modeling the error probabilities as functions of a fewer number of parameters is found to be all die more essential.

9.4 BIVARIATE BINARY REGRESSION MODELS WITH TWO TYPES OF ERROR

Here we discuss bivariate binary regression models with measurement errors in the covariates and classification errors in the responses. Let $y_i = (y_{i1}, y_{i2})^T$ denote the ith observation $(i = 1, 2, \ldots n)$ on the bivariate binary response vector. The response depends on a set of $(m \times 1)$ covariates $x_i = (x_{i1}^T, x_{i2}^T)^T$, where X_{ij} is the $(m_j \times 1)$ predictor of $y_{ij} (j = 1, 2; m = m_1 + m_2)$. Let us define Gaussian latent variables $w_i = (w_{i1}, w_{i2})^T$ such that $w_i \sim N_2(\beta_{01} + \beta_1^T x_{i1}, \beta_{02} + \beta_2^T x_{i2}, 1, 1, \rho)$. The binary response y_{ij} is related to the latent variable as follows:

$$y_{ij} = 1 \quad \text{if } w_{ij} > 0; \quad \text{otherwise } y_{ij} = 0 (\dot{j} = 1, 2). \tag{9.8}$$

Simple probability calculation yields

$$P(y_{i1} = 1, y_{i2} = 1 \mid x_i) = F_2(\beta_{01} + \beta_1^T x_{i1}, \beta_{02} + \beta_2^T x_{i2}), \tag{9.9}$$

where $F_2(.,.)$ is the cdf of a bivariate normal distribution with parameters $(0, 0, 1, 1, \rho)$. The model represented by (9.10) will be called the "naive model" (M_{N}). In this case the x_i terms are not observable; however, observations on the surrogate $z_i = (z_{i1}^T, z_{i2}^T)^T$ are available. It is further assumed that

$$x_i \mid z_i \sim N_m(z_i, \Sigma), \tag{9.10}$$

where Σ is a matrix consisting of the diagonal blocks Σ_{11}, Σ_{22} and the off-diagonal blocks Σ_{12}, Σ_{21}. For the purpose of identifiability, Σ is completely specified. Assuming nondifferential measurement errors, we obtain

$$P(y_{i1} = 1, y_{i2} = 1 \mid z_i) = \int P(y_{i1} = 1, y_{i2} = 1 \mid x_i) f(x_i \mid z_i) dx_i$$

$$= F_2(\gamma_{01} + \gamma_1^T z_{i1}, \gamma_{02} + \gamma_2^T z_{i2}). \tag{9.11}$$

Here $F_2(.,.)$ is the cdf of a bivariate normal distribution with parameters $(0, 0, 1, 1, \rho^*)$

$$\gamma_{0j} = \beta_{0j}(1 + \beta_j^T \Sigma_{jj} \beta_j)^{-0.5}; \quad \gamma_j^T = \beta_j^T (1 + \beta_j^T \Sigma_{jj} \beta_j)^{-0.5} \quad (j = 1,2) \tag{9.12}$$

and

$$\rho^* = \frac{\rho + \beta_1^T \Sigma_{12} \beta_2}{(1 + \beta_1^T \Sigma_{11} \beta_1)^{0.5} (1 + \beta_2^T \Sigma_{22} \beta_2)^{0.5}}. \tag{9.13}$$

The model represented by (9.11) incorporates measurement errors. For estimating the parameters, the MLEs of $\gamma_{0j}, \gamma_j^T, \rho^*$ are first derived, and then the MLEs of $\beta_{0j}, \beta_j^T, \rho$ are obtained by inverting the relationships (9.12) and (9.13). However, the MLEs of (β_{0j}, β_j^T) exist provided that $\gamma_j^T \Sigma_{jj} \gamma_j < 1$ and that of ρ exists, provided in turn that

$$\frac{-1 + \beta_1^T \Sigma_{12} \beta_2}{(1 + \beta_1^T \Sigma_{11} \beta_1)^{0.5} (1 + \beta_2^T \Sigma_{22} \beta_2)^{0.5}} < \rho^* < \frac{1 + \beta_1^T \Sigma_{12} \beta_2}{(1 + \beta_1^T \Sigma_{11} \beta_1)^{0.5} (1 + \beta_2^T \Sigma_{22} \beta_2)^{0.5}}.$$

It is clear from (9.12) that ignoring measurement error causes attenuation of the regression coefficients. However, the effect of measurement error on the correlation coefficient does not follow any pattern. It can be observed that in the case of a scalar β, if $\beta_1 = \beta_2 > 0$ (or <0), then the naive estimate of ρ becomes inflated.

In this case, in addition to measurement error in covariates, the data may also be contaminated with classification errors in binary responses. Suppose that $\tilde{y}_i = (\tilde{y}_{i1}, \tilde{y}_{i2})^T$ denotes the ith manifest response corresponding to the true response y_i. The probabilities of misclassification are given by

$$P(\tilde{y}_{i1} = j, \tilde{y}_{i2} = k \mid y_{i1} = l, y_{i2} = m) = \varepsilon(j, k \mid l, m), \quad \text{say}, \quad j, k, l, m \in \{0, 1\}, \tag{9.14}$$

where $j \neq l$ or $k \neq m$. The misclassification probabilities in (9.14) are treated as unknown constants and, to keep the treatment simple, are assumed to be independent of the true covariates x_i. Clearly, there are 12 misclassification probabilities corresponding to different choices of (j, k) and (l, m). Moreover, $(j, k) = (l, m)$ gives a correct classification. To avoid excessive error probabilities, one meaningful assumption may be that of conditional independence. More specifically, it is assumed that

$$P(\tilde{y}_{i1} = j, \tilde{y}_{i2} = k \mid y_{i1} = l, y_{i2} = m) = P(\tilde{y}_{i1} = j \mid y_{i1} = l) \times P(\tilde{y}_{i2} = k \mid y_{i2} = m). \tag{9.15}$$

This model involves only four unknown classification errors that are relatively easy to estimate along with the regression parameters. Now, the conditional probability of the manifest response given the surrogates is given by

$$P(\tilde{y}_{i1} = j, \tilde{y}_{i2} = k \mid z_i) = \sum_{(i,m)} \varepsilon(j, k \mid l, m) P(y_{i1} = l, y_{i2} = m \mid z_i), \tag{9.16}$$

where $(j, k) \in \{(1, 1), (1, 0), (0, 1), (0, 0)\}$. Note that Equation (9.16) gives the model incorporating both measurement errors and classification errors and will be denoted by M_{MC}.

A simulation study is carried out with $\beta_1 = \beta_2 = 1.0$ (no intercept terms are involved), $\rho = 0.6$, and for varying choices of measurement error variances and misclassification probabilities. In particular, the measurement error distribution is chosen to be univariate normal with mean equal to the value of the surrogate and for some prefixed choices of the measurement error variance, such as $\sigma^2 = 0.01, 0.5, 1.0$. It is further assumed that

$$P\{\tilde{y}_{i1} = 1 \mid y_{i1} = 0\} = P\{\tilde{y}_{i2} = 1 \mid y_{i2} = 0\} = \varepsilon_0,$$
$$P\{\tilde{y}_{i1} = 0 \mid y_{i1} = 1\} = P\{\tilde{y}_{i2} = 0 \mid y_{i2} = 1\} = \varepsilon_1,$$

where ε_0 and ε_1 are some prefixed numbers. Samples of size $n = 10000$ were selected, and the simulation was repeated 500 times. The average of the estimated values of the parameters was obtained along with the standard errors calculated from the inverse of the Fisher information matrix (given in parentheses). The results show that ignoring classification errors results in attenuation of the estimates of the regression parameters as well as that of ρ. However, no theoretical justification could be given for this phenomenon. As noted before, ignoring measurement errors attenuates the estimates of the regression coefficients. However, for the choice of the parameters considered in the simulation, ignoring measurement errors causes inflation of the estimate of ρ. Thus, in the presence of both these errors, the effect of ignoring these errors on the estimate of ρ works in opposite directions. As a result, we might chance upon a situation where the estimate of ρ is close to the true value under M_N. For instance, when $\sigma^2 = 0.5$, the estimate of ρ is 0.511 and 0.599 under models M_N and M_{MC}, respectively. Throughout the simulation it is observed that classification errors dominate small or moderate measurement errors. For completeness, we list the results in Table 9.2 corresponding to a specific choice of $(\sigma^2, \varepsilon_0, \varepsilon_1)$ and for the models M_N and M_{MC}. For further details, we refer the reader to Roy [42].

Table 9.2 $(\sigma^2, \varepsilon_0, \varepsilon_1) = (0.5, 0.05, 0.05)$

Estimate	Model M_N	Model M_{MC}
$\hat{\beta}_1$	0.546 (0.009)	1.002 (0.099)
$\hat{\beta}_2$	0.547 (0.009)	1.008 (0.120)
$\hat{\rho}$	0.511 (0.017)	0.599 (0.031)
$\hat{\varepsilon}_0$	—	0.050 (0.003)
$\hat{\varepsilon}_1$	—	0.950 (0.004)

9.5 MODELS FOR ANALYZING MIXED MISCLASSIFIED BINARY AND CONTINUOUS RESPONSES

In this formulation, let $y_i = (y_{1i}, y_{2i})^T$ denote the bivariate response, where y_{1i} is binary and y_{2i} is continuous. Suppose that y_{1i}^* is the unobserved latent variable associated with the binary response y_{1i} such that

$$y_{1i} = 1 \text{ if } y_{1i}^* > 0; \quad \text{otherwise} \quad y_{1i} = 0. \tag{9.17}$$

Associated with each observation there is a $p_1 \times 1$ covariate vector x_{1i} thought to predict y_{1i}^* (and hence y_{1i}) and a $p_2 \times 1$ covariate vector x_{2i} believed to predict y_{2i}. Further, it is assumed that

$$y_{1i}^*, y_{2i} \,|\, x_i = (x_{1i}^T, x_{2i}^T)^T \sim N_2(\beta_{01} + \beta_1^T x_{1i}, \beta_{02} + \beta_2^T x_{2i}, 1, \sigma_2^2, \rho). \tag{9.18}$$

In the case of measurement errors in covariates, let $z_{1i}^T(1 \times p_1)$ and $z_{2i}(1 \times p_2)$ be the surrogates for x_{1i}^T and x_{2i}^T, respectively. With the strength of Equation (9.18), and assuming that $(x_{1i}, x_{2i}) \,|\, (z_{1i}, z_{2i}) \sim N_p\{(z_{1i}, z_{2i}), \Sigma\}$, where $p = p_1 + p_2$ and Σ is a completely specified matrix consisting of the blocks $\Sigma_{11}, \Sigma_{12}, \Sigma_{21} = \Sigma_{12}^T$ and Σ_{22}, it can be shown that

$$y_{1i}^*, y_{2i} \,|\, z_i \sim N_2(\beta_{01} + \beta_1^T z_{1i}, \beta_{02} + \beta_2^T z_{2i}, 1 + \beta_1^T \Sigma_{11} \beta_1, \sigma_2^2 + \beta_2^T \Sigma_{22} \beta_2, \rho^*), \tag{9.19}$$

where

$$\rho^* = \frac{\rho \sigma_2 + \beta_1^T \Sigma_{12} \beta_2}{(1 + \beta_1^T \Sigma_{11} \beta_1)^{0.5} (\sigma_2^2 + \beta_2^T \Sigma_{22} \beta_2)^{0.5}}. \tag{9.20}$$

In this situation, in addition to measurement errors in covariates, the binary response y_{1i} is subject to classification errors and hence is not observable. Suppose that y_{1i} denotes the ith manifest response corresponding to the true response y_{1i}. We assume a simple probability model in terms of misclassification probabilities given by

$$\begin{aligned} P(\tilde{y}_{1i} = 1 \,|\, y_{1i} = 0, y_{2i}, x_i) = P(\tilde{y}_{1i} = 1 \,|\, y_{1i} = 0) = \varepsilon_0, \\ P(\tilde{y}_{1i} = 0 \,|\, y_{1i} = 1, y_{2i}, x_i) = P(\tilde{y}_{1i} = 0 \,|\, y_{1i} = 1) = \varepsilon_1. \end{aligned} \tag{9.21}$$

We treat the misclassification probabilities in (9.21) as unknown constants independent of the true covariates x_i and the continuous response y_{2i}. Now, some simple probability calculations yield

$$P(\tilde{y}_{1i} = 1 \mid y_{2i}, x_i) = (1 - \varepsilon_1)\Phi\left\{\frac{\mu_{1i}}{(1 - \rho^2)^{0.5}}\right\} + \varepsilon_0\left[1 - \Phi\left\{\frac{\mu_{1i}}{(1 - \rho^2)^{0.5}}\right\}\right]$$

$$= \varepsilon_0 + (1 - \varepsilon_0 - \varepsilon_1)\Phi\left\{\frac{(\mu_{1i})}{(1 - \rho^2)^{0.5}}\right\} = \pi_{2i} \text{ (say)},$$

$$P(\tilde{y}_{1i} = 1 \mid y_{2i}, z_i) = \varepsilon_0 + (1 - \varepsilon_0 - \varepsilon_1)$$

$$\times \Phi\left\{\frac{\gamma_{01} + \gamma_1^T z_{1i}}{(1 - \rho^{*2})^{0.5}} + \frac{\rho^*}{\sigma_2^*(1 - \rho^{*2})^{0.5}}(y_{2i} - \beta_{02} - \beta_2^T z_{2i})\right\} \quad (9.22)$$

$$= \pi_i \text{ (say)}, \text{ where } \rho^* \text{ is given by (9.20) and}$$

$$\gamma_{01} = \frac{\beta_{01}}{(1 + \beta_1^T \Sigma_{11} \beta_1)^{0.5}}, \gamma_1^T = \frac{\beta_1^T}{(1 + \beta_1^T \Sigma_{11} \beta_1)^{0.5}}. \quad (9.23)$$

Also, σ_2^* is given by

$$\sigma_2^{*2} = \sigma_2^2 + \beta_2^T \Sigma_{22} \beta_2. \quad (9.24)$$

Thus, the joint distribution of the manifest binary response $\tilde{y}_{1i}$ and the continuous response y_{2i} given the surrogates z_i is

$$f(\tilde{y}_{1i}, y_{2i} \mid z_i) = f(\tilde{y}_{1i} \mid y_{2i}, z_i) f(y_{2i} \mid z_{2i})$$

$$= \pi_i^{\tilde{y}_{1i}} (1 - \pi_i)^{1 - \tilde{y}_{1i}} \frac{1}{\sigma_2^*(2\pi)^{1/2}} \exp\left\{-\frac{1}{2}\frac{(y_{2i} - \beta_{02} - \beta_2^T z_{2i})^2}{\sigma_2^{*2}}\right\}. \quad (9.25)$$

The bivariate model given in (9.25) incorporates both the classification errors in binary responses and measurement errors in covariates.

Using the observed data $d_0 = \{(\tilde{y}_{1i}, y_{2i}, z_{1i}^T, z_{2i}^T)^T, i = 1, 2, \ldots, n\}$, the MLEs of the parameters $\xi_1 = (\gamma_{01}, \gamma_1^T, \rho^*)^T, \xi_2 = (\beta_{02}, \beta_2^T, \sigma_2^{*2})^T, \varepsilon_0$, and ε_1 are obtained. Let us also define $\theta_1 = (\beta_{01}, \beta_1^T, \rho)^T$ and $\theta_2 = (\beta_{02}, \beta_2^T, \sigma_2^2)^T$. As a consequence of the invariance property of MLEs, the MLE of θ_1 and θ_2 are obtained from $\hat{\xi}_1$ and $\hat{\xi}_2$ by using the relations (9.20), (9.23), and (9.24). It is to be noted that the first $p_2 + 1$ components of θ_2 and ξ_2 are the same. Hence β_{02} and β_2 remain unchanged because of the incorporation of measurement error. However, the scale parameter of the continuous response is affected. According to the assumption that $\sigma_2^{*2} > \beta_2^T \Sigma_{22} \beta_2$, it follows from (9.24) that ignoring measurement error inflates the estimate of σ_2^2. The estimates of β_{01} and β_1 are obtained from (9.23), assuming that $\gamma_1^T \Sigma_{11} \gamma_1 < 1$. Also equation (9.23) clearly shows that ignoring measurement errors causes attenuation of the estimates of the regression coefficients. Equation (9.20) shows that the estimate of the correlation coefficient is also affected; however, the effect of measurement errors in this case does not follow any clear pattern.

It is significant that for fixed θ_2, ignoring classification error causes attenuation of the estimates of θ_1. To be specific, the estimates of β_{01}, β_1, and ρ are attenuated under the naive model

Table 9.3 $(\sigma^2, \varepsilon_0, \varepsilon_1) = (0.5, 0.10, 0.10)$

Estimates	M_1	M_2
$\hat{\beta}_{01}$	0.0014 (0.0147)	00.0019 (0.0438)
$\hat{\beta}_1$	0.4305 (0.0081)	1.2336 (0.0692)
$\hat{\beta}_{02}$	-0.0002 (0.0124)	-0.0002 (0.0124)
$\hat{\beta}_2$	0.9997 (0.0054)	0.9997 (0.0054)
$\hat{\rho}$	0.3191 (0.0147)	0.5781 (0.0312)
$\hat{\sigma}_2^2$	1.5013 (0.0226)	1.0016 (0.0230)
$\hat{\varepsilon}_0$	—	0.1000 (0.0061)
$\hat{\varepsilon}_1$	—	0.1000 (0.0055)

M_{N}. The effect of classification errors on the estimates of, β_{02}, β_2, and σ_2^2 is not quite clear. However, in the special but important case of $x_{1i} = x_{2i} = x_i$ (say), the estimates of θ_2 remain unaffected because of the incorporation of classification errors.

An extensive simulation study was carried out for varying choices of the misclassification probabilities and measurement error variance such as $\Sigma = \sigma^2 = 0.01, 0.5, 1.0$. Here the sample size is chosen to be 10,000, and the simulation is repeated 500 times. The true values of the parameters are as follows: $\beta_{01} = \beta_{02} = 0, \beta_1 = \beta_2 = 1.0, \rho = 0.6$, and $\sigma_2^2 = 1.0$. The average values of the estimates of the parameters along with their standard errors calculated from the inverse of the Fisher information matrix (shown within parentheses) are reported for a specific choice of the measurement error variance ($\sigma^2 = 0.5$) and misclassification probabilities (0.10, 0.10). The results are listed in Table 9.3 for the naive model (M_{N}) and the model incorporating measurement errors and classification errors (M_{MC}). The findings support the theoretical justifications given above. For further details, we refer the reader to Roy [42], and Roy and Banerjee [43].

9.6 ATOM BOMB DATA ANALYSIS

About 5 years after the dropping of atomic bombs on Hiroshima and Nagasaki, a lifespan study (LSS) was begun at the behest of the Radiation Effect Research Foundation (RERF) that led to the establishment of a fixed study cohort of survivors who have been followed since October 1950. A major purpose of the study was to assess the effect of radiation exposure on cancer mortality. This cohort of 86,520 survivors includes both an exposed group and a nonexposed group, distinguished by distance (<2 km, $2-10$ km) from the bursting locations of the bombs. For those in the exposed group, interviews and other efforts were made to determine the survivor location and shielding at the time of explosion. On the basis of this information, elaborate physical computations were made to estimate the individual radiation exposures by using DS86 dosimetry [41,19]. The true dose for a person is represented by the absorbed radiation, measured in gray units (Gy), to his/her intestine at the time of exposure. For our purpose it is useful to think of the dosimetry system as a formula that would provide negligible error for a survivor's radiation exposure given the exact location and shielding condition. Thus the sources of error in dose measurements result not only from errors in location and shielding information but also from the fact that the two individuals receiving the same amount of intestinal exposures may absorb different amounts of radiation, possibly due to biological factors. For illustrative purpose here, it is assumed that the latter is the cause of measurement error.

Table 9.4 $(\sigma^2, \varepsilon_0, \varepsilon_1) = (0.5, 0.22, 0.035)$

Estimates	Model M_N	Model M_{MC}
$\hat{\beta}_0$	-0.778 (0.000071)	-0.688 (0.000114)
$\hat{\beta}_1$	0.291 (0.000778)	0.381 (0.001470)

Thus the true dose (x) for a person is a function of the estimated dose (z) obtained by using DS86. In other words, z is considered as a surrogate to x. It is assumed that given the surrogate (z), the distribution of the true dose (x) is normal with mean equal to z and variance equal to cz^2, where c is a known number representing the constant coefficient of variation. The choice of this model is motivated partly by the study of Pierce et al. [38]. However, the normal model is one of the several possible choices for measurement error distribution.

A close inspection of the data [48] shows an unexpected pattern in the proportion of cancer deaths. It suddenly decreases in the last two dose categories, contrary to what is expected. Shimizu et al. [46] pointed out that the observed dose response in noncancer mortality was due to the misclassification of cancer deaths as noncancer deaths on death certificates. Sposto et al. [48] estimated the misclassification probabilities by using a validation dataset obtained from a subset of deaths in the cohort for which autopsies were carried out. They found the overall crude misclassification rate of cancer deaths to be 22% and of noncancer deaths, 3.5%. From the study they concluded that the misclassification rates do not change significantly with change in dose categories although they significantly depend on age.

Thus the data are contaminated by both measurement error in covariate and classification error in response. In this analysis, for all practical purposes, it is assumed that the misclassification probabilities are all known and equal to the overall misclassification rates estimated by Sposto et al. [48]. The analysis is carried out with the data for different choices of c. It is found, for values of c lying between 0.1 and 0.8, that the estimates of neither the regression parameters norits standard error change significantly. The results are furnished below in Table 9.4, for $c = 0.5$ and for the two models, namely, the naive model (M_N) and the model incorporating measurement errors and classification errors (M_{MC}). The figures in parentheses indicate the standard errors. The study reveals that the presence of measurement error does not affect the results significantly. On the other hand, the results of the analysis in presence of response misclassification in the data significantly affect the estimates of regression parameters. This is expected, as the misclassification probabilities are quite high. Ignoring misclassification probabilities may result in significant underestimation of the regression parameters.

9.7   CONCLUDING REMARKS

This study considers the effect of measurement errors in binary regression models where the binary responses are subject to classification errors. The concepts are extended to correlated binary outcomes and mixed binary–continuous outcomes. It is possible to extend the work to random effects probit and logit models [22] as well as to ordered probit models when the responses are ordinal [14]. Finally, in binary regression models, in addition to misclassified responses, some of the responses may be missing as well. The responses depend on a set of covariates, discrete or continuous, which are subject to measurement error or classification error. In such a situation, investigating the joint effect of the missing mechanism, the missing proportions, the classification errors, and the measurement errors on the estimates of regression coefficients through some efficiency criteria is worth considering.

ACKNOWLEDGMENT

The authors thank two anonymous referees for their constructive comments that led to an improvement of the earlier version.

REFERENCES

1. Amemiya, Y., Instrumental variable estimator for the nonlinear errors-in-variables model, *J. Econometr.* **28**, 273–289 (1985).

2. Amemiya, Y., Two-stage instrumental variable estimator for the non-linear errors-in-variables model, *J. Econometr.* **44**, 311–332 (1990).

3. Amemiya, Y., Instrumental variable estimation of the nonlinear measurement error model, in *Statistical Analysis of Measurement Error Models and Applications*, Brown, P. J. and Fuller, W. A., eds., American Mathematical Society, Providence, RI, 1990, pp. 147–156.

4. Ashford, J. R. and Sowden, R. R., Multivariate probit analysis, *Biometrics* **26**, 535–546 (1970).

5. Armstrong, B., Mesurement error in generalized linear models, *Commun. Statist. Theor. Meth.* **14**, 529–544 (1985).

6. Berry, S. M., Carroll, R. J., and Ruppert, D., Bayesian smoothing and regression splines for measurement error problems, *J. Am. Statist. Assoc.* **97**, 160–169 (2002).

7. Buonaccorsi, J. P., A modified estimating equation approach for correcting for measurement error in regression, *Biometrika* **83**, 433–440 (1996).

8. Burr, D., On errors-in-variables in binary regression—Berkson case, *J. Am. Statist. Assoc.* **83**, 739–743 (1988).

9. Carroll, R. J. and Stefanski, L. A., Approximate quasi-likelihood estimation in models with surrogate predictors, *J. Am. Statist. Assoc.* **85**, 652–663 (1990).

10. Caroll, R. J. and Wand, M. P., Semiparametric estimation in logistic measurement error models, *J. Roy. Statist. Soc. Ser. B* **53**, 573–585 (1991).

11. Carroll, R. J., Spiegelman, C. H., Lan, K. K. G., Bailey, K. G., and Abbott, R. D., On errors-in-variables for binary regression models, *Biometrika* **70**, 19–25 (1984).

12. Carroll, R. J., Ruppert, D., and Stefanski, L. A., *Measurement Error in Non Linear Models*, Chapman & Hall, New York, 1995.

13. Carroll, R. J., Ruppert, D., Stefanski, L. A., and Crainiceanu, M., *Measurement Error in Non Linear Models: A Modern Perspective*, Chapman & Hall, New York, 2006.

14. Daykin, A. R. and Moffatt, P. G. Analyzing ordered responses: A review of the ordered probit model, *Understand. Statist.* **1**(3), 157–166 (2002).

15. Dellaportas, P. and Stephens, D. A., Bayesian analysis of errors-in-variables regression models, *Biometrics* **51**, 1085–1095 (1995).

16. Delaigle, A., Hall, P., and Qiu, P., Nonparametric methods for solving the Berkson errors-in-variables problems, *J. Roy. Statist. Soc. Ser. B* **68**(2), 201–220 (2006).

17. Ekholm, A. and Palmgrem, J., A model for a binary response with misclassification, in *GLIM 82: Proc. Int. Conf. Generalized Linear Models*, Gilchrist, R., ed., Springer-Verlag, New York, 1982, pp. 128–143.

18. Fan, J. and Truong, Y. K., Nonparametric regression with errors in variables, *Ann. Statist.* **21**, 1900–1925 (1993).

19. Fujita, S., Version of DS86, RERF Update 1, 3, 1989.

20. Fuller, W., *Measurement Error Models*, Wiley, New York, 1987.

21. Gelfand, A. E., Mallick, B. K., and Polasek, W., Broken bilogical size relationships: A truncated semiparametric regression approach with measurement error, *J. Am. Statist. Assoc.* **92**(43), 836–845 (1997).

22. Gibbons, R. D. and Hedeker, D., Random-effects probit and logistic regression models for three-level data, *Biometrics* **53**, 1527–1537 (1997).

23. Gustafson, P., Le, N. D., and Vallee, M., A Bayesian Approach to Case-Control Studies with Errors in Covariables, *Biostatistics* **3**(2), 229–243 (2002).

24. Gustafson, P., *Measurement Error and Misclassification in Statistics & Epidemiology: Impacts and Bayesian Adjustments*, Chapman & Hall/CRC Press, Boca Raton, FL, 2003.

25. Gustafson, P., Measurement error modelling with an approximate instrumental variable (preprint) (2006).

26. Hanfelt, J. J. and Liang, K. Y., Approximate likelihoods for generalized linear errors-in-variables models, *J. Roy. Statist. Soc. Ser. B* **59**(3), 627–637 (1997).

27. Jeffrey B. G., Ashima, M., Cheng, Q., and Gilberto, C., Increased risk of respiratory disease and diarrhea in children with pre-existing mild vitamin A deficiency, *Am. J. Clin Nutr.* **40**, 1090–1095 (1984).

28. Johnson, N. L. and Kotz, S., *Distributions in Statistics*, Houghton-Miffin, Boston, 1970, Vol. 2.

29. Kendall, M. G. and Stuart, A., *The Advanced Theory of Statistics*, 4th ed., Hafner Press, New York, 1979, Vol. 2.

30. Kmenta, J., *Elements of Econometrics*, Macmillan, New York, 1990.

31. Lin, X. and Carrol, R. J., SIMEX variance component tests generalized linear mixed measurement error models, *Biometrics* **55**, 613–619 (1999).

32. Mallick, B. K. and Gelfand, A. E., Semiparametric, errors-in-variables models: A Bayesian approach, *J. Statist. Plan. Infer.* **52**, 307–321 (1996).

33. Madansky, A., The fitting of straight lines when both variables are subject to error, *J. Am. Statist. Assoc.* **54**, 173–205 (1959).

34. Monahah, H. and Stefanski, L. A., Normal scale mixture approximation to and computation of the logistic-normal integral, in *Handbook of the Logistic Distribution*, Balakrishnan, N., ed., Marcel Dekker, New York, 1992, pp. 529–540.

35. Nakamura, T., Corrected score function for errors-in-variables models: Methodology and application to generalized linear models, *Biometrika* **77**, 127–137 (1990).

36. Neuhaus, J. M., Bias and efficiency loss due to misclassified responses in binary regression, *Biometrika* **86**(4), 843–855 (1999).

37. Pepe, M. S. and Fleming, T. R., A nonparametric method for dealing with mismeasured covariate data, *J. Am. Statist. Assoc.* **86**, 108–113 (1991).

38. Pierce, D. A., Stram, D. O., Vaeth, M., and Schafer, D. W., The errors-in-variables problem: Consideration provided by radiation dose-response analysis of the A-bomb survivor data, *J. Am. Statist. Assoc.* **87**, 351–359 (1992).

39. Richardson, S. and Gilks, W. R., A Bayesian approach to measurement error problems using conditional independence models, *Am. J. Epidemiol.* **138**, 430–442 (1993).

40. Richardson, S. and Gilks, W. R., Conditional independence models for epidemiological studies with covariate measurement error, *Statist. Med.* **12**, 1703–1722 (1993).

41. Roesch, W. C., ed., *U.S.-Japan Joint Reassessment of Atomic Bomb Radiation Dosimetry in Hiroshima and Nagasaki: Final Report*, Radiation Effects Research Foundation, Hiroshima, Japan, 1987.

42. Roy, S., *Measurement Error Models-Related Inference Problems*, Ph.D. thesis, Dept. Statistics, Univ. Calcutta, 2004.

43. Roy, S. and Banerjee, T., Analysis of mixed outcomes: Misclassified binary responses and measurement error in covariates (unpublished communication) (2006).

44. Roys, S., Banerjee, T., and Maiti, T., Measurement error model for misclassified binary responses, *Statist. Med.* **24**, 269–283 (2005).

45. Schafer, D. W., Covariate measurement error in generalized linear models, *Biometrika* **74**, 385–391 (1987).

46. Shimizu, Y., Kato, H., and Schull, W. J., Studies of mortality of A-bomb survivors: 9. Mortality, 1950–1985: Part 2. Cancer mortality based on recently revised doses (DS86), *Radiat. Res.* **121**, 120–141 (1990).

47. Spiegelman, D., Rosner, D. L., and Logan, R., Estimation and inference for logistic regression with covariate misclassification and measurement error in main study/validation study designs, *J. Am. Statist. Assoc.* **95**(449), 51–61 (2000).

48. Sposto, R., Preston, D. L., Shimizu, Y., and Mabuchi, K., The effect of diagnostic misclassification on non cancer and cancer mortality dose response in A-bomb survivors, *Biometrics* **48**, 605–617 (1992).

49. Stefanski, L. A., The effects of measurement error in parameter estimation, *Biometrika* **72**, 385–389 (1985).

50. Stefanski, L. A. and Carroll, R. J., Covariate measurement error in logistic regression, *Ann. Statist.* **13**, 1335–1351 (1985).

51. Stefanski, L. A. and Carroll, R. J., Conditional scores and optimal scores for generalized linear measurement-error models, *Biometrika* **74**, 703–716 (1987).

52. Stefanski, L. A. and Buzas, J. S., Instrumental variable estimation in binary regression measurement error models, *J. Am. Statist. Assoc.* **90**(430), 541–550 (1995).

53. Thoresen, M. and Laake, P., A simulation study of measurement error correction methods in logistic regression, *Biometrics* **56**(3), 868–872 (2000).

54. Tosteson, T. D., Stefanski, L. A., and Schafer, D. W., A measurement error model for binary and ordinal regression, *Statist. Med.* **8**, 1139–1147 (1989).

55. Wang, C. Y. and Pepe, M. S., Expected estimating equations to accommodate covariate Measurement error, *J. Roy. Statist. Soc. Ser. B* **62**, 509–524 (2000).

56. Schafer, D. W., Likelihood analysis for probit regression with measurement errors, *Biometrika* **80**, 899–904 (1993).

57. Schafer, D. W. and Purdy, K. G., Likelihood analysis for errors-in-variables regression with replicate measurements, *Biometrika* **83**, 813–824 (1996).

58. Cook, J. and Stefanski, L. A., A simulation extrapolation method for parametric measurement error models, *J. Am. Statist. Assoc.* **89**, 1314–1328 (1994).

59. Stefanski, L. A. and Cook, J., Simulation extrapolation: The measurement error jackknife. *J. Am. Statist. Assoc.* **90**, 1247–1256 (1995).

60. Carroll, R. J., Maca, J. D., and Ruppert, D., Nonparametric regression in the presence of measurement error. *Biometrika* **86**, 541–554 (1999).

61. Liang, K.-Y. and Liu, X., Estimating equations in generalized linear models with measurement error. in *Estimating Functions*, Godambe, V. P. ed., New York: Oxford University Press, 1991.

PART III

Survival Analysis

CHAPTER 10

Semiparametric Maximum-Likelihood Inference in Survival Analysis

Michael R. Kosorok

Department of Biostatistics, University of North Carolina, Chapel Hill, North Carolina

10.1 INTRODUCTION

Survival analysis techniques are widely used in biostatistics, econometrics, and many other areas where time-to-event data occur. Semiparametric versions of survival models have proved to be extremely useful in practice because of their meaningful blending of both interpretability (through the parametric regression component) and flexibility (through the nonparametric nuisance component). There are many classic references to this area that can help provide a thorough introduction (see, e.g., Refs. [24,3, and 26]). We will assume that the reader has had exposure to the basic semiparametric models and estimation techniques used in survival analysis.

Semiparametric survival models have both a parametric index θ and a nonparametric index η. Often, inference about θ is the primary interest and η is a nuisance parameter. The prototypic example is the Cox [11] regression model, where the components in θ are the hazard ratios for the covariates (the regression parameter vector) and η is the baseline hazard function. A broader example is the class of transformation models [49], which includes the odds-ratio family [12,46], the proportional odds model [39], and the Cox model as special cases. The parameter defining the odds-ratio transformation can also be considered unknown, resulting in the univariate proportional hazards frailty regression model family [28]. In all these settings, the parameter of interest θ is usually the regression parameter (but may also include other parameters), while the nuisance parameter η is usually related to the baseline survival function but may also include nonparametric covariate effects.

Statistical Advances in the Biomedical Sciences, edited by Atanu Biswas, Sujay Datta, Jason P. Fine, and Mark R. Segal

159

We use the adjective "regular" to designate a parameter that is $\sqrt{n}$-consistently estimable. Note that in many of our examples one or more of the nuisance parameters may also be regular (as happens, e.g., with the baseline hazard in the Cox model for right censoring). In addition to right censoring, other kinds of censoring may be involved, including current status data [21,34] or panel data [56,5], although the use of time-dependent covariates may not be feasible in some cases. Other models for survival data include additive hazards regression models [1,33], accelerated failure-time models [50,55], time-varying coefficient models [45], and other complex models for addressing departures from proportionality [4,19]. Correlated failure times may also be involved, as happens with multivariate frailty models both without correlation [37,38] and with correlation [42,57].

For simplicity of exposition, this chapter focuses on certain transformation models for independent and identically distributed univariate failure-time data under either right censoring or current status censoring (case 1 interval censoring), although the techniques we develop can be extended to more general situations.

We begin the chapter by a presenting several examples that will be used to illustrate the main inferential techniques. We will then briefly review some basic asymptotic theory needed for our results. The bootstrap inferential technique, both the nonparametric and weighted versions, will be presented next. We will then present the profile sampler, followed by the piggyback bootstrap. At that point, we will briefly review other inferential techniques, and the chapter will conclude with a brief discussion.

10.2 EXAMPLES OF SURVIVAL MODELS

We now introduce the key examples that will be used for illustration. We note that this is far from an exhaustive list; it is intended only for illustration. Here are the examples.

Example 10.1: The Cox Model for Right-Censored Data. A single observation consists of $X = (U, \delta, Z)$, where $U = T \wedge C$ is the minimum of a failure time T and censoring time C, $\delta = 1\,\{T \leq C\}$ is the indicator of observing a failure time, and Z is a covariate vector in $\Re^d$. We assume that T and C are independent given Z and that censoring is uninformative. The survival function for T given the possibly time-dependent covariates Z has the form

$$S_Z(t) = \exp\left(-\int_0^t e^{\beta' Z(s)}\, dA(s)\right),\tag{10.1}$$

where A is a continuous, unknown increasing function with $A(0) = 0$. We will assume for simplicity that the time-dependent covariates are external (see Sec. 6.3.1 of Ref. 24). This model has been widely studied.

Example 10.2: The Proportional Odds Model for Right-Censored Data. The data have the same form as in Example 10.1, but the survival function for T has the form

$$S_Z(t) = \left(1 + \int_0^t e^{\beta' Z(s)}\, dA(s)\right)^{-1}\tag{10.2}$$

Efficient estimation for this model was studied in Murphy et al. [39].

Example 10.3: The Odds-Ratio Model for Right-Censored Data. The data are the same as in the previous two examples, but the model has the following form for the survival function given the possibly time-dependent covariates Z

$$S_Z(t) = \left(1 + \gamma \int_0^t e^{\beta'Z(s)} \, dA(s)\right)^{-1/\gamma}, \tag{10.3}$$

where $\gamma \geq 0$ is a specified constant. Taking the limit $\gamma \downarrow 0$ results in the Cox model of Example 10.1, while setting $\gamma = 1$ results in the proportional odds model of Example 10.2. Efficient estimation for this family of models was considered in Scharfstein et al. [46].

Example 10.4: The Odds-Ratio Model for Right-Censored Data, with γ Known and a Change Point Based on a Covariate Threshold. A single observation consists of $X = (U, \delta, Z, W)$, where (U, δ, Z) is as in the previous examples, but $W \in \mathcal{R}$ is a time-independent covariate (which may or may not be a component of Z). The survival function in this case is the same as (10.3) but with $\beta'Z(s)$ replaced by

$$r_{\beta,\alpha,\zeta}(s) = \beta'Z(s) + \mathbf{1}\{W > \zeta\}\alpha'Z(s), \tag{10.4}$$

where $\alpha \in \mathcal{R}^d$ and $\zeta \in \mathcal{R}$. Here $\theta = (\beta, \alpha)$ are the regular parameters and $\eta = (\zeta, A)$ are the "nuisance" parameters. The special case of this model with $\gamma \downarrow 0$ (the Cox model case) was considered in Pons [44]. A more general version for general transformation models was considered in Kosorok and Song [29].

Example 10.5: The Odds-Ratio Model for Right-Censored Data with γ Unknown. For this model, $\theta = (\gamma, \beta')'$ are the regular parameters of interest. A slightly more general version of this model was considered in Kosorok et al. [28].

Example 10.6: The Cox Model for Current Status Data. A single observation consists of $X = (U, \delta, Z)$, where U is the random current status time, $\delta = \mathbf{1}\{T \leq U\}$ is the event status indicator at U, and $Z \in \mathcal{R}^d$ is a time-independent covariate. T and U are assumed to be independent given Z. The survival function of T is assumed to have the form given in (10.1). Since Z is time-independent, the survival function simplifies to

$$S_Z(t) = \exp\left(-e^{\beta'Z} A(t)\right). \tag{10.5}$$

This model was considered by Huang [21] (see also Ref. 20).

Example 10.7: The Partly Linear Cox Model for Current Status Data. A single observation consists of $X = (U, \delta, Z, W)$, where U, δ and Z are as defined above but $W \in \mathcal{R}$ is a single additional time-independent covariate. Both T and U are assumed to be independent given both Z and W. The survival function has the form

$$S_Z(t) = \exp\left(-e^{\beta'Z + h(W)} A(t)\right), \tag{10.6}$$

where h is an unknown smooth function assumed to be in a Sobolev space [34] and A is a baseline integrated hazard. In this case the nuisance parameter is the composite $\eta = (h, A)$. A generalization of this example for transformation models was studied in Ma and Kosorok [34].

10.3 BASIC ESTIMATION AND LIMIT THEORY

For Examples 10.1–10.5, semiparametric maximum-likelihood estimation (SPMLE) involves a hazard function (the derivative a of A). Maximizing over this a results in assigning mass to each observed failure time, and the resulting maximizer is no longer a continuous hazard as assumed in model (10.3). This is a well-known issue [39], and the solution is to utilize an empirical likelihood that replaces a with ΔA and assigns mass only at observed failure times. The maximizer for $A(t)$ is then just the sum of the maximizers for $\Delta A(s)$ for all $s \leq t$. The profile likelihood $pL_n(\theta)$, obtained by profiling the empirical likelihood over A, is used for estimation in these examples, although additional profiling over ζ is needed in Example 10.4. For Example 10.1, this profiling results in the celebrated partial likelihood, which does not involve A at all. The full MLE in this instance consists of $\hat{\theta}_n$ obtained from the partial likelihood and the Breslow estimator

$$\hat{\eta}_n(t) = \int_0^t \frac{\sum_{i=1}^n dN_i(s)}{\sum_{i=1}^n \tilde{Y}_i(s) e^{\hat{\theta}_n' Z_i(s)}},$$

where $N_i(t) = \mathbf{1}\{U_i \leq t\}\delta_i$ and $\tilde{Y}_i(t) = \mathbf{1}\{U_i \geq t\}$, for the sample observations $i = 1, \ldots, n$. The remaining MLEs are more complex.

In general, what is required for Examples 10.2–10.7 is to maximize over the infinite-dimensional nuisance parameter A before maximizing over the other parameters. For Examples 10.2–10.5, this maximization leads to a stationary point equation that can be solved iteratively to obtain an estimator $\hat{A}_\theta$ depending on θ. For Example 10.2, this stationary point equation has the form

$$\hat{A}_\theta(t) = \int_0^t \left(\mathcal{P}_n \left[\frac{\tilde{Y}(s) e^{\theta' Z(s)} (1 + \delta)}{1 + \int_0^U e^{\theta' Z(v)} d\hat{A}_\theta(v)} \right] \right)^{-1} \mathcal{P}_n[dN(s)], \tag{10.7}$$

where $\mathcal{P}_n$ is the empirical measure of the sample, for instance, $\mathcal{P}_n h(\tilde{Y}, Z, \delta, N) = n^{-1} \sum_{i=1}^n h(\tilde{Y}_i, Z_i, \delta_i, N_i)$. For Examples 10.6 and 10.7, profiling over A is still needed but is accomplished by using an iterative convex minorant algorithm [17,21]. For Example 10.7, additional complexity is present since some form of penalization is required to ensure proper maximization of the likelihood over h [34].

This maximizer $\hat{A}_\theta$ is used to compute the profile likelihood $pL_n(\theta)$, which is then further maximized to obtain the SPMLE estimates $\hat{\theta}_n$ and $\hat{\eta}_n$. In Example 10.7, as mentioned earlier, this may require penalized maximization. In all of these examples, the estimators have been shown to be consistent. In the case of the infinite-dimensional nonregular parameters, this consistency may be in terms of an L_1 norm or some other nonuniform norm. All regular parameters—all parameters in Examples 10.1–10.5 except for the threshold parameter ζ, and the β parameter in Examples 10.6 and 10.7—have been shown to be $\sqrt{n}$-consistent, asymptotically normal, and fully efficient (even for the infinite-dimensional regular parameters).

For Example 10.4, $\hat{\zeta}_n$ is n-consistent (more rapidly converging than the regular parameters) and converges to the argmax of an interesting Poisson process that is asymptotically independent of the other parameter estimates. For Examples 10.6 and 10.7, A and h are $n^{1/3}$-consistent.

We note that establishing consistency, existence, and rates of convergence are typically the most technically challenging steps in developing inference for semiparametric models. While we are not dwelling on these steps in this chapter, we acknowledge their importance and complexity. Establishing existence and consistency is especially challenging for infinite-dimensional parameter estimators, as highlighted in Murphy and van der Vaart [41], although the general approach used in Murphy [37] can often be successfully adapted for right-censored survival data settings. Establishing rates of convergence for nonregular parameter estimators, regardless of whether they are infinite-dimensional, is also quite difficult. While there are some general results and strategies available to help, highly specific methods are often needed for each new situation.

The focus of the remainder of the chapter is on inference for the regular parameters based on the limiting normal distributions. For Example 10.1, simultaneous inference for the regression parameter can be accomplished by using closed-form variance estimators that have by now become standard (for an overview, see Sec. 4.3 of Ref. 16). Because the limiting variances from the remaining examples involve complex operators, direct estimation of the variances in these settings is seldom not feasible, and inference can be quite challenging. We will not discuss inference for the nonregular parameters, except briefly in Section 10.8, although some progress has been made in this area (see, e.g., the discussion on inference for the threshold parameter in Ref. 29).

10.4 THE BOOTSTRAP

The bootstrap has a long and successful history as a general method of statistical inference. The use of the bootstrap for infinite-dimensional regular parameters or for finite-dimensional regular parameters in the presence of nonregular parameters is a more recent phenomenon. In both cases, empirical process theory plays an important role. The reason for this is that all the estimators from all our examples can be expressed as smooth functionals of an empirical process. We now briefly review empirical processes and the associated bootstrap results. As part of this, we will introduce a useful alternative to the nonparametric bootstrap, the weighted bootstrap. (For additional information on the empirical process bootstrap, see Sec. 3.6 of Ref. 54) and Ch. 10 of Ref. 27.)

The empirical probability measure $\mathcal{P}_n$ was introduced in Section 10.310.3 (above). Let $\mathcal{X}$ be the sample space for the random observation X. Then, for any measurable function $f : \mathcal{X} \mapsto \mathfrak{R}, \mathcal{P}_n f = n^{-1} \sum_{i=1}^{n} f(X_i)$ (this is sometimes called the empirical "expectation" of f). We let P denote the true probability distribution of X and define $Pf = \int_{\mathcal{X}} f(x)P(dx)$. Also let $\mathcal{G}_n f = \sqrt{n}(\mathcal{P}_n - P)f$. General empirical process theory involves a collection $\mathcal{F}$ of measurable functions $f : \mathcal{X} \mapsto \mathfrak{R}$. We say that $\mathcal{F}$ is P-Glivenko–Cantelli if $\sup_{f \in \mathcal{F}} |(\mathcal{P}_n - P)f| \to 0$, outer almost surely, where the "outer" here invokes a high level of generality (see Sec. 1.2 of Ref. 54) that is quite useful in survival analysis. We say that $\mathcal{F}$ is P-Donsker if $\mathcal{G}_n$ converges weakly to a tight Gaussian process $\mathcal{G}$ uniformly over all $f \in \mathcal{F}$, that is, all $\mathcal{G}_n f \rightsquigarrow \mathcal{G}f$ with an appropriate level of continuity over $f \in \mathcal{F}$. We drop the prefix P in P-Glivenko–Cantelli and P-Donsker if the choice of P is contextually clear.

Consider Example 10.2. Much of the theory we now review comes from Lee [30], which differs some from the approach in Murphy et al. [39]. Using the form of $\hat{A}_\theta$ given in

Equation 10.7, it is not difficult to establish that the profile empirical likelihood is composed of several empirical expectations. Thus the SPMLE is a functional of an empirical process over a suitable choice of $\mathcal{F}$. Consistency will essentially follow from $\mathcal{F}$ being P-Glivenko–Cantelli combined with the identifiability of the proportional odds model, although some verification of the existence of the estimators involved is also needed. The SPMLEs can also be expressed as the solution $(\hat{\beta}_n, \hat{A}_n)$ of the Z-estimating equation $P_n U(\beta, A)(h) = 0$, where $U(\beta, A)(h) =$

$$\int_0^\tau [Z'(s)h_1 + h_2(s)]dN(s) - (1+\delta)\frac{\int_0^\tau \tilde{Y}(s)e^{\beta'Z(s)}[Z'(s)h_1 + h_2(s)]dA(s)}{1 + \int_0^\tau \tilde{Y}(s)e^{\beta'Z(s)}dA(s)}, \tag{10.8}$$

$h = (h_1, h_2)$ ranges over $H = \mathcal{R}^d \times BV_1 [0,\tau]$, $BV_1 [0,\tau]$ is the space of functions on $[0,\tau]$ with total variation bounded by 1, and τ is the fixed upper limit of the censoring times.

Once we have consistency of $(\hat{\beta}_n, \hat{A}_n)$, we can usually obtain asymptotic normality provided the class of functions $\mathcal{F} = \{U(\beta)(h) : \beta \in B_0, A \in \mathcal{A}_0, h \in H\}$, where B_0 and A_0 are open neighborhoods of the true parameter values β_0 and A_0, respectively, is P-Donsker. (The basic theory for Z estimators of this kind can be found in Sec. 3.3 of Ref. 54). It turns out that once a class of functions is determined to be Glivenko–Cantelli or Donsker, there is an automatic corresponding validity of the bootstrap. This makes ensuring that the bootstrap estimator is consistent (as an estimator) and that its conditional distribution is also consistent (as an estimator of the limiting probability distribution) both somewhat automatic. Before we explain this in more detail, we need to define the bootstraps that we are interested in. The nonparametric bootstrap is obtained by sampling without replacement from the sample $X_1, \ldots, X_n$ to obtain the bootstrapped sample $X_1^*, \ldots, X_n^*$. It is not difficult to see that the empirical measure of the resulting bootstrapped sample is $P_n^* f = n^{-1} \sum_{i=1}^n W_{in} f(X_i)$, where the $W_n = (W_{1n}, \ldots, W_{nn})$ are independent multinomial vectors with n categories (one for each sample value) and probability n vector $(1/n, \ldots, 1/n)'$. Thus the nonparametric bootstrap of a functional of the empirical distribution can be viewed as the functional of a weighted empirical distribution.

An alternative bootstrap is the weighted bootstrap. This also can be expressed as the functional of a weighted empirical but with different weights. Let $\xi_1, \ldots, \xi_n$ be an independent identically distributed (i.i.d.) collection of positive "preweights" independent of the data with $0 < E[\xi_1] = \mu < \infty$, $0 < \text{Var}[\xi_1] = \sigma^2 < \infty$, and $\int_0^\infty \sqrt{P[\xi_1 > u]}du < \infty$. The latter moment condition is a little stronger than the existence of a second moment but not as strong as requiring $E[|\xi_1|^{2+\in}] < \infty$ for some $\in > 0$. Let $\bar{\xi}$ be the sample mean of these preweights. The proposed weighted bootstrap is then based on the weighted empirical measure $P_n^{**} f = n^{-1} \sum_{i=1}^n (\xi_i/\bar{\xi})f(X_i)$. Let $\mathcal{G}_n^* = \sqrt{n}(P_n^* - P_n)$ and $\mathcal{G}_n^{**} = \sqrt{n}(\mu/\sigma)(P_n^{**} - P_n)$. We have the following useful Glivenko–Cantelli result.

Theorem 10.1. Let $\mathcal{F}$ be a class of measurable functions $f : \mathcal{X} \mapsto \mathcal{R}$. The following are equivalent:

1. $\mathcal{F}$ is P-Glivenko–Cantelli.
2. $\text{Sup}_{f \in \mathcal{F}} |(P_n^* - P)f| \to 0$ outer almost surely and $\sup_{f \in \mathcal{F}} |f - Pf| < \infty$.
3. $\text{Sup}_{f \in \mathcal{F}} |(P_n^{**} - P)f| \to 0$ outer almost surely and $\sup_{f \in \mathcal{F}} |f - Pf| < \infty$.

The proof follows from Theorem 10.8 and Corollary 10.2 of Kosorok [27].   □

We also have a similar result for Donsker classes. First, however, we need to define what it means for the conditional bootstrap distributions to be consistent for the true limiting distribution. This is accomplished through bounded Lipschitz classes of functions. Specifically, let $BL_1(\mathcal{F})$ be the collection of all Lipschitz continuous functions $h : \ell^\infty(\mathcal{F}) \mapsto \Re$ with $\|h\|_\infty \leq 1$ and $|h(f) - h(g)| \leq \|f - g\|_\infty$, where $\|\cdot\|_\infty$ is the uniform norm. Let $\mathcal{G}$ be a mean zero Gaussian process indexed by $f \in \mathcal{F}$ such that the correlation between $\mathcal{G}f$ and $\mathcal{G}g$ is $P(fg) - PfPg$. We say that the bootstrapped process $\mathcal{G}_n^*$ is consistent for $\mathcal{G}$ if $\sup_{h \in BL_1(\mathcal{F})} |E_W h(\mathcal{G}_n^*) - Eh(\mathcal{G})| \to 0$ in outer probability, where E_W denotes expectation over the weights W given the sample data. This reduces to the standard consistency definition for the bootstrap when $\mathcal{F}$ is finite-dimensional.

The following theorem tells us that bootstrap conditional distribution consistency is an automatic consequence of the class $\mathcal{F}$ being Donsker in the first place.

Theorem 10.2. Let $\mathcal{F}$ be a class of measurable functions $f : X \mapsto \Re$. The following are equivalent:

1. $\mathcal{F}$ is P-Donsker.
2. $\sup_{h \in BL_1(\mathcal{F})} |E_W h(\mathcal{G}_n^*) - E(\mathcal{G})| \to 0$ in probability and $h(\mathcal{G}_n^*)$ is asymptotically measurable for all $h \in BL_1(\mathcal{F})$.
3. $\sup_{h \in BL_1(\mathcal{F})} |E_\xi h(\mathcal{G}_n^{**}) - E(\mathcal{G})| \to 0$ in probability and $h(\mathcal{G}_n^{**})$ is asymptotically measurable for all $h \in BL_1(\mathcal{F})$.

Note that the asymptotic measurability condition in Theorem 10.2 is discussed in van der Vaart Wellner [54, Sec. 1.3]; however, it poses no difficulties for the examples that we are considering and for most other practical survival analysis examples (and can be essentially ignored by the reader). The proof of Theorem 10.2 follows from Theorem 2.6 of Kosorok [27]. There is also a continuous mapping result for the bootstrap that converts the above results to validity of the bootstrapped estimators.

We will now apply these results to two settings where inference focuses on regular parameters. In the first setting, if there are any nonregular parameters, we assume that they converge at a rate faster than $\sqrt{n}$. In the second setting, inference focuses on the finite-dimensional parameter θ but the nuisance parameter estimators are allowed to converge at a slower than $\sqrt{n}$ rate.

10.4.1 The Regular Case

For Examples 10.1–10.3 and 10.5, all parameters are regular. In these settings, Theorems 10.1 and 10.2 enable verification of bootstrap validity to follow almost automatically from consistency of $(\hat{\theta}_n, \hat{\eta}_n)$ and asymptotic normality of $\sqrt{n}[(\hat{\theta}_n, \hat{\eta}_n) - (\theta_0, \eta_0)]$. The proof of Corollary 1 in Kosorok et al. [28] illustrates this principle for (our) Example 10.5. For Example 10.4, the only nonregular parameter is the change point ζ. Since the MLE for ζ, $\hat{\zeta}_n$, converges at the n rate, one can hold the value of ζ fixed at $\hat{\zeta}_n$ while maximizing over the other parameters for each bootstrap realization. The resulting bootstrap is valid for all the regular parameters [29]. We note that confidence bands for infinite-dimensional parameters, such as the baseline hazard, cannot be obtained from simply knowing the covariance structure of the involved limiting Gaussian processes, except for very simple processes such as Brownian motion or standard Brownian bridges, and that some sort of sampling is required.

10.4.2   When Slowly Converging Nuisance Parameters are Present

The validity of the bootstrap when the nuisance parameters converge at a rate slower than $\sqrt{n}$ is significantly more difficult to establish. The issue is that the convergence of the bootstrapped estimators $\hat{\theta}_n^*$ or $\hat{\theta}_n^{**}$ (for the nonparametric and weighted bootstraps, respectively) must be established in essentially the same fashion as the weak convergence of $\hat{\theta}_n$. Usually, establishing the asymptotic normality of $\sqrt{n}(\hat{\theta}_n - \theta_0)$ in this setting requires calculations of the entropy of the log-likelihood components from the full model. For the bootstrap calculations to be valid, these entropy calculations must be preserved in the bootstrap likelihood. This is quite complex to evaluate for the nonparametric bootstrap since the weights are dependent. However, since the weights $\xi_1, \ldots, \xi_n$ are not dependent, this issue is more easily evaluated for the weighted bootstrap. In some cases, it may also be necessary to require the ξ_i terms to be bounded. The general theory for this approach is given in Ma and Kosorok [35] and can be shown to apply to inference on θ in both Examples 10.6 and 10.7. This means that the theory is applicable to penalized nonparametric maximum-likelihood estimation. It turns out to be generally valid for obtaining inference for the θ component in general semiparametric M-estimators, including least-squares, least-absolute-deviation, and misspecified likelihood estimation, in addition to correctly specified likelihood estimation and penalized likelihood estimation.

10.5   THE PROFILE SAMPLER

The material for this section comes mostly from Lee et al. [31]. As mentioned earlier, there are special cases where the profile likelihood for θ does not involve η, as occurs with Example 10.1. Unfortunately, most often the form of the profile likelihood is quite complicated and η is not easily eliminated. Inferences about θ have been studied for specific survival analysis models, including Example 10.6 [21] and Murphy and van der Vaart [40] have provided a general justification for such practices. Under mild structural conditions, the profile likelihood for θ has an asymptotic quadratic expansion that resembles that of a parametric likelihood. Furthermore, the maximum profile likelihood estimator $\hat{\theta}_n$ for θ is asymptotically normal with mean θ_0, the true value of θ, and covariance matrix n^{-1} times the inverse of the efficient Fisher information matrix $\tilde{I}_0$, which is corrected for the presence of the infinite-dimensional nuisance parameter [9,53].

Inferences about θ may be obtained without $\hat{\theta}_n$. The quadratic expansion of the profile likelihood permits the construction of confidence sets by inverting the log-likelihood ratio. Translating this elegant theory into practice has been limited by computational difficulties. Even if the log profile likelihood ratio can be successfully inverted for a multivariate parameter, this inversion does not enable the construction of confidence intervals for each parameter subcomponent separately, as is standard practice in data analysis. For such confidence intervals, it would be necessary to further profile over all remaining components in θ. A related problem for which inverting the log likelihood is not adequate is the construction of rectangular confidence regions for θ, such as minimum volume confidence rectangles [13] or rescaled marginal confidence intervals. For many practitioners, rectangular regions are preferable to ellipsoids, for ease of interpretation.

In principle, having an estimator of θ and its variance simplifies these inferences considerably. However, the computation of these quantities using the semiparametric likelihood poses stiff challenges relative to those encountered with parametric models. Finding the maximizer of the profile likelihood is done implicitly and typically involves numerical approximations. When the nuisance parameter is not $\sqrt{n}$-estimable, nonparametric functional estimation of η

for fixed θ may be required, which depends heavily on the proper choice of smoothing parameters. Even when η is estimable at the parametric rate, and without smoothing, $\tilde{I}_0$ does not ordinarily have a closed form. When it does have a closed form, it may include linear operators that are difficult to estimate well, and inverting the estimated linear operators may not be straightforward. The validity of these variance estimators must be established on a case-by-case basis.

The bootstrap is a possible solution to some of these problems, but, as mentioned in the 10.4previous section, theoretical justification is not guaranteed for semiparametric models where the nuisance parameter is not $\sqrt{n}$-consistent. The results in van der Vaart and Wellner [54] apply only to estimators converging at the parametric rate. Even when the bootstrap can be shown to be valid, the computational burden is quite substantial, since maximization over both θ and η is needed for each bootstrap sample. A different approach to variance estimation may be based on Corollary 3 of Murphy and van der Vaart [40], which demonstrates that the curvature of the profile likelihood near $\hat{\theta}_n$ is asymptotically equal to $\tilde{I}_0$. In practice, one can perform second-order numerical differentiation by evaluating the profile likelihood on a hyperrectangular grid of 3^p equidistant points centered at $\hat{\theta}_n$, taking the appropriate differences, and then dividing by $4h^2$, where p is the dimension of θ and h is the spacing between grid points. While the properties of h for the asymptotic validity of this approach are well known, there are no clear-cut rules on choosing the grid spacing in a given dataset. Thus, it would seem difficult to automate this technique for practical usage.

Prior to the paper by Lee et al. [31], there does not appear to exist in the statistical literature a general theoretically justified and automatic method for approximating $\tilde{I}_0$. They [31] propose an application of Markov chain Monte Carlo to the semiparametric profile likelihood. The method involves generating a Markov chain $\{\theta^{(1)}, \theta^{(2)}, \dots\}$ with stationary density proportional to $p_{\theta, n}(\theta) = pL_n(\theta)q(\theta)$, where $q(\theta) = Q(d\theta)/(d\theta)$ for some prior measure Q. This can be accomplished by using, for example, the Metropolis–Hastings algorithm [36,18]. Begin with an initial value $\theta^{(1)}$, for the chain. For each $k = 2, 3, \dots$, obtain a proposal $\vartheta^{(k+1)}$ by random walk from $\theta^{(k)}$. Compute $\hat{\eta}_{\vartheta}^{(k+1)}$ and $p_{\vartheta}(k+1)$, $n^{(\vartheta^{(k+1)})}$, and decide whether to accept $\vartheta^{(k+1)}$ by evaluating the ratio $p_{\vartheta}(k+1)$, $n^{(\vartheta^{(k+1)})}/p_{\theta}(k) n^{(\theta^{(k)})}$ and applying an acceptance rule. After generating a sufficiently long chain, one may compute the mean of the chain to estimate the maximizer of $pL_n(\theta)$ and the variance of the chain to estimate $\tilde{I}_0^{-1}$. The output from the Markov chain can also be directly used to construct various confidence sets, including minimum volume confidence rectangles.

Part of the computational simplicity of this procedure is that $pL_n(\theta)$ does not need to be maximized; it only needs to be evaluated. As mentioned earlier, the profile likelihood is fairly easy to compute as a consequence of algorithms such as the stationary point algorithm for maximizing over the nuisance parameter. In Example 10.2, as a case in point, Equation (10.7) can be iteratively solved by starting with an initial guess on the right side, obtaining $\hat{A}_{\theta}$ on the left, and then plugging this in on the right and repeating until the change in value is below a prespecified threshold. The procedure's validity is established in Theorem 1 of Lee et al., and the arguments rest on a careful analysis of the stationary distribution of the chain, which involves an extension of the theory of Murphy and van der Vaart [40]. This extension enables the quadratic expansion of the log likelihood around $\hat{\theta}$ to be valid in a fixed, bounded set, rather than only in a shrinking neighborhood. The conclusion of these arguments is that the "posterior" distribution of the profile likelihood with respect to a prior on θ is asymptotically equivalent to the distribution of $\hat{\theta}_n$.

One requirement for the profile sampler to be useful is for the profile likelihood to be reasonable easy to compute. When this is not the case, the numerical differentiation method mentioned previously may be advantageous since it requires fewer evaluations of the profile likelihood. However, numerical evidence in Lee et al. [31] seems to indicate that, at least for moderately small samples, numerical differentiation does not perform as well as the profile sampler. This observation is supported by theoretical work on the profile sampler in Cheng and Kosorok [10], which indicates that the profile sampler yields frequentist inference that is second-order accurate.

Note that inferences about θ might also be based on the marginal posterior of θ from the full likelihood with respect to a joint prior on (θ, η). Shen [48] has shown that this approach yields valid inferences for $\hat{\theta}_n$ when θ is estimable at the parametric rate. The profile likelihood sampler greatly simplifies the theory and computations, since a prior is not explicitly specified for η. At the very least, the profile sampler is a useful alternative to fully Bayesian computations when η is strictly a nuisance parameter. It may also enable an exact Bayesian inference that complements the asymptotic frequentist inference, if one accepts the use of the profile likelihood for Bayesian analysis.

It is not difficult to verify the theoretical assumptions for the validity of the profile sampler for Examples 10.1–10.3, 10.5, and 10.6. Lee et al. [31] explicitly verify the assumptions for Examples 10.5 and 10.6. The validity of Examples 10.1–10.3 follow since these are essentially special cases of Example 10.5. Data analysis and simulation studies in their paper [31] verify that this method works well for moderate sample sizes. Note that the profile sampler is widely applicable for semiparametric models in general, not just for survival models. It is, however, unclear how the procedure will work when parameters faster than root n are involved, as in Example 10.4, but it is probably the case that, as with the bootstrap, the change point parameter $\hat{\zeta}_n$ can be held at its MLE value and then the algorithm can proceed as though ζ were known. Unfortunately, it is known that the profile sampler cannot be applied to many penalized maximum likelihoods [35], such as is used for Example 10.7 [34].

10.6 THE PIGGYBACK BOOTSTRAP

The material presented in this section comes mostly from Dixon et al. [15]. The focus in this section is on survival analysis settings where the MLEs $\hat{\theta}_n$ and $\hat{\eta}_n$ are both $\sqrt{n}$ consistent, as is the case for Examples 10.1–10.3 and 10.5. The difficulty is that $\sqrt{n}(\hat{\eta}_n - \eta_0)$, where a zero subscript denotes the true value, usually converges weakly to an infinite-dimensional Gaussian process, and constructing confidence bands usually requires the ability to sample from good approximations of this limiting process. As mentioned above in Section 10.4.1, having a uniformly consistent estimate of the covariance will seldom lead to a shortcut, except when the form of the covariance is extremely simple.

Bootstrap methods can circumvent this difficulty by using the information from a sample of size n to generate random draws that accurately approximate the desired Gaussian process. Valid bootstrap draws are realizations of random variables θ_n and η_n that satisfy the following asymptotic property: $\sqrt{n}(\theta_n - \hat{\theta}_n, \eta_n - \hat{\eta}_n)$ converges weakly, given the sample data, to the same distribution that $\sqrt{n}(\hat{\theta}_n - \theta_0, \hat{\eta}_n - \eta_0)$ does unconditional on the sample data, as $n \to \infty$. A challenge with the bootstrap is that for each set of bootstrap weights, one must maximize the likelihood over the parametric and nonparametric components. Thus both the nonparametric and weighted bootstraps are computationally intense.

We now introduce an alternative to these bootstrap methods, the "piggyback bootstrap". Let $L_n(\theta, \eta)$ be the full log likelihood associated with the profile likelihood $pL_n(\theta) = L_n(\theta, \hat{\eta}_\theta)$, where $\hat{\eta}_\theta = \mathrm{argmax}_\eta L_n(\theta, \eta)$. As mentioned earlier, the required computations are frequently facilitated by the existence of fixed-point algorithms for computing $\hat{\eta}_\theta$. We assume in this section that draws for the parametric component θ_n are readily available and that $\sqrt{n}(\theta_n - \hat{\theta}_n)$, given the sample data, converges in distribution to the unconditional limiting distribution of $\sqrt{n}(\hat{\theta}_n - \theta_0)$. Such draws can be achieved, for example, via the profile sampler discussed in the 10.5previous section. A key feature of the piggyback bootstrap, however, is that it doesn't matter how the draws θ_n are obtained, provided they have the appropriate conditional limiting distribution. Let $L_n^*(\theta, \eta)$ be the bootstrapped log likelihood. Then, for each parametric draw θ_n, the piggyback bootstrap draw is $\eta_n = \text{argmax}_\eta L_i^*(\theta_n, \eta)$, resulting in the pair (θ_n, η_n). Hence, given θ_n, only one maximization over η is required. This approach results in a manyfold decrease in computational intensity over the full bootstrap, since the full bootstrap requires simultaneous maximization over both θ and η for each set of bootstrap weights.

The proposed approach is useful for several survival analysis models, including Examples 10.1–10.3 and 10.5 as mentioned earlier. In the case of clustered survival data, the procedure also applies to the shared frailty model and the correlated gamma frailty model. The method also applies to the Cox model for doubly censored survival data. There are also a number of applications not arising in survival analysis, such as certain biased sampling models, for which this procedure works. For several of these examples, there are existing methods to simplify the computations. For the Cox proportional hazards model, Kim and Lee [25] propose a novel Bayesian method for obtaining asymptotically valid random draws. In the proportional odds model, Hunter and Lange [23] provide an accelerated bootstrap algorithm for maximization of the likelihood so that an ordinary or weighted bootstrap can be employed with relatively low computational cost. For the proportional hazards random-effects regression model, Vaida and Xu [51] consider using the EM algorithm to obtain maximum-likelihood estimates. A disadvantage of these procedures is that they are applicable only to certain families of models, whereas the piggyback bootstrap applies in general to a fairly large class of semiparametric efficient estimators. The proposed piggyback bootstrap has the additional advantage of reducing the dimension of the set over which maximization is needed.

We will now present the piggyback bootstrap in greater detail. The main idea is to first obtain valid random draws for the parametric component of the model. Usually, it is possible to do this in a manner that is computationally much less intense than maximizing the profile likelihood, such as is the case, for example, with the profile sampler. The second step is to piggyback the draws for the nonparametric component onto the parametric draws, by plugging the parametric draws into a bootstrapped likelihood and maximizing over the nonparametric component holding the parametric part fixed; that is, for each $\theta_n^{(k)}$ drawn, $k = 1, \ldots, m$, we generate i.i.d. random bootstrap weights $\xi_1, \ldots, \xi_n$, and compute $\hat{\eta}^* \theta_n^{(k)} = \text{argmax}_\eta L_n^*(\theta_n^{(k)}, \eta)$, where L_n^* is the bootstrapped log likelihood using the given bootstrap weights. We assume that these bootstrap weights are nonnegative, with mean and variance 1 and with $\int_0^\infty \sqrt{P[\xi_1 > x]}\, dx < \infty$. As with the weighted bootstrap of Section 10.4 10.4, some variations in the mean and variance of the weights are possible after suitable adjustments. We also define $A_n \approx B_n$ to mean that A_n has a limit law conditional on the data equal to the limit law of B_n. The following theorem, where we let θ_n denote a representative from $\theta_n^{(1)}, \ldots, \theta_n^{(m)}$, establishes the validity of the new approach.

Theorem 10.3. Under regularity conditions, $\sqrt{n}(\theta_n - \hat{\theta}_n, \hat{\eta}_{\theta_n}^* - \hat{\eta}_{\hat{\theta}_n}) \approx \sqrt{n}(\hat{\theta}_n - \theta_0, \hat{\eta}_{\hat{\theta}_n} - \eta_0)$

The regularity conditions and proof are given in Dixon et al. [15].

A key assumption is that both $\hat{\theta}_n$ and $\hat{\eta}_n$ are efficient estimators. This implies that the efficient score for θ (adjusted for not knowing η) is uncorrelated with the score for η, where the score for η is computed under the assumption that θ is known. This property yields a simple expression for the joint distribution of $\hat{\theta}_n$ and $\hat{\eta}_n$ that is utilized in the piggyback bootstrap. Fortunately, establishing efficiency is rarely a problem for semiparametric MLEs. Furthermore, obtaining the MLE of η corresponding to a specific θ is simplified in many of the examples considered in Dixon, including Examples 10.1–10.3 and 10.5 through the use of a fixed-point algorithm.

Before utilizing this result, it is necessary to obtain draws $\theta_n^{(k)}$, $k = 1, \ldots, m$, that have the right conditional distribution. Because $\hat{\theta}_n$ is efficient, $\sqrt{n}(\hat{\theta}_n - \theta_0)$ is asymptotically zero-mean normal with variance $\tilde{I}_0^{-1}$, where $\tilde{I}_0$ is the efficient Fisher information for θ. Thus one way to obtain the desired draws is to estimate $\tilde{I}_0^{-1}$ with a consistent estimator $\hat{V}_0$, and then let $\theta_n^{(k)} = \hat{\theta}_n + n^{-1/2}\hat{V}_0^{1/2}Z^{(k)}$, $k = 1, \ldots, m$, where the $Z^{(k)}$ are independent standard normal vectors of length p, where p is the dimension of θ. In some settings, such as Example 10.1, a consistent estimator of $\hat{V}_0$ is not difficult to construct, but in many other settings finding such an estimator can be quite challenging. An alternative is to utilize the quadratic expansion of the profile log likelihood $pL_n(\theta)$ given in Murphy and van der Vaart [40] as mentioned earlier in this chapter. As we have mentioned several times, the profile sampler is also an extremely useful and computationally efficient approach to accomplishing this, and it is the approach we most readily recommend.

10.7   OTHER APPROACHES

Important alternative approaches to the above procedures include the m within n bootstrap [8] and subsampling [43]. Since $\sqrt{n}(\hat{\theta}_n - \theta_0)$ is known to have a continuous limiting distribution $\mathcal{L}$, Theorem 2.1 of Politis and Romano [43] yields that the m out of n subsampling bootstrap converges—conditionally on the data—to the same distribution $\mathcal{L}$, provided $m/n \to 0$ and $m \to \infty$ as $n \to \infty$. Because of the requirement that $m \to \infty$ as $n \to \infty$, the subsampling bootstrap potentially involves many calculations of the estimator. Fortunately, the asymptotic linearity of the SPMLE $\hat{\theta}_n$ [as asserted, e.g., in expression (5) of Ref. 40] can be used to formulate a computationally simpler alternative as described in Ma and Kosorok [34].

Let $\tilde{\theta}_n$ be any asymptotically linear estimator of a parameter $\theta_0 \in \mathcal{R}^d$, based on an i.i.d. sample $X, \ldots, X_n$, having square-integrable influence function ϕ for which $E[\phi\phi^T]$ is nonsingular. Let m be a fixed integer $>d$, and, for each $n \geq m$, define $k_{m,n}$ to be the largest integer satisfying $mk_{m,n} \leq n$. Also define $N_{m,n} \equiv mk_{m,n}$. For the data $X_1, \ldots, X_n$, compute the estimator $\tilde{\theta}_n$ and randomly sample $N_{m,n}$ out of the n observations without replacement, to obtain $X_1^*, \ldots, X_{N_{m,n}}^*$. Note that we are using the notation $\tilde{\theta}_n$ rather than $\hat{\theta}_n$ to remind ourselves that this estimator is a general, asymptotically linear estimator and not necessarily an SPMLE. For $j = 1, \ldots, m$, let $\tilde{\theta}_{n,j}^*$ be the estimate of θ based on the observations $X_1^*, \ldots, X_{N_{m,n}}^*$ after omitting X_j^*, X_{m+j}^*, X_{2m+j}^*, $\ldots, X_{(k_{m,n}-1)m+j}^*$. Compute $\bar{\theta}_n^* \equiv m^{-1}\sum_{j=1}^{m} \tilde{\theta}_{n,j}^*$ and $S_n^* \equiv (m-1)k_{m,n} \sum_{j=1}^{m} (\tilde{\theta}_{n,j}^* - \bar{\theta}_n^*)(\tilde{\theta}_{n,j}^* - \bar{\theta}_n^*)^T$. The following lemma provides a method of obtaining asymptotically valid confidence ellipses for θ_0.

Lemma 10.1.   Let $\tilde{\theta}_n$ be an estimator of $\theta_0 \in \mathcal{R}^d$, based on an i.i.d. sample $X_1, \ldots, X_n$ which satisfies $n^{1/2}(\tilde{\theta}_n - \theta_0) = \sqrt{n}\mathcal{P}_n\phi + o_p(1)$, where $E[\phi\phi^T]$ is nonsingular. Then $n(\tilde{\theta}_n - \theta_0)^T[S_n^*]^{-1} (\tilde{\theta}_n - \theta_0)$ converges weakly to $d(m-1)F_{d,m-d}/(m-d)$, where $F_{r,s}$ has an F distribution with degrees of freedom r and s.

The key to the proof of Lemma 10.1 is the simultaneous validity of the asymptotic linearity expansion for all the jackknife estimates. The details of the proof are given in Ma and Kosorok [34].

The fact that m remains fixed as $n \to \infty$ in the proposed approach results in a potentially significant computational savings over subsampling that requires m to grow increasingly large as $n \to \infty$. A potential challenge for the proposed approach is in choosing m for a given dataset. The larger m is, the larger the denominator degrees of freedom in $F_{d, m-d}$ and the tighter the confidence ellipsoid. On the other hand, m cannot be so large that the required asymptotic linearity does not hold simultaneously for all jackknife components. The need to choose m makes this approach somewhat less automatic than the profile sampler. Nevertheless, that fact that this "block jackknife" procedure requires fewer assumptions than does the profile sampler makes it a potentially useful alternative.

We have already mentioned fully Bayesian alternatives to the frequentist approaches that we have discussed. This is an important area of research with much current activity. Of particular interest are those methods that have demonstrated good frequentist properties such as the general results of Shen [48] and the special results of Kim and Lee [25] for Example 10.1 that we mentioned earlier and that we now briefly review. The basic idea is to use a specially designed prior for both parameters. This prior, when applied to the empirical likelihood discussed previously for the Cox model, results in a posterior with a very convenient form. Kim and Lee [25] prove that the output of this sampling scheme has the desired asymptotic properties. Unfortunately, this technique relies on the special structure of the Cox partial likelihood, a feature not shared by other semiparametric survival models.

Another general approach when all parameters are regular is to accurately estimate the influence function ϕ with some $\hat{\phi}_n$ satisfying $n^{-1} \sum_{i=1}^{n} \| \phi(X_i) - \hat{\phi}_n(X_i) \|_{\infty}^2 = o_P(1)$. One then samples from $n^{-1/2} \sum_{i=1}^{n} Z_i \hat{\phi}_n(X_i)$, where $Z_1, \ldots, Z_n$ are i.i.d. standard normals, to construct confidence intervals. This is essentially the approach taken by Lin, et al. [32] for Example 10.1. They [32] utilize the nice structure of the SPMLE for the full Cox model to obtain a nice estimate of the full influence function. A key challenge of this approach for the other examples we have discussed is that the influence function generally involves complex operators and is thus not practical to estimate in many situations. There are a number of other important specialty approaches to inference in survival analysis that apply to specific situations but do not appear to be widely generalizable. Recall, for example, the accelerated bootstrap of Hunter and Lange [23] and the modified EM algorithm of Vaida and Xu [51] mentioned earlier.

10.8 CONCLUDING REMARKS

In this chapter, we have endeavored to present general methods for semiparametric survival analysis inference based on nonparametric maximum likelihood estimation. While the methods we have discussed apply to many survival analysis models, there are many addition models, such as the correlated gamma frailty model [42], which we have not examined here but for which the methods presented are applicable [14]. It is important to note that empirical processes are very important in all this development, and we encourage the interested reader to become well acquainted with empirical process theory and techniques.

We also note that the greatest challenges appear to occur when one or more of the parameters converge at a rate different than $\sqrt{n}$. Most open research questions appear to be in this direction. The use of sieved and penalized log likelihoods can be very useful in some of these situations, although inference based on these procedures can still be technically challenging as mentioned previously regarding penalization for Example 10.7 (see also Ref. 34). Huang [22] applies sieves to estimation in the partly linear Cox model for right-censored data, while Shen [47] studies sieved estimation for Example 10.2. A deeper discussion of sieved and penalized estimation is beyond the scope of the present chapter, but a useful introduction to these approaches can be found in of van de Geer [52, Ch. 10].

Note that we have not even started to discuss inference for nonregular parameters. This area is very challenging and is also a very active area of research. An interesting development in this area is the asymptotically pivotal distribution results for certain cube-root-consistent estimators as described in Banerjee [6]. This approach has been applied successfully to inference for the survival function evaluated at a chosen timepoint in the current status data setting of example 6 [7]. On the other hand, the problem of constructing uniform confidence bands for the survival function in this setting remains unsolved.

Yet another challenging, open area in survival analysis is inference under model misspecification. Some results in this direction for regular parameters can be found in Kosorok et al. [28], and research on inference under misspecification in the presence of nonregular parameters is currently underway.

ACKNOWLEDGMENT

This research was supported in part by National Cancer Institute grant CA075142. The author thanks Editor Jason Fine and two anonymous reviewers for very helpful suggestions for improving the manuscript.

REFERENCES

1. Aalen, O. O., Nonparametric inference for a family of counting processes, *Ann. Statist.* **6**, 701–726 (1978).

2. Aalen, O. O., A model for nonparametric regression analysis of counting processes, in *Mathematical Statistics and Probability Theory, Lecture Notes in Statistics*, Springer, New York, 1980, Vol. 2, pp. 1–25.

3. Andersen, P. K., Borgan, Ø, Gill, R. D., and Keiding, N., *Statistical Models Based on Counting Processes*, Springer, New York, 1993.

4. Bagdonavičius, V. B. and Nikulin, M. S., Generalized proportional hazards models based on modified partial likelihood, *Lifetime Data Anal.* **5**, 329–350 (1999).

5. Balshaw, R. F. and Dean, C. B., A semiparametric model for the analysis of recurrent-event panel data, *Biometrics*, **58**, 324–331 (2002).

6. Banerjee, M., Likelihood based inference for monotone response models, *Ann. Statist.* (2006).

7. Banerjee, M., Biswas, P., and Ghosh, D., A semiparametric binary regression model involving monotonicity constraints, *Scand. J. Statist.* **33**(4), 673–697 (2006).

8. Bickel, P. J., Götze, F., and van Zwet, W. R., Resampling fewer than *n* observations: Gains, losses, and remedies for losses, *Statistica Sinica* **7**, 1–31 (1997).

9. Bickel, P. J., Klaassen, C. A. J., Ritov, Y., and Wellner, J. A., *Efficient and Adaptive Estimation for Semiparametric Models*, Springer-Verlag, New York, 1997.

10. Cheng, G. and Kosorok, M. R., Higher order semiparametric frequentist inference with the profile sampler, *Ann. Statist.* Tentatively accepted.

11. Cox, D. R., Regression models and life-tables (with discussion), *J. Roy. Statist. Soc. Ser. B* **34**, 187–220 (1972).

12. Dabrowska, D. M. and Doksum, K. A., Estimation and testing in a two-sample generalized odds-rate model, *J. Am. Statist. Assoc.* **83**, 744–749 (1988).

13. Di Bucchiano, A., Einmahl, J. H. J., and Mushkudiani, N. A., Smallest nonparametric tolerance regions, *Ann. Statist.* **29**, 1320–1343, (2001).

14. Dixon, J. R., *The Piggyback Bootstrap for Functional Inference in Semi Parametric Models*, Ph.D. dissertation, Dept. Statistics, Univ. Wisconsin—Madison, 2003.

15. Dixon, J. R., Kosorok, M. R., and Lee, B. L., Functional inference in semiparametric models using the piggyback bootstrap, *Ann. Inst. Statist. Math.* **57**, 255–277 (2005).

16. Fleming, T. R. and Harrington, D. P., *Counting Processes and Survival Analysis*, Wiley, New York, 1991.

17. Groeneboom, P. and Wellner, J. A., *Information Bounds and Nonparametric Maximum Likelihood Estimation*, Birkhäuser, Basel, 1992.

18. Hastings, W. K., Monte Carlo sampling methods using Markov chains and their applications, *Biometrika* **57**, 97–109 (1970).

19. Hsieh, F., On heteroscedastic hazards regression models: Theory and application, *J. Roy. Statist. Soc. Ser. B* **63**, 63–79 (2001).

20. Huang, J., *Estimation in Regression Models with Interval Censoring*, Ph.D. dissertation, Dept. Statistics, Univ. Washington, 1994.

21. Huang, J., Efficient estimation for the proportional hazard model with interval censoring, *Ann. Statist.* **24**, 540–568 (1996).

22. Huang, J., Efficient estimation of the partly linear additive Cox model, *Ann. Statist.* **27**, 1536–1563 (1999).

23. Hunter, D. R. and Lange, K., Computing estimates in the proportional odds model, *Ann. Inst. Statist. Math.* **54**, 155–168 (2002).

24. Kalbfleisch, J. D. Prentice, R. L., *The Statistical Analysis of Failure Time Data*, 2nd ed. Wiley, New York, 2002.

25. Kim, Y. and Lee, J., Bayesian bootstrap for proportional hazards models, *Ann. Statist.* **31**, 1905–1922 (2003).

26. Klein, J. P. and Moeschberger, M. L., *Survival Analysis: Techniques for Censored and Truncated Data*, Springer, New York, 1997.

27. Kosorok, M. R., *Introduction to Empirical Processes and Semiparametric Inference*, Springer, New York. (Forthcoming).

28. Kosorok, M. R., Lee, B. L., and Fine, J. P., Robust inference for univariate proportional hazards frailty regression models, *Ann. Statist.* **32**, 1448–1491 (2004).

29. Kosorok, M. R. and Song, R., Inference under right-censoring for transformation models with a change-point based on a covariate threshold, *Ann. Statist.* (Forthcoming).

30. Lee, B. L. *Efficient Semiparametric Estimation Using Markov Chain Monte Carlo*, Ph.D. dissertation, Dept. Statistics, Univ. Wisconsin—Madison, 2000.

31. Lee, B. L., Kosorok, M. R., and Fine, J. P., The profile sampler, *J. Am. Statist. Assoc.* **100**, 960–969 (2005).

32. Lin, D. Y., Fleming, T. R., and Wei, L. J., Confidence bands for survival curves under the proportional hazards model, *Biometrika*, **81**, 73–81 (1994).

33. Lin, D. Y. and Ying, Z., Semiparametric analysis of the additive risk model, *Biometrika* **81**, 61–71 (1994).

34. Ma, S. and Kosorok, M. R., Penalized log-likelihood estimation for partly linear transformation models with current status data, *Ann. Statist.* **33**, 2256–2290 (2005).

35. Ma, S. and Kosorok, M. R., Robust semiparametric M-estimation and the weighted bootstrap, *J. Multivar. Anal.* **96**, 190–217 (2005).

36. Metropolis, N., Rosenbluth, A. W., Rosenbluth, M. N., Teller, A. H., and Teller, E., Equation of state calculations by fast computing machines, *J. Chem. Phys.* **21**, 1087–1092 (1953).

37. Murphy, S. A., Consistency in a proportional hazards model incorporating a random effect, *Ann. Statist.* **22**, 712–731 (1994).

38. Murphy, S. A. Asymptotic theory for the frailty model, *Ann. Statist.* **23**, 182–198 (1995).

39. Murphy, S. A., Rossini, A. J., and van der Vaart, A., Maximum likelihood estimation in the proportional odds model, *J. Am. Statist. Assoc.* **92**, 968–979 (1997).

40. Murphy, S. A. and van der Vaart, A. W., On profile likelihood (with discussion), *J. Am. Statist. Assoc.* **95**, 449–485 (2000).

41. Murphy, S. A. and van der Vaart, A. W., Semiparametric mixtures in case-control studies, *J. Multivar. Anal.* **79**, 1–32 (2001).

42. Parner, E., Asymptotic theory for the correlated gamma-frailty model, *Ann. Statist.* **26**, 183–214 (1998).

43. Politis, D. N. and Romano, J. P., Large sample confidence regions based on subsamples under minimal assumptions, *Ann. Statist.* **22**, 2031–2050 (1994).

44. Pons, O., Estimation in a cox regression model with a change-point according to a threshold in a covariate, *Ann. Statist.* **31**, 442–463 (2003).

45. Sargent, D. J., A flexible approach to time-varying coefficients in the Cox regression setting, *Lifetime Data Anal.* **3**, 13–25 (1997).

46. Scharfstein, D. O., Tsiatis, A. A., and Gilbert, P. B., Semiparametric efficient estimation in the generalized odds-rate class of regression models for right-censored time-to-event data, *Lifetime Data Anal.* **4**, 355–391 (1998).

47. Shen, X., Proportional odds regression and sieve maximum likelihood estimation, *Biometrika* **85**, 165–177 (1998).

48. Shen, X., Asymptotic normality of semiparametric and nonparametric posterior distributions, *J. Am. Statist. Assoc.* **97**, 222–235 (2002).

49. Slud, E. V. and Vonta, F., Consistency of the NPML estimator in the right-censored transformation model, *Scand. J. Statist.* **31**, 21–41 (2004).

50. Tsiatis, A. A., Estimating regression parameters using linear rank tests for censored data, *Ann. Statist.* **18**, 354–372 (1990).

51. Vaida, F. and Xu, R., Proportional hazards model with random effects, *Statist. Med.* **19**, 3309–3324 (2000).

52. van de Geer, S. A., *Empirical Processes in M-Estimation*, Cambridge Univ. Press, Cambridge, UK, 2000.

53. van der Vaart, A. W., *Asymptotic Statistics*, Cambridge Univ. Press, Cambridge, UK, 1998.

54. van der Vaart, A. W., and Wellner, J. A., *Weak Convergence and Empirical Processes: With Applications to Statistics*, Springer, New York, 1996.

55. Wei, L. J., Ying, Z., and Lin, D. Y., Linear regression analysis of censored survival data based on rank tests, *Biometrika* **77**, 845–851 (1990).

56. Wellner, J. A. and Zhang, Y., Two estimators of the mean of a counting process with panel count data, *Ann. Statist.* **28**, 779–814 (2000).

57. Zeng, D., Lin, D. Y., and Yin, G., Maximum likelihood estimation in the proportional odds model with random effects, *J. Am. Statist. Assoc.* **100**, 470–483 (2005).

CHAPTER 11

An Overview of the Semi–Competing Risks Problem

Limin Peng
Emory University, Atlanta, Georgia

Hongyu Jiang
Harvard University, Boston, Massachusetts

Rick J. Chappell and Jason P. Fine
University of Wisconsin—Madison, Madison, Wisconsin

11.1 INTRODUCTION

The semi–competing risks problem was first introduced by Fine et al. [15] to refer to the situation in which an event time can be censored by another event time but not vice versa. It often occurs in chronic disease studies and clinical trials involving both terminating events and nonterminating events. A terminating event potentially censors a nonterminating event, but the nonterminating event does not prevent subsequent observation of the terminating event. In contrast, the classic competing risks setting allows only for the observation of the event that occurs first.

   Semi–competing risks data may arise in the following two scenarios. The first scenario involves two kinds of endpoints—time to morbidity and time to mortality. For example, in a multicenter clinical trial of allogeneic bone marrow transplants in patients with acute leukemia [10], the primary endpoint was death and the secondary endpoints were relapse and graft-versus-host disease (GVHD). Both relapse and GVHD may lead to death but death caused by GVHD is not directly leukemia-related. Since mortality is quite complicated, a good measure of the biological efficacy might be based on the endpoint of relapse. Relapse was observable if it occurred earlier, but only mortality was observable otherwise. In this example, time to relapse and time to death formed the special bivariate structure of semi–competing risks data.

Statistical Advances in the Biomedical Sciences, edited by Atanu Biswas, Sujay Datta,
Jason P. Fine, and Mark R. Segal

The second scenario is frequently encountered in clinical trials in which the primary outcome is some nonterminating event, including a case of a surrogate endpoint. Mortality is not an important issue because the death rate may be quite low during the course of study. However, there exists a terminating event: dropout. A good example comes from AIDS Clinical Trial Group (ACTG) 364 Study [2]. The first virologic failure (confirmed HIV RNA ≥ 200 copies/mL) is one intermediate endpoint of interest. Many patients withdrew before the end of the study for disease-related reasons such as complications or excessively high viral load. This setting falls into the semi–competing risks paradigm, because the occurrence of virologic failure did not prevent subsequent follow-up so that the time to dropout was still potentially observable.

An instructive graph of semi–competing risks data was provided by Jiang et al. [23] to compare these data with bivariate right-censored data and with classic competing risks data. In Figure 11.1, (T_1, T_2) denote a pair of event times. Both failure times can be observed in the whole quadrant for bivariate right-censored data, but the semi–competing risks data are observable only in the upper wedge. With competing risks data, T_1 and T_2 are never observed together and all the information lies on the diagonal line.

Inferences with semi–competing risks data are generally focused on the development of the nonterminating event and the association between the nonterminating event and the terminating event. There are two distinct types of approach to analyzing semi–competing risks data. One may use crude quantities, including cause-specific hazard and cumulative incidence functions, which account for the presence of the terminating event and are nonparametrically identifiable. The other type of approach is based on net quantities, such as the marginal distribution of the nonterminating event, which hypothesize the removal of the terminating event and are not identifiable without further assumptions [44]. In practice, these two kinds of approach aid in addressing different scientific questions. For example, in the ACTG 364 study, the marginal distribution of virologic failure corresponds to the setting where there is no dropout, while the cumulative incidence function does not remove the effect of dropout and reflects the observational process, not the underlying biology of the disease. The cumulative incidence of virologic failure is therefore less relevant than the corresponding marginal distribution. In the leukemia example, estimating the marginal distribution of time to relapse may be useful in describing the behavior of morbidity as a process distinct from mortality due to other causes. However, it posits a hypothetical setting in which censoring of relapse time by death does not exist. The appropriateness of the counterfactual interpretation of the marginal

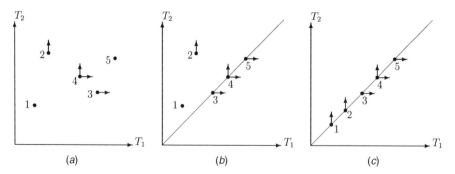

Figure 11.1 Illustration of semicompeting risks data; [a dot indicates that T_1 and T_2 are observed; an arrow in the direction of T_1 (T_2) means that T_1 (T_2) is censored]: (*a*) bivariate right-censored data; (*b*) semi–competing risks data; (*c*) competing risks data.

distribution has been scrutinized by the medical community [4, 5, 12, 6, 7]. In alike scenarios, analyses are often oriented to cause-specific hazard and cumulative incidence functions that characterize the progression of disease in the presence of death.

Because of lack of appropriate methodology, semi–competing risks data were previously analyzed as competing risks data, pretending as if only the time and cause of the first event had been recorded. The inferences involve either (1) restricting the joint distribution using either semiparametric or parametric models [29,14] or (2) performing a sensitivity analysis [37,42,26,49,40], in which bounds on the marginal distributions are obtained via estimation under various assumed dependence structures. Multistate modeling is an alternative approach, in which the data are viewed as a multistate process with finite state space, say $\{0,1,2\}$, with states 1 and 2 representing the occurrences of the nonterminating event and the terminating event, respectively. Transition probabilities between different states can be estimated by Aalen–Johansen estimators [1,3]. However, they provide limited information regarding the dependence structure as well as the distribution of the nonterminating event in the absence or presence of the terminating event.

The goal of this chapter is to provide an overview of the more recent methodology advances for semi–competing risks data. In Section 11.2 we first revisit the classic nonparametric inference based on crude quantities and outline nonparametric techniques for dealing with left truncation of the terminating event as often occurs in observational studies. Semiparametric one-sample inference based on net quantities are reviewed in Section 11.3, focusing on the work by Fine et al. [15] and its current extensions. In Section 11.4 we introduce regression methods for semi–competing risks data that have been developed under accelerated failure-time modeling and a class of functional regression models. Some concluding remarks are provided in Section 11.5.

Throughout this chapter, let T_1 be the time to the nonterminating event (e.g., morbidity, surrogate endpoint), and let T_2 be the time to the terminating event (e.g., mortality, dropout), which may dependently censor T_1. Let C be an independent censoring time for both T_1 and T_2, which often occur as the administrative censoring. Define $T = T_1 \wedge T_2$. In the semi–competing risks setting, observables include n i.i.d. replicates of $\{X = T \wedge C, \delta = I_{\{T_1 \leq T_2 \wedge C\}}, Y = T_2 \wedge C, \xi = I_{\{T_2 \leq C\}}\}$, denoted by $\{X_i, \delta_i, Y_i, \xi_i\}_{i=1}^n$. Here and in the sequel, $\wedge$ is the minimum operator and $I_A(\cdot)$ is the indicator function.

11.2   NONPARAMETRIC INFERENCES

Without considering covariates, the observed semi–competing risks data consist of n i.i.d. replicates of (X, Y, δ, ξ), denoted by $\{(X_i, Y_i, \delta_i, \xi_i), i = 1, \ldots, n\}$.

In the semi–competing risks setting, nonparametric analyses of identifiable quantities, including cause-specific hazard and cumulative incidence functions, have been widely adopted [38]. The cause-specific hazard and cumulative incidence functions for the nonterminating event are defined as

$$\lambda_1(t) = \lim_{h \to 0} h^{-1} \Pr(t \leq T_1 < t + h, T_2 > t | T_1 \geq t, T_2 \geq t),$$
$$F_1(t) = \Pr(T_1 \leq t, T_2 > T_1).$$

In words, $\lambda_1(t)$ and $F_1(t)$ represent the instantaneous rate and the distribution of the nonterminating event in the presence of the terminating event, respectively.

With semi–competing risks data, the estimation of $\Lambda_1(t) = \int_0^t \lambda_1(s)ds$ and $F_1(t)$ resembles that with classic competing risks data. To define the estimators, we use the counting process notation $Y_i(t) = I_{\{X_i \geq t\}}, N_{1,i}(t) = I_{\{X_i \leq t, \delta_i = 1\}}$, $N_{2,i}(t) = I_{\{X_i \leq t, \delta_i = 0, \xi_i = 1\}}$, $\bar{Y}(t) = \sum_{i=1}^n Y_i(t), \bar{N}_j(t)$ $= \sum_{i=1}^n N_{j,i}(t), j = 1, 2$. The Nelson–Aalen-type estimator for $\Lambda_1(t)$ is

$$\hat{\Lambda}_1(t) = \int_0^t \frac{d\bar{N}_1(s)}{\bar{Y}(s)} = \sum_{i=1}^n \frac{I_{\{X_i \leq t, \delta_i = 1\}}}{\sum_{l=1}^n I_{\{X_i \leq X_l\}}}.$$

Let $S_T(t) = \Pr(T > t)$ and $\hat{S}_T(t)$ be the Kaplan–Meier estimator for $S_T(t)$ based on $\{(X_i, I_{\{X_i < C_i\}}), i = 1, \ldots, n\}$; that is, $\hat{S}_T(t) = \prod_{X_i \leq t} [1 - \{d\bar{N}_1(t) + d\bar{N}_2(t)\}/\bar{Y}(X_i)]$. Using the fact that $F_1(t) = \int_0^t S_T(u^-)d\Lambda_1(u)$, an estimator for $F_1(t)$ is obtained as $\hat{F}_1(t) = \int_0^t \hat{S}_T(u^-)d\hat{\Lambda}_1(u)$.

In observational studies, complications may arise when T_2 is left truncated at time L. One example is the Denmark diabetes registry [3], consisting of insulin-dependent diabetes patients referred to the Steno Memorial Hospital in Greater Copenhagen between 1931 and 1988. The cumulative incidence of diabetic nephropathy, a common morbidity, is helpful in characterizing the disease progression. The analysis must account for the facts that time to death T_2 may dependently censor time to nephropathy T_1. Another complication is administrative left trunca-tion; specifically, only patients living long enough to enter the registry provided data.

With truncation, the observed data consist of n replicates of $(X^*, Y^*, \delta^*, \xi^*, L^*)$, denoted by $\{(X_i^*, Y_i^*, \delta_i^*, \xi_i^*, L_i^*)\}_{i=1}^n$, where $(X^*, Y^*, \delta^*, \xi^*, L^*)$ follows the conditional distribution of (X, Y, δ, ξ, L) given $Y \geq L$. It is assumed that (L, C) is independent of (T_1, T_2).

Andersen et al. [3] suggested using the Nelson–Aalen estimator with appropriately defined risk sets to estimate $\Lambda_1(t)$ and $F_1(t)$. To do so, the semi–competing risks data must be "forced" into a competing risks setup, where only $(X_i, I_{\{T_i \leq C_i\}})$ are observed conditional on $X_i \geq L_i, i = 1, \ldots, n$. The nonterminating event is then artificially truncated by L, the trun-cation time for the terminating event. Huang and Wang [22] give an account of estimation of $\Lambda_1(t)$ and $F_1(t)$ with independent left truncation and right censoring of T. Their estimators are essentially $\hat{\Lambda}_1(t)$ and $\hat{F}_1(t)$ with $\bar{N}_1(t), \bar{N}_2(t), \bar{R}(t)$ replaced by $\bar{N}_1^*(t) = \sum_{i=1}^n I_{\{L_i^* \leq X_i^* \leq t, \delta_i^* = 1\}}$, $\bar{N}_2^*(t) = \sum_{i=1}^n I_{\{L_i^* \leq X_i^* \leq t, \delta_i^* = 0, \xi_i^* = 1\}}, \bar{Y}^*(t) = \sum_{i=1}^n I_{\{L_i^* \leq t \leq X_i^*\}}$, respectively. The naive competing risk procedure employs data only with $X_i^* \geq L_i^*$. The removal of X_i^* terms that are smaller than the left truncation times for Y_i^* may incur considerable information loss.

Peng and Fine [34] proposed simple nonparametric estimators for $F_1(t)$ and $\Lambda_1(t)$ that better utilize semi–competing risks information. The strategy is to examine the connection between the transformed bivariate subsurvival function $F^*(x, y) = \Pr(X > x, Y > y, \xi = 1|L \leq Y)$ and the underlying bivariate survival function $\tilde{H}(x,y) = \Pr(T > x, T_2 > y)$. Weighting $F^*(dx, dy)$ inversely by the probability of observing complete uncensored $(T = x, T_2 = y)$ under a left truncation mechanism, one can transform F^* into $\tilde{H}$. Using F^* in place of $\tilde{H}$ facilitates the estimation of $F_1(t) = \int_0^t \int_v^\infty \tilde{H}(dv,du)$. Let $\tau > t$ be a constant defined to be slightly smaller than the upper bound of Y, and let $\hat{S}_{T_2}$ be the Lynden–Bell product limit estimator [30] for left-truncated right-censored data using $\{(Y_i^*, \eta_i^*, L_i^*), i = 1, \ldots, n\}$. Define $C_n(y) = n^{-1} \sum_{i=1}^n I_{\{L_i^* \leq y \leq Y_i^*\}}$ and $C_{2,n}(y, x) = n^{-1} \sum_{i=1}^n I_{\{L_i^* \leq y \leq Y_i^*, X_i^* > x\}}$. The estimators derived in [34] are given by

$$\breve{F}_1(t) = n^{-1} \sum_{i=1}^n \frac{\hat{S}_{T_2}(Y_i^{*-})}{C_n(Y_i^*)} I_{\{X_i^* \leq t, X_i^* < Y_i^* \leq \tau, \xi_i^* = 1\}} + \hat{S}_{T_2}(\tau)\left\{1 - \frac{C_{2,n}(\tau^+, t)}{C_n(\tau^+)}\right\}$$

and $\check{\Lambda}_1(t) = \int_0^t \{\check{R}(s)\}^{-1} d\check{F}_1(s)$, where $\check{R}(t)$ is an estimator of $\tilde{H}(t,t)$ and equals

$$n^{-1} \sum_{i=1}^n \left\{ \frac{\hat{S}_{T_2}(Y_i^{*-})}{C_n(Y_i^*)} I_{\{t \leq Y_i^* \leq \tau, X_i^* \geq t, \xi_i^* = 1\}} + \frac{\hat{S}_{T_2}(\tau)}{C_n(\tau)} I_{\{L_i^* \leq \tau < Y_i^*, X_i^* \geq t\}} \right\}.$$

Unlike $\hat{F}_1(t)$ and $\hat{\Lambda}_1(t)$, the estimators $\check{F}_1(t)$ and $\check{\Lambda}_1(t)$ do not require artificial truncation. Under mild regularity conditions, they are shown to be uniformly consistent and to weakly converge to tight zero-mean Gaussian processes. Given that $\check{F}_1(t)$ and $\check{\Lambda}_1(t)$ have simple closed forms and plug-in variance estimators are available, it would be reasonable to recommend them for practice use. In an analysis [34] of the Denmark diabetes registry, cumulative incidence rates were calculated separately for male and female patients who were diagnosed before age 31 and between 1933 and 1972. The variance estimates of $\hat{F}_1(t)$ are always larger than those of $\check{F}_1(t)$, with the variance reductions ranging from 24% to 67% for males and from 19% to 38% for females.

11.3 SEMIPARAMETRIC ONE-SAMPLE INFERENCE

With semi–competing risks data, studying the marginal distribution of the nonterminating event is plagued by the nonparametric nonidentifiability of the bivariate model for (T_1, T_2) [44]. The naive Kaplan–Meier estimator for $S_{T_1}(x) = \Pr(T_1 > x)$ based on $\{(X_i, \delta_i), i = 1, \dots, n\}$ is usually invalid because of the correlation between T_1 and T_2. For instance, relapse and death in the leukemia example are believed to be associated, while dropout in the AIDS example may be informative for the virologic endpoint. It is recognized that extra information on T_2 in the semi–competing risks is helpful in nonparametric estimation of the marginal distribution of T_2. However, the marginal distribution of T_1 cannot be identified without further assumptions. Methods that fully use the semi–competing risks data may better address these difficulties.

Fine et al. [15] exploited the special features of semi–competing risks data and developed inferences for a novel semiparametric model, avoiding extrapolations in the lower wedge of (T_1, T_2). Let $H(x, y) = \Pr(T_1 > x, T_2 > y)$ and $S_{T_j}(x) = \Pr(T_j > x), j = 1, 2$. The dependence structure of (T_1, T_2) is formulated via the gamma frailty model [8,33] in the upper wedge where $T_1 \leq T_2$. Thus, for $\theta \geq 1$ and $0 \leq x \leq y \leq \infty$, we obtain

$$H(x, y) = \begin{cases} \{S_{T_1}(x)^{1-\theta} + S_{T_2}(y)^{1-\theta} - 1\}^{1/(1-\theta)}, & \theta > 1; \\ S_{T_1}(x) S_{T_2}(y), & \theta = 1. \end{cases} \quad (11.1)$$

With the usual bivariate right-censored data, estimating the association parameter in copula models has been widely studied. For example, Shih and Louis [41] suggested estimating the association parameter on the basis of the likelihood function with plug-in marginal distribution estimators. The possibility of extending this two-stage estimation procedure in the presence of semi–competing risks seems very unlikely because of the nonidentifiability of S_{T_1}. Under model (11.1), Fine et al. [15] cleverly adapted Oakes' nonparametric estimator of predicative hazard ratio for bivariate survival data and proposed a closed-form estimator for θ without involving either S_{T_1} or S_{T_2} for semi–competing risks data. The main idea is that the concordance indicator $\Delta_{ij} = I_{\{(T_{1i} - T_{1j})(T_{2i} - T_{2j}) > 0\}}$ of two independent pairs of failure times (T_{1i}, T_{2i})

and (T_{1j}, T_{2j}) has expectation $\theta_0/(1 + \theta_0)$ under the assumed model, where θ_0 is the true value of θ, $i, j = 1, \ldots, n$. In the semi–competing risks setting, Δ_{ij} is determinable only when $\tilde{X}_{ij} < \tilde{Y}_{ij} < \tilde{C}_{ij}$, where $\tilde{X}_{ij} = T_{1i} \wedge T_{1j}, \tilde{Y}_{ij} = T_{2i} \wedge T_{2j}, \tilde{C}_{ij} = C_i \wedge C_j$. Let $D_{ij} = I_{\{\tilde{X}_{ij} < \tilde{Y}_{ij} < \tilde{C}_{ij}\}}$, $\tilde{X}'_{ij} = X_i \wedge Y_j$, and $\tilde{Y}'_{ij} = X_i \wedge Y_j$. A closed-form estimator for θ_0 is given by

$$\hat{\theta} = \frac{\sum_{i<j} W(\tilde{X}'_{ij}, \tilde{Y}'_{ij}) D_{ij} \Delta_{ij}}{W(\tilde{X}'_{ij}, \tilde{Y}'_{ij}) D_{ij} (1 - \Delta_{ij})},$$

where $W(u,v)$ is an appropriate random weight function. It has been shown that $\hat{\theta}$ is consistent for θ_0 and $n^{1/2}(\hat{\theta} - \theta_0)$ has a limiting normal distribution $N(0, \Sigma)$. The variance Σ can be consistently estimated by $\hat{\Sigma} = \hat{I}^{-2}\hat{J}$, where

$$\hat{I} = n^{-2} \sum_{i<j} W(\tilde{X}'_{ij}, \tilde{Y}'_{ij}) D_{ij} (1 + \hat{\theta})^{-2},$$

$$\hat{J} = 2n^{-3} \sum_{k<l<m} (\hat{Q}_{kl}\hat{Q}_{km} + \hat{Q}_{kl}\hat{Q}_{lm} + \hat{Q}_{lm}\hat{Q}_{km}),$$

and $\hat{Q}_{kl} = W(\tilde{X}'_{kl}, \tilde{Y}'_{kl}) D_{kl} \{\Delta_{kl} - \hat{\theta}/(1 + \hat{\theta})\}$.

A simple plug-in estimator of S_{T_1} can be constructed by plugging a consistent estimator for θ in model (11.1). A closed-form estimator is obtained as $\hat{S}_{T_1}(x) = \{\hat{S}_T(x)^{1-\hat{\theta}} - \hat{S}_{T_2}(x)^{1-\hat{\theta}} + 1\}^{1/(1-\hat{\theta})}$, where $\hat{S}_{T_2}$ and $\hat{S}T$ are the Kaplan–Meier estimators for S_{T_2} and S_T using $\{(Y_i, \xi_i), i = 1, \ldots, n\}$ and $\{(X_i, \delta_i + \xi_i - \delta_i\xi_i), i = 1, \ldots, n\}$. The continuous mapping theorem gives the uniform convergence of $\hat{S}_{T_1}(x)$ to $S_{T_1}(x)$. It is also shown that $n^{1/2}\{\hat{S}_{T_1}(x) - S_{T_1}(x)\}$ converges weakly to a Gaussian process. As a result of censoring, in small samples, $\hat{S}_T(t) \leq \hat{S}_{T_2}(t)$ may be violated and hence $\hat{S}_{t_1}(t)$ may not be monotone at all time-points. A simple variant $\hat{S}^*_{T_1}(t) = \min_{s \leq t}\hat{S}_{T_1}(s)$ is proposed that is always decreasing and is asymptotically equivalent to $\hat{S}_{T_1}(t)$.

Fine et al. [15] illustrated the semiparametric method with the leukemia example. Estimates, with standard errors in parentheses, for θ with $W(u, v) = 1$ and $W(u, v) = n^{-1} \sum_{i=1}^{n} I_{\{X_i \geq u, Y_i \geq v\}}$ are $\hat{\theta}_u = 8.79$ (2.15) and $\hat{\theta}_w = 8.61$ (2.15). Both estimates indicate that relapse is highly predictive of death. Figure 11.2 plots $\hat{S}^*_{T_1}(t)$ employing $\hat{\theta}_w$ and the Kaplan–Meier estimator using $\{(X_i, \delta_i), i = 1, \ldots, n\}$, along with the 0.95 confidence intervals for $\hat{S}_{T_1}(t)$. The naive Kaplan–Meier estimator is uniformly above the upper 0.95 limit of $\hat{S}^*_{T_1}(t)$, which may be explained by the substantial association between death and relapse.

General dependence structures were investigated by Wang [47] for semi–competing risks data. One extension is based on the predicative hazard ratio function $\theta(x, y) = \lambda_2(y | T_1 = x)/\lambda_2(y | T_1 > x)$ [11], where $\lambda_2(t | A)$ is the hazard function of T_2 given that event A occurs. In the upper wedge, $\theta(x, y)$ is parameterized as $\theta_{\alpha,\eta}(x, y)$, where α is a one-dimensional parameter of interest and η denotes the nuisance parameter. Note that the gamma frailty model is a special case in which $\theta(x, y)$ reduces to a constant. The other association model considered by Wang [47] is the general parametric copula model, in which $H(x, y) = C_\alpha\{S_{T_1}(x), S_{T_2}(y)\}$, where $C_\alpha(u, v) : [0, 1]^2 \rightarrow [0, 1]$ is a known copula function

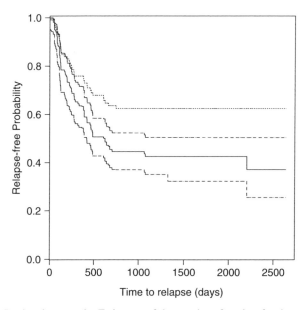

Figure 11.2 Leukemia example. Estimates of the survivor function for time to relapse. The solid line represents the point estimate from $\hat{S}_{T_1}^*$; the dashed lines, the limits of the corresponding 0.95 intervals; and the dotted line, the Kaplan–Meier estimate.

with an unknown parameter α. Two types of estimating equation are proposed for α, by generalizing the work by Day et al. [11] or utilizing the Doob–Meyer decomposition [17]. The resultant estimator under either dependence model is shown to be consistent and asymptotically normal. The asymptotic variance may be complicated but can be estimated via bootstrap-based approaches.

As the dependence structure is still formulated via the gamma frailty copula on the upper wedge, Jiang et al. [25] studied the estimation of S_{T_1} according to the principle of pseudo–self-consistency. The self-consistent estimator is superior to the simple closed-form estimator $\hat{S}_{T_1}(t)$, which may not be monotone and may jump at times other than the observed T_1. The idea of "self-consistency" was used by Efron to derive the Kaplan–Meier estimator under right censoring [13]. Similar techniques have also been used for interval censored data and for doubly censored data (e.g., see Refs. 46, 18, 31, and 43). The challenge in the current setting is that θ_0 is unknown. Under the full gamma frailty model, the self-consistency estimating equations have the following specific forms:

$$
\begin{aligned}
S_{T_1}(t) = \Bigg\{ &\sum_{i=1}^{n} I_{\{X_i > t\}} + \sum_{X_i \le t} (1 - \xi_i)\delta_i \left[\frac{S_{T_1}(t)^{1-\theta} + S_{T_2}(Y_i)^{1-\theta} - 1}{S_{T_1}(X_i)^{1-\theta} + S_{T_2}(Y_i)^{1-\theta} - 1} \right]^{\theta/(1-\theta)} \\
&+ \sum_{X_i \le t} (1 - \xi_i)(1 - \delta_i) \left[\frac{S_{T_1}(t)^{1-\theta} + S_{T_2}(Y_i)^{1-\theta} - 1}{S_{T_1}(X_i)^{1-\theta} + S_{T_2}(Y_i)^{1-\theta} - 1} \right]^{1/(1-\theta)} \Bigg\} \cdot n^{-1}
\end{aligned}
\qquad (11.2)
$$

$$S_{T_2}(t) = \left\{ \sum_{i=1}^{n} I\{Y_i > t\} + \sum_{Y_i \leq t} \xi_i(1 - \delta_i) \left[\frac{S_{T_1}(X_i)^{1-\theta} + S_{T_2}(t)^{1-\theta} - 1}{S_{T_1}(X_i)^{1-\theta} + S_{T_2}(Y_i)^{1-\theta} - 1} \right]^{\theta/(1-\theta)} \right.$$

$$\left. + \sum_{Y_i \leq t} (1 - \xi_i)(1 - \delta_i) \left[\frac{S_{T_1}(X_i)^{1-\theta} + S_{T_2}(t)^{1-\theta} - 1}{S_{T_1}(X_i)^{1-\theta} + S_{T_2}(Y_i)^{1-\theta} - 1} \right]^{1/(1-\theta)} \right\} \cdot n^{-1}. \qquad (11.3)$$

Given that $\hat{\theta}$ converges almost surely to θ_0, one can either first plug $\hat{\theta}, \hat{S}_{T_2}$, and an initial estimate of S_{T_1} in Equation (11.2) and update the estimator of S_{T_1} on the basis of Equation (11.2) iteratively until it converges, or estimate S_{T_1} and S_{T_2} simultaneously using both Equations (11.2) and (11.3) with $\theta = \hat{\theta}$. The theory in Jiang et al. [25] shows that the pseudo–self-consistent estimator for S_{T_1} exists and is robust to model misspecification in the lower wedge. Uniform consistency and weak convergence of the pseudo–self-consistent estimator are also established.

A conditional version of the gamma frailty model was considered by Jiang et al. [24] in the presence of left truncation of the terminating event. In the Denmark diabetes registry described in Section 11.2, the observations of time to nephropathy (T_1) and time to death (T_2) on each enrolled patient was left-truncated at the time of first contact at the Steno (L). The marginal distribution of T_1 may be used to evaluate the net effect of care on nephropathy, independent of its effects on other aspects of diabetes that lead to dependent censoring via mortality prior to nephropathy. With truncation, the unconditional distribution may not be identifiable, but one can typically estimate the conditional survival function given the event time greater than a certain timepoint a, where a satisfies $\Pr(L \leq a) > 0$. Define $H_a(x,y) = \Pr(T_1 > x, T_2 > y \mid T_2 > a), S_{T_1,a}(x) = \Pr(T_1 > x \mid T_2 > a)$, and $S_{T_2,a}(x) = \Pr(T_2 > x \mid T_2 > a)$. For $0 \leq \max(a,x) < y \leq \infty$, a conditional version of Clayton copula assumes

$$H_a(x,y) = \{ S_{T_1,a}(x)^{1-\theta} + S_{T_2,a}(y)^{1-\theta} - 1 \}^{1/(1-\theta)}. \qquad (11.4)$$

Note that the model (11.4) is satisfied if the gamma frailty model holds on the upper wedge. The conditional Clayton model preserves the nice interpretation of the association parameter θ as the predicative hazard ratio in the observable region. Under the model (11.4), θ can be estimated by using $\hat{\theta}$ with O_{ij} in place of D_{ij}, where $O_{ij} = I_{\{\max(\tilde{L}_{ij}, a, \tilde{X}_{ij}) < \tilde{Y}_{ij} < \tilde{C}_{ij}\}}$ and $\tilde{L}_{ij} = \max(L_i, L_j)$. The validity of this new estimator is ensured by the fact that the concordance probability is not affected by conditioning on event times larger than a. The closed-form estimator of $S_{T_1,a}$ can be constructed as in the semi–competing risks setting without truncation.

11.4 SEMIPARAMETRIC REGRESSION METHOD

In clinical studies involving semi–competing risks, covariate effects on the occurrence of the nonterminating event are often of scientific interest. For example, in the ACTG 364 study, the primary interest is in evaluating the effects of treatment and other baseline characteristics on the marginal distribution of virologic failure. The literature on regression analysis tailored to semi–competing risks data is rather limited, with such data typically analyzed as competing risks data. Heckman and Honore [20] established the identifiability of bivariate proportional hazards and accelerated lifetime models with competing risks data; however, the complexity of maximum-likelihood estimation has hindered their practical development in the competing risks setup. Fully utilizing the semi−competing risks data leads to more practical methodology.

Let Z denote a $p \times 1$ covariate vector of interest, and let $\{Z_i, i = 1, \ldots, n\}$ be n i.i.d. replicates of Z. In this section, two types of regression model are introduced and inference procedures with semi−competing risks data are outlined.

11.4.1 Functional Regression Modeling

In the semi−competing risks setting, $\{Y_i, \delta_i, Z_i\}_{i=1}^n$ can be viewed as independently right-censored data so that one can easily adopt standard censored regression models for T_2. The real challenge in analyzing semi−competing risks data with covariates is the inference on T_1, which needs to properly account for dependent censoring by T_2.

To formulate covariate effects on T_1, it is tempting to employ the popular proportional hazards model, that is

$$\lambda(t \mid Z) = \lambda_0(t) \exp(\beta_0^T Z),$$

where $\lambda(t \mid Z)$ denotes the hazard function of T_1 conditional on Z, $\lambda_0(t)$ is an unspecified baseline hazard function, and β_0 is a $p \times 1$ coefficient vector. In practice, restricting the hazard functions associated with two sets of covariates to be proportional over time may be unrealistic. Motivated by the fact that the proportional hazards model can be equivalently represented as

$$\Pr(T_1 > t \mid Z) = S(t \mid Z) = \exp[-\exp\{\log \Lambda_0(t) + \beta_0^T Z\}],$$

where $\Lambda_0(t) = \int_0^t \lambda_0(s)ds$, Peng and Fine [36] proposed accommodating time-varying covariate effects on the survival function of T_1 via a generalized functional linear model

$$\Pr(T_1 > t \mid Z) = g\{\theta_0(t)^T \tilde{Z}\}, \tag{11.5}$$

where $g(\cdot)$ is a known monotone function, $\tilde{Z} = (1, Z^T)^T$, and $\theta_0(t)$ is a $(p + 1) \times 1$ vector of unknown time-dependent coefficients. The parameter $\theta_0(t)$ is completely unspecified in t but is assumed to be "cadlag", that is, a right-continuous function with left-hand limits. This model defines a rich family of varying-coefficient regression models. Choosing $g = \exp\{-\exp(\cdot)\}$ and $g = \exp/(1 + \exp)$, the model (11.5) accommodates respectively the standard proportional hazards model and the proportional odds model. The survival-based functional regression modeling facilitates estimation without involving smoothing. It also renders straightforward interpretations of the time-varying parameter $\theta_0(t)$ via the generalized linear model representation, namely, $g^{-1}\{S(t \mid Z)\} = \theta_0(t)^T Z$. For example, with $g = \exp/(1 + \exp)$, the components of $\theta_0(t)$ are log odds ratios of surviving beyond t per unit change in the corresponding covariates.

With semi−competing risks, estimation of θ_0 requires a model for the dependence structure of (T_1, T_2), since T_2 may dependently censor T_1. Peng and Fine [36] proposed linking the joint distribution of (T_1, T_2) to its marginal distributions through a known time-independent copula function $C(u, v, w)$, where for fixed w, C satisfies the definition of a copula. It is assumed that in the observable region of the data

$$\Pr(T_1 > s, T_2 > t \mid Z) = C\{\Pr(T_1 > s \mid Z), \Pr(T_2 > t \mid Z), \alpha_0(s, t)\}, \quad \text{for}$$
$$0 \le s \le t,$$

where $\alpha_0(s, t)$ is an unknown time-varying parameter, which is also cadlag like $\theta_0(t)$. The model (11.6) generalizes the class of parametric copula models [9,21,33,19]. As an example, when $C(u, v, w) = [u^{1-w} + v^{1-w} - 1]^{1/(1-w)}$ and $\alpha_0(s, t) = \alpha^*$ for $0 \leq s \leq t$, the model (11.6) reduces to the gamma frailty copula restricted to the upper wedge. In general, $\alpha_0(s, t)$ can be interpreted as the standard odds ratio based on the binary random variables $I(T_1 > s)$ and $I(T_2 > t)$. This odds ratio is widely reported in biomedical studies for assessing association between two binary variables. Depending on the parameterization, larger values of $\alpha_0(s, t)$ generally correspond to either increasing positive or negative association defined by $\Pr(T_1 > s, T_2 > t)/\Pr(T_1 > s)\Pr(T_2 > t) > 1$ or < 1, respectively [32]. Unlike parameterizations based on hazard association measures (e.g., Ref. 42), the copula parameterization in (11.6) yields an explicit form for the joint distribution.

Since T_2 is subject to censoring only by C, the regression model for T_2 can be chosen among existing models for standard independently right-censored data. To simplify the developments, the model for $\Pr(T_2 > t \mid Z)$ is assumed to take the form

$$\Pr(T_2 > t \mid Z) = h\{\eta_0(t)^T \tilde{Z}\}, \tag{11.7}$$

where h is a known link function and $\eta_0(t)$ is estimable with existing methods. The estimator of $\eta_0(t)$ is denoted by $\hat{\eta}_0(t)$.

Under models (11.5)–(11.7), the covariate effects on T_1 and the dependence parameter can be estimated simultaneously on the basis of a set of nonlinear estimating equations, which adopts a "working independence" assumption across time [27]. Let $\alpha(t) = \alpha(t, t)$. The estimator $\{\hat{\alpha}(t), \hat{\theta}(t)\}$ is obtained as the solution of $U\{\alpha(t), \theta(t), \hat{\eta}(t), t\} = n^{-1/2} \sum_{i=1}^{n} A_i\{\alpha(t), \theta(t), \hat{\eta}(t), t\} = 0$, where $A_i\{\alpha(t), \theta(t), \eta(t), t\}$ equals $V_i\{\alpha(t), \theta(t), t\} D_i\{\alpha(t), \theta(t), \eta(t)\}$ $[I(X_i > t) - I(Y_i > t), \Psi\{\alpha(t), \theta(t)^T \tilde{Z}_i, \eta(t)^T \tilde{Z}_i\}]$, $D_i\{\alpha(t), \theta(t), \eta(t)\} = \partial \Psi\{\alpha(t), \theta(t)^T \tilde{Z}_i, \eta(t)^T \tilde{Z}_i\}/\partial \binom{\alpha(t)}{\theta(t)}$ and V_i is a scalar weight function, $i = 1, \ldots, n$. One can show that $\hat{\alpha}(t)$ and $\hat{\theta}(t)$ are step functions that jump only at observed failure and censoring times. The estimating equation needs to be solved only at finitely many timepoints.

Under certain regularity conditions including restrictions on $\hat{\eta}(t)$, as n approaches infinity, there exists a unique solution to $U\{\alpha(t), \theta(t), \hat{\eta}(t), t\} = 0$ in a neigborhood of (α_0, θ_0) that converges to $\binom{\alpha_0(t)}{\theta_0(t)}$ in probability, uniformly in $t \in [l, u]$. It is further shown that $n^{1/2}[\{\hat{\alpha}(t)^T, \hat{\theta}(t)^T\}^T - \{\alpha_0(t)^T, \theta_0(t)^T\}^T]$ converges weakly to a tight Gaussian process. The conditions on $\hat{\eta}(t)$ for validity of $\hat{\alpha}(t)$ and $\hat{\theta}(t)$ are verified under proportional hazards models.

On the basis of the desirable properties of $\{\hat{\alpha}(t), \hat{\theta}(t)\}$, nonparametric tests are developed for the null hypothesis $H_0 : C(t)\binom{\alpha_0(t)}{\theta_0(t)} = c(t)$, where $C(t)$ is a $r \times (p + 2)$ matrix and $c(t)$ is a $r \times 1$ vector. To explore the parametric forms of covariate effects and association parameters, parametric submodels for (α_0, θ_0) are also considered: $L(t)^T \{\alpha_0(t)^T, \theta_0(t)^T\}^T = q(\zeta_0, t)$, where $L(t)$ is a $(p + 2) \times 1$ vector, q is a known function, and ζ_0 is finite-dimensional parameter. The estimator of ζ_0, $\hat{\zeta}$, is defined as the minimizer of the least-squares criterion $\int_l^u \{L(t)^T \binom{\hat{\alpha}(t)}{\hat{\theta}(t)} - q(\zeta, t)\}^2 \tilde{\Xi}(t) dt$, where $\tilde{\Xi}$ is a nonnegative weight function. Under mild assumptions, $\hat{\zeta}$ is consistent and asymptotically normal. Goodness-of-fit tests for the assumed submodel are developed accordingly.

The functional regression method was applied to ACTG364 data with $h = g = \exp(-\exp)$ and $C(u, v, w) = \{u^{1-\exp(w)} + v^{1-\exp(w)}\}^{-1/(1-\exp(w))}$ [36]. Four covariates are considered: Z_1 and Z_2 are indicator variables for treatment arms EFV and NFV + EFV, respectively; Z_3 indicates whether the patient received 3TC as a new NRTI in the ACTG364 study and 0 otherwise; and Z_4 is

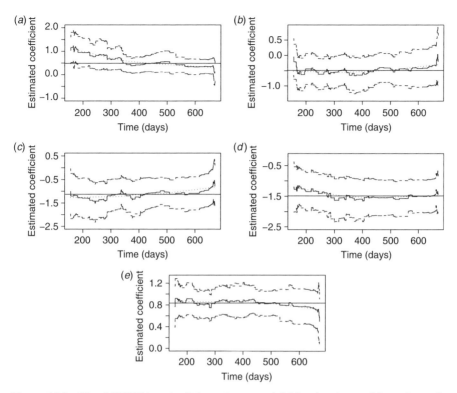

Figure 11.3 The ACTG364 study. Point estimates and 0.95 pointwise confidence intervals for time-varying copula parameter and covariate effects. The ragged solid lines represent the point estimates; the dashed lines, the 0.95 pointwise confidence intervals; the dotted lines, a lowess smoothing curve from the estimated parameters; and the horizontal solid lines, the fitted covariate effects from constant models. (a) Copula parameter; (b) EFV versus NFV; (c) EFV+NFV versus NFV; (d) New3TC; (e) $\log_{10}$ baseline RNA.

equal to $\log_{10}$ baseline RNA. Figure 11.3 displays estimates for time-varying copula parameter and covariate effects with 95% pointwise confidence intervals. It is observed that $\hat{\alpha}(t)$ decreases linearly to one year and then plateaus at 0.4 at later t. The goodness-of-fit test rejects the constant-dependence model and suggests that the association between virologic failure and dropout on the observable wedge is strong at early timepoints but noticeably diminished at later timepoints. It is suggested that time-independent coefficients may be adequate. Hypothesis testing on whether covariate effects differ from zero shows that the combination therapy may be superior to NFV, whereas EFV is the better of the two single-line treatments. Patients taking 3TC as a new NRTI have better prognoses then do patients with lower baseline RNA levels.

11.4.2 A Bivariate Accelerated Lifetime Model

Without loss of generality, suppose that T_1, T_2, and C are logarithm transformations of raw event times. The bivariate AFT model for (T_1, T_2) assumes that

$$T_1 = \theta_0^T Z + \epsilon^x, \quad T_2 = \eta_0^T Z + \epsilon^y, \tag{11.8}$$

where θ_0 and η_0 are $p \times 1$ vector of regression coefficients and $\epsilon = (\epsilon^x, \epsilon^y)^T$ has an unspecified joint survival function H not depending on Z. The inference on $\beta_0 = (\theta_0^T, \eta_0^T)^T$ discussed later requires only that the model hold in the upper wedge where $T_1 < T_2$. Formally, for $x < y$, $\Pr(X > x, Y > y \mid Z) = H(x - \theta_0^T Z, y - \eta_0^T Z)$.

Interestingly, it is more convenient to adjust for dependent censoring of T_1 by T_2 under accelerated failure-time assumptions. Lin et al. [28] first estimated the AFT model for T_2 using the standard rank procedures and then used an artificial censoring technique in rank estimation of the model for T_1. The estimating function for η_0 is a weighted log-rank test $S_n(\eta)$ based on $\tilde{\epsilon}_i^y(\eta) = Y_i - \eta^T Z_i$. An estimator $\hat{\eta}$ is obtained as the solution of $S_n(\eta) = 0$ [45,48]. Simply substituting $\{\tilde{\epsilon}_i^y(\eta), \xi_i\}$ with $\{\tilde{\epsilon}_i^x = X_i - \theta^T Z_i, \delta_i\}$ in S_n does not yield an unbiased estimator for θ_0 since the cause-specific hazard function may deviate from the net hazard function when ϵ_i^x and ϵ_i^y are correlated. Let $\beta = (\theta^T, \eta^T)^T$. The artificial censoring technique employed by Lin et al. [28] is to trim Y_i by a quantity $d(\beta)$ so that $(X_i - \theta^T Z_i) \wedge \{Y_i - \eta^T Z_i - d(\beta)\}$ can be viewed as censored analogs of ϵ_i^x and share a common distribution at $\beta = \beta_0$. A rank estimator $\tilde{\theta}$ is obtained by solving $\tilde{S}_n\{\binom{\theta}{\eta}\} = 0$, where $\tilde{S}_n(\beta)$ is a log-rank statistic constructed based on these residual analogs. The estimator $\tilde{\theta}$ has nice asymptotic properties such as consistency and asymptotic normality. However, when Z has several components or the components have wide ranges, $d(\beta)$ may be large and lead to excessive artificial censoring. Substantial rank information may be lost, and $\tilde{\theta}$ may be very inefficient.

The M-estimators studied by Robins [39] also permit dependent censoring and thus are valid under the model (11.8). An semiparametric efficient estimator was derived when X and Y are conditionally independent. Robins argued that this estimator is "nearly efficient" under heavy dependent censoring. However, nonparametric estimation of certain hazard functions is needed. The efficiency properties of the nearly efficient estimator are unclear with light/moderate dependent censoring.

Peng and Fine [35] developed a new artificial censoring technique using pairwise ranking. This approach avoids excessive artificial censoring by trimming separately within pairs of observations, $\{(X_i, Y_i, \delta_i, \xi_i, Z_i), (X_j, Y_j, \delta_j, \xi_j, Z_j)\}$. The data transformation within the (i, j) pair is $\{\tilde{X}_{i(j)}(\beta), \tilde{\delta}_{i(j)}(\beta); \tilde{X}_{j(i)}(\beta), \tilde{\delta}_{j(i)}(\beta)\}$, where

$$\tilde{X}_{i(j)}(\beta) = (T_{1i} - \theta^T Z_i) \wedge \{T_{2i} - \eta^T Z_i - d_{ij}(\beta)\} \wedge \{C_i - \eta^T Z_i - d_{ij}(\beta)\},$$

$$\tilde{\delta}_{i(j)}(\beta) = I_{(T_{1i} - \theta^T Z_i) \leq \{T_{2i} - \eta^T Z_i - d_{ij}(\beta)\} \wedge \{C_i - \eta^T Z_i - d_{ij}(\beta)\}},$$

and $d_{ij}(\beta) = \max\{0, (\theta - \eta)^T Z_i, (\theta - \eta)^T Z_j\}$. The choice of $d_{ij}(\beta)$ ensures that both $\tilde{X}_{i(j)}(\beta)$ and $\tilde{\delta}_{i(j)}(\beta)$ are determinable. Unlike the parameter in $\tilde{S}_n(\beta)$, the artificial censoring parameter is not fixed and a different value is determined for each pair using the covariate vectors Z_i and Z_j. Because $d_{ij}(\beta)$ is always $\leq d(\beta)$, large reductions in artificial censoring can be expected relative to that of Lin et al. [28].

Define $\psi_{ij}(\beta) = \tilde{\delta}_{i(j)}(\beta) I_{\{\tilde{X}_{i(j)}(\beta) \leq \tilde{X}_{j(i)}(\beta)\}} - \tilde{\delta}_{j(i)}(\beta) I_{\{\tilde{X}_{j(i)}(\beta) \leq \tilde{X}_{i(j)}(\beta)\}}$. Because $(T_{1i} - \theta_0^T Z_i, T_{2i} - \eta_0^T Z_i)$ and $(T_{1j} - \theta_0^T Z_j, T_{2j} - \eta_0^T Z_j)$ are independent and with common H on the upper wedge, $(Z_i - Z_j)\psi_{ij}(\beta)$ also has zero mean at the true value $\beta_0 = (\theta_0^T, \eta_0^T)^T$. It is observed that $(Z_i - Z_j)\psi_{ij}(\beta)$ is symmetric in i and j. This suggests a U-statistic-based estimating function

$$U_n(\beta) = 2\sqrt{n} \sum_{1 \leq i < j \leq n} (Z_i - Z_j)\psi_{ij}(\beta) / \{n(n-1)\}.$$

It can be shown that $E\{U_n(\beta_0)\} = 0$ by using the exchangeability of the observations. A reasonable estimator for θ_0, denoted by $\hat{\theta}_{PW}$, can be obtained by solving $U_n\{(\theta^T, \hat{\eta}^T)^T\} = 0$. Since U_n is discontinuous in θ, similar to S_n, an exact solution may not exist in practice. The estimator $\hat{\theta}_{PW}$ can alternatively be defined as $\operatorname{argmin}_\theta \|U_n\{ \begin{smallmatrix} \theta \\ \hat{\eta} \end{smallmatrix} \}\|$. The pairwise estimator $\hat{\beta} = (\hat{\theta}_{PW}^T, \hat{\eta}^T)^T$ is shown to be consistent and asymptotically normal under mild regularity conditions.

Simulation studies reported by Peng and Fine [35] showed that with realistic sample sizes, the pairwise estimator $\hat{\theta}_{PW}$ may achieve large reductions in artificial censoring (58–65%) and large efficiency gains over $\tilde{\theta}^T$. When T_1 and T_2 are conditionally independent, $\hat{\theta}_{PW}$ is slightly less efficient than Tsiatis' efficient estimator [45] using the known hazard of the residual for T_1, while the Tsiatis estimator may be somewhat less efficient than the pairwise estimator (15–20%) when the hazard is estimated, as is needed in practice. This may suggest that the pairwise estimator has better small-sample performance than does Robins' estimator [39]. When T_1 and T_2 are dependent with light to moderate censoring, there are large reductions in MSE with the pairwise approach, with threefold to fivefold improvements over the approach of Lin et al. [28] seen in some scenarios.

11.5 CONCLUDING REMARKS

Semi–competing risks data have received increased attention as distinct from classic competing risks data. Such data are frequently encountered in biomedical studies. In this chapter, we review major inferential techniques for semi–competing risks data, including nonparametric estimation, semiparametric one-sample inference, and semiparametric regression models.

In practice, analysis of semi–competing risks data should be planned carefully in order to better address scientific questions of interest and yield interpretable results. Analysis oriented to cause-specific hazard and cumulative incidence function may be more sensible when the elimination of the terminating event prior to the nonterminating event is not possible or is expected to alter the mechanism of the nonterminating event. Inference based on the marginal distribution of the nonterminating event is controversial under some circumstances but may have meaningful practical implications in settings such as the ACTG364 study. Our opinion takes the middle ground that the estimated marginal distribution of the nonterminating event is sometimes useful.

Future research on semi–competing risks data may be directed to the multivariate case rather than current bivariate structure. It corresponds to a more complicated but rather realistic situation involving multiple terminating and nonterminating events. The foremost issue in this complex setting is to determine which quantities are identifiable. Investigation of this identifiability problem may require delicate theoretical developments. Regarding the inference, multistate modeling [3] seems to be a natural approach attached to the multiendpoint scenario. The key issue with this approach is the translation of multistate transition intensities into relevant practical context. When an appropriate mapping between the estimable transition intensities and the quantities of interest is lacking, different joint modeling of terminating and nonterminating endpoints is warranted and merits future research.

REFERENCES

1. Aalen O. O. and Johansen S., An empirical transition matrix for non-homogeneous Markov chains based on censored observations, *Scand. J. Statist.* **5**, 141 (1978).

2. Albrecht M. A., Bosch R. J., Katzenstein D. A. et al., Nelfinavir, efavirenz, or both after the failure of nucleoside treatment of HIV infection, *New Engl. J. Med.* **345**, 398 (2001).

3. Andersen P. K., Borgan Ø., Gill R. D., and Keiding N., *Statistical Models Based on Counting Processes*, Springer-Verlag, New York, (1993).

4. Bentzen S. M., Vaeth M., Pederson D. E., and Overgaard J., Why actuarial estimates should be used in reporting late normal-tissue effects of cancer treatment . . . Now!, *Int. J. Radiat. Oncol. Biol. Phys.* **32**, 1531 (1995).

5. Caplan R. J., Pajak T. F., and Cox J. D., In response to Bentzen et al. IJROBP 32: 1531–1534, *Int. J. Radiat. Oncol. Biol. Phys.* **32**, 1547 (1995).

6. Caplan R. J., Pajak T. F., and Cox J. D., In response to Dr. Denham et al., *Int. J. Radiat. Oncol. Biol. Phys.* **35**, 1547 (1996).

7. Chappell R. J., Comment on the articles of Caplan et al., *Int. J. Radiat. Oncol. Biol. Phys.* **36**, 988 (1996).

8. Clayton D. G., A model for association in bivariate life tables and its application in epidemiological studies of familial tendency in chronic disease incidence, *Biometrika* **65**, 141 (1978).

9. Clayton D. G. and Cuzick J., Multivariate generalizations of the proportional hazards model (with discussion), *J. Roy. Statist. Soc. A* **148**, 82 (1985).

10. Copelan E. A., Biggs J. C., Tutschka P. J. et al., Treatment for acute myelocytic leukemia with allogeneic bone marrow transplantation following preparation with Bu/Cy2, *Blood* **78**, 838 (1991).

11. Day R., Bryant J., and Lefkopoulou M., Adaptation of bivariate frailty models for prediction, with application to biological markers as prognostic indicators, *Biometrika* **84**, 45 (1997).

12. Denham J. W., Hamilton C. S., and O'Brien P., Regarding actuarial late effect analyses: Bentzen et al., IJROBP 32: 1531–1534; 1995 and Caplan et al., IJROBP 32: 1547; 1995, *Int. J. Radiat. Oncol. Biol. Phys.* **35**, 197 (1996).

13. Efron B., The two sample problem with censored data, *Proc. 5th Berkeley Symp. Mathematical Statistics and Probability*, Vol. **4**, p. 831, 1967.

14. Emoto S. E. and Matthews P. C., A weibull model for dependent censoring, *Ann. Statist.* **18**, 1556 (1990).

15. Fine J. P., Jiang H., and Chappell R. J., On semi-competing risks data, *Biometrika* **88**, 907 (2001).

16. Fine J. P., Yan J., and Kosorok M. R., Temporal process regression, *Biometrika* **91**, 683 (2004).

17. Flemming T. R. and Harrington D. P., *Counting Processes and Survival Analysis*, Wiley, New York, 1991.

18. Frydman H., Nonparametric estimation of a Markov "illness-death" process from interval-censored observations, with application to diabetes survival data, *Biometrika* **82**, 773 (1995).

19. Genest C. and Mackay J., The joy of copulas: Bivariate distributions with uniform marginals, *Am. Statist.* **40**, 280 (1986).

20. Heckman J. J. and Honore B. E., The identifiability of the competing risks model, *Biometrika* **24**, 135 (1989).

21. Hougaard P., Modelling multivariate survival, *Scand. J. Statist.*, **14**, 291 (1987).

22. Huang Y. and Wang M. C., Estimating the occurrence rate for prevalent survival data in competing risks models, *J. Am. Statist. Assoc.* **90**, 1406 (1995).

23. Jiang H., Chappell R. J., and Fine J. P., Estimating the distribution of nonterminal event time in the presence of mortality or informative dropout, *Control. Clin. Trials* **24**, 135 (2003).

24. Jiang H., Fine J. P., and Chappell R. J., Semiparametric analysis of survival data with left truncation and dependent right censoring, *Biometrics* **61**, 567 (2005).

25. Jiang H., Fine J. P., Kosorok M. R., and Chappell R. J., Pseudo self-consistent estimation of a copula model with informative censoring, *Scand. J. Statist.* **32**, 1 (2005).

26. Klein J. P. and Moeschberger M. L., Bounds on net survival probabilities for dependent competing risks, *Biometrics* **44**, 529 (1988).

27. Liang K. Y. and Zeger S. L., Longitudinal data analysis using generalized linear models, *Biometrika* **73**, 13 (1996).

28. Lin D. Y., Robins J. M., and Wei L. J., Comparing two failure time distributions in the presence of dependent censoring, *Biometrika*, **83**, 381 (1996).

29. Link W. A., A model for informative censoring, *J. Am. Statist. Assoc.* **84**, 749 (1989).

30. Lynden-Bell D., A method of allowing for known observational selection in small samples applied to 3CR quasars, *Monthly Notices Roy. Astron. Soc.* **155**, 95 (1971).

31. Mykland P. A. and Ren J. J., Algorithms for computing self-consistent and maximum likelihood estimators with doubly censored data, *Ann. Statist.* **24**, 1740 (1996).

32. Nelsen R. B., *An Introduction to Copulas*, Springer, New York, 1999.

33. Oakes D., Bivariate survival models induced by frailties, *J. Am. Statist. Assoc.* **84**, 487 (1989).

34. Peng L. and Fine J. P., Nonparametric estimation with left truncated semi-competing risks data, *Biometrika* **93**, 367 (2006).

35. Peng L. and Fine J. P., Rank estimation of accelerated lifetime models with dependent censoring, *J. Am. Statist. Assoc.* **101**, 1085 (2006).

36. Peng L. and Fine J. P., Regression modeling of semi-competing risks data, *Biometrics* (in press).

37. Peterson A. V., Bounds for a joint distribution with subdistribution functions: application to competing risks, *Proc. Nat. Acad. Sci. USA* **73**, 11 (1976).

38. Prentice R. L., Kalbfleisch J. D., Peterson Jr. A. V., Flournoy N., Farewell V. T., Breslow N. E., The analysis of failure times in the presence of competing risks, *Biometrics* **34**, 541 (1978).

39. Robins J. M., An analytic method for randomized trials with informative censoring: Part II, *Lifetime Data Anal.* **1**, 417 (1995).

40. Scharfstein D. O. and Robins J. M., Estimation of the failure time distribution in the presence of informative censoring, *Biometrika* **89**, 617 (2002).

41. Shih J. H. and Louis T. A., Inferences on the association parameter in copula models for bivariate survival data, *Biometrics* **51**, 1384 (1995).

42. Slud E. V. and Rubinstein L. V., Dependent competing risks and summary survival curves, *Biometrika* **70**, 643 (1983).

43. Sun J., Self-consistency estimation of distributions based on truncated and doubly censored survival data with applications to AIDS cohort studies, *Lifetime Data Anal.* **3**, 305 (1997).

44. Tsiatis A. A., A nonidentifiability aspect of the problem of competing risks, *Proc. Nat. Acad. Sci.* **72**, 20 (1975).

45. Tsiatis A. A., Estimating regression parameters using linear rank tests for censored data, *Ann. Statist.* **18**, 354 (1990).

46. Turnbull B. W., Nonparametric estimation of a survivorship function with doubly censored data, *J. Am. Statist. Assoc.* **69**, 169 (1974).

47. Wang W., Estimating the association parameter for copula models under dependent censoring, *J. Roy. Statist. Soc. B* **65**, 257 (2003).

48. Wei L. J., Ying Z., and Lin D. Y., Linear regression analysis of censored survival data based on rank tests, *Biometrika* **77**, 845 (1990).

49. Zheng M. and Klein J. P., Estimates of marginal survival for dependent competing risks based on an assumed copula, *Biometrika* **82**, 127 (1995).

CHAPTER 12

Tests for Time-Varying Covariate Effects within Aalen's Additive Hazards Model

Torben Martinussen

Department of Natural Sciences, The Royal Veterinary and Agricultural University, Frederiksberg, Denmark

Thomas H. Scheike

Department of Biostatistics, University of Copenhagen, Copenhagen, Denmark

12.1 INTRODUCTION

The Aalen additive hazards model [4] is a useful alternative to the Cox model when analyzing survival data. A particularly useful aspect of the Aalen additive hazards model is that it allows for time-varying covariate effects. In many biomedical applications there will often be important time-varying effects. A typical example is a treatment effect that varies over time, and two important examples are that (1) treatment efficacy fades away over time, due to, for example, tolerance developed by the patient, or in the case of infectious diseases due to resistance developed by the targeted bacteria; and (2) a treatment with many side effects will lead to an initial adverse effect that is compensated by a beneficial effect for those surviving the initial phase. Below we analyze data on time to death for acute myocardial infarction (AMI) heart patients and for these data some risk predictors show highly time-varying effects. The strongest time-varying effect was found for patients with ventricular fibrillation that lead to an increased risk only for the first 30 days after AMI. In this period, however, the risk was strongly increased and the excess risk was at a level of ~ 5 on the intensity in years.

Statistical Advances in the Biomedical Sciences, edited by Atanu Biswas, Sujay Datta, Jason P. Fine, and Mark R. Segal

Even for a relatively small dataset it is possible to say something about such time-varying effects, and inferential procedures that account for these effects have been developed [8] on the basis of the semiparametric version of the Aalen model [9]; where the intensity $\alpha(t)$ has the specific form

$$\alpha(t) = X^T \beta(t) + Z^T \gamma, \tag{12.1}$$

where X and Z are p-dimensional and q-dimensional covariate vectors, respectively. The time-varying regression function $\beta(t)$ is a vector of locally integrable functions, and γ is a q vector of unknown parameters. Apart from its use for testing time-varying effects, this model is useful in its own right because it is then possible to make a sensible bias-variance tradeoff, where effects that are almost constant can be summarized as such and effects that are strongly time-varying can be described as such. Lin and Ying [7] considered a special case of (12.1), where only the intercept term is allowed to depend on time.

In this chapter we develop and study two types of test of time-varying effects within Aalen's additive model. The first test uses only information from one model fit of the semiparametric additive model and is simple to calculate and implement. The second test applies information from two semiparametric additive model fits and is more difficult to calculate. For both test statistics we give an asymptotically i.i.d. representation, which is used to approximate relevant limit distributions by applying a certain resampling technique [6].

In the next section, we describe the inferential procedures and give the i.i.d. representations of the test processes. In Section 12.3, we report the results of a simulation study and provide an illustration with real data from a study concerning myocardial infarction. Some remarks follow in Section 12.4 and some technical details are given in Appendix 12A (at the end of this chapter).

12.2 MODEL SPECIFICATION AND INFERENTIAL PROCEDURES

Let $\tilde{T}$ be the survival time of interest with conditional hazard function $\alpha(t; X, Z)$ given the covariate vectors X and Z. In practice $\tilde{T}$ may be right-censored by U so that we observe $((T = \tilde{T} \wedge U, \Delta = I(\tilde{T} \leq U), X, Z))$. Let $(T_i, \Delta_i, X_i, Z_i)$ be n i.i.d. replicates so that the ith counting process $N_i(t) = I(T_i \leq t, \Delta_i = 1)$ has intensity

$$\lambda_i(t) = Y_i(t)\left[X_i^T \beta(t) + Z_i^T \gamma\right],$$

where $Y_i(t) = I(t \leq T_i)$ is the at-risk indicator. Usually, the X_i will have 1 as its first component allowing for an intercept in the model. The intensity $\lambda_i(t)$ models the risk of a jump in the ith counting process $N_i(t)$ at time t. We assume that all counting processes are observed in the time interval $[0, \tau]$, where $\tau < \infty$. Each counting process has compensator $\Lambda_i(t) = \int_0^t \lambda_i(s)ds$ such that $M_i(t) = N_i(t) - \Lambda_i(t)$ is a martingale. Define the n-dimensional counting process $N = (N_1, \ldots, N_n)^T$ and the n-dimensional martingale $M = (M_1, \ldots, M_n)^T$. Let also $X = (Y_1 X_1, \ldots, Y_n X_n)^T$, $Z = (Y_1 Z_1, \ldots, Y_n Z_n)^T$. We assume the set of conditions listed in Appendix 12A.

The (unweighted) estimators of $\{B(t) = \int_0^t \beta(s)\, ds, \gamma\}$ [9] are

$$\hat{\gamma} = \left(\int_0^\tau Z^T H Z\, dt\right)^{-1} \int_0^\tau Z^T H dN(t),$$

$$\hat{B}(t) = \int_0^t X^- dN(t) - \int_0^t X^- Z\, dt\hat{\gamma},$$

where X^- denotes the generalized inverse $(X^T X)^{-1} X^T$ and $H = I - XX^-$. We assume that the required inverses exist. It is quite obvious that these simple and direct estimators will work. First, considering $\hat{\gamma}$, the counting process integral can be written as

$$\int_0^\tau Z^T H \, dN(t) = \int_0^\tau Z^T H \, d\{X \, dB(t) + Z\gamma \, dt\} + \int_0^\tau Z^T H \, dM(t)$$

$$= \int_0^\tau Z^T HZ \, dt\gamma + \int_0^\tau Z^T H \, dM(t)$$

since $HX = 0$, and therfore $\hat{\gamma}$ is an essentially unbiased estimator of γ. Similarly, if we consider $\hat{B}(t)$, we find that the first term can be written as

$$\int_0^t X^- \, dN(t) = \int_0^t X^- d\{X \, dB(t) + Z\gamma \, dt\} + \int_0^t X^- \, dM(t)$$

$$= B(t) + \int_0^t X^- Z \, dt\gamma + \int_0^t X^- \, dM(t)$$

since $X^- X = I_p$. The second term of this expression is estimated by the $\hat{\gamma}$ term of the estimator of $\hat{B}(t)$, and we therefore see that $\hat{B}(t)$ is a sensible estimator.

The limit distributions of the estimators are

$$n^{1/2}\{\hat{\gamma} - \gamma\} = C^{-1}(\tau) n^{-1/2} \int_0^\tau Z^T H \, dM(t),$$

$$n^{1/2}\{\hat{B}(t) - B(t)\} = n^{1/2} \int_0^\tau X^- \, dM(t) - P(t) n^{1/2}\{\hat{\gamma} - \gamma\},$$

where

$$C(t) = n^{-1} \int_0^t Z^T HZ \, ds, \quad P(t) = \int_0^t X^- Z \, dt.$$

It is useful to note that these limit distributions may be written as sums of essentially i.i.d. terms:

$$n^{1/2}\{\hat{\gamma} - \gamma\} = n^{-1/2} \sum_{i=1}^n \varepsilon_i^\gamma + o_P(1),$$

$$n^{1/2}\{\hat{B}(t) - B(t)\} = n^{-1/2} \sum_{i=1}^n \varepsilon_i^B(t) + o_P(1),$$

where

$$\varepsilon_i^\gamma = c^{-1}(\tau) \int_0^\tau \{Z_i - (z^T x)(x^T x)^{-1} X_i\} \, dM_i(t),$$

$$\varepsilon_i^B(t) = \int_0^\tau (x^T x)^{-1} X_i \, dM_i(t) - p(t)\varepsilon_i^\gamma,$$

where $c(t)$ and $p(t)$ denote the limits in probability of $C(t)$ and $P(t)$, respectively. Also, $x^T x$ is used as notation for the limit in probability of $n^{-1} X^T X$, and similarly with $z^T x$. The limit distributions may be simulated as described by Lin et al. [6] in a Cox model setting (see Appendix 12A for some details on this)

$$n^{1/2}\{\hat{\gamma} - \gamma\} \sim n^{-1/2} \sum_{i=1}^{n} \hat{\varepsilon}_i^{\gamma} G_i, \quad n^{1/2}\{\hat{B}(t) - B(t)\} \sim n^{-1/2} \sum_{i=1}^{n} \hat{\varepsilon}_i^{B}(t) G_i,$$

where $G_1, \ldots, G_n$ are independent standard normals and $\hat{\varepsilon}_i^{\gamma}$ is obtained from ε_i^{γ} by replacing deterministic quantities with their empirical counterparts and by replacing $M_i(t)$ with $\hat{M}_i(t)$, $i = 1, \ldots, n$, and similarly with $\hat{\varepsilon}_i^{B}(t)$. The result is that, conditional on the data

$$\left(n^{-1/2} \sum_{i=1}^{n} \hat{\varepsilon}_i^{\gamma} G_i, n^{-1/2} \sum_{i=1}^{n} \hat{\varepsilon}_i^{B}(t) G_i \right)$$

will have the same limit distribution as

$$\left(n^{1/2}\{\hat{\gamma} - \gamma\}, n^{1/2}\{\hat{B}(t) - B(t)\} \right).$$

We thus let $\sim$ indicate that two quantities have the same limit distribution.

We wish to develop inferential procedures for the following hypothesis

$$H_0 : \beta_p(t) = \beta_p,$$

focusing without loss of generality on the pth regression coefficient. The hypothesis may be reformulated in terms of the cumulative regression function $B_p(t) = \int_0^t \beta_p(s)\,ds$ as

$$H_0 : B_p(t) = \beta_p \cdot t.$$

Martinussen and Scheike [8] studied the test process

$$V_n(t) = n^{1/2}\left(\hat{B}_p(t) - \hat{B}_p(\tau)\frac{t}{\tau} \right),\qquad(12.2)$$

which is very easy to compute, and considered the test statistics

$$\sup_{t \leq \tau} |V_n(t)|.$$

It is clear that the test will be consistent, but it cannot be expected to be optimal against any alternative since the estimator itself is not efficient. One might also take the variance of $V_n(t)$ into account. Note that $V_n(\tau) = 0$ by construction. Under H_0, we have

$$V_n(t) = n^{1/2}\left\{ (\hat{B}_p(t) - B_p(t)) - (\hat{B}_p(\tau) - B_p(\tau))\frac{t}{\tau} \right\}.$$

Clearly, the limit distribution of $V_n(t)$ cannot be a martingale because of the term $\hat{B}_p(\tau)$, but one may use the resampling technique of [6] to approximate its limit distribution. Alternatively, one may also simulate the limit distribution by a more classical bootstrap, and it is not clear which of these approaches lead to the best approximation. We find, however, that the conditional multiplier approach tends to have quite good small-sample properties. The limit distribution of $V_n(t)$ may be approximated by

$$\hat{V}_n(t) = n^{-1/2} \sum_{i=1}^{n} \left[\{\hat{\varepsilon}_i^B(t)\}_p - \{\hat{\varepsilon}_i^B(\tau)\}_p \frac{t}{\tau} \right] G_i,$$

where v_k is the kth element of a given vector v and where we are fixing the data.

An alternative test process is

$$W_n(t) = n^{1/2}(\hat{B}_p(t) - \hat{\beta}_p \cdot t), \tag{12.3}$$

where $\hat{\beta}_p$ is the estimator of β_p under the null. Again, under H_0, we have

$$W_n(t) = n^{1/2}\left\{ (\hat{B}_p(t) - B_p(t)) - (\hat{\beta}_p - \beta_p) \cdot t \right\}.$$

Contrary to $V_n(t)$, we do not have $W_n(\tau) = 0$. To write down the simulation technique for this test process, we need some notation for the design and the parameters under the null. Let $\tilde{X}$ and $\tilde{Z}$ be the design matrices under the null, that is, $\tilde{X}$ and $\tilde{Z}$ have ith row $Y_i \, (X_{i1}, \ldots, X_{ip-1})$ and $Y_i \, (Z_{i1}, \ldots, Z_{iq}, X_{ip})$, respectively, and let $\tilde{\beta}(t) = (\beta_1(t), \ldots, \beta_{p-1}(t))^T$ and $\tilde{\gamma} = (\gamma_1, \ldots, \gamma_q, \beta_p)$. Then we have (ignoring lower-order terms)

$$W_n(t) = n^{-1/2} \sum_{i=1}^{n} \left[\{\varepsilon_i^B(t)\}_p - \{\varepsilon_i^{\tilde{\gamma}}\}_{q+1} \cdot t \right],$$

where $\varepsilon_i^{\tilde{\gamma}}$ are the i.i.d. terms corresponding to $n^{1/2}\{\hat{\tilde{\gamma}} - \tilde{\gamma}\}$ and computed under the null. The limit distribution of $W_n(t)$ may thus be simulated by generating samples from

$$\hat{W}_n(t) = n^{-1/2} \sum_{i=1}^{n} \left[\{\hat{\varepsilon}_i^B(t)\}_p - \{\hat{\varepsilon}_i^{\tilde{\gamma}}(t)\}_{q+1} \cdot t \right] G_i$$

while fixing the data.

12.2.1 A Pseudo–Score Test

In this section we consider a test statistic that is based directly on the underlying estimating equation, and we need some additional notation in this case. We start by partitioning X into X_1 and X_2 of dimensions $n \times (p-1)$ and $n \times 1$, respectively. Similarly, we write $B(t) = (B_1(t), B_2(t))$ with $B_1(t)$ a $(p-1)$ vector and $B_2(t)$ a scalar, and let $V(t) = (X_2, Z)$ and $\theta = (\beta_p, \gamma)$ and $H_1 = I - X_1 \, (X_1^T X_1)^{-1} X_1^T$. Note that this partitioning of X may be completely general, but for simplicity we let X_2 be one-dimensional.

First, under the model

$$\lambda_i(t) = Y_i(t)[X_i^T \beta(t) + Z_i^T \gamma],$$

the estimating function for $B(t) = \int_0^t \beta(s)\, ds$ is given by

$$U(t) = \int_0^t X^T (dN - X\, dB - Z\gamma\, dt) = \int_0^t X^T dM, \quad t \in [0, \tau],$$

thus giving the solution

$$dB = X^-(dN - Z\gamma\, dt),$$

as pointed out above.

Under the null hypothesis $H_0 : \beta_p(t) = \beta_p$, the model can then be written as

$$dN(t) = X_1\, dB_1(t) + V\theta\, dt + dM(t).$$

Then the estimating function for $B(t)$, computed under the null, becomes

$$\tilde{U}(t) = \int_0^t X^T (dN - X_1\, d\hat{B}_1 - V\hat{\theta}\, dt)$$

$$= \int_0^t X^T H_1 (dN - V\hat{\theta}\, dt).$$

Since $H_1 X_1 = 0$, it follows that $X^T H = (0, X_2)^T H_1$ and the first $(p-1)$ components of $\tilde{U}(t)$ is zero, we need to consider only the pth component of the estimating function, which we denote as $\tilde{U}_p(t)$. We obtain

$$\tilde{U}_p(t) = \int_0^t X_2^T H_1 (dN - V\hat{\theta}\, dt)$$

$$= \int_0^t X_2^T H_1\, dM - P_2(t)C_2 \int_0^t V^T H\, dM,$$

where

$$P_2(t) = \int_0^t X_2 H_1 V dt, \quad C_2 = \left(\int_0^\tau V^T H_1 V dt \right)^{-1}.$$

We suggest using the test statistic $\sup_{t \leq \tau} |\tilde{U}_p(t)|$. This estimating function or pseudo–score function, may then be resampled as in the previous section by making an i.i.d. decomposition of $\tilde{U}_p(t)$.

12.3 NUMERICAL RESULTS

12.3.1 Simulation Studies

To evaluate the finite sample performance of the proposed tests processes (12.2) and (12.3), we did a simulation study, investigating whether the correct nominal level is attained and whether the power of the tests based on the two test processes differs. In addition to the simple supremum tests

$$\sup_{t \leq \tau} |V_n(t)|, \quad \sup_{t \leq \tau} |W_n(t)|,$$

we also computed the integrated squared errors over the following time interval:

$$\int_0^\tau V_n^2(t)dt, \quad \int_0^\tau W_n^2(t)dt.$$

In Table 12.1 we denote these tests as sup V_n and $\int V_n^2$ and similarly with W_n. The simple test statistic based on V_n has the advantage of being insensitive to the increased variation at the end of the time interval. In order to remedy this problem for the $W_n(t)$ test process, we also computed a weighted version of the test process

$$\tilde{W}_n(t) = t \cdot (\tau - t) \cdot W_n(t),$$

which also has the property that it starts in 0 and ends in 0 in the same way as does $V_n(t)$.

Finally, we also computed the pseudo–score test

$$\sup_{t \leq \tau} |\tilde{U}_p(t)|,$$

which we denote as PS in the tables below. Note that the pseudo–score test also starts and ends in 0.

Table 12.1 **Performance of Two Estimators of γ for Different Censoring Times (3, 5 and 8)[a]**

n	Censor Time	Mean Censoring	$\hat{\gamma}$		$\tilde{\gamma}$	
			Mean	Emp SE	Mean	Emp SE
100	3	0.43	0.191	0.166	0.220	0.122
200	3	0.43	0.194	0.118	0.202	0.080
100	5	0.23	0.198	0.145	0.208	0.114
200	5	0.23	0.198	0.092	0.201	0.081
100	8	0.10	0.195	0.162	0.275	0.364
200	8	0.10	0.194	0.100	0.203	0.100

[a]Mean of estimates (mean) and empirical standard error of estimates (emp SE) for $\hat{\gamma}$ and $\tilde{\gamma}$ (see text).

We generated data from the additive hazards model:

$$\alpha_i(t) = 0.2 + 0.2 \cdot X_{i1} + (0.2 + \theta \cdot (t < 1)) \cdot X_{i2}.$$

The effect of X_2 is increased by θ in the first time unit (values of 0.2 and 0.4 of θ are considered). The covariates are drawn as independent standard uniform variables.

We generated samples of 100 and 200 and censored the data at three different points in time (3, 5, and 8).

The performance of the tests depend on how well behaved the cumulative regression coefficient is toward the end of the period. We start by considering one direct consequence of this. Recall that $\hat{\gamma}$ is the estimate of the constant effect under the semiparametric model, and denote similarly the estimator $\tilde{\gamma}_j = B_j(\tau)/\tau$, an estimator based on the additive Aalen model. This estimator is also discussed in Martinussen and Scheike [8], and it is more unstable because the matrix inverse (ZHZ) needs to be computed for all timepoints, in contrast to the estimator $\hat{\gamma}$, where only the matrix inverse of $\int ZHZ\,dt$ needs to be computed.

In Table 12.1 we compare the performance of these two estimators for estimating the effect of X_2 in the model where this effect is equal to 0.2.

Note that $\tilde{\gamma}$ is as good as and even slightly better than $\hat{\gamma}$ when we do not use the unstable part of the tail of the cumulative regression coefficient. When the estimate becomes unstable, as is the case for censoring time 8 and with only 100 observations, $\hat{\gamma}$ is clearly more precise than $\tilde{\gamma}$.

Table 12.2   Observed Power for 1000 Repetitions for Tests for Time-Varying Effects[a]

Test	n	X_1	X_2
sup V_n	100	0.07	0.09
int V_n	100	0.06	0.09
sup W_n	100	0.06	0.08
int W_n	100	0.07	0.08
sup $\tilde{W}_n$	100	0.07	0.09
int $\tilde{W}_n$	100	0.06	0.07
PS	100	0.05	0.09
sup V_n	200	0.05	0.12
int V_n	200	0.05	0.11
sup W_n	200	0.05	0.10
int W_n	200	0.06	0.09
sup $\tilde{W}_n$	200	0.06	0.11
int $\tilde{W}_n$	200	0.05	0.10
PS	200	0.05	0.10
sup V_n	400	0.05	0.12
int V_n	400	0.05	0.12
sup W_n	400	0.06	0.11
int W_n	400	0.05	0.13
sup $\tilde{W}_n$	400	0.06	0.12
int $\tilde{W}_n$	400	0.05	0.13
PS	400	0.05	0.16

[a]See text for details. Test computed for data censored at time 3. The nonlinear effect was given by 0.2.

We now carried out a test with significance level at 5% for constant effect of X_1 and X_2 for the different censoring times. In Tables 12.2 and 12.3 we show the results based on censoring time 3. Censoring time 5 led to somewhat similar results, but the level was a bit too high for the lower sample sizes. Censoring at time 8 led to tests where the level was severely skewed and much too high due to the instability toward the end of the time period for the tests to work, and this is reflected in Table 12.1. It is noteworthy, however, that the pseudo–score test was unaffected by the censoring times and in fact improved only when more of the data was used (see Table 12.4).

We first considered $\theta = 0.2$. The test for constant effect of X_1 attained the nominal 5% level for all tests. The nonconstant effect of covariate X_2 led to powers that were quite low, $\sim 8-9\%$ for all tests. When the sample size increased to 200 and 400, we found that the PS test had the best power improvement, and all other tests behaved similarly, with the simple V_n being competitive. Note, however, that the time-varying effect is quite minor and present only on the first part of the time interval, and the low powers simply reflect that the time-varying effect is really difficult to observe.

We then increased the time-varying effect by setting $\theta = 0.4$. This improved the performance of all tests, and we found that the simple V_n now showed a power that increased from 0.14 to 0.32 over the sample sizes. W_n showed an increased power compared to its weighted version. Again, PS showed the best improvement in power over the sample sizes.

The pseudo–score test is as good as any of the other tests and showed the best performance overall. The test also showed an ability to work even in the tail of the data where there is little

Table 12.3 Observed Power from 1000 Repetitions for Tests of Time-Varying Effects[a]

Test	n	X_1	X_2
sup V_n	100	0.07	0.14
int V_n	100	0.06	0.12
sup W_n	100	0.05	0.12
int W_n	100	0.05	0.12
sup $\tilde{W}_n$	100	0.06	0.14
int $\tilde{W}_n$	100	0.06	0.13
PS	100	0.05	0.13
sup V_n	200	0.07	0.22
int V_n	200	0.06	0.20
sup W_n	200	0.05	0.20
int W_n	200	0.06	0.22
sup $\tilde{W}_n$	200	0.07	0.20
int $\tilde{W}_n$	200	0.06	0.20
PS	200	0.05	0.23
sup V_n	400	0.06	0.32
int V_n	400	0.05	0.32
sup W_n	400	0.06	0.34
int W_n	400	0.05	0.38
sup $\tilde{W}_n$	400	0.06	0.33
int $\tilde{W}_n$	400	0.05	0.34
PS	400	0.05	0.40

[a]See text for details. Test computed for data censored at time 3. The nonlinear effect was given by 0.4.

Table 12.4   Observed Power for 1000 Repetitions for Pseudo–Score Test Time-Varying Effects^a

Test	n	θ	X_1	X_2
		Censoring at Time 5		
PS	100	0.2	0.06	0.09
PS	200	0.2	0.06	0.11
PS	400	0.2	0.06	0.17
PS	100	0.4	0.07	0.14
PS	200	0.4	0.05	0.26
PS	400	0.4	0.05	0.46
		Censoring at Time 8		
PS	100	0.2	0.07	0.08
PS	200	0.2	0.06	0.11
PS	400	0.2	0.05	0.18
PS	100	0.4	0.05	0.15
PS	200	0.4	0.05	0.25
PS	400	0.4	0.05	0.51

^{a}Power for censoring at times 5 and 8.

information. We illustrate this in Table 12.4, where we show the observed levels for censoring times 5 and 8.

We see that the PS test improves when more of the data is used to investigate the time-varying effects of the covariates. This is in contrast to all other tests that did not perform well when regions with little information were used.

The pseudo–score test gave the highest power and also was unaffected by the censoring time, and is thus to be preferred when testing for time-varying effects. If the esimates are stable, however, all tests considered lead to similar performance levels.

12.3.2   Trace Data

The TRACE study group [5] studied the prognostic importance of various risk factors on mortality for approximately 6600 patients with acute myocardial infarction (AMI). In this illustration we consider 1000 of these patients who were randomly selected. The data are part of the Glostrup cohort, and consist of consecutive admissions of patients with AMI to one hospital between 1979 and 1983. Additional details about the data are given in Martinussen and Scheike [8].

It was expected that ventricular fibrillation (VF) had a strongly time-varying effect and that other covariates such as clinical hear failure (CHF) and diabetes might have smaller time-varying effect. All covariates are measured at time 0, and as such it is expected that covariates such as VF, which give a condition that is very specific for time 0, will wear off with time. This may be in contrast to covariates such as diabetes, which refer to a more chronic condition of the patient. The VF covariate, for example, refers to a specific condition of the heart. Here we consider covariates for diabetes, sex, VF, CHF, and age.

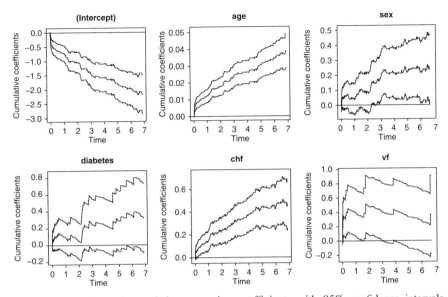

Figure 12.1 Estimated cumulative regression coefficients with 95% confidence intervals (solid lines).

We first fitted the nonparametric additive Aalen model and obtained the cumulative estimates shown in Figure 12.1 with 95% pointwise confidence bands and 95% Hall–Wellner bands (dotted lines).

The tests for time-varying effects based on V_n and PS yielded the p values listed in Table 12.5. This suggests that diabetes and sex have effects that may be constant. The P-values are based on the resampling processes, and Figure 12.2 shows the test processes for diabetes and VF along with 50 realizations under the hypothesis of a constant effect.

We simplified the model by successively testing the hypothesis of time-invariant covariate effect. In the model where the time-varying effect is allowed in all covariates, we conclude that the effect of diabetes may be constant ($p = 0.48$). The model where all effects are allowed to be time-dependent except for the effect of diabetes, which is taken to be constant, now forms the basis for further testing to determine whether the allowed time-varying effects could be taken as constant. This is acceptable for sex ($p = 0.17$). Finally, in this simplified model all

Table 12.5 **P-Values for Testing Time-Constant Effect Based on V_n Test Process and Pseudo–Score Test[a]**

Parameter	sup V_n	PS	sup V_n	PS	sup V_n	PS
Age	0.006	0.000	0.002	0.000	0.002	0.000
Sex	0.218	0.252	0.170	0.315	—	—
Diabetes	0.484	0.434	—	—	—	—
CHF	0.098	0.024	0.024	0.022	0.032	0.020
VF	0.012	0.004	0.006	0.007	0.002	0.007

[a]Successive testing leading to stepwise model reduction to model with constant effects [indicated by dashes (—) in blank tabular cell].

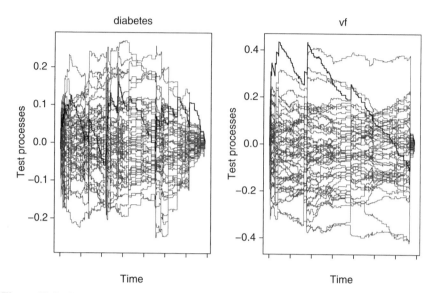

Figure 12.2   Test process for test for constant effect (solid dark line) and 50 resample realizations reflecting performance of text process under the null.

remaining effects are significantly time-varying. The p values for constant effects of VF, CHF, and age within this simplified model are 0.002, 0.032, and 0.002, respectively. This leads to a model in which the effects of VF, CHF, and age are time-varying and the other effects are summarized by constant excess risk (se): sex 0.0412 (0.0178), and diabetes 0.0617 (0.0350). The p values for constant effect of VF, CHF, and age within this simplified model are 0.002, 0.032, and 0.002, respectively.

12.4   CONCLUDING REMARKS

We have shown how the flexible additive hazards model may be used as a starting point for model reduction techniques, which, in the first step considered here, attempt to reduce the time-varying nonparametric regression effects to a parametric form (referred to as the *constant effect* here). This leads to a step-by-step model reduction that allows flexibility in the model if this is found in the data, and on the other hand gives a precise summary if this is reasonable according to the data. Another advantage of stepwise modeling is that the degrees of freedom at each test is kept low.

The reported simulations indicate that the tests based on the simple test process V_n performed reasonably well, but that the pseudo–score tests has the best performance and highest power. All tests but the pseudo–score test are highly sensitive to unstable behavior toward the end of the time-period.

12.5   SUMMARY

A useful alternative to the Cox model is the Aalen additive hazards model [1], which can easily accommodate time-varying covariate effects. For this model it is possible to test whether covariate effects significantly vary with time. If not all effects are time-varying, it is of interest to find the most useful model for summarizing the data. In this chapter we study different types of tests for time-varying effects within the additive hazards model. Large-sample results are obtained and a resampling technique for evaluating limit distributions is developed. The finite-sample properties of the proposed inference procedures are assessed through a simulation study. The methods are further applied to a dataset concerning myocardial infarction.

ACKNOWLEDGMENT

A major part of this work was done while the authors visited the Center for Advanced Study, Oslo. We are grateful to two referees for their useful comments, which were based on a very careful reading of the manuscript.

APPENDIX 12A: UNDERLYING ASSUMPTIONS AND JUSTIFICATION OF RESAMPLING

The following set of conditions are assumed throughout the chapter:

1. $P(Y_i(t) = 1, \text{ for all } t \in [0, \tau]) > 0$.
2. The covariates are bounded.
3. $E(Y_i(t)X_i X_i')$ is nonsingular for all $t \in [0, \tau]$.
4. As $n \to \infty$, $n^{-1}\tilde{C}(\tau)$ converges in probability to an invertible matrix.

Here we justify the resampling approach suggested in Section 12.2 following the line of Spiekerman and Lin of (Ref. 10, App. B). We consider only the case of $n^{1/2}\{\hat{\gamma} - \gamma\}$. By the martingale central-limit theorem it follows that $n^{1/2}\{\hat{\gamma} - \gamma\}$ converges in distribution toward a normal distribution. It also follows that $n^{1/2}\{\hat{\gamma} - \gamma\}$ is essentially a sum of n i.i.d. terms replacing terms like $(n^{-1} Z^T X) (n^{-1} X^T X)^{-1}$ by their limits in probability using the fact that we have uniform convergence due to condition 2. (Ranga Rao's SLL; see App. III of Ref. 3) and then applying Lenglart's inequality [2]. Thus

$$n^{1/2}\{\hat{\gamma} - \gamma\} = n^{-1/2} \sum_{i=1}^{n} \varepsilon_i^{\gamma} + o_P(1)$$

Now, by the conditional multiplier central-limit theorem [11], we obtain

$$n^{-1/2} \sum_{i=1}^{n} \varepsilon_i^{\gamma} G_i$$

converges weakly (in probability) toward the same limit distribution as $n^{1/2}\{\hat{\gamma} - \gamma\}$ conditional on the data. The argument is completed by showing that

$$|n^{-1/2} \sum_{i=1}^{n} \{\hat{\varepsilon}_i^{\gamma} - \varepsilon_i^{\gamma}\} G_i| \overset{P}{\Rightarrow} 0 \tag{12A.1}$$

We have that the left-hand side of (12A.1) is bounded above by

$$
\left| n^{-1/2} \sum_{i=1}^{n} G_i C^{-1}(\tau) \int_0^{\tau} Y_i \{ Z_i - (Z^T X)(X^T X)^{-1} X_i \} \right.
$$
$$
\times \{ (Z_i^T - X_i^T X^- Z)(\hat{\gamma} - \gamma) dt + X_i^T X^- dM(t) \} |
$$
$$
+ \left| n^{-1/2} \sum_{i=1}^{n} G_i \int_0^{\tau} \{ C^{-1}(\tau)(Z_i - (Z^T X)(X^T X)^{-1} X_i) \right.
$$
$$
\left. - c^{-1}(\tau)(Z_i - (z^T x)(x^T x)^{-1} X_i) \} dM_i \right| \tag{12A.2}
$$

The first term of (12A.2) is bounded above, using the triangular inequality, by

$$
\left| n^{-1/2} \int_{\theta}^{\tau} \sum_{i=1}^{n} f_i G_i \, dt (\tilde{\gamma}_1 - \gamma_1) \right| + \left| n^{-1/2} \int_{\theta}^{\tau} (n^{-1} \sum_{i=1}^{n} g_i G_i)(n^{-1} X^T X)^{-1} X^T \, dM \right|
$$

where the expressions for f_i and g_i are easily worked out. The first term of the latter display converges to 0 in probability by Ranga Rao's SLL, and since the G_i terms are independent of the data with $EG_i = 0$. The latter term also converges to 0 in probability using similar arguments and the inequality of Lenglart. Similarly the second term of (12A.2) is seen to converge to 0 in probability.

REFERENCES

1. Aalen, O. O., A model for non-parametric regression analysis of counting processes, in *Lecture Notes in Statistics—2: Mathematical Statistics and Probability Theory*, Klonecki, W., Kozek, A., and Rosinski, J., eds., Springer-Verlag, New York, 1980, pp. 1–25.

2. Andersen, P. K., Borgan, Ø., Gill, R. D., and Keiding, N., *Statistical Models Based on Counting Processes*, Springer-Verlag, New York, 1993.

3. Andersen, P. K. and Gill, R. D., Cox's regression model for counting processes: A large sample study, *Ann. Statist.* **10**, 1100–1120 (1982).

4. Cox, D. R., Regression models and life tables, *J. Roy. Statist. Soc. Ser. B* **34**, 406–424 (1972).

5. Jensen, G. V., Torp-Pedersen, C., Hildebrandt, P., Kober, L., Nielsen, F. E., Melchior, T., Joen, T., and Andersen, P. K., Does in-hospital ventricular fibrillation affect prognosis after myocardial infarction?, *Eur. Heart J.* **18**, 919–924 (1997).

6. Lin, D. Y., Wei, L. J., and Ying, Z., Checking the Cox model with cumulative sums of martingale-based residuals, *Biometrika* **80**, 557–572 (1993).

7. Lin, D. Y. and Ying, Z., Semiparametric analysis of the additive risk model, *Biometrika* **81**, 61–71 (1994).

8. Martinussen, T. and Scheike, T., *Dynamic Regression Models for Survival Data*, Springer-Verlag, New York, 2006.

9. McKeague, I. W. and Sasieni, P. D., A partly parametric additive risk model, *Biometrika* **81**, 501–514 (1994).

10. Spiekerman, C. F. and Lin, D. Y., Marginal regression models for multivariate failure time data, *J. Am. Statist. Assoc.* **93**, 1164–1175 (1998).

11. van der Vaart, A. W. and Wellner, J. A., *Weak Convergence and Empirical Processes: With Applications to Statistics*, Springer-Verlag, New York, 1996.

CHAPTER 13

Analysis of Outcomes Subject to Induced Dependent Censoring: A Marked Point Process Perspective

Yijian Huang

Department of Biostatistics, Emory University, Atlanta, Georgia

13.1 INTRODUCTION

For chronic diseases including cancer and HIV/AIDS, a time-to-event, such as overall survival, has been the typical primary outcome in clinical studies. However, such a single outcome is often inadequate to capture all the impacts that a treatment (and/or other covariates) might have on the disease process. For more comprehensive treatment assessment as being increasingly advocated, a number of secondary outcomes characterizing other features of the disease process toward the event of interest are often simultaneously evaluated. Examples include

1. *Lifetime Medical Cost.* Cost evaluation has become an accepted, and often required, adjunct to the standard effectiveness and safety assessment in today's medical research. This is due largely to the fact that demands on our healthcare system continue to outgrow the resources available. For example, lung cancer being the leading cause of cancer-related deaths in the United States, is estimated to cost the society $4.7 billion annually in direct medical costs [5]. The need to effectively control medical care cost becomes increasingly urgent.

2. *Quality-Adjusted Survival Time.* A treatment may affect not only the quantity but also the quality of life. Furthermore, tradeoffs may occur between these two aspects of life. The notion of quality-adjusted survival time offers a synthesis measure of the two, which has received great interest and attention in the health care community [18].

Statistical Advances in the Biomedical Sciences, edited by Atanu Biswas, Sujay Datta, Jason P. Fine, and Mark R. Segal
Copyright © 2008 John Wiley & Sons, Inc.

3. *Sojourn Times in Various Clinical Stages.* The course of a disease may comprise a
series of successive states representing progressive clinical stages. For example, longitudinal cancer studies often involve patients experiencing disease-free and disease-
relapse states before death. Sometimes more states are defined for better resolution of
the disease course; Gelber et al. [6] split the disease-free state further into two states,
with and without toxicity. The vector of state-specific sojourn times is clinically more
informative than the overall survival time.

Unfortunately, these secondary outcomes pose significant statistical challenges. Although there
are outcome-specific issues, one common difficulty arises from incomplete follow-up data as
typically obtained in clinical studies. Although censoring is nothing new in survival analysis,
the pattern associated with these endpoints turns out to be distinctive as being induced dependent censoring. In this chapter, we provide a review of a unified and effective approach to these
outcomes under the statistical framework of marked point process.

In Section 13.2, we elaborate on the phenomenon of induced dependent censoring and discuss
its associated identifiability issues. The concept of marked point process is introduced in Section
13.3, along with results on the one-sample nonparametric estimation. These results provide building blocks for the development of two-sample testing and regression analysis methods, which are
presented in Section 13.4. Section 13.5 concludes the chapter with final remarks.

13.2   INDUCED DEPENDENT CENSORING AND ASSOCIATED IDENTIFIABILITY ISSUES

Censoring occurs when a participant is lost to follow-up prior to the event of interest, due to
study termination or participant dropout. Denote time to the event by T, and the censoring
time by C. As a result of censoring, these underlying random variables are observed only
through follow-up time and censoring indicator

$$X = T \wedge C, \qquad \Delta = I(T \leq C),$$

where $\wedge$ is the minimization operator and $I(\cdot)$ is the indicator function. To make inference on T,
standard survival analysis techniques require a basic but critical assumption that T and C are
independent, possibly conditioning on covariates if applicable as in the regression setup.
Albeit not necessarily testable, this assumption is often plausible and widely accepted in
many practical situations. This is how the primary analysis on T is typically carried out.

Now, for a secondary endpoint described in Section 13.1, there exists an accumulation
process $U(\cdot)$ indexed by time. For lifetime medical cost, $U(t)$ is the accumulated cost at time
t. Similarly, in the case of quality-adjusted survival time, $U(t)$ corresponds to the accumulated
quality-adjusted life years (QALYs). These two examples involve scalar $U(\cdot)$. When state
sojourn times of a multistate process are under consideration, $U(t)$ is the vector of sojourn
times accumulated at time t. The secondary outcome of interest is $U \equiv U(T)$. However, in
the presence of censoring, it is observed through

$$W = U(X).$$

Given T as a measure of the event on the timescale, U may be viewed as a measure on a new
scale: cost to event, QALY to event, or state-specific sojourn times to event. This viewpoint
may suggest applying standard survival analysis techniques to such a new scale, that is,

analyzing the sample of $\{W, \Delta\}$ in a fashion similar to the standard survival analysis with that of $\{T, \Delta\}$. Unfortunately, such an analysis is inappropriate since the induced censoring pattern is in general dependent. This phenomenon has long been recognized; for example, see Glasziou et al. [8] and Lin et al. [15].

The induced dependence is due to the randomness nature of the accumulation process $U(\cdot)$, which is the transformation map from the timescale to the new scale. Since $U(\cdot)$ is nondecreasing, we obtain

$$W = U(T) \wedge U(C) = U \wedge U(C),$$

where $U(C)$ is the censoring measured on the new scale. If $U(\cdot)$ is deterministic, the independence between T and C leads to that between $U(T)$ and $U(C)$. But this is no longer the case in general for stochastic $U(\cdot)$, which becomes apparent with the following special cases.

Example 13.1: Lifetime Medical Cost or Quality-Adjusted Survival Time. To reveal the induced dependent censoring, we consider the special case where the accumulation rate, say, B, is time-independent. Nevertheless, B is a random variable. Thus, $U = BT$ and $U(C) = BC$ are not independent in general. Indeed, higher lifetime medical cost or quality-adjusted survival time tends to be censored later on cost or QALY scale, respectively.

Example 13.2: State Sojourn Times in a Progressive Multistate Process. For simplicity, consider a two-state process and denote the two underlying sojourn times by G_1 and G_2, with overall survival time $T = G_1 + G_2$. In this case

$$U(t) = \begin{pmatrix} t \wedge G_1 \\ (t - G_1)^+ \wedge G_2 \end{pmatrix},$$

where $a^+ = \max(a, 0)$. Clearly

$$W = \begin{pmatrix} C \wedge G_1 \\ (C - G_1)^+ \wedge G_2 \end{pmatrix}.$$

The censoring pattern for the two states is *serial* in that the censoring on G_1 would preclude the observation of G_2. The censoring pattern for the first state is still independent. However, unless G_1 and G_2 are independent, the censoring pattern on G_2 by $(C - G_1)^+$ becomes dependent. For a general progressive multistate process, the censoring pattern on any sojourn time beyond the first one is typically dependent.

As mentioned before, one implication of induced dependent censoring is that standard survival analysis techniques cannot be applied to the new scale. In addition and even more troublesome, the distribution of lifetime medical cost, quality-adjusted survival time, or any sojourn time but the first one in a progressive multistate process may be *nowhere identifiable*. In a typical clinical study, the duration is limited to, say, 3 years. This means that censoring time C is bounded by 3 years. In the cases of lifetime medical cost and quality-adjusted survival time, if a certain portion of the target population may incur little medical cost or QALYs within 3 years, then no information can be observed for their lifetime medical cost or quality-adjusted survival time. Therefore, the corresponding distribution function for the population is not identifiable away from 0. Similarly, for the second sojourn time in a progressive two-state process, its distribution is nowhere identifiable if the first sojourn time may exceed the study duration.

It is well known that limited study duration also causes identifiability issues in standard survival analysis. But the nonidentifiability is restricted only to the tail portion of the survival distribution, that is, beyond the study duration. In contrast, for outcomes subject to induced dependent censoring, nonidentifiability may be everywhere for the marginal distribution of interest.

13.3 MARKED POINT PROCESS

The issues described in the previous section plague the analysis of those secondary endpoints. To further complicate the analysis, the structure of observed data may vary not only from one endpoint to another but also from one study to another even for the same secondary endpoint. Taking lifetime medical cost as an example, the observation of $U(\cdot)$ on $[0, X]$ may exhibit a wide spectrum of patterns:

1. One less informative pattern corresponds to the observation of $U(X)$ only. Such data collection is often retrospective, that is, after the follow-up is complete.
2. The observation of $U(\cdot)$ may have an intermittent pattern on $[0, X]$. When cost data collection is driven by study visits, cost accumulated from the previous visit is recorded at each visit. Study participants may or may not share the same visit schedule.
3. Apparently the most informative pattern involves the continuous observation of $U(\cdot)$ on $[0, X]$. This is often the case when cost of interest is due to hospitalizations, by which the cost collection is driven.

Despite all these differences from one situation to another, there exists a common basic data structure with the observation of the following random variables [11]:

$$X, \qquad \Delta, \qquad Y = U\,I(T \le C).$$

Thus, in addition to $\{X, \Delta\}$ as in the standard survival data, the secondary endpoint U is observed among uncensored individuals. A random variable like U that is observed only on the occurrence of an event is termed the *mark* of the event, and the process counting the event along with the mark is referred to as the *marked point process* [2, Sec. II.4.1]. A simpler and better-known marked point process involves cause of death as a discrete mark. Both lifetime medical cost and quality-adjusted survival time are continuous marks of death. Meanwhile, a mark may also be a random vector as in the case of sojourn times in a multistate process.

The marked point process is a natural extension of the counting process. Not surprisingly, classical results and martingale theory with counting processes can be generalized to marked point processes [11]. In this section, we introduce some of them in the one-sample setup.

13.3.1 Hazard Functions with Marked Point Process

Write $F_{TU}(t, u) = \Pr(T \le t,\ U \le u)$ and its marginal for time to event $F_T(t) = 1 - S_T(t) = F_{TU}(t, \infty)$. The cumulative hazard function of T is given by

$$\Lambda_T(t) = \int_{[0,t]} \frac{F_T(ds)}{S_T(s-)}.$$

The one-to-one mapping between the survival function and the hazard function is a central idea in univariate survival analysis [1,7]

$$S_T(t) = \prod_{[0,t]} \{1 - \Lambda_T(ds)\},$$

where $\prod$ is product integral. Correspondingly, with marked point process, define the cumulative mark-specific hazard function

$$\Lambda_{TU}(t, u) = \int_{[0,t]} \frac{F_{TU}(ds, u)}{S_T(s-)}.$$

The following representation plays an important role in the estimation with marked point process

$$
\begin{aligned}
F_{TU}(t, u) &= \int_{[0,t]} S_T(s-)\Lambda_{TU}(ds, u) \\
&= \int_{[0,t]} \prod_{[0,s]} \{1 - \Lambda_{TU}(dv, \infty)\} \Lambda_{TU}(ds, u),
\end{aligned}
\tag{13.1}
$$

where the mapping from Λ_{TU} to F_{TU} is continuous and compactly differentiable.

13.3.2 Identifiability

To deal with censored data, we assume the random censorship mechanism; to be more specific, the pair $\{T, U\}$ is independent of C. In addition, we impose a standard assumption in the one-sample survival problem that the distributions of T and C do not have jump points in common [21]. Note that this assumption does not exclude the possibility that either function is discrete.

Let $F_{XY,\Delta=1}(t, u) = \Pr(X \le t, \ Y \le u, \ \Delta = 1)$, $S_X(t) = \Pr(X > t)$, and $S_C(t) = \Pr(C > t)$. Denote the maximum support of $S_X(t)$ by $\tau = \sup\{t : S_X(t) > 0\}$. Under the given assumptions, we obtain

$$F_{XY,\Delta=1}(dt, u) = F_{TU}(dt, u)S_C(t-), \qquad S_X(t) = S_T(t)S_C(t).$$

Therefore

$$\Lambda_{TU}(t, u) = \int_{[0,t]} \frac{F_{XY,\Delta=1}(t, u)}{S_X(s-)}, \qquad \forall t < \tau. \tag{13.2}$$

With this representation of Λ_{TU} in terms of the joint distribution of observed random variables $\{X, \Delta, Y\}$, Λ_{TU} and thus F_{TU} are identifiable up to the support $[0, \tau) \times (-\infty, \infty)$ as shown in Figure 13.1. This identifiability result on the joint distribution is important since the marginal distribution of U may be nowhere identifiable, which is due to the nonidentifiability of F_{TU} on $[\tau, \infty) \times (-\infty, \infty)$.

13.3.3 Nonparametric Estimation

The data consist of $\{X_i, \Delta_i, Y_i\}$, $i = 1, \ldots, n$, as n i.i.d. replicates of $\{X, \Delta, Y\}$. Define processes

$$N_i(t, u) = I(X_i \le t, Y_i \le u, \Delta_i = 1), \qquad R_i(t) = I(X_i \ge t),$$

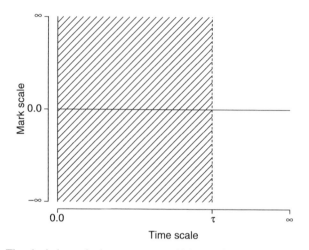

Figure 13.1   The shaded area is the support on which F_{TU} is identifiable in the one-sample problem.

where R_i is often referred to as an "at-risk process." Let $N(t, u) = \sum N_i(t, u)$ and $R(t) = \sum R_i(t)$. Note that N/n and R/n are the empirical versions of $F_{XY,\Delta=1}$ and $S_X(\cdot-)$, respectively, in identity (13.2). This motivates an estimator for Λ_{TU}.

$$\hat{\Lambda}_{TU}(t, u) = \int_{[0,t]} \frac{N(ds, u)}{R(s)},$$

which is a generalization of the Nelson–Aalen estimator. Indeed, $\hat{\Lambda}_T(\cdot) = \hat{\Lambda}_{TU}(\cdot, \infty)$ is the Nelson–Aalen estimator for Λ_T. Subsequently, with mapping of (13.1), a natural estimator of the joint distribution F_{TU} emerges as

$$\hat{F}_{TU}(t, u) = \int_{[0,t]} \prod_{[0,s)} \{1 - \hat{\Lambda}_{TU}(dv, \infty)\} \, \hat{\Lambda}_{TU}(ds, u).$$

Similar to the relationship between $\hat{\Lambda}_{TU}$ and the Nelson-Aalen estimator, $\hat{F}_{TU}$ is a generalization of the Kaplan–Meier estimator.

13.3.4   Martingales

In standard survival analysis, martingales prove to be powerful in not only finite- and large-sample investigations but also modeling with counting processes. It is not surprising that martingales can also be identified with marked point processes.

Define a filtration $\{\mathcal{F}_t : t \geq 0\}$ with

$$\begin{aligned}
\mathcal{F}_t = \sigma\{&I(X_i \leq s, \Delta_i = 1), I(X_i \leq s, \Delta_i = 0), \\
&Y_i I(X_i \leq s, \Delta_i = 1) : 0 \leq s \leq t, \qquad i = 1, \cdots, n\}.
\end{aligned} \tag{13.3}$$

For any given u

$$M_i(t, u) = N_i(t, u) - \int_{[0,t]} R_i(s)\Lambda_{TU}(ds, u)$$

is a martingale with respect to $\{\mathcal{F}_t : t \geq 0\}$. This facilitates the finite- and large-sample investigations of $\hat{\Lambda}_{TU}$ and $\hat{F}_{TU}$, similar to the studies of the Nelson–Aalen and Kaplan–Meier estimators [11].

Furthermore, define

$$N_i^{(m)}(t) = Y_i^m I(X_i \leq t)\Delta_i, \qquad \Lambda^{(m)}(t) = \int_{(-\infty, \infty)} u^m \Lambda_{TU}(t, du), \qquad m = 0, 1.$$

With respect to $\{\mathcal{F}_t : t \geq 0\}$

$$M_i^{(m)}(t) = N_i^{(m)}(t) - \int_{[0,t]} R_i(s)\Lambda^{(m)}(ds) \tag{13.4}$$

is a martingale for $m = 0, 1$. When $m = 0$, $N_i^{(0)}$ is a counting process, $\Lambda^{(0)}(t) = \Lambda_T(t)$, and the martingale result is well known. But the extension to $m = 1$ will be useful in the following section for testing and regression analysis involving a mark.

13.4 MODELING STRATEGY FOR TESTING AND REGRESSION

The secondary analyses mentioned in Section 13.1 are typically concerned with scientific questions such as

1. Does the intervention improve quality-adjusted survival time over the standard care?
2. How do the baseline demographics affect lifetime medical cost?
3. Are the cancer-free and cancer-relapse times in the intervention arm different from those in the standard arm?

Unfortunately, such issues may not be addressed on the basis of the marginal distribution of interest due to limited study duration. In light of the identifiability result in Section 13.3.2, one reasonable strategy is to formulate these questions in terms of the joint distribution with time to event. Such a joint distribution strategy appears to fit in well with the relationship between the primary and secondary analyses. In the following, we describe testing and regression analysis procedures with lifetime utility or cost, and a testing procedure with a progressive multistate process in line with this strategy. With each procedure, the marked point process data structure is considered and the one-sample results in Section 13.3 serve as a building block. The focus is on main ideas; details may be found in the references provided.

13.4.1 Two-Sample Test for Lifetime Utility or Cost

In many clinical trials, survival time is the primary endpoint and lifetime utility or cost is a secondary endpoint. One example is a randomized clinical trial conducted by the Southwest Oncology Group (SWOG), comparing vinorelbine plus cisplatin versus paclitaxel plus carboplatin therapies in earlier untreated patients with advanced non-small-cell lung cancer [13]. In addition to survival data, resource utilization data were collected for each participant, and comparing medical cost associated with the two treatments was a secondary objective. Such a comparison is not trivial not only because of the issues associated with incomplete follow-up data as discussed before but also because the two treatments may affect survival time differently.

With the two samples under consideration, we use the notation introduced in Section 13.3 for one sample and add an asterisk to indicate the other. Huang and Lovato [12] suggested testing the difference between U and U^* on the basis of the joint distributions of $\{T, U\}$ and $\{T^*, U^*\}$. It is acknowledged that T and T^* may not share the same marginal distribution, and their difference is a nuisance for the test under consideration. Therefore, a calibration of this potential and irrelevant difference is in order. For this purpose, an accelerated failure-time model may be adopted: $T \sim \exp(\beta)T^*$. Then, the null hypothesis is specified as

$$\mathcal{H}_0 : (T, U)^T \sim \{\exp(\beta)T^*, U^*\}^T.$$

Apparently, $U \sim U^*$ under $\mathcal{H}_0$. Writing $\mu(t) = E(U|T = t)$ and $\mu^*(t) = E(U^*|T^* = t)$, we consider the alternative hypothesis $\mathcal{H}_A : \mu(t) \geq \mu^*\{\exp(-\beta)t\}$ or $\mu(t) \leq \mu^*\{\exp(-\beta)t\}$, with strict inequality for at least some t. Clearly, $\mathcal{H}_A$ implies different marginal distributions of U and U^*.

For the test, we build on and extend the weighted log-rank statistics

$$\xi_m(b) = \int_{[0,\infty)} w_m(t, b) \left\{ \frac{dN^{(m)}(t)}{R(t)} - \frac{dN^{(m)*}\{\exp(-b)t\}}{R^*\{\exp(-b)t\}} \right\}, \qquad m = 0, 1,$$

where w_m is a nonnegative weight function. Note that ξ_0 is the familiar rank statistic for survival time, where $w_0(t, b) = R(t)R^*\{\exp(-b)t\}/[R(t) + R^*\{\exp(-b)t\}]$ and $w_0(t, b) = R(t)$ $R^*\{\exp(-b)t\}$ correspond to the log-rank and Gehan statistics, respectively. These statistics are justified by the martingales given in (13.4): $\xi_0(\beta)$ and $\xi_1(\beta)$ have mean 0 under $\mathcal{H}_0$. Of course, β is a nuisance parameter and unknown. Louis [17] and Wei and Gail [24] suggested using a consistent estimator of β, say, $\hat{\beta}$, as the zero-crossing of $\xi_0(b)$. Therefore, $\xi_1(\hat{\beta})$ may serve as a test statistic, which is asymptotically normal with mean 0 under $\mathcal{H}_0$.

13.4.2 Calibration Regression for Lifetime Medical Cost

In the previous section, we were concerned with the relationship between, say, lifetime medical cost and a binary covariate, such as the treatment indicator. More generally, the relationship between the lifetime medical cost U and the covariate vector, say, $\mathbf{Z}$, is of interest as in a regression problem. For example, with such a regression analysis in the SWOG study, one may address the treatment effect after adjustment for baseline patient characteristics and/or effects of baseline characteristics, on lifetime medical cost.

With censored data, it is necessary to model the covariate effect on the joint distribution of T and U, instead of the marginal distribution of U. Huang [10] generalized the univariate accelerated failure-time model for the joint distribution of $\{T, U\}$

$$\log\left(\frac{T}{U}\right) = \left(\frac{\alpha^T}{\beta^T}\right)\mathbf{Z} + \varepsilon,$$

where α and β are regression coefficient vectors and the bivariate error term ε has a completely unspecified bivariate distribution. Evidently this regression model is rooted in the two-sample model for the testing procedure in Section 13.4.1. Furthermore, the following conditional independence censoring mechanism is adopted

$$C \perp \{T, U\} \mid \mathbf{Z},$$

where $\perp$ represents independence.

The data consist of $\{X_i, Y_i, \Delta_i, \mathbf{Z}_i\}$, $i = 1, \ldots, n$, as n i.i.d replicates of $\{X, Y, \Delta, \mathbf{Z}\}$. To draw inference on α and β, estimating functions have been constructed. For α, the standard log-rank estimating function (or a weighted version of it) for the univariate accelerated failure-time model [20] is adopted

$$\mathbf{S}_0(\mathbf{a}) = \sum_{i=1}^{n} \int_{(-\infty, \infty)} \left\{ \mathbf{Z}_i - \frac{\sum_{j=0}^{n} \mathbf{Z}_j R_j(t; \mathbf{a})}{\sum_{j=0}^{n} R_j(t; \mathbf{a})} \right\} dN_i^{(0)}(t; \mathbf{a}),$$

where $R_i(t; \mathbf{a}) = I(\log X_i - \mathbf{a}^T\mathbf{Z}_i \geq t)$ is an at-risk process and $N_i^{(0)}(t; \mathbf{a}) = I(\log X_i - \mathbf{a}^T\mathbf{Z}_i \leq t)\Delta_i$ is a counting process. For the estimation of β, the above log-rank estimating function is generalized to the marked point process framework. Define process $N_i^{(1)}(t; \mathbf{a}, \mathbf{b}) = I(\log X_i - \mathbf{a}^T\mathbf{Z}_i \leq t)\Delta_i(\log Y_i - \mathbf{b}^T\mathbf{Z}_i)$, which differs from the counting process $N_i^{(0)}$ in jump size. An estimating function (or a weighted version of it) for α and β has been proposed as follows:

$$\mathbf{S}_1(\mathbf{a}, \mathbf{b}) = \sum_{i=1}^{n} \int_{(-\infty, \infty)} \left\{ \mathbf{Z}_i - \frac{\sum_{j=0}^{n} \mathbf{Z}_j R_j(t; \mathbf{a})}{\sum_{j=0}^{n} R_j(t; \mathbf{a})} \right\} dN_i^{(1)}(t; \mathbf{a}, \mathbf{b}).$$

Both $\mathbf{S}_0(\alpha)$ and $\mathbf{S}_1(\alpha, \beta)$ have mean 0, which is a result of the martingales given in (13.4). An estimator of α, $\hat{\alpha}$, can be obtained as a zero-crossing of $\mathbf{S}_0(\mathbf{a})$. Then, the zero-crossing of $\mathbf{S}_1(\hat{\alpha}, \mathbf{b})$ gives a reasonable estimator $\hat{\beta}$ for β.

13.4.3 Two-Sample Multistate Accelerated Sojourn-Time Model

We now turn to the problem of multistate process representing various clinical stages, and use the International Breast Cancer Study Group (IBCSG) trial V as a motivating example. This study investigated the effectiveness of short-duration (1 month) perioperative systemic treatment compared with long-duration adjuvant therapy (6 or 7 months) in node-positive breast cancer patients [22]. Cancer relapse and death were monitored, and the disease progression can be characterized as a progressive two-state process with cancer-free and cancer-relapse states. While treatments potentially have differential, sometimes even opposite, effects on different states, comparison with respect to individual-state sojourn times is of interest and importance.

Let $g = 0,1$ indicate the control and treatment groups. Suppose that K consecutive states end at times $T_{g1} \leq \ldots \leq T_{gK}$ from the start of follow-up. We assess the treatment effect on the K sojourn times $G_{g1} = T_{g1}$, $G_{g2} = T_{g2} - T_{g1}, \ldots, G_{gK} = T_{gK} - T_{g(K-1)}$. This process is subject to censoring by time C_g, and these underlying random variables are observed through

$$X_{gk} = T_{gk} \wedge C_g, \qquad \Delta_{gk} = I(T_{gk} \leq C_g), \qquad k = 1, \ldots, K.$$

The observed censored sojourn times are $H_{g1} = X_{g1}$, $H_{g2} = X_{g2} - X_{g1}, \ldots, H_{gK} = X_{gK} - X_{g(K-1)}$. Assume that, given $g = 0, 1$, C_g is independent of $\{T_{gk} : k = 1, \ldots, K\}$.

As discussed before, the marginal distribution of any sojourn time beyond the first state may be nowhere identifiable for each group. Therefore, a marginal model may not be identifiable. Huang [9] proposed the multistate accelerated sojourn-time model such that $(G_{11}, \ldots, G_{1K})^T$ has the same distribution as $\{\exp(\beta_1)G_{01}, \ldots, \exp(\beta_K)G_{0K}\}^T$ for some $\beta = (\beta_1, \ldots, \beta_K)^T$.

The estimation of β_1 can be carried out with standard methods for the univariate accelerated failure-time model including those proposed by Louis [17] and Wei and Gail [24]. However, the challenge lies in the estimation of β_k for $k \geq 2$. The result described in Section 13.3 is useful for this purpose. Given g, any function of $\{G_{gj} : j = 1, \ldots, k\}$ is a mark of the kth transition. In particular, we are interested in the linear function $\sum_{j=1}^k \exp(b_j)G_{gj}$ for some $\mathbf{b}_k = (b_1, \ldots, b_k)^T$. Define its joint distribution with T_{gk} as $F_{gk}(t, u; \mathbf{b}_k) = \Pr\{T_{gk} \leq t, \sum_{j=1}^k \exp(b_j)G_{gj} \leq u\}$, which is identifiable except for large t. Write $\beta_k = (\beta_1, \ldots, \beta_k)^T$. The multistate accelerated sojourn-time model implies

$$F_{1k}(t, u; -\beta_k) = F_{0k}(u, t; \beta_k), \qquad k = 2, \ldots, K.$$

This key identity, along with the estimation procedure for F_{gk} given in Section 13.3, leads to estimation equations for β.

13.5 CONCLUDING REMARKS

This chapter provides a review of the marked point process approach to the analysis of outcomes subject to induced dependent censoring. The marked point process is a natural extension of the counting process, and many classical results for counting process can be generalized. Meanwhile, this perspective reveals identifiability of the joint distribution of mark and time to event except for the tail region on the timescale. One-sample estimation, two-sample testing, and regression analysis procedures on the basis of the joint distribution are discussed.

Many investigations exist in the literature for the problems under consideration here, including those by Zhao and Tsiatis [26], Lin et al. [15], Lin [14], Zhao and Tian [25], Strawderman [20], and Bang and Tsiatis [3,4]. However, these methods differ from our approach in two aspects:

1. In order to address the identifiability issue, the time-restricted mark is considered in these investigations instead of the mark itself. For example, 3-year-restricted medical cost is the cost accumulated up to 3 years or death, whichever occurs earlier. If the maximum support of the censoring distribution is longer than 3 years, the marginal distribution for 3-year-restricted medical cost is identifiable. Nevertheless, such a time-restricted mark depends on the time limit, which is typically artificial. Consequently, results on a time-restricted mark may not be easy to interpret, especially in the two-sample and regression problems. In contrast, our approach in this chapter targets the mark of interest. To overcome the marginal identifiability issue, the strategy is to consider joint modeling with time to event.

2. The aforementioned investigations in the literature motivated their approach to induced dependent censoring by the Horvitz–Thompson method for missing data. The essence of this method is to weight each complete case inversely by its (estimated) probability of being complete, so as to eliminate the biased sampling due to missingness [16]. Indeed, censored data can be viewed as a special type of missing data. In the one-sample problem, perspectives from the Horvitz–Thompson approach and marked point process provide complementary insights. Specifically, the Horvitz–Thompson approach may be used to obtain $\hat{F}_{TU}$ [3]. On the other hand, the framework of marked point process may also facilitate the marginal estimation of a time-restricted mark. Note that the time-restricted mark is a mark associated with the event of interest and the time limit, whichever comes first. Nevertheless, the premise of the Horvitz–Thompson approach is that the estimated probability of being complete is readily available. In the regression setup, this requires a censoring mechanism more restrictive than the conditional independence censoring mechanism as considered in Section 13.4.2, as suggested by, for example, Lin [14] and Bang and Tsiatis [4]. In this respect, exploiting the marked point process may be advantageous.

We have approached these various problems by attempting a general solution on identifying a common data structure. When additional data are available, more efficient estimation might be possible. Robins et al. [19] developed general efficiency theories for estimation with missing data by augmenting the Horvitz–Thompson estimator. Along this line, Zhao and Tsiatis [27] and Bang and Tsiatis [3] identified the semiparametric efficiency bound and developed practical strategies for efficiency improvement when $U(\cdot)$ is continuously observed on $[0, X]$, in one-sample problems for time-restricted quality-adjusted survival time and lifetime medical cost. Extending these results to the estimation on the basis of the joint distribution of U and T requires further investigation.

REFERENCES

1. Aalen, O. O. and Johansen, S., An empirical transition matrix for non-homogeneous Markov chains based on censored observations, *Scand. J. Statist.* **5**, 141–150 (1978).

2. Andersen, P. K., Borgan, O., Gill, R. D., and Keiding, N., *Statistical Models Based on Counting Processes*, Springer-Verlag, New York, 1993.

3. Bang, H. and Tsiatis, A. A., Estimating medical costs with censored data, *Biometrika* **87**, 329–343 (2000).

4. Bang, H. and Tsiatis, A. A., Median regression with censored cost data, *Biometrics* **58**, 643–649 (2002).

5. Brown, M. L., Lipscomb, J., and Snyder, C., The burden of illness of cancer: Economic cost and quality of life, *Annu. Rev. Publ. Health* **22**, 91–113 (2001).

6. Gelber, R. D., Gelman, R. S., and Goldhirsch, A., A quality-of-life-oriented endpoint for comparing therapies, *Biometrics* **45**, 781–795 (1989).

7. Gill, R. D. and Johansen, S., A survey of product-integration with a view towards application in survival analysis, *Annal. Statist.* **18**, 1501–1555 (1990).

8. Glasziou, P. P., Simes, R. J., and Gelber, R. D., Quality adjusted survival analysis, *Statist. Med.* **9**, 1259–1276 (1990).

9. Huang, Y., Two-sample multistate accelerated sojourn times model, *J. Am. Statist. Assoc.* **95**, 619–627 (2000).

10. Huang, Y. Calibration regression of censored lifetime medical cost, *J. Am. Statist. Assoc.* **97**, 318–327 (2002) correction: **97**, 661.

11. Huang, Y. and Louis, T. A., Nonparametric estimation of the joint distribution of survival time and mark variables, *Biometrika* **85**, 785–798 (1998).

12. Huang, Y. and Lovato, L., Tests for lifetime utility or cost via calibrating survival time, *Statistica Sinica* **12**, 707–723 (2002).

13. Kelly, K., Crowley, J., Bunn, P. A., Jr., Presant, C. A., Grevstad, P. K., Moinpour, C. M., Ramsey, S. D., Wozniak, A. J., Weiss, G. R., Moore, D. F., Israel, V. K., Livingston, R. B., and Gandara, D. R., Randomized phase III trial of Paclitaxel plus Carboplatin versus Vinorelbine plus Cisplatin in the treatment of patients with advanced non-small-cell lung cancer: A Southwest Oncology Group Trial, *J. Clin. Oncol.* **19**, 3210–3218 (2001).

14. Lin, D. Y., Linear regression analysis of censored medical costs, *Biostatistics* **1**, 35–47 (2000).

15. Lin, D. Y., Feuer, E. J., Etzioni, R., and Wax, Y., Estimating medical costs from incomplete follow-up data, *Biometrics* **53**, 419–434 (1997).

16. Little, R. J. A. and Rubin, D. B., *Statistical Analysis with Missing Data*, 2nd ed., Wiley, New York, 2002.

17. Louis, T. A., Nonparametric analysis of an accelerated failure time model, *Biometrika* **68**, 381–390 (1981).

18. Olschewski, M. and Schumacher, M., Statistical analysis of quality of life in cancer clinical trials, *Statist. Med.* **9**, 749–763 (1990).

19. Robins, J. M., Rotnitzky, A., and Zhao, L. P., Analysis of semiparametric regression models for repeated outcomes in the presence of missing data, *J. Am. Statist. Assoc.* **90**, 106–121 (1995).

20. Strawderman, R. L., Estimating the mean of an increasing stochastic process at a censored stopping time, *J. Am. Statist. Assoc.* **95**, 1192–1208 (2002).

21. Stute, W. and Wang, J.-L., The strong law under random censorship, *Ann. Statist.* **21**, 1591–1607 (1993).

22. The Ludwig Breast Cancer Study Group, Combination adjuvant chemotherapy for node-positive breast cancer: Inadequacy of a single perioperative cycle, *New Engl. J. Med.* **319**, 677–683 (1988).

23. Tsiatis, A. A., Estimating regression parameters using linear rank tests for censored data, *Ann. Statist.* **18**, 354–372 (1990).

24. Wei, L. J. and Gail, M. H., Nonparametric estimation for a scale-change with censored observations, *J. Am. Statist. Assoc.* **78**, 382–388 (1983).

25. Zhao, H. and Tian, L., On estimating medical cost and incremental cost-effectiveness ratios with censored data, *Biometrics* **57**, 1002–1008 (2001).

26. Zhao, H. and Tsiatis, A. A., A consistent estimator for the distribution of quality adjusted survival time, *Biometrika* **84**, 339–348 (1997).

27. Zhao, H. and Tsiatis, A. A., Efficient estimation of the distribution of quality adjusted survival time, *Biometrics* **55**, 1101–1107 (1999).

CHAPTER 14

Analysis of Dependence in Multivariate Failure-Time Data

Li Hsu and Zoe Moodie

Biostatistics and Biomathematics Program, Public Health Sciences Division, Fred Hutchinson Cancer Research Center, Seattle, Washington

14.1 INTRODUCTION

Multivariate failure times, or times to response, are natural outcomes in many studies when each unit in the data is a cluster of multiple outcomes. Examples of such data include the time to multiple infections after bone marrow transplantation in a clinical trial, the ages at onset of some disease in a twin or family study, and the time to occurrences of multiple events for an individual in a cohort study. In these examples, the time to infection, age at onset, or the time to occurrence of an event is a failure time and is subject to independent right censoring. (By "right censoring" we mean that the individual had been under observation until a particular time and after that time we do not know what happened to that individual.) Moreover, the time to multiple infections on the same individual or the ages at onset of multiple individuals from the same family could be correlated with each other because they occur to the same individual or the same family.

When researchers are primarily interested in the effects of covariates on the marginal hazard function of an individual failure time, the dependence among these failure times is, to some degree, a nuisance even though it needs to be considered in order to draw valid statistical inference and/or increase the efficiency of parameter estimation. However, there are situations where dependences among correlated failure times are of main interest, too. For example, in a family study of a particular disease, if the parent–offspring or sibling–sibling age dependence at onset is greater than that between spouses, the pattern may suggest a possible genetic contribution in addition to environment to the etiology of the disease. In this case, the familial resemblance pattern sheds light on the disease etiology.

Statistical Advances in the Biomedical Sciences, edited by Atanu Biswas, Sujay Datta, Jason P. Fine, and Mark R. Segal

This chapter is concerned with the non- and semiparametric estimation of dependences among failure times. An important aspect of studying the dependences among multivariate failure times is by nonparametric estimation of its joint survivor distribution, which has the distinct advantage of not requiring any distributional assumptions. It is a useful quantity not only by itself but also for purposes such as identifying the appropriate parametric form and nonparametric estimation of summary measures of dependences over a finite follow-up region. Various approaches to the nonparametric estimation of the joint survival distribution are reviewed in this chapter.

A perhaps more common approach to estimation of dependence is through semiparametric modeling of the joint survival function in that the marginal hazard function is non- or semiparametric and the dependence is quantified by a finite number of parameters. The non- or semi-parametric marginal hazard functions allows for flexible modeling of an often skewed age-at-onset distribution. Because of the inclusion of an infinite-dimensional component in the marginal hazard function, the estimation of the dependence parameters is often termed *semiparametric* as well. This chapter will review two main approaches, frailty-model-based and partially specified model-based, to the semiparametric estimation of dependence parameters.

Two data examples will be used to illustrate the methods.

Example 14.1: Danish Twin Data. This example includes all twins born in Denmark between 1870 and 1910 and all same-sex twins born between 1911 and 1930, where both members were known to be alive at age 15 years. The data include follow-up of survival status up until January 1, 1980. Further details are found in Hauge [21]; analyses of these data can be found in Hougaard [24,25]. In this chapter we will use a subset of this data, selecting a random sample of 100 pairs from the 1366 monozygotic male twins in the computerized portion of the registry of same-sex twins born between 1911 and 1930 to illustrate the relative performance of bivariate survival function estimators. The methods can also be applied to the full dataset, but differences are best revealed in small to moderate-sized samples.

Example 14.2: Case–Control Family Data of Breast Cancer. This dataset includes two case–control studies conducted between 1983 and 1990 (white women only) and between 1990 and 1992 (all races) [37]. All women who were diagnosed before age 45 years were obtained through the Cancer Surveillance System, a member of the National Cancer Institute–sponsored Surveillance, Epidemiology, Endpoint and End Result program. Controls were selected through random digit dialing, and matched with cases on gender, age at diagnosis, and county of residence. In this chapter we will use a subset of these data— cases and controls and their mothers—to illustrate the semiparametric methods for estimating the dependence between paired ages at onset.

The rest of the chapter is organized as follows. The next section reviews nonparametric estimators of the joint survivor function mainly for the setting of bivariate survival data with a brief discussion of extension to multivariate survival data. The leading estimators are applied to a subset of the Danish Twin Data, as an illustrative example of their performances with real data. In Section 14.3, a class of weighted dependence measures that employs a nonparametric estimator of the bivariate survivor function is described. In the same section, two main approaches for semiparametric estimation of dependence parameters are reviewed. As an example, these procedures are applied to a case–control family study of breast cancer with a discussion of modifications of current methods to accommodate the retrospective sampling of the data. The chapter ends with a few final remarks.

14.2 NONPARAMETRIC BIVARIATE SURVIVOR FUNCTION ESTIMATION

In this section we will review various approaches to nonparametric estimation of the bivariate survivor function. Under univariate censoring, the estimation problem is relatively straightforward and several good estimators exist [36,63,59,58]. Each of these estimators is strongly consistent, weakly convergent to a Gaussian process whose covariance function can be estimated. The estimator proposed by Tsai and Crowley has the smallest asymptotic variance of all the path-dependent estimators, including that of Lin and Ying. However, at present there is no fully satisfactory solution to the estimation problem for bivariate right-censored data. In what follows, various approaches are reviewed for this situation. First, some notation are introduced. Let $(T_1, \ldots, T_K)$ and $(C_1, \ldots, C_K)$ be K-variate correlated failure times and censoring times, respectively. The corresponding failure and censoring survival functions are denoted by F and G, respectively. Furthermore define observed time $X_k = T_k \wedge C_k$ and disease status $\delta_k = I(T_k \leq C_k)$ for $k = 1, \ldots, K$, where $\wedge$ is minimum and $I(\cdot)$ is the indicator function. For bivariate failure times discussed here, $K = 2$.

14.2.1 Path-Dependent Estimators

One of the earliest approaches, path-dependent estimators, express the survivor function at a point (t_1, t_2) as a "path" of conditional and marginal survivor functions $P(T_1 > t_1, T_2 > t_2) = P(T_2 > t_2 | T_1 > t_1) P(T_1 > t_1)$. The Campbell–Földes estimator is one of the first examples of this type of approach [6]. Campbell and Földes showed that under certain smoothness conditions of the failure and censoring survivor functions, F and G, respectively (F, G continuous such that $-\log F$ is absolutely continuous with partial derivatives existing almost everywhere), the estimator is uniformly almost surely consistent with rate $O(\sqrt{(n^{-1}(\log \log n))})$ as n goes to infinity. Their estimator can be generalized to more than two dimensions with uniform almost sure consistency at the same rate. However, the estimator is not guaranteed to be monotonic, and its finite sample properties can vary depending on the selected path. It is also worth noting that interchanging the role of T_1 and T_2 results in a different estimator.

In general, the properties of estimators in this class are path-dependent. The estimators often have poor efficiency properties and make negative point mass assignments in the presence of censored observations.

14.2.2 Inverse Censoring Probability Weighted Estimators

Inverse censoring probability weighted estimators make direct use of estimates of the censoring probabilities by expressing the survivor function as

$$F(t_1, t_2) = F_1(t_1) + F_2(t_2) - 1 + \int_0^{t_1} \int_0^{t_2} \{G(u_1^-, u_2^-)\}^{-1} H^{uc}(du_1, du_2)$$

where $H^{uc}(du_1, du_2) = P(X_1 \in [u_1, u_1 + du_1), X_2 \in [u_2, u_2 + du_2), \delta_1 = 1, \delta_2 = 1)$ and $G(u_1, u_2) = P(X_1 \geq u_1, X_2 \geq u_2)$.

Since the Kaplan–Meier estimator for univariate data can be cast as an inverse probability weighted estimator, it was hoped that an extension of these methods might yield a good bivariate estimator. One of the earliest estimators in this class [4] is based on an adaptation of the Campbell–Földes estimator [6]; Burke imposed monotonicity while achieving the same rate of consistency.

His estimator has poor efficiency but is uniformly strongly consistent, computationally simple, and a monotonic estimator for F. Burke proposed that the survivor function be calculated from the marginal and joint distribution functions, $\hat{F}(t_1, t_2) = 1 - \underline{\hat{F}}(t_1, \infty) - \underline{\hat{F}}(t_2, \infty) + \underline{\hat{F}}(t_1, t_2)$; however, the use of marginal estimators over the whole plane is problematic given the limited range of the support of the data.

14.2.2.1 *Plug-In Estimators*   These explicit estimators express F as a function of the marginal survivor function, F_1 and F_2, and the bivariate (double failure) hazard function. The Dabrowska [9] and Prentice–Cai [49] estimators are the two most commonly cited plug-in estimators. Both estimators reduce to the usual empirical estimator in the absence of censoring and have similar asymptotic properties—strong consistency and asymptotic Gaussian distributions—and achieve nonparametric efficiency for the special case of complete independence of all failure and censoring times and continuity of the survivor function. Under more general conditions, however, the asymptotic efficiency is less than full [16]. Another drawback is the estimators' lack of monotonicity due to the incorporation of negative point mass. Pruitt studied the negative mass assignments of the Dabrowska estimator and found that although the amount of negative mass at each point decreases as $n \to \infty$, the number of points assigned negative mass actually increases at rate n^2 [52]. Also troubling, is the observation that the total amount of negative mass assigned does not decrease as $n \to \infty$, despite the estimator's strong consistency. The poor correspondence between the Kaplan–Meier marginal estimators and the hazard rate estimator is believed to be in part responsible for the negative properties of the Dabrowska and Prentice–Cai estimators [51]. The estimators are nevertheless appealing. They are easy to calculate, readily generalize to more than two dimensions, and have good performance in moderate-sized samples ($n \approx 50$) and under heavy censoring.

14.2.3   NPMLE-Type Estimators

In the univariate setting, the nonparametric MLE procedure yields the successful Kaplan–Meier estimator; however, difficulties are encountered in higher dimensions despite convexity of the likelihood function [5]. Censored observations often correspond to flat spots that prevent the definition of a unique NPMLE. Numerous attempts have been made to address this problem, primarily by reducing the number of parameters to be estimated. For example, smoothness assumptions can be imposed followed by kernel estimation [59,52,1]. Alternatively, data reduction can be done by coarsening or grouping the data [61,39]. NPMLE-type estimators are generally monotonic nonincreasing, not assigning any negative mass. Also, they do not use the Kaplan–Meier estimator for the marginal survivor functions and therefore may improve on this estimator when the failure times are correlated and censoring is heavy.

The nonparametric likelihood can be maximized by assigning mass at points of uncensored observations, at points of intersection of singly censored observations with later uncensored observations, and at points of intersections of doubly censored observations with either uncensored observations or singly censored observations. The NPMLE is unique for rare cases where the latter includes only points and not regions. The procedure does not specify how to optimally allocate mass along the half-lines or upper right quadrants resulting from the intersection of censored observations. In addition to the nonuniqueness issue, NPMLE procedures are computationally challenging, often involving high-dimensional iterative procedures. Mass assignments at grid points in the failure-time region are often mutually highly dependent.

Van der Laan's approach was to coarsen the data so that the mass redistribution along half-lines is specified by the likelihood function for the "reduced" or coarsened data. Singly censored observations (represented by half-lines) are interval-censored so that they intersect with later uncensored observations. The mass corresponding to each singly censored observation ($1/n$) can then be redistributed to later uncensored observations in the strip so that mass is placed at points rather than along half-lines. The censoring times are also discretized so that they lie on the same grid as the interval-censored failure times to recover orthogonality of the likelihood. Censoring times must be known or simulated if unknown. The EM algorithm [60,10] is iterated to find a solution to the self-consistency equation [11] for the reduced data:

$$
L = \prod_{i=0}^{I} \prod_{j=0}^{J} p_{ij}^{n_{ij}^{11}} \left(\sum_{\ell \in S_{1i}} \sum_{m > S_{2j}} p_{\ell m} \right)^{n_{ij}^{10}} \left(\sum_{\ell > S_{1i}} \sum_{m \in S_{2j}} p_{\ell m} \right)^{n_{ij}^{01}} \left(\sum_{\ell > S_{1i}} \sum_{m > S_{2j}} p_{\ell m} \right)^{n_{ij}^{00}} \tag{14.1}
$$

where $t_{11}, t_{12}, \ldots, t_{1I}$ denote the uncensored T_1 observations in a sample of size n and $t_{21}, t_{22}, \ldots, t_{2J}$, the uncensored T_2 observations. The total number of observations at (t_{1i}, t_{2j}) is denoted $n_{ij}^{\epsilon_1 \epsilon_2}$, where $\epsilon_k = 1$ indicates failure and $\epsilon_k = 0$ indicates censoring in the kth component; T_1 strip membership of an observation at t_{1i} is denoted by S_{1i}, where $S_{1i} = S_{1l}$ if t_{1i} and t_{1l} fall in the same strip. Similarly, T_2 strip membership is denoted by S_{2j}. Prentice et al. [51] showed that the estimator can also be obtained by Newton–Raphson iteration, which is simpler and noniterative for certain special cases. They also suggest likelihood-based variance and covariance estimators in lieu of bootstrap variance estimates.

Van der Laan's repaired NPMLE has many nice asymptotic properties: strongly consistent and convergent at $\sqrt{n}$ rate to a Gaussian process. In addition, the estimator is nonparametric efficient if the bandwidth goes to zero at a sufficiently slow rate ($<n^{-1/18}$) as the sample size increases [61,62]. Its practical performance, however, is less desirable. It is somewhat sensitive to the choice of bandwidth, and its performance is typically not as good as the simpler Dabrowska and Prentice–Cai estimators in moderate-sized samples [51]. It also suffers from nonuniqueness when there are interval censored observations that do not intersect with later uncensored observations, and the mass must be placed at some arbitrary later point (e.g., on the boundary of the truncated risk region).

Moodie and Prentice [39] proposed an adjustment to van der Laan's repaired NPMLE that improves the estimator's small-sample behavior, reduces sensitivity to bandwidth choice, and has smaller variance estimates when the bandwidth is large. The estimator is termed *reassigned NPMLE*. They argue that the procedure is closer to the conventional NPML procedure than the repaired NPMLE in that mass is assigned not only in the support of uncensored observations but also in the support of the singly censored observations along half-lines. For example, the conventional NPMLE procedure places mass along the half-line implied by a singly censored observation but does not specify the exact coordinate if there is no later uncensored observation along the half-line. The repaired NPMLE places the mass at the next (original or created) uncensored observation in the *strip*; if the bandwidth is large, this may result in mass placement far from the half-line along which the underlying failure time occurred. For the reassigned NPMLE, information is borrowed in the same manner as the repaired NPMLE but the mass is retained along the half-line. The different mass placements of the two procedures for calculating repaired and reassigned NPMLEs are illustrated in Figure 14.1.

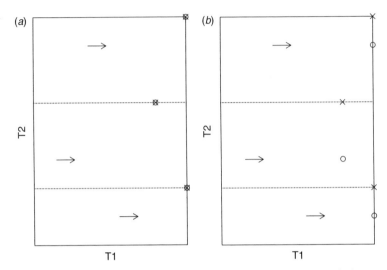

Figure 14.1 Van der Laan's repaired NPMLE (*a*) and Moodie and Prentice's reassigned NPMLE (*b*) approaches to the mass redistribution of singly censored observations (denoted by arrows). Crosses denote uncensored observations, circles denote the location of mass redistribution from singly censored observations, and dashed lines define the partition strips.

For fixed n, van der Laan's repaired NPMLE and the reassigned NPMLE are equivalent in the limit as bandwidth goes to zero, the area of the grid regions decreases and the amount of mass assigned to each region is the same for both estimators. Hence, if $h \to 0$ as $n \to \infty$, both estimators have the same asymptotic properties.

The reassigned NPMLE and van der Laan's repaired NPMLE can be generalized to higher dimensions as both can be expressed using the representation described in Prentice et al. [51]. The representation expresses the bivariate survivor function as a mapping of the hazard function for truncated failure-time variables and involves bivariate Peano series [15]. Extension to multivariate survival times follows in a straightforward manner with the definition of hyperrectangles for the partition grid elements and a multivariate hazard function over the hyperrectangles [51].

14.2.4   Data Application to Danish Twin Data

For illustration, the two plug-in estimators [9,49] and the two NPMLE-type estimators (van der Laan's repaired NPMLE [61] and Moodie and Prentice's reassigned NPMLE [39]) were applied to a subset of the Danish twin data with age at death as the outcome. Example 14.1 contains further details about the data. A random sample of 100 monozygotic male twins was selected, conditional on survival to 15 years of age to compare the performance of the four estimators in a moderate-sized sample. No discernible differences could be seen in the full dataset of 1366 twins (recall the estimators are all consistent). Censoring times were known for all twins in the study; the censoring pattern was typical of end-of-study censoring. The data were truncated at 84.72 years (30944 days, denoted by the solid lines in Fig. 14.2) to provide an appropriate risk region for the NPMLE-type estimators that require $F(\tau_1-, \tau_2-)G(\tau_1, \tau_2) > 0$. Few points (4%) fell outside the truncated risk region.

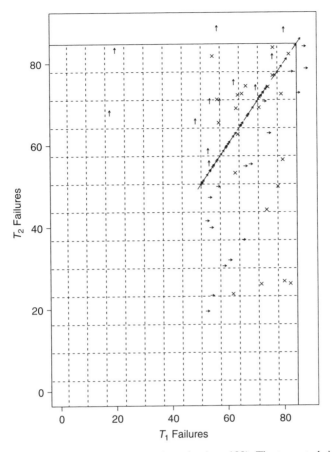

Figure 14.2 Danish twin study: monozygotic males ($n = 100$). The truncated risk region is delimited by the solid lines; dashed lines indicate the coarsening partition for the NPMLE-type estimators. Crosses denote uncensored observations, horizontal arrows denote T_1 singly censored observations, vertical arrows denote T_2 singly censored observations, and diagonal arrows denote doubly censored observations.

Plots of the bivariate survivor function estimators for fixed failure times of one coordinate are shown in Figure 14.3. The Dabrowska estimator is denoted by $\hat{F}_d$, the Prentice–Cai estimator by $\hat{F}_{pc}$, the reassigned estimator by $\hat{F}_{re}$, and van der Laan's estimator by $\hat{F}_{vdl}$. The two plug-in estimators were indistinguishable. The NPMLE-type estimators were very similar, although some minor differences could be noted. The same bandwidth ($h = 2500$) was used for both estimators. Other bandwidths ($h = 1000, 1500, 2000$) were considered but did not differ markedly from $h = 2500$. The joint survivor function estimates were slightly larger toward the tail of the distribution for the smaller bandwidth, and variance estimates were slightly smaller at some points. For the most part, all four estimators gave similar results, with the exception of the functions at 80 years of age, where estimated survival was higher for the plug-in estimators. By convention, the NPMLE-type estimators place the singly censored mass at the boundary of the risk region when the mass cannot be redistributed to later uncensored

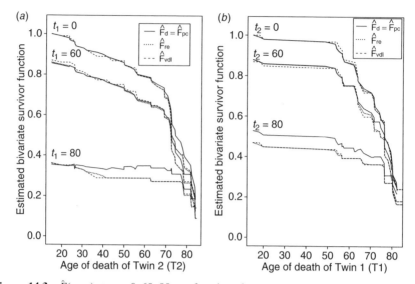

Figure 14.3 $\hat{F}(t_1, t_2)$ at $t_1 = 0, 60, 80$ as a function of t_2 (*a*) and at $t_2 = 0, 60, 80$ as a function of t_1 (*b*) for the plug-in estimators—Dabrowska estimator $\hat{F}_d$ and the Prentice–Cai estimator $\hat{F}_{pc}$; NPMLE-type estimators—reassigned estimator $\hat{F}_{re}$ and van der Laan's estimator $\hat{F}_{vdl}$.

observations. The plug-in estimators also encounter difficulties at these points; their nonmonotonicity is most pronounced here. The marginal survivor functions of the NPMLE-type estimators were similar to the Kaplan–Meier estimators used by the plug-in estimators, and virtually identical to each other.

Bootstrap variances were obtained for the joint and marginal survivor function estimators (Tables 14.1 and 14.2, respectively). The marginal survivor function estimators from the plug-in estimators are Kaplan–Meier estimators, which are denoted by $\hat{F}_{km}$ in Table 14.2. To ensure an appropriate risk region for the NPMLE-type estimators, the point $(\tau_1, \tau_2) = (84.72, 84.72)$ was sampled with probability 1 for each of the 500 bootstrap samples. Greenwood-like variance estimates (denoted by "Grn" in the table) for the NPMLE-type estimators are also provided; they are slightly smaller than the bootstrap variances. Larger sample sizes may be needed for good agreement between the bootstrap and Greenwood-like estimates ($n \approx 250$) [51].

Table 14.1 Bootstrap Variance Estimates ($\times 10^{-3}$) for Select Ages of Prentice–Cai Estimator $\hat{F}_{pc}$, Dabrowska Estimator $\hat{F}_d$, Moodie and Prentice reassigned NPMLE $\hat{F}_{re}$, van der Laan Repaired NPMLE $\hat{F}_{vdl}$ for Danish Twin Data, and Greenwood-Like Variance Estimates (Grn)

(t_1, t_2)	(40,40)	(40,60)	(40,80)	(60,40)	(60,60)	(60,80)
$\hat{V}(\hat{F}_{pc})$	1.062	1.977	5.901	1.732	2.340	5.375
$\hat{V}(\hat{F}_d)$	1.062	1.978	5.960	1.732	2.332	5.516
$\hat{V}(\hat{F}_{re})$	1.062	2.022	7.062	1.749	2.408	6.345
$\hat{V}(\hat{F}_{vdl})$	1.162	2.022	7.457	1.866	2.408	6.716
Grn	1.131	1.859	6.599	1.946	2.464	6.151

Table 14.2 Bootstrap Variance Estimates ($\times 10^{-3}$) for Select Ages of Marginal Survivor Functions Estimated by Prentice–Cai Estimator $\hat{F}_{pc}$, Dabrowska Estimator $\hat{F}_d$, Moodie and Prentice Reassigned NPMLE $\hat{F}_{re}$, and van der Laan Repaired NPMLE $\hat{F}_{vdl}$ for Danish Twin Data

Function	T_1					T_2				
	40	50	60	70	80	40	50	60	70	80
$\hat{V}(\hat{F}_{km})$	0.202	0.295	1.077	2.948	6.287	0.873	1.342	1.822	2.869	5.776
$\hat{V}(\hat{F}_{re})$	0.202	0.307	1.082	2.977	6.308	0.873	1.368	1.871	3.287	7.070
$\hat{V}(\hat{F}_{vdl})$	0.202	0.307	1.082	3.224	6.308	0.975	1.368	1.871	3.646	7.478
Grm	0.196	0.303	1.225	3.076	5.235	0.979	1.317	1.744	3.160	6.665

[a]The marginal survivor function estimates for $\hat{F}_{pc}$ and $\hat{F}_d$ are denoted by Kaplan–Meier estimates $\hat{F}_{km}$. The Greenwood-like variance estimates are also provided.

In short, all four of the estimators exhibited fairly similar performances. The plug-in estimators suffer from nonmonotonicity but have slightly smaller variance estimates than do the NPMLE-type estimators. There were no appreciable differences between the two NPMLE-type estimators in this dataset. The reassigned NPMLE did not considerably outperform van der Laan's repaired NPMLE for any of the bandwidths studied. This is likely due to the configuration of the singly censored and uncensored failure times. Many of the singly censored observations merely placed mass at the boundary given the absence of later uncensored observations in the strips. Therefore the choice of bandwidth did not have much effect on the mass redistribution. Computation time is not an issue for any of the estimators; for the twin data, each were calculated in less than a second in R. Computation is simpler and faster for the plug-in estimators in that a bandwidth needs not be chosen, but the NPMLE estimators have the potential of direct calculation of Greenwood-like variance estimates.

14.3   NON- AND SEMIPARAMETRIC ESTIMATION OF DEPENDENCE MEASURES

Various nonparametric dependence measures that require no assumption of joint survival distribution have been proposed in quantifying the degree of dependence between paired failure times. They can be used in characterizing the changes in the strength of dependence over the support for T_1 and T_2 in the cohort setting. Semiparametric modeling, in that the dependence is quantified by one or a few parameters while the marginal hazard function is non- or semiparametric, provides a flexible approach to estimation of dependence parameters under a variety of settings. In what follows, we will describe both approaches to dependence estimation.

14.3.1   Nonparametric Dependence Estimation

Global dependence measures assess the overall strength of dependence of failure times over $(0, \infty)$. Commonly used ones, such as the Kendall's τ [31] and Spearman's ρ [57], are typically rank-based because the failure-time distribution is skewed and the dependence is usually nonlinear. The generalization of the Kendall's τ to the censored survival data was studied by Brown et al. [3] and later by Oakes [45]. For non-rank-based measures such as the Pearson's correlation coefficient, the failure time can be first transformed by, for example, the cumulative hazard function. Under the Cox proportional hazards model, such transformation would yield unit exponential distributed random variables [29]. The correlation is then assessed for the transformed variables. The text by Hougaard [24] and references cited therein gave a comprehensive review of various global measures and compared them in real datasets, including the Danish twin data. While the global measures are intuitive to interpret and straightforward to calculate, they are not suitable for correlated failure times that have restricted support and are subject to right censoring.

Local dependence measures that take a snapshot of the dependence at a single timepoint can capture the varying degree of strength over time. One particular useful such measure is the *cross-ratio function* [7,46], which is defined as

$$C(t_1, t_2) = \frac{F^{(11)}(t_1, t_2)/F^{(01)}(t_1, t_2)}{F^{(10)}(t_1, t_2)/F(t_1, t_2)},$$

where $F^{(10)}(t_1, t_2), F^{(01)}(t_1, t_2)$, and $F^{(11)}(t_1, t_2)$ are partial derivatives with respect to the first variate, the second variate, and both, respectively. It is the ratio of conditional hazard rates for the first subject of the pair at time t_1 given the second subject failing at time t_2 and given the second subject disease-free until time t_2. When the cross-ratio function is expressible as a function of $F(t_1, t_2)$, the joint survival function $F(t_1, t_2)$ takes the Archimedean model with the functional form determined implicitly [46]. One can therefore directly focus on $C(t_1, t_2)$, or any function of $C(t_1, t_2)$, instead of $F(t_1, t_2)$ when assessing the strength of the dependence. For example, the conditional version of the commonly used Kendall's τ [46] for concordance measure given the componentwise minima can be equivalently written as $\{C(t_1, t_2) - 1\}/\{C(t_1, t_2) + 1\}$.

Fan et al. [14] proposed a class of weighted dependence measures based on $C(t_1, t_2)$, which can be written as

$$D(B_1, B_2) = \frac{\int_{B_1} \int_{B_2} \phi\{C(S_1, S_2)\} w(ds_1, ds_2)}{\int_{B_1} \int_{B_2} w(ds_1, ds_2)},$$

where ϕ is a continuous function, w is a weight function, and B_1 and B_2 are the intervals for the region of the bivariate failure times over which the dependence is assessed. The weight function can take the function of the joint survivor function and its partial differentiations. The nonparametric estimators of bivariate survival function described in Section 14.2 can be used to estimate the weight functions $w(t_1, t_2)$ and $C(t_1, t_2)$. The caveat, as pointed out by Fan et al. [14], is that $\phi\{C(S_1, S_2)\} w(ds_1, ds_2)$ and $w(ds_1, ds_2)$ should not have an unintegrated differential function in the denominator because such functions are nonzero only at the uncensored observations. In this situation, $\hat{D}(B_1, B_2)$ with plug-in nonparametric estimators $\hat{F}(t_1, t_2)$ is not consistent with the true dependence measure.

There is a wide choice in the functional form ϕ of $C(t_1, t_2)$ as well as the weight function $w(t_1, t_2)$. For example, one can choose the weight function $w(t_1, t_2)$ to give greater weight to early or late dependence measured by $\phi\{C(t_1, t_2)\}$. One also need not necessarily use only the cross-ratio function $C(t_1, t_2)$ for measuring the association. Yan and Fine [66] proposed a local odds ratio (LOR)

$$\text{LOR}(t_1, t_2) = \frac{F(t_1, t_2)\{1 - F(t_1) - F(t_2) - F(t_1, t_2)\}}{\{F(t_1) - F(t_1, t_2)\}\{F(t_2) - F(t_1, t_2)\}},$$

as a time-varying dependence measure. A class of weighted dependence measure can be constructed using $\text{LOR}(t_1, t_2)$ in the same fashion as in $D(B_1, B_2)$. Large-sample properties and caveats should follow similarly to $D(B_1, B_2)$.

The weighted measure $D(B_1, B_2)$ can be extended to include covariates by transforming the failure times T_1 and T_2 with the cumulative hazard function $\Lambda(\cdot)$ and then calculate $D(B_1, B_2)$ using the transformed failure times [13]. The estimators $\hat{\Lambda}(\cdot)$ can be obtained through the generalized estimating equations approach, which treats the dependency as nuisance and only adjusts it in the variance estimators [65].

Generally speaking, the nonparametric dependence estimation is appealing because the estimation procedure is noniterative and is distribution-free. However, it is rather restrictive in dealing with complex sampling situations such as the case–control design in the family study of breast cancer of Example 14.2. In this situation, it is not even clear whether a nonparametric bivariate survival function estimator is identifiable from case–control family data.

14.3.2 Semiparametric Dependence Estimation

In this section, we focus on the broad class of the copula model for the multivariate survival distribution. The copula model includes two elements: the marginal hazard function for each of the failure times in the cluster and the correlation function of these failure times. It is given by

$$F(t_1, \ldots, t_K) \equiv \Pr(T_1 > t_1, \ldots, T_K > t_K)$$
$$= g\{F(t_1), F(t_2), \ldots, F(t_K); \theta\},$$

where $g(\cdot)$ is a fixed function, $F(t_k)$ is the marginal survival function for the kth failure time in a cluster, and θ is a vector of a finite-dimensional parameter that governs the dependence among $F(t_1), F(t_2), \ldots$, and $F(t_K)$. When the marginal distributions $F(t_K), k = 1, \ldots, K$, include an unspecified function such as the unspecified baseline hazard function in the Cox proportional hazards model, the resulting joint survivor distribution is semiparametric. The estimation of dependence parameters is termed *semiparametric*, even though the dependence parameters themselves are of only finite dimension. This term is somewhat misleading and used here simply for conforming to the conventional usage in the literature. The copula model is a very general model and encompasses many commonly used multivariate survival models, such as the Clayton model [7], the positive stable model [22,23], and the more complicated survival model that accounts for multiple levels of clustering [2]. Further details about the copula model can be found in the books by Nelson [43] and Hougaard [24].

There are two main approaches to semiparametric modeling: the frailty model and the partially specified model. Before describing these approaches, we will first introduce some notation. Consider n independent clusters. Let T_{ik} and C_{ik} be the failure time and censoring time, respectively, for the kth member in the ith cluster, $k = 1, \ldots, K_i, i = 1, \ldots, n$. We assume that K_i is relatively small compared to n, and that $K = \max_i K_i$ is bounded. For each individual, one can observe only $X_{ik} = T_{ik} \wedge C_{ik}$ and $\delta_{ik} = I(t_{ik} \leq C_{ik})$. Let Z_{ik} be a $p \times 1$ vector of bounded covariates. For simplicity, we assume that each family has the same set of relatives K. The presence or absence of any particular type of relative is assumed to be independent of the observed data. The absence of these members can be denoted by setting $C_{ik} = 0$. Naturally, such members will not contribute to any of the likelihoods. In the counting process notation, $Y_{ik}(t) = I(X_{ik} \geq t)$ and $N_{ik}(t) = \delta_{ik} I(X_{ik} \leq t), t \in [0, \tau]$, where τ is the maximum follow-up time.

14.3.2.1 Frailty-Model-Based Approach

A frailty model, or a random-effects model for failure time, extends the univariate hazard function by allowing unobserved random variables or frailties in addition to the fixed effects of covariates. The frailties are used to describe part of the risk shared by the individuals within the cluster and reflect some unmeasured influences on the disease risk. In family studies, the frailties can be either common genetic material or a shared environment. The parameters in the frailty distribution therefore measure the strength of dependences among multivariate failure times.

A simple shared frailty model extends the Cox proportional hazards model by having frailty ω acting multiplicatively on the hazard function

$$\lambda^c(t; Z_{ik}, \omega_i) = \lambda_0^c(t) \exp\{(\beta^c)^T Z_{ik}\} \omega_i, \tag{14.2}$$

where $\lambda_0^c(t)$ is an unspecified conditional baseline hazard function, Z_{ik} is a vector of fixed covariates, β^c is the corresponding regression coefficient, and ω_i follows a frailty distribution $h_\theta(\cdot)$ with dependence parameter θ. The superscript T refers to *transpose*. Conditional on ω_i and all Z_{ik} terms, the failure times $T_{i1}, T_{i2}, \ldots, T_{ik}$ are independent of each other.

Clayton and Cuzick [8] were perhaps the first ones to consider the joint estimation of parameters in the hazard function and in the frailty distribution by using order statistics and imputed frailties. In the discussion that accompanied the Clayton–Cuzick paper, Richard Gill proposed a nonparametric maximum-likelihood method by using the expectation–maximization (EM) algorithm. In that the latent frailties are estimated given the current parameter estimates in the E-step and the M-step updates the parameter estimates by maximizing the likelihood with latent frailties substituted with estimated ones obtained in the E-step. The expectation and maximization steps are iterated until all parameter estimates converge. It was not, however, until 1992 that Nielsen et al. [44] and Klein [32] formalized the approach for the semiparametric gamma frailty model. Murphy [41,42] provided consistency and asymptotic normality for this model without covariates. Parner [47] further extended the asymptotic results to a more general correlated gamma frailty model with covariates. He also presented a consistent estimator of the limiting covariance matrix of the estimator, based on inverting a discrete observed information matrix. Since the dimension of the observed information matrix is the number of the regression coefficients plus the number of observed failure times, inverting the matrix is practically infeasible for a large dataset with many distinct failure times. He proposed another covariance estimator based on solving a discrete version of a second-order Sturm–Liouville equation. This covariance estimator requires substantially less computational effort, but is still difficult to implement.

Another approach to estimating the parameter is by using the stochastic intensity process for the counting process $N_{ik}(t)$ conditional on the preceding observed information for all members within the ith cluster up to time t [53]. The observed history for all members up to time t denoted by filtration $\mathcal{F}_t$. The $\mathcal{F}_t$ intensity of $N_{ik}(t)$ can be obtained by replacing ω_i with its conditional expectation with respect to $\mathcal{F}_t$:

$$\lambda_{ik}^{\mathcal{F}}(t) = Y_{ik}(t)\lambda_0^c(t)\exp\{(\beta^c)^T Z_{ik}\}E(\omega_i|\mathcal{F}_t)$$

A natural estimator of $\Lambda_0^c(t)$ is a Nelson–Aalen-type estimator, where the jump size at time t is the ratio of the number of failures at time t and the summation of $\exp\{(\beta^c)^T Z_{ik}E(\omega_i|\mathcal{F}_t)\}$ over at-risk individuals at time t. The estimators $\hat{\beta}^c$ and $\hat{\theta}$ can be obtained by maximizing the likelihood function with plug-in $\hat{\Lambda}_0^c(t)$ estimator [20]. Gorfine et al. [20] showed that the parameter estimates were consistent and asymptotically normal with a consistent covariance estimator that could be easily computed. Since a plug-in estimator $\hat{\Lambda}_0^c(t)$ is used for parameter estimation, some efficiency loss is expected. However, the simulation study [20] showed that there was little efficiency loss compared to the nonparametric MLE. The correlations of the estimates between the two methods for 1000 simulated datasets were ≥ 0.95 for the situations considered.

14.3.2.2 *Partially Specified Model*

A partially specified model assumes only the selected features of the joint distribution. The selected features are typically the first few moments of the data. In survival analysis, this could simply be the marginal survival functions and/or pairwise dependences of failure times. For example, one can use the widely used Cox proportional hazards model to postulate the effects of covariates on the hazard function, which can be written as

$$\lambda^m(t; Z_{ik}) = \lambda_0^m(t) \exp\{(\beta^m)^T Z_{ik}\}, \tag{14.3}$$

where $\lambda_0^m(t)$ is an unspecified marginal hazard function and β^m is a vector of regression coefficients. Note that the superscript "m" refers to marginal to distinguish from the conditional hazard function (14.2) given the latent frailty ω, where the superscript "c" refers to conditional. The interpretation for β^m and β^c are different in that β^m measures the population-averaged effects of the covariates, whereas β^c measures the cluster-specific effects of the covariates. Wei et al. [65] proposed an approach for estimating $\Lambda_0^m(t)$ and β^m assuming the independence among the correlated failure times and incorporated the correlation in the covariance estimators of the parameter estimates by a "sandwich" type of estimator. In this approach, the parameter estimates are solution to the equations

$$\sum_{i=1}^n \sum_{k=1}^K \int_0^\tau Z_{ik}^T [N_{ik}(dt) - Y_{ik}(t) \exp\{(\beta^m)^T Z_{ik}\} \hat{\Lambda}_0(dt)] = 0,$$

where $\hat{\Lambda}_0(dt) = \sum_i \sum_k N_{ik}(dt) / [\sum_i \sum_k Y_{ik}(t) \exp\{(\beta^m)^T Z_{ik}\}]$. An intuitive view of the left side of this equation is that it is the weighted difference between the observed events and the expected events under the assumed Cox proportional hazards model with the weight being the individual covariate vector.

Prentice and Hsu [50] generalized this approach in a similar manner for estimating pairwise dependencies, in that the estimating function $U(\theta)$ takes the differences between the empirical pairwise conditional covariance rate or the A function defined in Prentice and Cai [49] and the corresponding conditional covariance rate function under an assumed copula model. The A function is an analog of the marginal hazard rate function of the univariate failure time in the bivariate failure times for measuring the dependence function. It is a rate of concurrence of events for both members of the pair conditional on the pair being at risk. Similar to the procedure due to Wei et al. [65], the dependence parameter estimators can be obtained by setting the estimating function equal to 0, $U(\theta) = 0$.

Another approach to estimating dependence parameters in a copula model is to construct a likelihood for θ assuming the marginal survival functions to be known

$$L_1 = \prod_{i=1}^n g^{(\delta_{i1}\delta_{i2}\cdots\delta_{iK})}\{F(X_{i1}), F(X_{i2}), \ldots, F(X_{iK}); \theta\}, \tag{14.4}$$

where $g^{(\delta_{i1}\delta_{i2}\cdots\delta_{iK})}$ takes the corresponding partial derivatives of g with respect to the components for which $\delta_{ik} = 1, k = 1, \ldots, K$. Since $F(X_{ik})$ involves unknown cumulative marginal hazard function $\Lambda_0^m(X_{ik})$ and β, the generalized estimating equation estimators proposed by Wei et al. [65] may be used to substitute these unknown parameters [55].

The cluster $\{1, \ldots, K\}$ can be further decomposed into subsets within which the members share the same dependence. For example, in family studies, one may decompose a family into subsets of spouse, parent–offspring, and sibling–sibling to examine the patterns of familial resemblance in age at onset. It is worth noting that the dependences between subsets need not be different. The decomposition of a large cluster into smaller units can be merely for the simplicity of calculations and for the robustness against the misspecification of a higher-dimensional joint distribution. One may construct a pseudolikelihood function by taking the product of all subset contributions, each of which takes the form of (14.4). Glidden [17] established a large-sample theory for such maximum pseudolikelihood estimators.

14.3.2.3 Frailty versus Partly Specified Models. The two types of model differ in several aspects:

1. The interpretation of parameters in the partly specified model is population-averaged, whereas the parameter interpretation in the frailty model is cluster-specific. In family studies, the marginal hazard function $\lambda^m(t; Z)$ is the averaged hazard of developing a disease for an individual with covariates Z. In the frailty model, the conditional hazard function $\lambda^c(t; Z, \omega)$ is the hazard function for an individual with covariates Z given the family-specific frailty ω.

2. The partly specified model-based approach is more robust to model misspecification because the higher-order dependences are often left unspecified. On the other hand, the frailty model assumes a distribution for the frailties and can be restrictive for the higher-order dependences. Our and others' experiences [19] are that the parameter estimates are actually quite robust to the misspecification of the frailty distribution.

3. While the partly specified approach is more robust to model specification, it may suffer some efficiency loss in parameter estimates compared to the frailty approach. The extent of loss likely depends on the strength of the dependences and the cluster sizes: the stronger the strength and/or the larger the size, the greater the loss.

4. The frailty model provides a natural framework for estimating the individualized disease risk, as in the counseling situation. The frailty for a particular family can be calculated from the posterior distribution conditional on the data from the family. The predicted disease risk incorporates the familial heterogeneity through the latent frailty, as the failure time of a member is completely independent of that of others when conditional on the frailty and covariates. Under the partly specified model, one may similarly obtain the family-specific disease risk by calculating the survival function conditional on the observed failure times and disease status of all other members in the same family under a copula model. However, the complexity of fitting the joint distribution, including all the partial derivatives, in calculating the conditional distribution increases exponentially with the number of the family size, posing significant feasibility issues in implementation.

Despite these differences, one can obtain the marginal survival function estimates by integrating $\int_0^\infty \lambda^c(t; Z_{ik}, \omega_i) h_\theta(\omega_i) d\omega_i$. The marginal survival function after integrating out the frailty depends on the dependence parameters and may not be desirable if the main interest is in the marginal hazard function. In this situation, in order to take advantage of the frailty-model-based approach, one can derive the frailty model such that the marginalized hazard function, after integrating out the latent frailty, follows a model that is free of dependence parameters. Glidden and Self [18] and Pipper and Martinussen [48] developed statistical methods for linking the marginal hazard function to the frailty model. For example, under a gamma frailty model, the conditional hazard function

$$\lambda_{ik}^c(t; Z_{ik}, \omega_i) = \lambda_0^m(t) \exp\{(\beta^m)^T Z_{ik} + \theta \exp\{(\beta^m)^T Z_{ik}\} \Lambda_0^m(t-)\} \omega_i \qquad (14.5)$$

yields a marginalized hazard function that follows the Cox proportional hazards model (14.3).

In this formulation, an added bonus is that one can obtain both the population-averaged $\lambda^m(t)$ and the cluster-specific hazard function $\lambda^c(t)$ simultaneously.

14.3.3 An Application to a Case–Control Family Study of Breast Cancer

In this section, we illustrate the semiparametric dependence estimation methods by analyzing a subset of a case–control family study of early-onset breast cancer (see Example 14.2 for description of the study). We will first describe the aspects that are specific to retrospective correlated failure times and the modifications of current methods for accommodating such a design.

14.3.3.1 Methods for Retrospective Correlated Failure Times. There are two assumptions that may be unique to retrospective correlated failure times arising from family studies: (1) the baseline hazard function is common for all family members and (2) family members are interchangeable with respect to dependences. The interchangeability assumption could be relaxed by specifying relation-specific frailty such as mother–daughter, sister–sister, or grandmother–granddaughter. However, the common baseline hazard function for all family members cannot be relaxed as it is the basis for using the information from the relatives of cases and controls to estimate the baseline hazard function. Otherwise, the baseline hazard function would be confounded with the dependence parameters. This is different from the prospective correlated data, in which each individual in a cluster could have a different baseline hazard function.

Let the correlated age-at-onset data from the ith ($i = 1, \ldots, n$) family be denoted by $\mathbf{X}_i = (X_{i1}, \ldots, X_{iK}), \delta_k = (\delta_{i1}, \ldots, \delta_{iK})$, and $\mathbf{Z}_i = (Z_{i1}, \ldots, Z_{iK})$ for the relatives, and $\{X_{i0}, \delta_{i0}, Z_{i0}\}$ for the case or control of the ith family. The likelihood function for case–control family data can be written as follows:

$$L_2 = \prod_{i=1}^{n} f(\mathbf{X}_i, \delta_i, \mathbf{Z}_i, Z_{i0} \mid X_{i0}, \delta_{i0}). \tag{14.6}$$

The extension of the frailty-based approach to accommodate this design is straightforward because, conditional on the frailty and all other covariates, the failure times of the relatives are independent among themselves as well as from that of cases and controls. The parameter estimates $\{\hat{\beta}^c, \hat{\Lambda}_0^c(t)\}$ can simply be obtained from the relatives by using the standard software routine for the Cox proportional hazards model with an offset term of log (frailty). Since the frailties are unknown for all families, they are estimated by the posterior expectations conditional on the observed data, for which the retrospective case–control design is naturally accommodated. Dependence parameter estimates $\hat{\theta}$ are updated by maximizing the profile likelihood function L in (14.6) with plug-in estimates for $\hat{\beta}^c$ and $\hat{\Lambda}_0^c(t)$. This procedure is iterated among all of these steps until all estimators $\hat{\beta}^c$, $\hat{\Lambda}_0(t)$, and $\hat{\theta}$ converge. For further details of this method, readers are referred to Hsu et al. [27] and Hsu and Gorfine [28].

To use the partially specified model-based approach for case–control family data, the key is to obtain a proper estimate of the baseline hazard function in the presence of oversampling cases. Shih and Chatterjee [54] proposed a Nelson–Aalen-type estimator for the baseline hazard function

$$\hat{\Lambda}_0^m(t) = \int_0^t \frac{\sum_{i=1}^{n} \sum_{k=1}^{K} N_{ik}(ds)}{\sum_{i=1}^{n} \sum_{k=1}^{K} Y_{ik}(s) r_{ik}(\beta, s) h_{ik}\{F_{i0}(X_{i0}), \delta_{i0}, F_{ik}(s)\}} \tag{14.7}$$

where $r_{ik}(\beta, s) = \exp[(\beta^m)^T Z_{ik} - \Lambda_0^m(s-) \exp\{(\beta^m)^T Z_{ik}\}]$, and

$$h_{ik}(u_1, \delta_{i0}, u_2) = \left\{\frac{g^{(11)}(u_1, u_2)}{g^{(10)}(u_1, u_2)}\right\}^{\delta_{i0}} \left\{\frac{g^{(01)}(u_1, u_2)}{g(u_1, u_2)}\right\}^{1-\delta_{i0}}$$

under an assumed copula function g. Note that the function $h_{ik}(\cdot)$ involves the marginal survival function $F_{i0}(X_{i0})$ at the case or control observational time X_{i0}. It requires looking ahead of the baseline hazard function at X_{i0} if the observational time X_{i0} for the case or control is greater than time s. An iterative procedure is thus needed to obtain the estimate $\hat{\Lambda}_0^m(t)$. Once an estimate for the cumulative baseline hazard function is obtained, a likelihood function can be constructed for β and θ by plugging in $\hat{\Lambda}_0^m(t)$. Standard maximum-likelihood theory and numerical routines can be applied to the likelihood function, yielding the maximum-likelihood estimators $\hat{\beta}$ and $\hat{\theta}$. It is worth noting that $\hat{\Lambda}_0^m(t)$ is a function of both the joint distribution $g(\cdot)$ and the dependence parameter θ, which is quite different from the generalized estimating equation approach [65] for prospective correlated failure-time data. In that sense, the partially specified approach is no different from the frailty approach as both need to make an assumption of a joint distribution in order to identify $\Lambda_0^m(t)$. The difference between the two approaches lies in that the partially specified approach only needs to assume pairwise joint distributions, whereas the frailty approach would need to specify the full joint distribution of all family members.

14.3.3.2   *Generalization to Multivariate Dependencies*   In family studies, family members have different strength of dependences in ages at onset because they share varying degrees of genes and environment. Such patterns have been useful for providing first clues to the etiology of the disease. In what follows we will discuss possible generalizations for the frailty and the partially specified approaches, respectively.

The frailty-model-based approach provides a natural framework to dissect genetic and environmental contributions to the observed aggregation of disease occurrence in families. For example, Li and Thompson [33] and Siegmund and McKnight [56] introduced an extended Cox proportional hazards model that included two random-effects components: individual-specific and family-specific. For individual-specific random effects, they assume that a single major Mendelian diallelic locus governs the disease susceptibility, whereas the family-specific random effect is used to accommodate risk heterogeneity over families that may be due to shared environmental risk factors or disease loci or mutations other than the major locus considered in the individual-specific random effects. This modeling makes it possible, at least in theory, to separate the major gene effect from the shared family-specific random effect and thus provides important evidence for future gene discovery efforts through the linkage analysis [34].

There are also other approaches for generalizing to include multivariate dependences among family members. Yashin et al. [67] and Wienke et al. [64] studied a correlated gamma frailty model. Under this model, individual family members can carry different, possibly dependent, random effects. Bandeen-Roche and Liang [2] proposed a recursive nesting of univariate frailty-type distribution through which Archimedean copula forms were determined for all bivariate margins. Unfortunately, because the joint distribution is fully specified, some of these approaches may impose constraints on the dependences. For example, under Bandeen-Roche and Liang's nested model, between-subcluster associations are not allowed to exceed the associations within subclusters.

The partially specified model, on the other hand, is flexible in assessing the aggregation pattern because it only needs to specify the lower-dimensional joint distribution among a subset of family members. For example, in case–control family data, family members could be grouped by their relations (e.g., father, mother, sibling) to the case or control, or simply by their individual relation to the case or control. For each subset, one could assume a joint distribution that involves only the relatives within each subset. The dependence parameters can be different for different relationships to account for varying strength of dependences among family members. Furthermore, even the joint distributions can differ as long as the marginal survival function of the case or control remains the same regardless of which subset that individual is in. Obviously such simplification gains robustness against the misspecification of a higher-dimensional joint distribution, albeit with the price of efficiency loss in some situations. Moreover, it lacks a proper full joint distribution that permits arbitrary specification of distributions for the subclusters. One possible remedy for this problem is to use a multivariate normal transformation model [35]. In this model, a failure time T is transformed marginally to a standard normal variate $W = \Phi^{-1}\{1 - e^{-\Lambda(T)}\}$, where Φ is the cumulative distribution function for $N(0, 1)$. Then the transformed $(W_1, \ldots, W_K)$ for a family are assumed to follow a K-variate normal distribution $N(0, \Sigma)$, where the off-diagonal elements of Σ measure the correlations of family members. The correlation structure may be modeled by either following a genetic model [26] or setting the correlations same for all those having the same relationship. The most general form of Σ is to let each paired relatives have a different correlation coefficient, leaving Σ completely arbitrary.

Other than modeling different dependences among family members, efforts have also been made to generalize the joint survival distribution to accommodate the often complex dependence over time. These include, but are not limited to, the power variance function [24], piecewise constant cross-ratio function [30,42], and compound Poisson distribution [38].

In principle, estimation procedures for the dependence parameters in these distributions could largely follow the procedures described in this section. However, these procedures are complicated and often require special programming to implement them. This, unfortunately, limits their usage in a general data setting.

14.3.3.3 Analysis Results

Because of the lack of software for conducting a comprehensive analysis, we will illustrate the methods using only paired relatives, specifically, mother–daughter pairs. In this subset of the dataset, there are 820 cases, 942 controls, and their mothers. Among these, 110 (13.4%) case mothers have breast cancer with mean age at diagnosis about 52 years, and 45 (14.8%) control mothers have breast cancer with mean about 56 years.

In order to facilitate the comparison of the two approaches, we modify the partially specified approach to estimate $\hat{\Lambda}_0^c(t)$ and $\hat{\beta}^c$ by using the conditional hazard function (14.2), deriving the bivariate distribution, and obtaining corresponding baseline hazard function estimator similar to (14.7). The gamma frailty model, equivalently, the Clayton–Oakes model, is assumed for the joint survivor distribution. Since there is no covariate in the breast cancer data, Table 14.3 shows only the estimates of $\hat{\Lambda}_0^c(t)$ and $\hat{\theta}$. Fifty bootstrap samples with families as resampling units were used to estimate the standard errors of these parameter estimates. One can see that both methods yield very similar parameter estimates and the standard error estimates. In these data, the mothers of cases have an ~ 2.7-fold increase in the risk of developing breast cancer compared to the mothers of controls.

We also modify the frailty-model-based approach to estimate marginal $\Lambda_0^m(t)$ and β^m instead of conditional cumulative hazard function and regression coefficients. Under the

Table 14.3   Summary of Conditional Cumulative Hazard Function Estimate $\hat{\Lambda}_0^c(t)(\times 10^{-2})$ and Dependence Parameter Estimate $\hat{\theta}(\times 10^{-2})$ for Frailty-Model-Based and Partially Specified Model-Based Approaches

Function	Frailty		Partially Specified	
	Estimate	SE	Estimate	SE
$\Lambda_0^c(40)$	0.815	0.193	0.820	0.194
$\Lambda_0^c(50)$	2.348	0.397	2.354	0.397
$\Lambda_0^c(60)$	4.108	0.639	4.115	0.639
$\Lambda_0^c(70)$	7.249	1.240	7.257	1.240
θ	1.715	0.560	1.711	0.559

[a]In these estimates, age index $t = 40, 50, 60$, and 70 years; also, 50 bootstrap samples were used to calculate bootstrap standard deviations as standard error (SE) estimates.

Table 14.4   Summary of Conditional Cumulative Hazard Function Estimate $\hat{\Lambda}_0^c(t)(\times 10^{-2})$ and Dependence Parameter Estimate $\hat{\theta}(\times 10^{-2})$ for Frailty-Model-Based and Partially Specified Model-Based Approaches

Function	Frailty		Partially Specified	
	Estimate	SE	Estimate	SE
$\Lambda_0^m(40)$	0.749	0.157	0.749	0.157
$\Lambda_0^m(50)$	1.983	0.315	1.984	0.315
$\Lambda_0^m(60)$	3.742	0.587	3.745	0.587
$\Lambda_0^m(70)$	5.732	1.050	5.738	1.050
θ	1.898	0.589	1.894	0.594

[a]In these estimates, age index $t = 40, 50, 60$, and 70 years; also, 50 bootstrap samples were used to calculate bootstrap standard deviations as standard error (SE) estimates.

gamma frailty model, a closed form such as that described in (14.5) for the conditional hazard function would yield a marginalized hazard function that obeys the Cox proportional hazards model and is free of the dependence parameter θ. Table 14.4 presents the results for $\hat{\Lambda}_0^m$ and $\hat{\theta}$ under the Cox proportional hazards model with gamma frailty distribution. Again both approaches yield very similar parameter estimates with comparable efficiency. The mothers of cases again have about a 2.9-fold increased risk of developing breast cancer compared to the mothers of controls. Using the estimates from the frailty model approach, the probability of a woman developing breast cancer by age 70 is $\exp(-0.05732) = 5.6\%$ with the 95% confidence interval (3.6%, 7.5%).

14.4   CONCLUDING REMARKS

Nonparametric estimation of the bivariate survivor function has proved difficult; relatively few estimators have been proposed since the early 1980s. The first viable options were the plug-in estimators of Dabrowska and Prentice–Cai. Despite their theoretical asymptotic inefficiency, they offer good practical performance with variance estimators that are hard to beat in small to moderate-sized samples [62,51]. They are, however, nonmonotonic. Prentice and Cai [49] suggested a simple correction to impose a monotonicity requirement, but asymptotic properties of the resulting estimator have not been proved. In theory, the NPMLE-type estimators offer significant advantages (monotonic and efficient under a condition on the bandwidth); however, the sample sizes may need to be very large for the estimators to improve on the simply calculated plug-in estimators [62,51]. The reassigned estimator shows some advantage over van der Laan's repaired NPMLE, but the improvement is not dramatic. With the exception of large sample sizes and very strong dependence between the two failure times, the Dabrowska and Prentice–Cai estimators may be preferred.

There has been considerable progress in the methodologic and theoretical development in the semiparametric estimation of dependencies of the multivariate failure time data since the mid-1980s. Both frailty-model-based and partially specified model-based approaches have stable numerical properties in the finite sample sizes. The frailty model describes a cluster-specific hazard function, whereas the partially specified model typically focuses on the population-averaged hazard function. One can, however, use the frailty-model-based approach to estimate the population-averaged hazard function and the partially specified model-based approach to estimate cluster-specific hazard function. The choice of which approach to use will depend on the actual application and the primary quantities of interest.

Applications of these methods to real datasets often require substantial modifications, extensions, and theoretical developments. This chapter describes one particular application to case–control family data, where the non-cohort-based sampling scheme needs to be adjusted in order to yield consistent parameter estimates. In this situation, even though the methods appeared to work well in both the simulated and real datasets, the large sample theory for these methods is largely unavailable. Further work in this area is warranted.

REFERENCES

1. Akritas, M. G. and Keilegom, I., Estimation of bivariate and marginal distributions with censored data, *J. Roy. Statist. Soc. Ser. B Methodol.* **65**, 457–471 (2003).

2. Bandeen-Roche, K. J. and Liang, K.-Y., Modelling failure-time associations in data with multiple levels of clustering, *Biometrika* **83**, 29–39 (1996).

3. Brown, W. B., Hollander, M., and Korwar, R. M., Nonparametric tests of independence for censored data with applications to heart transplant studies, in *Reliability and Biometry: Statistical Analysis of Lifelength*, Proschan, F. and Serfling, R. G., eds., SIAM, Philadelphia, 1974. pp. 327–354.

4. Burke, M. D., Estimation of a bivariate distribution function under random censorship, *Biometrika* **75**, 379–382 (1988).

5. Campbell, G., Nonparametric bivariate estimation with randomly censored data, *Biometrika* **68**, 417–422 (1981).

6. Campbell, G. and Földes, A., Large-sample properties of nonparametric bivariate estimators with censored data, in *Nonparametric Statistical Inference* (in 2 vols.), 1982, pp. 103–122.

7. Clayton, D. G., A model for association in bivariate life tables and its application in epidemiological studies of familial tendency in chronic disease incidence, *Biometrika* **65**, 141–152 (1978).

8. Clayton, D. and Cuzick, J., Multivariate generalization of the proportional hazards model (with discussion), *J. Roy. Statist. Soc. A* **148**, 82–117 (1985).

9. Dabrowska, D. M., Kaplan-Meier estimate on the plane, *Ann. Statist.* **16**, 1475–1489 (1988).

10. Dempster, A. P., Laird, N. M., and Rubin D. B., Maximum likelihood from incomplete data via the EM algorithm, *J. Roy. Statist. Soc. Ser. B Methodol.* **39**, 1–22 (1977).

11. Efron, B., The two sample problem with censored data, *Proc. 5th Berkeley Symp. Mathematical Statistics*, 1967, Vol. IV, pp. 831–853.

12. Fan, J., Hsu, L., and Prentice, R. L., Dependence estimation over a finite bivariate failure time region, *Lifetime Data Anal.* **6**, 343–355 (2000).

13. Fan, J. and Prentice, R. L., Covariate-adjusted dependence estimation on a finite bivariate failure time region, *Statistica Sinica* **12**, 689–705 (2002).

14. Fan, J., Prentice, R. L., and Hsu, L., A class of weighted dependence measures for bivariate failure time data, *J. Roy. Statist. Soc. Ser. B Methodol.* **62**, 181–190 (2000).

15. Gill, R. D. and Johansen, S., A survey of product-integration with a view toward application in survival analysis, *Ann. Statist.* **18**, 1501–1555 (1990).

16. Gill, R. D., van der Laan, M. J., and Wellner J. A., Inefficient estimators of the bivariate survival function for three models, *Ann. Inst. Henri Poincare Sect. B Calcul Probabil. Statistiques* **31**, 545–597 (1995).

17. Glidden, D. V., A two-stage estimator of the dependence parameter for the Clayton-Oakes model, *Lifetime Data Anal.* **6**, 141–156 (2000).

18. Glidden, D. V. and Self, S. G., Semiparametric likelihood estimation in the Clayton-Oakes failure time model, *Scand. J. Statist.* **26**, 363–372 (1999).

19. Glidden, D. V. and Vittinghoff, E., Modeling clustered survival data from multicenter clinical trials, *Statist. Med.* **23**, 369–388 (2004).

20. Gorfine, M., Zucker, D., and Hsu, L., Prospective survival analysis with a general semiparametric shared frailty model—a pseudo full liklihood approach, *Biometrika* **93**, 735–741 (2006).

21. Hauge, M., *Prospective Longitudinal research: An Empirical Basis for Primary Prevention of Psychosocial Disorders*, Oxford Univ. Press (1981).

22. Hougaard, P., Survival models for heterogeneous populations derived from stable distributions, *Biometrika* **73**, 387–396 (1986).

23. Hougaard, P., A class of multivariate failure time distributions, *Biometrika* **73**, 671–678 (1986).

24. Hougaard, P., *Analysis of Multivariate Survival Data*, Springer-Verlag, New York, 2000.

25. Hougaard, P., Harvald, B., and Holm, N., Measuring the similarities between the lifetimes of adult Danish twins born between 1881–1930, *J. Am. Statist. Assoc.* **87**, 17–24 (1992).

26. Houwing-Duistermaat, J. J., Derkx, B. H. F., Rosendaal, F. R., and van Houwelingen, H. C., Testing familial aggreation, *Biometrics* **51**, 1292–1301 (1995).

27. Hsu, L., Chen, L., Gorfine, M., and Malone, K., Semiparametric estimation of marginal hazard function from case-control family studies, *Biometrics* **60**, 936–944 (2004).

28. Hsu, L. and Gorfine, M., Multivariate survival analysis for case-control family data, *Biostatistics* (e-publication ahead of print) (2006).

29. Hsu, L. and Prentice, R., On assessing the strength of dependency between failure time variates, *Biometrika* **83**, 491–506 (1996).

30. Hsu, L., Prentice, R. L., Zhao, L. P., and Fan, J. J., On dependence estimation using correlated failure time data from case-control family studies, *Biometrika* **86**, 743–53 (1999).

31. Kendall, M. G., *Rank Correlation Methods*, 3rd ed., Griffin, London, 1962.

32. Klein, J. P., Semiparametric estimation of random effects using the Cox model based on the EM algorithm, *Biometrics* **48**, 795–806 (1992).

33. Li, H. and Thompson, E., Semiparametric estimation of major gene and family-specific random effects for age of onset, *Biometrics* **53**, 282–293 (1997).

34. Li, H. and Zhong, X., Multivariate survival models induced by genetic frailties, with application to linkage analysis, *Biostatistics* **3**, 57–75 (2002).

35. Li, Y. and Lin, X., Semiparametric normal transformation models for spatially correlated survival data, *J. Am. Statist. Assoc.* **101**, 591–603 (2006).

36. Lin, D. Y. and Ying, Z., A simple nonparametric estimator of the bivariate survival function under univariate censoring, *Biometrika* **80**, 573–581 (1993).

37. Malone, K. E., Daling, J. R., Weiss, N. S., McKnight, B., White, E., and Voigt, L. F., Family history and survival of young women with invasive breast carcinoma, *Cancer* **78**, 1417–1425 (1996).

38. Moger, T. A. and Aalen, O. O., A distribution for multivariate frailty based on the compound poisson distribution with random scale, *Lifetime Data Anal.* **11**, 41–59 (2005).

39. Moodie, Z. and Prentice, R. L., An adjustment to improve the bivariate survivor function repaired NPMLE, *Lifetime Data Anal.* **11**, 291–307 (2005).

40. Murphy, S. A., Consistency in proportional hazards model incorporating a random effect, *Ann. Statist.* **22**, 721–731 (1994).

41. Murphy, S. A., Consistency in a proportional hazards model incorporating a random effect, *Am. Statist.* **22**, 712–731 (1995).

42. Nan, B., Lin, X., Lisabeth, L. D., and Harlow, S. D., Piecewise constant cross-ratio estimator for association of age at marker event and age at menopause, *J. Am. Statist. Assoc.* **101**, 65–77 (2006).

43. Nelsen, R. B., *An Introduction to Copulas* (lecture notes in statistics), Springer-Verlag, New York, 1998.

44. Nielsen, G. G., Gill, R. D., Andersen, P. K., and Soerensen, T. I. A., A counting process approach to maximum likelihood estimation in frailty models, *Scand. J. Statist.* **19**, 25–43 (1992).

45. Oakes, D., A concordance test for independence in the presence of censoring, *Biometrics*, 451–455 (1982).

46. Oakes, D., Bivariate survival models induced by frailties, *J. Am. Statist. Assoc.* **84**, 487–493 (1989).

47. Parner, E., Asymptotic theory for the correlated gamma-frailty model, *Ann. Statist.* **26**, 183–214 (1998).

48. Pipper, C. B. and Martinussen, T., An estimating equation for parametric shared frailty models with marginal additive hazards, *J. Roy. Statist. Soc. B* **66**, 207–220 (2004).

49. Prentice, R. L. and Cai, J., Covariance and survivor function estimation using censored multivariate failure time data, *Biometrika* **79**, 495–512 (1992).

50. Prentice, R. L. and Hsu, L., Regression on hazard ratios and cross ratios in multivariate failure time analysis, *Biometrika* **84**, 349–363 (1997).

51. Prentice, R. L., Moodie, F. Z., and Wu, J., Hazard-based nonparametric survivor function estimation, *J. Roy. Statist. Soc. B* **66**, 305–319 (2004).

52. Pruitt, R. C., On negative mass assigned by the bivariate Kaplan-Meier estimator, *Ann. Statist.* **19**, 443–453 (1991).

53. Self, S. G. and Prentice, R. L., Asymptotic distribution theory and efficiency results for case-cohort studies, *Ann. Statist.* **16**, 64–81 (1986).

54. Shih, J. H. and Chatterjee, N., Analysis of survival data from case-control family studies, *Biometrics* **58**, 502–509 (2002).

55. Shih, J. H. and Louis, T. A., Inferences on the association parameter in copula models for bivariate survival data, *Biometrics* **51**, 1384–1399 (1995).

56. Siegmund, K. and McKnight, B., Modeling hazard functions in families, *Genetic Epidemiol.* **15**, 147–171 (1998).

57. Spearman, C., The proof and measurement of correlation between two things, *Am. J. Psychiatr.* **15**, 72–101 (1904).

58. Tsai, W. and Crowley, J., A note on nonparametric estimators of the bivariate survival function under univariate censoring, *Biometrika* **85**, 573–580 (1998).

59. Tsai, W., Leurgans, S., and Crowley, J., Nonparametric estimation of a bivariate survival function in the presence of censoring, *Annals Statist.* **14**, 1351–1365 (1986).

60. Turnbull, B. W., The empirical distribution with arbitrary grouped censored and truncated data, *J. Roy. Statist. Soc., Ser. B Methodol.* **38**, 290–295 (1976).

61. van der Laan, M. J., Efficient estimation in the bivariate censoring model and repairing NPMLE, *Ann. Statist.* **24**, 596–627 (1996).

62. van der Laan, M. J., Nonparametric estimators of the bivariate survival function under random censoring, *Statist. Neerland.* **51**, 178–200 (1997).

63. Wang, W. and Wells, M. T., Nonparametric estimators of the bivariate survival function under simplified censoring conditions, *Biometrika* **84**, 863–880 (1997).

64. Wienke, A., Holm, N. V., Christensen, K., Skytthe, A., Vaupel, J. W., and Yashin, A. I., The heritability of cause-specific mortality: A correlated gamma-frailty model applied to mortality due to respiratory diseases in Danish twins born 1870–1930, *Statist. Med.* **22**, 3873–3887 (2003).

65. Wei, L. J., Lin, D. Y., and Weissfeld, L., Regression analysis of multivariate incomplete failure time data by modeling marginal distributions, *J. Am. Statist. Assoc.* **84**, 1065–1073 (1989).

66. Yan, J. and Fine, J. P., Functional association models for multivariate survival processes, *J. Am. Statist. Assoc.* **100**, 184–196 (2005).

67. Yashin, A. I., Vaupel, J. W., and Iachine, I., Correlated individual frailty: An advantageous approach to survival analysis of bivariate data, *Math. Popul. Stud.* **5**, 145–159 (1995).

CHAPTER 15

Robust Estimation for Analyzing Recurrent-Event Data in the Presence of Terminal Events

Rajeshwari Sundaram

Biometry and Mathematical Statistics Branch Division of Epidemiology,
Statistics and Prevention Research National Institute of Child Health
and Human Development, National Institutes of Health
Rockville, Maryland

15.1 INTRODUCTION

In many biomedical studies, subjects may experience multiple failure events of the same type; outcomes of this type are referred to as *recurrent events*. Recurrent events occur in varied fields of study, ranging from medical studies and actuarial studies to sociology, and reliability engineering. For example, patients with cerebrovascular disease may experience repeated transient ischemic attacks [11]; HIV patients may experience recurrent opportunistic infections [16]. In addition, patients often die during the study period as a result of repeated occurrences of severe diseases. In these studies, the investigators are often interested in assessing the effects of covariates on the recurrent-event process.

In analyzing recurrent failure-time data, the majority of the work has focused on Anderson and Gill's [1] intensity-based models. Under the Anderson–Gill model, a nonhomogeneous Poisson process structure is assumed for the counting process denoting the recurrent-event process; time-dependent covariates are used to model the dependence between recurrent events. Various generalizations of the Anderson–Gill model to more arbitrary counting processes have been studied by, for example, Pepe and Cai [23], Lawless et al. [15], and Lin et al. [19]. More recently, a general class of models have been suggested by Peña and Hollander [22] that also incorporates the intervention effect. An extensive survey of different models used in recurrent-event analysis can be found in Cook and Lawless [5]. In these

Statistical Advances in the Biomedical Sciences, edited by Atanu Biswas, Sujay Datta,
Jason P. Fine, and Mark R. Segal

models, the effect of covariates is taken to be multiplicative on the baseline rate or the baseline mean function. However, in certain instances such an assumption may not be valid.

Lin et al. [20] introduced semiparametric transformation models for modeling the mean function of the recurrent-event process. These semiparametric transformation models offer great flexibility in modeling the effect of covariates on the mean function of the recurrent-event processes without specifying the underlying stochastic structure, and the effect of covariate need not be assumed multiplicative. Under the transformation model, the conditional mean $m_Z(t)$ of the recurrent events process $N^*(t)$ given the covariates Z satisfies

$$m_Z(t) = g(\mu_0(t)e^{\beta'Z}), \tag{15.1}$$

where the link function $g(\cdot)$ is a twice continuously differentiable and strictly increasing function, $\mu_0(\cdot)$ is an increasing function, and β is the unknown regression parameter of interest. Here, $m_0(t) = E(N^*(t)|Z = 0) = g(\mu_0(t))$ is the unknown baseline means function. The model introduced in (15.1) includes a very rich class of models through the link function $g(\cdot)$. A very important special case occurs when $g(x) = x$, which yields $m_Z(t) = \mu_0(t)e^{\beta'Z}$ This is known as the *proportional means model* and has been studied in the context of time to event as well as recurrent event data by several authors. Some other common choices for the link function g are the Box–Cox transformations:

$$g(x) = ((1 + x)^\rho - 1)/\rho, \rho > 0, \quad \log(1 + x), \rho = 0.$$

See Lin et al. [20] for more examples. The choice of time-dependent covariates $Z(\cdot)$ should be restricted to external covariates, that is, those covariates whose value may influence the rate of recurrence over time, but its future path up to any time $s > t$ is not affected by the occurrence of a recurrent event at time t [12, p. 196]. The choice of time-dependent covariates is also restricted to be monotonic to ensure that $m_Z(\cdot)$ is nondecreasing.

Much of the effort in recurrent-event analysis has focused on independent censoring. Attention is now focusing on dealing with the regression analysis of recurrent events in the presence of death [9,10,26]. In the case of the transformation model for the recurrent-event processes, Lin et al. [20] have adapted the generalized estimation equation procedure developed by Liang and Zeger [17] and have focused only on independent censoring. In this chapter, we propose a class of (easily computable) minimum L_2 distance estimators for the regression parameter β in (15.1) and extend the methodology to deal with terminal events. Such an estimating procedure appears to be novel in dealing with recurrent-event data.

Minimum-distance (MD) estimation is well known to be "automatically robust"; that is, MD estimates remain consistent even when the model is only approximately valid and have good finite sample property [6,7]. Donoho and Liu further demonstrated certain additional advantages of using Cramér–von Mises L_2 distance. These estimators are usually consistent and $n^{1/2}$-consistent under minimal conditions in the valid model. Also, the minimum L_2 distance estimators are highly efficient in the case of some familiar parametric models for proper choices of the integrating measures [14,13,21]. Yang and Prentice [29] and Sundaram [27] have studied these estimators for various semiparametric models with censored data.

The chapter is organized as follows. In the next section, we develop the estimation procedure for a class of robust minimum-distance estimators for the regression parameter β and the baseline function $\mu_0(t)$. In Section 15.3, the large-sample properties of the proposed estimators are discussed. In Section 15.4, results from extensive numerical studies are reported.

We conclude with an illustration of the proposed methods to the rhDNase data discussed in Therneau and Hamilton [28] and to the well-known bladder cancer data [3].

15.2 INFERENCE PROCEDURES

Let $N^*(t)$ be the number of recurrent events in the time interval $[0, t]$ pertaining to a subject, and let D be the survival time of the subject. Obviously, subjects who die cannot experience any further recurrent events, so $N^*(\cdot)$ is unobservable after D. In practice, the subject is usually followed for a limited period of time, so $N^*(\cdot)$ is typically right-censored. Let C denote the follow-up or censoring time. So, in the presence of censoring C and terminal event D, we get to observe only $N(t) = N^*(t \wedge X)$, where $X = D \wedge C$. Let $\delta = I\{D \leq C\}$. We assume that C is independent of $N^*(\cdot)$, and is independent of D. However, we do not make any such assumption regarding the dependence structure between the recurrent events and survival time. Let $Z(\cdot)$ denote a p-dimensional covariate process. For a random sample of n subjects, the observable data consist of $\{X_i, \delta_i, N_i(t), Z_i(t), t \leq X_i\}$ $(i = 1, \cdots, n)$.

15.2.1 Estimation in the Presence of Only Independent Censoring (with All Censoring Variables Observable)

Define

$$M_i(t) = Y_i(t)[N_i(t) - g(\mu_0(t)e^{\beta Z_i(t)})], \quad i = 1, \ldots, n, \tag{15.2}$$

where $Y_i(t) = I\{C_i \geq t\}$. Under the transformation model (15.1), $M_i(t)$ are zero mean stochastic processes. Let $w_{i0}(\cdot) \equiv 1$ and $w_{ij}(t) = f_j(Z_{ij}, t), j = 1, \ldots, p$ be nonnegative, subject-specific weights that may depend on the covariates through adequate functions f_j. We observe that

$$E(w_{ij}(t)M_i(t)) = 0, \quad j = 0, 1, \ldots, p. \tag{15.3}$$

For instance, under the assumption that the covariates $Z(\cdot) \geq 0, w_{ij}(\cdot) = Z_{ij}(\cdot)^\alpha, \alpha > 0$ is a common choice in regression.

Thus, consulting (15.3), for a given β one may obtain estimates $\hat{\mu}_{0j}, j = 0, 1, \ldots, p$ of $\mu_0(t)$ by solving the equation

$$\sum_{i=1}^{n} w_{ij}(t)Y_i(t)[N_i(t) - g(\hat{\mu}_{0j}(t)e^{\beta'Z_i(t)})] = 0, \quad j = 0, 1, \ldots, p. \tag{15.4}$$

When $w_{\cdot j} \equiv 1$, this type of estimating equation has been used by Lin et al. [20] to estimate $\mu_0(\cdot)$. Here, we have a class of estimators for μ_0, one for every j. We can show, analogous to A.1 of Lin et al. [20], that (15.4) has a unique solution for $\hat{\mu}_{0j}$. In the case of proportional means model [i.e., $g(x) = x$], $\hat{\mu}_{0j}$ has an explicit formula:

$$\hat{\mu}_{0j}(t; \beta) = \frac{\sum_{i=1}^{n} w_{ij}(t)Y_i(t)N_i(t)}{\sum_{i=1}^{n} w_{ij}(t)Y_i(t)e^{\beta'Z_i(t)}}, \quad j = 0, 1, \ldots, p.$$

Observe that this estimator is the well-known Aalen–Breslow estimator when $w_{ij} \equiv 1$. So, one can view the estimator displayed above as a (covariate) weighted version of the Aalen–Breslow estimator. In general, one can solve (15.4) for every given t and β relatively easily since the equation reduces to a scalar equation with a negative slope. So, any root solving method can be used.

Next, we outline our estimation procedure for the unknown regression parameter β in (15.1). As mentioned before, equation (15.4) yields a unique solution $\hat{\mu}_{0j}(t, b)$ for any b. We will show in the next section that under certain regularity conditions, $\hat{\mu}_{0j}(\cdot, b)$ converges (in supnorm) to a nonrandom function $\mu_{0j}(\cdot, b)$ almost surely. Moreover, $\mu_{0j}(\cdot, b) = \mu_0(\cdot) \, \forall j = 0, 1, \ldots, p$ if and only if $b = \beta$. Motivated by this, we can obtain an estimator for β by minimizing the L_2 distance between $\hat{\mu}_{0j}$ and $\hat{\mu}_{00}$. In other words, our proposed estimator $\hat{\beta}$ of β is the minimizer of the Cramér–von Mises distance

$$D_n(b) = \int_0^{\infty} \sum_{j=1}^{p} (\hat{\mu}_{0j}(t; b) - \hat{\mu}_{00}(t; b))^2 \Phi_n(dt; b). \tag{15.5}$$

Here, Φ_n is a finite, compactly supported integrating measure that may be data-dependent. In fact, by varying the integrating measure Φ_n, we can get a class of estimators for the regression parameter.

15.2.2 Estimation in the Presence of Terminal Events

Under this situation, C_i is unknown if the ith subject dies before being censored. Consequently, $Y_i(t) = I\{C_i \geq t\}$, and hence (15.4) is not calculable. In fact, $Y_i(t \wedge D_i)$ can only be calculated.

One can then modify (15.4) by replacing $Y_i(t)$ by a quantity $\hat{Y}_i(t)$, which can be computed according to the observed sample and has the same expectation as $Y_i(t)$ Observe that $E(Y_i(t \wedge D_i)) = E(E(Y_i(t \wedge D_i)|D_i))$, and note that $Y_i(t \wedge D_i)G(t)/G(t \wedge D_i)$ has the same expectation as $Y_i(t)$, where $G(t) = P(C_i \geq t)$. As G is unknown, we can replace G by the Kaplan–Meier estimator $\hat{G}$. This gives us the choice $\hat{Y}_i(t) = Y_i(t \wedge D_i)\hat{G}(t)/\hat{G}(t \wedge D_i)$, which is the same as $Y_i(t \wedge D_i)\hat{G}(t)/\hat{G}(t \wedge X_i)$. This method is similar to the inverse probability of censoring weighing scheme suggested by Robins and Rotnitzky [25], which has been used successfully in various contexts by Lin and Ying [18], Cheng et al. [4], Fine and Gray [8], and Ghosh and Lin [9], among others. In fact, Ghosh and Lin [9] have used the same weight in dealing with terminal events in their context.

Hence, in the presence of terminal events, for a given β one may obtain estimates $\hat{\mu}_{0j}, j = 0, 1, \dots, p$ of $\mu_0(t)$ by solving the equation

$$\sum_{i=1}^{n} w_{ij}(t)\hat{Y}_i(t)[N_i(t) - g(\hat{\mu}_{0j}(t)e^{\beta' Z_i(t)})] = 0, \quad j = 0, 1, \dots, p. \tag{15.6}$$

The estimation method for the regression coefficients remains the same as before.

15.3 LARGE-SAMPLE PROPERTIES

We begin by introducing some notation and state some sufficient assumptions under which the large-sample properties are established. Recall that $N_i^*(t)$ denotes the number of events that occur by time t, and C_i and D_i respectively denote the censoring time and the survival time for the ith individual. Also, recall that

$$X_i = C_i \wedge D_i, \delta_i = I\{D_i \leq C_i\}, N_i(t) = N_i^*(t \wedge X_i), \quad Y_i(t) = I\{C_i \geq t\}.$$

Denote by $\dot{g}(x) = [dg(x)/dx]$.

We prove our large-sample results under some standard regularity assumptions (15A.1)–(15A.4), presented in a rigorous form in the end-of-chapter appendix. The assumptions include the independence and identical distribution of the recurrent event process and censoring and survival times. Our results hold when the recurrent event process is independent of the censoring variable. The results can be shown to hold true even when the recurrent-event process is independent of the censoring variable, conditionally on the covariates when the individual is subject only to independent censoring. We further assume that the covariates are bounded and sufficiently smooth. The integrating measure Φ is assumed finite over an interval $[0, \tau]$, where τ is such that the recurrent-event process has finite second moment and that $P(C > \tau) > 0$ to avoid tail instability under right censoring.

Let

$$s_j^{(1)}(t; b) = G(t)E(w_{1j}(t)Z_{1j}(t)e^{b' Z_1(t)}\dot{g}(\mu_{0j}(t; b)e^{b' Z_1(t)})),$$

$$s_j(t; b) = G(t)E(w_{1j}(t)e^{b' Z_1(t)}\dot{g}(\mu_{0j}(t; b)e^{b' Z_1(t)})) \tag{15.7}$$

for $j = 0, 1, \dots, p$. Define the p-dimensional column vectors $s^{(1)}(t; b), s(t; b)$ by

$$s^{(1)}(t; b) = [s_1^{(1)}(t; b) \cdots s_p^{(1)}(t; b)]', \quad s(t, b) = [s_1(t; b) \cdots s_p(t; b)]'.$$

Also, for a column vector a, let $a^{\otimes 2} = aa'$. Let the $p \times p$ matrix $\Psi(t) = ((\psi_{kj}(t)))$, where

$$\psi_{kj}(t) = \frac{\partial}{\partial \beta_k}(\mu_{0j}(t; \beta) - \mu_{00}(t; \beta)), \quad j, k = 1, \dots, p. \tag{15.8}$$

Also we will denote the column vector $[\psi_{1j} \cdots \psi_{pj}]'$ by $\Psi_{\cdot j}$. We further denote

$$A = \left(\int_0^\tau \Psi(t)'\Psi(t)\Phi(dt;\beta) \right). \tag{15.9}$$

We first establish the strong consistency of the proposed estimators $\hat{\mu}_{0j}, j = 0, 1, \ldots, p$, and $\hat{\beta}$. The proof of strong consistency of $\hat{\mu}_{0j}$ is similar to the discussion in A.1 of Lin et al. [20]. Hence

$$\hat{\mu}_{0j}(t;b) \rightarrow \mu_{0j}(t;b) \tag{15.10}$$

almost surely and uniformly in $t \in [0,\tau]$ and $b \in N(\beta)$, where $\mu_{0j}(t;b)$ is the unique solution of the equation

$$E\left(w_{1j}(t)\left[g(\mu_0(t)e^{\beta'Z(t)}) - g(\mu_{0j}(t;b)e^{b'Z(t)}) \right] \right) = 0, t \leq \tau$$

and $\mu_{0j}(t;b) = \mu_0(t)$ when $b = \beta$. Using assumption (15A.4), we obtain

$$D_n(b) = \int_0^\infty (\hat{\mu}_{0j}(t;b) - \hat{\mu}_{00}(t;b))^2 \Phi_n(dt;b) \rightarrow D(b) = \int_0^\infty (\mu_{0j}(t;b) - \mu_{00}(t;b))^2 \Phi(dt;b),$$

almost surely and uniformly in $b \in N(\beta)$ and that $D(\beta) = 0$. So, if

$$D(b) \neq 0 \quad \text{for} \quad b \in N(\beta)\backslash\{\beta\}, \tag{15.11}$$

any minimizer of $D_n(b)$ is strongly consistent.

Theorem 15.1. Suppose that assumptions (15A.1)–(15A.4) hold. Let $\hat{\beta}$ be the minimum L_2 distance estimator of β defined in (15.5). Then, under (15.11), with probability 1, we obtain

$$\hat{\beta} \rightarrow \beta, \quad \text{as} \quad n \rightarrow \infty. \tag{15.12}$$

Remark 15.1: In case of the proportional means model [i.e., $g(x) = x$], (15.11) can be restated as follows. For all $b \in N(\beta)\backslash\{\beta\}$, there exists a $j \in \{1, \ldots, p\}$ such that

$$\frac{E[w_{1j}(t)e^{\beta'Z_1(t)}]}{E[w_{1j}(t)e^{b'Z_1(t)}]} \neq \frac{E[e^{\beta'Z_1(t)}]}{E[e^{b'Z_1(t)}]}$$

for t in a subset of $[0,\tau]$ with positive Φ measure. In the general transformation model, one can show that (15.11) holds true in the k-sample setup. It can also be verified that (15.11) holds true for time-dependent covariates that are step functions with random jumps and/or random jump points. These types of time-dependent covariates are quite general.

The weak convergence of $\sqrt{n}(\hat{\beta} - \beta)$ is established in Appendix 15A through the following steps. In the first step, we establish the asymptotic linearity of $\hat{\mu}_{0j}(t; b)$ near β. Next, we establish the asymptotic distribution of $\sqrt{n}(\hat{\mu}_{0j}(t; \beta) - \mu_0(t))$. Combining these two, we further show that the distance $D_n(b)$ can be approximated by a quadratic function of b. This quadratic function has a unique minimizer $\sqrt{n}(\hat{\beta} - \beta)$, and that its asymptotic distribution is normal with covariance Σ that can be consistently estimated.

Theorem 15.2. Assume that assumptions (15A.1)–(15A.4) and (15.11) are satisfied, and let $\hat{\beta}$ be the estimator of β defined in (15.5). Then

$$\sqrt{n}(\hat{\beta} - \beta) \longrightarrow N(0, \Sigma),$$

where Σ is

$$\Sigma = A^{-1}E\left[\int_0^\tau M(t) \sum_{j=1}^p \Psi_{\cdot j}(t; \beta)\left(\frac{w_{1j}(t)}{s_j(t; \beta)} - \frac{1}{s_0(t; \beta)}\right)\Phi(dt; \beta)\right]^{\otimes 2} A^{-1}. \tag{15.13}$$

Remark 15.2: Consistent Estimator for the Asymptotic Variance of $\hat{\beta}$. The covariance matrix Σ can be consistently estimated by replacing all unknown quantities by their empirical counterparts. The resulting estimator is easily computable.

Remark 15.3: Choice of Weight Functions w_j, Φ. In defining the estimating equations for μ_0, the weights w_{ij} have been taken to be positive, smooth functions of the covariate. In practice the weights should be chosen in an optimal way in the sense that it leads to minimum asymptotic variance for the proposed estimators. The choice of weight functions as the covariate process, specifically, $w_{ij}(t) = Z_{ij}(t)$, often results in minimum asymptotic covariance [14].

The performance of the minimum-distance estimator also depends on the integrating measure Φ_n. The role of the integrating measure Φ_n is similar to that of the weight function for the weighted log-rank statistics. Ideally, Φ_n should be chosen in an optimal way, for example, by minimizing the variance of $\hat{\beta}$. However, in general it is difficult to find such an optimal integrating measure analytically as the asymptotic variance is a function of the covariate, as well as the underlying distribution of the censoring variable and event times. We will consider some integrating measures in the simulation study in the next section that perform reasonably well.

We also compared the asymptotic variance of the proposed estimators with those of the existing estimators [20] for the case of independent censoring. In fact, by choosing $w_{ij} = Z_{ij}$ and making appropriate choices of the integrating measure Φ_n, one may achieve exactly the same asymptotic variance as that of $\hat{\beta}_{\text{LWY}}$ (where the subscript "LWY" denotes Lin–Wei–Ying). To see this, restricting our attention to a single covariate, the asymptotic variance of $\hat{\beta}_{\text{LWY}}$ is given by

$$\Sigma_{\text{LWY}} = A_{\text{LWY}}^{-1}E\left(\int_0^\tau M(t)\left(Z_1(t) - \frac{s_1(t; \beta)}{s_0(t; \beta)}\right)dH(t)\right)^{\otimes 2} A_{\text{LWY}}^{-1},$$

where $H(\cdot)$ is any integrating measure satisfying assumption (15A.2), and

$$A_{\mathrm{LWY}} = \int_0^{\tau} \left(\frac{s_1^{(1)}(t;\beta)s_0(t;\beta) - s_1^2(t;\beta)}{s_0(t;\beta)} \right) m_0(t)dH(t).$$

It is easy to verify that for the choice

$$\Phi(t) = \int_0^t \frac{s_1(s;\beta)}{\Psi_{11}(s;\beta)} dH(s), \tag{15.14}$$

the asymptotic variance of the proposed minimum-distance estimators $\hat{\beta}$ is precisely the same as that of $\hat{\beta}_{\mathrm{LWY}}$. It is conceivable that by optimizing the choice of w and/or Φ, one may obtain an estimator $\hat{\beta}$ in the class of estimators proposed here with asymptotic variance lower than that of $\hat{\beta}_{\mathrm{LWY}}$. However, such an optimal choice does not seem to be analytically tractable. One may instead use resampling techniques to obtain an optimal choice within a large (finite) class of estimators.

15.4 NUMERICAL RESULTS

15.4.1 Simulation Studies

Simulation studies were conducted to examine the finite sample properties of the proposed minimum-distance estimators. Our primary interest was in investigating their performance at different levels of correlation between recurrent event times and survival time within an individual. In absence of terminal events, we also compared them with the estimators proposed by Lin et al. [20] for proportional as well as nonproportional means models. (Henceforth, the Lin–Wei–Ying estimators will be referred to as LWY estimators.) The proportional means model (model 1) was generated under the following scheme. We generated gap times between recurrences from the following model:

$$\lambda(t|Z,\psi) = \psi\lambda_0(t)e^{\beta'Z},$$

where ψ is a gamma variable with mean 1 and variance σ^2 and Z is a $0/1$ treatment indicator (0 for placebo, 1 for treatment). When $\sigma^2 = 0$, recurrent events for an individual are independent and nonzero values of σ^2 induces correlation between the recurrent times. It can be shown that this regression model implies a proportional means model:

$$m_Z(t) = \mu_0(t)e^{\beta'Z}.$$

We set $\lambda_0(t) = 1$, $\beta = 0.5$, and $\sigma^2 = 0, 0.25, 0.5, 1$. The censoring variable C was generated from uniform $[2,3]$, which results in an average of 3.3 recurrences per individual. These parametric values were used in simulation studies in Lin et al. [20]. The simulation results for model 1 are presented in Table 15.1.

Next, we also compared the performance of the proposed estimators with $\hat{\beta}_{\mathrm{LWY}}$ in nonproportional means model. The events times $T_1 < T_2 < \cdots$ were generated from a unit

Table 15.1 Simulation Results for Model 1

		MDEs[a]		LWY Estimators		
σ^2	Φn	Bias	SE	Bias	SE	H
0.0	$\hat{\Phi}_{nt}$	−0.0011	0.1307	0.0004	0.1312	H_t
	$\hat{\Phi}_{nF}$	−0.0007	0.1338	0.0007	0.1343	H_F
	$\hat{\Phi}_{nG}$	−0.0004	0.1269	0.0002	0.1272	H_G
0.25	$\hat{\Phi}_{nt}$	−0.0008	0.1681	0.0008	0.1688	H_t
	$\hat{\Phi}_{nF}$	−0.0002	0.1696	0.0013	0.1703	H_F
	$\hat{\Phi}_{nG}$	−0.0011	0.1728	0.0005	0.1733	H_G
0.5	$\hat{\Phi}_{nt}$	−0.0004	0.1937	0.0008	0.1966	H_t
	$\hat{\Phi}_{nF}$	−0.0006	0.1942	0.0011	0.1949	H_F
	$\hat{\Phi}_{nG}$	−0.0014	0.2032	0.0004	0.2041	H_G
1.0	$\hat{\Phi}_{nt}$	−0.0011	0.2462	0.0010	0.2477	H_t
	$\hat{\Phi}_{nF}$	−0.0008	0.2458	0.0012	0.2469	H_F
	$\hat{\Phi}_{nG}$	−0.0002	0.2626	0.0029	0.2641	H_G

[a]Minimum-distance estimators.

intensity Poisson process, and

$$N^*(t) = \sum_{k\geq 1} I\{T_k \leq \psi g(\mu_0(t)e^{\beta Z})\}.$$

Hence, $m_Z(t) = g(\mu_0(t)e^{\beta Z})$, as ψ is a gamma variable with mean 1 and variance σ^2. The following parametric values were chosen (model 2):

$$g(\cdot) = \{1 + \cdot\}^2 - 1, \quad \mu_0(t) = t, \quad \beta = 0.5, \quad \sigma^2 = 0, 0.25, 0.5, 1, \quad C \sim U(0.6, 0.9).$$

Table 15.2 Simulation Results for Model 2

		MDEs		LWY Estimators		
σ^2	Φ_n	Bias	SE	Bias	SE	H
0.0	$\hat{\Phi}_{nt}$	−0.0015	0.1183	0.0030	0.1189	H_t
	$\hat{\Phi}_{nF}$	−0.0004	0.1189	0.0031	0.1195	H_F
	$\hat{\Phi}_{nG}$	−0.0006	0.1093	0.0019	0.1095	H_G
0.25	$\hat{\Phi}_{nt}$	0.0005	0.1432	0.0034	0.1440	H_t
	$\hat{\Phi}_{nF}$	0.0010	0.1439	0.0042	0.1441	H_F
	$\hat{\Phi}_{nG}$	0.0012	0.1395	0.0039	0.1400	H_G
0.5	$\hat{\Phi}_{nt}$	−0.0025	0.1653	0.004	0.1665	H_t
	$\hat{\Phi}_{nF}$	0.0000	0.1652	0.0046	0.1661	H_F
	$\hat{\Phi}_{nG}$	0.0017	0.1618	0.0033	0.1625	H_G
1.0	$\hat{\Phi}_{nt}$	−0.0040	0.2065	0.0042	0.2079	H_t
	$\hat{\Phi}_{nF}$	−0.0010	0.2057	0.0046	0.2068	H_F
	$\hat{\Phi}_{nG}$	−0.0017	0.2081	0.0057	0.2091	H_G

The choice of parameters for the censoring variable were such that they resulted in an average of 3.5 recurrences per individual. Table 15.2 lists the simulation results for model 2. Lin et al. [20] have suggested the following choices of integrating measures:

$$H_t(t) = t, \quad H_F = \frac{\sum\limits_{i=1}^{n} N_i(t)}{n}, \quad H_G = \frac{\sum\limits_{i=1}^{n} I\{C_i \leq t\}}{n}.$$

Recall from the previous section that the choice of the integrating measure Φ of the form (15.14) leads to an asymptotic variance of the minimum-distance estimators equal to that of the LWY estimators. We replace the terms s_1 and Ψ_{11} by their estimates as follows:

$$S_1(t;b) = \sum_{i=1}^{n} Z_i \hat{Y}_i(t) e^{bZ_i}$$

$$\hat{\Psi}_{11}(t;b) = \left(\frac{\sum\limits_{i=1}^{n} Z_i^2 \hat{Y}_i(t) e^{bZ_i}}{\sum\limits_{i=1}^{n} Z_i \hat{Y}_i(t) e^{bZ_i}} - \frac{\sum\limits_{i=1}^{n} Z_i \hat{Y}_i(t) e^{bZ_i}}{\sum\limits_{i=1}^{n} \hat{Y}_i(t) e^{bZ_i}} \right) \hat{\mu}_{00}(t;b).$$

Using these estimates, we choose the following integrating measures

$$\Phi_{nt}(t;b) = \int_0^t \frac{S_1(s;b)}{\hat{\Psi}_{11}(s;b)} dH_t(s), \quad \Phi_{nF}(t;b) = \int_0^t \frac{S_1(s;b)}{\hat{\Psi}_{11}(s;b)} dH_F(s),$$

$$\Phi_{nG}(t;b) = \int_0^t \frac{S_1(s;b)}{\hat{\Psi}_{11}(s;b)} dH_G(s)$$

to make our comparisons. Recall that our proposed class of estimators for β is obtained by minimizing the distance D_n defined in (15.5). In practice, this optimization is very easy to implement as any standard optimization method can be used. The results reported in Tables 15.1 and 15.2 are based on 5000 replications and the sample size $n = 100$. The bias is the average of the difference between estimated $\hat{\beta}$ and the true β, and SE is the sampling standard deviations. The bias of the minimum distance estimators are negligible across all combinations of the parameters investigated here. In fact, the bias of the minimum distance estimators are much smaller than that of Lin et al. [20] for nonproportional means model. The standard errors of the proposed minimum-distance estimators are comparable to (slightly lower than) those of Lin et al. [20], thus confirming the appropriateness of the empirical approximation of (15.14). The performance of the minimum-distance estimators seems to be comparable for the integrating measures considered here with $\hat{\Phi}_{nG}$ performing a little better in terms of the standard error, and $\hat{\Phi}_{nF}$ performing better for the power models considered in the simulation studies reported here.

We further conducted simulations to investigate the performance of the proposed minimum-distance estimators in the presence of terminal events.

In both models (1 and 2), we introduced the survival time D with a hazard rate of $\lambda_D(t) = 0.25\psi$ and in the presence of D, the recurrent events process is observed only up to $X = C \wedge D$ instead of C. Under this setup, we are only specifying the marginals of D and $N^*(\cdot)$ and the dependence between the recurrence times and the survival time is captured

Table 15.3 Simulation Results for Model 1 in Presence of Terminal Event

σ^2	Φ_n	$n = 100$				$n = 200$			
		Bias	SE	SEE	CP	Bias	SE	SEE[a]	CP[b]
0.0	$\hat{\Phi}_{nt}$	0.0039	0.1602	0.1573	0.941	0.0057	0.1106	0.1111	0.960
	$\hat{\Phi}_{nF}$	0.0042	0.1667	0.1634	0.939	0.0063	0.1153	0.1165	0.958
	$\hat{\Phi}_{nG}$	0.0044	0.1525	0.1512	0.939	0.0049	0.1064	0.1081	0.961
0.25	$\hat{\Phi}_{nt}$	0.0068	0.2140	0.2010	0.922	0.0027	0.1454	0.1439	0.943
	$\hat{\Phi}_{nF}$	0.0074	0.2167	0.2058	0.930	0.0027	0.1521	0.1472	0.938
	$\hat{\Phi}_{nG}$	0.0056	0.2122	0.1977	0.927	0.0033	0.1438	0.1419	0.943
0.5	$\hat{\Phi}_{nt}$	−0.0028	0.2491	0.2358	0.928	−0.0039	0.1693	0.1689	0.950
	$\hat{\Phi}_{nF}$	−0.0015	0.2505	0.2404	0.936	−0.0025	0.1727	0.1719	0.946
	$\hat{\Phi}_{nG}$	−0.0036	0.2499	0.2340	0.929	−0.0034	0.1683	0.1681	0.951
1.0	$\hat{\Phi}_{nt}$	−0.0003	0.3011	0.2898	0.929	−0.0042	0.2041	0.2087	0.943
	$\hat{\Phi}_{nF}$	0.0011	0.3050	0.2943	0.935	−0.0027	0.2142	0.2116	0.947
	$\hat{\Phi}_{nG}$	−0.0014	0.3013	0.2894	0.935	−0.0041	0.2132	0.2089	0.943

[a]Standard error, estimated.
[b]Coverage probability.

Table 15.4 Simulation Results for Model 2 in Presence of Terminal Event

σ^2	Φ_n	$n = 100$				$n = 200$			
		Bias	SE	SEE	CP	Bias	SE	SEE	CP
0.0	$\hat{\Phi}_{nt}$	0.0077	0.1378	0.1328	0.934	−0.0026	0.0963	0.0943	0.95
	$\hat{\Phi}_{nF}$	0.0087	0.1398	0.1352	0.939	−0.0021	0.0978	0.0961	0.949
	$\hat{\Phi}_{nG}$	0.0089	0.1272	0.1221	0.935	−0.0020	0.0891	0.0872	0.954
0.25	$\hat{\Phi}_{nt}$	−0.0006	0.1661	0.1652	0.942	0.0055	0.1194	0.1174	0.944
	$\hat{\Phi}_{nF}$	0.0015	0.1679	0.1672	0.942	0.0062	0.1225	0.1190	0.948
	$\hat{\Phi}_{nG}$	−0.0011	0.1616	0.1570	0.940	0.0091	0.1134	0.1123	0.944
0.5	$\hat{\Phi}_{nt}$	0.0090	0.2038	0.1928	0.933	0.0055	0.1345	0.1372	0.951
	$\hat{\Phi}_{nF}$	0.0126	0.2068	0.1951	0.936	0.0068	0.1363	0.1380	0.952
	$\hat{\Phi}_{nG}$	0.0110	0.1945	0.1856	0.936	0.0069	0.1292	0.1317	0.951
1.0	$\hat{\Phi}_{nt}$	0.0106	0.2503	0.2356	0.935	0.0058	0.1714	0.1690	0.945
	$\hat{\Phi}_{nF}$	0.0157	0.2505	0.2384	0.947	0.0090	0.1743	0.1725	0.944
	$\hat{\Phi}_{nG}$	0.0118	0.2476	0.2293	0.924	0.0073	0.1711	0.1678	0.943

through the frailty term ψ. Under this setup, 30% of individuals were observed only until the terminal event D. Tables 15.3 and 15.4 report the simulation results based on 1000 replications for sample sizes $n = 100$ and $n = 200$. In addition to bias and standard error, these tables also report estimated values of the standard error of $\hat{\beta}$ and the observed coverage probabilities of 95% confidence intervals. The results indicate that the minimum-distance estimators have very small bias. The standard error estimators are very close the standard error of the replications indicating the appropriateness of the proposed estimators for the asymptotic standard deviation of the parameter estimators. Also, the coverage probabilities of the considered 95% confidence intervals are very good, with the performance improving with increasing sample size.

15.4.2 rhDNase Data

We now illustrate the proposed methods used in the rhDNase trial analyzed in Therneau and Hamilton [28]. Cystic fibrosis patients often suffer from repeated exacerbations of respiratory symptoms. The randomized clinical trial was conducted to assess the efficacy of rhDNase, a

Table 15.5 Summary of Recurrent Pulmonary Exacerbations in the rhDNase Trial

	Number of Recurrent Events						
Treatment	0	1	2	3	4	5	Total
rhDNase	218	65	30	6	3	0	322
Placebo	186	97	24	13	4	1	325

Table 15.6 Summary of Regression Analysis of rhDNase Trial

Model: $g(\mu_0(t)e^{\beta Z})$	$\hat{\Phi}_n$	$\hat{\beta}$	SE($\hat{\beta}$)	$\hat{\beta}$/SE($\hat{\beta}$)	p-Value
Model 1: $\mu_0(t)\,e^{\beta'Z}$	$\hat{\Phi}_{nt}$	-0.330	0.136	-2.42	0.016
	$\hat{\Phi}_{nF}$	-0.331	0.137	-2.42	0.016
	$\hat{\Phi}_{nG}$	-0.305	0.129	-2.36	0.018
Model 2: $\log(\mu_0(t)e^{\beta'Z} + 1)$	$\hat{\Phi}_{nt}$	-0.389	0.162	-2.40	0.016
	$\hat{\Phi}_{nF}$	-0.390	0.162	-2.40	0.016
	$\hat{\Phi}_{nG}$	-0.373	0.159	-2.35	0.018
Model 3: $\log(\log(\mu_0(t)e^{\beta'Z} + 1) + 1)$	$\hat{\Phi}_{nt}$	-0.482	0.199	-2.42	0.016
	$\hat{\Phi}_{nF}$	-0.482	0.199	-2.42	0.016
	$\hat{\Phi}_{nG}$	-0.459	0.194	-2.37	0.018
Model 4: $\log(0.3\log(\mu_0(t)e^{\beta'Z} + 1) + 1)$	$\hat{\Phi}_{nt}$	-0.714	0.297	-2.40	0.016
	$\hat{\Phi}_{nF}$	-0.709	0.295	-2.40	0.016
	$\hat{\Phi}_{nG}$	-0.782	0.328	-2.38	0.017

highly purified recombinant enzyme, in treating such patients. There were 647 patients in the study, and most of them were followed for about 170 days. Out of the 647 patients, 322 received rhDNase and the remaining 325 received placebo. By the end of the study, 104 of the 322 patients on rhDNase and 139 of the 325 placebo patients had developed pulmonary exacerbations. In fact, many of them had experienced multiple exacerbations. The recurrent events are summarized in Table 15.5.

These data were analyzed using the treatment as the only covariate based on four models: the proportional means model and certain log transformation models advocated by Lin et al. [20]. In Table 15.6 we report the point estimate of the regression parameter signifying the effect of treatment and estimates for the asymptotic standard deviation and the p value for the two-sided test $\beta = 0$ of the regression parameter.

Under all the four models, the regression analysis indicates that the effect of treatment is significant in reducing the recurrence of pulmonary exacerbation events in cystic fibrosis patients. This is in agreement with the findings of previous analysis of Lin et al. [20].

Now, to choose an appropriate model, we propose the following sum of squared residuals as a measure of overall lack of fit:

$$S_n = \sum_{j=1}^{p} \sum_{l=1}^{m} \left(\frac{\hat{\mu}_{0j}(t_l; \hat{\beta})}{\max_l \hat{\mu}_{0j}(t_l; \hat{\beta})} - \frac{\hat{\mu}_{00}(t_l; \hat{\beta})}{\max_l \hat{\mu}_{00}(t_l; \hat{\beta})} \right)^2 .$$

This is motivated by the fact that the proposed estimating procedure gives a class of estimators $\hat{\mu}_{0j}(\cdot; \beta), j = 0, 1, \ldots, p$ for the baseline function $\mu_0(\cdot)$. So, if the model is appropriate, the $\hat{\mu}_{0j}(\cdot; \hat{\beta})$ terms estimate the same function $\mu_0(\cdot)$. Hence, their squared differences can be used as residuals indicating departure from the proposed model.

The values of S_n based on minimum-distance (MD) estimators with respect to integrating measure Φ_{nt} for the four models are 1.2901, 1.9300, 1.0645, and 0.1749 for the four models considered. The value of S_n is smallest for model 4 (last model listed in Table 15.5), which agrees with the findings of Lin et al. [20]. In fact, one can plot the residuals $S_n(t)$ as a function of t (Fig. 15.1) for a graphical check for validity of the model for the data presented

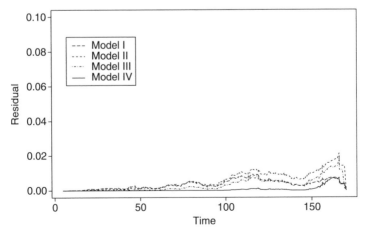

Figure 15.1 Plot of $S_n(t)$ against time.

Table 15.7   Summary of Regression Analysis of Bladder Tumor Trial

Variable	Estimate	SE	p Value
Treatment	0.5804	0.1321	<0.0001
Number	0.1658	0.1048	0.0568
Size	0.005	0.0545	0.4635

above. As can be seen, the last model in Table 15.6 is a good fit for the rhDNase data analyzed here. Another point to be noted from Table 15.6 is that the p values for the regression coefficient under the four models are close. This attests to the well-known robustness of the MD estimation procedure, namely, that this procedure performs well even when the model is only approximately valid.

15.4.3   Bladder Tumor Data

We also applied our proposed method to a well-known cancer trial conducted by the Veterans Administration Cooperative Urological Research Group [3]. In this trial, 117 patients were randomized to placebo, pyridoxine (Vitamin B_6), or intravesical Thio-Tepa (triethylenetriphosphamide) and were followed for subsequent recurrences of superficial bladder tumors. Following previous authors, we will focus on the 85 patients on Thio-Tepa and placebo. During the study, 11 of the patients on placebo and 12 patients on Thio-Tepa died. So, we have the presence of terminal events in addition to the usual censoring. We will consider a proportional means model with the treatment type and the number of tumors and the size of the largest tumors measured at the baseline as the time-invariant covariates. It has been shown in the literature that the proportionality assumption in the proportional means model is a reasonable assumption. In addition, the baseline covariates do not significantly affect the censoring distribution [9]. Hence it is appropriate to use our weight $\hat{Y}_i$ for Y_i, which involves estimating the censoring distribution. For the purpose of illustrating our method, we used the integrating weight $\Phi(t) = t$. In Table 15.7 we report the point estimate of the regression parameters: treatment (covariate coded as Thio-Tepa—1 vs. placebo—0), number of tumors at baseline, size of the largest tumor, estimate for the asymptotic standard deviation, and the p value for the two-sided test $\beta = 0$ of the regression parameters.

The findings of this regression analysis are in agreement with those of previous authors [12, p. 293] that Thio-Tepa significantly (highly in this case) reduces the mean number of recurrences.

15.5   CONCLUDING REMARKS

We have proposed a new method for estimation based on a marginal means model for recurrent events. As evidenced in the numerical study section, the estimators have performed very competitively. The numerical methods used to perform these studies involved random search techniques such as simulated annealing and/or a combination of grid search and the box-constrained quasi-Newton method for optimization, which are all available in popularly used software such as S-Plus/R and MATLAB.

In the presence of terminal events, we have proposed a method for estimation that entails estimation of the censoring distribution. This is appropriate when the censoring distribution is independent of the covariates. As the transformation model for the mean recurrent process is inherently marginal, this approach cannot discriminate between (for example) a treatment that reduces the recurrent event mean through reducing the rate of events while the subject is alive or by decreased survival probability. To address this issue, one may estimate the survival function via modeling the death hazard and use a weight similar to Inverse Probability of Survival Weighting (IPSW) approach used in Ghosh and Lin [9]. This approach is worth investigating in the future. The models studied here are inherently marginal; however, it may be worthwhile also to study the recurrence process and mortality simultaneously to properly ascertain the effect of covariates on these two phenomena.

APPENDIX 15A

We begin by first presenting the assumptions under which our asymptotics hold true.

Assumptions

15A.1 $\{N_i^*(\cdot), C_i(\cdot), D_i, Z_i(\cdot)\}_{i=1}^n$ are independent and identically distributed, where $Z_i = [Z_{i1} \cdots Z_{ip}]'$ denotes the p-dimensional column vector of covariates.

15A.2 $Z_{ij}(0) + \int_0^\tau |dZ_{ij}(t)| < M < \infty$ almost surely, for $j = 1, \ldots, p$, that is, $Z_{ij}(\cdot)$ is uniformly bounded and is of uniform bounded variation on $[0, \tau]$. Here, τ is a prespecified positive constant. Also, the weights $w_{ij}(\cdot) = f_j(Z_{ij}(\cdot), \cdot)$ are assumed to be uniformly bounded and of uniform bounded variation.

15A.3 Let τ be such that $E(N_1^{*2}(\tau)) < \infty$, $G(\tau) = P(C > \tau) > 0$ and

$$\inf_{0 \le t \le \tau, b \in N(\beta)} E(w_{1j}(t)e^{b'Z_1(t)}) > 0, \quad j = 0, 1, \ldots, p,$$

where $N(\beta)$ is a compact neighborhood containing β.

15A.4 $\Phi(t; b) = \lim_n \Phi_n(t; b)$ exists uniformly for all b in a compact set $N(\beta)$ and t in $(0, \tau]$ and where Φ_n is as in (15.5) with support in $(0, \tau]$.

Proof of Theorem 2: Weak Convergence of $\sqrt{n}(\hat{\beta} - \beta)$ In the first step, we prove the weak convergence of $\sqrt{n}(\hat{\mu}_{0j}(t; \beta) - \mu_0(t))$. Using the mean value theorem for $g(\hat{\mu}_{0j}(t; \beta))$ at $\mu_0(t)$ and assuming that $\hat{\mu}_{0j}(t; \beta)$ is the unique solution of (15.4), one can express

$$\hat{\mu}_{0j}(t; \beta) - \mu_0(t) = \frac{\sum_{i=1}^n w_{ij}(t)M_i(t) + \sum_{i=1}^n w_{ij}(t)(\hat{Y}_i(t) - Y_i(t))[N_i(t) - g(\mu_0(t)e^{\beta Z_i(t)})]}{\sum_{i=1}^n w_{ij}(t)\hat{Y}_i(t)\dot{g}(\mu_j^*(t)e^{\beta'Z_i(t)})e^{\beta'Z_i(t)}},$$

where $\mu_j^*(t)$ lies between $\hat{\mu}_{0j}(t, \beta)$ and $\mu_{0j}(t)$. Furthermore, by the uniform strong law of large numbers, the almost sure convergence of $\hat{\mu}_{0j}(\cdot; \beta)$ to $\mu_0(\cdot)$, the strong consistency of the

Kaplan–Meier estimator and the continuity of $\dot{g}(\cdot)$ ensure that

$$n^{-1} \sum_{i=1}^{n} w_{ij}(t)(\hat{Y}_i(t) - Y_i(t)) \to 0, \quad j = 0, 1, \ldots, p, \tag{15A.1}$$

and

$$n^{-1} \sum_{i=1}^{n} w_{ij}(t)\hat{Y}_i(t)\dot{g}(m_j^*(t)e^{\beta'Z_i(t)})e^{\beta'Z_i(t)} \to s_j(t; \beta), \quad j = 0, 1, \ldots, p, \tag{15A.2}$$

almost surely and uniformly in $t \in [0, \tau]$, where $s_j(t; \beta)$ is as in (15.7).

Consequently, recalling $w_{i0} \equiv 1$, for $i = 1, \ldots, n$, we have

$$\sqrt{n}(\hat{\mu}_{0j}(t; \beta) - \hat{\mu}_{00}(t; \beta)) = \sqrt{n} \sum_{i=1}^{n} \left(\frac{w_{ij}(t)}{\sum_{i=1}^{n} w_{ij}(t)\hat{Y}_i(t)\dot{g}(\mu_j^*(t)e^{\beta'Z_i(t)})e^{\beta'Z_i(t)}} \right.$$

$$\left. - \frac{1}{\sum_{i=1}^{n} \hat{Y}_i(t)\dot{g}(\mu_0^*(t)e^{\beta'Z_i(t)})e^{\beta'Z_i(t)}} \right) M_i(t) \tag{15A.3}$$

$$= \sqrt{n} \sum_{i=1}^{n} \left(\frac{w_{ij}(t)}{s_j(t; \beta)} - \frac{1}{s_0(t; \beta)} \right) M_i(t) + o_P(1)$$

where the second equality follows from the almost sure convergence of (15A.1), (15A.2), and the fact that $\sum_{i=1}^{n} M_i(t) = O_P(n^{1/2})$. Note the first term in (15A.3) is sum of n independent and identically distributed (i.i.d.) terms for every t. By the multivariate central-limit theorem, it converges in finite-dimensional distributions to a zero mean Gaussian process W_Z. Obviously $M_i(t)$ is the difference between two monotone functions in t. By assumption (15A.4) and the fact that $w_{ij}(\cdot)$ is of bounded variation, it follows that $w_{ij}(t)/s_j(t; \beta)$, $j = 0, 1, \ldots, p$ is of bounded variation. Hence, it can be expressed as a difference of two monotonically increasing functions in t. Combining this with the fact that $M_i(t)$ can be written as a difference between two monotonic functions in t, each summand of (15A.3) can be expressed as the sum of monotonic functions in t. Hence, they have pseudodimension 1. This in turn implies that they are manageable [see Lemma A.2 in Ref. 2]; hence $\sqrt{n}(\hat{\mu}_{0j}(t; \beta) - \hat{\mu}_{00}(t; \beta))$ is tight. By the functional central-limit theorem [24, p. 53], we have

$$\sqrt{n}(\hat{\mu}_{0.}(t; \beta) - \hat{\mu}_{00}(t; \beta)\mathbf{1}) \Longrightarrow W_Z(t; \beta),$$

where W_Z is a zero mean Gaussian process with covariance matrix $\Gamma(s, t) = ((\Gamma_{jk}(s, t)))$ given by

$$\Gamma_{kj}(s, t) = E\left[M(s)M(t)\left(\frac{w_{1j}(s)}{s_j(s; \beta)} - \frac{1}{s_0(s; \beta)} \right)\left(\frac{w_{1k}(t)}{s_k(t; \beta)} - \frac{1}{s_0(t; \beta)} \right)' \right] \tag{15A.4}$$

for $1 \le k, j \le p$, and $0 \le s, t \le \tau$. In the convergence displayed above, $\hat{\mu}_0 = [\hat{\mu}_{01} \ldots \hat{\mu}_{0p}]'$ and $\mathbf{1} = [1 \ldots 1]'$.

To establish the asymptotic normality of the MD estimators $\hat{\beta}$, in the next step we exhibit the asymptotic linearity of $\hat{\mu}_{0j}(t; b)$ near β. Recall from (15.4) that $\hat{\mu}_{0j}(t; b)$ is the unique solution to the equation

$$\sum_{i=1}^{n} w_{ij}(t)\hat{Y}_i(t)\left(N_i(t) - g(\hat{\mu}_{0j}(t; b)e^{b'Z_i(t)})\right) = 0.$$

So, differentiating this equation with respect to b, we obtain

$$\frac{\partial}{\partial b}\hat{\mu}_{0j}(t; b) = -\frac{\sum_{i=1}^{n} w_{ij}(t)\hat{Y}_i(t)Z_i(t)\dot{g}(\hat{\mu}_{0j}(t; b)e^{b'Z_i(t)})}{\sum_{i=1}^{n} w_{ij}(t)\hat{Y}_i(t)\dot{g}(\hat{\mu}_{0j}(t; b)e^{b'Z_i(t)})e^{b'Z_i(t)}}\hat{\mu}_{0j}(t; b). \tag{15A.5}$$

Furthermore, by the almost sure convergence of $\hat{\mu}_{0j}$, the strong consistency of the Kaplan–Meier estimator $\hat{G}$, the continuity of $\dot{g}$, and the uniform strong law of large numbers, we obtain

$$\frac{\partial}{\partial b}\hat{\mu}_{0j}(t; b) \to \frac{\partial}{\partial b}\mu_{0j}(t; b) = -\frac{s_j^{(1)}(t; b)}{s_j(t; b)}\mu_{0j}(t; b), \tag{15A.6}$$

almost surely and uniformly in $t \in [0, \tau]$, $b \in \mathbf{N}(\beta)$. For $\|b_n\| \to 0$, using once again the mean value theorem for $\hat{\mu}_{0j}(t, \beta + b_n)$ at β, we have

$$\hat{\mu}_{0j}(t; \beta + b_n) - \hat{\mu}_{0j}(t; \beta) = \frac{\partial}{\partial\beta}\hat{\mu}_{0j}(t; b^*)b_n,$$

where b^* is on the line segment joining β and $\beta + b_n$. Hence, using (15A.6), as $n \to \infty$, we obtain

$$\hat{\mu}_{0j}(t; \beta + b_n) - \hat{\mu}_{00}(t; \beta + b_n) = (\hat{\mu}_{0j}(t; \beta) - \hat{\mu}_{00}(t; \beta))$$
$$+ \left[\frac{\partial}{\partial\beta}(\mu_{0j}(t; \beta) - \mu_{00}(t; \beta)) + o(1)\right]b_n, \tag{15A.7}$$

almost surely and uniformly in $t \in [0, \tau]$. Recall that $\Psi_{\cdot j}$ is the column vector $[\psi_{1j} \dots \psi_{pj}]'$, where Ψ_{kj} is as in (15.8). Define $\tilde{D}(b)$ as

$$\tilde{D}(b) = \int_0^\tau \sum_{j=1}^{p} (\hat{\mu}_{0j}(t; \beta) - \hat{\mu}_{00}(t; \beta) + \Psi_{\cdot j}(t)'b)^2 \Phi(dt; \beta).$$

Observe that $\tilde{D}(b)$ is quadratic in b and has the minimizer

$$A^{-1}\int_0^\tau \sum_{j=1}^{p} \Psi_{\cdot j}(t; \beta)(\hat{\mu}_{0j}(t; \beta) - \hat{\mu}_{00}(t; \beta))\Phi(dt; \beta), \tag{15A.8}$$

where A is as in (15.9). Furthermore, for some $0 < B < \infty$, we have the following:

$$\sup_{\|b\| \leq B} |D_n(\beta + n^{-1/2}b) - \tilde{D}(n^{-1/2}b)| = o_P(1). \tag{15A.9}$$

This follows from some algebra, the asymptotic linearity of $\hat{\mu}_{0j}(t; b)$ near β given in (15A.7), and the convergence of the measures Φ_n to Φ in assumption (15A.4).

Hence, the minimizer of $D_n(\beta + n^{-1/2}b)$ with respect to b is given by

$$\sqrt{n}(\hat{\beta} - \beta) = A^{-1}\sqrt{n}\int_0^\tau \sum_{j=1}^p \Psi_j(t; \beta)(\hat{\mu}_{0j}(t; \beta) - \hat{\mu}_{00}(t; \beta))\Phi(dt; \beta) + o_P(1)$$

$$= A^{-1}\sqrt{n}\int_0^\tau \sum_{j=1}^p \Psi_j(t; \beta)\left(\frac{w_{ij}(t)}{s_j(t; \beta)} - \frac{1}{s_0(t; \beta)}\right)M_i(t)\Phi(dt; \beta) + o_P(1), \tag{15A.10}$$

where the first equality follows from (15A.9) and the fact that the minimizer of $\tilde{D}$ is given by (15A.8) and the second equality follows from (15A.3). Observe that the first term of (15A.10) is a sum of n i.i.d. terms. Using the multivariate central-limit theorem, it now follows that $\sqrt{n}(\hat{\beta} - \beta)$ converges in distribution to a zero mean normal vector with covariance matrix Σ as defined in (15.13).

REFERENCES

1. Andersen, P. K. and Gill, R. D., Cox's regression model for counting processes: A large sample study, *Ann. Statist.* **10**, 1100–1120 (1982).

2. Billias, Y., Gu, M., and Ying, Z., Towards a general asymptotic theory for Cox model with staggered entry, *Ann. Statist.* **25**, 662–682 (1997).

3. Byar, D. P., The Veterans Administration study of chemoprophylaxis for recurrent stage I bladder tumors: Comparisons of placebo, pyridoxine and topical thiotepa, in *Bladder Tumors and Other Tumors in Urological Oncology*, Pavonne-Macaluso, M., Smith, P. H., and Edsmyr, F., eds., Plenum Press, New York, 1980, pp. 363–370.

4. Cheng, S. C., Wei, L. J., and Ying, Z., Analysis of transformation models with censored data, *Biometrika* **82**, 835–845 (1995).

5. Cook, R. J. and Lawless, J. F., Analysis of repeated events, *Statist. Meth. Med. Res.* **11**, 141–166 (2002).

6. Donoho, D. L. and Liu, R. C., The "automatic" robustness of minimum distance functionals, *Ann. Statist.* **16**, 552–586 (1988).

7. Donoho, D. L. and Liu, R. C., Pathologies of some minimum distance estimators, *Ann. Statist.* **16**, 587–608 (1988).

8. Fine, J. P. and Gray, R. J., A proportional hazards model for the subdistribution of a competing risk, *J. Am. Statist. Assoc.* **94**, 496–509 (1999).

9. Ghosh, D. and Lin, D. Y., Marginal regression models for recurrent and terminal events, *Statistica Sinica* **12**, 663–688 (2002).

10. Ghosh, D. and Lin, D. Y., Semiparametric analysis of recurrent events data in the presence of dependent censoring, *Biometrics* **59**, 877–885 (2003).

11. Hobson, R. W., Weiss, D. G., Fields, W. S., Goldstone, J., Moore, W. S., Towne, J. B., Wright, C. B., and the Veteran Affairs Cooperative Study Group, Effect of carotid endarterectomy for asymptotmatic carotid stenosis, *New Engl. J. Med.* **328**, 221–227 (1993).

12. Kalbfleisch, J. D. and Prentice, R. L., *The Statistical Analysis of Failure Time Data*, Wiley, New York, 2002.

13. Koul, H. L. and DeWet, T., Minimum distance estimation in a linear regression model, *Ann. Statist.* **11**, 921–932 (1983).

14. Koul, H. L., Minimum distance estimation in multiple linear regression, *Sankhya Ser. A* **47**(Part I), 57–74 (1985).

15. Lawless, J. F., Nadeau, C., and Cook, R. J., Analysis of mean and rate functions for recurrent events, in *Proc. 1st Seattle Survival Analysis Symp.* Lin, D. Y. and Fleming, T. R., eds., Springer-Verlag, New York, 1997, pp. 137–149.

16. Li, Q. H. and Lagakos, S. W., Use of the Wei-Lin-Weissfeld method for the analysis of a recurring and a terminating event, *Statist. Med.* **16**, 925–940 (1997).

17. Liang, K. Y. and Zeger, S. L., Longitudinal data analysis using generalized linear models, *Biometrika* **73**, 13–22 (1986).

18. Lin, D. Y. and Ying, Z., A simple nonparametric estimator of the bivariate survival function under univariate censoring, *Biometrika* **80**, 573–581 (1993).

19. Lin, D. Y., Wei, L. J., Yang, I., and Ying, Z., Semiparametric regression for the mean and rate events, *J. Roy. Statist. Soc. Ser. B* **62**, 711–730 (2000).

20. Lin, D. Y., Wei, L. J., and Ying, Z., Semiparametric transformation models for point processes, *J. Am. Statist. Assoc.* **96**, 620–628 (2001).

21. Parr, W. C. and Schucany, W. R., Minimum distance and robust estimation, *J. Am. Statist. Assoc.* **75**, 616–624 (1980).

22. Pena, E. and Hollander, M., Models for recurrent phenomena in survival analysis and reliability, in *Mathematical Reliability: An Expository Perspective*, Mazzuchi, T., Singpurwalla, N., and Soyer, R., eds., 2004, pp. 105–123.

23. Pepe, M. S. and Cai, J., Some graphical displays and marginal regression analysis for recurrent failure times and time-dependent covariates, *J. Am. Statist. Assoc.* **88**, 811–820 (1993).

24. Pollard, D., *Empirical Processes: Theory and Applications*, Institute of Mathematical Statistics, Hayward, CA, 1990.

25. Robins, J. and Rotnitzky, A., Recovery of information and adjustment for dependent censoring using surrogate markers, in *AIDS Epidemiology-Methodological Issues*, Jewell, N., Dietz, K., and Farewell, V., eds., Birkhauser, Boston, 1992, pp. 297–331.

26. Strawderman, R., Estimating the mean of an increasing stochastic process at a censored stopping time. *J. Am. Statist. Assoc.* **95**, 1192–1208 (2000).

27. Sundaram, R., Semiparametric inference for proportional odds model with time-dependent covariates, *J. Statist. Plan. Infer.* **36**, 320–334 (2006).

28. Therneau, T. M. and Hamilton, S. A., rhDNase as an example of recurrent event analysis, *Statist. Med.* **16**, 2029–2047 (1997).

29. Yang, S. and Prentice, R. L., Semiparametric inference in the proportional odds regression model, *J. Am. Statist. Assoc.* **94**, 125–136 (1999).

CHAPTER 16

Tree-Based Methods for Survival Data

Mousumi Banerjee and Anne-Michelle Noone

Department of Biostatistics, University of Michigan, Ann Arbor, Michigan

16.1 INTRODUCTION

Tree-based methods have become one of the most flexible, intuitive, and powerful data analytic tools for exploring complex data structures. The applications of these methods are far-reaching. The best documented, and arguably most popular uses of tree-based methods are in biomedical research, where classification is a central issue. For example, a clinician may be very interested in whether a patient with chest pain is suffering from a heart attack or simply has a strained muscle [15]. To answer this question, information on the patient must be collected, and a good diagnostic test utilizing such information must be in place. Tree-based methods provide one solution for constructing such diagnostic tests. Some interesting applications of tree-based methods are described by Zhang et al. [45] and Segal et al. [38].

Original tree-based methods, introduced by Morgan and Sonquist [31], were used in classification and regression. Advances in the practical and theoretic aspects of tree-based methods were developed by Breiman et al. [5] in their monograph on classification and regression trees. Generally, tree-based methods recursively partition the covariate space into disjoint regions and the corresponding data into groups (nodes). For each node to be split, some measure of separation in the response distribution between the two daughter nodes resulting from a split is calculated. All possible splits for each covariate are evaluated, and the variable and corresponding split point that best separates the daughter nodes is chosen. The same procedure is applied recursively to increase the number of nodes until each contains only a few subjects. The resulting model can be represented as a binary tree. After a large tree is grown, there are rules for pruning and for readjusting the size of the tree.

Statistical Advances in the Biomedical Sciences, edited by Atanu Biswas, Sujay Datta, Jason P. Fine, and Mark R. Segal

265

Interest in tree-based methods for survival data naturally came from the need of clinical researchers to define interpretable prognostic classification rules both for understanding the prognostic structure of data (by forming a small number of groups of patients with differing prognoses) and for designing future clinical trials. Several authors have studied extensions of original tree-based methods in the setting of censored survival data [16,10,11,35,12,26, 27,21,43]. Some applications of tree-based survival analyses are given by Albain et al. [1,2], Banerjee et al. [3,4], Freedman et al. [13], and Katz et al. [23].

In this chapter, we discuss general methodological aspects of tree-based modeling for survival data. Although several splitting criteria have been proposed in the literature for survival data, the choice of an appropriate criterion is not obvious. Thus, we focus on comparing and contrasting different splitting criteria. On the basis of a simulation study and analyses of a clinical dataset, we compare five different splitting rules that use either a measure of within-node error or between-node separation.

Another exciting more recent development is the expansion of trees into forests or ensemble of trees [6,7]. Growing an ensemble of trees and aggregating is a way to improve predictive performance and address the problem of instability that is recognized to be inherent in a single tree. In this chapter, we present a method for growing survival forests by using the null deviance residuals from a Cox proportional hazards model as the outcome variable for growing trees in the forest. This approach is easy to implement, and circumvents the complexity induced by censoring. Ensemble predictions are computed by aggregating across different trees in the forest.

The chapter is organized as follows. In Section 16.2, we present a review of classification and regression of trees (CART). Section 16.3 describes algorithms for growing and pruning trees in the survival data setting. In Section 16.4 we describe the design and results of the simulation study to compare different splitting rules. Section 16.5 presents analyses of data from a cohort study of breast cancer, based on single tree methods employing various splitting rules. Section 16.6 presents the methodology for growing a survival forest, and survival forest analyses of the breast cancer data are presented in Section 16.6.1. Finally, Section 16.7 contains some concluding remarks.

16.2 REVIEW OF CART

The literature on tree-based methods dates from work in the social sciences by Morgan and Sonquist [31] and Morgan and Messenger [30]. In statistics, Breiman et al. [5] had a seminal influence in both bringing the work to the attention of statisticians and proposing new algorithms for constructing trees. At around the same time decision tree induction was beginning to be used in the field of machine learning and in engineering.

The terminology of trees is graphic; a tree T has a *root* that is the top node, and observations are passed down the tree, with decisions made at each *node* (also called "daughters") until a *terminal node* or *leaf* is reached. Each nonterminal node (also called *internal node*) contains a question on which a split is based. The terminal nodes of a tree T are collectively denoted by $\tilde{T}$, and the number of terminal nodes is denoted by $|\tilde{T}|$. Each terminal node contains the class label (for a classification problem) or an average response (for a least-squares regression problem). The branch T_t that stems from node t includes t itself and all its daughters. A *subtree* of T is a tree with root a node of T; it is a *rooted subtree* if its root is the root of T.

In the CART paradigm, the covariate space is partitioned recursively in a binary fashion. The partitioning is intended to increase within-node homogeneity, where homogeneity is determined by the dependent variable in the problem. There are three basic elements for constructing a tree under the CART paradigm: (1) tree growing, (2) finding the "right-sized tree," and (3) testing. The first element is aimed at addressing the question *how and why a parent node is split into daughter nodes*. CART uses binary splits, phrased in terms of the covariates, that partition the predictor space. Each split depends on the value of a single covariate. For ordered (continuous or categorical) covariates X_j, only splits resulting from questions of the form "Is $X_j \le c$?" for $c \in$ domain(X_j) are considered, thereby allowing at most $n - 1$ splits for a sample of size n. For nominal covariates no constraints are imposed on possible subdivisions. Thus, for a nominal covariate with M categories, there are $2^{M-1} - 1$ splits to examine.

Using the covariates univariately entails that all splits are orthogonal to the coordinate axes. Methods extending the allowable splits to (1) linear combinations of covariates and (2) Boolean combinations of binary covariates have been proposed in the literature. The price for this improved flexibility is reduced interpretability and a greater computational burden. In particular, use of linear combination splits can be very computer-intensive. Thus, at least for regression, CART advises against use of linear combination splits.

The question that logically comes next is: How do we select one or several preferred splits from the pool of allowable splits? Before selecting the best split, one must define the goodness of split. The objective of splitting is to make the two daughter nodes as homogeneous as possible. Therefore, the goodness of a split must weigh the homogeneities in the two daughter nodes. Extent of node homogeneity is measured quantitatively using an impurity function. Potential splits are evaluated for each covariate, and the covariate and split value resulting in the greatest reduction in impurity is chosen.

Corresponding to a split s at node t into left and right daughter nodes t_L and t_R, the reduction in impurity is given by

$$\Delta I(s, t) = i(t) - P(t_L)i(t_L) - P(t_R)i(t_R),$$

where $i(t)$ is the impurity in node t and $P(t_L)$ and $P(t_R)$ are the probabilities that a subject falls in nodes t_L and t_R, respectively. For classification problems, $i(t)$ is measured in terms of entropy or Gini impurity. For regression problems, $i(t)$ is typically the mean residual sum of squares. The probabilities $P(t_L)$ and $P(t_R)$ are estimated via corresponding sample proportions. The splitting rule that maximizes $\Delta I(s, t)$ over the set S of all possible splits is chosen as the best splitter for node t.

A useful feature of CART is that of growing a large tree and then *pruning* it back to find the right-sized tree. During the early development of recursive partitioning, stopping rules were proposed to quit the partitioning process before the tree becomes too large. For example, the *automatic interaction detection* (AID) program proposed by Morgan and Sonquist [31] declared a terminal node based on the relative merit of its best split to the quality of the root node.

Breiman et al. [5] argued that depending on the stopping threshold, the partitioning tends to end too soon or too late. Accordingly, they made a fundamental shift by introducing a second step, called *pruning*. Instead of attempting to stop the partitioning, they propose to let the partitioning continue until it is saturated or nearly so. Beginning with this generally large tree, they prune it from the bottom up. The point is to find a subtree of the saturated tree that is most "predictive" of the outcome and least vulnerable to the noise in the data.

Let $c(t)$ be the misclassification cost of a node t. Now define $C(T)$ as the misclassification cost of the entire tree T: $C(T) = \sum_{t \in \tilde{T}} P(t)c(t)$. Note that $C(T)$ is a measure of the quality of the tree T. The purpose of pruning is to select the best subtree of an initially overgrown (or saturated) tree, such that $C(T)$ is minimized. In this context, an important concept introduced by Breiman et al. [5] is the concept of tree cost–complexity. It is defined as

$$C_\alpha(T) = C(T) + \alpha|\tilde{T}|,$$

where α (≥ 0) is a penalty parameter for the complexity of the tree. The total number of terminal nodes $|\tilde{T}|$ is used as a measure of tree complexity. Note that the total number of nodes in a tree T (i.e., its size) is twice the number of its terminal nodes minus 1. Thus, *tree complexity* is really another term for the size of the tree. The difference between $C_\alpha(T)$ and $C(T)$ as a measure of tree quality resides in that $C_\alpha(T)$ penalizes a large tree.

For any tree, there are many subtrees, and therefore many ways to prune. The challenge is how to prune, that is, which subtrees to cut first. Breiman et al. [5] showed that (1) for any value of the penalty parameter α, there is a unique smallest subtree of T that minimizes the cost–complexity; and (2) if $\alpha_1 > \alpha_2$, the optimal subtree corresponding to α_1 is a subtree of the optimal subtree corresponding to α_2. The use of tree cost–complexity therefore allows one to construct a sequence of nested optimal subtrees from any given tree T. This is done by recursively pruning the branch(es) with the weakest link; that is, the node t with the smallest value of α such that $C_\alpha(t) \leq C_\alpha(T_t)$.

Having obtained a nested sequence of pruned optimal subtrees, one is left with the problem of selecting a *best* tree from this sequence. Using the learning sample (resubstitution) estimate of misclassification cost results in selecting the largest tree. Breiman et al. [5] suggest using a test sample or cross-validation to obtain honest estimates of $C(T)$. The subtree with the smallest estimate of misclassification cost is chosen as the final tree. Details of the cross-validation method are described in Breiman et al. [5] and Zhang and Singer [44].

16.3 TREES FOR SURVIVAL DATA

Consider the usual setting for censored survival data that includes a measurement of time under observation and covariates that are potentially associated with the survival time. Specifically, an observation from a sample of size n consists of the triple $(y_i, \delta_i, \mathbf{X}_i)$, $i = 1, \ldots, n$, where y_i is the time under observation for individual i, δ_i is the event indicator for individual i [i.e., $\delta_i = 1$ if the ith observation corresponds to an event ("failure"), and $= 0$ if the ith observation is censored], and $\mathbf{X}_i = (X_{i1}, \ldots, X_{ip})$ is the vector of p covariates for the ith individual. For simplicity, we will assume that there are no tied events.

Several authors have proposed extensions of CART in the setting described above, [16,10,11,35,12,26,27,43]. Algorithms for growing trees for survival data can be broadly classified under two general approaches. One approach is to measure the within-node homogeneity with a statistic that measures how similar the subjects in each node are and choose splits that minimize the within-node error. The alternative is to summarize the dissimilarity in survival experiences between two groups induced by a split and choose splits that maximize this difference.

16.3.1 Methods Based on Measure of Within-Node Homogeneity

Tree growing and pruning based on measures of within-node homogeneity adopt the CART algorithm directly, since the measures defined are all subadditive, allowing comparisons between subtrees. Gordon and Olshen [16] presented the first extension of CART to censored survival data, which involved a distance measure (the Wasserstein metric) between Kaplan–Meier curves and certain point masses. Their approach amounts to assuming a piecewise exponential model with one data-determined knot. When L_2 Wasserstein distances are used, the homogeneity corresponds to the variance of the Kaplan–Meier estimate. Another method proposed by Ciampi et al. [10] is based on a parametric model and likelihood ratio statistics. Below we briefly describe methods based on within-node homogeneity that are most commonly used in practical settings.

16.3.1.1 Using Martingale Residuals

Therneau et al. [40] proposed using the null martingale residuals from a Cox proportional hazards model as the outcome variable in a regression tree. In the absence of time-dependent covariates these residuals are given by

$$\widehat{M}_i = \delta_i - \hat{\Lambda}_0(y_i),$$

where $\hat{\Lambda}_0(\cdot)$ is the Breslow estimator [9] of the baseline cumulative hazard. Since this transforms the censored data into uncensored values in the form of the martingale residuals, they can be used directly as continuous outcome in CART without modification to the regression tree algorithm. The same authors further established that improved results could be attained by including the covariate of interest in the Cox model prior to obtaining martingale residuals as opposed to using null martingale residuals [17].

A drawback of this approach is that the use of martingale residuals does not provide an easily interpretable summary measure for the terminal nodes. As is standard with regression trees, node summaries are simply averages. The root node average in this case is zero by definition. But it is not possible to confer any meaning to the terminal node averages from a survival analysis perspective. This can be easily remedied, though, by simply obtaining survival-based plots and summaries for the terminal nodes. However, an additional drawback of using the martingale residuals is that there is no guarantee that minimizing the residual sums of squares improves fit to the survival data [41].

16.3.1.2 Likelihood-Based Methods

A likelihood-based splitting criterion was proposed by Davis and Anderson [12], who assumed that the survival function in a node is exponential with a constant hazard. The measure for within-node homogeneity is based on the negative log likelihood of the exponential model at a node; for node h, this is given by

$$R(h) = D_h \left[1 - \log\left(\frac{D_h}{y_h}\right) \right],$$

where $D_h = \sum_{i \in h} \delta_i$ is the total number of events and $y_h = \sum_{i \in h} y_i$ is the sum of observation times for all subjects in node h. An advantage of this method is that each terminal node can be summarized by the hazard ratio of that node. This is a meaningful and easily interpretable description of the subgroups identified by the tree. A disadvantage, however, is the assumption of a constant underlying hazard in each node, which may not hold true and may degrade the performance of the method.

LeBlanc and Crowley [26] developed a splitting method based on the popular semiparametric proportional hazards model, where the hazard at time y for an individual i with covariate vector x_i is the product of a baseline hazard that depends only on time and a structural component that depends on the individual through that person's covariates. Consequently, their splitting criterion is also based on the assumption that the hazard functions in each daughter node are proportional. LeBlanc and Crowley define the within-node homogeneity measure based on the deviance residual. For this, the full likelihood function under the proportional hazards model for a tree T must be constructed. Under the assumption that the proportional hazards model

$$\lambda_h(y) = \lambda_0(y)\theta_h$$

is true, where θ_h is a structural nonnegative parameter that depends on an individual's covariates and λ_0 is the baseline hazard, the full likelihood can be written as

$$L = \prod_{h \in \tilde{T}} \prod_{i \in h} (\lambda_0(y_i)\theta_h)^{\delta_i} e^{-\Lambda_0(y_i)\theta_h},$$

where $\tilde{T}$ is the set of terminal nodes and $\Lambda_0(\cdot)$ is the baseline cumulative hazard function. The full tree likelihood must be computed every time a node is split, and since all possible splits are under consideration, the computation quickly becomes burdensome. Furthermore, the baseline cumulative hazard must be estimated for each node. LeBlanc and Crowley [26] proposed using the Nelson [32] estimator for Λ_0. This is the Breslow [9] estimator in a Cox model without covariates. LeBlanc and Crowley [26] referred to this as their "one-step estimator," denoted by $\hat{\Lambda}_0^1$. In growing the tree, they propose splitting based on the deviance residual

$$d_i = 2\left[\delta_i \log\left(\frac{\delta_i}{\hat{\Lambda}_0^1(y_i)\hat{\theta}_h^1}\right) - \left(\delta_i - \hat{\Lambda}_0^1(y_i)\hat{\theta}_h^1\right)\right],$$

where

$$\hat{\theta}_h^1 = \frac{\sum_{i \in h} \delta_i}{\sum_{i \in h} \hat{\Lambda}_0^1(y_i)},$$

which can be interpreted as the number of failures divided by the expected number of failures in node h under the assumption of no structure in survival times. The deviance for a node h is $R(h) = \sum_{i \in h} d_i$, which is the log-likelihood ratio test statistic when the null is the saturated model at h. This method of splitting provides meaningful and interpretable node summaries. Specifically, the node summary is the ratio of observed to expected events in that node under the proportional hazards model.

For the likelihood-based splitting rules, improvement for split s at a node h into left and right daughter nodes h_L and h_R is given by

$$\Delta R(s, h) = R(h) - [R(h_L) + R(h_R)],$$

where $R(h)$ is the node deviance under an exponential or proportional hazards model. The tree is split by the variable at s so as to lead to the largest value $\Delta R(s, h)$.

16.3.1.3 Weighted Impurity Function

Zhang [43] proposed a method in which the node impurity is determined by the observed times and the proportion of censored versus uncensored observations in the node. He argues that a homogeneous node should consist of subjects whose event times are close and who are either mostly censored or mostly uncensored. Since this within-node homogeneity is based on the observed times and the censoring status, the impurity at node h can simply be written as a weighted combination of the impurity in these two quantities

$$i(h) = w_1 i_y(h) + w_2 i_\delta(h),$$

where w_1 and w_2 are weights and $i_y(h)$ and $i_\delta(h)$ are the impurities for the observed time and censoring, respectively. The impurity for the observation time y is given by

$$i_y = \frac{\sum_{i \in h} (y_i - \bar{y}(h))^2}{\sum y_i^2},$$

where $\bar{y}(h)$ is the average of y in node h. The sum in the denominator can be over the subjects in node h (adaptive normalization), or over the entire sample (global normalization). The impurity of the censoring indicator $i_\delta(h)$ is given by

$$i_\delta(h) = -p_h \log(p_h) - (1 - p_h) \log(1 - p_h),$$

where p_h is the proportion of censored subjects in node h.

In simulations performed by Zhang [43], three different pairs of weights (1:2, 1:1, and 2:1) with each normalization method were tested. Adaptive normalization outperformed global normalization, and the most reliable weight choice for this was 1:2. As noted by Zhang, since global normalization is easier to implement and retains the subadditivity it should still be considered. Under global normalization, simulation results revealed that equal weighting (1:1) was better than the other two weighting schemes.

Benefits of this approach are that it provides an intuitive impurity measure and is easily implemented. Once the weighted sum is calculated, it can be used as a continuous outcome in a regression tree using CART engineering. A potential drawback of this approach is that it may not perform well under heavy censoring.

16.3.2 Methods Based on Between-Node Separation

A different approach to splitting is to recursively partition the data by maximizing the dissimilarity of the two daughter nodes resulting from a split [35]. One such algorithm was proposed by LeBlanc and Crowley [27], who use the two-sample log-rank statistic to measure the separation in survival times between two daughter nodes. The two-sample log-rank statistic was chosen because of its extensive use in the survival analysis setting, and also because it is an appropriate measure of dissimilarity in survival between two groups. The numerator of the log-rank statistic can be expressed as a weighted difference between estimated hazard functions

$$G = \int_0^\infty w(u) \frac{n_1(u) n_2(u)}{n_1(u) + n_2(u)} (d\hat{\Lambda}_1(u) - d\hat{\Lambda}_2(u)),$$

where $w(\cdot) = 1$, $n_1(u)$, and $n_2(u)$ are the number of subjects at risk in each group at time u and $\hat{\Lambda}_1$ and $\hat{\Lambda}_2$ are the Nelson cumulative hazard estimators for each group. In general, other weights could be chosen to have greater sensitivity to early or late differences in the hazards of the two groups. LeBlanc and Crowley [27] propose using the ratio of G squared divided by an estimate of its variance as the splitting statistic. Partitioning at node h involves finding the split s among all variables that maximizes the standardized two-sample log-rank statistic.

16.3.3 Pruning and Tree Selection

In the survival data setting, two general approaches have been proposed for pruning and sub-sequent tree selection. As stated earlier, methods that are based on a measure of within-node homogeneity directly adopt the CART algorithm. This includes cost–complexity pruning, which efficiently yields subtrees that perform best in terms of residual error (deviance for the likelihood-based splitting) for their size. In addition, cross-validation can be used to select the final tree from the sequence of pruned subtrees. Thus, the entire CART engineering of Breiman et al. can be adopted in this approach.

For the log-rank-based splitting aimed at maximizing between-node separation, LeBlanc and Crowley [27] developed an optimal pruning algorithm analogous to the cost–complexity pruning algorithm of CART. Their algorithm uses a measure of the tree's performance according to the dissimilarity in survival between daughter nodes in the tree. LeBlanc and Crowley [27] define the split complexity of a tree as

$$G_\alpha(T) = G(T) - \alpha|S|,$$

where $S = T - \tilde{T}$ is the set of internal nodes of the tree T and $G(T)$ is the sum over the standardized splitting statistics $G(h)$ in the tree T, specifically $G(T) = \sum_{h \in S} G(h)$, and $\alpha \geq 0$ is the complexity parameter. One can interpret $G(T)$ as the amount of prognostic structure in the tree T. A tree T_1 is an optimally pruned subtree of T for complexity parameter α if $G_\alpha(T_1) = \max_{T' \preceq T} G_\alpha(T')$, where the symbol "$\preceq$" means "is a subtree of." Furthermore, T_1 is the smallest optimally pruned subtree if $T_1 \preceq T'$ for every optimally pruned subtree T' of T. The algorithm repeatedly prunes off branches with smallest average log-rank statistics in the branch. Thus the pruning algorithm borrows the idea of weakest link cutting from the cost–complexity algorithm of CART.

Having obtained a sequence of optimally pruned subtrees, the next step is to select the best tree from this sequence. Since the same data are used to select the split point and variable, as well as to calculate the statistic, LeBlanc and Crowley [27] suggest using a bias-corrected version of the split complexity for the final tree selection, using the bootstrap method to estimate the bias.

16.4 SIMULATIONS FOR COMPARISON OF DIFFERENT SPLITTING METHODS

Several authors have compared different splitting methods for growing survival trees. In particular, Keles and Segal [24] provided an analytic relationship between log-rank and martingale residual-based splitting. Zhang [43] compared the weighted-impurity-based splitting [43] with the splitting rules proposed by Davis and Anderson [12], Gordon and Olshen [16], and Segal [35]. LeBlanc and Crowley [26] compared their one-step full-likelihood-based splitting method with the methods proposed by Davis and Anderson [12].

In this section, we describe findings from our simulation experiments to compare the various splitting rules discussed in Section 16.3. Our interest lies in contrasting the splitting rules, as opposed to entire tree architectures derived from repeated splitting. Thus, we considered only a single covariate (x) generated from a uniform(0,1) distribution. Survival distributions were Weibull—$S(t; \lambda, \alpha) = \exp(-\lambda t^\alpha)$—with the following choices of the shape parameter: (1) $\alpha = 0.5$, (2) $\alpha = 1$ (exponential survival), and (3) $\alpha = 2$. We considered three different survival models for λ: (a) $\lambda = 1$; (b) $\lambda = 1$ if $x > 0.6$, and $= \exp(1)$ $(=2.72)$ if $x \leq 0.6$; and (c) $\lambda = \exp(4x)$. We assumed uniform censoring with censoring proportions of 0%, 25%, 50%, and 75%. For each simulation design scenario, we generated samples of size $n = 250$. We calculated the following five split statistics for all possible split values: martingale-residual-based split statistic (M), Davis and Anderson's [12] exponential-model-deviance-based splitting (ED), LeBlanc and Crowley's [26] proportional hazards model (based on one-step full likelihood) splitting (PH), Zhang's [43] weighted-impurity-based splitting (WI), and log-rank statistic-based splitting (LR).

Figure 16.1 presents line graphs of the five split statistics versus covariate cutpoint in the 25% censoring scenario. For the null case (model A), the line graphs of M, ED, PH, and LR exhibit strong tracking for all three survival distributions. The correlations between the split statistics M, ED, PH, and LR range from 0.91 to 0.99. Although local features for these four split statistics were very similar, the maxima did not always coincide, since there were competing splits that had comparable values of the split statistics. The line graph of WI, on the other hand, exhibits only moderate tracking with the line graphs of the other four splitting statistics in the $\alpha = 0.5$ scenario, but poor tracking for the other two survival distributions. All five statistics exhibit end-cut preference. Although not presented here, the results obtained in the 50% and 75% censoring scenarios are consistent with the general patterns mentioned above. For 0% censoring, as expected, all five statistics exhibit strong tracking with correlations between split statistics ranging from 0.83 to 0.99.

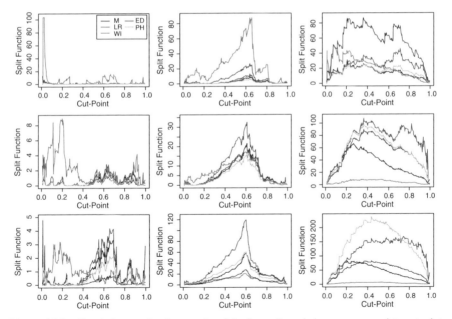

Figure 16.1 Simulation results: line graphs of the five split statistics versus covariate cutpoint in the 25% censoring scenario. The first column corresponds to model A, the second column corresponds to model B, and the third column corresponds to model C for λ. The first row corresponds to $\alpha = 0.5$, the second row corresponds to $\alpha = 1$, and the third row corresponds to $\alpha = 2$ in the Weibull model.

For model B, the line graphs of all five split statistics track very closely for all three survival distributions, with correlations between split statistics ranging from 0.92 to 0.99. Given that tree methods are especially adept at picking the covariate threshold-type structure such as in model B, this result is reassuring. It is interesting to note that, for the Weibull survival distribution with $\alpha = 0.5$, M, ED, PH, and LR were all optimized by the same split (0.64) and WI was optimized at 0.66. For the exponential survival distribution, M, ED, PH, and LR were optimized at 0.61 and WI was optimized at 0.63. For the Weibull survival distribution with $\alpha = 2$, M and LR were optimized at 0.6 and WI, ED, and PH were optimized at 0.59. Thus the split statistics M, ED, PH, and LR were largely comparable in their ability to pick the target cutpoint of 0.6. Once again, the results obtained (not shown here) in the 0%, 50%, and 75% censoring scenarios are consistent with the general patterns mentioned above, although the concordance between WI and the other four split statistics seems to somewhat decline with higher censoring.

For model C, the line graphs of the split statistics fail to show clear optimal cutpoints for all three survival distributions. This is because tree methods are not particularly suited for uncovering smooth covariate survival association such as in model C. Correlations between the split statistics are generally in the range 0.12–0.63.

16.5 EXAMPLE: BREAST CANCER PROGNOSTIC STUDY

As an illustrative example, we present tree-based analyses of data from a cohort study of breast cancer patients. Women eligible for this study were newly diagnosed patients with stage I, II, or III breast cancer, diagnosed between January 1990 and December 1996 at the Harper Hospital in Detroit, Michigan. Detailed demographic, clinical, pathological, treatment, and follow-up data were obtained from the Surveillance, Epidemiology, and End Results (SEER) database, hospital, and clinic records. Recurrence-free survival (RFS) was the primary endpoint of the study, defined as the interval between diagnosis and documented regional/local or distant recurrence. The goals of the study were to analyze the relative contributions of patient and tumor-related prognostic factors on RFS, and to identify patient subgroups with homogeneous RFS within a group but different RFS between groups (i.e., prognostic grouping of patients).

The analysis cohort consisted of 1055 patients. A total of 10 covariates were considered for the analysis. These included sociodemographic variables (age, race, marital status, and socioeconomic status), factors characterizing tumor [tumor size, number of positive lymph nodes, tumor differentiation, estrogen receptor (ER), and progesterone receptor (PR) status], and body mass index (BMI) as a comorbid factor. Patients were classified as obese if their BMI was > 30, per the standard guideline recommended by the World Health Organization. Number of positive lymph nodes was categorized as: 0, 1–3, 4–9, and >10 positive nodes. Tumor differentiation was categorized as well, moderate, and poor. Estrogen and progesterone receptors are binary categorical variables (positive/negative).

Figure 16.2 shows the survival tree based on log-rank splitting. At each level of the tree, we show the best splitter (covariate with cutpoint), and the corresponding LR split statistic. The permutation sampling method was used to add an approximate p value to each split conditional on the tree structure above the split to facilitate the interpretation of individual splits. Circles denote terminal nodes in the tree. Within each terminal node, n denotes the number of patients, R denotes the (crude) number of recurrences, and $5Yr$ is the 5-year RFS rate. Competitor splits (i.e., covariate with cutpoint that had the second largest value of the LR split statistic) were also generated at each step of the tree to assess the relative strength of the chosen best splitter. Knowledge of such splits also enables the elucidation of alternate, competing models.

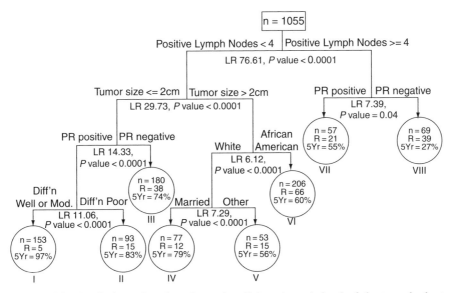

Figure 16.2 Survival tree based on log-rank splitting. At each level of the tree, the best splitter (covariate with cutpoint), along with the corresponding LR split statistic, and permutation p value are presented. Circles denote terminal nodes in the tree.

The root node was split by the number of positive nodes, with the best cutoff fewer than four versus at least four positive nodes (LR = 76.61; permutation $p < 0.0001$). The competitor at this step was tumor size (≤ 2 cm vs. >2 cm), and the corresponding LR statistic was 50.39. The subgroup with at least four positive nodes was next split by PR status (LR = 7.39; $p = 0.04$), and the competitor at this step was differentiation (well or moderately vs. poorly differentiated; corresponding LR = 6.3), followed by ER status (LR = 3.33). Patients with positive PR status had significantly better outcome than did the PR-negative patients (estimated 25th percentile RFS, 31 months and 16 months, respectively). None of these subgroups were further split and formed terminal nodes VII and VIII in the tree. Of note, PR-negative patients with at least four positive nodes (i.e., terminal subgroup VIII) had the worst prognosis among all subgroups. On the opposite side of the tree, the subgroup with fewer than four positive nodes was next split by tumor size (best cutoff ≤ 2 vs. >2 cm; LR = 29.73; $p < 0.0001$). The competitor was differentiation (well or moderately vs. poorly differentiated; LR = 20.3). The subgroup with fewer than four positive nodes and tumor size ≤ 2 cm was subsequently split by PR status (LR = 14.33; $p < 0.0001$), and the competitor was ER status (LR = 8.7). Patients with PR-positive tumors had significantly better outcome compared with the PR-negative patients. The latter formed terminal node III in the tree; the 25th percentile RFS of patients in this group was 57 months. The subgroup with fewer than four positive lymph nodes, tumors ≤ 2 cm, and positive PR status was further split by tumor differentiation (well or moderately vs. poorly differentiated; LR = 11.06; $p < 0.0001$). The competitor at this step was tumor size (≤ 1 cm vs. 1–2 cm), and the corresponding LR was 3.7. None of the resulting subgroups had any further split and formed terminal nodes I and II in the tree. Notably, patients with fewer than four positive nodes, tumor size ≤ 2 cm, positive PR status, and well or moderately differentiated tumors (i.e., terminal node I) had the best prognosis, with a 5-year RFS of 97%.

The patient subgroup with fewer than four positive nodes and tumor size >2 cm was then split by race (LR $= 6.12$; $p < 0.0001$). The competitor at this step was age (<50 vs. ≥ 50 years), and the corresponding LR was 4.1. White patients had significantly better RFS than did African-American patients. The latter formed terminal node VI in the tree; the 25th percentile RFS of these patients was 30 months. The white subgroup with fewer than four positive nodes and tumor size >2 cm was further split by marital status (LR $= 7.29$; $p < 0.0001$), with married patients having better prognosis. The competitor at this step was socioeconomic status (LR $= 3.7$). None of these subgroups had any other significant split and formed terminal nodes IV and V in the tree.

We also constructed survival trees based on the other splitting criteria discussed earlier (Figs. 16.3–16.5). The purpose was to illustrate the similarities and differences in survival trees grown on the basis of the other split statistics. Using WI, the root node splitter was <10 versus ≥ 10 positive nodes (Fig. 16.3). The resulting daughter nodes were terminal nodes, yielding a simple tree structure with only two terminal nodes. The competitor for the root node splitter was tumor size ≤ 2 cm versus >2 cm. Using M, the root node was split by <4 versus ≥ 4 positive nodes (Fig. 16.4). The resulting left daughter node was split by tumor size ≤ 2 cm versus >2 cm, and the right daughter node was split by SES. All the resultant nodes were declared terminal nodes, thereby resulting in a tree with four terminal nodes. Structurally, the trees grown using PH (Fig. 16.5) and LR (Fig. 16.2) as splitting statistics were most similar. In fact, the optimal splitters (covariate with cutpoint) at each level of the

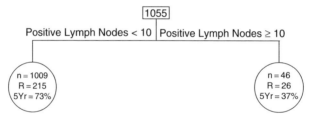

Figure 16.3   Survival tree based on weighted impurity (WI).

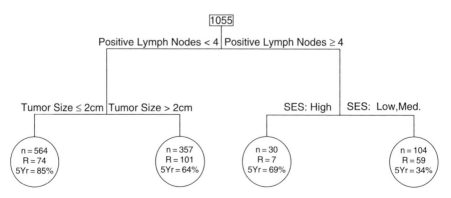

Figure 16.4   Survival tree based on martingale residual (M).

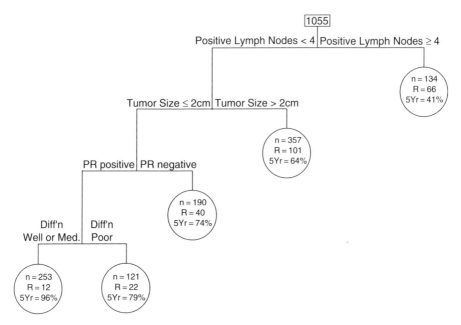

Figure 16.5 Survival tree using proportional hazards model – based splitting (PH).

PH- and LR-based trees were identical. The number of terminal nodes in the PH-based tree was five, yielding once again a smaller sized tree than the LR-based tree.

Prognostic grouping of the patients was based on further amalgamation of the terminal nodes in Figure 16.2. Since only a small number of prognostic groups was of interest, further amalgamation of the terminal nodes with similar prognosis was performed. We

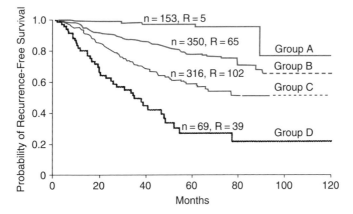

Figure 16.6 Recurrence-free survival for the four prognostic groups derived from amalgamation of terminal nodes in Figure 16.2. Group A (best prognostic group) corresponds to terminal node I; group B corresponds to terminal nodes II, III, and IV; group C corresponds to terminal nodes V, VI, and VII; and group D (worst prognostic group) corresponds to terminal node VIII.

chose the 5-year RFS rate as a measure of prognosis, and ranked each terminal node in the tree according to that measure. After ranking the nodes, there are several options for amalgamation [25]. One option would be to grow another tree on the ranked nodes and only allow the second tree to select three or four nodes. A second option would be to divide the nodes according to the quantiles of the data. Yet another option would be to evaluate all possible amalgamations of nodes into K groups and choose the partition that yields the largest partial likelihood or largest K sample log-rank test statistic. The result of amalgamation to yield the largest partial likelihood for a four-group combination of the breast cancer tree given in Figure 16.2 is presented in Figure 16.6. Terminal nodes II, III, and IV in Figure 16.2 were amalgamated to form prognostic group B; and nodes V, VI, and VII were amalgamated to form prognostic group C. These, together with subgroups I and VIII, resulted in four distinct prognostic groups for RFS, designated A–D in Figure 16.6. The 5-year RFS rates for patients in groups A, B, C, and D were 97%, 78%, 58%, and 27%, respectively ($p = 0.0001$).

16.6   RANDOM FOREST FOR SURVIVAL DATA

The mechanism of selecting a best split and the recursive partitioning of data leads to smaller and smaller datasets. This can lead to instability [6] in the tree structure, whereby small changes in the data and/or algorithm inputs can have dramatic effects on the nature of the solution (variables and splits selected). Another major shortcoming of tree-based methods is their modest prediction performance, attributable to algorithm greediness and constraints that, while enhancing interpretability, reduce flexibility of the fitted functional forms. Growing an ensemble of trees and aggregating is a way to fix these problems. The advantage in growing many trees and using an aggregated estimate is that it is a way to reduce variance [7]. It also leads to classifiers and predictors that are drawn from a richer class of models [18]. Ensemble methods such as bagging [6,33], boosting [14,33], and random forest [7] yield substantial performance improvement over a single tree, and are known to be stable.

Bagging involves bootstraping the training data. A large number of pseudodatasets are generated by resampling the original observations with replacement, and a tree grown on each pseudodataset. This results in an ensemble of trees. In boosting, instead of random resampling, the data are iteratively reweighted. The algorithm alternates between fitting a tree and reweighting the data. The weights are adaptively chosen, with more weight given to observations that the tree models poorly. Again, an ensemble of trees result. The simple mechanism whereby bagging and boosting reduce prediction error, is well understood in terms of variance reduction resulting from averaging [18]. Such variance gains can be enhanced by reducing the correlation between the quantities being averaged. It is this principle that motivates random forest.

Random forest [7] is an ensemble of unpruned classification or regression trees, induced from bootstrap samples of the training data, using random feature selection in the tree induction process. Correlation reduction is achieved by the random feature selection. Instead of determining the optimal split of a given node of a tree by evaluating all allowable splits on all covariates, as is done with growing a single tree, a subset of the covariates drawn at random is employed. Prediction is made by aggregating (majority vote for classification or averaging for regression) the predictions of the ensemble. Random forests demonstrate exceptional prediction accuracy [7], comparable to artificial neural networks and support vector machines.

The published literature on ensemble techniques for survival data is sparse owing to the difficulties induced by censoring. Hothorn et al. [19] studied an aggregation scheme for bagging survival trees. Breiman [8] introduced a software implementation of a random forest variant for survival data; however, it does not come with a formal description of the methodology. Ishwaran et al. [22] proposed a method that combines random forest methodology with survival trees grown using Poisson likelihoods. In a more recent article, Hothorn et al. [20] proposed a unified and flexible framework for ensemble learning in the presence of censoring.

In this section, we present an adaptation of Breiman's [7] random forest methodology to the survival data setting. The strategy involves substituting suitably chosen residuals for the survival endpoint, and enabling inheritance of the random forest algorithm applicable to continuous outcomes, thereby bypassing difficulties that result from censoring. This general strategy has been employed to adapt additive (Cox) models [17,37], multivariate adaptive regression splines (MARS) [28], regression trees [26,24], and least-angle regression–lasso [39] to censored survival outcomes.

For growing random forest in the survival data setting, we propose using the null deviance residuals from a Cox proportional hazards model as the outcome variable in the random forest algorithm. Therneau et al. [40] had advocated the use of null martingale residuals from a Cox proportional hazards model in growing a regression tree for censored data. However, Therneau and Grambsch [41] later described pitfalls surrounding the minimization of sums of squared martingale residuals. Moreover, it has been shown in other contexts [39] that it is via use of deviance residuals that methods devised for uncensored outcomes are best extended to survival settings.

For growing the trees in the forest, we substitute the null deviance residuals from a Cox proportional hazards model for the survival endpoint, and adopt the random forest algorithm directly. This approach is easy to implement, and circumvents the complexity induced by censoring. Ensemble predictions are computed by aggregating across different trees in the forest. This reduces variance and avoids the instability of working with a single tree. We illustrate this approach using data from the breast cancer prognostic study.

Following Breiman [7], the idea is to grow trees by injecting two types of randomness into the process. To grow the trees in the forest, the following steps are recommended:

1. Bootstrap the training data. Grow each tree on an independent bootstrap sample using null deviance residuals from a Cox proportional hazards model as the outcome variable.
2. At each node, randomly select m covariates out of all M possible covariates. Find the best split on the selected m covariates.
3. Grow the tree to maximal depth under the restriction of minimum node size = 5 (i.e., splitting is stopped when a node has fewer than five subjects). No pruning is performed.
4. Repeat for each bootstrap sample.
5. Average the trees to get predictions.

Steps 1 and 2 introduce randomness. To ensure that random forests have good prediction properties, it is important to ensure that the correct amount of randomization has been introduced. This means that we need to determine an appropriate number of randomly selected covariates m to be used in step 2 of the procedure. If we select too few covariates, the trees might be too sparse, and the ensemble estimator will have suboptimal properties. Choosing too many covariates will make the trees highly correlated, which can also degrade performance. As discussed in Breiman [7], one method for assessing the accuracy of a forest is through its generalization error. As m increases, the strength of a tree increases, which contributes to a lower forest generalization error; at the same time, however, the correlation between residuals increases, which increases error.

An estimate of the prediction error rate is obtained, based on the training data, as follows:

1. At each bootstrap iteration, predict the data not in the bootstrap sample (Breiman calls these "out-of-bag data") using the tree grown with the bootstrap sample.
2. Average the out-of-bag predictions. Calculate the error rate, and call it the "out-of-bag estimate of error rate."

Given that enough trees have been grown, the out-of-bag estimate of error rate is an accurate estimate of test set prediction error rate [7].

In addition to excellent prediction performance, random forests possess a number of advantages. These include the distinction of forests from so-called blackbox methods (e.g., neural nets), and accurate, internal estimates of test set prediction error. Furthermore, a byproduct of forests is a collection of variables that are frequently used in the forests, and the frequent uses are indicative of the importance of these variables. Zhang et al. [46] examined the frequencies of the variables in a forest and used them to rank the variables. We illustrate these in our analysis of the breast cancer data.

16.6.1 Breast Cancer Study: Results from Random Forest Analysis

In view of the potential improvement in predictive performance afforded by random forest over a single tree, we performed a random forest analysis of the breast cancer data discussed in Section 16.5. We used the random forest software, available as an R interface [29], to grow regression forests, based on using the null deviance residuals from a Cox proportional hazards model as the outcome variable. Null deviance residuals are obtained simply in R by specifying residual type and zero iterations in the call to `coxph()`. The same 10 covariates used in growing a single tree in Section 16.5 were used for the random forest analysis as well: age, race, marital status, socioeconomic status, tumor size, number of positive lymph nodes, tumor differentiation, ER, PR status, and BMI. We grew 500 trees in a forest. The size of the individual trees constituting the forest is controlled by a tuning parameter, which specifies the number of cases in a node below which the tree will not split. This was set to the default value of 5, which is claimed to give generally good results.

Table 16.1 shows the prediction errors from the forest as a function of varying the primary tuning parameter m. The entries are out-of-bag estimates of prediction error variance. For the single tree, the estimate of prediction error variance is based on 10-fold cross validation.

Table 16.1 Tree and Forest Prediction Error Variances for Breast Cancer Data

Method	m	Prediction Error Variance
Random forest	2	697.51
	3	734.13
	4	754.07
	5	769.19
Single tree (unpruned)		743.92
Single tree (pruned)		735.39

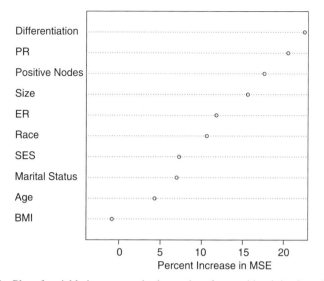

Figure 16.7 Plot of variable importance in the random forest with minimal prediction error.

Note that the best prediction error achieved by $m = 2$ among the range of forests examined is only a marginal improvement over the prediction error attained from a single pruned tree. A possible reason for this is the strong correlation between the covariates in this study. Roughly 65% of all possible pairwise correlations between covariates were significant. This could have potentially hindered the effectiveness of the random forest variance reduction strategy.

Variable important factors corresponding to the forest with minimal prediction error are depicted in Figure 16.7. Note the consistency with the tree results in terms of the prominence of differentiation, PR status, number of positive lymph nodes, and tumor size.

16.7 CONCLUDING REMARKS

In this chapter, we discussed methodological and practical aspects of tree-based modeling for survival data. Our focus was on comparing and contrasting different splitting criteria that affect tree growth. The splitting criteria based on M, ED, PH, and LR were found to be largely comparable on the basis of the simulations, as well as the data example. In all our simulations, Zhang's weighted impurity criterion demonstrated at best only moderate concordance with the four split statistics described above, and the concordance seemed to decline with higher censoring. The weighted impurity criterion decouples the link between survival time and censoring indicator. In our simulations, censoring was assumed to be uniform and independent of survival. However, for the Cox proportional hazards model, under conditional (on covariates) independence, the weighted impurity criterion may break down.

In our simulations, we considered only a single covariate since our focus was on contrasting the splitting functions. However, with multiple covariates, ultimately the interest may lie on how the various splitting functions affect the entire tree architecture derived from repeated splitting. In that context, all that matters is the upper ranks of the split functions, so the extent of tracking or concordance may be excessively stringent measures. We also did not consider the subsequent pruning procedure in our simulations. Because tree growing and pruning are two independent processes, the existing pruning procedures can be applied to any large tree that is grown by any of the splitting rules discussed in this chapter. Further simulations are warranted to study the abovementioned aspects in terms of their effects on the tree architecture.

It is well recognized that the mechanism of selecting a best split and the recursive partitioning of data leading to smaller and smaller datasets to be considered can lead to instability in the tree structure. Ensemble methods such as random forests can be useful in understanding the stability of tree structures and improving predictive performance. However, the published literature on ensemble techniques for survival data is sparse because of the difficulties induced by censoring. Breiman's [8] software implementation of a random forest variant for survival data comes with no formal description of the methodology, and needs to be evaluated before it can be widely used.

In this chapter, we presented a simple method for growing survival forests based on using the null deviance residuals from a Cox proportional hazards model. Our approach is easy to implement, and circumvents the difficulties induced by censoring. Variable importance summaries derived from the forest can be used to assess the relative importance of the covariates. However, the predictions obtained are not amenable to easy interpretation in the survival analysis setting. Methods that directly control the censoring are therefore warranted. For boosting-based ensemble approaches, such direct control has been described by Hothorn et al. [20]. An area of future investigation involves handling of time-dependent covariates under ensemble methods.

There are many versions of freewares implementing tree-based methods for survival data. Most of these are available through Statlib (http://lib.stat.cmu.edu) and CRAN (http://cran. r-project.org/). In particular, the RPART program [42] could be used to implement the method of LeBlanc and Crowley [26], the martingale-residual-based method of Therneau et al. [40], and Zhang's [43] weighted impurity approach. Software programs implementing log-rank-statistic-based splitting [38,27] are available from the first authors. Free, open-source code for random forests is available from http://www.stat.berkeley.edu/users/breiman/RandomForests. There is also an R implementation of random forests [29]. Finally, a commercial version is available through Salford Systems.

ACKNOWLEDGMENT

Dr. Banerjee's research was supported by the grant P30-CA46592-05 from the NCI.

REFERENCES

1. Albain, K. S., Crowley, J. J., LeBlanc, M., and Livingston, R. B., Determinants of improved outcome in small-cell lung cancer: An analysis of the 2,580-patient Southwest Oncology Group data base, *J. Clin. Oncol.* **8**, 1563–1574 (1990).

2. Albain, K. S., Green, S., LeBlanc, M., Rivkin, S., O'Sullivan, J., and Osborne, C. K., Proportional hazards and recursive partitioning and amalgamation analyses of the Southwest Oncology Group node-positive adjuvant CMFVP breast cancer data base: A pilot study, *Breast Cancer Res. Treat.* **22**, 273–284 (1992).

3. Banerjee, M., Biswas, D., Sakr, W., and Wood, D. P., Jr., Recursive partitioning for prognostic grouping of patients with clinically localized prostate carcinoma, *Cancer* **89**, 404–411 (2000).

4. Banerjee, M., George, J., Song, E. Y., Roy, A., and Hryniuk, W., Tree-based model for breast cancer prognostication, *J. Clin. Oncol.* **22**, 2567–2575 (2004).

5. Breiman, L., Friedman, J. H., Olshen, R. A., and Stone, C. J., *Classification and Regression Trees*, Wadsworth, Belmont, CA, 1984.

6. Breiman, L., Bagging predictors, *Machine Learn.* **24**, 123–140 (1996).

7. Breiman, L., Random forests, *Machine Learn.* **45**, 5–32 (2001).

8. Breiman, L., *How to Use Survival Forests,* http://www.stat.berkeley.edu/users/breiman/, 2002.

9. Breslow, N., Contribution to the discussion of the paper by D. R. Cox, *J. Roy. Statist. Soc. Ser. B* **34**, 216–217 (1972).

10. Ciampi, A., Chang, C.-H., Hogg, S., and McKinney, S., Recursive partitioning: A versatile method for exploratory data analysis in biostatistics, in MacNeil, I. B. and Umphrey, G. J., eds., *Biostatistics*, Reidel, Dordrecht, 1987, pp. 23–50.

11. Ciampi, A., Hogg, S., McKinney, S., and Thiffault, J., RECPAM: A computer program for recursive partitioning and amalgamation for censored survival data, *Comput. Meth. Programs Biomed.* **26**, 239–256 (1988).

12. Davis, R. B. and Anderson, J. R., Exponential survival trees, *Statist. Med.* **8**, 947–961 (1989).

13. Freedman, G. M., Hanlon, A. L., Fowble, B. L., Anderson, P. R., and Nicoloau, N., Recursive partitioning identifies patients at high and low risk for ipsilateral tumor recurrence after breast-conserving surgery and radiation, *J. Clin. Oncol.* **20**, 4015–4021 (2002).

14. Freund, Y. and Schapire, R. E., Experiments with a new boosting algorithm, in *Proc. 13th Int. Conf. Machine Learning,* Bari, Italy, 1996, pp. 148–156.

15. Goldman, L., Cook, F., Johnson, P., Brand, D., Rouan, G., and Lee, T., Prediction of the need for intensive care in patients who come to emergency departments with acute chest pain, *New Engl. J. Med.* **334**, 1498–1504 (1996).

16. Gordon, L. and Olshen, R., Tree-structured survival analysis, *Cancer Treat. Rep.* **69**, 1065–1069 (1985).

17. Grambsch, P. M., Therneau, T. M., and Fleming, T. R., Diagnostic plots to reveal functional form for covariates in multiplicative intensity models, *Biometrics* **51**, 1469–1482 (1995).

18. Hastie, T., Tibshirani, R., and Friedman, J., *The Elements of Statistical Learning*, Springer, New York, 2001.

19. Hothorn, T., Lausen, B., Benner, A., and Radespiel-Tröger, M., Bagging survival trees, *Statist. Med.* **23**, 77–94 (2004).

20. Hothorn, T., Buhlmann, P., Dudoit, S., Molinaro, A., and van der Laan, M. J., Survival ensembles, *Biostatistics* **7**, 355–373 (2006).

21. Intrator, O. and Kooperberg, C., Trees and splines in survival analysis, *Statist. Meth. Med. Res.* **4**, 237–261 (1995).

22. Ishwaran, H., Blackstone, E. H., Pothier, C. E., and Lauer, M. S., Relative risk forests for exercise heart rate recovery as a predictor of mortality, *J. Am. Statist. Assoc.* **99**, 591–600 (2004).

23. Katz, A., Buchholz, T. A., Thames, H., Smith, C. D., McNeese, M. D., Theriault, R., Singletary, S. E., and Strom, E. A., Recursive partitioning analysis of locoregional recurrence patterns following mastectomy: Implications for adjuvant irradiation, *Int. J. Radiat. Oncol. Biol. Phys.* **50**, 397–403 (2001).

24. Keles, S. and Segal, M. R., Residual-based tree-structured survival analysis, *Statist. Med.* **21**, 313–326 (2002).

25. LeBlanc, M., Tree-based methods for prognostic stratification, in *Handbook of Statistics in Clinical Oncology*, Crowley, J., ed., Marcel Dekker, New York, 2001, pp. 457–472.

26. LeBlanc, M. and Crowley, J., Relative risk trees for censored survival data, *Biometrics* **48**, 411–425 (1992).

27. LeBlanc, M. and Crowley, J., Survival trees by goodness of split, *J. Am. Statist. Assoc.* **88**, 457–467 (1993).

28. LeBlanc, M. and Crowley, J., Adaptive regression splines in the Cox model, *Biometrics* **55**, 204–213 (1999).

29. Liaw, A. and Wiener, M., Classification and regression by random-Forest, *R News* **2**, 18–22 (2002).

30. Morgan, J. N. and Messenger, R. C., THAID: *A Sequential Search Program for the Analysis of Nominal Scale Dependent Variables*, Technical Report, Institute for Social Research, Univ. Michigan, Ann Arbor, 1973.

31. Morgan, J. N. and Sonquist, J. A., Problems in the analysis of survey data and a proposal, *J. Am. Statist. Assoc.* **58**, 415–434 (1963).

32. Nelson, W., On estimating the distribution of a random vector when only the coordinate is observable, *Technometrics* **12**, 923–924 (1969).

33. Quinlan, J., Bagging, boosting, and C4.5, in *Proc. 13th American Association for Artificial Intelligence National Conf. Artificial Intelligence,* Menlo Park, CA, AAAI Press, 1996, pp. 725–730.

34. R Development Core Team, *R: A Language and Environment for Statistical Computing*, R Foundation for Statistical Computing, Vienna, Austria, 2005 (ISBN 3-900051-07-0, URL http://www.R-project.org).

35. Segal, M., Regression trees for censored data, *Biometrics* **44**, 35–48 (1988).

36. Segal, M., Extending the elements of tree-structured regression, *Statist. Meth. Med. Res.* **4**, 219–236 (1995).

37. Segal, M., James, I. R., French, M. A. H., and Mallal, S., Statistical issues in the evaluation of markers of HIV progression, *Int. Statist. Rev.* **63**, 179–197 (1995).

38. Segal, M., Barbour, J. D., and Grant, R. M., Relating HIV-1 sequence variation to replication capacity via trees and forests, *Statist. Appl. Genet. Molec. Biol.* **3**(1) Art. 2 (2004).

39. Segal, M., Microarray gene expression data with linked survival phenotypes: diffuse large B-cell lymphoma revisited, *Biostatistics* **7**, 268–285 (2006).

40. Therneau, T. M., Grambsch, P. M., and Fleming, T. R., Martingale-based residuals for survival models, *Biometrika* **77**, 147–160 (1990).

41. Therneau, T. M. and Grambsch, P. M., *Modeling Survival Data: Extending the Cox Model*, Springer, New York, 2000.

42. Therneau, T. M. and Atkinson, B., R port by Brian Ripley ⟨ripley@stats.ox.ac.uk⟩. rpart: Recursive Partitioning. R package version 3.1–27, 2005.

43. Zhang, H., Splitting criteria in survival trees, in *Statistical Modelling, Proc. 10th Int. Workshop on Statistical Modelling,* Innsbruck, Austria, July 1995, Seeber, G. U. H., Francis, B. J., Hatzinger, R., and Steckel-Berger, G., eds., Springer, New York, 1995, pp. 305–314.

44. Zhang, H. and Singer, B., *Recursive Partitioning in the Health Sciences*, Springer, New York, 1999.

45. Zhang, H., Yu, C. Y., Singer, B., and Xiong, M. M., Recursive partitioning for tumor classification with gene expression microarray data, *Proc. Natl. Acad. Sci. USA* **98**, 6730–6735 (2001).

46. Zhang, H., Yu, C. Y., and Singer, B., Cell and tumor classification using gene expression data: Construction of forests, *Proc. Natl. Acad. Sci. USA* **100**, 4168–4172 (2003).

CHAPTER 17

Bayesian Estimation of the Hazard Function with Randomly Right-Censored Data

Jean-François Angers

Department of Mathematics and Statistics, University of Montreal, Canada

Brenda MacGibbon

Department of Mathematics, University of Quebec at Montreal, Canada

17.1 INTRODUCTION

For survival analysis in medical research, it is useful to have clear summaries of the data for clinicians, and, as advocated by Efron [16], this can often be achieved by a graphical presentation of the hazard function. The data may consist solely of observed survival times, or with each time there may be associated a vector of covariates. In the latter case the hazard is often modeled as a product of a hazard function that depends only on time and a function of the covariates that is presumed independent of the time. In analyzing such data the two components are often analayzed separately, and such an analysis is called *semiparametric*. For an excellent review of Bayesian semiparametric analysis for even more complex models, we cite Sinha and Dey [55]. Here we prefer to concentrate on the problem of estimating the hazard function without covariates, although indications will be given on how the methodology can be extended to the semiparametric model.

   Our purpose here is threefold: (1) to give an overview of the relevant literature concerning both Bayesian and frequentist nonparametric estimation of the hazard rate, mainly in the case without covariates; and (2) to introduce a new nonparametric Bayesian method using monotone wavelets; and (3) to compare our estimators with both a frequentist estimator and a Bayesian nonparametric estimator.

Statistical Advances in the Biomedical Sciences, edited by Atanu Biswas, Sujay Datta, Jason P. Fine, and Mark R. Segal

We first briefly review the frequentist approaches to this problem. Various authors have used splines for estimating the survival function and the hazard rate in the random right-censorship model. We cite, in particular, Bloxom [9], Klotz and Yu [33], Whittemore and Keller [58], Efron [16], O'Sullivan [46], Jarjoura [30], Kooperberg and Stone [34], and Kooperberg et al. [35]. Senthilselvan [54] proposed penalized likelihood methods, and Loader [42] used local likelihood methods for hazard rate estimation with censored data. Kernel estimation of the hazard rate has also proved to be a useful method [50,52,53,27,21]. For frequentist estimation of a monotone hazard rate with randomly right-censored data, we cite the original work of Grenander [20] and that of Prakasa Rao [49] for uncensored data and for censored data, that of Padgett and Wei [47], Huang and Wellner [25], and MacGibbon et al. [44], which is based on least concave majorants (greatest convex minorants).

Some researchers have previously used orthogonal series methods; in particular, Patil [48] used orthogonal wavelet methods for hazard rate estimation in the uncensored case, and Antoniadis et al. [5] in the random right-censorship model. We also cite the theoretical research on wavelet density and hazard estimation by Li [37,38] and Liang et al. [39].

Early Bayesian research in survival analysis concentrated mainly on the estimation of the survival function. Susarla and Van Ryzin [56] used Dirichlet priors [17] to estimate the survival function with censored data. Ferguson and Phadia [18] extended this work to include prior distributions that are neutral to the right, previously studied by Doksum [14]. Kalbfleisch [31] used a gamma process prior for survival function estimation. Kuo and Smith [36] found Bayes estimators of the survival function with censored data using the Gibbs sampler. We also cite other interesting Bayesian research related to hazard rate estimation such as Arjas and Liu [7] and Berger and Sun [8].

Among the first to estimate the hazard directly were Dykstra and Laud [15] and Broffitt [11]. Dykstra and Laud [15] defined an appropriate prior stochastic process called an *extended gamma process* whose sample paths are hazard rates, and obtained the posterior distribution of the hazard rates for both exact and censored data. Bayesian nonparametric hazard function estimation methodology in Dykstra and Laud [15] was generalized in different ways by Ammann [1], who used conditional Laplace transforms, and by Thompson and Thavaneswaran [57]. Hjort [22] used beta process priors to estimate the cumulative hazard rate process. Further generalizaitons by Lo and Weng [41], Ho and Lo [24], and James [28,29], culminated in the characterization given by Ho [23] of the posterior distribution of the mixture hazard model of a monotone hazard rate via a finite mixture of S paths.

One of the most interesting methods, perhaps, for estimating the hazard rate, influenced by Dykstra and Laud [15] is that proposed by Arjas and Gasbarra [6]. Using a hierarchical model structure, they modeled the hazard rate nonparametrically as a jump process having a martingale structure with respect to the prior distribution. They describe an algorithm that generates sample paths from the posterior by a dynamic Gibbs sampler and illustrate the method in simulated examples. We have chosen here to compare our proposed method with theirs.

Angers and MacGibbon [3] developed a Bayesian adaptation of the Antoniadis et al. [5] method by employing Bayesian nonparametric estimation techniques with Fourier series methods in order to obtain a procedure that is easier to implement. Their method did not perform as well as the method of Antoniadis et al. [5] for estimation of the subdensity, but in simulations, it was shown to be as good or superior to the method of Antoniadis et al. [5] for estimation of the hazard rate. Here we propose the use of monotone wavelet approximation introduced by Anastassiou and Yu [2] to estimate the subdensity and hazard function.

We proceed as in Antoniadis et al. [5] to estimate the number of events and the survival functions separately. In order to describe our method here, in Section 17.1.1 we follow the description given by Antoniadis et al. [5]. For ease of presentation, Section 17.1.2 is devoted to recalling the Bayesian approach to linear models. In Section 17.2, the Bayesian model using monotone wavelet approximation is introduced. In Section 17.3 we develop our method of Bayesian functional estimation for the hazard rate problem with right-censored data. Section 17.4 contains a simulation study and the comparison of our results with those of Antoniadis et al. [5]. Section 17.5 presents an application of our method as well as those of Antoniadis et al. [5] and Arjas and Gasbarra [6] to a bone marrow transplantation dataset and to the Standford heart data. Section 17.6 consists of some concluding remarks.

17.1.1 The Random Right-Censorship Model

Survival analysis is usually based on study of a group of individuals of size n for which we assume their failure times, where the nonnegative random variables $T_1, \ldots, T_n$ are independent and identically distributed (i.i.d.) with distribution function $F(t)$, survival function $S(t) = 1 - F(t)$, and density $f(t)$. However, one feature that distinguishes the analysis of survival data from classical statistical analysis is the possibility that the data may be incomplete; that is, some individuals may not be observed until failure. For example, some patients will survive to the end of a clinical trial, and thus their failure times cannot be observed. If this happens in a random fashion, then this type of incompleteness is modeled by assuming that there exist $C_1, \ldots, C_n$, i.i.d. random variables with distribution function G and density g representing the censoring mechanism. Instead of observing the complete data $T_1, \ldots, T_n$, we observe $X_i = \min(T_i, C_i)$, $i = 1, \ldots, n$ and an indicator function $\delta_i = 1$ if $T_i \leq C_i$ and $= 0$ if not.

Since the density function of T exists, the hazard rate function can also be defined a

$$\lambda(t) = \frac{f(t)}{1 - F(t)}, \quad F(t) < 1.$$

With T_j, C_j, δ_j defined as above, the observed random variables are then X_j and δ_j. Henceforth we assume that

1. $T_1, T_2, \ldots, T_n$ are nonnegative and i.i.d. with distribution function F and density f.
2. $C_1, C_2, \ldots, C_n$ are nonnegative, i.i.d. with distribution function G and density g.
3. The T and C terms are independent.

In the censored case, if $G(t) < 1$, we have

$$\lambda(t) = \frac{f(t)\{1 - G(t)\}}{\{1 - F(t)\}\{1 - G(t)\}}. \quad F(t) < 1.$$

If we let $L(t) = P(X_i \leq t)$, then

$$1 - L(t) = \{1 - F(t)\}\{1 - G(t)\}.$$

Letting

$$f^*(t) = f(t)\{1 - G(t)\}$$

be the subdensity of those observations that are still to fail, clearly

$$\lambda(t) = \frac{f^*(t)}{1 - L(t)}, \qquad L(t) < 1.$$

17.1.2   The Bayesian Model

The estimator for the hazard rate proposed in the next section is obtained by writing the estimation problem using a Bayesian linear model. Hence, for ease of presentation, we first recall the Bayesian linear model as found in Lindley and Smith [40] and Robert [51]. Let

$$Y = X\theta + \epsilon,$$

where $Y = n \times 1$ vector of observations
$X = n \times p$ known matrix
$\theta = p \times 1$ vector of regression coefficients
$\epsilon \sim N_n(0, \sigma^2 \, \mathbf{I}_n)$

Note that X is assumed to be of full rank, but even if X is singular, the theory holds. Furthermore, σ^2 might be known or unknown. If σ^2 is unknown, it will also be considered as a random variable.

Given this model, the likelihood function is given by

$$\ell(\theta, \sigma^2) = \frac{1}{(2\pi\sigma^2)^{n/2}} \exp\left\{ -\frac{1}{2\sigma^2}(Y - X\theta)'(Y - X\theta) \right\}.$$

The loss function typically used is

$$L(\theta, \hat{\theta}) = (\theta - \hat{\theta})'\mathbf{Q}(\theta - \hat{\theta}), \qquad (17.1)$$

where $\mathbf{Q}$ is a positive-definite matrix.

Letting

$$\theta_{\mathrm{LS}} = (X'X)^{-1}X'Y$$
$$S = (Y - X\theta_{\mathrm{LS}})'(Y - X\theta_{\mathrm{LS}}),$$

the likelihood function can be rewritten as

$$\ell(\theta, \sigma^2) \propto \frac{1}{(\sigma^2)^{n/2}} \exp\left\{ -\frac{1}{2\sigma^2}[(\theta - \theta_{\mathrm{LS}})'X'X(\theta - \theta_{\mathrm{LS}}) + S] \right\}$$
$$= \left(\frac{1}{(\sigma^2)^{p/2}} \exp\left\{ -\frac{1}{2\sigma^2}(\theta - \theta_{\mathrm{LS}})'X'X(\theta - \theta_{\mathrm{LS}}) \right\} \right)$$
$$\times \left(\frac{1}{(\sigma^2)^{(n-p)/2}} \exp\left\{ -\frac{S}{2\sigma^2} \right\} \right)$$
$$= [\theta \mid \sigma^2, Y \sim N_p(\theta_{\mathrm{LS}}, \sigma^2(X'X)^{-1})]$$
$$\times [\sigma^2 \mid Y \sim I\Gamma((n-p-2)/2, S/2)], \qquad (17.2)$$

where $N_p(\mu, \Sigma)$ denotes the multivariate normal density with mean μ and covariance matrix Σ while $I\Gamma(a, b)$ represents the inverse gamma density with shape parameter a and scale parameter b. Note that if $\lambda \sim I\Gamma(a, b)$, then

$$\pi(\lambda) = \begin{cases} \dfrac{b^a}{\Gamma(a)\lambda^{a+1}} e^{-b/\lambda} & \text{if } \lambda > 0, \\ 0 & \text{otherwise.} \end{cases}$$

Now a conjugate prior for (θ, σ^2) is given by

$$\theta \mid \sigma^2 \sim N_p(\eta, \sigma^2 C), \tag{17.3}$$

$$\sigma^2 \sim I\Gamma(\alpha/2, \gamma/2), \tag{17.4}$$

where η, C, α, and γ are assumed to be known.

With the prior model given in Equations (17.3) and (17.4) and the likelihood function given in Equation (17.12), the posterior density of (θ, σ^2) and the marginal of the least-squares estimator θ_{LS} are

$$\theta \mid \sigma^2, Y \sim N_p(\theta_*, \sigma^2 C_*), \tag{17.5}$$

$$\sigma^2 \mid Y \sim I\Gamma\left(\frac{n+\alpha}{2}, \frac{\gamma_*}{2}\right),$$

$$\theta_{LS} \sim T_p\left(n + \alpha - p, \theta_0, \frac{S+\gamma}{n+\alpha-p} A_*\right),$$

where

$$\theta_* = \theta_{LS} - C_* C^{-1}(\theta_{LS} - \eta),$$

$$C_* = (X'X + C^{-1})^{-1} = C - C(C + (X'X)^{-1})^{-1}C,$$

$$\gamma_* = S + \gamma + (\theta_{LS} - \eta)' A_*^{-1}(\theta_{LS} - \eta),$$

$$A_* = (X'X)^{-1} + C.$$

Then, under the loss function given in Equation (17.1), the Bayes estimator of θ is given by

$$\hat{\theta} = \mathbb{E}[\theta \mid Y] = \theta_{LS} - C_* C^{-1}(\theta_{LS} - \eta).$$

Another intersting loss function is given by

$$L((\theta, \sigma^2), (\hat{\theta}, \hat{\sigma}^2)) = (\theta - \hat{\theta})' Q(\theta - \hat{\theta}) + (\sigma^2 - \hat{\sigma}^2)^2.$$

The associated Bayes estimators of β and σ^2 are given by

$$\hat{\theta} = \theta_* = \theta_{LS} - (X'X + C^{-1})^{-1}C^{-1}(\theta_{LS} - \theta_0),$$

$$\hat{\sigma}^2 = \frac{\gamma_*}{n + \alpha - 2} = \frac{S + \gamma + (\theta_{LS} - \theta_0)'A_*^{-1}(\theta_{LS} - \theta_0)}{n + \alpha - 2}.$$

Remark 17.1. If we use the reference prior $\pi(\theta, \sigma^2) \propto 1/\sigma^2$, the posterior densities become

$$\theta \mid \sigma^2, Y \sim N_p(\theta_{LS}, \sigma^2(X'X)^{-1}),$$

$$\sigma^2 \mid Y \sim I\Gamma\left(\frac{n}{2}, \frac{S}{2}\right).$$

(The reference prior is the limit case of Equations (17.3) and (17.4) with $C^{-1} \to 0$, $\alpha \to 0$ and $\gamma \to 0$.)

17.2   BAYESIAN FUNCTIONAL MODEL USING MONOTONE WAVELET APPROXIMATION

We first consider the estimation of the cumulative distribution function $L(t)$ and $F^*(t)$, which are monotone functions. Let us consider the general case and assume that $H(t)$ is a monotone nondecreasing function. [The term $H(t)$ stands for $L(t)$ or $F^*(t)$ depending on the observations considered.] Now, we can develop this Bayesian functional estimation model using the Bayesian linear model described in the previous section as a basis. As many authors including Antoniadis et al. [5] have indicated, wavelet estimators are ideal for estimating functions with inhomogeneous spatial smoothness. This is often the case with hazard functions. Here we introduce the terminology from Anastassiou and Yu [2].

Let $\varphi(x)$ denote a bounded right-continuous function on $\mathbb{R}$ with compact support, that is, supp $\varphi(x) \subseteq [-a, a]$, $0 < a < +\infty$ and define

$$\varphi_{kj}(x) := 2^{k/2}\varphi(2^k x - j) \quad \text{for} \quad k, j \in Z.$$

If H is continuous, then define

$$B_k(H)(x) := \sum_j H(2^{-k}j)\varphi_{kj}(x) \quad \text{for} \quad k \in Z. \tag{17.6}$$

Since $\varphi(x)$ is compactly supported, for any fixed $x \in \mathbb{R}$ the summation in (17.6) involves only a finite number of terms (see Appendix 17A). So $B_k(H)(x)$ is well defined on $\mathbb{R}$, that is

$$B_k(H)(x) = \sum_{j=j_0}^{j_1} H(2^{-k}j)\varphi_{kj}(x).$$

Theorem 6 of Anastassiou and Yu [2] states that if $\varphi(x)$ satisfies conditions C1–C4 given below and if $H(x) \in C(\mathbb{R})$ is a nondecreasing function, then the linear wavelet operator $B_k(H)(x)$ given by Equation (17.6) is also nondecreasing on $\mathbb{R}$ satisfying

$$|B_k(H)(x) - H(x)| \leq C\omega_2(H, 2^{-k+1}a) \quad \text{for} \quad x \in \mathbb{R}, k \in \mathbb{Z},$$

where C is an absolute constant and

$$\omega_2(H, \delta) = \sup_{h < \delta} \sup_x |H(x + 2h) - 2H(x + h) + H(x)|.$$

The conditions on $\varphi(x)$ are

C1: $\sum_{j \in \mathbb{Z}} \varphi(x - j) = 1 \ \forall x \in \mathbb{R}$.
C2: There exists a number b such that $\varphi(x)$ is nondecreasing if $x \leq b$ and is non-increasing if $x \geq b$.
C3: $\int_{-\infty}^{\infty} \varphi(x)\,dx = 1$.
C4: $\sum_{j \in \mathbb{Z}} j\varphi(x - j) = x \quad \forall x \in \mathbb{R}$.

Note that if $a = 1$, then conditions C1 and C4 can be written as

C1:

$$\begin{cases} \varphi(x) + \varphi(x + 1) = 1 & \text{if } -1 \leq x \leq 0, \\ \varphi(x) + \varphi(x - 1) = 1 & \text{if } 0 < x \leq 1, \end{cases}$$

C4:

$$\begin{cases} x + \varphi(x + 1) = 0 & \text{if } -1 \leq x \leq 0, \\ x - \varphi(x - 1) = 0 & \text{if } 0 < x \leq 1. \end{cases}$$

It can be easily shown that the only function satisfying these two conditions is

$$\varphi(x) = \begin{cases} 1 + x & \text{if } -1 \leq x \leq 0, \\ 1 - x & \text{if } 0 < x \leq 1 \end{cases} \tag{17.7}$$

Since $H(\cdot)$ is unknown, we cannot compute $H(2^{-k}j)$ directly. Consequently, let $\{\theta_j\}_{j=j_0}^{j_1}$ be a sequence of real numbers. Hence, renumbering the θ_j, $B_k(H)$ can be written as

$$B_k(H)(x) = \sum_{j=j_0}^{j_1} \theta_j \varphi_{kj}(x)$$

$$= \sum_{j=j_0}^{j_1} \theta_j 2^{k/2} \varphi(2^k x - j) \tag{17.8}$$

$$= 2^{k/2} \sum_{l=0}^{j_1 - j_0} \theta_l \varphi(2^k x + j_0 - l).$$

However, since $H(x)$ is a nondecreasing function, the θ_j terms should also be non-decreasing.

Hence, given θ_0, $\theta_l(l = 1, 2, \ldots, j_1 - j_0)$ can be written as

$$\theta_1 = \theta_0 + \zeta_1,$$

$$\vdots$$

$$\theta_l = \theta_0 + \zeta_1 + \cdots + \zeta_l$$

$$= \sum_{q=0}^{l} \zeta_q,$$

where $\zeta_0 = \theta_0$ and $\zeta_l \geq 0$ for $l = 1, 2, \ldots, j_1 - j_0$. Consequently, Equation (17.8) can be written as

$$
\begin{aligned}
B_k(H)(x) &= 2^{k/2} \sum_{l=0}^{j_1-j_0} \left[\sum_{q=0}^{l} \zeta_q \right] \varphi(2^k x + j_0 - l) \\
&= 2^{k/2} \sum_{q=0}^{j_1-j_0} \zeta_q \left[\sum_{l=q}^{j_1-j_0} \varphi(2^k x + j_0 - l) \right] \\
&= 2^{k/2} \sum_{q=0}^{j_1-j_0} \zeta_q \Phi_q(2^k x + j_0),
\end{aligned}
\tag{17.9}
$$

where

$$\Phi_q(2^k x + j_0) = \sum_{l=q}^{j_1-j_0} \varphi(2^k x + j_0 - l).$$

Using standard techniques, we can obtain a linear model as in Equation (17.2) with

$$Y = (H(x_1), H(x_2), \ldots, H(x_n))',$$

$$(X)_{i,j} = 2^{k/2} \Phi_{j-1}(2^k x_i) \quad \text{for} \quad i = 1, 2, \ldots, n \quad \text{and} \quad j = 1, 2, \ldots, j_1 - j_0 + 1,$$

$$\zeta = (\zeta_0, \zeta_1, \ldots, \zeta_{j_1-j_0})'.$$

However, the prior on ζ is different from that in equation (17.3). To account for the non-negativity of ζ_q for $j = 1, 2, \ldots, j_1 - j_0$, the prior is then

$$\zeta_0 \sim N(\eta_0, \sigma^2/n_0),$$

$$\zeta_q \sim N(\eta_q, \sigma^2/n_0) I_{[0,\infty)}(\zeta_q) \quad \text{for } q = 1, 2, \ldots, j_1 - j_0,$$

$$\sigma^2 \sim I\Gamma(\alpha/2, \gamma/2),$$

where $I_{[0,\infty)}(\zeta_q)$ represents the indicator function of the set $[0, \infty)$. The posterior density of ζ is similar to that in Equation (17.5) but we have to account for the nonnegativity of

$\zeta_1, \zeta_2, \ldots, \zeta_{j_1 - j_0}$. The Bayes estimator of ζ is then given by

$$\hat{\zeta} = \int_0^\infty \int_0^\infty \cdots \int_0^\infty \int_{-\infty}^\infty \zeta \times [\zeta \,|\, \sigma^2, Y \sim N_{j_1 - j_0 + 1}(\zeta_*, \sigma^2 C_*) \prod_{q=1}^{j_1 - j_0} I_{[0,\infty)}(\zeta_q)]$$

$$\times \left[\sigma^2 \,|\, Y \sim I\Gamma\left(\frac{n+\alpha}{2}, \frac{\gamma_*}{2}\right)\right] d\zeta_0 d\zeta_1 \cdots d\zeta_{j_1 - j_0} d\sigma^2. \tag{17.10}$$

17.3 ESTIMATION OF THE SUBDENSITY F^*

To obtain the estimator of $f^*(t)$, we start by estimating the cumulative distribution function (cdf) F^* using Equation (17.10) based only on the uncensored observations (values of i such that $\delta_i = 1$). The vector Y, described at the beginning of Section 17.1.2, will then be based on the empirical cdf of the uncensored observations; that is

$$Y = \frac{1}{n_o + 1}(1, 2, \ldots, n_o - 1, n_o)',$$

where n_o represents the number of uncensored observations. Referring to Equation (17.9), it is clear that the estimator of Y is also an estimator of the subdistribution, which we denote as $\hat{F}^*(x)$ given by

$$\hat{F}^*(x) = 2^{k/2} \sum_{q=0}^{j_1 - j_0} \hat{\zeta}_q \Phi_q(2^k x + j_0),$$

where $\hat{\zeta}$ is defined by Equation (17.10). To obtain the estimator of $f^*(t)$, we proceed as follows:

$$\hat{f}^*(x) = \frac{\partial}{\partial x} \hat{F}^*(x)$$

$$= 2^{3k/2} \sum_{q=0}^{j_1 - j_0} \hat{\zeta}_q \Phi'_q(2^k x + j_0).$$

To estimate $L(x) = P(X_i \leq x)$, we proceed as for $F^*(x)$ in order to obtain $\hat{L}(x)$, but this time all the observations (censored and uncensored) are used. The estimator of the hazard function is then given by

$$\hat{\lambda}(x) = \frac{\hat{f}^*(x)}{1 - \hat{L}(x)}.$$

Remark 17.2. This method is easy to adopt for hazard estimation in more general models such as the Cox proportional risk model

$$\lambda(t) = \lambda_0(t)\exp\{-\boldsymbol{\beta}'\boldsymbol{x}\},$$

where x is a vector of covariates, β a vector of parameters, and $\lambda_0(t)$ is the baseline hazard function, and the usual assumption of noninformative censoring [19] is made. We proceed with the usual semiparametric Bayesian approach to find an estimator $\hat{\beta}$ of β. The vector Y, described in Section 17.2, will now be based on the empirical cumulative hazard function given by

$$\hat{\Lambda}(t) = \sum_{t_i \leq t} d_i \Big/ \sum_{k \in R(t_i)} \exp\left\{ -\sum_{m=1}^{p} \hat{\beta}_m x_{km} \right\},$$

where t_i represent the observed "event" times and d_i, the observed number of events occurring at time t_i and $R(t_i)$, the risk set associated with t_i [32].

Again, referring to Equation (17.9), the estimator of Y is also an estimator $\hat{\Lambda}$ of the cumulative hazard function Λ, the survival function is estimated by $\exp\{-\hat{\Lambda}(t)\}$, and an estimator of the density function is found by differentiation. Proceeding in a manner analogous to that described above, we obtain an estimator of the baseline hazard function $\hat{\lambda}_0(t)$ and consequently of $\lambda(t)$.

Remark 17.3. We conjecture that in an analogous way, the methodology presented here can be extended to include more general situations such as some of those mentioned in Sinha and Dey [55]. In particular, we feel that our methodology can be extended to such models as a Cox model with informative censoring or one with a cure fraction as in Ibrahim et al. [26].

17.4   SIMULATIONS

We consider here the first simulation study proposed by Antoniadis et al. [5]. Samples of size n, $\{T_i, 1 \leq i \leq n\}$, from the gamma distribution with shape parameter 5 and scale parameter 1, denoted by f_1, and an independent sample $C_i, 1 \leq i \leq n$, from the exponential distribution with mean 6 (the mean was chosen to yield $\sim 50\%$ censoring) were generated. The performance measure used to compare the different estimators is the average mean-squared error obtained by averaging the mean-squared errors given by

$$\text{ASE}(f^*) = n_*^{-1} \sum_{i=1}^{n_*} [\hat{f}^*(x_i) - f^*(x_i)]^2,$$

$$\text{ASE}(\lambda) = n_*^{-1} \sum_{i=1}^{n_*} [\hat{\lambda}(x_i) - \lambda(x_i)]^2,$$

where n_* represents the number of observations with $x_i \leq 6$. Two values of n are considered, that is, $n = 200$ and $n = 500$.

The simulation results are given in Table 17.1. The true function along with the proposed estimator and the one given in Antoniadis et al. [5] are given in Figure 17.1. Because $\varphi'(x)$ is a step function [see Eq. (17.7)], the estimator of the subdensity is also a step function. Hence, we did a linear interpolation based on the center of each interval to smooth $\hat{f}^*$. In Table 17.1, the results for this interpolation is denoted by linear Bayes, that is, "Lin–Bayes" in the different figures.

Table 17.1 **Average Mean-Squared Errors ($\times 10^{-5}$ for subdensity and $\times 10^{-3}$ for Hazard Function) in First Simulation Setup of Antoniadis et al. [5] Based on 200 Repeitions**

Source	f_1^*		λ_1	
	$n = 200$	$n = 500$	$n = 200$	$n = 500$
Antoniadis	[14.6; 20.5]	[5.2; 13.6]	[2.5; 5.8]	[1.6; 5.9]
Bayes	50.1	38.6	11.6	7.0
Linear Bayes	33.4	21.9	3.9	2.0

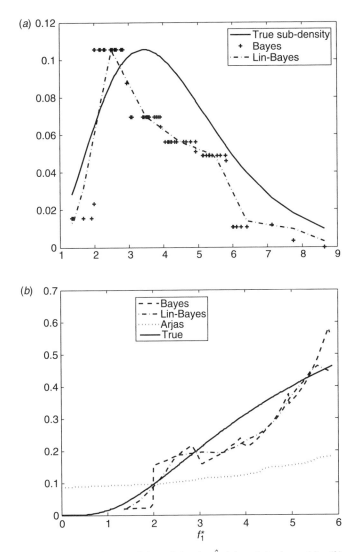

Figure 17.1 Estimate of the subdensity $\hat{f}_1$ (a) and the hazard λ_1 (b).

From the table, it can be seen that our proposed estimators are not as efficient as the one proposed [5] for the subdensity. However, the linear interpolation performs as well as the Antoniadis estimator for the hazard function.

17.5 EXAMPLES

In this section, two real datasets are considered and the subdensities along with the corresponding hazard functions are estimated using the monotone wavelet estimator described here as well as the estimators proposed by Antoniadis et al. [5] and Arjas and Gasbarra [6].

We have chosen to illustrate our method on a dataset consisting of a follow-up study of acute leukemia patients after allogenic bone marrow transplantion. The survival times are given in months. The dataset consist of 162 patients (including 63 deaths). For a more complete description of these data, see Brochstein et al. [10].

Various subsets of these data have been used [45,4] to illustrate different change points methods. Here, as a preliminary step in a more complete data analysis to be pursued elsewhere, we have chosen to use the complete dataset with death due to any cause as the endpoint and leukemic relapse or end of study as the censoring mechanism.

Figure 17.2 illustrates our estimate of the subdensity along with the estimator proposed in Antoniadis et al. [5]. [This estimator has been computed using Rice Wavelet Toolbox, version 2.4, with 64 bins and hard thresholding, resolution set at $k = 6$, and threshold level chosen to yield the smoother graph of f^* (threshold level at 0.0032) and λ (0.023).] Figure 17.3 gives our estimate of the actual hazard rate for this example, the one using wavelets proposed by Antoniadis et al. [5] and the Bayesian one of Arjas and Gasbarra [6].

From Figure 17.2 it can be seen that the linearized Bayesian estimator is similar to the one obtained using the Antoniadis et al. [5] approach, although theirs gives more weight to shorter survival times. However, there is much less data manipulation required in order to

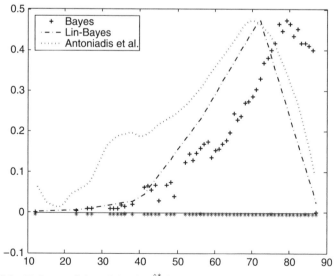

Figure 17.2 Estimate of the subdensity $\hat{f}^*$ for the bone marrow transplantation example.

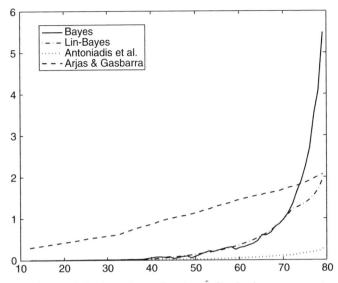

Figure 17.3 Estimate of the hazard rate function $\hat{\lambda}$ for the bone marrow transplantation example.

obtain our estimator. From Figure 17.3, all estimators of the hazard function except the one from Arjas and Gasbarra [6], are similar for $x \leq 50$. Starting from this point, the direct Bayes estimator is similar to that of Antoniadis et al. [5], while the linearized version yields a smaller hazard.

The second dataset that we consider here is the February 1980 version of the Stanford heart transplant data, published in Cox and Oakes [12]. This dataset has been analyzed by many

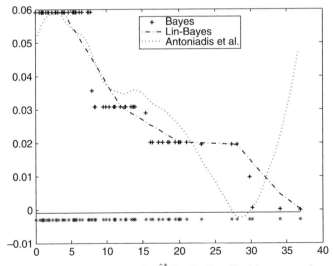

Figure 17.4 Estimate of the subdensity $\hat{f}^*$ for the Standford heart transplant example.

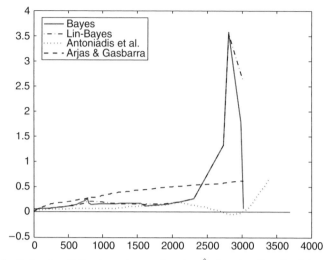

Figure 17.5 Estimate of the hazard rate function $\hat{\lambda}$ for the Standford heart transplant example.

authors, including Loader [42] and Antoniadis et al. [4], who considered a change point model for it. Figure 17.4 shows the graphs of the subdensity functions given by our methods (Bayes and linear Bayes) and that of Antoniadis et al. [5]. It should be noted that the subdensity function of Antoniadis et al. [5] is negative over a small interval and then increases dramatically after that. The linear Bayes graph seems a more reasonable representation of a subdensity. Figure 17.5 compares the four different hazard function estimators: those of Arjas and Gasbarra [6], and Antoniadis et al. [5], and our Bayes and linear Bayes estimators. Up to $t = 2400$, the four estimators are similar with the Bayes and the linear Bayes representing a compromise between Antoniadis et al. [5] and Arjas and Gasbarra [6]. The estimator of Antoniadis et al. [5] does have the disadvantage of being negative for long survival times, while ours have a sharp peak between 2500 and 3000. The estimator of Arjas and Gasbárra [6] is the most stable.

17.6 CONCLUDING REMARKS

Our objective here, in addition to reviewing some of the relevant literature on frequentist and Bayesian nonparametric hazard estimation, was to use Bayesian functional estimation techniques combined with the monotone wavelet approximation methods of Anastassiou and Yu [2] to estimate the hazard rate with randomly right-censored data by a relatively easy method to implement. This has been accomplished.

Although our model is not as effective in the simulation study as is the frequentist method of Antoniadis et al. [5] for the subdensity estimation, the performance of the linear Bayes estimator for the hazard function is comparable, and ours is much easier to implement and extremely flexible. Because of the monotonicity of the wavelet approximation, our estimators of the subdensity are theoretically always positive. This is not the case for the estimator of Antoniadis et al. [5]. In fact, in one of the real data examples (see Fig. 17.4), it is negative.

We have also chosen the Bayesian nonparametric method of Arjas and Gasbarra [6] for purposes of comparison. Arjas and Gasbarra [6] have an excellent Bayesian nonparametric method that performs very well on the examples here. It is more stable than ours or that of Antoniadis et al. [5]. However, it does not provide an estimate of the subdensity, which we consider an interesting function in its own right. We, therefore conclude that ours, that is, the linear Bayes one, is a good compromise method between the fully frequentist wavelet one and that of Arjas and Gasbarra [6]. It is our hope, however, to somehow combine ours with the latter and have a fully Bayesian nonparametric method that combines the good qualities of both.

ACKNOWLEDGMENT

The authors wish to acknowledge the support of the Natural Sciences and Engineering Research Council of Canada and to thank the referees for their helpful comments, which improved this manuscript. They are also grateful to Dr. R. J. O'Reilly of the Sloan Kettering Memorial Cancer Center and to Dr. Susan Groshen of the University of Southern California, Keck School of Medicine for making available to us the bone marrow transplant data [10] that we used to illustrate our method here.

APPENDIX 17A: CHOICE OF RESOLUTION LEVEL

In this appendix, we will discuss the choice of k and the bounds j_0 and j_1. Suppose that the observed times are $0 < x_1 \leq x_2 \leq \cdots \leq x_n < T$. Since the support of φ is $[-a; a]$, then, for a fixed k, we obtain

$$\varphi_{k,j}(t) = 0 \Leftrightarrow 2^{k/2}\varphi(2^k x - j) = 0$$

$$\Leftrightarrow 2^k x - j < -a \text{ or } 2^k x - j > a$$

$$\Leftrightarrow 2^k x + a < j \text{ or } j < 2^k x - a.$$

Since $x \in [0, T]$, then

$$\varphi_{k,j}(t) = 0 \Rightarrow j \notin [2^k x - a, 2^k x + a]$$

$$\Rightarrow j \notin [-a, 2^k T + a].$$

Hence $j_0 = -a$ and $j_1 = 2^k T + a$.

To choose k, we proceed as follows. For each value of j, we want a series of m observations such that $\varphi_{k,j}(x_i) > 0$, $\varphi_{k,j}(x_{i+1}) > 0$, ..., $\varphi_{k,j}(x_{i+m-1}) > 0$. Since the support of φ is $[-a; a]$, k should be such that

$$\frac{j - a}{2^k} \leq x_i \leq x_{i+1} \leq x_{i+m-1} \leq \frac{j + a}{2^k}.$$

Let $\Delta_m = \max_i (x_{i+m-1} - x_i)$. This condition is satisfied if

$$\Delta_m \leq \frac{j+a}{2^k} - \frac{j-a}{2^k}$$

$$\Rightarrow \Delta_m \leq \frac{a}{2^{k-1}}$$

$$\Rightarrow 2^{k-1} \leq \frac{a}{\Delta_m}$$

$$\Rightarrow k \leq 1 + \frac{\log(a/\Delta_m)}{\log(2)}$$

$$\Rightarrow k \leq 1 + \log_2\left(\frac{a}{\Delta_m}\right).$$

REFERENCES

1. Ammann, L. P., Conditional Laplace transforms for Bayesian nonparametric inference in reliability theory, *Stochast. Process. Appl.* **20**, 197–212 (1985).

2. Anastassiou, G. A. and Yu, X. M., Monotone and probabilistic wavelet approximation, *Stochast. Anal. Appl.* **10**, 251–264 (1992).

3. Angers, J.-F. and MacGibbon, B., Bayesian fuctional estimation of the hazard rate for randomly right censored data using Fourier series methods, in *Statistical Modeling and Analysis for Complex Data Problems*, Duchesne, P. and Remillard, B., eds., Kluwer, Academic Publishers, Boston, 2004, pp. 53–69.

4. Antoniadis, A., Gijbels, I., and MacGibbon, B., Nonparametric estimation for the location of a change-point in an otherwise smooth hazard function under random censoring, *Scand. J. Statist.* **27**, pp. 501–519 (2000).

5. Antoniadis, A., Grégoire, G., and Nason, G., Density and hazard rate estimation for right-censored data by using wavelet methods, *J. Roy. Statist. Soc. Ser. B* **61**, 63–84 (1999).

6. Arjas, E. and Gasbarra, D. Nonparametric Bayesian inference from right censored survival data, using the Gibbs sampler, *Statistica Sinica* **4**, 505–524 (1994).

7. Arjas, E. and Liu, L. Assessing the losses caused by an industrial intervention: a hierarchical Bayesian approach, *Appl. Statist.* **44**, 357–368 (1995).

8. Berger, J. O. and Sun, D., Bayesian inference for a class of poly-Weibull distributions, in *Bayesian Analysis in Statistics and Econometrics*, Berry, D. A., Chaloner, K. M., and Geweke, J. K., eds., Wiley, New York, 1996, pp. 101–113.

9. Bloxom, B., A constrained spline estimator of a hazard function, *Psychometrika* **50**, 301–321 (1985).

10. Brochstein, J. A., Kernan, N. A., Groshen, S., Cirrincione, C., Shank, B., Emanual, D., Laver, J., and O'Reilly, R. J., Allogenic bone marrow transplantation after hyperfractionated total body irradiation and cyclophosphamide in children with acute leukaemia, *New Engl. J. Med.* **317**, 1618–1624 (1987).

11. Broffitt, J. D., A Bayes estimator for ordered parameters and isotonic Bayesian graduation, *Scand. Actuarial J.* **67**, 231–247 (1984).

12. Cox, D. R. and Oakes, D., *Analysis of Survival Data*, Chapman & Hall, London, (1984).

13. Digital Signal Processing Group, *Rice Wavelet Toolbox*, version 2.4, Rice Univ. (http://www.dsp.rice.edu/software/RWTl).

14. Doksum, K., Tailfree and neutral random probabilities and their posterior distributions, *Ann. Probabil.* **2**, 183–201 (1974).

15. Dykstra, R. L. and Laud, P. W., A Bayesian nonparametric approach to reliability, *Ann. Statist.* **9**, 356–367 (1981).

16. Efron, B., Logistic regression, survival analysis, and the Kaplan-Meier curve, *J. Am. Stastist. Assoc.* **83**, 414–425 (1988).

17. Ferguson, T. S., A Bayesian analysis of some nonparametric problems, *Ann. Statist.* **1**, 209–230 (1973).

18. Ferguson, T. S. and Phadia, E. G., Bayesian nonparametric estimation based on censored data, *Ann. Statist.* **7**, 163–186 (1979).

19. Fleming, T. R. and Harrington, D. P., *Counting Processes and Survival Analysis*, Wiley, New York, (1991).

20. Grenander, U., On the theory of mortality measurement. Part II, *Scand. navisk Aktuarietidskrift* **39**, 125–153 (1956).

21. Hall, P., Huang, L.-S., Gifford, J. A., and Gijbels, I., Nonparametric estimation of hazard rate under the constraint of monotonicity, *J. Comput. Graph. Statist.* **10**, 592–614 (2001).

22. Hjort, N. L., Nonparametric Bayes estimators based on beta processes in models for life history data, *Ann. Statist.* **18**, 1259–1294 (1990).

23. Ho, M.-W., A Bayes method for a monotone hazard rate via S-Paths, *Ann. Stastist.* **34**, 820–836 (2006).

24. Ho, M.-W. and Lo, A. Y., Bayesian nonparametric estimation of a monotone hazard rate, in *System and Bayesian Reliability: Essays in Honor of Professor Richard E. Barlow on His 70th Birthday*, Hayakawa, Y., Irony, T., and Xie, M., eds., World Scientific, Singapore, 2001, pp. 301–314.

25. Huang, J. and Wellner, J. A., Estimation of a monotone density or monotone hazard under random censoring, *Scand. J. Statist.* **22**, 3–33 (1995).

26. Ibrahim, J. G., Chen, M.-H., and Sinha, D., Bayesian semiparametric models for survival data with a cure fraction, *Biometrics* **57**, 383–388 (2001).

27. Izenman, A. J. and Tran, L. T., Kernel estimation of the survival function and hazard rate under weak dependence, *J. Statist. Plan. Infer.* 233–247 (1990).

28. James, L. F., Bayesian calculus for Gamma processes with applictions to semiparametric models, *Sankhya* **65**, 159–206 (2003).

29. James, L. F., Bayesian Poisson process partition calculus with an application to Bayesian Lévy moving averages, *Ann. Statist.* **33**, 1771–1799 (2005).

30. Jarjoura, D., Smoothing hazard rates with cubic splines, *Commun. Statist. Simul. Comput.* 377–392 (1988).

31. Kalbfleisch, J. D., Nonparametric Bayesian analysis of survival time data, *J. Roy. Statist. Soc. Ser. B* **40**, 214–221 (1978).

32. Klein, J. P. and Moeschberger, M. L., *Survival Analysis: Techiniques for Censored and Truncated Data*, Springer-Verlag, New York, (1997).

33. Klotz, J. and Yu, R.-Y., Small sample relative performance of the spline smooth survival estimator, *Commun. Statist. Simul. Comput.* **15**, 815–818 (1986).

34. Kooperberg, C. and Stone, C. J., Logspline density estimation for censored data, *J. Comput. Graph. Statist.* **1**, 301–328 (1992).

35. Kooperberg, C., Stone, C. J., and Truong, Y. K., The L_2 rate of convergence for hazard regression, *Scand. J. Statist.* **22**, 143–157 (1995).

36. Kuo, L. and Smith, A. F. M., Bayesian computations in survival models via the Gibbs sampler, in *Survival Analysis: State of the Art*, Klein, J. P. and Goel, P., eds., Kluwer Academic Publishers, Boston, 1992, pp. 11–24.

37. Li, L., Hazard rate estimation for censored data by wavelet methods, *Commun. Statist. Theor. Meth.* **31**, 943–960 (2002).

38. Li, L. Y., On the minimax optimality of wavelet estimators with censored data, *J. Statist. Plan. Infer.* (2006).

39. Liang, H.-Y., Mammitzsch, V., and Steinebach, J., Nonlinear wavelet density and hazard rate estimation for censored data under dependent observations, *Statist. Decisions,* **23**, 161–180 (2005).

40. Lindley, D. V. and Smith, A. F. M., Bayes estimates for the linear model, *J. Roy. Statist. Soc. Ser. B* **34**, 1–41 (1972).

41. Lo, A. Y. and Weng, C. S., On a class of Bayesian nonparametric estimates. II. Hazard rates estimates, *Ann. Inst. Statist. Math.* **41**, 227–245 (1989).

42. Loader, C., Inference for a hazard rate change point, *Biometrika* **74**, 301–209 (1991).

43. Loader, C., *Local Regression and Likelihood*, Statistics and Computing series, Springer-Verlag, New York, (1999).

44. MacGibbon, B., Lu, J., and Younes, H., Limit theorems for asymptotically minimax estimation of a distribution with increasing failure rate under a random mixed censorship/truncation model, *Commun. Statist. Theor. Meth.* **31**, 1309–1333 (2002).

45. Mueller, H. G. and Wang, J. L., Nonparametric analysis of changes in hazard rates for censored survival data: An alternative to change-point models, *Biometrika* **77**, 305–314 (1990).

46. O'Sullivan F., Fast computation of fully automated log-density and log-hazard estimators, *SIAM J. Sci. Statist. Comput.* **9**, 363–379 (1988).

47. Padgett, W. J. and Wei, L. J., Maximum likelihood estimation of a distribution function with increasing failure rate based on censored observations, *Biometrika* **67**, 470–474 (1980).

48. Patil, P., Nonparametric hazard rate estimation by orthogonal wavelet methods, *J. Statist. Plan. Inf.* **60**, 53–168 (1997).

49. Prakasa Rao, B. L. S., Estimation of distributions with monotone failure rate, *Ann. Math. Statist.* **41**, 507–519 (1970).

50. Ramlau-Hansen, H., Smoothing counting process intensities by means of kernel functions, *Ann. Statist.* **11**, 453–466 (1983).

51. Robert, C. P., *The Bayesian Choice*, 2nd ed., Springer, New York, 2001.

52. Roussas, G. G., Hazard rate estimation under dependence conditions, *J. Statist. Plan. Infer.* **22**, 81–93 (1989).

53. Roussas, G. G., Asymptotic normality of the kernel estimate under dependence conditions: Application to hazard rate, *J. Statist. Plan. Infer.* **25**, 81–104 (1990).

54. Senthilselvan, A., Penalized likelihood estimation of hazard and intensity functions, *J. Roy. Statist. Soci. Ser. B* **49**, 170–174 (1987).

55. Sinha, D. and Dey, D. K., Semiparametric Bayesian analysis of survival data, *J. Am. Statist. Assoc.* **92**, 1195–1212 (1997).

56. Susarla, V. and Van Ryzin, J., Empirical Bayes estimation of a distribution (survival) function with right censored observations, *Ann. Statist.* **6**, 740–751 (1978).

57. Thompson, M. E. and Thavaneswaran, A., On Bayesian nonparametric estimation for stochastic processes, *J. Statist. Plan. Infer.* **33**, 131–141 (1992).

58. Whittemore, A. S. and Keller, J. B., Survival estimation using splines. *Biometrics* **16**, 1–11 (1986).

PART IV

Bioinformatics

CHAPTER 18

The Effects of Intergene Associations on Statistical Inferences from Microarray Data

Kerby Shedden

Department of Statistics, University of Michigan, Ann Arbor, Michigan

18.1 INTRODUCTION

Gene expression measurements obtained from microarrays provide a detailed picture of relative messenger RNA (mRNA) abundances in a cell population, one important facet of biological state [7]. While it is now technically possible to measure the expression levels for tens of thousands of genes, the amount of information present in such large measurement sets is limited by the degree of statistical correlation among the gene expression levels. To the extent that the number of analyzed mRNA transcript types exceeds the number of independently varying cellular characteristics, it is possible to substantially overstate the level of evidence in the data supporting apparent associations between gene expression and clinical or other high-level biological characteristics.

For example, use of microarrays in clinical studies often focuses on identifying prognostic or diagnostic markers (e.g., see Ref. 9). Such markers are identified by testing large numbers of mRNA candidates in order to identify a subset of transcripts showing an apparent correlation between their expression levels, and the medical trait or outcome of interest. This analysis often involves directly testing each mRNA for an association with the trait. As will be demonstrated below, intergene correlations can greatly exaggerate the evidence supporting a candidate marker. Later on, we will argue that if the association is identified in a timecourse experiment, or is supported through the use of meta-analysis to combine the results of several studies, the problem of overstating significance levels is accentuated.

Several physical mechanisms for intergene correlations can be proposed. One important mechanism is that the target on a microarray designed to detect a single type of mRNA probe

Statistical Advances in the Biomedical Sciences, edited by Atanu Biswas, Sujay Datta,
Jason P. Fine, and Mark R. Segal

may actually respond to multiple mRNA types, a phenomenon known as *cross-hybridization*. In addition, most microarrays contain replicate targets for the same gene, or for closely related mRNA transcripts. This replication is rarely accounted for explicitly in analysis. At a more biological level, the control elements governing changes in gene expression often sit at the end of signaling pathways that have wide-ranging effects on a cell. Critical factors such as whether the cells are actively growing, or are responding to environmental stress are also likely to have widespread effects on many mRNA types.

While the systematic mapping of the pathways and control mechanisms regulating changes in gene expression is a major long-term scientific goal [1,19], the effects of intergene correlations on the analysis of microarray measurements in routine clinical and biomedical research are less well appreciated. As a simple illustration, suppose that 10^4 genes are assessed individually using some statistical test, and 130 are deemed interesting based on having p values less than 0.01. This is 30% more than the expected number of $10^4 \times 10^{-2} = 100$, supposing that the individual tests are properly calibrated. The question of whether any of these 130 genes are actually responding to the treatment reduces to asking whether it is likely to obtain 130 significant tests by chance. If the genes are independent, the standard deviation of the number of affirmative tests is around 10, so the observed value 130 is well beyond the 99th null percentile for the number of successful tests. However, suppose, as a simple illustration, that there are 10^3 gene clusters containing 10 genes each, and the clusters respond identically to a change in cellular state. This is an extreme form of the *block correlation* structure used by Storey [17], and in many other investigations. In this setting, the expected number of affirmative tests remains 100, but the standard deviation for the number of significant 10-gene blocks is around 3.15. The standard deviation for the number of affirmative tests is more than 30. Observing 130 affirmative tests is no longer inconsistent with replicate observations of a constant system.

In this chapter, we will explore the effects of intergene correlations on several simple analysis strategies that are commonly used in routine biomedical and clinical research using microarrays. First, in Section 18.2 we introduce a summary measure to empirically quantify the level of intergene correlation in a cell population. Sections 18.3, 18.4, and 18.5 look at three common data analysis settings, considering the effects of intergene correlation in each. Section 18.6 summarizes the chapter, and more difficult issues are briefly discussed.

18.2   INTERGENE CORRELATION

As noted above, it is expected that pairwise correlations between mRNA transcripts measured with a microarray will show a wider range of intergene correlations than will an equivalent number of paired data vectors drawn from a pairwise independent population. This is demonstrated for two experimental datasets in Figure 18.1. The distribution of Pearson correlation coefficients between randomly selected gene pairs across 40 independent samples is shown. Results are given for two solid tumor datasets, denoted "OV" and "UVA" (further discussion of the data and references are given below). The distribution of correlation coefficients between simulated independent Gaussian "genes" is shown as broken lines. Both experimental sets show a relative increase of correlations at both the positive and negative end of the correlation scale compared to independent, identically distributed (i.i.d.) data. The OV set shows a greater number of positive associations, whereas the UVA set is balanced between positive and negative associations. This is notable since some of the technical sources of intergene correlation, such as cross-hybridization, can give rise to positive but not negative associations.

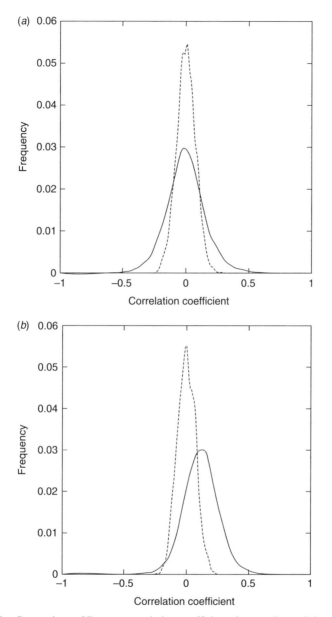

Figure 18.1 Comparison of Pearson correlation coefficients in experimental data with those in simulated i.i.d. data. The empirical distribution of Pearson correlation coefficients in two real datasets (solid lines) are shown compared to the distribution of Pearson correlation coefficients in simulated i.i.d. sets of the same size (broken lines). Plot (*a*) shows the UVA data; plot (*b*), the OV data (see text for details).

An elementary way to quantify the presence of intergene correlations in a dataset is through the average pairwise population correlation coefficient

$$R = \sum_{i<j} \rho_{ij}^2 / \binom{n}{2},$$

where n is the number of measured genes and ρ_{ij} is the population correlation coefficient between the expression levels of genes i and j. This is far from a complete description. For example, it fails to capture whether the most highly correlated gene pairs tend to contain a few key genes in common, a situation that has been exploited to identify gene coexpression "modules" [21]. Nevertheless, it captures one important component of the correlation structure, and in Section 18.5, we will see that it contains all relevant correlation information for an important inference problem. To estimate R, we can use the bias-corrected average of sample pairwise correlation coefficients. Suppose that p independent experimental conditions are studied (the dependent case will be considered below). Since

$$E \sum_{i<j} \hat{\rho}_{ij}^2 / \binom{n}{2} \approx \sum_{i<j} (\rho_{ij}^2 + \mathrm{var}\hat{\rho}_{ij}) / \binom{n}{2} \approx \sum_{i<j} \rho_{ij}^2 / \binom{n}{2} + 1/p,$$

where the first approximation is due to the small bias in $\hat{\rho}_{ij}$ relative to ρ_{ij} and the second is due to the use of the asymptotic sampling variance for $\hat{\rho}_{ij}$. It follows that

$$\hat{R} \equiv \sum_{i<j} \hat{\rho}_{ij}^2 / \binom{n}{2} - 1/p$$

can serve as an estimate of R. In practice the average over all pairs will be replaced with an average over 10^4 randomly sampled pairs.

Table 18.1 contains the value of $\hat{R}$ for seven experimental datasets. These sets were obtained using several microarray platforms, and include studies of both human cell lines and human

Table 18.1   Calculated $\hat{R}$ Values for Eight Experimental Datasets[a]

Dataset	$\hat{R}$
UVA	0.019
OV	0.020
CO	0.027
LU	0.021
NCI60$_a$	0.011
NCI60$_b$	0.018
NCI60$_c$	0.015
NCI60$_d$	0.014
Null	0.002 $(-0.001, 0.005)$

[a]This study included 2000 genes and 40 samples randomly selected from each set. The "Null" line gives the average and 95% range for $\hat{R}$ values calculated from i.i.d. Gaussian data of the same size.

tissue sample biopsies. To improve comparability, 2000 genes and 40 samples were randomly selected from each dataset. The set denoted "UVA" [18] contains 141 solid tumors of various types assayed on Affymetrix HG-U95Av2 microarrays. The three sets denoted "OV" (ovary), "CO" (colon), and "LU" (lung) contain tumor samples assayed on Affymetrix HuGeneFL arrays (see Ref. 8 for more details and primary references). The four datasets denoted "NCI60" measure gene expression in the human tumor cell line screening panel used by the Developmental Therapeutics Program at the National Cancer Institute for studying small molecule growth inhibitory activity. The $NCI60_a$ [15] experiment used the Affymetrix HuGeneFL array, while sets $NCI60_b$, $NCI60_c$, and $NCI60_d$ (available from dtp.nci.nih.gov/mtargets/download.html) are replicated measurements on Affymetrix U95A arrays contributed by the Novartis Foundation. For the eight experimental datasets, the data were log-transformed and genes with sample standard deviation on the log_2 scale less than 10^{-4} were removed.

The $\hat{R}$ values for experimental datasets in Table 18.1 are substantially similar across multiple array platforms and specimen types. There is some trend for the tissue samples to have higher levels of intergene correlation than the cell lines, which would be consistent with correlation arising from certain signaling pathways operating more efficiently in tissue compared to cultured cells. However, since the cell line experiments all represent the same cell line panel, this phenomenon may have other explanations. The values of $\hat{R}^{1/2}$, which is on the correlation scale, range from 0.10 to 0.16, suggesting that most gene pairs are only weakly associated. Nevertheless, we will see below that this level of intergene correlation can have substantial effects on the sampling behavior of important statistics. The line labeled "Null" shows the result of calculating $\hat{R}$ for 1000 independent Gaussian datasets, each with 2000 "genes" and 40 "samples" (all simulated values are i.i.d.). The interval is the 95% empirical range of the resulting $\hat{R}$ values. Clearly the experimental sets have $\hat{R}$ values that are inconsistent with a complete lack of intergene correlation.

We conclude this section by pointing out an interesting characteristic of the $\hat{R}$ values reflecting the extensiveness of intergene correlations in microarray data. Table 18.2 lists the calculated $\hat{R}$ values for 40 samples randomly selected from each of the nine experimental sets. The values were calculated based on different numbers of randomly selected genes (500, 1000, 2000, and 4000). Notably, the $\hat{R}_n$ values (where n denotes the number of sampled genes) are nearly constant within each dataset across the different numbers of sampled genes. To better understand the possible implication of this, consider the grossly simplistic model in which the n genes can be partitioned into q mutually independent blocks, such that the $r = n/q$ genes within each

Table 18.2 Estimated Values of R for Nine Experimental Datasets Based on Four Different Values for Number of Genes Considered

Dataset	Number of Genes			
	500	1000	2000	4000
UVA	0.021	0.020	0.019	0.018
OV	0.021	0.020	0.020	0.019
CO	0.027	0.027	0.027	0.024
LU	0.023	0.024	0.021	0.022
$NCI60_a$	0.015	0.014	0.011	0.011
$NCI60_b$	0.017	0.017	0.018	0.018
$NCI60_c$	0.014	0.015	0.015	0.014
$NCI60_d$	0.011	0.013	0.014	0.012

block are exchangeably correlated with correlation coefficient ρ. A simple calculation shows that the value of R based on n genes (denoted R_n) varies with q according to

$$R_n \equiv \sum_{i \neq j} \rho_{ij}^2 \bigg/ \binom{n}{2} \sim \rho/q.$$

Under this simple model, the observation that R_n is constant in n suggests that after 500 genes have been sampled, additional genes fit into already-seen blocks, rather than founding new blocks. Because of the simplicity of the model used to characterize the pattern in the $\hat{R}_n$ values, we must be cautious in overstating the implications of this analysis. But these findings are consistent with the number of "effective degrees of freedom" in a gene expression dataset being far smaller than the number of measured genes.

18.3 DIFFERENTIAL EXPRESSION

Identifying genes that are differentially expressed between two experimentally controlled or naturally observed conditions is often a major goal of analysis. Calculating a test statistic T_i for each measured gene and ranking the genes accordingly is a starting point, but fails to characterize the evidence level for reproducible differential expression in the top scoring genes. Aiming to control the familywise error rate using Bonferroni or stepdown procedures, while sometimes useful, is not consistent with the goals of a screening study in which a moderate proportion of false-positive calls is tolerable and expected.

A major development was the formalization of the concept of "false discovery rate" (FDR) (e.g., see Refs. 4 and 6). Several definitions have been proposed. For simplicity we will use the definition

$$\text{FDR} = \frac{E_0 N(t)}{N(t)},$$

where $N(t)$ is the number of measured genes with test statistics greater than or equal to the number t, and E_0 is the expected number of positive calls when all the null hypotheses are true. This definition is most suitable when it is very likely that at least one null hypothesis is false, but the number of false null hypotheses is a small fraction of the total number of hypotheses considered.

The fact that FDR is defined in terms of single-gene expectations minimizes the effect of intergene correlations. However, FDR is not completely insensitive to intergene correlations, as has been explored in several papers [2,3,10].

A matter of practical importance is that the standard definitions of FDR do not account for the possibility that by chance, the number of actual false positives can be much greater than $E_0 N(t)$. The null variance of $N(t)$ is influenced by intergene correlations; therefore, the probability of a large excess of false positives occurring by chance is also increased. To quantify this type of event, we will consider conservative false discovery rates of the form

$$\text{FDR}_q = \frac{Q_0^q N(t)}{N(t)},$$

where $Q_0^q N(t)$ is the qth percentile of the null distribution. Specifically, we will consider $q = 0.75$ and $q = 0.9$. The ratio FDR_q/FDR depends only on the null hypothesis, so it

**Table 18.3 Ratio Between Conservative and Standard FDR, for Four
Threshold Levels of the Test Statistic**

$|T|$ Threshold:	Q_{75}/Q_{50}				Q_{90}/Q_{50}			
	2	2.5	3	3.5	2	2.5	3	3.5
UVA	1.3	1.5	1.7	1.7	1.7	2.0	2.4	2.9
OV	1.5	1.9	2.3	2.7	2.4	3.3	4.7	6.0
CO	1.3	1.5	1.7	1.8	1.8	2.4	2.9	3.3
LU	1.3	1.4	1.5	1.8	1.8	2.0	2.3	2.8
NCI60$_a$	1.2	1.3	1.3	1.5	1.4	1.6	1.8	2.0
NCI60$_b$	1.3	1.5	1.6	1.7	1.8	2.2	2.6	3.0
NCI60$_c$	1.3	1.4	1.6	1.8	1.6	1.9	2.3	3.0
NCI60$_d$	1.3	1.4	1.5	1.7	1.6	1.9	2.4	2.9
Null	1.0	1.0	1.1	1.2	1.1	1.1	1.2	1.4

can be estimated using permutations of the samples in the eight experimental datasets
studied above, along with the i.i.d. Gaussian set. Table 18.3 shows these ratios based on
various thresholds for a test of differential expression based on the two-sample Z statistic.

In most practical settings, a test statistic threshold between 2.5 and 3.5 will be used.
According to Table 18.3, the actual FDR could easily be $1.5-2$ times higher than that given
by the standard definition ($FDR_{0.75}$), and could possibly be $2-3$ times higher than the standard
definition ($FDR_{0.9}$). The multipliers are fairly consistent across the datasets, with the exception
of the OV dataset. Since that set did not have an unusual $\hat{R}$ value, there may be some unusual
pattern of correlations in the OV set that is not detected by $\hat{R}$. Also of note is that the "null"
(i.i.d. Gaussian) data have much smaller multipliers than do those in the experimental datasets.
The discrepancy is the result of intergene correlations.

In many cases, an increase by a factor of $2-3$ in the FDR is tolerable. For example, if
FDR < 0.05 is used as the gene selection criterion, a true FDR of 0.15 is generally worth pursuing. However, examples exist where the goal is to identify genes predictive of a subtle trait,
for example, differential treatment response to two fairly similar treatments. In such cases, it is
not uncommon to be forced to consider genes at the $0.1-0.2$ FDR level. Genes at a true FDR of
0.2 may sometimes be of genuine interest, but clearly a threefold inflation of the FDR value in
this situation is not tolerable. This discussion is related to the more general issue of variance
and sampling behavior of the various empirical FDR measures.

18.4 TIMECOURSE EXPERIMENTS

Many important questions in biology rest on the pattern of gene expression changes over time.
In a typical experiment, a cell culture is treated in a certain way, then cells are removed from the
culture at a sequence of timepoints for measurement on microarrays. The key question for inference is the probability that the pattern of interest (e.g., a cycling pattern, or a particular timing
of peak expression) occurs by chance. In most timecourse experiments, a single cell culture is
maintained, and a small subset of cells are removed from the culture at each timepoint for
analysis. Therefore, small perturbations such as changes in temperature or humidity may

affect all subsequent measurements. The slowly varying dynamical patterns deriving from such perturbations can increase the chances of a spurious match to the pattern of interest.

To be concrete, we focus here on the detection of genes showing coordinated gene expression with the cell cycle (e.g., see Ref. 13). In this type of experiment, a synchronized cell culture is obtained either by arresting the cells at a common cell cycle stage and then releasing them to grow in unison, or selecting cells from an unsynchronized culture that are at a similar size. Gene expression is then measured at a set of timepoints, usually over two generations of growth. Typically 5–10 timepoints per generation (10–20 measurements overall) are used. A cell-cycle-coordinated gene will show defined peaks of expression that are reproduced in both observed passages through the cell cycle. One way to quantify the presence of such a pattern is to look at the proportion of variance explained (PVE) when regressing the measured timecourse of gene expression on a three-component basis (constant, sine wave, cosine wave), where the sine and cosine waves have period equal to the doubling time in the experiment (which is known).

The goal of this section is to look at how serial correlations and intergene correlations jointly affect the sampling properties of statistics such as the PVE. We will begin by quantitatively assessing the degree of serial correlation in microarray data from timecourse experiments and then consider how this affects the sampling properties of the statistic $\hat{R}$ discussed earlier. Finally, we will demonstrate how permutation approaches may be misleading in this type of problem.

We will work with the fission yeast data described in Rustici et al. [13]. Around 5000 genes (essentially the entire yeast genome) were measured using two-color cDNA arrays. Cells were synchronized and then assayed on arrays every 15 min following release for 300 min. This gives data for two complete generations, 10 experimental points per generation. The entire process was replicated 3 times independently.

Our first goal is to quantify the autocorrelation in the data due to factors other than cell cycle coordination, that is, the level of autocorrelation that would be observed in an unperturbed, exponentially growing cell population. Since a gene that truly is coordinated with the cell cycle will yield a smooth pattern under the dense sampling scheme used in the experiment, such genes must be excluded from this analysis as they would upwardly bias the autocorrelation measure. We removed all genes with PVE > 0.3 and calculated the first-order autocorrelation coefficient for the remaining genes. The PVE threshold 0.3 was selected since this gives negligible visible appearance of cell-cycle-coordinated expression. The results of this calculation for the three replicates are $0.30(0.25)$, $0.20(0.25)$, and $0.19(0.25)$, where the average over all selected genes is given, followed by the standard deviation.

The occurrence of three similar average autocorrelation coefficients $(0.30, 0.20, 0.19)$ in triplicated experiments strongly suggests that positive serial correlation is a reproducible feature of the "background signal" of this system—apart from any correlation induced by cycling genes. The standard deviations are somewhat more difficult to interpret. A simple question is whether the observed standard deviation is consistent with a constant population value of the autocorrelation coefficient being shared by all genes. The probability of getting a large average autocorrelation coefficient by chance is increased by intergene correlation, but intuitively the standard deviation should not be affected. To address this question, we considered a simple model with 1000 simulated genes over 20 "timepoints." We considered the situation in which expression levels at a given timepoint were identical within each of n blocks containing m genes. Smaller values of n give greater intergene correlation.

Table 18.4 gives the results of this analysis, showing several characteristics of the null sampling distribution of the average autocorrelation coefficient over all genes on the array. The average value (Avg) over replicates is slightly negative owing to the small bias in the sample autocorrelation coefficient. As expected, the 97.5 null percentile increases with

Table 18.4   Characteristics of the Null Sampling Distribution of the Average First-Order Autocorrelation Coefficient[a]

n	m	Avg	2.5%	97.5%	σ
1000	1	−0.05	−0.06	−0.04	0.21
100	10	−0.05	−0.09	−0.01	0.20
50	20	−0.05	−0.11	0.01	0.20
10	100	−0.05	−0.18	0.08	0.19

[a]The expected value (Avg) and 2.5 and 97.5 percentiles are given, along with the standard deviation of first-order autocorrelation coefficients across all genes on the array. The rows correspond to population structures with different levels of intergene correlation; see text for details.

increasing intergene correlation, but remains well below the experimental values. The standard deviation σ of sample autocorrelation coefficients across genes on the array is nearly invariant with intergene correlation, as expected, and is slightly below the experimental value. Thus there is a small extra dispersion of the sample autocorrelation coefficients in the experimental data, indicating that most genes are similar in their serial correlation levels, with some degree of small differences.

The statistic R used above to quantify intergene correlation must be modified if serial correlations are present. Specifically, the bias correction must be larger since it is easier to get large sample $\hat{\rho}_{ij}$ values in the presence of autocorrelation. The approximately unbiased estimate of R is given by

$$\sum_{i<j} \rho_{ij}^2 \Big/ \binom{n}{2} \approx \sum_{i\neq j} \hat{\rho}_{ij}^2 \Big/ \binom{n}{2} - 1/p - 2p^{-1}\tau^2 \Big/ (1-\tau^2).$$

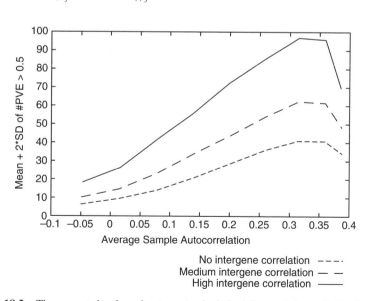

Figure 18.2 The expected value plus two standard deviations of the null distribution of the statistic N, given for a range of different intergene correlation levels and serial autocorrelation levels.

The values of this statistic for the three datasets under consideration are 0.11, 0.04, and 0.03, where the estimated autocorrelation values $\tau = 0.3, 0.2, 0.2$ are substituted into the bias term. All of these values are substantially greater than the values for eight datasets of independent samples shown in Table 18.1, with the value 0.11 for the first replicate being grossly out of place. Given that these experiments were carried out on a different organism, using two-color cDNA arrays, it is unclear whether these results are anomalous.

Next we consider how intergene and serial correlations interact in their effects on the sampling properties of the statistic N, defined as the number of genes with PVE greater than a given threshold. To generate data with controlled levels of intergene and serial correlation, first simulate independent "cluster centers" $X_i(t)$ $i = 1, \ldots, n$, $t = 1, \ldots, p$ as stationary Gaussian processes with mean zero and correlation structure given by $\mathrm{cor}(X_i(t), X_i(t + d)) = r^d$. Next, simulate m correlated versions of each cluster center using

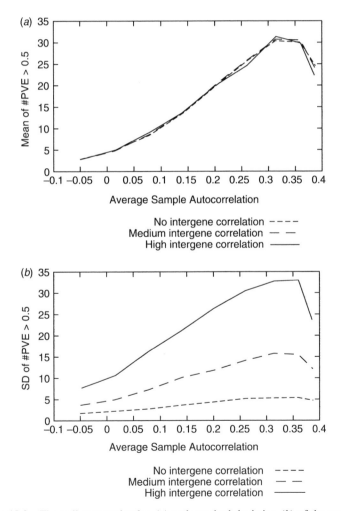

Figure 18.3   The null expected value (*a*) and standard deviation (*b*) of the statistic N.

$Z_{m(i-1)+j}(t) = X_i(t) + \epsilon_{ij}(t)$, $j = 1, \ldots, m$, where the $\epsilon_{ij}(t)$ are i.i.d. $N(0, \sigma^2)$ and mn is the number of observed genes. Values of r, σ, m, and n were adjusted so that the distribution of intergene correlations in the resulting simulated data matched the distribution in the NCI60$_b$ dataset discussed above. Quantile–quantile plots of the sample intergene correlation distribution showed a nearly perfect match.

Figure 18.2 shows the null expected value of N plus two standard deviations (which we will refer to as Q) under different levels of serial correlation and intergene correlation. Increasing intergene correlation increases Q up to a point, after which it declines. The reason for this is that a perfect sinusoidal pattern that maximizes the PVE has autocorrelation around 0.3 under the sampling scheme used here. Since the experimental autocorrelation is in the range 0.2–0.3, we are operating near the most unfavorable point in this range. The value of Q also increases with intergene correlation. The "medium" level of intergene correlation in the plot is close to the experimentally observed level.

Variation in the value of Q can be decomposed in terms of the effect on the mean and the effect on the standard deviation, which are shown separately in Figure 18.3. As expected, intergene correlation has no effect on the mean, but has a substantial effect on the standard deviation. Serial correlation affects both the expected value and standard deviation. Notably, the slopes in Figure 18.3b become steeper as intergene correlation increases. This indicates a synergy between serial and intergene correlation in which their joint effects are greater than the sum of their individual contributions.

As a concluding comment, we note that it is not straightforward to use permutation approaches for calibrating the statistic N. The standard approach for independent samples would be to randomize whole arrays over the time points. This preserves intergene correlation but destroys serial correlations. From Figure 18.2, it is clear that this substantially understates the minimal value of N needed to conclude that at least some genes are cell cycle regulated. A reasonable threshold from Figure 18.2 would be $N = 35$ (an autocorrelation value of 0.2 with "medium" intergene correlation). The permutation method forces the serial correlation to be zero, hence gives a threshold of around $N = 12$ for "medium" intergene correlation.

18.5 META-ANALYSIS

When similar treatments are applied in different experimental settings, it is natural to ask whether the same genes are affected. A common example occurs when the expression of a gene of interest, say, G, is rendered experimentally controllable in a genetically modified cell line. The genes that respond to changes in the expression of G are putative targets of G (commonly G is a transcription factor or an upstream regulator of one or more transcription factors). For comparison, one can consider a large collection of tissue samples in which variation in the expression of G is uncontrolled. If the cell line model is realistic, and if the action of G on its targets is not confounded with the actions of other regulatory elements, then the set of genes having expression correlated with G in tissue samples should be similar to the set of genes that are differentially expressed in the cell line model as the levels of G are experimentally varied.

Concretely, suppose that S_{ik}, $k = 1, 2$ are test statistics applied to measurements of gene i in either the cell line model ($k = 1$) or the set of tissue samples ($k = 2$), quantifying the association between the levels of genes G and i. Since gene G is assumed directly measured in both cases, the test statistic can be a simple correlation coefficient or Z statistic. We will also assume that the S_{ik} are transformed so that under the null hypothesis for gene i in experiment k, the test statistic's expected value is zero and its variance is one. This would be true of the Z statistics, and of the Fisher transformed correlation coefficient. The association between test statistics in the two experiments can be quantified by

$$T = n^{-1} \sum_i S_{i1} S_{i2}.$$

Under the null hypothesis, T estimates the correlation coefficient between S_{i1} and S_{i2}.

Rejecting the null hypothesis $ET = 0$ in favor of $ET > 0$ is weak support for the claim that the cell line model captures the biological effects of variation in G in the tissue environment. Nevertheless it is a starting point. Moreover, the test can be inverted to form a confidence interval for ET. If ET is greater than, say, 0.2 with 95% confidence, claims that the cell line model is valid are justified.

The variance of T is easily seen to be

$$\operatorname{var} T = \frac{1}{n^2} \sum_{ij} ES_{i1} S_{j1} \cdot ES_{i2} S_{j2}.$$

For test statistics that are linear in the data, $\operatorname{cor}(S_{i1}, S_{j1})$ is equal to ρ_{ij} (the correlation between expression measurements for the two genes). Consider either a two-sample Z-test statistic of the expression levels of gene i (comparing samples at the high and low experimental levels of gene G) or the correlation coefficient between the expression levels of gene i and the naturally varying expression levels of gene G. Both of these statistics have a numerator that is linear in the data, and a standard deviation term in the denominator that also involves the data. Simulation studies (not shown) indicate that if the sample size is not too small, the relationship $\operatorname{cor}(S_{i1}, S_{j1}) \approx \rho_{ij}$ holds for these statistics. Furthermore, using standardized statistics yields

$$\begin{aligned} ES_{ik} S_{jk} &= \operatorname{cov}(S_{ik}, S_{jk}) \\ &= \operatorname{cor}(S_{ik}, S_{jk}) \\ &\approx \rho_{ij}. \end{aligned}$$

Thus

$$\operatorname{var} T \approx \sum_{i<j} \rho_{ij}^{(1)} \rho_{ij}^{(2)} \Big/ \binom{n}{2},$$

where $\rho_{ij}^{(k)}$ denotes the population correlation coefficient between genes i and j in either the cell line data ($k = 1$) or the tissue data ($k = 2$). If we are willing to assume that these intergene correlations are equal, then $\operatorname{var} T \approx R$, as defined earlier. Taking two standard deviations as the significance threshold, we find that for the eight human datasets studied here, the value of T would need to exceed a threshold of $0.2-0.32$ to be deemed significant. By contrast, if the genes were independent, the significance threshold would be around 0.03. This 10-fold discrepancy on the standard deviation scale corresponds to a 100-fold discrepancy on the variance scale. Since the variance scale is inverse linear with sample size (the number of genes), the data are behaving as if the sample size were 100 times smaller than the number of measured genes.

We conclude this section with a note about permutation tests. A natural way to obtain a significance level for T would be to permute samples within the cell line and tissues datasets separately, recalculate all the S_{ik} values, and then recalculate T. A set of T values obtained in this way could be considered as an empirical null distribution for T. However, it is important to note that under this permutation scheme, not only is there no systematic trend between the test statistics in the cell line and tissue experiments; in addition, there is no systematic signal within either of the experiments—that is, each S_{ik} has mean zero (all the null hypotheses are true). This is a strong assumption, since it is reasonable to assume that even if G has completely different effects in cell lines and in tissue, it is likely to have separate systematic effects in one or both environments. It can be easily shown [14] that under weak conditions the approximation var $T \approx R$ holds whenever there is no association between the test statistic values in the two experiments, even if within one or both experiments there are nonnull genes.

18.6 CONCLUDING REMARKS

A number of papers have appeared in the literature utilizing intergene expression associations to understand pathways and networks involving gene expression (e.g., see Refs. 6 and 20). Ultimately, the structures of these pathways and networks encode the "degrees of freedom" inherent in the variation of cellular gene expression. Here we have provided some evidence, consistent with other published reports, that the degrees of freedom in gene expression is rather limited, especially in comparison to the number of genes measured.

For some of our simulation work, and also for conceptual illustration, we have used a "blockwise" correlation pattern among the genes as a working model. Such correlation structures have been previously used in similar work (e.g., Ref. 17). An alternative view has been developed more recently [11,12], claiming that intergene correlations arise from long chains consisting of hundreds or even thousands of genes that are correlated serially. To the extent that an estimable quantity such as the "R statistic," discussed in Section 18.2, can be used to determine sampling distributions, the precise nature of intergene correlations (blockwise, serial, or neither) does not directly impinge on the inference problems discussed here. As shown in Section 18.5, this is the case in at least one type of meta-analysis. However, it is possible that a pairwise statistic such as our R is not sufficient to address many important inferential questions, in which case the higher-order pattern of intergene correlation may become critical.

It is already well known that inferences involving the relationship between gene expression and other biological traits are sensitive to intergene correlations [16,10]. In the case of false discovery rates, various mixing conditions have been invoked to justify the validity of the empirical FDR in the presence of intergene correlations. As noted above, since FDR is defined in terms of expectations rather than variances, the impact of intergene correlations in that setting is limited. On the other hand, the effects of intergene correlations on other types of analyses including timecourse and metagene analyses have not been as widely studied. Since permutation methods preserve intergene correlations, it is generally assumed that they are safe to use in this regard. As demonstrated here, if there is additional structure in the experiment that is not respected when permuting whole arrays, problems can potentially arise. On the positive side, we have shown in one example that a test statistic can be constructed that is easily calibrated on the basis of an aggregated measure of the intergene correlation structure of the data. Whether this is a particularly powerful test has not been addressed, but would be important to determine before the test can be considered for routine use.

REFERENCES

1. Barabasi, A. L. and Oltvai, Z. N., Network biology: Understanding the cell's functional organization, *Nature Rev. Genet.* **5**(2), 101–115 (2004).

2. Benjamini, Y., Hochberg, T., and Klint, Y., *False Discovery Rate Control in Multiple Hypotheses Testing Using Dependent Test Statistics*, Research Paper 97-1, Dept. Statistics and Operations Research, Tel Aviv Univ., Israel, 1997.

3. Benjamini, Y. and Yekutieli, D., The control of the false discovery rate in multiple testing under dependency, *Ann. Statist.* **29**, 1165–1188 (2001).

4. Efron, B., Tibshirani, R., Storey, J. D., and Tusher, V., Empirical Bayes analysis of a microarray experiment, *J. Am. Statist. Assoc.* **96**, 1151–1160 (2001).

5. Efron, B., *Correlation and Large-Scale Simultaneous Significance Testing*, Technical Report, Stanford Univ. 2006.

6. Friedman, N. et al., Using Bayesian networks to analyze expression data, *J. Comput. Biol.* **7**, 601–620 (2000).

7. Gibson, G., Microarray analysis, *PLoS Biol.* **1**(1) (2003).

8. Giordano, T. J., Shedden, K. A., Schwartz, D. R., Kuick, R., Taylor, J. M., Lee, N., Misek, D. E., Greenson, J. K., Kardia, S. L., Beer, D. G., Rennert, G., Cho, K. R., Gruber, S. B., Fearon, E. R., and Hanash, S., Organ-specific molecular classification of primary lung, colon, and ovarian adenocarcinomas using gene expression profiles, *Am. J. Pathol.* **159**, 1231–1238 (2001).

9. Huang, E., Cheng, S. H., Dressman, H. et al., Gene expression predictors of breast cancer outcomes, *Lancet* **361**, 1590–1596 (2003).

10. Owen, A. B., Variance of the number of false discoveries, *J. Roy. Statist. Soc. Ser. B* **67**, 411 (2005).

11. Qiu, X., Brooks, A. I., Klebanov, L., and Yakovlev, A., The effects of normalization on the correlation structure of microarray data, *BMC Bioinformatics* **6**, Art. 120 (2005).

12. Qiu, X., Klebanov, L., and Yakovlev, A., Correlation between gene expression levels and limitations of the empirical bayes methodology in microarray data, *Statist. Appl. Genet. Molec. Biol.* **4**(1), Art. 3 (2005).

13. Rustici, G., Mata, J., Kivinen, K., Lio, P., Penkett, C. J., Burns, G., Hayles, J., Brazma, A., Nurse, P., and Bahler, J., Periodic gene expression program of the fission yeast cell cycle. *Nature Genet.* **36**(8), 809–817 (2004).

14. Shedden, K., Confidence levels for the comparison of microarray experiments, *Statist. Appl. Genet. Molec. Biol.* **3**(1) (2004).

15. Staunton, J. E., Slonim, D. K., Coller, H. A., Tamayo, P., Angelo, M. J., Park, J., Scherf, U., Lee, J. K., Reinhold, W. O., Weinstein, J. N., Mesirov, J. P., Lander, E. S., and Golub, T. R., Chemosensitivity prediction by transcriptional profiling, *Proc. Natl. Acad Sci. USA* **98**, 10787–10792 (2001).

16. Storey, J. D., A direct approach to false discovery rates, *J. Roy. Stat. Soc. Ser. B* **64**, 479–498 (2002).

17. Storey, J. D., Comment on "Resampling-based multiple testing for DNA microarray data analysis" by Ge, Dudoit and Speed, *Test* **12**, 1–77 (2003).

18. Su, A. I., Welsh, J. B., Sapinoso, L. M., Kern, S. G., Dimitrov, P., Lapp, H., Schultz, P. G., Powell, S. M., Moskaluk, C. A., Frierson, H. F. Jr., and Hampton, G. M., Molecular classification of human carcinomas by use of gene expression signatures, *Cancer Res.* **61**, 7388–7393 (2001).

19. van Steensel, B., Mapping of genetic and epigenetic regulatory networks using microarrays, *Nature Genet.* **37**, S18–S24 (2005).

20. Woolf, P. J., Prudhomme, W., Daheron, D., Daley, G. Q., and Lauffenburger, D. A., Bayesian network analysis of signaling networks governing mouse embryonic stem cell self-renewal, *Bioinformatics* **21**(6), 741–753 (March 2005).

21. Zhang, B. and Horvath, S., *Statist. Appl. Genet. Molec. Biol.* **4**, Art. 17 (2005).

CHAPTER 19

A Comparison of Methods for Meta-Analysis of Gene Expression Data

Hyungwon Choi and Debashis Ghosh

Department of Biostatistics, University of Michigan, Ann Arbor, Michigan

19.1 INTRODUCTION

With the development of microarray technology, it has become possible to globally monitor the biochemical activity of populations of cells. Microarrays are being increasingly used in medical and scientific research and have allowed for characterization of biological activity on a high-throughput basis. A major resource that has made the use of microarray technology feasible is large-scale genome sequencing projects, such as the Human Genome Project [14,32]. Having such sequence data available allows for the characterization of the probes on the microarray.

While transcript mRNA microarrays have received much attention in the literature, there has been work on other types of microarrays. Examples include chromatin–immunoprecipitation (ChIP-chip) microarrays, which measure transcription factor–DNA binding expression [21] and methylation microarrays Yan et al. [35], which assess DNA methylation on a global scale. In addition, there has also been much attention on high-throughput assays that measure protein–protein interactions, such as yeast two-hybrid systems [31]. Because of the various large-scale datasets that are being generated, there is much interest in attempting to integrate them in order to provide a more complete understanding of the biological mechanisms that are at play. This type of analysis has been termed *systems biology* in the bioinformatics literature [13].

For the statistician, this area brings many interesting and challenging problems. The particular problem we will be addressing in this chapter is meta-analysis of microarray data. We are

Statistical Advances in the Biomedical Sciences, edited by Atanu Biswas, Sujay Datta, Jason P. Fine, and Mark R. Segal

325

dealing with data from multiple studies in which the same biological activity is measured using microarrays, including transcript mRNA activity. While the term *meta-analysis* is familiar among most statisticians [22], the term takes a very different meaning here. The situation that statisticians are familiar with involves attempting to combine information from relatively homogeneous data structures from multiple similar experiments. There are technical challenges that make meta-analysis of gene expression data more challenging; we discuss them in Sections 19.2 and 19.3. The number of available microarray studies is the most for gene expression, but we anticipate that there will be increasing studies done for assessing epigenetic and copy number using microarrays in the future. The methods described in the chapter will potentially be applicable to meta-analysis of other types of microarray.

Our goal here is to outline the major issues involved in such analyses and describe some solutions that have been proposed. It is not our intent to provide an up-to-date listing of all methodologies that have been used, as the literature is constantly changing. Given the dynamic nature of the field, an important component will be benchmarking of methods to see which should be used in practice.

19.2 BACKGROUND

19.2.1 Technology Details and Gene Identification

In this chapter, we consider multiple microarray studies in which the same comparison was considered. For measuring gene expression, different technologies might be used. The two dominant technologies currently being used are the two-color microarrays and Affymetrix arrays. We now provide a brief description of each type of technology. For the spotted array, each spot represents a cDNA (complementary DNA) fragment, which represents the complement of a messenger RNA molecule that has been transcribed from an individual's DNA. The length of a cDNA is between 500 and 5000 bases.

Affymetrix arrays for measuring gene expression employs 11–20 different and sometimes overlapping 25-mer oligonucleotides per gene. In addition to perfect-match oligonucleotides, Affymetrix microarrays also contain mismatch oligonucleotides. The mismatch oligonucleotides carry a mutation at position 13 of the 25 mers. While their initial purpose was to serve as negative controls, there has been much controversy as to what is actually being measured with the mismatches. Many preprocessing algorithms avoid using this information (e.g., the robust multisample average method of Irizarry et al. [16]), and Affymetrix is planning to make arrays in the future that do not have any mismatch probes.

There are several issues that must be considered when attempting such an analysis. First, one must consider the problem of study-specific artifacts, such as collection of available samples, variations in experimental protocols, and differences in laser scanners. However, there are two bigger issues in the analysis of such data. The first is that of matching genes from studies that use different platforms. This is where the availability of large-scale genomic data is important. Each spot on a microarray corresponds to a DNA sequence. What one can do is to match up each spot to a putative gene in the National Centers of Biotechnology Information (NCBI) Database. The LocusLink identifier from NCBI can then be used to identify common genes across multiple datasets. Such a task can be done for Affymetrix chips from their Website (http://www.netaffx.com/) or for two-color cDNA microarrays using the SOURCE tool at Stanford University [8].

Some work has been done on using the sequence data for the spots on microarrays and reperforming database searches in order to determine the genetic identity of spots [7]. These authors found a variety of annotation issues with the probe sequences on the Affymetrix

GeneChip arrays and estimate between 30% and 50% discrepancy between the genes identified by their method versus those given by the Affymetrix website. Dai et al. [7] point out that many probes may not be grouped into the appropriate probeset or that probesets may not map to the "correct" gene. Another sequence-based method was proposed by Carter et al. [4]. In their approach, the short oligoprobes of the Affymetrix platform were mapped to the cDNA clones of the Stanford microarray platform. Affymetrix probes were reassigned to redefined probe sets if they mapped to the same cDNA clone sequence, regardless of the original manufacturer-defined grouping. Even when the probes across platforms are appropriately mapped to the same gene, probes in one platform may hybridize to the 5′ end, where probes in a different platform may hybridize at the 3′ end. Discordance between the two platforms may then be a result of alternative splicing, where the different probes bind to different isoforms (translation variants) of a gene. Another approach to understanding the behavior across different platforms is through studies of the same samples [28,16].

While determining the genetic identity of the spots is important, there is still the issue of attempting to combine expression measurements from diverse data platforms. Because of the differing technologies, expression values from different microarray platforms are on different scales. For example, an expression value of 200 from a cDNA two-color microarray is very different from an expression value of 200 measured on an Affymetrix array. The two-color microarrays typically use a reference sample for one of the color channels so that the measurement represents relative hybridization of two samples. By contrast, the Affymetrix microarrays contain only one sample per slide, so the intensity measurement represents an absolute expression of the sample. Thus, it is important to have tools that can enhance comparability across arrays of different platforms.

One such technique that has proved useful as a filtering device is known as the *integrative correlation coefficient* or *correlation of correlation coefficients* [21,24]. Integrative correlation coefficients presume that while raw expression values vary from study to study, the intergene correlations do not vary as much. Thus, one would consider combining genes that have similar intergene correlations across the studies. A nice of example of this technique was presented in the study by Lee et al. [21]. In this study, Affymetrix and two-color microarrays were used to profile the National Cancer Institute 60 cell line data. While the correlation between expression between the two technologies was reported to be poor in a previous study [18], by using the integrative correlation of correlation, Lee et al. [21] found that a subset of genes has correlation above 0.8 between the two technologies. A caveat of the technique is that the filtered set of genes may not be a random sample of genes on the microarray so that the generalizability of resulting gene signatures may be limited.

19.2.2 Analysis Methods

In terms of meta-analysis methods put forward, many have been based on the fact that the standardized effect size is combinable across studies. This is the approach advocated by Parmigiani et al. [24] after filtering based on the integrative correlation coefficient. In Rhodes et al. [25], the t statistic was transformed into a p value, a transformation of which was combined across multiple studies. By contrast, in Ghosh et al. [11], the t statistic was combined directly. A large-scale comprehensive meta-analysis of 40 independent datasets (>3700 array experiments) was performed by Rhodes et al. [26], who found a universal profile of 67 genes that could differentiate cancer versus noncancer tissue for a variety of cancers. In addition, they determined 36 cancer-specific signatures for determining a tissue-specific cancer. The signatures also demonstrated good discrimination performance on three independent datasets.

Another approach more in line with classification or supervised learning analyses is to build a classifier or find a gene expression signature on one dataset and to see how well it predicts in an independent microarray dataset. Such approaches were taken by Beer et al. [1], Wright et al. [34], and Jiang et al. [17]. An alternative method using hierarchical clustering, which is an unsupervised learning procedure, was taken by Sorlie et al. [30], who found a gene expression signature that defined molecular subtypes in breast cancer; they found through interrogation of other datasets that the subtypes were present there as well. Given the increasing availability of publicly available large-scale gene expression datasets, it is increasingly important that results found by one investigator on a particular dataset be validated using other datasets as well.

Much of the meta-analysis methods have studied differential expression across multiple studies. A notable exception is the study by Lee et al. [20], in which intergene correlations across multiple studies was considered. The authors sought pairs of genes that were consistently coexpressed across several datasets.

19.3   EXAMPLE

We now discuss the application of various meta-analysis methodologies to a study looking at metastatic cancer. Generally speaking, cancer is either nonmetastatic (nonaggressive) or metastatic (aggressive). The latter type of cancer tends to be much more lethal. It is thus important from the point of view of treatment to determine whether a cancer detected at early stage will become indolent or metastatic. As one step toward this goal, we would like to determine whether there is a gene signature that can discriminate metastatic cancer from nonmetastatic cancer. We will use these data to illustrate the various methodologies that are available for meta-analysis of microarray data.

While the purpose of the chapter is to illustrate the meta-analysis methods, it should also be noted that the goal of the analysis, classification of metastasis, is a tricky issue. First, there are many statistical limitations and misconceptions in developing a classifier; a reference on these issues is the paper by Simon [29]. Moreover, metastasis is a clinically detectable event, and it is rather a dichotomization of an underlying process that is likely to be a continuum.

The datasets were obtained from publicly available data sources [19,10,5]. They were selected according to the presence of both localized and metastatic samples, as well an overlap of a sufficiently large number of genes. In general, only a small number of metastatic samples are profiled in all datasets. We also note that the organ sites are different. We are postulating that there is a common profile separating localized from metastatic cancer across the three sites. Similar evidence for this type of hypothesis has been suggested before [26]. The LaTulippe et al. data include 23 nonmetastatic and 9 metastatic samples; the Garber et al. data, 54 nonmetastatic and 6 metastatic; and the Chen et al. data, 99 nonmetastatic and 9 metastatic samples. The LaTulippe study used Affymetrix arrays, while the other two studies used two-color microarrays.

The microarray platforms differ by studies, so we mapped clone/probeset IDs to Unigene cluster IDs (UG ID) in its most recent build through SOURCE [8]. When multiple clones were mapped to the same UG ID, we averaged the expression values over the clones within each sample. Such a mapping produced 1633 common UG IDs.

Note that the sample size of metastatic tumors is relatively small in the three studies. For this reason, it is important to validate study-specific metastasis signatures externally, or perform a meta-analysis across multiple experiments that reported features of cancer metastasis, which would result in a more robust set of genes relevant to metastasis.

Table 19.1 Misclassification Rates Using Classifiers Constructed from Individual Studies[a]

Signature from/Error in	Chen	Garber	LaTulippe
Chen [5] 50 genes	0% (0/0)	15% (10/67)	9% (13/0)
Garber [10] 30 genes	1% (0/11)	21% (13/67)	6% (4/11)
LaTulippe [19] 80 genes	6% (6/11)	15% (7/50)	3% (4/0)

[a]Table entries are error rates using classifier from study defined in column 1. The error rates in parentheses are percentage of primary tumors and metastatic tumors misclassified, respectively.

19.4 CROSS-COMPARISON OF GENE SIGNATURES

In this section, we describe how to select optimal gene signatures for classification of metastatic tumors versus primary tumors in each study separately. We first obtain study-specific signatures and estimate the misclassification error rate for each signature in all three studies. Misclassification errors were estimated for all tumors and for each type of tumor (primary/metastatic) separately as well. We emphasize that metastatic tumors often represent only a small fraction of the samples in individual studies; thus the gene signature specific to a single study generally requires external validation. The signature selection was based on the "leave-one-out cross-validation" (LOOCV) risk index method that was used in Shen et al. [27]. We extract the top k^* genes that are significant in univariate logistic regression, where k^* is the optimal sample size. The size of the optimal signatures varies between the studies. The signature sizes are tabulated in Table 19.1.

Using the risk index approach, each study-specific signature predicts the tumor types in the other studies, as can be seen in Table 19.1. The Garber et al. lung study profiles gene expression in blocks of different subclasses of adenocarcinoma; hence the classification between metastatic and primary tumors is not noticeably clear.

Although the classification performance is acceptable when study-specific signatures are applied to independent datasets, one drawback is the poor overlap between signatures. Only 3 genes are shared by all three signatures, and 10 genes are shared between the signatures from the Chen et al. [5] liver study and the LaTulippe et al. [19] prostate study. No genes are shared by the other two pairs of studies. This suggests that using study-specific analyses fails to identify generalizable genes. Therefore, we consider more integrative approaches for signature selection. In the following sections, we apply three meta-analytic methods: best common mean difference (BCMD) [33], effect size (ES) [6], and probability of expression assimilation (POEA) [27]. We describe their application and compare the results.

19.5 BEST COMMON MEAN DIFFERENCE METHOD

The BCMD method involves calculation of differential expression across multiple studies based on a classic calculation of best common mean estimator that minimizes the variance among all linear estimators [3]. The estimated mean difference in individual studies is weighted by the reciprocal of the variance and combined into a single statistic that asymptotically follows standard normal distribution. The BCMD approach was applied to the multiple microarray study context by Wang et al. [33]. Before forming the univariate statistic, they stabilize the mean and variance relationship of the gene expression values using locally weighted regression.

We applied BCMD to our three datasets to select differentially expressed genes. Genes with a false discovery rate (FDR)–corrected p value less than 0.05 were used to form the signature. FDR correction on the p value was done using the step-down procedure of Benjamini and Hochberg [2]. The final number of genes selected was 286. Assigning significance based on permutation versus the normal-based null did not alter the signature profoundly.

The gene signature has some overlap with the study-specific signatures (23/50 in Chen et al. [5], 11/30 in Garber et al. [10], 46/80 in LaTulippe et al. [19]). See Figure 19.1 for the heatmap of the signature. A drawback of this signature is that there is discordance in the direction of differential expression for some genes in the signature. The horizontal rectangular

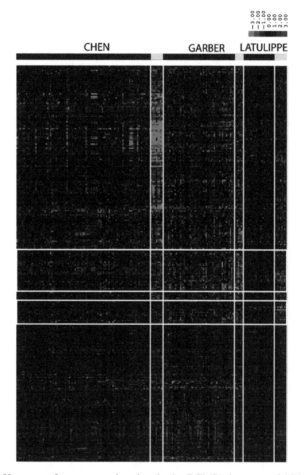

Figure 19.1 Heatmap of raw expression data in the BCMD signature of 286 genes. Within each study, arrays were scale-normalized and then genes were centered at the median of primary tumors. After centering, all three datasets were juxtaposed for visualization. Black and white tabs on the top denote primary and metastatic tumors, respectively.

boxes in Figure 19.1 show examples of such genes. For many of these cases, a large difference in one direction for one study contrasts with the direction for the effects in the other two studies.

Annotation of the 20 genes includes general molecular functions such as apoptosis (MYBL2, RAD21), cell cycle (CCNB1), and cell differentiation (INHBB). When the signature was used in each study, the corresponding risk index classifiers reduced classification error relative to the study-specific signatures described in Table 19.1. In the Garber study, the signature yielded an overall misclassification error of 9%; it gave perfect classification in the other two studies.

19.6 EFFECT SIZE METHOD

The effect size (ES) method entails a direct application of traditional meta-analytic methods [12] to microarray data. This approach resembles BCMD in that it pools the standardized t statistic for differential expression of a single gene from all studies. The ES approach directly models expression levels with gene-specific mean and variance parameters. Since the models are gene-specific, there is no information pooled across all genes as in BCMD. Choi et al. [6] suggested modeling the effect size using a Bayesian framework based on incorporating the study effect as random effects. Such an approach might potentially improve the performance of the ES procedure by pooling information across genes, information on which is specified through the prior distribution.

Applying the ES method to the three datasets results in a 90-gene signature that has a 60-gene overlap with the BCMD-derived signature. See Figure 19.2 for the heatmap of the signature. The selection criterion is that all the genes must have an estimated empirical FDR of <0.05. This signature has few genes in common with the study-specific signatures: 4/50 with the Chen, 3/30 with the Garber, and 9/80 with LaTulippe. Classification using this signature shows performance worse than that of BCMD in the individual studies.

However, it is clear from Figure 19.2 that the genes in the signature show more consistent directionality of differential expression between the two tumor types than do those found using

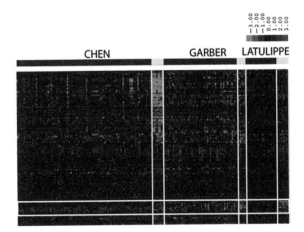

Figure 19.2 Heatmap of raw expression data in the ES signature of 90 genes. Note that the directionality of differential expression is consistent across all studies.

BCMD. If one wishes to look for genes that show concordance in expression across studies, the ES approach finds more genes with this behavior than does BCMD.

19.7 POE ASSIMILATION METHOD

In the POEA method, expression data are combined after the raw expression has been transformed into the probability of over/underexpression, making it possible to treat the transformed expression as independent observations regardless of the source of the data. The estimation procedure borrows strength from all genes in each study.

The concept of probability of expression (POE), proposed in a single-study setting by Parmigiani et al. [23], is to treat expression of a sample relative to a baseline group as a scale-free quantity using a latent variable model. This leads to a three-component mixture model for the preprocessed microarray data, from which one transforms the data to a POE scale. POE is then the difference between the posterior probability of overexpression minus that of underexpression. We refer the reader to Shen et al. [27] for mathematical and computational details of POE.

The transformation sets the mean gene expression values of one class (primary tumors in our example) close to zero and normalizes the expression values of the other class (metastatic tumor) relative to the distribution of the former. This is done using a scale within $[-1,1]$. It helps single out over- or underexpressed genes in metastatic tumors based on a relative strength of differential expression of one class to the other. The resulting transformation produces an expression profile where the contrast of between the two tumor types is more noticeable than in that using expression values.

To illustrate the behavior of the POE transformation, two examples are graphically presented in Figures 19.3 and 19.4. Figure 19.3 shows the distribution of POE values for the gene Hs.69771, a complement factor that localizes to major histocompatibility complex class III on chromosome 6. The three panels are histograms of raw expression in three studies, and the last panel is the histogram of POE values. This gene is the top ranked gene using POEA but is ranked 83rd in BCMD and does not appear in the ES signature. Figure 19.4 is the same plot for the gene Hs.79088, RCN2. This gene is ranked 82nd in the POEA signature but is 2nd in the ES signature and 19th in the BCMD signature.

After the POE transformation had been applied to each study separately, the datasets were combined into a single dataset. To construct an optimal cutoff point and signature size, we used leave-one-out cross-validation. Supplementary Figure 19.1, available from the companion Website, shows a contour plot of misclassification errors for a wide range of signature size and risk index cutoff pairs in the risk index approach. The error rate is lowest at several points; we chose a signature of size 110 with risk index dichotomized at the 60th percentile was deemed to be the optimal choice.

The POEA signature shares 90 out of 110 genes with BCMD signature, and 19 genes with ES signature. Like the BCMD signature, it tends to select genes that have discordant differential expression across studies. Supplementary Figure 19.2 shows three blocks of genes that show discordance with respect to differential expression. A possible remedy is to filter out the genes whose expression pattern is inconsistent with respect to the other genes in each study using the integrative correlation coefficient [21,24].

Regarding the classification performance of this signature, Figure 19.5 shows the risk index for all samples.

The classifier does not achieve good classification results in the Garber et al. study but successfully identifies the tumor types in the other two studies.

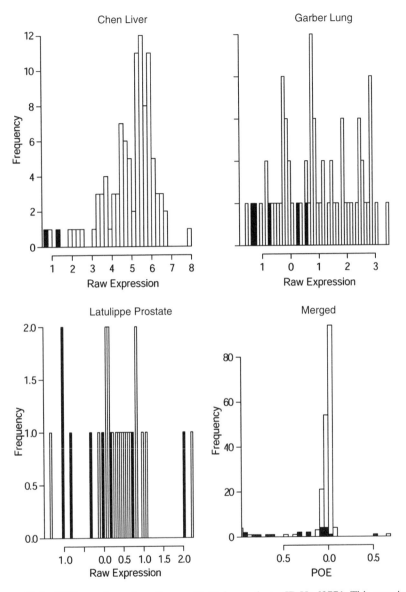

Figure 19.3   POE transformation of gene with Unigene cluster ID Hs.69771. This gene is the top-ranked gene in POEA signature. Unfilled and filled bars mark the expression of primary and metastatic tumors for this gene, respectively.

One novel feature of the POEA method is that it shows the contrast between the two tumor types to be less strong in the Garber et al. data than it is in the other two studies. The ES signature still misclassifies 50% of the metastatic samples in the Garber et al. data (Table 19.2). However, it classifies samples worse in the other two studies relative to the the BCMD and POEA signatures.

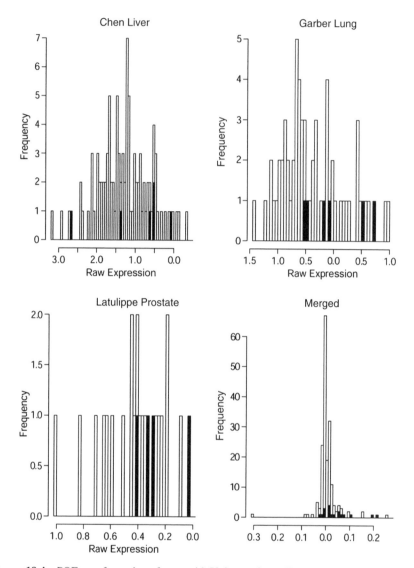

Figure 19.4 POE transformation of gene with Unigene cluster ID Hs.79088. This gene is the second-ranked gene in the ES signature.

19.8 COMPARISON OF THREE METHODS

In this section we characterize the signatures obtained by each method and compare them. Supplementary Figures 19.3–19.5 visualize the signatures using a heatmap. For the BCMD and ES signatures, the raw expression of each gene was centered at the median of primary tumors to highlight the contrast between the two tumor types. For POEA, the modal expression values of primary tumors are centered near zero automatically by the estimation procedure.

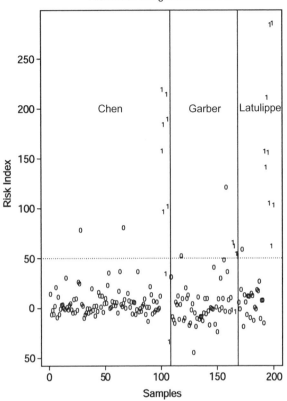

Figure 19.5 Misclassification performance of metasignature of size 110 dichotomizing at the 60th percentile.

19.8.1 Signatures

The BCMD method by Wang et al. [33] has the largest signature size. The ES signature was constructed based on the same rule of FDR being 0.05 or less, which led to a signature size of 90 genes. Finally, all 110 genes in POEA signature are significant predictors in univariate logistic regression with p value < 0.0001 and will thus still be strongly significant after FDR correction.

In addition to the varying sizes of the signatures, the variability of gene expression profiles for the genes included in the signatures is also diverse. In our three-dataset example, the results from the POEA method resemble those of BCMD more than those of the ES method. POEA and BCMD select more genes that display discordant differential expression.

19.8.2 Classification Performance

The classification results do not vary widely among the three methods. All classifiers applied to the Chen et al. and LaTulippe et al. datasets have nearly perfect separation of the two tumor types. In those two studies, ES signature produces relatively weaker separation of risk

Table 19.2   Misclassification Rates Using Classifiers Obtained from BCMD, ES, and POEA[a]

Method	Chen (99/9)	Garber (54/6)	LaTulippe (23/9)	Remarks
BCMD (286 genes)	0/0	4/3	0/0	Raw scale
ES (90 genes)	8/0	3/3	0/2	Raw scale
POEA (110 genes)	2/0	1/6	0/0	POE scale

[a]Each entry lists the number of misclassified primary tumors/metastatic tumors.

indices (plot not shown). Table 19.2 tabulates the misclassification of each signature across the three studies.

19.8.3   Directionality of Differential Expression

Although the ES method selects a signature with genes displaying concordant differential expression, it is highly sensitive to the heterogeneity of of expression levels across studies. The heterogeneity is quantified in Cochran's Q statistic, which follows an approximately χ^2_{k-1} distribution under the null hypothesis of absence of heterogeneity, with k being the number of studies included. A large Q statistic indicates substantial heterogeneity in expression level and variability across studies.

The ES signature keeps concordant differentially expressed genes at the cost of losing highly significant genes with discordant expression, while the BCMD and POEA signatures appear to behave in an opposite manner. One remedy is to apply the integrative correlation coefficient [21,24] as a filter. One can use it to filter genes out in the beginning of the analysis, or apply it on the chosen signature. Since throwing out genes will reduce information for the POE transformation, we take the former approach and obtain new signatures for the POEA method. We eliminated the genes whose average integrative correlation across three studies is less than 0.1 and reanalyzed the data using all three methods. Using this criterion leaves 884 out of 1633 genes.

The correlation filter results in better agreement among the three methods. Many of the genes with discordant differential expression across studies disappear from the signatures of the three methods. This can be seen in Supplementary Figures 19.3–19.5. Note that there remains a large block of discordant genes in the BCMD signature. The classification performance of the POEA signature is much better than the previous signature in the Garber et al. study, which shows the impact of integrative correlation filter.

19.9   CONCLUSIONS

In this chapter, we have compared three approaches to the analysis of microarray data from multiple studies. One criterion for their use is ease of implementation. The ES method was implemented using the GeneMeta package available at the Bioconductor Project Website (http://www.bioconductor.org). The BCMD and POEA methods are available in a package that we developed, called MetaArray, which is also available through Bioconductor.

For the example presented here, we found that all methods gave very similar classification performance. The major reason for this is the fact that the phenotype under consideration (metastatic vs. nonmetastatic) is very visibly different. In addition, the number of candidate genes

(1633) is much larger than the sample size (200). Thus, because of the gross phenotypic difference, several gene signatures will provide good classification. This was seen in breast cancer recently by Ein-Dor et al. [9]. However, the genes selected by the various methods had some variation. In particular, there was more concordance between the BCMD and POEA methods than with the ES signature. Concordance increased with the use of the integrative correlation coefficient. We do expect that for more subtle phenotypic changes, the BCMD and POEA will provide better classification than will the ES approach. However, we should also note that because metastatic cancers are heterogeneous, the nonoverlapping gene lists might also be more a reflection of the biological heterogeneity in the samples.

More broadly, we have attempted to describe a problem in genomic data analysis that we expect to become more common in the future, that of genomic data integration. With the development of new large-scale functional genomic technologies in full bloom, how to combine these datasets is an important analytical problem. This is an enterprise that will require substantial statistical input and is an area open to further development.

ACKNOWLEDGMENT

This research has been supported by grant GM72007 from the Joint DMS/DBS/NIGMS Biological Mathematics Program.

REFERENCES

1. Beer, D. G., Kardia, S. L., Huang, C. C., Giordano, T. J., Levin, A. M., Misek, D. E., Lin, L., Chen, G., Gharib, T. G., Thomas, D. G., Lizyness, M. L., Kuick, R., Hayasaka, S., Taylor, J. M., Iannettoni, M. D., Orringer, M. B., and Hanash, S., Gene-expression profiles predict survival of patients with lung adenocarcinoma, *Nature Med.* **8**, 816–824 (2002).

2. Benjamini, Y. and Hochberg, Y., Controlling the false discovery rate: A practical and powerful approach to multiple testing. *J. Roy. Statist. Soc. B* **57**, 289–300 (1995).

3. Box, G. E. P. and Tiao, G. C., *Bayesian Inference in Statistical Analysis*, Wiley, New York, (1973).

4. Carter, S. L., Eklund, A. C., Mecham, B. H., Kohane, I. S., and Szallasi, Z., Redefinition of Affymetrix probe sets by sequence overlap with cDNA microarray probes reduces cross-platform inconsistencies in cancer-associated gene expression measurements, *BMC Bioinformatics* **6**, 107 (2005).

5. Chen, X., Cheung, S. T., So, S., Fan, S. T., Barry, C., Higgins, J., Lai, K. M., Ji, J., Dudoit, S., Ng, I. O., Van De Rijn, M., Botstein, D., and Brown, P. O., Gene expression patterns in human liver cancers, *Molec. Bio. Cell* **13**, 1929–1939 (2002).

6. Choi, J. K., Yu, U., Kim, S., and Yoo, O. J., Combining multiple microarray studies and modeling interstudy variation, *Bioinformatics* **19**, 84–90 (2003).

7. Dai, M., Wang, P., Boyd, A. D., Kostov, G., Athey, B., Jones, E. G., Bunney, W. E., Myers, R. M., Speed, T. P., Akil, H., Watson, S. J., and Meng, F., Evolving gene/transcript definitions significantly alter the interpretation of GeneChip data, *Nucleic Acids Res.* **33**, e175 (2005).

8. Diehn, M., Sherlock, G., Binkley, G., Jin, H., Matese, J. C., Hernandez-Boussard, T., Rees, C. A., Cherry, J. M., Botstein, D., Brown, P. O., and Alizadeh, A. A., SOURCE: A unified genomic resource of functional annotations, ontologies, and gene expression data, *Nucleic Acids Research* **31**, 219–223 (2003).

9. Ein-Dor, L., Kela, I., Getz, G., Givol, D., and Domany, E., Outcome signature genes in breast cancer: Is there a unique set?, *Bioinformatics* **21**, 171–178 (2005).

10. Garber, M. E., Troyanskaya, O. G., Schluens, K., Petersen, S., Thaesler, Z., Pacyna-Gengelbach, M., van de Rijn, M., Rosen, G. D., Perou, C. M., Whyte, R. I., Altman, R. B., Brown, P. O., Botstein, D., and Petersen, I., Diversity of gene expression in adeno-carcinoma of the lung, *Proc. Nat. Acad. Sci. USA* **98**, 13784–13789 (2001).

11. Ghosh, D., Barette, T. R., Rhodes, D., and Chinnaiyan, A. M., Statistical issues and methods for meta-analysis of microarray data: A case study in prostate cancer, *Funct. Integrat. Genom.* **3**, 180–188 (2003).

12. Hedges, L. V. and Olkin, I., *Statistical Method for Meta-Analysis*, Academic Press, New York, 1985.

13. Ideker, T., Galitski, T., and Hood, L., A new approach to decoding life: Systems biology. *Annual Review of Genom. Hum. Genet.* **2**, 343–372 (2001).

14. International Human Genome Sequencing Consortium, Initial sequencing and analysis of the human genome, *Nature* **409**, 860–921 (2001).

15. Irizarry, R. A., Hobbs, B., Collin, F., Beazer-Barclay, Y. D., Antonellis, K. J., Scherf, U., and Speed, T. P., Exploration, normalization, and summaries of high density oligonucleo-tide array probe level data, *Biostatistics* **4**, 249–264 (2003).

16. Irizarry, R. A., Warren, D., Spencer, F., Kim, I. F., Biswal, S., Frank, B. C., Gabrielson, E., Garcia, J. G., Geohegan, J., Germino, G., Griffin, C., Hilmer, S. C., Hoffman, E., Jedlicka, A. E., Kawasaki, E., Martinez-Murillo, F., Morsberger, L., Lee, H., Petersen, D., Quackenbush, J., Scott, A., Wilson, M., Yang, Y., Ye, S. Q., and Yu, W., Multiple-laboratory comparison of microarray platforms. *Nature Meth.* **2**, 345–350 (2005).

17. Jiang, H., Deng, Y., Chen, H. S., Tao, L., Sha, Q., Chen, J., Tsai, C. J., and Zhang, S., Joint analysis of two microarray gene-expression data sets to select lung adenocarcinoma marker genes, *BMC Bioinformatics* **5**, 81 (2004).

18. Kuo, W. P., Jenssen, T. K., Butte, A. J., Ohno-Machado, L., and Kohane, I. S., Analysis of matched mRNA measurements from two different microarray technologies, *Bioinformatics* **18**, 405–412 (2002).

19. LaTulippe, E., Satagopan, J., Smith, A., Scher, H., Scardino, P., Reuter, V., and Gerald, W. L., Comprehensive gene expression analysis of prostate cancer reveals distinct tran-scriptional programs associated with metastatic disease, *Cancer Res.* **62**, 4499–4506 (2002).

20. Lee, H. K., Hsu, A. K., Sajdak, J., Qin, J., and Pavlidis, P., Coexpression analysis of human genes across many microarray data sets, *Genome Res.* **14**, 1085–1094 (2004).

21. Lee, J. K., Bussey, K. J., Gwadry, F. G., Reinhold, W., Riddick, G., Pelletier, S. L., Nishizuka, S., Szakacs, G., Annereau, J. P., Shankavaram, U., Lababidi, S., Smith, L. H., Gottesman, M. M., and Weinstein, J. N., Comparing cDNA and oligonucleotide array data: Concordance of gene expression across platforms for the NCI-60 cancer cells, *Genome Bio.* **4**, R82 (2003).

22. Normand, S. L., Meta-analysis: Formulating, evaluating, combining and reporting, *Statist. Med.* **18**, 321–359 (1999).

23. Parmigiani, G., Garrett, E. S., Anbazhagan, R., and Gabrielson, E., A statistical framework for molecular-based classification in cancer, *J. Roy. Statist. Soc. Ser. B* **64**, 717–736 (2002).

24. Parmigiani, G., Garrett-Mayer, E. S., Anbazhagan, R., and Gabrielson, E., A cross-study comparison of gene expression studies for the molecular classification of lung cancer, *Clin. Cancer Res.* **10**, 2922–2927 (2004).

25. Rhodes, D., Barrette, T. R., Rubin, M. A., Ghosh, D., and Chinnaiyan, A. M., Meta-analysis of microarrays: Interstudy validation of gene expression profiles reveals pathway dysregulation in prostate cancer, *Cancer Res.* **62**, 4427–4433 (2002).

26. Rhodes, D. R., Yu, J., Shanker, K., Deshpande, N., Varambally, R., Ghosh, D., Barrette, T., Pandey, A., and Chinnaiyan, A. M., Large-scale meta-analysis of cancer microarray data identifies common transcriptional profiles of neoplastic transformation and progression, *Proc. Nat. Acad. Sci. USA* **101**, 9309–9314 (2004).

27. Shen, R., Ghosh, D., and Chinnaiyan, A. M., Prognostic meta-signature of breast cancer developed by two-stage mixture modeling of microarray data, *BMC Genom.* **5**, 94 (2004).

28. Shi, L., Tong, W., Fang, H., Scherf, U., Han, J., Puri, R. K., Frueh, F. W., Goodsaid, F. M., Guo, L., Su, Z., Han, T., Fuscoe, J. C., Xu, Z. A., Patterson, T. A., Hong, H., Xie, Q., Perkins, R. G., Chen, J. J., and Casciano, D. A., Cross-platform comparability of microarray technology: Intra-platform consistency and appropriate data analysis procedures are essential, *BMC Bioinformatics* **6** Suppl 2, S12 (2005).

29. Simon, R., When is a genomic classifier ready for prime time?, *Nature Clin. Practice Oncol.* **1**, 4–5 (2004).

30. Sorlie, T., Tibshirani, R., Parker, J., Hastie, T., Marron, J. S., Nobel, A., Deng, S., Johnsen, H., Pesich, R., Geisler, S., Demeter, J., Perou, C. M., Lonning, P. E., Brown, P. O., Borresen-Dale, A. L., and Botstein, D., Repeated observation of breast tumor subtypes in independent gene expression data sets, *Proc. Natl. Acad. Sci. USA* **100**, 8418–8423 (2003).

31. Uetz, P., Giot, L., Cagney, G., Mansfield, T. A., Judson, R. S., Knight, J. R., Lockshon, D., Narayan, V., Srinivasan, M., Pochart, P., Qureshi-Emili, A., Li, Y., Godwin, B., Conover, D., Kalbfleisch, T., Vijayadamodar, G., Yang, M., Johnston, M., Fields, S., and Rothberg, J. M., A comprehensive analysis of protein-protein interactions in Saccharomyces cerevisiae, *Nature* **403**, 623–627 (2000).

32. Venter, J. C., Adams, M., Myers, E. W., Li, P. W. et al., The sequence of the human genome, *Science* **291**, 1304–1351 (2001).

33. Wang, J., Coombes, K. R., Highsmith, W. E., Keating, M. J., and Abruzzo, L. V., Differences in gene expression between B-cell chronic lymphocytic leukemia and normal B cells: A meta-analysis of three microarray studies, *Bioinformatics* **20**, 3166–3178 (2004).

34. Wright, G., Tan, B., Rosenwald, A., Hurt, E. H., Wiestner, A., and Staudt, L. M., A gene expression-based method to diagnose clinically distinct subgroups of diffuse large B cell lymphoma, *Proc. Nat. Acad. Sci. USA* **100**, 9991–9996 (2003).

35. Yan, P. S., Chen, C. M., Shi, H., Rahmatpanah, F., Wei, S. H., Caldwell, C. W., and Huang, T. H., Dissecting complex epigenetic alterations in breast cancer using CpG island microarrays, *Cancer Res.* **61**, 8375–8380 (2001).

C H A P T E R 20

Statistical Methods for Identifying Differentially Expressed Genes in Replicated Microarray Experiments: A Review

Lynn Kuo, Fang Yu, and Yifang Zhao
Department of Statistics, University of Connecticut, Storrs, Connecticut

20.1 INTRODUCTION

A common task in microarray analysis is to compare the expression levels of thousands of genes in samples collected under two different conditions to determine which genes are upregulated, downregulated, or unchanged in their expression levels. Research on the statistical methods for microarray analysis has been very active recently. Too many genes and not enough replications have also complicated the problem. Give the huge body of literature, we list only a few of the most relevant works in our study. Dudoit et al. [12], Cui et al. [7], and Pan [28] provide easy overviews on some of the methods. Tusher et al. [39] and Storey and Tibshirani [36] discuss significance analysis of microarrays (SAM). Benjamini and Hochberg [2], Westfall and Young [42], and Dudoit et al. [11] discuss issues on multiple comparisons. More in-depth developments in multiple testing can be found elsewhere in the literature [10,40,41] Efron and Tibshirani [17] provide an empirical Bayes justification for the Benjamini–Hochberg procedure of controlling the false discovery rate (FDR). Efron [13–16] provides a gene-specific measure called *local FDR* to bound the global FDR and to estimate the false negative rate. He also introduces the concept of empirical null distribution that may be more dispersed than the usual theoretical null distribution. It is particularly applicable when there are correlations across arrays or across genes. Lönnstedt and Speed [25] provide an alternative empirical Bayes method that combines information across genes based on odds ratios for detecting differentially expressed (DE) genes. Wright and Simon [45]

Statistical Advances in the Biomedical Sciences, edited by Atanu Biswas, Sujay Datta,
Jason P. Fine, and Mark R. Segal
Copyright © 2008 John Wiley & Sons, Inc.

propose a model where the gene variances are drawn from an inverse gamma distribution, whose parameters are estimated across all genes. Microarray ANOVA methods and linear regression methods can be found in Kerr et al. [22], Wolfinger et al. [44], Smyth [31], and Gottardo et al. [19,20]. McLachlan et al. [26] propose a simple two-component normal mixture model for a transformed score for detecting DE genes. Yu et al. [51] propose calibrated Bayes factor methods for selecting DE genes. The books by Speed [33], Wit and McClure [43], and Lee [24] provide more details on microarray data analysis.

Historically, the fold change method was used by Schena et al. [30], DeRisi et al. [8], Chen et al. [4], and Draghici [9] for identifying DE genes. The method calls for a gene to be upregulated (down-regulated) if, for example, the mean fold change ratio for the treatment over control is at least 2 (at most $\frac{1}{2}$). This method is not adopted any more, because it ignores the variability of the expression levels over replicates, it is biased toward genes with low expression levels, and it provides no assessment of false-positive rates. Numerous alternative methods have been studied. They include various versions of the two-sample t test, two-sample nonparametric test, hierarchical and Bayesian versions of the two-sample comparison, ANOVA, generalized linear models, and principal-component reduction techniques. Practitioners are often faced with the question of which method to adopt. In this chapter, we focus on five methods: the Benjamini–Hochberg (BH) procedure [1] applied to the two-sample t tests (abbreviated BH-T), the significance analysis of microarray (SAM) by Tusher et al. [39], the semiparametric hierarchical (SPH) method by Newton et al. [27], the linear models for microarray (LIMMA) by Smyth [31], and the microarray ANOVA (MAANOVA) by Cui et al. [6]. The primary reason for selecting these five methods is the ease in implementing them. The software programs are available via the Internet, and the methods can be implemented and interpreted fairly easily by practitioners. We will review each method in detail, and evaluate the performance of these methods with reference to two simulated data cases.

The statistical procedures here consist of determining a test statistic, a rejection region (a cutoff value or values), calling all the genes in the rejection region to be upregulated (downregulated, or expressed). Finally, the procedures assess the rate of false discovery (FDR).

For each gene g, we are testing $H_{g,0}$: gene g is EE (equivalently expressed), versus $H_{g,a}$: gene g is DE. Similarly, to discover upregulated (downregulated) genes, the $H_{g,0}$ versus $H_{g,a}$ are changed to nonupregulated (nondownregulated) versus upregulated (downregulated) for gene g. For the rejection region, let us first define two concepts: (1) the familywise error rate (FWER) that is the probability of at least one false positive in the simultaneous testing of G genes; and (2) the false discovery rate (FDR) that is the false-positive rate among the rejected hypotheses. It is a less stringent criterion than the FWER. In Kuo et al. [23], we review two FWER controlling methods: one uses the Westfall–Young [42] stepdown max-T permutation adjusted p values; the other, called the cyber-T method, uses a Bayesian version of the two-sample t procedure and the Bonferroni procedure for adjusting p values. In this chapter, we are focusing on the five methods mentioned earlier that restrict the estimated FDR to be $\leq\alpha$. The BH-T method estimates this by using Benjamini and Hochberg adjusted p values computed from two-sample t statistics. SAM controls the median of the estimated FDR, and MAANOVA controls the estimated FDR. Both achieve this by bootstrap methods via the null distribution that the gene is hypothesized to be EE. The SPH method calculates a Bayesian version of the FDR based on the posterior distribution. The LIMMA procedure estimates the FDR using the BH adjustment of its p value calculated from its modified t test statistics.

On the test statistics, BH-T uses the two-sample t statistics for each gene. The SAM method uses a modified two-sample t statistics for each gene with the standard error corrected by a more stable variance estimate incorporating information from all the genes; SPH is an empirical Bayes procedure with a hierarchical structure. The level 1 model consists of an individual gene-specific gamma model that is used to explain replicated experiments for each gene. The level 2 model is the common population average model for each unknown latent pair of mean intensities with the first coordinate for the control and the second coordinate for the treatment. Newton et al. [27] assume a mixture distribution with three components (corresponding to upregulated, downregulated, and nonregulated genes) with each component consisting of a nonparametric bivariate density. Both mixing weights and nonparametric densities are estimated from the data. The test statistics are the empirical Bayes estimates of the posterior probabilities for each gene to be upregulated, downregulated, or nonregulated. Both MAANOVA and LIMMA use generalized linear models on simultaneously testing many contrasts to select DE genes. The LIMMA, SPH, and MAANOVA methods are all hierarchical in nature, each with a baseline density assumed for all the latent mean intensities for each gene. Therefore, they allow a mechanism for sharing information among genes. The other two methods are not hierarchical: BH-T makes decisions from the marginal information, that is, just the information for each gene; and the SAM "fudge" correction factor is determined from all the genes, so the link that combines information across genes is subtle, not as direct as the hierarchical structures in the other three methods. The only method restricted to two independent samples for comparison is SPH; all the other four methods allow paired-t comparisons. Moreover, MAANOVA and LIMMA allow for more complicated designs, including loop designs.

Performance evaluation for the five methods is based on simulated data, and we use the receiver operating characteristics (ROC) curves that plot sensitivity versus (1-specificity) for each method. The sensitivity is the probability of correctly calling a DE gene. The specificity is the probability of correctly calling an EE gene. A good procedure should have large partial area under the curve, and it should have large sensitivity for small (1- specificity) up to a bounded limit. The partial area is desired because investigators are interested only in genes exhibiting strong evidence of differential expression. In addition to the ROC curves that evaluate the five methods using frequentist measures, we evaluate them in terms of their predictive power. As an analog to the ROC curve, for each method, we plot the positive predictive value (PPV) against the negative predictive value (1-NPV; NPV scenario in this case). The better method should have larger partial area under the curve. The PPV is the conditional probability of a gene being DE given that it is claimed to be DE, and NPV is the conditional probability of a gene being EE given that it is claimed to be EE.

We simulated two data cases. Both mimic a four-array, two-channel dye-swap experiment comparing 5000-gene expressions in control and treatment conditions. The array effect on the log intensity is a random draw from a normal distribution. So is the dye effect. The two simulation scenarios differ mainly in how the treatment effects are simulated and which distribution is used to describe the variation of intensities. For gene expression intensities, scenario 1 uses gamma distributions while scenario 2 utilizes lognormal distributions. Regarding generation of treatment means, scenario 1 uses inverse gamma distributions for the raw treatment means. Scenario 2 simulates the $\log_2$ treatment means through its relationship with the log ratio intensities M and the log average intensities A, with M and A simulated from two uniform distributions, respectively. In each case, we consider both 5% and 10% of DE prevalence; that is, we simulate 250 and 500 DE genes. Among the four F-like statistics in the MAANOVA package, we use F_S only, because it is most robust and has a good power. In the first simulation study, we compare the five methods by varying target FDR levels from 0.005 to 0.655 with an increment of 0.05. The results in both ROC and predictive power comparison given in Figures 20.3–20.3 indicate that the direct or indirect information-sharing methods, SPH, SAM, LIMMA, and F_S, are superior to the BH-T procedure. In scenario 1, SPH performs slightly better than do SAM, LIMMA, and F_S, but vice versa for scenario 2.

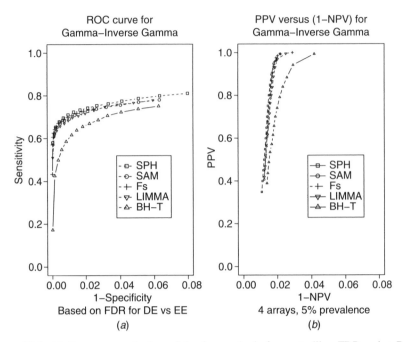

Figure 20.1 Performance evaluation of the five methods for controlling FDR, using ROC curves and predictive power curves for scenario 1 simulated data with 5% prevalence of DE genes.

Instead of controlling FDR, the second simulation study evaluates these methods by fixing the same number of claimed DE genes, for instance, 50–250 with an increment of 50, and 300–1000 with an increment of 100 for claimed DE genes. In this case, we are essentially controlling cFDR [38], which is defined as the FDR conditioning on the observed number of rejections. The BH-T method reduces to the two-sample t tests and all methods are performed without the FDR adjustments. Figures 20.7 and 20.8 display the evaluations for the two simulated data cases with 5% DE genes. We observe results very similar to those for controlling FDR. The procedure of using two-sample t tests to rank and select DE genes is inferior to those in the other four methods, because the latter methods take the advantage of hierarchical structures directly or indirectly to borrow information across genes.

We also provide mean MA plots in Figures 20.4–20.6 to give a snapshot of selected DE genes for each controlling FDR method. As we increase the FDR, more DE genes are called by each method. The BH-T method is conservative in that it tends to select the smallest number of DE genes.

20.2   NORMALIZATION

There are several computational issues associated with microarray analysis. The first is to process the data to produce a result that reflects either absolute amounts of transcripts in cells or the ratios of these amounts under two different experimental conditions. This process is called the *normalization* that adjusts data to remove bias from dye labeling or other instrumental errors. The second process involves interpretation of the intensity data to

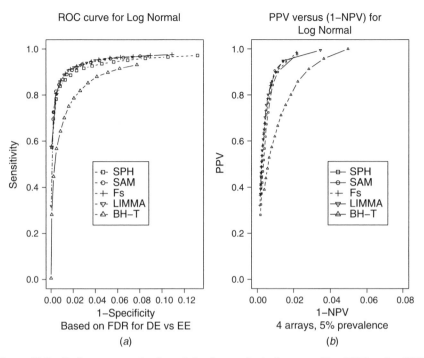

Figure 20.2 Performance evaluation of the five methods for controlling FDR, using ROC curves and predictive power curves for scenario 2 with 5% prevalence of DE genes.

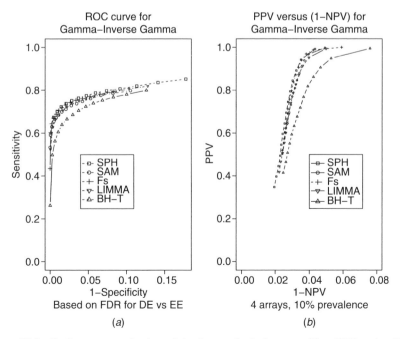

Figure 20.3 Performance evaluation of the five methods for controlling FDR, using ROC curves and predictive power curves for scenario 1 with 10% prevalence of DE genes.

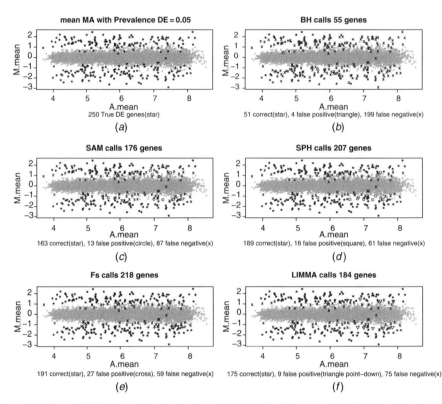

Figure 20.4 Display of simulated truly DE genes and selected DE genes by each method for controlling FDR at 0.05 for scenario 2 with 5% prevalence of DE genes: (*a*) black stars represent truly 250 DE genes; (*b–f*) black stars indicate correctly selected genes; X symbols denote false negative; other symbols, false positive.

provide biological insight. It includes statistical methods for selecting DE genes, clustering, principal-component analysis, and prediction. We will discuss normalization first in this section, and then some statistical methods for selecting DE genes in the next section.

Suppose that we have equal concentrations of each species and no dye bias or other systematic errors; we would expect a plot of Cy5 intensity versus Cy3 intensity for spots in a microarray to fall on a straight line having a slope of 1.0. In fact, Cy5 intensities are systematically lower than Cy3 intensities when equal amounts of sample are present. If we indicate Cy5 intensities by R ("red") and Cy3 intensities by G ("green"), a regression of R against G would produce a slope k below 1.0. If the regression is linear, then $R = kG$, for $k < 1.0$. So, to correct the R values, we would multiply R by $1/k$; that is, $R_{corrected} = k^{-1}R$. This is called a *global normalization*. The constant k can be estimated by linear regression of R against G for all the genes. This is valid because most genes are not DE, and among the small number of genes that are DE, roughly the same number of genes will be upregulated as downregulated.

However, it often occurs that the dye bias in $\log_2(R/G)$ is not constant, but varies as a function of intensity. An *MA* plot [49] that plots $M = \log_2(R/G) = \log_2 R - \log_2 G$ versus $A = (\log_2 R + \log_2 G)/2$ would usually show this. So it calls for an intensity-dependent

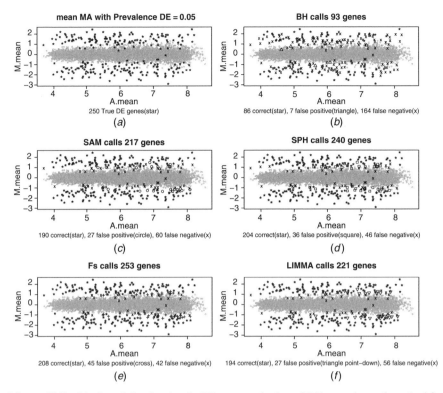

Figure 20.5 Display of simulated truly DE genes and selected DE genes by each method for controlling FDR at 0.01 for scenario 2 with 5% prevalence of DE genes: (*a*) black stars represent truly 250 DE genes; (*b–f*) black stars indicate correctly selected genes; X symbols denote, false negative; other symbols, false positive.

normalization. We can use locally weighted scatterplot smoother (lowess) or locally polynomial scatterplot smoother (loess) to correct the *M* values. After the intensity-dependent normalization, we would expect the average value of the *M* values to be 0.0 for all pairs of intensities with the same *A* value.

The dependence of log ratios on fluorescent intensities is often depicted in *MA* plots as either a curvature trend or a large variance of *M* values at the low- or high-intensity regions. MAANOVA includes curvature-adjusting methods such as shift and intensity-based lowess, linlog (log-linear) transformation for stabilizing variance, linlogshift (log-linear with shift) transformation for both straightening curvature, and stabilizing variance as discussed in Kerr et al. [22] and Cui et al. [6]. The shift method adds a constant to the intensities of one channel and subtracts the same constant from the intensities in the other channel prior to the logarithmic transformation. This constant is chosen so that the absolute deviation of each log ratio from the median log ratio of the array is minimized. The intensity-based lowess method corrects the *M* values by using the residuals from the fitted values of lowess or loess on an *MA* plot. Since linear transformation works better for low intensities and logarithmic transformation is more appropriate for high intensities, linlog transformation combines them. Below the channel-specific intensity cutoff, the linlog transformation is linear, while

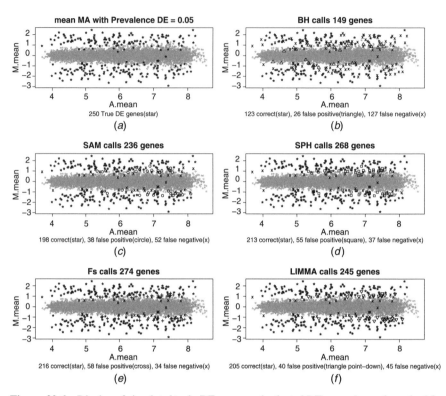

Figure 20.6 Display of simulated truly DE genes and selected DE genes by each method for controlling FDR at 0.15 for scenario 2 with 5% prevalence of DE genes: (*a*) black stars represent truly 250 DE genes; (*b–f*) black stars indicate correctly selected genes; in star, X symbols denote false negative; other symbols, false positive.

above this cutoff it is logarithmic. The value of the cutoff is chosen to minimize the absolute deviation of the variance of M values in each bin from the median variance in an MA plot. Linlogshift is a combination of linlog and shift methods. MAANOVA also incorporates the joint-lowess method to remove spatial bias due to lack of uniformity in array printing or hybridization. The joint-lowess transformation modifies the intensity-based lowess method by adding the grid locations as predictors in the lowess or loess fit on an MA plot.

Different from MAANOVA, LIMMA provides normalization not only within an array but also between arrays. The normalization within arrays will normalize the expression log ratios M so that the M values average to zero within each array. The normalization between arrays will render the expression data comparable across different arrays.

There are two different between-array normalization options in the LIMMA package: the scale normalization [49,50,32] and the quantile normalization [3,50]. The scale normalization stems from the idea of scaling the M values to ensure the same median absolute deviation (MAD) across arrays. The quantile normalization normalizes the M values from each array to end up with the same distribution.

The LIMMA package also includes numerous methods for normalization within arrays [49,50,32]. The median normalization subtracts the weighted median from the M values for

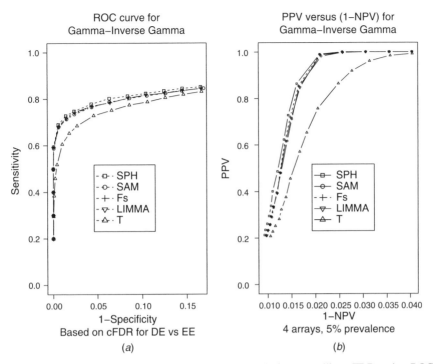

Figure 20.7 Performance evaluation of the five methods for controlling cFDR, using ROC curves and predictive power curves for scenario 1 with 5% prevalence of DE genes.

each array. The loess normalization, the print-tip loess normalization, and the composite normalization are three different loess-based normalization methods, which use the residuals from the fitted M values of the loess curves. The loess normalization works similarly to the intensity-based lowess normalization under the MAANOVA package. The print-tip loess normalization fits loess curves for each print-tip group. The composite normalization uses a compromise between the loess curves fit for each print-tip group and the global MSP (specially designed microarray sample pool) titration series curve. The purpose of the composite loess is to conduct normalization based on the global MSP curves rather than the individual tip-group curves when the intensities are low and unreliable.

The BH-T, SAM, and SPH methods assume that data have been normalized and converted to gene expression matrix with foci on selecting DE genes.

20.3 METHODS FOR SELECTING DIFFERENTIALLY EXPRESSED GENES

In this section, we review the five statistical methods for determining DE genes.

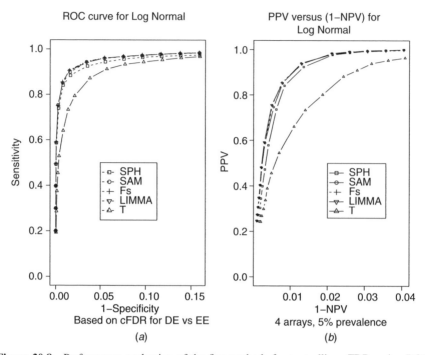

Figure 20.8   Performance evaluation of the five methods for controlling cFDR, using ROC curves and predictive power curves for scenario 2 with 5% prevalence of DE genes.

20.3.1   BH-T

Let $\mathbf{y}_{g,c} = (y_{g,1}, \ldots, y_{g,m})$ denote the normalized log intensities of the m replicates for gene g in the control condition, and let $\mathbf{y}_{g,t} = (y_{g,m+1}, \ldots, y_{g,m+n})$ represent that of the n replicates of the treatment condition. Typically, $m = n$, and $(y_{g,j}, y_{g,m+j})$ for $j = 1, \ldots, m$ is the pair of (R, G) or (G, R) discussed earlier. So the gth row of the gene expression matrix consists of $\mathbf{y}_{g,c}$ and $\mathbf{y}_{g,t}$, a vector of $m + n$ dimension of the log intensities for the replicated experiment. We compute the t statistic from each row:

$$t_g = \frac{\bar{y}_{g,t} - \bar{y}_{g,c}}{\left(\frac{s_{g,t}^2}{n} + \frac{s_{g,c}^2}{m}\right)^{1/2}},\tag{20.1}$$

where $\bar{y}_{g,c}$ denotes the sample mean of $\mathbf{y}_{g,c}$ and $s_{g,c}^2$ denotes the sample variance of $\mathbf{y}_{g,c}$. Similarly for $\mathbf{y}_{g,t}$.

To select DE (versus EE) genes, we first evaluate the p value associated with test statistics $|t_g|$ obtained from (20.1) for the two-sided t test. For discovering the up-regulated genes, we evaluate the p value by $\Pr(T > t_g)$. Similarly, for discovering the downregulated genes, we evaluate the p value by $\Pr(T < t_g)$.

Now we describe the BH procedure. We first order the p values so that $p_{r_1} \leq p_{r_1} \cdots \leq p_{r_G}$. Let k be the largest integer i for which $p_{r_i} \leq \frac{i}{G}\alpha$ for all i. Then we declare all the genes with labels $r_1, \ldots, r_k$ to be DE.

Note that the value $p_{r_i}\frac{G}{i}$ for each i is referred to as the *adjusted p value*. Benjamini and Hochberg [1] used a sequential p-value method so that on the average FDR $\leq \alpha$ for some pre-specified α. However, the BH method controls FDR at a level too low by a factor of G_0/G, where G_0 and G are the number of truly EE genes and the total number of genes, respectively. The adaptive BH procedure [2] incorporates an estimate of G_0 into their 1995 procedure [1], and uses $\alpha^* = \alpha(\hat{G}_0/G)$ instead of α to gain more power. So the adaptive procedure performs better only by using the difference between G_0/G and 1. As prevalence of DE increases, the adaptive procedure will offer a more detectable advantage.

Let us note that both the BH and adaptive BH methods can be applied to certain dependent test statistics and p values, including the modified t statistics in LIMMA and the F statistics in MAANOVA.

20.3.2 SAM

The input to SAM [5] is a set of gene expression measurements from a set of microarray experiments, as well as a response variable from each experiment. The response variable may be a grouping such as untreated, treated (either unpaired or paired), a multiclass grouping (e.g., breast cancer, lymphoma, colon cancer), a quantitative variable (e.g., blood pressure), or a possibly censored survival time. The SAM method computes a statistics d_g for each gene g. It uses repeated permutations of the group labels to determine whether the expression of any gene is significantly related to the response. The cutoff for the significance is determined by a tuning parameter Δ chosen by the user on the basis of the false-positive rate. The SAM procedure can also choose a fold change parameter to ensure that the called genes are with fold-changes at least at a prespecified amount. In this chapter, we are considering only the unpaired two-group comparison.

For each gene, we compute the d statistics

$$d_g = \frac{\bar{y}_{g,t} - \bar{y}_{g,c}}{s_g + s_0}, \tag{20.2}$$

with
$$s_g = \left(\frac{1}{m} + \frac{1}{n}\right)^{1/2} \left(\frac{(m-1)s_{g,c}^2 + (n-1)s_{g,t}^2}{m+n-2}\right)^{1/2},$$

where s_0 is a "fudge" factor that is chosen as a percentile of the s_g values in order to make the coefficient of variation (CV) of d_g approximately constant as a function of s_g. It has the effect of dampening large values of d_g that arise from genes whose expressions are near zero.

The SAM method can be summarized in the following steps:

1. Compute the order statistics $d_{r_1} \leq d_{r_2} \leq \cdots \leq d_{r_G}$.
2. Take B sets of permutation of the expression levels, where each permutation b consists of relabeling the two conditions for the columns of the gene expression matrix. Then compute the corresponding order statistics $d_{r_1}^b \leq d_{r_2}^b \leq \cdots \leq d_{r_G}^b$. From the set of B permutations, estimate the expected order statistics by $\bar{d}_{r_g} = \frac{1}{B}\sum_b d_{r_g}^b$ for $g = 1, 2, \ldots, G$.
3. Plot d_{r_g} versus $\bar{d}_{r_g}$ for each g. For a fixed threshold Δ, moving up to the right, finding the first index such that $d_{r_g} - \bar{d}_{r_g} > \Delta$. Then call all the genes with larger d to be upregulated. This d is called the *cutup point*, denoted by C_u. Similarly, move down to the

left, find the first index such that $\bar{d}_{r_g} - d_{r_g} > \Delta$. Then call all the genes with smaller d to be downregulated. This d is called the *cutdown point*, denoted by C_d. If $C_d > C_u$, then set $C_d = C_u = 0$.

4. Estimate the FDR in two steps by the permutation method. For a fixed grid of Δ values, SAM first computes the total number of DE genes and the median number of falsely called genes by computing the median number of values among each of the B sets of d^b that fall above the C_u or below the C_d. This estimated FDR number is computed treating all the genes as not expressed, so SAM further adjusts this number downward by estimating the total number of genes that are truly not expressed.

20.3.3   SPH

Unlike the other methods modeling the log intensities, the SPH method models the intensities on the raw scale by gamma distributions. This is referred to as a *level 1 gene-specific model* that explains the variations within the replicated experiments for each gene. Specifically, the replicates for the control $\mathbf{y}_{g,c} = (y_{g,1}, \ldots, y_{g,m})$ are considered as a random sample from a gamma distribution with shape parameter a_c and rate parameter $\frac{a_c}{\mu_{g,c}}$, so that $E(y_{g,j}) = \mu_{g,c}$ for $j = 1, \ldots, m$. Likewise, the replicates in the treatment, $\mathbf{y}_{g,t} = (y_{g,m+1}, \ldots, y_{g,m+n})$, form a random sample from a gamma distribution with shape a_t and rate $\frac{a_t}{\mu_{g,t}}$. The SPH method classifies the latent gene expression profiles by the following three hypotheses: $H_{g,0} : \mu_{g,c} = \mu_{g,t}$, $H_{g,1} : \mu_{g,c} < \mu_{g,t}$, and $H_{g,2} : \mu_{g,c} > \mu_{g,t}$, respectively. They are respectively called equivalently expressed, underexpressed, and overexpressed in the first condition (for control). The joint density f of $(\mu_{g,c}, \mu_{g,t})$ for all g, referred to as the *baseline density*, is assumed to be

$$f(\mu_{g,c}, \mu_{g,t}) = p_0 f_0(\mu_{g,c}, \mu_{g,t}) + p_1 f_1(\mu_{g,c}, \mu_{g,t}) + p_2 f_2(\mu_{g,c}, \mu_{g,t}),$$

where f_0, f_1, and f_2 are the densities of μ values within the preceding hypotheses $H_{g,0}, H_{g,1}$, and $H_{g,2}$, respectively. Scalars p_0, p_1, and p_2 are the marginal proportions of genes satisfying each of the three hypotheses. So the parameters $(p_0, p_1, p_2, f_0, f_1, f_2)$ with $p_0 + p_1 + p_2 = 1.0$ describe the level 2 population-average model that describes the variation of mean intensities among genes. Newton et al. [27] make further assumptions on the nonparametric joint densities f_0, f_1, and f_2 using just a univariate density π for estimability. In their setup, π represents a probability vector on the support of equally spaced finite grids of log mean expressions. To generate the μ values of gene g, two independent draws, U_g and V_g, arise from π. A three-sided die with the outcome of Z_g being 0, 1, or 2 with probabilities (p_0, p_1, p_2) is tossed once. If the die comes up with $Z_g = 0$, then $\mu_{g,c} = \mu_{g,t} = U_g$ and V_g is ignored. If the die comes up with $Z_g = 1$, then $\mu_{g,c} = \min(U_g, V_g)$ and $\mu_{g,t} = \max(U_g, V_g)$ and vice versa if the die comes up with $Z_g = 2$. Consequently

$$f_0(\mu_{g,c}, \mu_{g,t}) = \pi(\mu_{g,c}) 1[\mu_{g,c} = \mu_{g,t}]$$
$$f_1(\mu_{g,c}, \mu_{g,t}) = 2\pi(\mu_{g,c})\pi(\mu_{g,t}) 1[\mu_{g,c} < \mu_{g,t}]$$
$$f_2(\mu_{g,c}, \mu_{g,t}) = 2\pi(\mu_{g,c})\pi(\mu_{g,t}) 1[\mu_{g,c} > \mu_{g,t}].$$

Regarding estimation, the common shape parameter within the control in the gamma distribution is estimated by the method of moments, where a_c is estimated as $\hat{a}_c = 1/\text{mean}(\text{CV}^2_{g,c})$;

the mean is over g, where $\mathrm{CV}_{g,c} = \mathrm{SD}(\mathbf{y}_{g,c})/\mathrm{mean}(\mathbf{y}_{g,c})$. Similarly, we estimate a_t for the treatment. A nonparametric EM algorithm using all the data is employed to estimate the baseline mixture distribution indexed by the mixing proportions (p_0, p_1, p_2) and the density π. Then gene-specific posterior probabilities for each of the three hypotheses are computed by the Bayes rule using the baseline mixture distribution and the expression for the gth gene. The details are given below.

1. EM starts with equally spaced grids on the log intensities as initial values for $\pi(u)$, where u denotes the grid values. Use equal probabilities $\left(\frac{1}{3}, \frac{1}{3}, \frac{1}{3}\right)$ as crude guesses for (p_0, p_1, p_2)

2. For iteration $t = 0, 1, 2, \ldots$:

    a. *E-step*: We need to evaluate the expected value of the missing log likelihood given the current parameter estimates. Therefore we need to evaluate the conditional expectations of the gene-specific indicator functions $1[Z_g = j]$ for each $j = 0, 1, 2$ and $1[U_g = u] + 1[V_g = u]$ given the observed data ($\mathbf{y}_{g,c}, \mathbf{y}_{g,t}$), respectively. Let $\phi = (p_0, p_1, p_2, \pi(u))$ and $\phi^{(0)}$ denote the initial values. So for each j, we need to evaluate

$$E_{\phi^{(t)}}\{1[Z_g = j|\mathbf{y}_{g,c}, \mathbf{y}_{g,t}]\} = P_{\phi^{(t)}}(Z_g = j|\mathbf{y}_{g,c}, \mathbf{y}_{g,t})$$

and

$$E_{\phi^{(t)}}\{1[U_g = u|\mathbf{y}_{g,c}, \mathbf{y}_{g,t}] + 1[V_g = u|\mathbf{y}_{g,c}, \mathbf{y}_{g,t}]\}$$
$$= P_{\phi^{(t)}}(U_g = u|\mathbf{y}_{g,c}, \mathbf{y}_{g,t}) + P_{\phi^{(t)}}(V_g = u|\mathbf{y}_{g,c}, \mathbf{y}_{g,t})$$

Note that by Bayes' rule, we obtain

$$P_{\phi^{(t)}}(Z_g = j|\mathbf{y}_{g,c}, \mathbf{y}_{g,t}) \propto p_j^{(t)} P_{\phi^{(t)}}(\mathbf{y}_{g,c}, \mathbf{y}_{g,t}|Z_g = j)$$

The quantity $p_{\phi^{(t)}}(\mathbf{y}_{g,c}, \mathbf{y}_{g,t}|Z_g = j)$ is the predictive probability whose value involves calculating $\pi^{(t)}(u)$. Similarly, $P_{\phi^{(t)}}(U_g = u|\mathbf{y}_{g,c}, \mathbf{y}_{g,t}) + P_{\phi^{(t)}}(V_g = u|\mathbf{y}_{g,c}, \mathbf{y}_{g,t})$ is proportional to

$$p_0^{(t)} \pi^{(t)}(\mathbf{y}_{g,c}, \mathbf{y}_{g,t})[\pi(u|\mathbf{y}_{g,c}, \mathbf{y}_{g,t}) + \pi^{(t)}(u)]$$
$$+ 2p_1^{(t)} \pi^{(t)}(\mathbf{y}_{g,c})\pi^{(t)}(\mathbf{y}_{g,t})[\pi^{(t)}(u|\mathbf{y}_{g,t})\pi^{(t)}(U_g \leq u|\mathbf{y}_{g,c})$$
$$+ \pi^{(t)}(u|\mathbf{y}_{g,c})\pi^{(t)}(V_g > u|\mathbf{y}_{g,t})]$$
$$+ 2p_2^{(t)} \pi^{(t)}(\mathbf{y}_{g,c})\pi^{(t)}(\mathbf{y}_{g,t})[\pi^{(t)}(u|\mathbf{y}_{g,c})\pi^{(t)}(V_g \leq u|\mathbf{y}_{g,t})$$
$$+ \pi^{(t)}(u|\mathbf{y}_{g,t})\pi^{(t)}(U_g > u|\mathbf{y}_{g,c})]$$

where the marginal probability $\pi(\mathbf{y}_{g,c}) = \sum_{u \in \mathrm{grid}} p(\mathbf{y}_{g,c}|u)\pi(u)$, and the posterior $\pi(u|\mathbf{y}_{g,c}) = p(\mathbf{y}_{g,c}|u)\pi(u)/\pi(\mathbf{y}_{g,c})$. Similar relationship exists for $\pi(\mathbf{y}_{g,t})$ and $\pi(u|\mathbf{y}_{g,t})$.

b. *M-step*: The maximum-likelihood estimates on the $(t+1)$th iteration are derived as

$$\hat{p}_j^{(t+1)} = \frac{\sum_g P_{\phi^{(t)}}(Z_g = j|\mathbf{y}_{g,c}, \mathbf{y}_{g,t})}{G}, \quad j = 0,1,2$$

$$\hat{\pi}^{(t+1)}(u) = \frac{\sum_g [P_{\phi^{(t)}}(U_g = u|\mathbf{y}_{g,c}, \mathbf{y}_{g,t}) + P_{\phi^{(t)}}(V_g = u|\mathbf{y}_{g,c}, \mathbf{y}_{g,t})]}{2G}$$

where G is the number of genes. Thus, in forming the estimates of (p_0, p_1, p_2) and $\pi(u)$, there is a contribution from each gene intensity.

The EM algorithm iterates between the E-step and M-step until convergence. Using the estimates obtained by the EM algorithm, the posterior probabilities of the three hypotheses, $H_{g,j}$ for $j = 0,1,2$, are calculated respectively as

$$P(H_{g,j}|\mathbf{y}_{g,c}, \mathbf{y}_{g,t}) = P(Z_g = j|\mathbf{y}_{g,c}, \mathbf{y}_{g,t}) = p_j p(\mathbf{y}_{g,c}, \mathbf{y}_{g,t}|Z_g = j)/p(\mathbf{y}_{g,c}, \mathbf{y}_{g,t}).$$

These genewise marginal posterior probabilities measure the evidence of differential expression for each gene. More specifically, a gene manifests stronger evidence of being DE if it has smaller $P(H_{g,0}|\mathbf{y}_{g,c}, \mathbf{y}_{g,t})$. Therefore, genes will be declared as DE if they are located at the top of the ordered gene list obtained by ranking from smallest to largest by $P(H_{g,0}|\mathbf{y}_{g,c}, \mathbf{y}_{g,t})$. SPH decides on the size of the DE gene-list by controlling FDR through a direct posterior probability approach. To report a list of upregulated genes, genes are ranked according to the increasing values of β_g with $\beta_g = 1 - P(H_{g,1}|\mathbf{y}_{g,c}, \mathbf{y}_{g,t})$. β_g is interpreted as the conditional probability that placing gene g on the upregulated list creates a type I error. The expected FDR given the data is estimated by the average of the conditional probabilities β_g in the reported list. To bound FDR at the target level of α is equivalent to ensuring the expected FDR less than or equal to α. Similarly, a list of downregulated genes is selected with now $\beta_g = 1 - P(H_{g,2}|\mathbf{y}_{g,c}, \mathbf{y}_{g,t})$.

20.3.4 LIMMA

The LIMMA method [31] (linear models for microarray data), constructs hierarchical linear models for the intensity data in microarray experiments with an arbitrary number of treatments. In this section, we focus only on the situation with two conditions: control and treatment. Let the column vector $\mathbf{y}_g$ represent the normalized log intensities with the first m replicates from the control sample and the remaining n replicates from the treatment sample. Assume that $\mathbf{y}_g$ has expectation $E(\mathbf{y}_g) = \mathbf{X}\alpha_g$ and variance $\text{var}(\mathbf{y}_g) = \mathbf{W}_g \sigma_g^2$. The matrix $\mathbf{X}$ here is a design matrix of full rank. In our case, we have:

$$\mathbf{X}^T = \begin{pmatrix} 1 & \cdots & 1 & 0 & \cdots & 0 \\ 0 & \cdots & 0 & 1 & \cdots & 1 \end{pmatrix}_{2*(m+n)}.$$

The vector $\alpha_g^T = (\alpha_{g1}, \alpha_{g2})$, where α_{g1} and α_{g2} represent the effect from the control and the treatment group, respectively, for gene g. The term $\mathbf{W}_g$ is a known non-negative definite weight matrix. LIMMA conducts a test of the hypotheses that the difference $\alpha_{g2} - \alpha_{g1}$ is zero. Let $\mathbf{c}^T = (-1,1)$ and $\beta_g = \mathbf{c}^T \alpha_g$; then LIMMA tests $H_0 : \beta_g = 0$ vs. $H_a : \beta_g \neq 0$.

The LIMMA method fits the linear model on each gene to obtain the coefficient estimate $\hat{\alpha}_g$, the variance estimate s_g^2 for σ_g^2, and the covariance matrix $\widehat{\mathrm{var}}(\hat{\alpha}_g) = s_g^2 \mathbf{V}_g$, where $\mathbf{V}_g$ is a positive-definite matrix not depending on s_g^2. Define $v_g = \mathbf{c}^T \mathbf{V}_g \mathbf{c}$, then the contrast estimator $\hat{\beta}_g = \mathbf{c}^T \hat{\alpha}_g = \hat{\alpha}_{g2} - \hat{\alpha}_{g1}$ and its estimated variance $\mathrm{var}(\hat{\beta}_g) = v_g s_g^2$. With the normal assumption in the linear model, the distribution of the parameters can be summarized as

$$\hat{\beta}_g | \beta_g, \sigma_g^2 \sim N(\beta_g, v_g \sigma_g^2) \quad \text{and} \quad s_g^2 | \sigma_g^2 \sim \frac{\sigma_g^2}{d_g} \chi_{d_g}^2,$$

where d_g is the residual degree of freedom for the linear model on gene g.

Since a large number genes share the same latent structure, LIMMA proposes prior distributions (level 2 information) on the parameters β_g and σ_g^2 as follows:

$$P(\beta_g \neq 0) = p, \quad \beta_g | \sigma_g^2, \beta_g \neq 0 \sim N(0, v_0 \sigma_g^2) \quad \text{and} \quad \frac{1}{\sigma_g^2} \sim \frac{1}{d_0 s_0^2} \chi_{d_0}^2.$$

Under this hierarchical model, the posterior mean of σ_g^2 given s_g^2 is

$$\tilde{s}_g^2 = \frac{d_0 s_0^2 + d_g s_g^2}{d_0 + d_g}.$$

The LIMMA method tests the hypotheses using a moderated t statistic defined as

$$t_g = \frac{\hat{\beta}_g}{\tilde{s}_g \sqrt{v_g}}.$$

Note that under the null hypothesis, $t_g \sim t_{d_0 + d_g}$, while under the alternative hypothesis $t_g \sim \left(1 + \frac{v_0}{v_g}\right)^{1/2} t_{d_0 + d_g}$. According to the distribution of t_g under each hypothesis, LIMMA calculates the p value and the adjusted p value using the BH method. LIMMA selects the genes with adjusted p values less than some prespecified level α to be DE, so that it can control the FDR at the targeted α level. LIMMA also calculates the posterior log odds ratio B_g for each gene. The user can also select the genes with larger B_g to be DE. The results based on the adjusted p values or the posterior odds ratios B_g agree with each other when there are no missing values in the observed intensities.

20.3.5 MAANOVA

The MicroArray ANalysis Of VAriance (MAANOVA) package provides a collection of functions that visualize gene expression quality, transform intensities through various normalization methods to diminish extra noise, discover DE genes, and cluster genes by bootstrapping. In this subsection, we focus on how MAANOVA finds DE genes. It applies the following gene-specific general linear mixed model structure on the normalized log-transformed intensities for gene g, denoted by $\mathbf{y}_g$:

$$\mathbf{y}_g = \mathbf{X}\beta_g + \mathbf{Z}u_g + \epsilon_g,$$

where the matrix $\mathbf{X}$ denotes the design matrix of fixed effects β_g, $\mathbf{Z}$ is the design matrix of random effects u_g, and ϵ_g is the residual vector. Since this linear mixed model is fit to the expression data one gene at a time, the design matrices, $\mathbf{X}$ and $\mathbf{Z}$ remain the same for all genes. In the cDNA experiments, for example, the array effect and biological replicates of samples, if any, which account for inherent biological variation, are usually considered as random factors. The cellular conditions (treatment effect) and dye effect are viewed as fixed factors. Depending on experimental designs, covariates can also be built into the fixed-effect part of the model to reduce the estimate of the systematic errors. The flexibility of the generalized linear model enables MAANOVA to adapt to multifarious experimental designs such as simple two-array dye-swap experiments or elaborate loop designs. In mixed-effect models, the variance components are estimated by restricted maximum-likelihood method (REML). Generalized least-square techniques are applied for estimation of the fixed-effect coefficients and prediction of random effects.

Identification of a list of DE genes is achieved by testing linear combinations of treatment effects for each gene. For instance, in a four-array dye-swap experiment in which per gene is singled-spotted within a slide to compare control versus treatment samples, let A_i, D_j, and T_k represent array, dye, and treatment effects, respectively. Then the fixed-effect vector can be written as $\beta_g = (\mu, D_1, D_2, T_1, T_2)^T$ and the random-effect vector as $u_g = (A_1, A_2, A_3, A_4)^T$. Note that the array effects are not necessarily always assumed to be random; they can be modeled as fixed effects in some experiments if the variation of array effects are not significant. Detecting DE genes involves testing $H_0 : T_1 = T_2$ versus $H_1 : T_1 \neq T_2$ for each gene. For this hypothesis testing, there are four types of F statistics in MAANOVA. Let Δ_g denote the estimate of the treatment effect for gene g. Use the notation $rss_{g,0}$ for the residual sum of squares for the null model that gene g is EE across cellular conditions and df_0 for the corresponding degrees of freedom. Similarly, $rss_{g,1}$ and df_1 represent the alternative model of gene g being DE across conditions. The design matrices, $\mathbf{X}$ and $\mathbf{Z}$, are the same across genes, and so is the number of replicates. Consequently, the degrees of freedom are equal across genes, and we can suppress the subscript g for them. Then $\Delta_g = (rss_{g,0} - rss_{g,1})/(df_0 - df_1)$. Let $\hat{\sigma}_g^2 = rss_{g,1}/df_1$ be the residual mean square error for gene g, and let $\hat{\sigma}_{pool}^2$ be the mean of $\hat{\sigma}_g^2$ across genes, which estimates the common variance. The four types of F statistics are as follows:

$$F_1 = \frac{\Delta_g}{\hat{\sigma}_g^2}$$

$$F_2 = \frac{\Delta_g}{\frac{1}{2}(\hat{\sigma}_g^2 + \hat{\sigma}_{pool}^2)}$$

$$F_3 = \frac{\Delta_g}{\hat{\sigma}_{pool}^2}$$

$$F_S = \frac{\Delta_g}{\tilde{\sigma}_g^2},$$

where $\tilde{\sigma}_g^2$ is the shrinkage estimator for σ_g^2. The derivation of $\tilde{\sigma}_g^2$ involves first obtaining the James–Stein (JS) estimator [21] for $\ln(\sigma_g^2)$, then transforming it back to the original scale for estimating σ_g^2. In deriving the JS estimator for $\ln(\sigma_g^2)$, it is assumed that the ratios of the residual sum of squares to the true variance $rss_{g,1}/\sigma_g^2$, are independent, each following a χ^2 distribution with v degrees of freedom; that is, $rss_{g,1} \sim \sigma_g^2 \chi^2(v)$. Taking the natural logarithmic transformation on $\mathrm{rss}_{g,1}$ transfers the problem to that of estimating a common location parameter. We have

$$\ln\frac{\mathrm{rss}_{g,1}}{\nu} \sim \ln(\sigma_g^2) + \ln\frac{\chi^2(\nu)}{\nu}. \tag{20.3}$$

Both sides of Equation (20.3) are centered by m, the mean of $\ln(\chi_\nu^2/\nu)$. Let X'_g be $\ln(\mathrm{rss}_{g,1}/\nu) - m$. Using $\mathrm{var}(\ln(\chi^2(\nu)/\nu)) \approx \mathrm{var}(\chi^2(\nu)/\nu) = 2/\nu$, which is a direct result of the first order Taylor expansion of $\ln(\chi^2(\nu)/\nu)$, the JS estimator shrinks X'_g toward the common mean $\bar{X}' = \sum_{g=1}^{G} X'_g/G$. It was argued in [6] that the shrinkage estimator performs best in the construction of the test statistics.

The null distributions of the F statistics are obtained through permutation. MAANOVA provides residual shuffling and sample shuffling. Residual shuffling applies to fixed-effect models only because of its assumption of homogeneous variance. In residual shuffling, the residuals obtained from fitting the null hypothesis model are shuffled globally among all genes to get a new dataset, then we compute the test statistics for the new dataset. The permuted p values are attained by repeating these steps of fitting models, generating new datasets, and computing the test statistics. In mixed-effect models, variation of a gene's expression consists of residual error and variances of random components such as array and biological replicates. Therefore the sample-shuffling method should be applied in order to construct the test statistics based on the proper variation. There are two key aspects of sample shuffling. One is to ensure that the shuffle base is the random term nested within the test term. For instance, in the Paigen's experiment in the MAANOVA manual [47], data were collected on mice of three strains that were administered low- or high-fat diets. Two mice were selected from each strain–diet combination, and the biological replicates (mice) are nested within the strain–diet interaction. The interaction is the test term, and the shuffle base is the biological replicates. If there are more than one random terms nested within the test term, the shuffle base should be the random term with the lowest level. The other is to preserve the array structure specified in the model. However, as pointed out in the MAANOVA manual, the number of possible permutations will be limited if the experiment size is small. In this case, users have to rely on the tabulated p values, which are obtained using F distributions to approximate the null distributions of F_1 and F_S, with χ^2 distributions for F_2 and F_3.

The four F statistics hence differ solely in the estimation of the true variance. The statistic F_1 is simply the ordinary F statistic, which uses the mean-squared error, $\hat{\sigma}_g^2$, to estimate the gene-specific variance. However, because of the usual case of small sample size per gene, F_1 is not a very good choice. The statistic F_3 works best in the situation of homogeneous variance, which does not occur often in microarray experiments with many sources of noise. As a hybrid of F_1 and F_3, F_2 gives $\hat{\sigma}_g^2$ and $\hat{\sigma}_{\mathrm{pool}}^2$ an equal (uniform) weight of $1/2$. The shrinkage estimator $\tilde{\sigma}_g^2$ enhances the performance of F_S by borrowing strength from all genes. By varying the CV of residuals and ν in their simulation studies, Cui et al. [6] show that F_S is robust to changes in CV values and has a good power of detecting DE genes. We therefore focus only on F_S in our simulation studies. To adjust for multiple comparisons, MAANOVA also provides Šidák's [34] stepdown procedure, the BH [1] stepup procedure, and their adaptive procedure [2].

20.4 SIMULATION STUDY

To assess the performance of these five methods, we conduct simulation studies consisting of two simulation scenarios. In both scenarios, we mimic a four-array, two-channel dye-swap experiment comparing 5000-gene expressions under control and treatment conditions. Since it is reasonable to assume that the experimental environment is independent of genes, in both scenarios, we simulate the array effect on the $\log_2$ intensities A_i from a $N(0, 0.6^2)$ density for $i = 1, 2, 3, 4$. Similarly, the dye effect D_j for $j = 1, 2$, is randomly drawn from $N(0, 0.3^2)$ for each gene.

In scenario 1, let $Z_{g,k}$ represent the raw expressions attributed to the treatment effect for gene g. We suppose that $Z_{g,k}$ are random draws from $\Gamma(20, \mu_{g,k}/20)$ so that the expected raw expressions are $\mu_{g,k}$, where $k = 1$ represents control and $k = 2$, treatment. Suppose that we set the prevalence of DE genes at 5%. For the first 125 genes ($g = 1, \ldots, 125$), the upregulated ones, $\mu_{g,1} = \min(U_g, V_g)$ and $\mu_{g,2} = \max(U_g, V_g)$ and vice versa for the downregulated genes ($g = 126, \ldots, 250$). For EE genes ($g = 251, \ldots, 5000$), we set $\mu_{g,1} = \mu_{g,2} = U_g$ and $V_g = 0$. Here U_g and V_g are randomly drawn from $\Gamma^{-1}(2, 1/200)$. Let $Y_{g,ijk}$ denote the $\log_2$ intensity of gene g from array i, dyed in color j and under the cellular condition k. Then $Y_{g,ijk} = A_i + D_j + \log_2(Z_{g,k})$. This indicates that the array and dye effects influence the expected raw intensity through a multiple of μ_k.

In scenario 2, we assume that the expected value of $Y_{g,ijk}$ is a linear combination of array, dye, and treatment effects. Let $T_{g,k}$ denote the treatment effect on the $\log_2$ scale. We simulate $T_{g,k}$ through its relationship with log intensity ratio $M_g = T_{g,2} - T_{g,1}$ and log average intensity $A_g = 1/2(T_{g,1} + T_{g,2})$. Let $A_g \sim U(4, 8)$ for all genes, and

$$
M_g \begin{cases} \sim U(1, 2) & \text{for upregulated genes} \\ \sim U(-2, -1) & \text{for downregulated genes} \\ = 0 & \text{for EE genes.} \end{cases}
$$

In other words, we assume that the DE genes have at least one- to two fold change due to the treatment effect. Then

$$
Y_{g,ijk} \sim N(\mu = A_i + D_j + T_{g,k}, \qquad \sigma = (A_i + D_j + T_{g,k})/12)
$$

We set σ this way so that CV values within one cellular condition remain constant across genes, which is a crucial assumption for the SPH method.

Each method presumes normalized expressions as the input. We thus apply the quantile normalization provided in the LIMMA package. This method ensures the same empirical distribution of intensities across arrays and across channels. Since MAANOVA factors array and dye effects into the modeling, we use the original data as its input while other methods use normalized data. In MAANOVA, users need to specify the general linear model according to their experiments. In our simulation studies, we fit a fixed-effect model by considering array, dye, and treatment as fixed factors, because we observe that the variance of array effects is very small relative to that of the residuals in the variance component plots. We have also explored fitting the mixed-effect model by treating the array as a random effect. The performance of the MAANOVA is not satisfactory. Both MAANOVA and LIMMA allow us to fit more complex models. However, caution is also needed in fitting the correct model. Finally let us note that the appropriateness of SPH relies on the assumption of constant CV in each condition across genes as pointed out by its authors. So we simulate our data to mimic this condition. Our following comparison results are based on this condition. In other cases, we have seen situations where SPH may not be as competitive as SAM, LIMMA, and MAANOVA. Therefore, it is important to caution users to check that the constant CV conditions are satisfied before using the SPH.

20.4.1 Results of Simulation Studies

We apply the five methods to the simulated data, one is for FDR control and the other, for cFDR control. In both studies, we consider both the 5% and the 10% DE prevalence rates and run 10 simulations for each scenario. We measure the classification accuracy of these five methods using ROC curves. The predictive power is also evaluated through the plots of PPV versus (1-NPV).

While SPH and BH-T propose their own method of controlling FDR, the LIMMA and MAANOVA packages incorporate some popular FDR-controlling procedures and leave the options to users. We use the BH [1] procedure to adjust the p values for the moderated t statistic in LIMMA, and the adaptive BH procedure [2] for the permuted p values of F_S. The SAM package offers estimated median FDR, 90th percentile FDR, and local FDR. It gives q values as a measure of pFDR as well. To put the methods on equal ground for comparison, we estimated the median FDR for SAM. Figures 20.1–20.3 evaluate these five methods by controlling FDR at 0.005–0.655 with increments of 0.05 and using both ROC curves and predictive curves. Figure 20.1 shows that in scenario 1, SPH has a slight edge over SAM, LIMMA, and F_S regarding classification accuracy, while these four methods are very comparable in terms of predictive power. As shown in Figure 20.2 of scenario 2 with uniform data generation mechanisms, these three methods slightly surpass SPH. In both scenarios, BH-T has the lowest sensitivity and PPV, and its setback in predictive power is more dramatic. This results from the fact that SPH, SAM, LIMMA, and F_S combine all information among genes, whereas BH-T is merely a gene-specific classifier with no borrowing information among genes. As the prevalence of DE genes increases, the adaptive BH procedure [2] will offer a more detectable advantage. So at 10% DE prevalence of scenario 1 as shown in Figure 20.3, at the same set of α values, the p values of the F_S statistic adjusted by the adaptive BH procedure indicate more DE genes than at 5% prevalence. Figure 20.3 reflects the fact that the ROC curve for F_S does not terminate as soon as that in Figure 20.1. Moreover, it shows that the superiority in predictive power of SPH and F_S relative to SAM and LIMMA becomes slightly more pronounced at 5% prevalence.

In order to observe the selected DE genes from each method, we draw a mean MA plot for a simulation data of scenario 2 with 250 genes being truly DE. The mean MA plot basically plots the average of M values and the average of A values for replicates of M and A for the same gene. As shown in Figure 20.4, when FDR is controlled at 0.05, SPH, SAM, F_S and LIMMA identify a large proportion of truly DE genes. However, BH-T correctly selects only $\sim$20% of truly DE genes. Moreover, it does not always prefer genes with large log intensity ratios; on the contrary, it picks up some genes having small log ratios and yet very small standard deviations. Combined with Figures 20.5 and 20.6, as α increases to 0.1 and 0.15, BH-T correctly calls genes less than 50% of the times. This suggests that the BH procedure is conservative. We also note that the other four methods tend to select genes with large M values.

In the FDR control studies we have discussed so far, we fix the acceptable FDR α beforehand and estimate a significance threshold to obtain this target FDR. The true but unknown data-generating distribution affects the quantity that measures the evidence against a gene being EE given all genes involved, such as adjusted p values. Often, biologists are interested in finding a list of DE genes to investigate, but are not as concerned as finding the strength of evidence against being EE. So we investigate the methods in terms of ranking genes in the second simulation study. We compare the methods by letting them declare the same number of DE genes. So the evaluation is equivalent to controlling the conditional false discovery rate (cFDR) at the same level. By varying the cFDR at different levels, we do not need to obtain the adjusted p values in the BH method. Consequently, we indeed compare the two-sample t test with the other four methods without FDR adjustments. Both Figures 20.7 and 20.8 evaluate the five methods by choosing the same number of claimed DE genes with 5% prevalence of DE genes for the two simulated data cases. Figure 20.7 shows that SPH performs moderately better than SAM, LIMMA, and F_S with respect to both accuracy and predictive power, but vice versa for scenario 2, as shown in Figure 20.8. The two-sample t tests perform worst in both scenarios.

20.4.2   Other Considerations

In addition to accuracy and predictive power, experimental designs and computation time should be considered as comparison criteria as well. Note that SPH and two-sample t tests manage two independent conditions only, while SAM can extend to multiple conditions. Both MAANOVA and LIMMA can manage loop designs. In addition, covariates can be incorporated in MAANOVA to reduce the residual errors. Both SAM and LIMMA can analyze timecourse data. Since MAANOVA uses permutation to obtain the null distributions and REML for variance components estimation in mixed effects models, the computing time becomes much longer than the other methods as the number of genes increases. The SNOW parallel computing package [29,37] alleviates this problem, however, it is not so easily set up by users.

20.5   CONCLUDING REMARKS

We have reviewed some computational methods associated with microarray analysis. We first briefly described the normalization procedure to reduce instrumental bias. Then we reviewed five statistical methods for selecting and ranking differentially expressed (DE) genes in replicated microarray experiments. These five methods are the Benjamini–Hochberg procedure for controlling the false discovery rate (FDR) applied to two-sample t tests, the significance analysis of microarray, the semiparametric hierarchical method, the linear models for microarray data, and the microarray analysis of variance. We simulated two data cases. Then we evaluated the performance of these methods controlling the FDR at the same levels using (1) the receiver operating characteristic curves and (2) the predictive powers. Our simulation results show that the first method of applying the Benjamini–Hochberg procedure to two-sample t tests is inferior to the other four methods. The other four methods are comparable with the advantage of sharing information across genes. Moreover, we also compared the five methods with the same numbers of claimed DE genes using the same criteria. In this study, the first method reduces to using two-sample t tests without the FDR adjustment. We have essentially compared the five methods on their abilities in ranking genes without the FDR adjustments. Our simulation study revealed much the same results as before.

ACKNOWLEDGMENT

Research was supported by NIH/NIGMS P20 GM5764-01. The authors wish to thank Drs. Dong-Guk Shin, David Rowe, and Winfred Krueger, and many members of the bioinformatics core at the University of Connecticut for discussions.

REFERENCES

1. Benjamini, Y. and Hochberg, Y., Controlling the false discovery rate: A practical and powerful approach to multiple testing, *J. Roy. Staitsit. Soc. B* **57**, 289–300 (1995).

2. Benjamini, Y. and Hochberg, Y., On the adaptive control of the false discovery rate in multiple testing with independent statistics, *J. Educ. Behav. Statist.* **25**, 60–83 (2000).

3. Bolstad, B. M., Irizarry, R. A., Astrand, M., and Speed, T. P., A comparison of normalization methods for high density oligonucleotide array data based on bias and variance, *Bioinformatics* **19**, 185–193 (2003).

4. Chen, Y., Dougherty, E. R., and Bittner, M. L., Ratio-based decisions and the quantitative analysis of cDNA micaroarray images, *J. Biomed. Opt.* **2**, 364–374 (1997).

5. Chu, G., Narasimhan, B., Tibshirani, R., and Tusher, V., SAM "Significance analysis of microarrays" users guide and technical document. (http://www-stat.stanford.edu/~tibs/SAM/), 2006.

6. Cui, X., Hwang, J. T., Qiu, J., Blades, N. J., and Churchill, G. A., Improved statistical tests for differential gene expression by shrinking variance components estimates, *Biostatistics* **6**(1), 59–75 (2005).

7. Cui, X., Kerr, M. K., and Churchill, G. A., Transformations for cDNA microarray data, *Statist. Appl. Genet. Molec. Biol.* **2**(1), Art. 4 (2003).

8. DeRisi, J. L., Iyer, V. R., and Brown, P. O., Exploring the metabolic and genetic control of gene expression on a genomic scale, *Science* **278**, 680–686 (1997).

9. Draghici, S., Statistical intelligence: Effective analysis of high-density microarray data, *Drug Discov. Today* **7**, S55–S63 (2002).

10. Dudoit, S., van der Laan, M., and Pollard, K., Multiple testing. Part I. Single-step procedures for control of general type I error rate, *Statist. Appl. Genet. Molec. Biol.* **3**(1), Art. 13 (2004).

11. Dudoit, S., Shaffer, J., and Boldrick, J. C., Multiple hypothesis testing in microarray experiments, *Statist. Sci.* **18**, 71–103 (2003).

12. Dudoit, S., Yang, Y. H., Callow, M. J., and Speed, T. P., Statistical methods for identifying differentially expressed genes in replicated cDNA microarray experiments, *Statistica Sinica* **12**, 111–139 (2002).

13. Efron, B., Large-scale simultaneous hypothesis testing: The choice of a null hypothesis, *J. Am. Statist. Assoc.* **99**, 96–104 (2004).

14. Efron, B., *Local False Discovery Rates*, Technical Report, Dep. Statistics, Stanford Univ., 2005.

15. Efron, B., *Correlation and Large-Scale Simultaneous Significance Testing*, Technical Report, Dep. Statistics, Stanford Univ., 2006.

16. Efron, B., Size, Power, and False Discovery Rates, Technical Report, Dep. Statistics, Stanford Univ., 2006.

17. Efron, B. and Tibshirani, R., Empirical Bayes methods and false discovery rates for microarrays, *Gene. Epidemiol.* **23**, 70–86 (2002).

18. Efron, B., Tibshirani, R., Storey, J. D., and Tusher, V., Empirical Bayes analysis of a microarray experiment, *J. Am. Statist. Assoc.* **96**, 1151–1160 (2001).

19. Gottardo, R., Raftery, A. E., Yeung, K. Y., and Bumgarner, R. E., Robust estimation of cDNA microarray intensities with replicates, *J. Am. Statist. Assoc.* **101**, 30–40 (2006).

20. Gottardo, R., Raftery, A. E., Yeung, K. Y., and Bumgarner, R. E., Bayesian robust inference for differential gene expression in cDNA microarrays with multiple samples, *Biometrics* **62**, 10–18 (2006).

21. James, W. and Stein, C., Estimation with quadratic loss, in *Proc. 3rd Berkeley Symp. Mathematics Statististics and Probability*, Univ. California Press, Berkeley, 1960, vol. 1, 361–380.

22. Kerr, M. K., Martin, M., and Churchill, G. A., Analysis of variance for gene expression microarray data, *J. Comput. Biol.* **7**, 819–837 (2000).

23. Kuo, L., Yu, F., and Zhao, Y., On Statistical Methods for Identifying Differentially Expressed Genes in Replicated Microarray Experiments, Technical Report 04-32, Statistics Dep., Univ. Connecticut, Storrs, 313–314 (2004).

24. Lee, M. T., *Analysis of Microarray Gene Expression Data*, Kluwer Academic Publishers, Boston, (2004).

25. Lönstedt, I. and Speed, T., Replicated microarray data, *Statistica Sinica* **12**, 31–46 (2002).

26. McLachlan, G. J., Bean, R., and Ben-Tovim Jones, L., A simple implementation of a normal mixture approach to differential gene expression in multiclass microarrays, *Bioinformatics* **22**, 1608–1615 (2006).

27. Newton, M. A., Noueiry, A., Sarkar, D., and Ahlquist, P., Detecting differential gene expression with a semiparametric hierarchical mixture method, *Biostatistics* **5**, 155–176 (2004).

28. Pan, W., A comparative review of statistical methods for discovering differentially expressed genes in replicated microarray experiments, *Bioinformatics* **18**, 546–554 (2002).

29. Rossini, A., Tierney, L., and Li, N., Simple parallel statistical computing in R, *UW Biostatistics working Paper Series*, Paper 193, Univ. Washington, Seattle, (2003).

30. Schena, M., Shalon, D., Heller, R., Chai, A., Brown, P. O., and Davis, R. W., Parallel human genome analysis: Microarray-based expression monitoring of 1000 genes, *Proc. Nat. Acad. Sci. USA* **93**, 10614–10619 (1996).

31. Smyth, G. K., Linear models and empirical Bayes methods for assessing differential expression in microarray experiments, *Statist. Appl. Genet. Molec. Biol.* **3**(1), Art. 3 (2004).

32. Smyth, G. K. and Speed, T. P., Normalization of cDNA microarray data, *Methods* **31**(4), 265–273 (2003).

33. Speed, T., Statistical Analysis of Gene Expression Microarray Data, Chapman & Hall/ CRC, Boca Raton, FL, 2003.

34. Šidák, Z., Rectangular confidence regions for the means of multivariate normal distributions, *J. Am. Statist. Assoc.* **62**, 626–633 (1967).

35. Storey, J. D., A direct approach to false discovery rates, *J. Roy. Statist. Soc. B* **64**, 479–498 (2002).

36. Storey, J. D. and Tibshirani, R., SAM thresholding and false discovery rates for detecting differential gene expression in DNA microarrays, in *The Analysis of Gene Expression Data: Methods and Software*, Parmigani, G., Garrett, E. S., Irizarry, R. A., and Zeger, S. L., eds., Springer, New York, 2003.

37. Tierney, L., Rossini, A., Li, N., and Sevcikova, H., *The Snow Package: Simple Network of Workstations*, version 0.2–1 (2004).

38. Tsai, C. A., Hsueh, H. M., and Chen, J. J., Estimation of false discovery rates in multiple testing: Application to gene microarray data, *Biometrics* **59**, 1071–1081 (2003).

39. Tusher, V. G., Tibshirani, R., and Chu, G., Significance analysis of microarrays applied to the ionizing radiation response, *Proc. Natl. Acad. Sci. USA* **98**, 5116–5121 (2001).

40. van der Laan, M., Dudoit, S., and Pollard, K., Multiple testing part II. Step-down procedures for control of the family-wise error rate, *Statist. Appl. Genet. Molec. Biol.* **3**(1), Art. 14 (2004).

41. van der Laan, M., Dudoit, S., and Pollard, K., Augmentation procedures for control of the generalized family-wise error rate and tail probabilities for the proportion of false positives, *Statist. Appl. Genet. Molec. Biol.* **3**(1), Art. 15 (2004).

42. Westfall, P. H. and Young, S. S., *Resampling-Based Multiple Testing: Examples and Methods for P-Value Adjustment*, Wiley, New York, 1993.

43. Wit, E. and McClure, J., *Statistics for Microarrays: Design, Analysis, and Inference*, Wiley, The Atrium, UK, 2004.

44. Wolfinger, R. D., Gibson, G., Wolfinger, E. D., Bennett, L., Hamadeh, H., Bushel, P., Afshari, C., and Paules, R. S., Assessing gene significance from cDNA microarray expression data via mixed models, *J. Comput. Biol.* **8**(6), 625–637 (2001).

45. Wright, G. W. and Simon, R. M., A random variance model for detection of differential gene expression in small microarray experiments, *Bioinformatics* **19**, 2448–2455 (2003).

46. Wu, H., Kerr, M. K., Cui, X. Q. and Churchill, G. A., MAANOVA: A software package for the analysis of spotted cDNA microarray experiments in the analysis of gene expression data, in *The Analysis of Gene Expression Data: Methods and Software*, Parmigiani, G. Garret, E. S. Irizarry, R. A. and Zeger, S. L., eds, Springer (London), 313–341 (2003).

47. Wu, H. and Churchill, G. A., R/MAANOVA: An extensive R environment for the analysis of microarray experiments (http://www.jax.org-/staff/churchill/labsite/software/Rmaanova/maanova.pdf), 2005.

48. Yang, Y. H., Dudoit, S., Luu, P., and Speed, T. P., Normalization for cDNA microarray data, in *Microarrays: Optical Technologies and Informatics*, Bittner, M. L., Chen, Y., Dorsel, A. N., Dougherty, E. R., eds., *Proc. SPIE* **4266**, 141–152 (2001).

49. Yang, Y. H., Dudoit, S., Luu, P., Lin, D. M., Peng, V., Ngai, J., and Speed, T. P., Normalization for cDNA microarray data: A robust composite method addressing single and multiple slide systematic variation, *Nucleic Acids Res.* **30**(4), 15 (2002).

50. Yang, Y. H. and Thorne, N. P., Normalization for two-color cDNA microarray data, in Goldstein, D. R., ed., Science and statistics: A Festschrift for Terry Speed, *IMS Lecture Notes-Monograph Series* **40**, 403–418 (2003).

51. Yu, F., Chen, M. H., and Kuo, L., Screening for Differentially Expressed Genes Using Bayes Factors, Technical Report 05-22, Dep. Statistics, Univ. Connecticut, Storrs, 2005.

CHAPTER 21

Clustering of Microarray Data via Mixture Models

Geoffrey J. McLachlan and Angus Ng
Department of Mathematics, University of Queensland,
Brisbane, Australia

Richard W. Bean
Institute for Molecular Bioscience, University of Queensland,
Brisbane, Australia

21.1 INTRODUCTION

The widespread use of DNA microarray technology [15] to perform experiments on thousands of gene fragments in parallel has led to an explosion of expression data. A variety of multivariate analysis methods have been used to explore these data for relationships among the genes and the tissue samples. Cluster analysis has been one of the most frequently used methods for these purposes. It is an exploratory technique that attempts to find groups of observations that have similar values on a set of variables. Sometimes emphasis is placed on the distinction between the search for naturally occurring clusters and the division of the entities into a given number of groups, where there is no implication that the resulting groups are in any sense a natural division of the data; see, for example, Hand and Heard [23]. But often there is no emphasis, particularly as most methods for finding natural clusters are also useful for segmenting the data.

Agglomerative hierarchical clustering (encompassing single-, complete-, and average-linkage variants), k-means clustering, and self-organizing maps (SOMs) have been the most widely used methods. Eisen et al. [14] was the first to apply cluster analysis to microarray data, using an agglomerative hierarchical method using average linkage with a correlation-based metric, or equivalently, the Euclidean metric after standardization of the data.

Statistical Advances in the Biomedical Sciences, edited by Atanu Biswas, Sujay Datta,
Jason P. Fine, and Mark R. Segal

More recently, increasing attention has focused on model-based methods of clustering microarray data (e.g., see Refs. 21,60,39, and 43).

A useful way to think about the different clustering procedures is in terms of the shape of the clusters produced [53]. Many clustering methods assume that the appropriate distance function (metric) is known (e.g., they may use Euclidean distance). But clearly, it would be more appropriate to use a metric that depends on the shape of the clusters. As pointed out by Coleman et al. [8], the difficulty is that the shape of the clusters is not known until the clusters have been found, and the clusters cannot be effectively identified unless the shapes are known. The majority of the existing clustering methods assume that a similarity measure or metric is known a priori; often the Euclidean metric is used. In particular, k means effectively uses the Euclidean metric, as it can be viewed as being a "hard" version of the mixture clustering procedure based on a mixture in equal proportions of multivariate normal components with a common spherical covariance matrix. In the absence of any prior knowledge on the metric, it is reasonable to adopt a clustering procedure that is invariant under affine transformations of the data; that is, invariant under transformations of the data $\mathbf{y}$ of the form

$$\mathbf{y} \to \mathbf{Cy} + \mathbf{a}, \tag{21.1}$$

where $\mathbf{C}$ is a nonsingular matrix. One attractive feature of adopting mixture models with elliptically symmetric components such as the normal or t densities is that the implied clustering is invariant under affine transformations of the data (i.e., under operations relating to changes in location, scale, and rotation of the data). Thus the clustering process does not depend on irrelevant factors such as the units of measurement or the orientation of the clusters in space. If the clustering of a procedure is invariant under (21.1) for only diagonal $\mathbf{C}$, then it is invariant under change of measuring units but not rotations. But as mentioned by Hartigan [25], this form of invariance is more compelling than affine invariance.

In this chapter, we shall focus on a model-based approach to the clustering of microarray data using mixtures of normal distributions, which are commonly used in statistics; see, for example, Ganesalingam and McLachlan [19], McLachlan and Basford [38], Banfield and Raftery [2], Fraley and Raftery [16,17], and McLachlan and Peel [42]. As noted by Aitkin et al. [61], "Clustering methods based on such mixture models allow estimation and hypothesis testing within the framework of standard statistical theory." Previously, Marriott [35, p. 70] had noted that the mixture likelihood-based approach "is about the only clustering technique that is entirely satisfactory from the mathematical point of view. It assumes a well-defined mathematical model, investigates it by well-established statistical techniques, and provides a test of significance for the results." More recently, Yeung et al. [60] noted that "in the absence of a well-grounded statistical model, it seems difficult to define what is meant by a 'good' clustering algorithm or the 'right' number of clusters."

The normal mixture model-based approach is to be applied here in a nonhierarchical manner, as there is no reason why the clusters of tissues or genes should be hierarchical in nature. It is true that if there is a clear, unequivocal grouping, with little or no overlap between the groups, any method will reach this grouping. But as pointed out by Marriott [35], "hierarchical methods are not primarily adapted to finding groups." For instance, if the division into $g = 2$ groups given by some hierarchical method is optimum with respect to some criterion, then the subsequent division into $g = 3$ groups is unlikely to be so. This is due to the restriction that one of the groups must be the same in both the $g = 2$ and $g = 3$ clusterings. As explained by Marriott [35], this restriction is not a natural one to impose if the purpose is to find a natural grouping of the data. As advocated by Marriott [35, p. 67], "it is better to consider the clustering problem *ab initio*, without imposing any conditions."

Another attractive feature in using mixture models for clustering is that the number of clusters can be formulated in terms of a criterion or a test for the smallest number of components in the mixture model compatible with the data. One such criterion is the Bayesian information criterion (BIC) of Schwarz [55], while a test can be carried out on the basis of the likelihood ratio statistic λ.

One potential drawback with the normal mixture-model-based approach to clustering is that normality is assumed for the cluster distributions. However, this assumption would appear to be reasonable for the clustering of microarray data after appropriate normalization.

In practice, the problem of relatively large local maxima that occur as a consequence of a fitted component having a very small (but nonzero) variance for univariate data or generalized variance (the determinant of the covariance matrix) for multivariate data deserves consideration. Such a component corresponds to a cluster containing a few data points either relatively close together or almost lying in a lower-dimensional subspace in the case of multivariate data. There is thus a need to monitor the relative size of the fitted mixing proportions and of the component variances for univariate observations, or of the generalized component variances for multivariate data, in an attempt to identify these spurious local maximizers. One situation where an apparent spurious solution would be of practical interest is where one (or more) of the fitted components correspond to a small number of points that are distant from the remaining points.

21.2 CLUSTERING OF MICROARRAY DATA

There are two distinct but related clustering problems with microarray data: (1) clustering of tissues on the basis of genes and (2) clustering of genes on the basis of tissues. This duality is quite common. One may be interested in grouping tissues (patients) with similar expression values or in grouping genes on patients with similar types of tumors or similar survival rates.

In clustering microarray data, the clusters of tissues can play a useful role in the discovery and understanding of new subclasses of diseases. The clusters of genes obtained can be used to search for genetic pathways or groups of genes that might be regulated together. Also, in problem 1, we may wish first to summarize the information in the very large number of genes by clustering them into groups (of hyperspherical shape), which can be represented by some metagenes, such as the group-sample means. We can then carry out the clustering of the tissues in terms of these metagenes. As noted by Pollard and van der Laan [52], in most research these two problems have been considered separately rather than simultaneously. They [52] propose a statistical framework for two-way clustering; see also Getz et al. [20] and the references cited therein for earlier approaches on this problem.

We first consider the clustering of tissue samples, using the EMMIX-GENE procedure of McLachlan et al. [39]. For the clustering of gene profiles, we shall describe a mixture model with random effects, EMMIX-WIRE (*EM*-based *mix*ture analysis *wi*th *r*andom *e*ffects), as developed in 2006 by Ng et al. [46]. More information on these programs can be found at the Web addresses http://www.maths.uq.edu.au/~gjm/emmix-gene/ and http://www.maths.uq.edu.au/~gjm/emmix/emmix.html.

21.3 NOTATION

Although biological experiments vary considerably in their design, the data generated by microarray experiments can be viewed as a matrix of expression levels. For M microarray experiments (corresponding to M tissue samples), where we measure the expression levels of N genes in each experiment, the results can be represented by the $N \times M$ matrix. For each tissue, we can consider the expression levels of the N genes, called its *expression signature*. Conversely, for each gene, we can consider its expression levels across the different tissue samples, called its *expression profile*. The M tissue samples might correspond to each of M different patients or, say, to samples from a single patient taken at M different timepoints.

The expression levels are taken to be the measured (absolute) intensities for Affymetrix oligonucleotide arrays, whereas for the spotted arrays (cDNA or oligonucleotide arrays) are taken to be the ratios of sample versus control intensities, represented by the Cy5-channel (red) and Cy3-channel (green) images (see, e.g., Ref. 12). It is assumed that one starts the clustering process with preprocessed (relative) intensities, such as those produced by RMA (for Affymetrix data), loess-modified log ratios, or differences of logged/generalized-logged data; see, for example, Parmigiani et al. [50], Huber et al. [28], Irizarry et al. [30], Rocke and Durbin [54], and Speed [57]. The $N \times M$ matrix is portrayed in Figure 21.1, where each sample represents a separate microarray experiment and generates a set of N expression levels, one for each gene.

In the sequel, we shall use the vector $\mathbf{y}_j$ to represent the measurement (feature observation) on the jth entity to be clustered. In the context of the classification of the tissues on the basis of the gene expressions, we can represent the $N \times M$ matrix $\mathbf{A}$ of gene expressions as

$$\mathbf{A} = (\mathbf{y}_1, \ldots, \mathbf{y}_M), \tag{21.2}$$

where the feature vector $\mathbf{y}_j$ (the *expression signature*) contains the expression levels on the N genes in the jth experiment ($j = 1, \ldots, M$). The latter is a nonstandard problem in parametric cluster analysis because the dimension of the feature space (the number of genes) is typically much greater than the number of observations (the number of tissues).

In the context of the clustering of the genes on the basis of the tissues, we can represent the transpose of the matrix $\mathbf{A}$ in terms of the feature vectors as

$$\mathbf{A}^T = (\mathbf{y}_1, \ldots, \mathbf{y}_N), \tag{21.3}$$

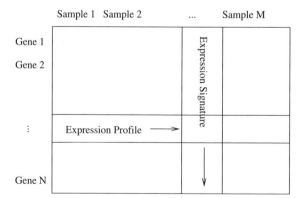

Figure 21.1 Gene expression data from M microarray experiments represented as a matrix of expression levels with the N rows corresponding to the N genes and the M columns to the M tissue samples.

where the feature vector $\mathbf{y}_j$ (the *expression profile*) contains the expression levels on the M tissues on the jth gene ($j = 1, \ldots, N$). For this clustering problem, the number of observations (the number of genes) is very large relative to the dimension of the feature space (the number of tissues), and so in this sense it falls in the standard framework. However, it is not really a standard problem, as not all the genes are independently distributed.

21.4 CLUSTERING OF TISSUE SAMPLES

In the standard setting of a model-based cluster analysis, the n observations $\mathbf{y}_1, \ldots, \mathbf{y}_n$ to be clustered are taken to be independent realizations where the sample size n is much larger than the dimension p of each vector $\mathbf{y}_j$:

$$n >> p. \tag{21.4}$$

It is also assumed that the sizes of the clusters to be produced are sufficiently large relative to p to avoid computational difficulties with near-singular estimates of the within-cluster covariance matrices.

In the cluster analysis of the M tissue samples on the basis of the N genes, we have $n = M$ and $p = N$. Thus the sample size n will be typically small relative to the dimension p, causing estimation problems under the normal mixture model

$$f(\mathbf{y};\ \Psi) = \sum_{i=1}^{g} \phi(\mathbf{y};\ \boldsymbol{\mu}_i, \Sigma_i), \tag{21.5}$$

where $\phi\ (\mathbf{y};\ \boldsymbol{\mu}_i,\ \Sigma_i)$ denotes the p-dimensional normal density function with mean $\boldsymbol{\mu}_i$ and covariance matrix Σ_i and Ψ is the vector of unknown parameters. This is because the g-component normal mixture model (21.5) with unrestricted component–covariance matrices is a highly parameterized model with $\frac{1}{2}p(p + 1)$ parameters for each component–covariance matrix $\Sigma_i (i = 1, \ldots, g)$.

An obvious way to handle the very large number of genes is to perform a principal-component analysis (PCA) and carry out the cluster analysis on the basis of the leading components. The shortcomings of a PCA in such a context is that the leading components need not necessarily reflect the direction in the feature space best for revealing the group structure of the tissues. This is because it is concerned with the direction of maximum variance, which is composed of variance both within and between the clusters. If the latter are relatively large, then the leading components may not be so useful for the purposes of cluster analysis. But with the analysis of microarray data, this problem is compounded by the very large number of genes and their associated noise. Thus artificial directions can result from noisy genes and highly correlated ones. Consequently, a potential problem with a PCA is the determination of an appropriate number of principal components (PCs) useful for clustering. A common practice is to choose the first few leading components. But it may not be clear where to stop and whether some of these components are caused by some artifact or noises in the data. An excellent account of these problems may be found in Liu et al. [33]. They have developed a Bayesian approach to model-based clustering that after an initial PCA simultaneously clusters the observations and selects "informative" variables or components for the cluster analysis.

21.5   THE EMMIX-GENE CLUSTERING PROCEDURE

The EMMIX-GENE procedure handles the problem of a high-dimensional feature vector by using mixtures of factor analyzers whereby the component correlations between the genes are explained by their conditional linear dependence on a small number q of latent or unobservable variables specific to each component. In practice we may wish to work with a subset of the available genes, particularly as the fitting of a mixture of factor analyzers will involve a considerable amount of computation time for an extremely large number of genes. Indeed, the simultaneous use of too many genes in the cluster analysis may serve only to create noise that masks the effect of a smaller number of genes. Also, the intent of the cluster analysis may not be to produce a clustering of the tissues on the basis of all the available genes, but rather to discover and study different clusterings of the tissues corresponding to different subsets of the genes; see the papers of Pollard and van der Laan [52] and Friedman and Meulman [18] on this point. As explained in Belitskaya-Levy [3], the tissues (cell lines or biological samples) may cluster according to cell or tissue type (e.g., cancerous or healthy) or according to cancer type (e.g., breast cancer or melanoma). However, the same samples may cluster differently according to other cellular characteristics, such as progression through the cell cycle, drug metabolism, mutation, growth rate, or interferon response, all of which have a genetic basis.

Therefore, the EMMIX-GENE procedure has two optional steps before the final step of tissue clustering. The first step considers the selection of a subset of relevant genes from the available set of genes by screening the genes on an individual basis to eliminate those that are of little use in clustering the tissue samples in terms of the likelihood ratio test statistic. The second step clusters the retained genes N_o into groups on the basis of Euclidean distance so that highly correlated genes are clustered into the same group. The third and final step of the EMMIX-GENE procedure considers the clustering of the tissues by fitting mixtures of normal distributions or factor analyzers. It can be implemented by considering the groups of genes either simultaneously on the basis of their means or individually on the basis of all or a subset of the genes in a given group. We now describe these three steps in more detail.

21.5.1   Step 1.   Screening of Genes

In step 1 of EMMIX-GENE, we screen the genes by attempting to delete those genes that individually are of little use in clustering the tissue samples into two groups. This screening is undertaken in the absence of tissue samples that are of known classification. The relevance of a gene for clustering the tissue samples can be assessed on the basis of the value of $-2\log\lambda$, where λ is the likelihood ratio statistic for testing $g = 1$ versus $g = 2$ components in the mixture model. In order to reduce the effect of atypically large observations on the value of λ, we fit mixtures of t components with their degrees of freedom inferred from the data. However, the use of t components in place of normal components still does not eliminate the effect of outliers on inference of the number of groups in the tissue samples. For example, suppose that for a given gene there is no genuine grouping in the tissues, but that there are a small number of gross outliers. Then a significantly large value of λ might be obtained, with one component representing the main body of the data (and providing robust estimates of their underlying distribution) and the other representing the outliers. In other words, although the t mixture model may provide robust estimates of the underlying distribution, it does not provide a robust assessment of the number of groups in the data.

In light of the above, the EMMIX-GENE software automatically assesses the relevance of each of the N genes by fitting one- and two-component t mixture models to the expression data over the M tissues for each gene considered individually. If $-2 \log \lambda$ is greater than a specified threshold b_1

$$-2 \log \lambda > b_1, \qquad (21.6)$$

then the gene is taken to be relevant provided that

$$s_{\min} \geq b_2, \qquad (21.7)$$

where $s_{\min}$ is the minimum size of the two clusters implied by the two-component t mixture model and b_2 is a specified threshold. If (21.6) holds but (21.7) does not for a given gene, then the three-component t mixture model is fitted to the tissue samples on this gene, and the value of $-2 \log \lambda$ calculated for the test of $g = 2$ versus $g = 3$. If (21.6) holds for this value of $-2 \log \lambda$, the gene is selected as being relevant (provided at least two of the three clusters implied by the $g = 3$ solution have sizes not less than b_2). Although the null distribution of $-2 \log \lambda$ for $g = 2$ versus $g = 3$ is not the same as for $g = 1$ versus $g = 2$ components, it would appear to be reasonable here to use the same threshold (21.6). The null distribution of $-2 \log \lambda$ for the test of the null hypothesis $H_0 : g = g_0$ versus the alternative hypothesis $H_1 : g = g_1$ is unknown (for finite sample sizes) for normal or t components [42, Ch. 6]. In our applications of EMMIX-GENE, we have taken

$$b_1 = b_2 = 8. \qquad (21.8)$$

The majority of genes in microarray datasets tend to exhibit near-constant expressions across samples [11], and so many methods preselect genes by eliminating those with small variance. For example, the gene-shaving methodology of Hastie et al. [26] is concerned with the identification of small, homogeneous subsets of genes that have maximal variance across the tissue samples. As noted by Pollard and van der Laan [52], genes with low variance can be equally interesting biologically, and so their two-way clustering procedure using hierarchical PAM (partitioning around medoids) is aimed at identifying clusters of genes with both low and high variance across tissues. The gene selection procedure in EMMIX-GENE aims to identify genes whose distributions are not consistent with a single normal distribution, and so it can identify potentially valuable genes for clustering that can have both small and high variances across the tissues.

21.5.2 Step 2. Clustering of Genes: Formation of Metagenes

Concerning the end problem of clustering the tissue samples on the basis of the genes considered simultaneously, we could examine the univariate clusterings provided by each of the selected genes taken individually. But this would be rather tedious when a large number of genes have been selected. Thus with the EMMIX-GENE approach, there is a second (optional) stage for clustering the genes into a user-specified number (N_o) of groups by fitting a mixture in equal proportions of $g = N_o$ normal distributions with covariance matrices restricted to being equal to a multiple of the $(M \times M)$ identity matrix. That is, if the mixing proportions were fixed at 0.5, then it would be equivalent to using a soft version of k means and grouping the genes in terms of the Euclidean distance between them. Since the gene profiles have been normalized, they lie on the surface of the unit hypersphere. Thus, after each M-step of the EM algorithm, we normalize the updated estimates of the component means so that they lie on the surface of the unit hypersphere. More precisely, we could fit mixtures of von Mises–Fisher distributions as in Banerjee et al. [1].

Each group (cluster) of genes can be represented by one or more M-dimensional profile vectors over the M tissues. We follow Huang et al. [27] in referring to these cluster representatives as *metagenes*. In EMMIX-GENE, we take the sample mean of the genes within a cluster to be the metagene representing the cluster. This strategy of using a linear combination of the genes within a cluster to represent it and thereby reducing the dimension of the feature (gene) space also helps smooth out gene-specific noise through the aggregation within a cluster.

The groups of genes are ranked in terms of the likelihood ratio statistic calculated on the basis of the fitted mean of a group over the tissues for the test of single versus two t components. This is provided that the minimum cluster size is greater than a specified threshold. Otherwise, such a group of genes would be placed at the end of the list.

A heatmap of genes in a group versus the tissues is provided for each of the groups where, in each group, the tissues can be left in their original order or rearranged according to their cluster membership obtained by fitting a univariate t mixture model on the basis of the group mean. Alternatively, one could cluster the tissues by fitting a two-component mixture of factor analyzers on the basis of the genes within the group. Heatmaps present a grid of colored points where each color represents a gene expression value for a gene in the tissue sample. They are used here primarily to exhibit similarities between groups or clusters of the tissue samples. Thus they are most effective in this role when the tissue samples have been grouped according to their group (cluster) memberships. Of course, heatmaps are also useful in revealing similarities between the genes.

21.5.3   Step 3. Clustering of Tissues

If a clustering is sought on the basis of the totality of the genes, then it can be obtained by fitting a mixture model to these group means. However, it may be that the number of group means N_o is too large to fit a normal mixture model with unrestricted component–covariance matrices. In this circumstance EMMIX-GENE has the option on the third step that allows for the fitting of mixtures of factor analyzers. The use of mixtures of factor analyzers reduces the number of parameters by imposing the assumption that the correlations between the genes can be expressed in a lower space by the dependence of the tissues on q ($q < N$) unobservable factors. In addition to clustering the tissues on the basis of all the genes, there may be interest in seeing whether the different groups of genes lead to different clusterings of the tissues when each is considered separately. For example, a subset of the genes may be all that is required to identify certain subtypes of the cancer being studied.

It can be seen from above that with the EMMIX-GENE procedure, the genes are being treated anonymously; that is, we do not incorporate existing biological information on the function of genes into the selection procedure. Spang [56] infuses some biological context into an otherwise unsupervised learning task. He structures the feature space by using a functional grid provided by the gene ontology annotations.

21.6   CLUSTERING OF GENE PROFILES

In the remainder of this chapter, we consider the clustering of gene profiles with or without replication across some experimental conditions of interest. For this clustering problem, the number of observations n to be clustered is the number of genes ($n = N$), which will usually be very large relative to the dimension p of the feature space ($p = M$). In this sense it falls in the standard framework. However, this clustering problem is not straightforward as the profiles of the genes are not all independently distributed and the expression levels may have been obtained from an experimental design involving replicated arrays. Thus the standard normal mixture model (21.5) cannot directly be applied to cluster the gene profiles. This is because in unmodified form, this approach does not incorporate experimental design information such as disease status of the tissue samples in which the genes are measured in cross-sectional studies, covariate information such as the time ordering of the gene measurements in time-course studies, or the structure of the replicated data as in longitudinal studies. Pan [48] has proposed to incorporate known gene functions as prior probabilities in model-based clustering. But there is a need to develop further clustering procedures that are applicable to data from a wide variety of experimental designs. For example, microarray experiments are now being carried out with replication for capturing either biological or technical variability in expression levels to improve the quality of inferences made from experimental studies [32,51]. Replicated measurements from each tissue sample (subject) are often interdependent and tend to be more alike in characteristics than are data chosen at random from the population as a whole. Similarly, in timecourse studies [59], where expression levels are measured under various conditions or at different timepoints, gene expressions obtained from the same condition (subject) are correlated.

Ng et al. [46] have developed a random-effect model that provides a unified approach to the clustering of genes with correlated expression levels measured in a wide variety of experimental situations. Their model is an extension of the normal mixture model (21.5) to account for the correlations between the gene profiles and to enable incorporation of covariate information into the clustering process. Hence the model is applicable to longitudinal studies with or without replication, for example, timecourse experiments by using time as a covariate, and to cross-sectional experiments by using categorical covariates to represent the different experimental classes. Ng et al. [46] have shown that their random-effect model EMMIX-WIRE can be fitted by maximum likelihood via the expectation–maximization (EM) algorithm for which the E- and M-steps can be implemented in closed form. Hence their model can be fitted deterministically without the need for time-consuming Monte Carlo approximations.

In related work, Ng et al. [47] have applied this method of clustering to two real timecourse datsets from the budding yeast (*Saccharomyces cerevisiae*) genome. They showed that the proposed method provided clusters of cell-cycle-regulated genes that are supported by existing gene function annotations, and hence enables inference on regulatory interactions for the genetic network. Their approach was to search for regulatory control elements (activators and inhibitors) shared by the clusters of coexpressed genes, based on time-lagged correlations.

As noted by Bryan [5] with the clustering of gene profiles, any clustering structure found may not be directly reflective of biological realities, but might be more due to the preprocessing of the data, which can create sparsely populated areas in the profile space as an artifact. In such situations, the clustering may still be of interest in terms of which genes are put together in the same cluster for various choices of the number of clusters.

21.7 EMMIX-WIRE

The EMMIX-WIRE procedure of Ng et al. [46] formulates a (multilevel) linear mixed-effect model (LMM) for the mixture components in which covariate information can be incorporated. It can be used for the clustering of correlated genes, based on expression microarray data obtained from various experimental designs such as repeated measurement data and timecourse data. Their proposed general random-effect model is formulated by incorporating both "gene" effects and "tissue" effects in the mixture modeling of the microarray data. This is in contrast to the mixed-effect models approaches in Celeux et al. [6], Luan and Li [34], and McLachlan et al. [40] that involve only gene-specific random effects. Their methods thus require the independence assumption for the genes, which, however, will not hold in practice for all pairs of genes [40].

With the EMMIX-WIRE procedure, it is assumed that the observed M-dimensional vectors $\mathbf{y}_1, \ldots, \mathbf{y}_N$ are assumed to have come from a mixture of a finite number, say, g, of components in some unknown proportions $\pi_1, \ldots, \pi_g$, which sum to one. Conditional on its membership of the hth component of the mixture, the vector $\mathbf{y}_j$ for the jth gene follows the model

$$\mathbf{y}_j = \mathbf{X}\boldsymbol{\beta}_h + \mathbf{U}\mathbf{b}_{hj} + \mathbf{V}\mathbf{c}_h + \boldsymbol{\varepsilon}_{hj}, \tag{21.9}$$

where the elements of $\boldsymbol{\beta}_h$ (an M-dimensional vector) are fixed effects (unknown constants) modeling the conditional mean of $\mathbf{y}_j$ in the hth component and $\mathbf{b}_{hj}$ (a q_b-dimensional vector) and $\mathbf{c}_h$ (a q_c-dimensional vector) represent the unobservable gene- and cluster-specific random effects, respectively. The random effects $\mathbf{b}_{hj}$ and $\mathbf{c}_h$, and the measurement error vector $\boldsymbol{\varepsilon}_{hj}$, are assumed to be mutually independent. In (21.9), $\mathbf{X}$, $\mathbf{U}$, and $\mathbf{V}$ are known design matrices of the corresponding fixed or random effects. The specification of (21.9) covers many general random-effect models for the clustering of correlated gene expression data arising from various microarray experiments, including those with replications. For example, let t be the number of distinct tissues in the experiment. We are given for the jth gene a feature vector $\mathbf{y}_j = (\mathbf{y}_{1j}^T, \ldots, \mathbf{y}_{tj}^T)^T$, where $\mathbf{y}_{lj} = (y_{l1j}, \ldots, y_{lrj})^T$ contains the r replications on the jth gene from the lth tissue ($l = 1, \ldots, t$). With respect to (21.9), $\boldsymbol{\beta}_h$ is a M-dimensional vector ($M = t$) modeling the conditional mean of $\mathbf{y}_j$ in the hth component. Moreover, conditional on membership of the hth component, it is assumed that the random effects are shared among the repeated measurements of expression on the same gene from the same tissue [$\mathbf{b}_{hj}$ in (21.9) with $q_b = t$], along with the random effects that are shared among gene expressions from the same tissue [$\mathbf{c}_h$ in (21.9) with $q_c = M = tr$]. The component-specific effects $\mathbf{c}_h$ for the tissues induce dependence among the gene expression levels of genes from the same component and from the same tissue (correlated genes). By allowing the expression levels of the genes in a cluster to have their own and cluster-specific random-effect terms, greater individual and collective variation, respectively, can be exhibited by the genes in the same cluster than would otherwise be possible under a fixed-effects model without gene- and cluster-specific random effects.

With the LMM, the distributions of $\mathbf{b}_{hj}$ and $\mathbf{c}_h$ are taken to be multivariate normal, $N_{q_b}(\mathbf{0}, \theta_{bh} \mathbf{I}_{q_b})$ and $N_{q_c}(\mathbf{0}, \theta_{ch} \mathbf{I}_{q_c})$, respectively, where $\mathbf{I}_{q_b}$ and $\mathbf{I}_{q_c}$ are identity matrices with dimensions being specified by the subscripts. The measurement error vector $\boldsymbol{\varepsilon}_{hj}$ is also taken to be multivariate normal $N_M(\mathbf{0}, \mathbf{D}_h)$, where $\mathbf{D}_h = \text{diag}(\mathbf{W}\boldsymbol{\phi}_h)$ is a diagonal matrix constructed from the vector $(\mathbf{W}\boldsymbol{\phi}_h)$ with $\boldsymbol{\phi}_h = (\sigma_{h1}^2, \ldots, \sigma_{hq_e}^2)^T$ and $\mathbf{W}$ a known $M \times q_e$ zero–one design matrix. Thus, we allow the hth component variance to be different among the M microarray experiments.

21.8   MAXIMUM-LIKELIHOOD ESTIMATION VIA THE EM ALGORITHM

We let $\boldsymbol{\Psi} = (\boldsymbol{\psi}_1^T, \ldots, \boldsymbol{\psi}_g^T, \pi_1, \ldots, \pi_{g-1})^T$ be the vector of all the unknown parameters, where $\boldsymbol{\psi}_h$ is the vector containing the unknown parameters $\boldsymbol{\beta}_h$, θ_{bh}, θ_{ch}, and $\boldsymbol{\phi}_h$ of the hth component density ($h = 1, \ldots, g$). Ng et al. [46] showed that the estimation of Ψ can be obtained by maximum-likelihood (ML) via the EM algorithm of Dempster et al. [9]. The implementation of the E-step is straightforward for mixture models provided that the data can be treated as being independently distributed. In their model (21.9), the gene profile vectors $\mathbf{y}_j$ are not all independently distributed as genes within the same cluster (i.e., from the same component in the mixture model) are allowed to be dependent due to the presence of the random-effect term $\mathbf{c}_h$ for the hth component in (21.9). However, this problem can be circumvented by proceeding conditionally on the random cluster effects $\mathbf{c}_h$, as, given these terms, the gene profile vectors $\mathbf{y}_j$ are all conditionally independent. In this way, Ng et al. [46] showed that the E- and M-steps can be carried out in closed form. In particular, we do not have to approximate the E-step by carrying out time-consuming Monte Carlo approximations.

Within the EM framework, each $\mathbf{y}_j$ is conceptualized to have originated from one of the g components. We let $\mathbf{z}_1, \ldots, \mathbf{z}_N$ denote the unobservable component indicator vectors, where the hth element z_{hj} of $\mathbf{z}_j$ is taken to be one or zero according to whether $\mathbf{y}_j$ originates from the hth component or not given $\mathbf{c}$, where $\mathbf{c} = (\mathbf{c}_1^T, \ldots, \mathbf{c}_g^T)^T$. We let $\mathbf{y} = (\mathbf{y}_1^T, \ldots, \mathbf{y}_N^T)^T$ denote the observed data and, correspondingly, put $\mathbf{z}^T = (\mathbf{z}_1^T, \ldots, \mathbf{z}_N^T)$. The ML estimation of the normal mixture of LMMs via the EM algorithm can be formulated by treating the unobservable component indicator variables $\mathbf{z}$ and the random effects $\mathbf{b} = (\mathbf{b}_1^T, \ldots, \mathbf{b}_g^T)^T$ and $\mathbf{c}$ as missing data in the EM framework [45], where $\mathbf{b}_h = (\mathbf{b}_{h1}^T, \ldots, \mathbf{b}_{hN}^T)^T$ for $h = 1, \ldots, g$. Let $\varepsilon_h = (\varepsilon_{h1}^T, \ldots, \varepsilon_{hn}^T)^T$ for $h = 1, \ldots, g$. With

$$(\mathbf{y}^T, \mathbf{z}^T, \mathbf{b}^T, \mathbf{c}^T)^T$$

taken to be the complete data, it follows that the complete data log likelihood is given, apart from an additive constant, by

$$\log L_c(\Psi) = \sum_{h=1}^{g} \left[\sum_{j=1}^{n} z_{hj} \log \pi_h - \frac{1}{2} \left\{ \sum_{j=1}^{n} z_{hj} q_b \log \theta_{bh} + q_c \log \theta_{ch} \right. \right.$$
$$\left. \left. + \sum_{j=1}^{n} z_{hj} \log |A_h| + \frac{\mathbf{b}_h^T \mathbf{b}_h}{\theta_{bh}} + \frac{\mathbf{c}_h^T \mathbf{c}_h}{\theta_{ch}} + \varepsilon_h^T \Omega_h \varepsilon_h \right\} \right], \qquad (21.10)$$

where

$$\mathbf{b}_h^T \mathbf{b}_h = \sum_{j=1}^{n} z_{hj} \mathbf{b}_{hj}^T \mathbf{b}_{hj}$$

and

$$\Omega_h = \mathbf{I}_n \otimes A_h^{-1}$$

for $h = 1, \ldots, g$, and hence

$$\varepsilon_h^T \Omega_h \varepsilon_h = \sum_{j=1}^{n} z_{hj} \varepsilon_{hj}^T A_h^{-1} \varepsilon_{hj}.$$

In the above, the sign $\otimes$ denotes the Krönecker product of two matrices. By consideration of (21.10), Ng et al. [46] showed that the E- and M-steps can be implemented in closed form.

To effect a probabilistic or an outright clustering of the genes into g components, we can condition on the cluster random-effect vector $\mathbf{c}_h$. As the latter is unobservable, we use its estimated conditional expectation given the observed data

$$\hat{\mathbf{c}}_h = E_{\hat{\Psi}}(\mathbf{c}_h|\mathbf{y}), \tag{21.11}$$

where $E_{\hat{\Psi}}$ denotes taking expectation using the ML estimate $\hat{\Psi}$ for the vector Ψ of unknown parameters. Since the genes within a cluster are independently distributed given $\mathbf{c}_h$, it suffices to effect a clustering with each gene considered individually in terms of its estimated posterior probabilities of component membership given its profile vector and $\mathbf{c}_h$, for $h = 1, \ldots, g$ and $j = 1, \ldots, n$. Using Bayes' theorem, the posterior probability that the jth gene belongs to the hth component given $\mathbf{y}_j$, and $\mathbf{c}$, $\tau(\mathbf{y}_j, \mathbf{c}; \Psi)$ can be expressed as

$$
\begin{aligned}
\tau(\mathbf{y}_j, \mathbf{c}; \Psi) &= \Pr\{Z_{hj} = 1 | \mathbf{y}_j, \mathbf{c}\} \\
&= \frac{\pi_h f(\mathbf{y}_j|z_{hj} = 1, \mathbf{c}_h; \psi_h)}{\sum_{i=1}^{g} \pi_i f(\mathbf{y}_j|z_{ij} = 1, \mathbf{c}_i; \psi_i)},
\end{aligned} \tag{21.12}
$$

where $f(\mathbf{y}_j \mid z_{hj} = 1, \mathbf{c}_h; \psi_h)$ denotes the hth component density of $\mathbf{y}_j$ given the random effect $\mathbf{c}_h$. The log of this density is given by

$$
\begin{aligned}
\log f(\mathbf{y}_j|z_{hj} = 1, \mathbf{c}_h; \psi_h) = -\tfrac{1}{2}\Big\{ & \log |\mathbf{B}_h| \\
& + (\mathbf{y}_j - \mathbf{X}\beta_h - \mathbf{V}\mathbf{c}_h)^T \mathbf{B}_h^{-1}(\mathbf{y}_j - \mathbf{X}\beta_h - \mathbf{V}\mathbf{c}_h)\Big\},
\end{aligned}
$$

apart from an additive constant, is the log of the hth component density of $\mathbf{y}_j$ conditional on $\mathbf{c}_h$, where $\mathbf{B}_h = \mathbf{A}_h + \theta_{bh} \mathbf{U}\mathbf{U}^T$.

21.9 MODEL SELECTION

Specification of the random-effect components in the model (21.9) needs careful consideration. An identifiability problem could arise if the random-effect model is specified so that the design matrix $\mathbf{V}$ for the random effects $\mathbf{c}_h$ is the same as the $\mathbf{X}$ for the fixed effects β_h. In their study, Ng et al. [46] were concerned with situations where the emphasis is on the grouping of the genes rather than the number of clusters and their link with externally existing groups; that is, they were concerned primarily in finding which genes are put together in the same cluster for plausible choices of the number of components g in the mixture model. A guide to plausible values of g can be obtained using the Bayesian information criterion (BIC) of Schwarz [55], whereby the number g of components in the mixture model is taken to minimize $-2 \log L(\hat{\Psi}) + d \log n$, where d denotes the number of parameters in the model. In the EM framework, $L(\Psi)$ is the incomplete-data likelihood function for Ψ. However, as the gene profile vectors $\mathbf{y}_j$ are not all independently distributed, this likelihood function $L(\Psi)$ is unable to be calculated directly by taking the product of the (marginal) densities of the $\mathbf{y}_j$. Ng et al. [46] suggested that $L(\Psi)$ be approximated by forming it as if all the $\mathbf{y}_j$ were independent. Another approach would be to use resampling methods [13,37,41].

21.10 EXAMPLE: CLUSTERING OF TIMECOURSE DATA

To illustrate the EMMIX-WIRE approach to the clustering of gene profiles, Ng et al. [46] applied it to three representative datasets, each arising from different kinds of microarray experiments: timecourse data as in the yeast cell cycle study of Spellman et al. [58], data with repeated measurements as in the yeast galactose study of Ideker et al. [29], and finally cross-sectional data involving two groups of tissues (tumuor and normal) as in the study of human colorectal carcinomas of Muro et al. [44].

We report here their first example. By analyzing cDNA microarrays from yeast cultures synchronized by three independent methods over approximately two cell cycle periods, Spellman et al. [58] identified 800 yeast genes that meet an objective minimum criterion for cell cycle regulation. In their study, Ng et al. [46] considered the 18α-factor (pheromone) synchronization where the yeast cells were sampled at 7-min intervals for 119 min. They worked with a subset of 612 genes that had no missing expression data across any of the 18 timepoints. Their aim was to cluster the cell-cycle-regulated genes on the basis of the microarray expression data matrix of $N = 612$ rows (genes) and $M = 18$ columns (timepoints). They then analyzed the clusters so formed for common regulatory elements, as described by Spellman et al. [58]. With reference to (21.9), they took the design matrix $\mathbf{X}$ to be an 18×2 matrix with the $(l + 1)$th row ($l = 0, \ldots, 17$)

$$\left(\cos\left(\frac{2\pi(7l)}{\omega} + \Phi \right), \ \sin\left(\frac{2\pi(7l)}{\omega} \right) + \Phi \right),$$

where ω is the period of the cell cycle and Φ is the phase offset. They adopted here the least-squares estimation approach considered by Booth et al. [4] to obtain the cell cycle period $\omega = 53$ and the initial phase $\Phi = 0$ from the dataset. For the design matrices of the random-effect parts, they took $\mathbf{U} = \mathbf{1}_{18}$ and $\mathbf{V} = \mathbf{I}_{18}$; that is, it is assumed that there exist random gene effects b_{hj} with $q_b = 1$ and random temporal effects $(c_{h1}, \ldots, c_{hq_c})$ with $q_c = m = 18$. The latter introduce interdependence among expression levels within the same cluster obtained from the same timepoint. Also, they took $\mathbf{W} = \mathbf{1}_{18}$ and $\phi_h = \sigma_h^2$ ($q_e = 1$) so that the component variances were common among the $m = 18$ experiments. The mixture model of LMMs was fitted to the data with $g = 4$ to $g = 15$ components. The number of components g was determined using BIC for model selection. These experiments indicated that there are 12 clusters.

The clustering results for $g = 12$ as obtained by Ng et al. [46] are given in Figure 21.2, where the expression profiles for genes in each cluster are presented. From Figure 21.2, it can be seen that the genes have very similar expression patterns within each cluster, except in clusters 4 and 7, where there is greater individual variation in some of the genes. This clustering result is different from Spellman's clustering, which was based on time of peak expression only.

For clusters 1, 3, 10, 11, and 12, which show clear periodic expression patterns, Ng et al. [46] searched through the 700-bp (basepair) upstream region of the start codon of each gene for the presence of binding site sequences for any known yeast cell cycle transcription factors such as MBF, SBF, Mcm1p-containing factors, and Swi5p factors. The results are summarized in Table 21.1. They found that the majority of the genes in these clusters share common promoter elements; furthermore, they correspond to known cell cycle transcription factor binding sites relevant to the time of peak expression.

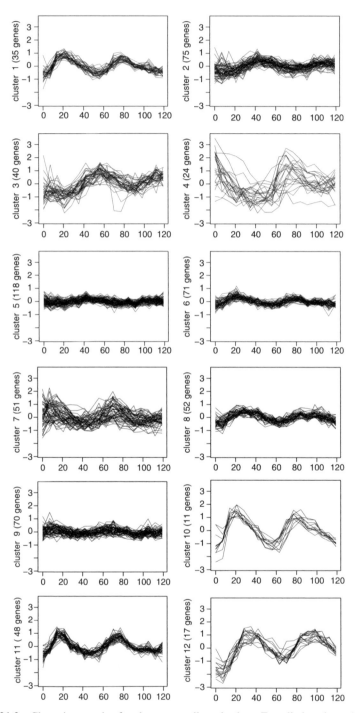

Figure 21.2   Clustering results for the yeast cell cycle data. For all the plots, the *x* axis (abscissa) is the timepoint and the *y* axis (ordinate) is the gene expression level.

Table 21.1 Promoter Elements (Yeast Cell Cycle Data)

Cluster	Number of Genes	Binding Site	Regulator	Peak Expression
1	35	ACGCGT	MBF, SBF	G1
3	40	MCM1 + SFF	Mcm1p + SFF	G2/M
10	11	ACGCGT	MBF, SBF	G1
11	48	Unknown	Unknown	G1
12	17	ATGCGAAR	Unknown	S

21.11 CONCLUDING REMARKS

As an increasing number and variety of high-throughput datasets become available, cluster analysis is playing an ever-increasing role in the analysis of these biological data. Hierarchical methods have been the primary clustering tool employed to date. The hierarchical algorithms have been applied mainly heuristically to these cluster analysis problems. Also, there is no reason why the clusters of tissues (or genes) should belong to a hierarchy such as in the evolution of species. Further, a major limitation of these methods is their inability to determine the number of clusters. Thus there is a need for a model-based approach to this clustering problem. Concerning the clustering of tissue samples, a clustering of some tumors, for example, will reveal whether tumors that have traditionally been lumped together as one type should be divided into a number of distinct subtypes, and whether these subtypes have different prognoses and respond differently to specific therapies. For this clustering problem, we have described the EMMIX-GENE procedure, which is a model-based approach to the clustering of high-dimensional independent observations.

The EMMIX-GENE procedure fits a mixture of multivariate normals without regression structure on the component means and without constraints on the covariance matrices that arise in experimental designs with structure, including replications taken over time. Thus it is not directly applicable to the other clustering problem of grouping the gene profile vectors as in longitudinal or cross-sectional studies. This problem arises where, say, the interest is in studying the changes in gene expression of entire groups of (correlated) genes as a means to finding possible functional relationships among them, the identification of transcription factor binding sites, and the elucidation of biological pathways. The biological rationale underlying the clustering of the gene profiles is the fact that often many coexpressed genes are also coregulated, which is supported by an immense body of empirical observations as well as a detailed mechanistic explanation [62]. However, it has been observed that genes with similar profiles sometimes do not share biological similarity [7,22,10]. Thus clustering does not provide proof of relationships between the genes, but it does provide suggestions that help to direct further research. The idea is that we can establish a guilt by association—that is, genes with similar expression patterns are more likely to have similar biological function. For this clustering problem, we have described the EMMIX-WIRE procedure, which provides a unified approach to the clustering of genes with correlated expression levels measured in a wide variety of experimental situations. This procedure is applicable to longitudinal studies with or without replication, for example, timecourse experiments by using time as a covariate, and to cross-sectional experiments by using categorical covariates to represent the different experimental classes.

Most clustering algorithms require that one gene be assigned to one cluster, adding an arbitrary element to the analysis. Mixture modeling provides one way to reduce this arbitrariness and to handle the clustering of the borderline cases. It gives a probabilistic or "soft" clustering through the posterior probabilities of component membership of each gene. An overlapping clustering can be obtained by making a hard assignment of each gene to one or more of the components (clusters) using a threshold on the posterior probabilities of component membership; for example, the jth gene with profile vector $\mathbf{y}_j$ belongs to the hth component if its posterior probability of membership of the hth component is greater than some specified threshold c.

REFERENCES

1. Banerjee, A., Dhillon, I. S., Ghosh, J., and Sra, S., Clustering on the unit hypersphere using von Mises-Fisher distributions, *J. Machine Learn. Res.* **6**, 1345–1382 (2005).

2. Banfield, J. D. and Raftery, A. E. Model-based Gaussian and non-Gaussian clustering, *Biometrics* **49**, 803–821 (1993).

3. Belitskaya-Levy, I., A generalized clustering problem, with application to DNA microarrays, *Statist. Appli. Genet. Molec. Biol.* **5**, Art. 2, (2006).

4. Booth, J. G., Casella, G., Cooke, J. E. K., and Davis, J. M., *Statistical Approaches to Analysing Microarray Data Representing Periodic Biological Processes: A Case Study Using the Yeast Cell Cycle*, Technical Report, Dept. Biological Statistics and Computational Biology, Cornell Univ., 2004.

5. Bryan, J., Problems in gene clustering based on gene expression data, *J. Multivar. Anal.* **90**, 44–66 (2004).

6. Celeux, G., Martin, O., and Lavergne, C., Mixture of linear mixed models for clustering gene expression profiles from repeated microarray experiments, *Statist. Model.* **5**, 243–267 (2005).

7. Clare, A. and King, R. D., How well do we understand the clusters in microarray data?, *In Silico Biol.* **2**, 511–522 (2002).

8. Coleman, D., Dong, X., Hardin, J., Rocke, D. M., and Woodruff, D. L., Some computational issues in cluster analysis with no a priori metric, *Comput. Statist. Data Anal.* **31**, 1–11 (1999).

9. Dempster, A. P., Laird, N. M., and Rubin, D. B. Maximum likelihood from incomplete data via the EM algorithm (with discussion), *J. Roy. Statist. Soc. B* **39**, 1–38 (1977).

10. DeRisi, J. L., Iyer, V. R., and Brown, P. O., Exploring the metabolic and genetic control of gene expression on a genomic scale, *Science* **278**, 680–686 (1997).

11. Dudoit, S. and Fridlyand, J., A prediction-based resampling method for estimating the number of clusters in a dataset, *Genome Biol.* **3**, research0036.1–0036.21 (2002).

12. Dudoit, S., Yang, Y. H., Callow, M. J., and Speed, T. P., Statistical methods for identifying differentially expressed genes in replicated cDNA microarray experiments, *Statistica Sinica* **12**, 111–139 (2002).

13. Efron, B. and Tibshirani, R., *An Introduction to the Bootstrap*, Chapman & Hall, London, 1993.

14. Eisen, M. B., Spellman, P. T., Brown, P. O., and Botstein, D., Cluster analysis and display of genome-wide expression patterns, *Proc. Natl. Acad. Sci. USA* **95**, 14863–14868 (1998).

15. Eisen, M. B. and Brown, P. O., DNA Arrays for analysis of gene expression, *Meth. Enzymol.* **303**, 179–205 (1999).

16. Fraley, C. and Raftery, A. E., How many clusters? Which clustering method? Answers via model-based cluster analysis, *Comput. J.* **41**, 578–588 (1998).

17. Fraley, C. and Raftery, A. E., Model-based clustering, discriminant analysis, and density estimation, *J. Am. Statist. Assoc.* **97**, 611–631 (2002).

18. Friedman, J. H. and Meulman, J. J., Clustering objects on subsets of attributes (with discussion), *J. Roy. Statist. Soc. B* **66**, 815–849 (2004).

19. Ganesalingam, S. and McLachlan, G. J., The efficiency of a linear discriminant function based on unclassified initial samples, *Biometrika* **65**, 658–662 (1978).

20. Getz, G., Levine, E., and Domany, E., Coupled two-way clustering analysis of gene microarray data, *Cell Biol.* **97**, 12079–12084 (2000).

21. Ghosh, D. and Chinnaiyan, A. M., Mixture modelling of gene expression data from microarray experiments, *Bioinformatics* **18**, 275–286 (2002).

22. Gibbons, F. D. and Roth, F. P., Judging the quality of gene expression-based clustering methods using gene annotation, *Genome Res.* **12**, 1574–1581 (2002).

23. Hand, D. J. and Heard, N. A., Finding groups in gene expression data, *J. Biomed. Biotechnol.* **2005**, 215–225 (2005).

24. Goldstein, H., *Multilevel Statistical Models*, 2nd ed., Arnold, London, 1995.

25. Hartigan, J. A., Statistical theory in clustering, *J. Classification* **2**, 63–76 (1975).

26. Hastie, T., Tibshirani, R., Eisen, M. B., Alizadeh, A., Levy, R., Staudt, L., Chan, W. C., Botstein, D., and Brown, P., "Gene shaving" as a method for identifying distinct sets of genes with similar expression patterns, *Genome Biol.* **1**, research0003.1–0003.21 (2000).

27. Huang, E., Cheng, S. H., Dressman, H., Pittman, J., Tsou, M.-H., Horng, Ch.-F., Bild, A., Iversen, E. S., Liao, M., Chen, C.-M., West, M., Nevins, J. R., and Huang, A. T., Gene expression predictors of breast cancer outcomes, *Lancet* **361**, 1576–1577 (2003).

28. Huber, W., von Heydebreck, A., Sueltmann, H., Poustka, A., and Vingron, M., Parameter estimation for the calibration and variance stabilization of microarray data, *Statist. Appl. Genet. Molec. Biol.* **2**(1), Art. 3 (2003).

29. Ideker, T., Thorsson, V., Siegel, A. F., and Hood, L. E., Testing for differentially-expressed genes by maximum-likelihood analysis of microarray data, *J. Comput. Biol.* **7**, 805–817 (2000).

30. Irizarry, R. A., Hobbs, B., Collin, F., Beazer-Barclay, Y. D., Antonellis, K. J., Scherf, U., and Speed, T., Exploration, normalization, and summaries of high density oligonucleotide array probe level data, *Biostatistics* **4**, 249–264 (2003).

31. Kettenring, J. R., The practice of cluster analysis, *J Classification* **23**, 3–30 (2006).

32. Lee, M. L. T., Kuo, F. C., Whitmore, G. A., and Sklar, J., Importance of replication in microarray gene expression studies: statistical methods and evidence from repetitive cDNA hybridizations, *Proc. Natl. Acad. Sci. USA* **97**, 9834–9838 (2000).

33. Liu, J. S., Zhang, J. L., Palumbo, M. J., and Lawrence, C. E., Bayesian clustering with variable and transformation selections, in Bayesian Statistics, Bernardo, J. M., Bayarri, M. J., Berger, J. O., Dawid, A. P., Heckerman, D., Smith, A. F. M., and West, M., eds., Oxford Univ. Press, Oxford, UK, 2003, vol. 7, pp. 249–275.

34. Luan, Y. and Li, H., Clustering of time-course gene expression data using a mixed-effects model with B-splines, *Bioinformatics* **19**, 474–482 (2003).

35. Marriott, F. H. C., *The Interpretation of Multiple Observations*, Academic Press, London, 1974.

36. McCulloch, C. E. and Searle, S. R., *Generalized, Linear, and Mixed Models*, Wiley, New York, 2001.

37. McLachlan, G. J., On bootstrapping the likelihood ratio test statistic for the number of components in a normal mixture, *Appl. Statist.* **36**, 318–324 (1987).

38. McLachlan, G. J. and Basford, K. E., *Mixture Models: Inference and Applications to Clustering*, Marcel Dekker, New York, (1988).

39. McLachlan, G. J., Bean, R. W., and Peel, D., A mixture model-based approach to the clustering of microarray expression data, *Bioinformatics* **18**, 413–422 (2002).

40. McLachlan, G. J., Do, K. A., and Ambroise, C., *Analyzing Microarray Gene Expression Data*, Wiley, Hoboken, NJ, (2004).

41. McLachlan, G. J. and Khan, N., On a resampling approach for tests on the number of clusters with mixture model-based clustering of tissue samples, *J. Multivar. Anal.* **90**, 90–105 (2004).

42. McLachlan, G. J. and Peel, D., *Finite Mixture Models*, Wiley, New York, 2000.

43. Medvedovic, M. and Sivaganesan, S., Bayesian infinite mixture model based clustering of gene expression profiles, *Bioinformatics* **18**, 1194–1206 (2002).

44. Muro, S., Takemasa, I., Oba, S., Matoba, R., Ueno, N., Maruyama, C., Yamashita, R., Sekimoto, M., Yamamoto, H., Nakamori, S., Monden, M., Ishii, S., and Kato, K., Identification of expressed genes linked to malignancy of human colorectal carcinoma by parametric clustering of quantitative expression data, *Genome Biol.* **4**(5), Art. R21 (2003).

45. Ng, S. K., Krishnan, T., and McLachlan, G. J., The EM algorithm, in *Handbook of Computational Statistics*, Gentle, J., Hardle, W., and Mori, Y., eds, Springer-Verlag, New York, Vol. 1, pp. 137–168 (2004).

46. Ng, S. K., McLachlan, G. J., Wang, K., Ben-Tovim, L., and Ng, S. W., A mixture model with random-effects components for clustering correlated gene-expression profiles, (in press).

47. Ng, S. K., Wang, K., and McLachlan, G. J., Multilevel modelling for inference of genetic regulatory networks, in *Proc. SPIE 2005, Complex Systems in the Int. Symp. Microelectronics, MEMS, and Nanotechnology*, Bender, A., ed., International Society for Optical Engineering, Bellingham, WA, 2006, Vol. 6039, pp. 60390S-1–60390S-12.

48. Pan, W. Incorporating gene functions as priors in model-based clustering of microarray gene expression data, *Bioinformatics* **22**(7) 795–801 (2006).

49. Pan, W., Lin, J., and Le, C. T., Model-based cluster analysis of microarray gene-expression data, *Genome Biol.* **3**, research0009.1–0009.8 (2002).

50. Parmigiani, G., Garrett, E. S., Irizarry, R. A., and Zeger, S. L., eds., *The Analysis of Gene Expression Data*, Springer-Verlag, New York, 2003.

51. Pavlidis, P., Li, Q., and Noble, W. S., The effect of replication on gene expression microarray experiments, *Bioinformatics*, **19**, 1620–1627 (2003).

52. Pollard, K. S. and van der Laan, M. J., Statistical inference for simultaneous clustering of gene expression data, *Math. Biosci.* **176**, 99–121 (2002).

53. Reilly, C., Wang, C., and Rutherford, R., A rapid method for the comparison of cluster analyses, *Statistica Sinica* **15**, 19–33 (2005).

54. Rocke, D. M. and Durbin, B., Approximate variance-stabilizing transformations for a gene-expression microarray data, *Bioinformatics* **19**, 966–972 (2003).

55. Schwarz, G., Estimating the dimension of a model, *Ann. Statist.* **6**, 461–464 (1978).

56. Spang, R., Diagnostic signatures from microarrays: A bioinformatics concept for personalized medicine, *Biosilico* **1**, 64–68 (2003).

57. Speed, T., ed., *Statistical Analysis of Gene Expression Microarray Data.* Chapman & Hall/CRC, Boca Raton, FL, 2003.

58. Spellman, P., Sherlock, G., Zhang, M. Q., Iyer, V. R., Anders, K., Eisen, M. B., Brown, P. O., Botstein, D., and Futcher, B., Comprehensive identification of cell cycle-regulated genes of the yeast Saccharomyces cerevisiae by microarray hybridization, *Molec. Biol. Cell* **9**, 3273–3297 (1998).

59. Storey, J. D., Xiao, W., Leek, J. T., Tompkins, R. G., and Davis, R. W., Significance analysis of time course microarray experiments, *Proc. Natl. Acad. Sci. USA* **102**, 12837–12842 (2005).

60. Yeung, K. Y., Fraley, C., Murua, A., Raftery, A. E., and Ruzzo, W. L. Model-based clustering and data transformations for gene expression data, *Bioinformatics* **17**, 977–987 (2001).

61. Aitkin, M., Anderson, D., and Hinde, J., Statistical modeling of data on teaching styles, *J. Roy. Statist. Soc.* **144**(4), 419–461 (1981).

62. Boutros, P. C. and Okey, A. B., Unsupervised pattern recognition: An introduction to the whys and wherefores of clustering microarray data, *Br. Bioinf.* **6**(4), 331–343 (2005).

CHAPTER 22

Censored Data Regression in High-Dimensional and Low-Sample-Size Settings for Genomic Applications

Hongzhe Li

Department of Biostatistics and Epidemiology, University of Pennsylvania School of Medicine, Philadelphia, Pennsylvania

22.1 INTRODUCTION

High-throughput technologies generate many types of high-dimensional genomic and proteomics data. Important examples include DNA microarray technology, which permits simultaneous measurements of expression levels for thousands of genes, array-based comparative genomic hybridization (aCGH) data that measure the change of DNA copy numbers, array-based single-nucleotide polymorphism (SNP) data, and mass spectrometry data to measure protein expression levels. Such high-throughout genomic data offer the possibility of a powerful, genomewide approach to the genetic basis of different types of tumors and can be used for molecular classification of cancers, for studying varying levels of drug responses in the area of pharmacogenomics, and for predicting different patients' clinical outcomes. The problem of cancer class prediction using the gene expression data, which can be formulated as predicting binary or multicategory outcomes, has been studied extensively and has demonstrated great promise [21,42]. There has also been active research of methods development in relating gene expression profiles to other phenotypes, such as quantitative continuous phenotypes or censored survival phenotypes such as time to cancer recurrence or time to death. Because of the wide variability in time to certain clinical events such as cancer recurrence among cancer patients, and in age of onset of many complex diseases, studying possibly censored survival phenotypes can be more informative than treating the phenotypes as binary or categorical variables.

Statistical Advances in the Biomedical Sciences, edited by Atanu Biswas, Sujay Datta, Jason P. Fine, and Mark R. Segal

The goal of linking genomic data to censored survival data is twofold: to identify genes that are involved in the risk of a clinical event and to build a predictive model for future patients' survival based on both genomic data and patient-specific covariates. These two goals are related but not equivalent, although a good predictive model often implies that the variables used in the model are relatively important or predictive. Owing to the problem of censoring, survival analysis models are obviously relevant to this problem. The Cox regression model [7] is the most popular method in regression analysis for censored survival data. Alternately, one can consider the accelerated failure-time (AFT) model [6,48] and the additive hazard model [34]. For a given censored data regression model, because of the very high-dimensional space of the predictors (i.e., where the genes with expression levels measured by microarray experiments), the standard estimation method cannot be applied directly to obtain the parameter estimates. Besides the high-dimensionality, the expression levels of some genes are often highly correlated, which creates the problem of high collinearity. Finally, we should also expect complex interactions between genes to affect the risk of survival. To deal with these problems, Li and Luan [31], Li and Gui [30], Li and Li [33], Gui and Li [22,23], and Li and Luan [32] were the first to investigate the use of penalized estimation procedures for the Cox model in the high-dimensional and low-sample-size settings. These regularized estimation methods were subsequently extended for the AFT models and the additive hazard models by Huang et al. [26] and Ma and Huang [35,36] by using appropriately defined loss functions.

The focus of this review is to present some more recently developed statistical and computational methods for relating high-throughout genomic data to censored survival outcomes, including both the methods for identifying genes related to such survival outcomes and the methods for building predictive models for future patient survival. The remainder of the chapter is organized as follows. We first review some commonly used censored data regression models. We then present a class of penalized estimation procedures for various models. We also present ensemble boosting methods for censored data regression models and briefly mention methods based on dimension reduction and Bayesian variable selection. We present a comparison of some of these methods using a real dataset of diffuse large B-cell lymphoma (DLBCL) survival times and gene expression data [40]. Finally, we briefly discuss the methods and present several important problems for future research.

22.2 CENSORED DATA REGRESSION MODELS

Suppose that we have a sample size of n from which to estimate the relationship between the survival time T and the gene expression levels $X = \{X_1, \ldots, X_p\}$ of p genes. In addition, let Z be the vector of other patient-specific covariates. As a result of censoring, for $i = 1, \ldots, n$, the ith datum in the sample is denoted by $(t_i, \delta_i, x_{i1}, x_{i2}, \ldots, x_{ip}, z_i)$, where δ_i is the censoring indicator and t_i is the survival time if $\delta_i = 1$ or censoring time if $\delta_i = 0$, and $x_i = \{x_{i1}, x_{i2}, \ldots, x_{ip}\}'$ is the vector of the gene expression level of p genes for the ith sample. In this section, we briefly review the three most commonly used censored data regression models, including the Cox proportional hazards model, the AFT model, and the additive hazards model.

22.2.1 The Cox Proportional Hazards Model

The Cox proportional hazards model is the most commonly used censored data regression model in survival analysis. The model assumes the following hazard function for cancer recurrence or death at time t

$$\lambda(t|X, Z) = \lambda_0(t) \exp(F(X, Z))$$
$$= \lambda_0(t) \exp(\beta_1 X_1 + \beta_2 X_2 + \cdots + \beta_p X_p + \gamma' Z)$$
$$= \lambda_0(t) \exp(\beta' X + \gamma' Z), \tag{22.1}$$

where $\lambda_0(t)$ is an unspecified baseline hazard function and $F(X, Z)$ is the function that links the (X, Z) to the hazard function. If this function is assumed to be linear, then $\beta = \{\beta_1, \ldots, \beta_p\}$ is the vector of the regression coefficients related to the p genomic data, and $X = \{X_1, \ldots, X_p\}$ is the vector of gene expression levels with the corresponding sample values of $x_i = \{x_{i1}, \ldots, x_{ip}\}$ for the ith sample. Finally, γ is the risk ratio parameter associated with covariate vector Z.

Using the available sample data, the Cox's partial likelihood [7] can be written as

$$L(\beta, \gamma) = \prod_{r \in D} \frac{\exp(\beta' x_r + \gamma' z_r)}{\sum_{j \in R_r} \exp(\beta' x_j + \gamma' z_j)},$$

where D is the set of indices of the events (e.g., deaths) and R_r denotes the set of indices of the individuals at risk at time $t_r - 0$. Note that when $p > n$, there is no unique β to maximize this partial likelihood function, and therefore some regularization is required (see Sections 22.3.1–22.3.3). Even when $p \leq n$, some regularization may still be required in order to reduce the variances of the estimates and to improve the prediction performance.

22.2.2 Accelerated Failure-Time Model

Let T be the random variable of time to event. For the ith individual, let t_i be the respective random variable. Let c_i be the censoring times, assumed to be i.i.d and following a survival function $G(t) = \Pr(c_i > t)$. The linear AFT model assumes

$$g(T) = \alpha + \beta' X + \gamma' Z + \varepsilon, \tag{22.2}$$

where g is some prespecified monotone function (e.g., log function) and ε is (heteroscedastic) unobservable error, assumed to be independent with zero means and bounded variances across n individuals. Because of censoring, for $i = 1, \ldots, n$, the ith datum in the sample is denoted by $(y_i, \delta_i, x_{i1}, x_{i2}, \ldots, x_{ip})$, where δ_i is the censoring indicator and y_i is g transformation of the survival time if $\delta_i = 1$ or g transformed of the censoring time if $\delta_i = 0$:

$$y_i = \min(g(t_i), g(c_i)), \quad \delta_i = I[t_i \leq c_i], \quad i = 1, \ldots, n.$$

Wei [48] discussed some advantages of using such AFT models over the popular Cox regression model, including easy interpretation of the model parameters and better fits for some datasets. One approach for estimating the parameter β is the Buckley–James (BJ) [6] procedure. In Section 22.3.4, we present simple modification of the BJ procedure to deal with the problem of large p. Alternatively, one can estimate β by minimizing the inverse probability of censoring weighted (IPCW) loss function introduced in Robins and Rotnitsky [39]. On the basis of this loss function, one can develop regularized estimation procedures for β and extend the random forests and boosting procedure to censored survival data (see Section 22.4.2). For simplicity, we consider only model (22.3.2) without covariate Z.

22.2.3   Additive Hazard Regression Models

The additive risk model as described in Lin and Ying [34] assumes the following conditional hazard at time t

$$\lambda(t|X, Z(\cdot)) = \lambda_0(t) + \beta'X + \gamma'Z(t), \tag{22.3}$$

given a p-dimensional vector of genomic data X and patient-specific covariate $Z(\cdot)$, which can be time-dependent. Here β, γ, and $\lambda_0(t)$ denote the unknown regression parameter and the unknown baseline hazard function. In the following discussion, we simply assume that there is no covariate Z in the model (22.3). Denote $\{N_i(t) = I(t_i \leq t, \delta_i = 1); t \geq 0\}$ and $\{Y_i(t) = I(t_i \leq t); t \geq 0\}$ as the observed event process and the at-risk process. Lin and Ying [34] proposed the following estimation equation for β

$$U(\beta) = \sum_{i=1}^{n} \int_0^\infty X\{dN_i(t) - Y_i(t)d\Lambda_0(\beta, t) - Y_i(t)\beta'X\,dt\} = 0,$$

where

$$\Lambda_0 = \sum_{i=1}^{n} \int_0^t \frac{\{dN_i(u) - Y_i(u)\beta'X_i\,du\}}{\sum_{i=1}^n Y_i(u)}$$

is the estimate of the baseline hazard function. As noted by Lin and Ying [24] and Ma and Huang [35,36], the resulting estimation of β is obtained by solving the equation

$$\left[\sum_{i=1}^{n} \int_0^\infty Y_i(t)\{X - \bar{X}(t)\}^{\oplus 2}dt\right]\beta = \left[\sum_{i=1}^{n} \int_0^\infty \{X - \bar{X}(t)\}dN_i(t)\right], \tag{22.4}$$

where $\bar{X}(t) = \sum_{i=1}^n Y_i(t)X_i / \sum_{i=1}^n Y_i(t)$. As noted by Ma and Huang [35,36], the estimate of β by this equation is equivalent to minimizing a loss function of β (see Section 22.3.6). On the basis of this loss function, the lasso or threshold gradient descent procedure can be developed for estimating the β in the additive hazard model (22.10).

22.3   REGULARIZED ESTIMATION FOR CENSORED DATA REGRESSION MODELS

In this section, we review several regularized estimation procedures for estimating the censored data regression models reviewed in previous sections. Most of these procedures are based on extensions of the procedures developed for linear regression and classification, with appropriate definitions of the loss functions.

22.3.1 L_2 Penalized Estimation of the Cox Model Using Kernels

Since the dimension of the x_i vector is usually far exceeds that of sample size n, standard methods such as the Cox partial likelihood for estimating the unspecified function f are unfeasible. In addition, to deal with the problem of collinearity, the most popular approach is to use the penalized partial likelihood, including the L_2 penalized estimation, which is often called the ridge regression. Li and Luan [31] investigated the L_2 penalized estimation of the Cox model in the high-dimensional low-sample-size settings and applied their method to relate the gene expression profile to survival data. To avoid the inversion of large matrix, they used the kernel tricks to reduce the computation to involve only the inversion of a matrix of the size of the sample size. In the model (22.1) with no covariate Z, a regularized formulation of the Cox regression is considered as a variational problem in reproducing kernel Hilbert space H

$$\min_{f \in H} R_{\text{reg}}(f) = \frac{1}{n} \sum_{i=1}^{n} V(t_i, \delta_i, f(x_i)) + \xi \|f\|_H^2,$$

where $V(t_i, \delta_i, f(x_i))$ is the loss function which is a function of f depending on only the values of $f(x)$ at the data points, $\{f(x_i)\}_{i=1}^{n}$. For the general Cox model (22.1), we propose to use the negative log partial likelihood as the loss function and reformulate the problem as finding function $f(x)$ such that

$$R_{\text{reg}} = -\frac{1}{n} \sum_{i=1}^{n} \delta_i \left[f(x_i) - \log \left\{ \sum_{j \in R_i} \exp\left(f(x_j)\right) \right\} \right] + \xi \|f\|_H^2 \tag{22.5}$$

is minimized, where $R_i = \{j = 1, \ldots, n, x_j \geq x_i\}$ is the set of individuals who were at risk at time x_i.

The solution to this problem was given by Kimeldorf and Wahba [29], and is known as the *representer theorem*. By this theorem, the optimal $f(x)$ has the form

$$f(x) = b + \sum_{i=1}^{n} a_i K(x, x_i) \tag{22.6}$$

where K is a positive definite reproducing kernel, which gives the inner product in the transform space. Since b can be absorbed into the baseline hazard function in model (22.1), we can omit b in the following discussion. For the simplest case of inner product kernel with $K(x_i, x_j) = \langle x_i, x_j \rangle$, the function $f(x)$ can be expressed as a linear function of x_i terms. In cases where the data are not linearly separable, one can choose a more general kernel such as the polynomial kernels with $K(x_i, x_j) = (\langle x_i, x_j \rangle + 1)^d$ or the Gaussian kernels with $K(x_i, x_j) = \exp(\|x_i - x_j\|/\sigma_d^2)$, where d and σ_d^2 are the kernel parameters. From the representer formula (22.6), it can be shown that minimizing equation (22.5) is equivalent to the finite-dimensional form

$$R_a = -\delta'(K_a a) + \delta' \log \left\{ \sum_{j \in R_i} \exp(K_a a) \right\} + \xi a' K_q a, \tag{22.7}$$

where $a' = (a_1, \ldots, a_n)$, the regressor matrix $K_a = [K(x_i, x_j)]_{n \times n}$, and the regularization matrix $K_q = K_a$. Here the matrix K_a is called the *kernel matrix*. One can use the Newton–Raphson method to minimize the loss function over a, which is n-dimensional.

This procedure can be simply modified to include other covariates Z. For example, we can estimate γ in model (22.1) by maximizing a profile partial likelihood.

22.3.2 L_1 Penalized Estimation of the Cox Model Using Least-Angle Regression

One limitation of the L_2 penalized estimation of the Cox model is that it uses all the genes in the prediction and does not provide a way of selecting relevant genes for prediction. However, from a biological perspective, one should expect that only a small subset of the genes is relevant to predicting the phenotypes. Including all the genes in the predictive model introduces noise and is expected to lead to poor predictive performance. Owing to the high-dimensionality, the standard variable selection methods such as stepwise and backward selection cannot be applied. The lasso method was proposed by Tibshirani [44] for variable selection for linear models and was further extended for variable selection for the Cox proportional hazard models [45]. For the model (22.1) with no covariate Z, let $l(\beta) = \log L(\beta)$ be the log of the partial likelihood function; then the lasso estimate of β [44,45] can be expressed as

$$\hat{\beta}(s) = \text{argmax } l(\beta), \text{ subject to } \sum_{j=1}^{p} |\beta_j| \leq s,$$

where s is a tuning parameter determining how many covariates with coefficients are zero.

Tibshirani [45] proposed the following iterative procedure for reformulating this optimization problem with constraint as a lasso problem for linear regression models. Specifically, let $\eta = \beta' X$, $\mu = \partial l / \partial \eta$, $A = -\partial^2 l / \partial \eta \, \eta^T$, and $z = \eta + A^- \mu$. With this reparameterization, a one-term Taylor series expansion for $l(\beta)$ has the form of

$$(z - \eta)^T A (z - \eta).$$

Although there are multiple choices of A^-, it is easy to show that if $\text{rank}(A) = n - 1$, for any A^- that satisfies $AA^- A = A$ and $z = \eta + A^- \mu$, $(z - \eta)^T A(z - \eta)$ is invariant to the choice of the generalized inverse of A. The iterative procedure of Tibshirani [45] involves the following four steps:

1. Fix s and initialize $\hat{\beta} = 0$.
2. Compute η, μ, A, and z on the basis of the current value of $\hat{\beta}$.
3. Minimize $(z - \beta' X)^T A(z - \beta' X)$ subject to $\sum |\beta_j| \leq s$.
4. Repeat steps 2 and 3 until $\hat{\beta}$ does not change.

Tibshirani [45] proposed using quadratic programming for solving step 3. However, in the high-dimensional and low-sample-size setting (i.e., where $p \gg n$), the quadratic programming algorithm cannot be directly applied. Gui and Li [22] proposed a simple modification of the LARS algorithm of Efron et al. [8] for step 3. Specifically, Gui and Li [22] apply the Choleski decomposition to obtain $T = A^{1/2}$ such that $T'T = A$; then step 3 of the iterative procedure can be rewritten as step 3: minimize $(y - \beta'\hat{X})^T(y - \beta'\hat{X})$ subject to $\sum|\beta_j| \leq s$, where $y = Tz$ and $\hat{X} = TX$. This can be efficiently solved by using the LARS–lasso procedure as presented in Efron et al. [8].

Segal [41] proposed using LARS to minimize a residual-based loss function for the Cox model, in which the IRWLS iterations are not required. Park and Hastie [38] proposed a generalization of the LARS algorithm for the Cox model using the predictor–corrector algorithm of convex optimization. Finally, for a given tuning parameter s, one can estimate γ in model (22.1) by maximizing a profile partial likelihood in γ. For a given γ and s, the LARS procedure can be used for estimating the β, denoted as $\beta(\gamma, s)$. Then we can maximize the partial likelihood over γ.

22.3.3 Threshold Gradient Descent Procedure for the Cox Model

Treating the negative log partial likelihood function $(-l(\beta))$ as the loss function, Gui and Li [22] presented a threshold gradient descent (TGD) regularization procedure for estimating the β in the Cox model following the key idea presented in Friedman and Popescu [17]. The main idea of the TGD is that during the gradient descent minimization, a thresholding is imposed to the absolute values of the gradients. Specifically, for any threshold value $0 \leq \tau \leq 1$, the threshold gradient descent algorithm for Cox model involves the following five steps:

1. $\beta(0) = 0, v = 0$.
2. Calculate η, μ, $g(v) = \partial l/\partial\beta$ for the current β.
3. $f_j(v) = I[|g_j(v)| \geq \tau \cdot \max_{0 \leq k \leq n}|g_k(v)|]$
4. Update $\beta(v + \Delta v) = \beta(v) + \Delta v \cdot g(v) \cdot f(v)$, $v = v + \Delta v$.
5. Repeat steps 2–4 until the β terms converge.

This procedure involves two tuning parameters τ and v, both of which control the sparsity of the estimates of β. Compared to the lasso estimate of the Cox model, this TGD procedure is computationally fast and does not involve matrix inversion. Simulations and applications to real datasets indicated that when $\tau = 1$, the TGD procedure performs very similarly to the Lasso procedure. Note that this procedure is quite general and can be applied to essentially any convex loss function. Finally, if covariate Z is included in model (22.1), one can estimate γ by maximizing a profile partial likelihood or by iteratively updating γ and β during the TGD iterations.

Although empirical evidence indicated that the TGD procedure works well in selecting variables and in building predictive models [22,26], it is not clear precisely what the corresponding penalty function is. The procedure, which is similar in spirit to the LARS, indeed provides a way to regularize the parameter estimates and is therefore expected to perform well in high-dimensional settings.

22.3.4 Regularized Buckley–James Estimations for the AFT Model

Buckley and James [6] used the transformation ϕ on the observed responses y_i, where $\phi(y_i) = \delta_i y_i + (1 - \delta_i) E(t_i|t_i \leq y_i)$ and proposed simultaneously updating $\phi(y_i)$ and β at each step and proceed iteratively:

1. Select an initial estimate β_0, and let $\tilde{t}_i = \beta_0 x_i$.

2. Compute the residuals $e_i = y_i - \tilde{t}_i$ and estimate transformation

$$\hat{\phi}(y_i) = \delta_i y_i + (1 - \delta_i)\left[\tilde{t}_i - \{\hat{S}_e(e_i)\}^{-1} \int_{e_i}^{\infty} s\, d\{\hat{S}_e(s)\}\right]$$

$$= \delta_i y_i + (1 - \delta_i)\left[\tilde{t}_i + \frac{\sum_k^u v_k \tilde{t}_k I(e_i < e_k)}{\hat{S}_e(e_i)^{-1}}\right],$$

where $\hat{S}_e(s)$ is the Kaplan–Meier estimator of the survival function based on $\{\epsilon_i, \delta_i\}_{i=1}^n$, v_k is the probability mass assigned to uncensored residual e_k, and $\sum^u$ denotes summation over uncensored values only.

3. Apply least-squares estimation to $\{\hat{\phi}(y_i), x_i\}$ and update β.
4. Stop if β converges or oscillates. Otherwise, go to step 2.

When $p > n$, one cannot implement the least-squares estimation in step 3 of the BJ procedure. However, one can perform LARS–lasso or the threshold gradient descent procedure for step 3 to obtain a regularized estimation of β. Alternatively, one can perform L_2 penalization or partial least-squares (PLS) methods for estimating the β in step 3 [25], which provides a PLS procedure for linear models with censoring on the responses.

22.3.5 Regularization Based on Inverse Probability of Censoring Weighted Loss Function for the AFT Model

If there are no censoring in the data, the most commonly used method for estimating the model (22.3) is by minimizing a quadratic loss function

$$l(\beta) = \sum_{i=1}^n (y_i - \beta' x_i)^2$$

over β. However, such a loss function cannot be evaluated at the censored observations. One solution to this problem is to use the inverse probability of censoring weighted (IPCW) loss function introduced in Robins and Rotnitsky [39], who showed that for any loss function $l(T, F(x))$, one has

$$E(l(T, F(x))\Delta G(T|x)) = E(l(T, F(x))),$$

where the T is the random variable of time to event, $F(x)$ is an estimator, and $G(T|x)$ is the survival function of the censoring variable, which may be dependent on x. This suggests the use of the following loss function to estimate the AFT model (22.2)

$$l_{\text{ipcw}}(\beta) = \sum_{i=1}^n \left[(y_i - \beta x_i)^2 \frac{\delta_i}{G(y_i)}\right], \tag{22.8}$$

where $G(t)$ is the survival function of the censoring variable. This loss function can be regarded as the weighted squared loss function with weight $w_i = \delta_i / G(y_i)$ for the ith individual. In practice, $G(t)$ is, of course, unknown and must be estimated by the Kaplan–Meier estimator $G(t)$

from the observation (t_i, δ_i, x_i). Notice for the purpose of estimating $G(t)$, a $\delta_i = 0$ means a complete observation and $\delta_i = 1$ means a censored observation.

Alternatively, we can use the robust Huber [27] loss function as

$$l_{\text{ipcw}}^{\text{H}}(\beta) = \sum_{i=1}^{n} l_i^{\text{H}}(y_i, x_i; \beta) \frac{\delta_i}{G(y_i)}, \tag{22.9}$$

where $l_i^{\text{H}}(\cdot)$ is the Huber loss function for the ith observation defined as

$$l_i^{\text{H}}(y_i, x_i, \beta) = \begin{cases} (y_i - \beta x_i)^2/2 & |y_i - \beta x_i| < \tau \\ \tau(|y_i - \beta x_i| - \tau/2) & |y_i - \beta x_i| < \tau \end{cases},$$

where τ is the transition point, the value of which is often taken to be αth quartile of the current absolute residuals $\tau(\beta) = \text{quantile}_\alpha\{|y_i - \beta x_i|\}_{i \in D}$. Here $1 - \alpha$ is a specified fraction of the observations that are treated as outliers, subject to absolute loss.

On the basis of the loss function defined by Equation (22.8), Huang et al. [26] developed L_1 penalized estimation or lasso by using the LARS, namely, min $l_{\text{ipcw}}(\beta)$ subject to $\sum_{i=1}^{p} |\beta_i| < s$, and a TGD procedure. For the Huber version of the loss function (22.9), one can similarly perform a gradient boosting procedure or the threshold gradient descent procedure [14,17]. Finally, on the basis of loss functions defined in Equation (22.8) or (22.9), one can easily develop principal-components or partial least-square components analysis for the AFT models.

22.3.6 Penalized Estimation for the Additive Hazard Model

In the estimation Equation (22.4) for β in the additive hazard model (22.3), we denote $H^i = \int_0^\infty Y_i(t)\{X_i - \bar{X}(t)\}^{\oplus 2} dt$, and $R^i = \int_0^\infty \{X_i - \bar{X}(t)\} dN_i(t)$, and $H_{s,l}^i$ as the (s, l) element of H^i and the sth components of R^i and β as R_s^i and β_s, then Equation (22.4) is equivalent to the following p equations:

$$\left(\sum_{i=1}^{n} H_{s,1}^i\right)\beta_1 + \cdots + \left(\sum_{i=1}^{n} H_{s,p}^i\right)\beta_p = \sum_{i=1}^{n} R_s^i, i = 1, \ldots, p.$$

Ma and Huang [35,36] further note that the estimate defined by this equation is the same as minimizing the following loss function $L(\beta)$,

$$\beta = \text{argmin}_\beta \left\{ l(\beta) = \sum_{s=1}^{p} \left\{ \left(\sum_{i=1}^{n} H_{s,1}^i\right)\beta_1 + \cdots + \left(\sum_{i=1}^{n} H_{s,p}^i\right)\beta_p - \sum_{i=1}^{n} R_s^i \right\}^2 \right\}.$$

On the basis of this loss function, Ma and Huang [35,36] proposed using the lasso or TGD procedure to obtain regularized estimates of β.

22.3.7   Use of Other Penalty Functions

Besides the L_2 and L_1 penalty functions, one can also consider penalized estimation of various censored data survival models using other penalties. One attractive alternative is the smoothly clipped absolute deviation (SCAD) penalty proposed in Fan and Li [10]. The SCAD penalty $p_\xi(\beta)$ is defined by its derivative as

$$p_\xi'(\beta) = \xi\left\{\frac{I(\beta \leq \xi) + (a\xi - \beta)_+}{(a-1)\xi}I(\beta > \xi)\right\}$$

for some $a > 2$ and $\beta > 0$, where ξ is a tuning parameter. The corresponding penalty $p_\xi(\beta)$ is a nonconcave function and is continuously differentiable. This penalty function is constructed to ensure that the resulting estimator has unbiasedness, sparsity, and continuity properties. In addition to ensuring less biased estimates of the coefficients for large true values, the SCAD estimators also possess an oracle property for both the finite-dimensional cases [10] and the cases with a diverging number of parameters [11]. The oracle property of the SCAD estimators is very attractive for variable selection. The SCAD penalty has also been applied to the Cox's proportional hazards model [10] and can, of course, be applied to other censored data regression models in the finite-dimensional settings. Because of the computational difficulty in implementing the SCAD penalization in very high-dimensional settings, its application in linking genomic data to censored survival data has not been seen in the literature. However, research along this line should be promising.

22.4   SURVIVAL ENSEMBLE METHODS

Since the early 1990s, ensemble methods such as random forests [2] and boosting procedures [12–14,4,5] have gained much popularity in classification and linear regression analysis because of their superior predictive performances. In addition, Bühlmann [4] demonstrated the applicability of the boosting procedure in the high-dimensional settings. In this section, we first review two extensions of the gradient boosting procedure to the Cox model and the AFT model.

22.4.1   The Smoothing-Spline-Based Boosting Algorithm for the Nonparametric Additive Cox Model

Li and Luan [32] proposed to use the boosting procedure for estimating the function $F(X)$ in model (22.1) nonparametrically. Boosting essentially is an iterative procedure to update function estimators successively. Friedman [14] developed a novel general framework, called the "gradient boosting machine," to obtain additive expansions adapted to any fitting criterion. The framework is quite general and works for various models. For linear regression with no censoring, Bühlmann and Yu [5] show that the L_2 boosting achieves the optimal rate of convergence.

Following Friedman [14] and Bühlmann and Yu [5], Li and Luan [32] proposed a componentwise boosting procedure using cubic smoothing splines as base learner. At the kth boosting step, they obtain the estimate of the function $F^{(k)}(X)$, which is a nonparametric additive function of each component of X, some of which are identically zero. It should be noted that when the iteration k increases by 1, one more term is added to the fitted procedure; however, this term may have already been in the model. Owing to the dependence of this new term on the previous terms, the complexity of the fitted model is not increased by a constant amount. The final model provides an estimate of possible nonlinear effects of gene expression levels on the risk of an event.

22.4.2 Random Forests and Gradient Boosting Procedure for the AFT Model

Hothorn et al. [28] presented a random forests algorithm and a gradient boosting algorithm for the construction of prognostic and diagnostic AFT models by using the IPCW loss function (22.8). Using the IPC weights, they [28] proposed modifying the original random forests procedure of Breiman [2] in two ways: in the bootstrap (or bagging) step, the case samples are weighted by their IPC weights to obtain the case counts, and in the base learner step, the tree is built using the learner sample with case counts obtained from the bootstrap step. Similarly, using this IPCW loss function, the generic gradient descent boosting procedure of Friedman [14] can be directly applied to develop a boosting procedure for a linear model with censoring. The base learner can be a regression tree, univariate splines, or componentwise least squares. One benefit of using the componentwise least squares as base learner is that there is a closed-form definition of the AIC score, which can be used for selecting the boosting step.

22.5 NONPARAMETRIC-PATHWAY-BASED REGRESSION MODELS

For many complex diseases, especially for cancers, there are many types of metadata available that are related to biological pathways. Currently, information derived from metadata such as known biological knowledge has been used primarily to select promising candidates for genetic characterization and for studying gene–gene and gene–environment interactions. Such information has hardly been utilized in the modeling step for identifying such interactions or for identifying genes or pathways that are related to the phenotypes.

Wei and Li [49] proposed a pathways-based boosting procedure for estimating a nonparametric-pathway-based regression model. Suppose that we have K pathways whose activities may be related to the phenotype of interest. Assume that there are p_k genes involved in the kth pathway. We allow that some genes belong to multiple pathways and let p be the total number of genes involved in the K pathways and therefore $p < \sum_{k=1}^{K} p_k$. Suppose that we have n independent individuals and let $y_i = (t_i, \delta_i)$, where t_i is time to event or censoring and δ_i is an event indicator. Let $x_{ij}^{(k)}$ be the genomic measurement of the jth gene in the kth pathway for the ith patient, let $x_i^{(k)} = \{x_{i1}^{(k)}, \ldots, x_{ipk}^{(k)}\}$ be the vector of the genomic measures of the genes in the kth pathway for the ith patient, and let $x_i = (x_i^{(1)}, \ldots, x_i^{(K)})$ be the vector of the genomic measurements of all the p genes. Here the genomic measurements can be SNP data or gene expression data. Our goal is to relate the phenotype data Y to $X = \{X^{(1)}, \ldots, X^{(K)}\}$ in order to identify the pathways that are related to the phenotype and to identify genes and their interactions that determine the pathway activities.

Here we assume that the phenotype is related to the total activity level across multiple pathways through an additive pathway activity function

$$F(X) = \sum_{k=1}^{K} F_k(X^{(k)}), \tag{22.10}$$

where $F_k(X^{(k)})$ can be interpreted as the activity level associated with the kth pathway as determined by the genomic measurements of the p_k genes in this pathway. We assume that conditioning on the genes of the pathways, the pathway activities across the K pathways are additive. For the censored survival phenotype, we can assume that the hazard function at time t given the observed genomic data X is modeled as

$$\lambda(t \mid X, Z) = \lambda_0(t) \exp(F(X) + \gamma Z), \qquad (22.11)$$

where $\lambda_0(t)$ is the baseline hazard function, $F(X)$ is the pathway activity function as defined in (22.10), Z is a covariate vector, and γ is the corresponding risk ratio parameter. The main motivation of these models is that we aim to model complex interactions between genes within pathways nonparametrically, rather than assume particular parametric forms for functions $F_k(X^{(k)})$. We use the term "nonparametric pathway-based regression" (NPR) to particularly emphasize this point, that the genetic and pathways effects are modeled nonparametrically. It is obvious that without any constraints on the functions $F_k(X^{(k)})$, model (22.11) is not identifiable.

Wei and Li [49] proposed a general pathway-based gradient descent boosting procedure to identify such NPR models with the particular form of (22.11). The key idea of our proposed extension of the boosting procedure of Friedman [14] is that instead of performing gradient boosting over all the p genes, we perform gradient descent boosting over genes in each of the K pathways separately. We first consider the case where no other covariates are included in model (22.11). Let $L(y_i, F(x_i))$ be a loss function for the ith observation, which can be defined as negative of the partial likelihood based on model (22.11). During each boosting iteration, one pathway is selected that gives the best fit of the negative gradients using the base learner. This effectively utilizes the known pathway information and reduces the dimensionality from considering all the genes to considering only those genes in a given pathway. Then the functions are updated by adding the tree corresponding to the k^*th pathway selected. In order to model interactions between genes in a given pathway, Wei and Li [49] proposed using a J-terminal node regression tree [3] as the base learning procedure. The boosting procedure with regression trees as base procedures inherits the favorable characteristics of trees such as robustness and flexibility in modeling interactions [3]. In addition, trees tend to be quite robust against the addition of irrelevant input variables and therefore serve as internal feature selection [14,3]. J controls the size of the tree, which is often chosen to be small.

22.6 DIMENSION-REDUCTION-BASED METHODS AND BAYESIAN VARIABLE SELECTION METHODS

There have also been some attempts to generalize the dimension reduction procedures to censored survival data. Li and Gui [30] and Park et al. [37] generalized the partial least squares (PLS) method to the Cox model taking into account censoring. Li and Li [33] extended the sliced inverse procedure to censored data. Similarly, one can also develop PLS procedures for the AFT models [25] and the additive hazard models. One limitation of such extensions is that these procedures do not provide a rigorous way of selecting genes in the model. As we expect that only a small set of genes might be related to survival endpoints, these procedures may introduce too much noise to the estimation, and therefore may have relatively low predictive performance. Bair and Tibshirani [1] proposed a supervised principal-components analysis

(PCA), where genes are selected by univariate Cox regression analysis and the selected genes are used to define several principal components. The number of genes and the number of components used in the final model are selected by cross-validation. While this is a step forward beyond the use of PCA on all the genes, selecting genes by univariate analysis may not capture possible joint effects of genes.

Bayesian variable and model selection procedures for linear regression models and for clustering analysis [19,20,50] in high-dimensional settings can also be extended for censored data regression models (M. G. Tadesse, personal communication, 2006). However, we have seen publications of such extensions in the literature.

22.7 CRITERIA FOR EVALUATING DIFFERENT PROCEDURES

The goal of linking genomic data to censored survival outcomes is twofold: to identify the genes that are related to the outcome and to build a model for predicting future outcomes. Since genomic data such as microarray gene expression often include many highly correlated features, it should not be surprising that different methods may identify different sets of genes. Because of the complexity of biological systems, it is almost impossible for a method to identify all the "correct" or "relevant" genes. It is therefore crucial to link the genes identified to biological functions or pathways such as those listed in gene ontology [18]. If the genes identified by two different procedures are not exactly the same but most of these genes belong to the same functional groups, we can conclude that the two procedures perform similarly in terms of selecting genes.

In order to assess how well the model predicts the outcome, we applied the concept of the time-dependent receiver operating characteristics (ROC) curve for censored data and area under the curve (AUC) as our criteria. For a given score function $f(X)$, we can define time-dependent sensitivity and specificity functions as

$$\text{Sensitivity}(c, t|f(X)) = \Pr\{f(X) > c|\delta(t) = 1\},$$
$$\text{Specificity}(c, t|f(X)) = \Pr\{f(X) \leq c|\delta(t) = 0\},$$

and define the corresponding $\text{ROC}(t|f(X))$ curve for any time t as the plot of sensitivity $(c, t|f(X))$ versus $1 - \text{specificity}(c, t|f(X))$ with cutoff point c varying, and the AUC as the area under the $\text{ROC}(t|f(X))$ curve, denoted by $\text{AUC}(t|f(X))$. Here $\delta(t)$ is the event indicator at time t. Note that larger AUC at time t based on a score function $f(X)$ indicates better predictability of time to event at time t as measured by sensitivity and specificity evaluated at time t. The time-dependent AUCs should be compared in both the means and variances or even their complete distributions [47] using cross-validation analysis or independent test samples.

22.8 APPLICATION TO A REAL DATASET AND COMPARISONS

We demonstrate the utility of some of these procedures using a published dataset of DLBCL by Rosenwald et al. [40]. This dataset includes a total of 240 patients with DLBCL, including 138 patient deaths during the follow-ups with median death time of 2.8 years. Rosenwald et al. divided the 240 patients into a training set of 160 patients and a validation set or test set of

80 patients and built a multivariate Cox model. The gene expression measurements of 7399 genes are available for analysis.

For the Cox model, we applied several methods to build a predictive model using the training dataset and we used zero as a cutoff point of the risk scores and divided the test patients into two groups according to whether they had positive or negative risk scores. Using the L_1 penalized estimation, the two groups of patients showed very significant differences (p value = 0.0004) in overall survival between the high-risk and low-risk groups. We observe that the two risk groups defined by the LARS–Cox estimated model showed more significant differences in risk of death than did the groups defined by the other three models: p value of 0.0004 versus 0.003, 0.003, and 0.034 for the partial Cox regression method of Li and Gui [30], the L_2 penalized method of Li and Luan [31], and the supervised PCA method of Bair and Tibshirani [1], respectively Finally, the AUCs based on the risk scores estimated by the LARS–Cox procedure were also higher than those from the other three procedures.

Owing to computational difficulty with the penalized estimation procedures for the AFT model and the additive hazard model, 1656 genes out of 7399 with large correlation coefficients (with the uncensored event times) were chosen for the AFT and the additive hazard model analysis. The results are summarized as follows. For the AFT model, the modified Lasso identified 37 genes and resulted in a test set p value of 0.05, the TGD procedure identified 91 genes with a test set p value of 0.776; for the additive hazard model, the modified lasso selected 7 genes with a test set p value of 0.331, and the TGD procedure identified 10 genes with a test set p value of 0.13. These results indicate that at least for this particular dataset, the AFT or the additive hazard models did not provide as good predictive results as the Cox regression models.

Finally, analysis by Li and Luan [32] using the splines-based boosting procedure indicated that some genes indeed show strong nonlinear effects on the risk of death from lymphoma.

22.9 DISCUSSION AND FUTURE RESEARCH TOPICS

It is clinically relevant and very important to predict a patient's time to cancer relapse or time to death due to cancer after treatment using gene expression profiles of the cancerous cells prior to treatment. Powerful statistical methods for such prediction allow microarray gene expression data to be used most efficiently. Because of the high-dimensionality of the genomic data, standard estimation and test methods for various censored data regression models cannot be applied directly to analyze such data. In this chapter, we have reviewed the latest developed regularized estimation procedures for several classes of the most commonly used censored survival data regression models, including the Cox proportional hazards model, the accelerated failure-time model, and the additive hazard models. The methods reviewed include penalization estimation, threshold gradient-based regularization, and gradient boosting procedures. These methods have been evaluated by simulation studies and found effective in identifying relevant genes and in building predictive models [22,32,35,36].

Among the methods reviewed, most of the penalized estimation procedures were developed for censored data regression models with simple linear functional forms (i.e., $\beta'X$). The kernel-based L_2 penalization [31] and extensions of the boosting procedure or random forests to censored data regression [32,51] allow for nonlinear effects and potential gene–gene interaction effects on the risk of event. In general, in the published results in the papers we reviewed, we observed that the methods with variable selection often perform better in prediction than do the dimension-reduction-based procedures. In addition, we should expect that the ensemble methods such as boosting and random forests perform better in prediction than other methods, especially in high-dimensional and low-sample-size settings. However, we should not expect one model or method to always perform better than the others. One useful avenue of research is to comprehensively compare these methods by simulations and application to many different datasets. Besides empirical results, theoretical results are also required in order to gain insight into the methods and to provide theoretical basis for the methods proposed. Currently, little is known about the theoretical properties of these penalized estimators in high-dimensional and low-sample-size settings.

While the emphasis of this review is on the methods for identifying genes that are related to censored survival outcome and building predictive models for future patients' survival using gene expression, there are several other interesting topics related to censored data regression in the high-dimensional and low-sample-size settings that deserve further research. In the following section we present some of the problems and possible solutions and some possible extensions of the methods presented in this chapter.

22.9.1 Test of Treatment Effect Adjusting for High-Dimensional Genomic Data

Consider the clinical trial setting where a treatment effect is evaluated with time to clinical event as an endpoint. In standard analysis of the data obtained from the clinical trials, the treatment effect is often tested using the Cox model adjusting for other low-dimensional covariates. It is becoming common practice that high-dimensional genomic data are often collected for such clinical trials. How to adjust for the genomic heterogeneity when testing for the treatment effect deserves further research. For example, in model (22.1), where Z is the treatment indicator in randomized clinical trials, the null hypothesis is

$$\gamma = 0,$$

where the effect of genomic data β is treated as a high-dimensional nuisance parameter. A valid test for such a null hypothesis is required. A related problem is to identify a subset of patients who respond to treatment differently.

22.9.2 Development of Flexible Models for Gene–Gene and Gene–Environment Interactions

Most of the models and methods reviewed in this chapter assume a simple linear functional form to relate genomic data to the phenotypes. However, most phenotypes are expected to be affected by the interplay of different genes and environments, and therefore simple linear functional form cannot capture the complexity of the genomic effects on phenotypes. Ensemble methods using trees offer one way of modeling potential interactions between the variables. However, new methods are required for identifying and assessing such complex interactions. This is especially challenging in the high-dimensional and low-sample-size settings. For example, the patient rule induction method (PRIM) [15] provides an alternative to the tree method, which may capture the genomic interactions better.

22.9.3 Methods for Other Types of Genomic Data

The methods presented in this chapter are developed mainly for microarray gene expression data, where the data structures are relatively simple. Since genes usually function in coordinated modules, it is often observed that the expression levels of some genes are highly correlated. Methods that can account for such clusters of genes in the models are expected to predict well. In addition, special features of other types of genomic data such as aCGH data, mass spectrometry data and genomewide SNP/haplotype data need to be accounted for when building predictive models. For example, to build a predictive model using aCHG data, one needs to account for local dependence of the measurements. Similarly, to identify SNPs that are related to censored survival phenotypes, one has to account for linkage disequilibrium for the SNPs. Tibshirani et al. [46] proposed a fused-Lasso procedure, which provides a way of accounting for such local dependency.

22.9.4 Development of Pathway- and Network-Based Regression Models for Censored Survival Phenotypes

Since genes and proteins almost never work alone, they interact with each other and with other molecules in highly structured but incredibly complex ways. Understanding this interplay of human genome and environmental influences is crucial to developing a systems understanding of human health and disease. An important avenue for future research is to develop methods that can incorporate known biological knowledge such as pathways and networks into statistical modeling in order to limit the search space for gene–gene and gene–environment interactions. Wei and Li [49] attempted to incorporate known biological pathways and networks information into the censored data regression model in order to reduce the dimensionality of the problem. However, how to best identify genes and pathways that are related to censored survival phenotypes clearly deserves future research.

22.10 CONCLUDING REMARKS

High-throughput genomic and proteomic data provide a unique opportunity for dissecting genes and pathways that are related to risk of complex diseases or the responses to treatments. Because of the variation in disease onset or time to clinical event, studying censored survival data can significantly facilitate the identification of the genes and pathways involved. As user-friendly software packages implementing these methods become available, we should expect to see more applications of these methods in identifying genes and pathways involved in complex diseases. We should also expect more new method developments in this important area.

ACKNOWLEDGMENT

This research was supported by NIH grant R01-ES009911 and a Pennsylvania Department of Health grant. I thank my students and postdocs (Dr. Yihui Luan, Dr. Gui Jiang, and Zhi Wei) for implementing some of the ideas presented in this chapter, Dr. Shuange Ma for providing the analysis results using the AFT and the additive hazards model for the lymphoma dataset, and Mr. Edmund Weisberg, M.S., for his editorial help.

REFERENCES

1. Bair, E. and Tibshirani, R., Semi-supervised methods for predicting patient survival from gene expression papers, *PLoS Biol.* **2**, 5011–5022 (2004).

2. Breiman, L., Random forests, *Machine Learn.* **45**(1), 5–32 (1994).

3. Breiman, L., Friedman, J. H., Olshen, R. A., and Stone, C. J., *Classification and Regression Trees*, Wadsworth and Brooks/Cole, Monterey, CA, 1984.

4. Bühlmann, P., Boosting methods: Why they can be useful for high-dimensional data, in *Proc. 3rd Int. Workshop on Distributed Computing (DSC 2003)*, 2003.

5. Bühlmann, P. and Yu, B., Boosting with the L_2-Loss: Regression and classification, *J. Am. Statist. Assoc.* **98**, 324–339 (2003).

6. Buckley, J. and James, I., Linear regression with censored data, *Biometrika* **66**, 429–436 (1979).

7. Cox, D. R., Regression models and life-tables, *J. Roy. Statist. Soc. Ser. B* **34**, 187–220 (1972).

8. Efron, B., Johnston, I., Hastie, T., and Tibshirani, R., Least angle regression, *Ann. Statist.* **32**, 407–499 (2004).

9. Fan, J. and Li, R., Variable selection via nonconcave penalized likelihood and its oracle properties, *J. Am. Statist. Assoc.* **96**, 1348–1360 (2001).

10. Fan, J. and Li, R., Variable selection for Cox's proportional hazards model and frailty model, *Ann. Statist.* **30**, 74–99 (2002).

11. Fan, J. and Peng, H., Nonconcave penalized likelihood woth a diverging number of parameters, *Ann. Statist.* **32**, 928–961 (2004).

12. Freund, Y., Boosting a weak learning algorithm by majority, *Inform. Comput.* **121**, 256–285 (1995).

13. Freund, Y. and Schapire, R., A decision-theoretic generalization of on-line learning and an application to boosting, *J. Comput. Syst. Sci.* **55**, 119–139 (1997).

14. Friedman, J., Greedy function approximation: A gradient boosting machine, *Ann. Statist.* **29**, 1189–1232 (2001).

15. Friedman, J. and Fisher, N., Bump hunting in high dimensional data, *Statist. Comput.* **9**, 123–143 (1999).

16. Friedman, J., Hastie, T., and Tibshirani, R., Additive logistic regression: A statistical view of boosting (with discussion), *Ann. Statist.* **28**, 337–407 (2000).

17. Friedman, J. H. and Popescu, B. E., *Gradient Directed Regularization*, Technical Report, Stanford Univ., 2004.

18. The Gene Ontology Consortium, Gene ontology: Tool for the unification of biology, *Nature Genet.* **25**, 25–29 (2000).

19. George, E. I., The variable selection problem, *J. Am. Statist. Assoc.* **95**, 1304–1308 (2000).

20. George, E. I. and McCulloch, R. E., Variable selection via Gibbs sampling, *J. Am. Statist. Assoc.* **88**, 881–889 (1993).

21. Golub, T. R., Slonim, D. K., Tamayo, P., Huard, C., Gaasenbeek, M., Mesirov, J. P., Coller, H., Loh, M. L., Downing, J. R., Caligiuri, M. A., Bloomfield, C. D., and Lander, E. S., Molecular classification of cancer: Class discovery and class prediction by gene expression monitoring, *Science* **286**, 531–537 (1999).

22. Gui, J. and Li, H., Threshold gradient descent method for censored data regression, with applications in pharmacogenomics, *Pacific Symp. Biocomput.* **10**, 272–283 (2005).

23. Gui, J. and Li, H., Penalized Cox regression analysis in the high-dimensional and low-sample size settings, with applications to microarray gene expression data, *Bioinformatics* **21**, 3001–3008 (2005).

24. Heagerty, P. J., Lumley, T., and Pepe, M., Time dependent ROC curves for censored survival data and a diagnostic marker, *Biometrics* **56**, 337–344 (2000).

25. Huang, J. and Harrington, D., Penalized partial likelihood regression for right-censored data with bootstrap selection of the penalty parameter, *Biometrics* **58**, 781–791 (2002).

26. Huang, J., Ma, S., and Xie, H., *Regularized Estimation in the Accelerated Failure Time Model with High Dimensional Covariates*, Technical Report, Univ. Iowa, 2005.

27. Huber, P., Robust estimation of a location parameter, *Ann. Math. Statist.* **53**, 73–101 (1964).

28. Hothorn, T., Buhlmann, P., Dudoit, S., Molinaro, A. M., and van der Laan, M. J., Survival ensembles, *Biostatistics* **7**(3), 355–373 (2006).

29. Kimeldorf, G. S. and Wahba, G., A correspondence between Bayesan estimation on stochastic processes and smoothing by splines, *Ann. Math. Statist.* **2**, 495–502 (1971).

30. Li, H. and Gui, J., Partial Cox regression analysis for high-dimensional microarray gene expression data, *Bioinformatics* **20** (Suppl. 1), i208–i215 (2004).

31. Li, H. and Luan, Y., Kernel Cox regression models for linking gene expression profiles to censored survival data, *Pacific Symp. Biocomput.* **8**, 65–76 (2003).

32. Li, H. and Luan, Y., Boosting proportional hazards models using smoothing splines, with applications to high-dimensional microarray data, *Bioinformatics* **21**, 2403–2409 (2005).

33. Li, L. and Li, H., Dimension reduction methods for microarrays with application to censored survival data, *Bioinformatics* **20**, 3406–3412 (2004).

34. Lin, D. Y. and Ying, Z., Semiparametric analysis of the additive risk model, *Biometrika* **81**, 61–71 (1994).

35. Ma, S. and Huang, J., *Threshold Gradient Descent Regularization in the Additive Risk Model With High-Dimensional Covariates*, Technical Report 346, Univ. Iowa, (2005).

36. Ma, S. and Huang, J., *Lasso Methods for Additive Risk Models with High-Dimensional Covariates*, Technical Report 347, Univ. Iowa, (2005).

37. Park, P. J., Tian, L., and Kohane, I. S., Linking expression data with patient survival times using partial least squares, *Bioinformatics* **18**, S120–S127 (2002).

38. Park, M. Y. and Hastie, T., *An L1 Regularization-Path Algorithm for Generalized Linear Models*, Technical Report, Stanford Univ., 2006.

39. Robins, J. and Rotnitsky, A., Recovery of information and adjustment for dependent censoring using surrogate markers, in *AIDS Epidemiology, Methodological Issues*, Bikhauser, 1992.

40. Rosenwald, A., Wright, G., Chan, W., Connors, J. M., Campo, E., Fisher, R., Gascoyne, R. D., Muller-Hermelink, K., Smeland, E. B., and Staut, L. M., The use of molecular profiling to predict survival after chemotherapy for diffuse large-B-Cell lymphoma, *New Engl. J. Med.* **346**, 1937–1947 (2002).

41. Segal, M. R., Microarray gene expression data with linked survival phenotypes: Diffuse large-B-cell lymphoma revisited, *Biostatistics* **7**(2), 268–285 (2006).

42. Sorlie, T., Perou, C. M., Tibshirani, R., Aas, T., Geisler, S., Johnsen, H., Hastie, T., Eisen, M. B., van de Rijn, M., Jeffrey, S. S., Thorsen, T., Quist, H., Matese, I. C., Brown, P. O., Botstein, D., et al., Gene expression patterns of breast carcinomas distinguish tumor subclasses with clinical implications, *Proc. Natl. Acad. Sci.* **98**, 10869–10874 (2001).

43. Tadesse, M. G., Sha, N., and Vannucci, M., Bayesian variable selection in clustering high-dimensional data, *J. Am. Statist. Assoc.* **100**, 602–617 (2005).

44. Tibshirani, R., Regression shrinkage and selection via the Lasso, *J. Roy. Statist. Soc. B* **58**, 267–288 (1995).

45. Tibshirani, R., The Lasso method for variable selection in the Cox model, *Statist. Med.* **16**, 385–395 (1997).

46. Tibshirani, R., Saunders, M., Rosset, S., Zhu, J., and Knight, K., Sparsity and smoothness via the fused lasso, *J. Roy. Statist. Soc. Ser. B* **67**, 91–108 (2005).

47. Tian, L., Cai, T., Goetghebeur, E., and Wei, L. J., Model evaluation based on the distribution of estimated absolute prediction error, *Biometrika* (in press).

48. Wei, L. J., The accelerated failure time model: A useful alternative to the Cox regression model in survival analysis, *Statist. Med.* **11**, 1871–1879 (1992).

49. Wei, Z. and Li, H., Nonparametric pathway-based regression models for analysis of genomic data, *Biostatistics* (in press).

50. West, M., Bayesian factor regression models in the "large p, small n" paradigm, *Bayesian Statist.* **7**, 723–732 (2003).

51. Wei, L. J., A generalized Gehan and Gilbert test for a period observations that are subject to arbitrary right censorship, *J. Am. Stat. Assoc.* **75**, 634–637 (1980).

CHAPTER 23

Analysis of Case–Control Studies in Genetic Epidemiology

Nilanjan Chatterjee

Biostatistics Branch, Division of Cancer Epidemiology and Genetics, National Cancer Institute, National Institutes of Health, Rockville, Maryland

23.1 INTRODUCTION

Case–control studies retrospectively sample subjects on the basis of their disease status to obtain an efficient way of collecting covariate information for epidemiological studies of rare diseases. The standard method for estimating the odds-ratio parameters from this design involves prospective logistic regression analysis of the data, ignoring the retrospective nature of the design. The validity of this approach relies on the classic results by Cornfield [3], who showed the equivalence of prospective and retrospective odds ratios. The efficiency of the approach was established in two other classic papers by Anderson [1] and Prentice and Pyke [13], who showed that the prospective logistic regression of case–control data yields the proper maximum-likelihood estimates of the odds-ratio parameters under a "semiparametric" model that allows the covariate distribution to remain completely unrestricted. More recently, Rabinowitz [14] and Breslow et al. [2] used modern semiparametric theory to show that the prospective logistic regression analysis of case–control data is efficient in the sense that it achieves the variance lower bound of the underlying semiparametric model.

Case–control studies are now increasingly used to study the role of genetic susceptibility and gene–environment interactions in the etiology of rare complex diseases. A special feature for studies in genetic epidemiology is that it often may be reasonable to assume certain parametric or semiparametric models for the population distribution of covariates of interest. The assumptions of Hardy–Weinberg equilibrium (HWE) and gene–environment

Statistical Advances in the Biomedical Sciences, edited by Atanu Biswas, Sujay Datta, Jason P. Fine, and Mark R. Segal
Copyright © 2008 John Wiley & Sons, Inc.

independence are examples of such models. The HWE model, which specifies the simple relationship between *allele*[1] and *genotype*[2] frequencies at a given chromosomal locus, is a natural law for a randomly mating large stable population in the absence of new genetic mutation, inbreeding, and selective survivorship among genotypes [see, e.g., Ref. 6, Ch. 3]. Often, it is also natural to assume that a subject's genetic susceptibility, a factor that is determined at birth, is independent of that person's subsequent environmental exposures. Standard logistic regression analysis, which is the semiparametric maximum-likelihood (ML) solution for the problem that allows an arbitrary covariate distribution, clearly remains a valid option for analyzing case–control data under these assumptions. The method, however, may not be efficient because it fails to exploit the natural model constraints.

Piegorsch et al. [12] proposed a method for exploiting gene–environment independence in case–control studies. Assuming gene–environment independence and rare disease, the authors noted that the interaction odds ratio between a genetic and an environmental exposure can be simply estimated as the association odds ratio between these two factors in cases alone. Estimates of interactions obtained from this method can be much more efficient than those obtained from standard logistic regression analysis, which uses both cases and controls. This case-only analysis, however, is limited. It discards all the information from controls and hence loses the ability to estimate the main effect parameters of the logistic regression model that are required for deriving various alternative scientific parameters of interest. Assuming rare disease and categorical exposures, Umbach and Weinberg [23] showed that under gene–environment independence, the ML estimates of all the parameters of a logistic regression model can be obtained in a fairly general setting by fitting a suitably constrained log-linear model to the case-control data. For a rich model, with a large number of covariates, however, the log–linear modeling approach can easily become cumbersome. Moreover, the method cannot handle continuous covariates.

This chapter reviews some more recent developments for ML analysis of case–control studies of genetic epidemiology. Section 23.2 assumes a setting where complete information on both genetic and environmental exposures of interest is available on all subjects in the study. Section 23.3 considers the problem of haplotype-based analysis of genetic data where haplotype-phase information may be missing on certain study subjects. In each section, two classes of methods, namely, "prospective" and "retrospective," are presented. The connections and the differences between the alternative methods are pointed out to shed light on their relative merits. Applications of the methodologies are illustrated using numerical examples. In Section 23.4, the chapter concludes with a short discussion on potential pitfalls of the novel methodologies and some practical recommendations are for their applications.

23.2 MAXIMUM-LIKELIHOOD ANALYSIS OF CASE–CONTROL DATA WITH COMPLETE INFORMATION

Here we discuss ML analysis of case–control data in the case where complete information on both genetic and environmental factors is available.

---

[1]Genetic variant at a given locus on a single chromosome.
[2]The pair of genetic variants possessed by an individual at a given locus on the paternally and maternally inherited chromosomes.

23.2.1 Background

Let D be the binary indicator of the presence $(D = 1)$ or absence $(D = 0)$ of a disease. Suppose that the prospective risk model for the disease given a subject's genetic risk factors G and environmental risk factors X is given by the logistic regression model $\Pr(D = 1 \mid G, X) = \mathcal{L}\{\beta_0 + m(G, X; \beta_1)\}$, where $\mathcal{L}(w) = \{1 + \exp(-w)\}^{-1}$ is the logistic distribution function and $m(\cdot)$ is a known but arbitrary function that parameterize the joint odds ratios for G and X in terms of the regression parameters β_1. Let $\mathcal{H}(g, x)$ denote the joint distribution of G and X in the underlying population. We assume that N_0 controls and N_1 cases are sampled from the conditional distributions $\Pr(G, X \mid D = 0)$ and $\Pr(G, X \mid D = 1)$, respectively, and let $(G_i, X_i)_{i=1}^{N_0+N_1}$ denote the corresponding covariate data of the $N_0 + N_1$ study subjects.

In the above model setting, the fundamental "retrospective" likelihood for the data is given by

$$L_1^R = \prod_{i=1}^{N_0+N_1} \Pr(G_i, X_i \mid D_i)$$

$$= \prod_{i=1}^{N_0+N_1} \frac{\Pr_{\beta_0, \beta_1}(D_i \mid G_i, X_i) d\mathcal{H}(G_i, X_i)}{\int \Pr_{\beta_0, \beta_1}(D_i \mid g, x) d\mathcal{H}(g, x)}$$

From the classic results of Prentice and Pyke [13], it is well known that neither the intercept parameter β_0 nor the nonparametric distribution function $\mathcal{H}(g, x)$ is uniquely identifiable from L_1^R. There is, however, a unique value of β_1 that maximizes L_1^R, and the corresponding estimate of β_1 can be obtained by simple maximization of the "prospective likelihood" of the data

$$L_1^P = \prod_{i=1}^{N_0+N_1} \Pr_{\beta_0, \beta_1}(D_i \mid G_i, X_i).$$

Notably, the estimate of the intercept parameter obtained by maximization of L_1^P yields unbiased estimate for, not β_0, but

$$\kappa = \beta_0 + \log(N_1/N_0) - \log\{\pi/(1 - \pi)\},$$

where $\pi = \Pr(D = 1)$ denotes the marginal probability of the disease in the underlying population. More recently, Roeder et al. [15] presented a number of elegant results regarding equivalence of the retrospective and prospective likelihoods in a very general setting.

An alternative angle to the prospective analysis is useful. Consider the "randomized recruitment" sampling approach for case–control studies [24], where subjects, given their disease status $D = d$, are individually randomized to be recruited or not with the recruitment/selection probability being proportional to $\psi_d = N_d/\Pr(D = d)$. Under this sampling scheme, if a total of $N = N_1 + N_0$ subjects are sampled into the study, then in expectation there will be $N_d (= N \times \Pr(D = d) \times \psi_d)$ subjects with $D = d$ in the sample. Let $R = 1$ denote the indicator of whether a subject is selected in the case–control sample. Then, under the case–control sampling scheme described above, the distribution of the disease given covariates can be derived as

$$\Pr_{\beta_0, \beta_1}(D = 1 \mid G, X, R = 1) = \frac{\Pr(R = 1 \mid D = 1)\Pr(D = 1 \mid G, X)}{\sum_{d=0,1} \Pr(R = 1 \mid D = d)\Pr(D = d \mid G, X)},$$

which, with some algebra, can be written in the logistic form

$$\frac{\exp\{\kappa + m(G, X; \beta_1)\}}{1 + \exp\{\kappa + m(G, X; \beta_1)\}}$$

In words, under case–control selection, the probability distribution of D given G and X maintains the same structure of the original logistic regression model, except that the original intercept parameter β_0 is replaced by κ. Thus, prospective analysis of case–control data essentially corresponds to maximization of the likelihood

$$L_1^* = \prod_{i=1}^{N_0+N_1} \Pr_{\beta_0, \beta_1}(D_i \mid G_i, X_i, R = 1).$$

23.2.2 Maximum-Likelihood Estimation Under HWE and Gene–Environment Independence

Assume that the joint distribution of G and X in the *underlying population* is given by the product form $\mathcal{H}(g, x) = Q(g) F(x)$, where Q and F are the marginal distribution functions for G and X, respectively. Also assume that for a subject the genetic factor G can take values in a fixed set $\{g_1, \ldots, g_J\}$. Thus the distribution Q can be parameterized by the corresponding probability masses $\{q_1, \ldots, q_J\}$. Moreover, using population genetics theory, in many situations the probabilities q_j, $j = 1, \ldots, J$, can be further modeled as $q_j = q_j(\theta)$, for some known functions q_j and some parameter vector θ. For example, if A/a denote the major/ minor allele at a given biallelic loci, then under HWE, the population frequency of three genotypes AA, Aa, and aa can be written as $f_{AA} = (1 - p_a)^2$, $f_{Aa} = 2p_a(1 - p_a)$, and $f_{aa} = p_a^2$, where p_a denotes the allele frequency of a. If no population genetics model assumptions are made, one can assume in the notation above that θ represents the vector of q_j values themselves. The covariate distribution $F(x)$ is left completely nonparametric and is allowed to have a mass point at each distinct value of X in the observed sample.

Under the new model, the retrospective likelihood of the case–control data is given by

$$L_2^R = \prod_{i=1}^{N_0+N_1} \Pr(G_i, X_i \mid D_i)$$

$$= \prod_{i=1}^{N_0+N_1} \frac{\Pr_{\beta_0, \beta_1}(D_i \mid G_i, X_i) q_\theta(G_i) dF(X_i)}{\int_x \sum_{g_j} \Pr_{\beta_0, \beta_1}(D_i \mid g_j, x) q_\theta(g_j) dF(x)}$$

Chatterjee et al. [4] considered joint maximization of L_2^R with respect to $\gamma = (\beta_0, \beta_1, \theta)$ and the nonparametric distribution function $F(\cdot)$. They showed that here, unlike in the classical setting of Prentice and Pyke [13], the intercept parameter of the logistic regression model β_0 is theoretically identifiable from the retrospective likelihood, except for some boundary parameter settings. They assumed that the nonparametric ML estimator of F can allow positive masses only within the set $\chi = \{x_1, \ldots, x_K\}$, which represents the unique values of X that are observed in the case–control sample of $N_0 + N_1$ study subjects. Thus, to obtain the ML estimator, it is sufficient to consider the class of discrete F that have support points within the set χ. Any F in this class can be parameterized with respect to the probability masses $\{\delta_1, \ldots, \delta_K\}$

that it assigns to the points $\{x_1, \ldots, x_K\}$. The authors then derived a profile likelihood of the data by maximizing the likelihood L_2^R with respect to δ, the probability masses associated with $F(\cdot)$, for fixed values of $\gamma = (\beta_0, \beta_1, \theta)$. If $\hat{\delta}(\gamma)$ denotes the value of δ that maximizes L_2^R for fixed γ, the profile log likelihood is then given by $L_2^R\{\gamma, \hat{\delta}(\gamma)\}$. They derived a number of key results to simplify the computation of the profile likelihood. Below is a lemma summarizing the results.

Lemma 23.1. Define the parameters $\mu_d = N_d/\{N\mathrm{Pr}(D=d)\}$ for $d = 0, 1$ and $\kappa = \mu_1/\mu_0$. Define a conditional distribution function for (D, G) given X as

$$P_{\gamma,\kappa}^*(D, G \mid X) = \frac{\mathrm{Pr}_{\beta_0,\beta_1}(D \mid G, X)\kappa^D q_\theta(G)}{\sum_{d=0,1} \sum_j \mathrm{Pr}_{\beta_0,\beta_1}(d \mid g_j, X)\kappa^d q_\theta(g_j)}.$$

The profile log likelihood $\log L_2^R\{\gamma, \hat{\delta}(\gamma)\}$ can be computed as $l^*\{\gamma, \hat{\kappa}(\gamma)\}$, where

$$l^*(\gamma, \kappa) = \sum_{i=1}^{N_0+N_1} \log P_{\gamma,\kappa}^*(D_i, G_i \mid X_i),$$

and $\hat{\kappa}(\gamma)$ is obtained by solving the score equations $\partial l^*/\partial\kappa = 0$ for each fixed value of γ.

The main consequence of Lemma 23.1 is that the profile likelihood $L_2^R\{\gamma, \hat{\delta}(\gamma)\}$ can be computed without having to maximize numerically the likelihood L_2^R with respect to the potentially high-dimensional nuisance parameter δ. Instead, the profile likelihood $L_2^R\{\gamma, \hat{\delta}(\gamma)\}$ can be obtained in closed form up to only one additional parameter κ, which in turn is defined as the solution of a single score equation. By recalling the definition of the sampling indicator variable $R = 1$ from Section 23.2.1, we further observe that

$$
\begin{aligned}
P_{DG}^*(E) &= \mathrm{Pr}(D, G \mid X, R = 1) \\
&= \mathrm{Pr}(D \mid G, X, R = 1) \times \mathrm{Pr}(G \mid X, R = 1).
\end{aligned}
\tag{23.1}
$$

Standard logistic regression analysis corresponds to maximization of $\mathrm{Pr}(D \mid G, X, R = 1)$, which is only the first part of the likelihood given in formula (23.1). Intuitively, if it is known that G and X are independently distributed in the population, then any association between G and X in the case–control sample, which is enriched by the diseased subjects, would indicate that both G and X are associated with D, under some mechanism or the other. Similarly, when HWE is known to hold in the population, departure of the case–control sample from HWE would be indicative of genetic association. Thus, in general, if certain model assumptions can be made to specify $\mathrm{Pr}(G \mid X)$ in the underlying population, then, under case–control sampling, the conditional distribution $\mathrm{Pr}(G \mid X, R = 1)$ contains information about the association parameters β_1. Prospective logistic regression analysis loses efficiency by ignoring this piece of information from the case–control data.

Chatterjee et al. [4] used the abovementioned result to derive the asymptotic theory for the ML estimator of the odds-ratio parameters in a "semiparametric" setting that allows the dimension of the nuisance parameter δ to potentially increase with sample size, as would be the case for dealing with continuous covariates. They showed how one can incorporate information on marginal probability of the disease in the population to further enhance the efficiency of the methodology. They described simplification of the methodology under a rare disease

Table 23.1   Odds-Ratio Estimates and 95% Confidence Intervals from the Israeli Ovarian Cancer Study

Parameter	Logistic Regression	ML Assuming $G \perp\!\!\!\perp X \mid S^a$	Case-Only
BRCA1-2	3.58 (2.27, 4.89)	3.15 (2.51, 3.79)	—
OC use[b]	−0.05 (−0.10, 0.00)	−0.05 (−0.10, 0.00)	—
Parity[c]	−0.06 (−0.12, 0.00)	−0.06 (−0.13, 0.00)	—
OC*BRCA1-2	0.06 (−0.15, 0.26)	0.09 (0.02, 0.15)	0.056 (−0.01, 0.12)
Parity*BRCA1-2	−0.20 (−0.63, 0.23)	−0.04 (−0.14, 0.07)	−0.014 (−0.12, 0.09)

[a]Stratification variables include age, ethnicity, personal history of breast cancer, and family history of breast/ovarian cancer.
[b]Years of oral contraceptive use.
[c]Number of children.

approximation. They extended the methodology for dealing with "population stratification" that may cause G and E to be related at a population level even when they are independent within subpopulations. A MATLAB program for implementing these methods is available under the software link from the Website www://dceg.cancer.gov/people/ChatterjeeNilanjan.html.

23.2.3   An Example

Chatterjee et al. [4] illustrated an application of their proposed methodology on a case–control study of ovarian cancer designed to investigate how the high-risk BRCA1-2 genetic mutations interact with various reproductive risk factors, such as oral contraceptive use and parity. Table 23.1 shows the analysis of the data using three different methods: (1) standard logistic regression, (2) the case-only method, and (3) ML assuming independence of BRCA1-2 mutations and the reproductive risk factor conditional on certain stratification factors. Clearly, the ML method assuming gene–environment independence yielded more precise estimates of the interaction parameters than did logistic regression. The case-only approach was also precise in estimating the interaction parameters, but could not yield estimates of the main effect parameters, which were key in addressing one of the main scientific questions of the study, that is, whether oral contraceptive (OC) use, a factor that is known to reduce the risk of ovarian cancer in the general population, decreases the risk of the disease among the high-risk women who carry BRCA1-2 mutations. Based on the maximum-likelihood method, the odds ratio for OC use among BRCA1-2 carriers can be computed as $\exp(-0.051 + 0.089) = 1.034$, with an associated 95% confidence interval of (0.977, 1.095). These calculations suggested that unlike noncarriers, the risk of ovarian cancer for carriers did not decrease with increasing OC use. The original study by Modan et al. [11], which used yet another alternative method, had reached a similar conclusion.

23.3   HAPLOTYPE-BASED GENETIC ANALYSIS WITH MISSING PHASE INFORMATION

23.3.1   Background

Often, the goal of a genetic epidemiologic study involves studying the association between a disease and a candidate genomic region of biologic interest. Typically, in such studies,

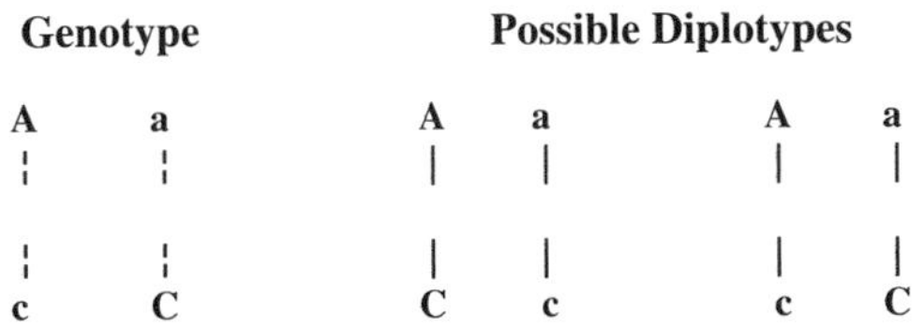

Genotype	Possible Diplotypes

Figure 23.1   An example of genotype and haplotype data for a subject involving three bial-lelic loci. In truth, the subject carries the haplotype (A-B-c) ad (a-B-C) on the two chromo-somes. When only locus-specific genotype data are observed, the phase information, that is, which combinations of alleles arise in the same chromosome, is lost. The dotted lines in the figure represent this loss of information. In the absence of the phase information, the genotype data cannot distinguish between two possible haplotype configurations: $\{(A\text{-}B\text{-}C), (a\text{-}B\text{-}c)\}$ and $\{(A\text{-}B\text{-}c), (a\text{-}B\text{-}C)\}$.

genotype information is obtained on multiple loci that are known to harbor genetic variations within the region(s) of interest. An increasingly popular approach for analysis of such multi-locus genetic data has been the haplotype-based regression methods where the effect of the genomic region on disease-risk is modeled through "haplotypes," the combinations of alleles (gene variants) at multiple loci along individual homologous chromosomes. It is believed that association analysis based on haplotypes, which can efficiently capture interloci interactions as well as "indirect association" due to *linkage disequilibrium*[3] of the haplotypes with unobserved causal variant(s), can be more powerful than the more traditional locus-by-locus methods for analysis of association [18].

A technical problem for haplotype-based regression analysis is that in traditional epidemio-logic studies the haplotype information for the study subjects is not directly observable. Instead, locus-specific genotype data are observed, which contain information on the pair of alleles that a subject carries on his/her pair of homologous chromosomes at each locus, but does not provide the "phase information," that is, which combinations of alleles appear across multiple loci along the individual chromosomes. Figure 23.1 shows an example of gen-otype and haplotype data for a subject involving three biallelic loci. In general, the genotype data of a subject will be phase-ambiguous whenever the subject is *heterozygous*[4] at two or more loci. Statistically, the lack of phase information can be viewed as a special missing data problem.

Suppose that there are M loci of interest within a genomic region. Let $H^{\text{di}} = (H_1, H_2)$ denote the corresponding diplotype status for an individual, that is, the two haplotypes that the indi-vidual carries in his/her pair of homologous chromosomes. Given the diplotype status H^{di} and environmental covariate X, assume that the risk of disease for a subject is given by the logistic regression model

$$\text{logit}\{\Pr(D = 1 | H^{\text{di}}, X)\} = \beta_0 + m(H^{\text{di}}, X; \beta_1). \tag{23.2}$$

Typically, the main interest is in estimating the regression parameters β_1. In the model pre-sented above, the effect of the diplotypes could be further specified in terms of the effect of

[3]Association among alleles at physically nearby loci of a chromosome due to their tendency of co-inheritance.
[4]Carries two different alleles on the given locus of a pair locus of a pair of homologus chromosomes.

the constituent haplotypes assuming different mode of effects such as *dominant*,[5] *recessive*,[6] or *additive*[7] [21]. These modeling assumptions may be necessary for identifiability purpose when only genotypes, and not the diplotypes, are directly observed [5]. Let $\mathbf{G} = (G_1, \ldots, G_M)$ denote the genotype data at M loci. As explained above, the same genotype data $\mathbf{G}$ could be consistent with multiple diplotypes. We denote by $\mathcal{H}_{\mathbf{G}}$ the set of all possible diplotypes that are consistent with the genotype data $\mathbf{G}$. Assume that H^{di} and environmental factors X are independent in the population, with a parametric form

$$\Pr(H^{di} = h^{di} \mid X) = \Pr(H = h^{di}) = q(h^{di}, \theta)$$

where the model $q(h^{di}; \theta)$, in turn, could be specified according to HWE or some of its extensions as considered by Satten and Epstein [16] and Lin and Zeng [9]. In particular, under HWE

$$\begin{aligned} \Pr_\theta\left\{H^{di} = (h_i, h_j)\right\} &= \theta_i^2 \quad \text{if} \quad h_i = h_j \\ &= 2\theta_i\theta_j \quad \text{if} \quad h_i \neq h_j, \end{aligned} \tag{23.3}$$

where θ_i denotes the population frequency for haplotype h_i. Let N_d, $d = 0, 1$ denote the number of controls and cases in the observed sample. Let D_i, $\mathbf{G}_i$, and X_i be the value of D, $\mathbf{G}$, and X for the ith subject in the sample.

In the more recent past, a number of researchers have developed methods for analysis of case–control data using the regression model (23.2). The methods can be broadly classified into two types: (1) prospective and (2) retrospective. A brief review of these methods follows.

23.3.2   Methods

Lake et al. [8] proposed jointly estimating the haplotype frequencies θ and the regression parameters $\beta = (\beta_0, \beta_1)$, by maximization of the prospective likelihood

$$\begin{aligned} L_{\text{haplo}}^P &= \prod_{i=1}^{N_0+N_1} \Pr(D_i \mid \mathbf{G_i}, X_i) \\ &= \prod_{i=1}^{N_0+N_1} \sum_{H^{di} \in \mathcal{H}_{\mathbf{G_i}}} \Pr_\beta(D_i \mid H^{di}, X_i) \Pr_\theta(H^{di}) \end{aligned}$$

via an expectation–maximization (EM) algorithm. Unfortunately, use of this purely prospective method for case–control data has not been justified on theoretical grounds. In fact, Spinka et al. [19] showed that the standard argument for unbiasedness of the prospective score functions under the case–control design may not hold in the current setting because of the constraints on the covariate distribution. Nevertheless, for practical purposes, the bias in estimates of the odds-ratio parameters from the purely prospective method tends to be quite small unless there is a large amount of phase ambiguity and the haplotype effects are large [20]. Zhao et al. [22] assumed a rare disease to propose a modification of the prospective

---

[5]Effect of the haplotype is the same whether one two copies are carried.
[6]Effect of the haplotype exists only if two copies are carried.
[7]Effect of carrying two copies of the haplotype is twice that of carrying one copy.

score equations that is theoretically unbiased under the case–control design. They proposed estimating θ in a separate step by using genotype data only from the controls. Spinka et al. [19] proposed an alternative modification of the prospective score functions that remain unbiased even without requiring the rare disease assumption.

Had there been no phase ambiguity, that is if H^{di} was directly observed, all the different prospective methods would be equivalent to the standard logistic regression analysis, which, as discussed in Section 23.2, does not rely on any covariate distributional assumptions. In the presence of phase ambiguity, however, estimation under a completely nonparametric model for the joint distribution of (H^{di}, X) is not possible because of the lack of parameter identifiability. Assumptions such as HWE or/and gene–environment independence are typically needed. The prospective methods, however, rely on those assumptions rather weakly in the sense that they utilize the covariate models only in the "expectation" steps of the respective EM-type algorithms to assign a probability distribution for the possible diplotypes of the subjects given their genotype, environmental covariates, and outcome data. Once these probabilities are assigned, the "maximization" steps of the methods do not rely on the assumptions; they involve simply fitting weighted prospective logistic regression model to the data. Thus, the prospective methods gain robustness against violations of HWE or/and gene–environment independence assumptions. Retrospective methods, on the other hand, can fully exploit the assumptions to gain efficiency.

Ignoring environmental covariates X, Epstein and Satten [5] described an algorithm for joint estimation of β and θ by maximization of the proper retrospective likelihood

$$L_{\mathrm{haplo}}^{R} = \prod_{i=1}^{N_0+N_1} \Pr(\mathbf{G_i} \mid D_i)$$

$$= \prod_{i=1}^{N_0+N_1} \sum_{H^{\mathrm{di}} \in \mathcal{H}_{\mathbf{G_i}}} \Pr_{\beta,\theta}(H^{\mathrm{di}} \mid D_i).$$

Assuming a rare disease, the authors approximated the diplotype distribution for the disease-free subjects as $\Pr(H^{\mathrm{di}} \mid D = 0) = \Pr_{\theta}(H^{\mathrm{di}})$. They further expressed $\Pr(H^{\mathrm{di}} \mid D = 1)$, the diplotype distribution for diseased subjects, in terms of $\Pr(H^{\mathrm{di}} \mid D = 0)$ and the odds-ratio parameters of the logistic regression model using formulas derived in Satten and Kupper [17]. An EM-type algorithm was described for maximization of the retrospective likelihood. Satten and Epstein [16] conducted extensive simulation studies to illustrate the efficiency advantage of the retrospective over the prospective methods.

Spinka et al. [19] described methods for incorporating environmental covariates X in the retrospective approach based on the likelihood

$$L_{\mathrm{haplo}}^{R} = \prod_{i=1}^{N_0+N_1} \Pr(\mathbf{G_i}, X_i \mid D_i)$$

$$= \prod_{i=1}^{N_0+N_1} \sum_{H^{\mathrm{di}} \in \mathcal{H}_{\mathbf{G_i}}} \Pr_{\beta,\theta}(H^{\mathrm{di}}, X_i \mid D_i).$$

Assuming a completely nonparametric form for $F(x)$, the authors obtained results similar to those described in Lemma 23.1 to show that estimates of β and θ that maximize the retrospective likelihood L_{haplo}^{R} can be obtained by the maximization of an alternative pseudolikelihood of the form

$$\log(L^*_{h\,aplo}) = \sum_{i=1}^{N_0+N_1} \log\{\Pr(D_i, \mathbf{G_i} \mid X_i, R_i = 1)\}$$

$$= \sum_{i=1}^{N_0+N_1} \log\left[\sum_{h^{di} \in \mathcal{H}_{G_i}} \Pr_{\beta,\theta}(D_i, H_i^{di} = h^{di} \mid X_i, R_i = 1)\right], \qquad (23.4)$$

where conditioning on the event $R = 1$ reflects the nonrandom "ascertainment" mechanism of the case–control design. The analytic formula for $\Pr(D_i, H_i^{di} \mid X_i, R_i = 1)$ is the same as that of P^*_{DG} shown in Lemma 23.1, with G simply replaced by H^{di}. Computationally, the maximization of L^*_{haplo} is relatively simple because it, unlike L^R_{haplo}, does not depend on the distribution of the environmental covariates X. When no environmental covariates are involved, Stram et al. [20] had previously proposed use of an "ascertainment corrected joint-likelihood" of the form $L^{asc}_{haplo} = \Pi_i \Pr(D_i, \mathbf{G}_i \mid R_i = 1)$. The representation of the pseudolikelihood L^*_{haplo} given in (23.4) suggests that when $F(x)$ is treated completely nonparametrically, the efficient retrospective ML estimator can be obtained by simply conditioning on X in the approach of Stram et al.

23.3.3   Application

Spinka et al. [19] illustrated the bias–efficiency tradeoffs between the retrospective and prospective methods using a simulated case–control study. The study involved a population consisting of two strata, with frequencies 0.40 ($S = 1$) and 0.60 ($S = 2$), which differ in their distribution of both haplotypes and environmental factors. They assumed a simple scenario involving four haplotypes constructed from two binary SNPs, with the haplotypes {(0,0), (0,1), (1,0), (1,1)} having frequencies (0.35, 0.30, 0.15, 0.20) and (0.35, 0.20, 0.30, 0.15) in strata 1 and 2, respectively. They generated the environmental covariate from a lognormal distribution with the mean and variance for the underlying normal distribution to be 0.67 and 1 for $S = 1$ and 0 and 1 for $S = 2$. Additionally, they assumed that the stratification variable S is a risk factor for the disease. In particular, the disease status for each subject was generated according to the model

$$\text{logit}\{\Pr(D|H^{di}, X, S)\} = \beta_0 + \beta_X + \beta_H N_2(H^{di}) + \beta_{HX} N_2(H^{di})X$$
$$+ \beta_S I(S = 2) + \beta_{HS} N_2(H^{di})I(S = 2),$$

where $N_2(H^{di})$ denotes the number of copies of $h_2 = (0,1)$ contained in H^{di}. The true value of $(\beta_0, \beta_X, \beta_H, \beta_{HX}, \beta_S, \beta_{HS})$ was $(-3.5, 0.1, 0.15, 0.20, 0.69, 1.10)$. The results given in Spinka et al. are replicated in Table 23.2.

Several key observations can be made. When the true model assumed that H^{di} and X are independent conditional on S, but the data were analyzed using the retrospective method that assumes H^{di} and (X, S) are independent in the entire population, substantial bias was introduced in estimating the parameters β_H, β_S, and β_{HS}. Neither the prospective method nor the retrospective method, which explicitly accounts for the conditional independence model, suffered from such bias. The prospective method had the largest variance of the three methods, while the retrospective method under the unconditional independence model had the smallest.

Table 23.2   Results from 500 Simulated Case–Control Studies^a

Parameter	(1) Unconditional RML		(2) Conditional RML		(3) Modified PSE	
	Bias	Empirical SE	Bias	Empirical SE	Bias	Empirical SE
β_X	−0.0050	0.0403	0.0012	0.0395	0.0047	0.0496
β_H	0.4951	0.2000	0.0209	0.2009	−0.0039	0.2370
β_{HX}	−0.0021	0.0389	0.0018	0.0388	0.0060	0.0567
β_S	0.1266	0.2155	0.0287	0.2299	−0.0377	0.2386
β_{HS}	−1.1509	0.1971	−0.0151	0.2395	−0.0214	0.2561

^{a}These results are from a population where HWE and the independence between haplotype (H) and environmental covariate (X) hold within strata defined by S. Each replicate contains 1000 cases and 1000 controls and is analyzed using (1) the unconditional retrospective maximum-likelihood (RML) method, assuming that HWE and $H - X$ independence hold in the entire population; (2) the conditional RML method assuming HWE and $H - X$ independence hold conditional on S; and (3) the modified proposed modified prospective score equation (PSE) method described in Spinks et al. [19].

The retrospective method, which assumed the correct conditional independence model, provided both small bias and relatively small variance.

23.4 CONCLUDING REMARKS

Case–control studies with modest sample size often have very little power for studying recessive genetic effects and gene–environment interactions using prospective logistic regression analysis. Efficiency of such investigations can be dramatically improved by use of retrospective ML methods that can exploit gene–environment independence and HWE assumptions. Caution, however, is needed because retrospective methods can produce large bias in estimates of odds-ratio parameters when the underlying covariate distributional assumptions are violated. The assumptions of HWE and gene–environment independence can be violated in a number of different ways. A common source for the problem could be a phenomenon called *population stratification*, which arises if there are underlying substrata in the population across which genotype or/and exposure distribution varies. In these situations, the assumptions of HWE and gene–environment independence are likely to hold within the substrata, but not in the entire population. Satten and Epstein [16] considered using the so-called *fixation index* model to account for violation of HWE toward excess homozygosity, a consequence of hidden population stratification (see, e.g., Ref. 6, Ch. 4). Chatterjee et al. [2] and Spinka et al. [19], on the other hand, proposed explicitly accounting for population stratification by using conditional models for HWE or/and gene–environment independence when information on the source of population stratification, such as ethnicity, is available in the study. Lin and Zeng [9] described methods for haplotype analysis of case–control data allowing for genotype–environment association with the constraint that haplotypes and environmental exposures are independent given the genotypes. For family-based case–control studies, Chatterjee et al. [2] have described methods that require the assumption of gene–environment independence to hold only within families, but not necessarily for the underlying population.

   Methods for exploiting the gene–environment independence assumption could be practically useful without concerns about bias in many important situations. For "randomized exposure" such as the treatment assigned in a randomized trial, the gene–environment independence assumption would be satisfied by definition of randomization. The phenomenon of independence due to randomization was utilized in a case-only study of interaction between BRCA1-2 mutation and tamoxifen, conducted within the Breast Cancer Prevention Trial [7]. The assumption of gene–environment independence is also very likely to be satisfied for external environmental agents, exposure to which is not directly controlled by an individual's own behavior. Some examples of such exposure are radiation exposure among the cohort of atom bomb survivors in Japan, carcinogenic exposure from a chemical factory to employees or nearby residents, and pesticide exposure in an agricultural community. When an exposure depends on a subject's individual behavior, on the other hand, the independence assumption should be used more cautiously. There could be spurious association between G and E for established risk factors such as smoking because family history of lung cancer, which is associated with G, may also influence a subject to change his/her smoking behavior. There could also be direct association. Genetic polymorphisms in the smoking metabolism pathway, for example, not only can modify a subject's risk from smoking but may also influence a subject's degree of addiction to smoking.

When violation of the gene–environment independence or/and HWE seems plausible, effort should be made to validate the assumption empirically. Tests for these assumptions within a given study, however, may have very little power and empirical evidence from external data sources should be investigated. Marcus et al. [10], for example, have demonstrated the independence of smoking behavior and NAT2 genotype that are believed to be involved in smoking metabolism using a meta-analysis approach that employs data on controls from a series of different case–control studies. When substantial uncertainty remains regarding the validity of the assumption because of lack of empirical data or for other reasons, positive findings based on the retrospective methodologic should be considered as a preliminary screen, which should be pursued with high priority in future studies.

ACKNOWLEDGMENT

This research was supported by the Intramural Program of the National Cancer Institute, National Institute of Health, USA.

REFERENCES

1. Andersen, J. B., Asymptotic properties of conditional maximum likelihood estimators, *J. Roy. Statist. Soc. Ser. B* **32**, 283–301 (1970).

2. Breslow, N. E., Robins, J. M., and Wellner, J. A., On the semi-parametric efficiency of logistic regression under case-control sampling, *Bernoulli* **6**, 447–455 (2000).

3. Cornfield, J., A statistical problem arising from retrospective studies, in *Proc. 3rd Berkeley Symp. Mathematical Statistics and Probability*, 1956, Vol. 4, pp. 135–148.

4. Chatterjee, N., Kalaylioglu, Z., and Carroll, R. J., Exploiting gene-environment independence in family-based case-control studies: Increased power for detecting associations, interactions and joint effects, *Genet. Epidemiol.* **28**, 138–156 (2005).

5. Epstein, M. P. and Satten, G. A., Inference on haplotype effects in case-control studies using unphased genotype data, *Am. J. Hum. Genet.* **73**, 1316–1329 (2003).

6. Hartl, D. L. and Clark, A. G., *Principles of Population Genetics*, Sinauer Associates Inc., Sunderland, MA, 1997.

7. King, M. C., Wieand, S., Hale, K., Lee, M., Walsh, T., Owens, K., Tait, J., Ford, L., Dunn, B. K., Costantino, J., Wickerham, L., Wolmark, N., and Fisher, B., National Surgical Adjuvant Breast and Bowel Project, Tamoxifen and breast cancer incidence among women with inherited mutations in BRCA1 and BRCA2: National Surgical Adjuvant Breast and Bowel Project (NSABP-P1) Breast Cancer Prevention Trial, *JAMA* **14**, 2251–2256 (2001).

8. Lake, S., Lyon, H., Silverman, E., Weiss, S., Laird, N., and Schaid, D., Estimation and tests of haplotype-environment interaction when linkage phase is ambiguous. *Hum. Hered.* **55**, 56–65 (2003).

9. Lin, D. Y. and Zeng, D., Likelihood based inference on haplotype effects in genetic association studies, *J. Am. Statist. Assoc.* **101**, 89–104 (2006).

10. Marcus, P. M., Vineis, P., and Rothman, N., NAT2 slow acetylation and bladder cancer risk: A meta-analysis of 22 case-control studies conducted in the general population, *Pharmacogenetics* **10**, 115–122 (2000).

11. Modan, M. D., Hartge, P., Hirsh-Yechezkel, G., Chetrit, A., Lubin, F., Beller, U., Ben-Baruch, G., Fishman, A., Menczer, J., Struewing, J. P., Tucker, M. A., and Wacholder, S., for the National Israel Ovarian Cancer Study Group, Parity, oral contraceptives and the risk of ovarian cancer among carriers and non-carriers of a BRCA1 or BRCA2 mutation, *New Engl. J. Med.* **345**, 235–240 (2001).

12. Piegorsch, W. W., Weinberg, C. R., and Taylor, J. A., Non-hierarchical logistic models and case-only designs for assessing susceptibility in population based case-control studies, *Statist. Med.* **13**, 153–162 (1994).

13. Prentice, R. L. and Pyke, R., Logistic disease incidence models and case-control studies, *Biometrika* **66**, 403–411 (1979).

14. Rabinowitz, D., A note on efficient estimation from case-control data, *Biometrika* **84**, 486–488 (1997).

15. Roeder, K., Carroll, R. J., and Lindsay, B. G., A nonparametric mixture approach to case-control studies with errors in covariables, *J. Am. Statist. Assoc.* **91**, 722–732 (1996).

16. Satten, G. A. and Epstein, M. P., Comparison of prospective and retrospective methods for haplotype inference in case-control studies, *Genet. Epidemiol.* **27**, 192–201 (2004).

17. Satten, G. A. and Kupper, L. L., Inferences about exposure-disease associations using probability-of-exposure information, *J. Am. Statist. Assoc.* **88**, 200–208 (1993).

18. Schaid, D., Evaluating associations of haplotypes with traits, *Genet. Epidemiol.* **27**, 348–364 (2004).

19. Spinka, C., Carroll, R. J., and Chatterjee, N., Analysis of case-control studies of genetic and environmental factors with missing genetic information and haplotype-phase ambiguity, *Genet. Epidemiol.* **29**, 108–127 (2005).

20. Stram, D., Pearce, C., Bretsky, P., Freedman, M., Hirschhorn, J., Altshuler, D., Kolonel, L., Henderson, B., and Thomas, D., Modeling and E-M estimation of haplotype-specific relative-risks from genotype data for case-control study of unrelated individuals, *Hum. Hered.* **55**, 179–190 (2003).

21. Wallenstein, S., Hodge, S., and Weston, A., A logistic regression model for analyzing extended haplotype data, *Genet. Epidemiol.* **15**, 173–181 (1998).

22. Zhao, L. P., Li, S. S., and Khalid, N., A method for the assessment of disease associations with single-nucleotide polymorphism haplotypes and environmental variables in case-control studies, *Am. J. Hum. Genet.* **72**, 1231–1250 (2003).

23. Umbach, D. M. and Weinberg, C. M., Designing and analyzing case-control studies to exploit independence of genotype and exposure, *Statist. Med.* **16**, 1731–1743 (1997).

24. Weinberg, C. R. and Sandler, D. P., Randomized recruitment in case-control studies, *Am. J. Epidemiol.* **134**, 431–432 (1991).

CHAPTER 24

Assessing Network Structure in the Presence of Measurement Error

Denise Scholtens

Department of Preventive Medicine, Northwestern University Medical School, Chicago, Illinois

Raji Balasubramanian

BG Medicine, Inc., Waltham, Massachusetts

Robert Gentleman

Program in Computational Biology, Fred Hutchinson Cancer Research Center, Seattle, Washington

24.1 INTRODUCTION

High-throughput experimental procedures such as affinity purification–mass spectrometry (AP-MS) and yeast two-hybrid (Y2H) systems have allowed the collection of genomewide protein interaction data. Such data typically indicate the presence or absence of specific protein–protein physical interactions and complex comemberships. One major emphasis in data-generated biological networks has been the description of their overall network topology. Many biological networks have been described as both small-world and scale-free (to be defined in Section 24.4). Proposals for the biological implications of these topologies include the relative robustness of biological graphs to random perturbations, but tremendous breakdown in response to targeted disruption of central hub nodes [16].

The data used to generate these graphs are experimentally obtained and hence are subject to a variety of errors and incompleteness. They are subject to experimental errors that can result in both false-positive (FP) and false-negative (FN) observations. Furthermore, the observed data are limited to the set of tested edges induced by the sampling scheme. The remaining untested edges are fundamentally different from tested edges, which yield a negative result. Standard

Statistical Advances in the Biomedical Sciences, edited by Atanu Biswas, Sujay Datta, Jason P. Fine, and Mark R. Segal

419

graph statistics are commonly applied to such data, and subsequent biological conclusions are often made without regard to the errors or incompleteness of the data on which they are based. There is therefore a need to assess the performance of these statistics, and the inference that can be drawn, when these data are imperfect.

In this chapter we consider the analysis of biological network data that are subject to measurement error. We identify particular issues that can profoundly affect inference on the true underlying graph structures. We also identify a number of specific network models and carry out simulation studies of the properties of global graph statistics and their biological interpretations in the presence of stochastic and systematic sources of measurement error.

24.2   GRAPHS OF BIOLOGICAL DATA

Graphs and networks have become foundational data structures for representing high-throughput genomics and proteomics data. Huber et al. [13] give a good review of graph types often used in computational biology, and Gentleman et al. [9] describe a few specific applications. Generally speaking, graphs consist of a set of nodes V and a set edges E can be represented as $G(V, E)$. Nodes represent objects of interest, and edges represent relationships between those objects. In this chapter, we focus on data pertaining to protein–protein relationships. Nodes represent proteins, and edges represent either protein complex comembership or physical interactions depending on the technology being discussed. A *multigraph* is a graph in which multiple edges between nodes are permitted. One useful type of multigraph is the super-position of graphs from different data sources in which each edge set represents a different sort of binary relationship. One can use a multigraph to combine binary interaction data together with protein complex comembership data while still allowing the different data types to be extracted and manipulated separately.

We will discuss two large-scale affinity purification–mass spectrometry (AP-MS) datasets published by Gavin et al. [7] and Ho et al. [12]. These two groups used slightly different techniques, and call their methods *tandem affinity purification* (TAP) and *high-throughput mass spectrometry–protein complex identification* (HMS-PCI), respectively. While different in analytical details, both of these AP-MS technologies gather data on protein complex comembership. In AP-MS experiments a set of bait proteins are specified, and each bait is used for a separate purification. In each purification, proteins that are comembers with the bait in at least one, but possibly more than one, complex are reported as hits for that bait. The set of hits for a particular bait may all belong to the same complex, or if the bait is involved in multiple complexes, the hits may represent these multiple complexes.

"Spoke" and "matrix" models are frequently used to represent complex comembership data detected by the AP-MS technology [2]. In "spoke" graphs, nodes represent proteins and edges are drawn from bait proteins to the hits that they detect as complex comembers in their respective purifications. In "matrix" graphs, additional edges connect all pairs of hits for a given bait protein. We will use the spoke representation since the matrix model often assumes edges between pairs of hit proteins for which complex comembership was never directly tested [22].

We will also discuss the physical interactions between proteins detected using the yeast two-hybrid (Y2H) system reported by Ito et al. [14] and Uetz et al. [28]. Y2H technology is also a bait–hit system in which the set of tested edges consists of those originating at bait proteins and extending to the set of proteins detected by each bait as physically interacting partners. Strictly speaking, if a common set of baits were used, Y2H data should be a subset of AP-MS data since complex comembers should bind to at least one other protein in the complex, but need not physically interact with all complex comembers. Furthermore, any proteins that physically interact under the experimental conditions form a complex and should be detected as complex comembers. The veracity of this relationship depends on similarity of experimental conditions and high sensitivity and specificity of the technology, none of which are guaranteed.

For accurate data representation, edges between proteins in both AP-MS and Y2H graphs should be directed from baits to hits. Directionality is useful in tracking the number of times each edge is tested, which edges result from common purifications, and potential reciprocity of the tested relationships. Generally only the underlying undirected versions of these graphs are used for analysis, and unfortunately, some of the resolution is lost.

We will also make use of protein complex estimates reported by Scholtens et al. [22]. These estimates use the statistical penalized likelihood approach described in Scholtens and Gentleman [21] to estimate protein complex membership using both the TAP and HMS-PCI data. This algorithm effectively estimates the presence or absence of untested edges given the set of tested edges, and results in the prediction of the true underlying set of protein complexes. In particular, we will focus on the set of estimated complexes that contain more than one bait protein and more than one edge; these are referred to as the *multibait–multiedge* (MBME) complexes and are believed to be the most reliable estimates.

24.2.1 Integrating Multiple Data Types

While we have chosen to focus on protein–protein interactions, graphs have also been used to model other systems biology data, including coregulation of gene expression [3], synthetic lethality [27], and metabolic and signal transduction networks [16,20,23]. The graph-theoretic framework has facilitated joint analyses of multiple biological data types. For example, Ge et al. [8] show that interacting proteins tend to be coregulated, and Balasubramanian et al. [3] translate these analyses into a graph-theoretic framework, providing permutation techniques for appropriate evaluation of statistical significance. Similar integrative approaches combine data from multiple sources to determine which experimentally observed edges are most likely to be true [15,30,6,25]. Wong et al. [32] use multiple data types to predict synthetic lethal interactions, and Ye et al. [33] note the significant overlap between genetic congruence graphs with protein interaction graphs. Qi et al. [19] merge AP-MS data from Gavin et al. [7] with the synthetic lethal data from Tong et al. [27] and develop a genetic interaction motif finding (GIMF) algorithm. Kelley and Ideker [17] combine several data types to form a loose definition of pathway, and investigate the relative abundance of inter- and intrapathway pairs of synthetic genetic interactions.

These analyses study the overlap of different graph types and often rely on the global graph statistics, discussed in this chapter, to make biological conclusions. All of these data are subject to the sources of measurement error discussed in this chapter, hence it will be of interest to explore, in the future, the extent to which the observations made here apply when simultaneously studying multiple graphs.

24.3 STATISTICS ON GRAPHS

Three of the most commonly discussed statistics on graphs are the average path length between any pair of nodes, the average clustering coefficient for all nodes in a graph, and the distribution of node degree. While not formally model-based from a statistical likelihood perspective, various values of these statistics have been associated with different graph types.

The average path length L between any pair of nodes in a graph is designed to measure global connectivity within the graph. For all pairs of nodes, the length of the shortest path from one node to the other is computed, and the lengths of all shortest paths are averaged together. In an undirected graph, the path length from node v_i to v_j will necessarily equal the path length from v_j to v_i; hence the path length for each pair is symmetric. In a directed graph, however, path lengths may differ depending on which node is the starting node since a path must always follow the direction of the edges. Theoretically, L pertains only to connected graphs in which a path exists between all pairs of nodes. If a graph is unconnected, then the path length between pairs of nodes that belong to separate connected components is reported by most algorithms to be infinite. Many report the global L statistic for unconnected graphs by ignoring the infinite values or by reporting values computed only for the largest connected component. Further research into appropriate analogs for L in unconnected graphs is warranted if this statistic is of general interest for all graph types.

Designed to measure local density of the network, the clustering coefficient C measures the extent to which neighbors of a node are also neighbors of each other. In an undirected graph, if a node v is connected to k_v other nodes, C_v is the fraction of $k_v(k_v - 1)/2$ possible edges between the k_v neighbors that exist, and C is the average of all C_v. The coefficient C tends to be reported only for undirected graphs, although its directed graph analog is straightforward to compute. Calculation of C is not complicated by a lack of connectivity of the entire graph.

Node degree distribution is generally plotted on a log–log scale and checked for linearity. Scale-free graphs (to be discussed in Section 24.4) are characterized by a node degree distribution that follows a power law with $f(x) \approx cx^{-(1+\alpha)}$, where x is degree, $0 < c < \infty$, and $\alpha > 0$. On a log–log scale, $f(x)$ should follow a straight line, hence scale-freeness is often diagnosed using log–log plots. Li et al. [18] note problems with inference made on network structure on the log–log scale for the probability density function (pdf) for node degree, and instead suggest diagnosis of linearity on a log–log scale for the complementary cumulative distribution function (ccdf). In this case, the ccdf is $1 - F(x)$, where $F(x)$ is the cumulative distribution function (cdf). Hence the ccdf is $1 - F(x) \approx cx^{-\alpha}$. A plot of x versus $1 - F(x)$ on the log–log scale should also be approximately linear. Li et al. [18] note that Erdös–Renyí random graphs can misleadingly look scale-free if the pdf is plotted on a log–log scale, whereas this is less likely using the ccdf. We will use the R^2 statistic to assess the fit of a straight line to the ccdf on the log–log scale.

24.4 GRAPH-THEORETIC MODELS

Perhaps the most widely studied family of graphs are the Erdös–Renyí random graphs (ER graphs). In this graph model, a prespecified number of undirected edges are randomly assigned to connect a prespecified number of nodes. An alternate construction begins with the prespecified number of nodes and edges generated at random with a fixed probability p. Two of the most widely cited characterizations of ER graphs in the biological world are the relatively short path lengths connecting any two pairs of nodes and the observation that node degree follows a Poisson distribution. Erdös–Renyí graphs often serve as the starting point for discussions of global network topologies as a contrast to the nonrandom behavior of cellular networks. Figure 24.1a illustrates an example of a random graph with 15 nodes and 29 edges.

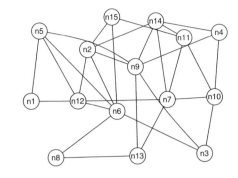

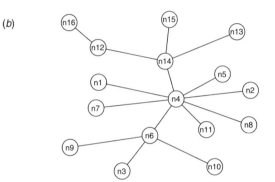

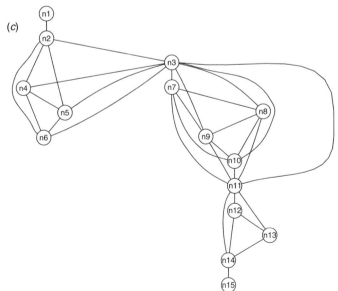

Figure 24.1 Examples of ER (*a*), scale-free (*b*), and overlapping (*c*), cluster graphs.

Watts and Strogatz [31] describe "small-world networks" as lying somewhere between the two extremes of random graphs and regular ring lattices. Regular ring lattices contain n nodes each with degree m. The nodes are arranged in a ring configuration and each node is connected to its m nearest neighbors. Nodes that are next to each other in the ring share $m - 2$ common neighbors. Two statistics, L and C, measure the small-world properties of networks. For small-world networks $L \approx L_{random}$ but $L \ll L_{lattice}$, where L_{random} and $L_{lattice}$ are the average path lengths for a random graph and a regular lattice with the same number of nodes and edges, respectively. In small-world networks, $C \gg C_{random}$ but $C \approx C_{lattice}$.

Barabási and Albert [4] and Amaral et al. [1] describe a particular class of small-world networks in which the distribution of node degree follows a power law. These graphs are termed "scale-free" since there is no 'typical' node characterizing all nodes in the sense that mean node degree could typify a Poisson degree distribution. Barabási and Albert [4] further characterize scale-free networks by noting that they can be generated by continuously adding new nodes to the network, and with each addition, connecting the new node preferentially to nodes that are already highly connected. This mechanism creates the hubs characteristic of scale-free graphs and confers linearity on the log–log plot of node degree versus frequency that is often used to diagnose scale-freeness. In Figure 24.1b, an example of a scale-free graph, node n4 is a highly connected hub node.

Our working definition of a scale-free graph will be the preferential attachment model of Barabási and Albert [4], but we do note that scale-freeness is not always a straightforward characterization of a graph. For example, Barabási and Oltvai [5] point out the contradiction that some biological networks that appear to be scale-free also contain modular features in which collections of nodes are highly connected to each other but loosely connected to other collections of nodes in the graphs. This contrasts with the scale-free concept in that node degree can be typified by an "average" with relatively little variability and contradicts the scale-free metric due to Li et al. [18]. Ravasz et al. [20] synthesize these concepts into a hierarchical graph model in which central nodes that connect to their own local neighbors also connect with centers of other local neighborhood clusters. Their specific algorithm, while limited in terms of cluster size and overall topological organization, does retain both scale-free and modular properties. Several diagnostic statistics confirm that this network structure is applicable at least to metabolic networks, if not other biological networks as well.

Keeping to the modular theme, we also consider cluster graphs that frequently arise in biological data, although their existence is not always recognized. Cluster graphs consist of sets of nodes that are connected to all other nodes within their set, and are not connected to any other nodes in the graph. Microarray expression data are frequently explored using clustering algorithms, and the sets of clusters then form mutually exclusive sets of genes that are often said to be differentially expressed. In a graph of all genes on the microarray, edges can be drawn between all pairs of nodes in the same expression cluster, resulting in a series of completely connected subgraphs. In fact, any partition on a set of objects can be represented in this manner. Balasubramanian et al. [3] demonstrate the uses of this type of cluster graph for assessing coregulation of interacting proteins.

Overlapping cluster graphs occur when the nodes are not strictly partitioned but can instead be members of more than one set. An example is the protein complex comembership graph induced by the MBME complex estimates. This graph consists of a series of clusters containing all proteins forming distinct complexes, some of which share common proteins as members and thus are connected to each other. Figure 24.1c shows an example of an overlapping cluster graph involving 15 nodes and 5 clusters.

24.5 TYPES OF MEASUREMENT ERROR

In this section we discuss two types of measurement error, namely, stochastic and systematic errors. While not formally treated, we also briefly describe sampling issues in networks and point out how misinterpretation of experimental design can lead to large-scale systematic error. In Section 24.7, we note the influence of various sources of measurement error on the commonly used graph statistics. Measurement error is a noted problem for biological network data and will be shown to greatly influence graph statistics; however, it is not generally accounted for in analyses of biological graphs.

24.5.1 Stochastic Error

Network data, like most biological data, are subject to stochastic measurement error. Stochastic error refers to mistakes made in the observation of relationships that are not attributable to a particular source, or "random noise." Both false-positive (FP) and false-negative (FN) observations may occur as a result of stochastic error mechanisms. As a simple example, Figure 24.2 shows the complex comemberships detected in three purifications in the TAP data for the baits Apl5, Apl6, and Apm3. Because protein–protein interactions should be reciprocated with perfectly sensitive and specific technology, the unreciprocated edge from Apl5 to Apm3 is either a FP or a FN observation between these two proteins. Either Apl5 errantly detected Apm3 as a hit, or Apm3 errantly failed to detect Apl5 as a hit.

The interpretation of FP and FN depends entirely on the relationship under investigation and must be treated with care, particularly when jointly analyzing related data types. For example, for AP-MS data, a FN will occur when an edge is observed as missing between two proteins that are in fact complex comembers, and a FP will occur when an edge is observed between two proteins that do not share complex membership. For Y2H-detected physical interactions, a FN AP-MS observation could be a true negative (TN) Y2H observation since a pair of proteins can

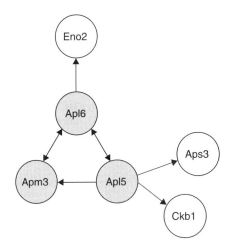

Figure 24.2 All directed edges represent detected complex comemberships and extend from the shaded bait proteins to the hits (either other baits or hit-only proteins) that they find.

be complex comembers without physically interacting. For the same reason, a FP Y2H observation could be a true positive (TP) AP-MS observation.

24.5.2 Systematic Error

Systematic error refers to errors that are consistently made for some particular reason. Both FN and FP observations can result from systematic error mechanisms. For example, AP-MS and Y2H FPs may arise systematically for certain bait proteins as a result of extra "stickiness" in the purification process. In this case, sticky baits will detect their neighbors as hits, as well as a set of FP hits at rates exceeding that of stochastic FPs. These extra hits may consist of a random sample of proteins in the cell, or they may tend to include neighbors of a sticky bait protein's true neighbors.

Similarly, FNs may arise systematically for AP-MS and Y2H data. Depending on the technology being used, some bait proteins experience enough conformational deformity when the experimental construct is attached to prevent interaction with their complete set of true neighbors. In this case, the rate of FNs for the deformed bait will exceed the rate of stochastic FN errors.

There is one important difference between systematic FPs and FNs. Systematic FPs may extend either globally to the entire set of proteins in the cell, or locally to the neighbors of a sticky bait's true neighbors. Systematic FNs, on the other hand, are restricted to the local set of relationships for the bait in question. While we have explained some sources of systematic errors for AP-MS and Y2H data, we note that similar errors can be made in the case of other technologies for unique biological and experimental reasons.

24.5.3 Sampling

Often, especially in high-dimensional settings, not all edges are tested, but rather only a subset are queried using the technology of interest. Both Y2H and AP-MS experiments use a "neighborhood sampling" scheme in which a subset of nodes is specified to be baits and all edges extending from those nodes are tested. Note that this implies that all edges that connect pairs of nodes that were not used as baits are in fact untested. This is different from a "subgraph sampling," scheme in which all edges in the subgraph induced by a subset of nodes are tested.

Misinterpretation of a sampling scheme can be the source of large-scale systematic error. In particular, Scholtens et al. [22] point out that most graph analyses errantly treat untested edges as missing, thereby inducing large-scale systematic FNs. This problem can be resolved by restricting attention only to the set of tested edges, but unfortunately, the global graph statistics are relatively underdeveloped for this purpose. Hakes et al. [10], Han et al. [11], and Stumpf et al. [24] deal with other important aspects of sampling in a graph-theoretic paradigm.

24.6 EXPLORATORY DATA ANALYSIS

Some very basic exploratory data analysis (EDA) techniques can diagnose the extent to which measurement error is a problem in network data. Here we discuss three very basic techniques involving reciprocity, sampling, and underlying network structure.

24.6.1 Reciprocity

In all bait–hit technologies, edges between pairs of baits are tested twice since each bait can potentially find the other as a hit. Edges between baits and hit-only proteins are only tested once since a hit-only protein is never given a chance to reciprocate the detected relationship or lack thereof. Simple inspection of the doubly tested relationships indicates very high levels of both FNs and FPs in AP-MS data. For the subgraph of all doubly tested edges in the TAP data induced by the 358 bait proteins that find at least one other bait as a hit, Figure 24.3 demonstrates the number of reciprocated and unreciprocated edges connected to each bait. If FNs and FPs did not exist, all edges would be reciprocated. Instead, only 43 of the 358 baits have exclusively reciprocated edges in the bait-induced subgraph. Of the 43, 30 have 1 reciprocated edge, 12 have 2, and 1 has 5. The lack of reciprocity is likely due to both stochastic and systematic FNs and FPs, although the exact mechanism and compilation of these errors are not obvious from this plot.

24.6.2 Sampling

As discussed previously, large-scale systematic error can arise when sampling of the network is misinterpreted. Specifically, when all untested edges are treated as missing, a substantial number of FNs may occur.

For the TAP data [7], of the 1364 reported proteins, 455 are baits and 909 are hit-only proteins. Consequently, in the graph restricted to all baits and observed hits and excluding self-edges, $455 \times 454/2 = 103,285$ edges were tested twice, $455 * 909 = 413,595$ edges were

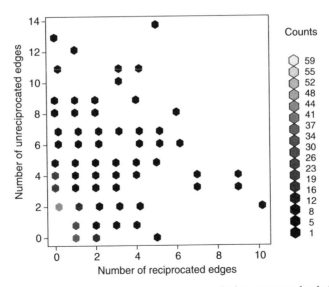

Figure 24.3 Summary of unreciprocated and reciprocated edges connected to bait proteins in the TAP data. The existence of unreciprocated edges indicates the presence of FP and FN observations.

tested once, and $909 * 908/2 = 412,686$—close to half of the edges in the graph—were never tested. Similar figures apply to the AP-MS data published by Ho et al. [12], in which 493 baits were used and 1085 proteins were found as hits but not used as baits.

Treating untested edges as missing forces FNs for all pairs of hit-only proteins that are indeed complex comembers. If the proportion of true edges in the untested set is consistent with the tested set, then in our case roughly half of the true positives would be treated as FNs. Simple calculation of the number of tested and untested edges, given the sampling scheme, highlights the potentially high number of systematic FNs resulting from misinterpretation of experimental design.

24.6.3   Underlying Network Structure

Comparison of data sources with other related data can also be very helpful in uncovering the extent of measurement error in a graph. In our case, AP-MS data can be very informative about error rates for Y2H data. In particular, we assume that a multiprotein complex is connected so that all complex comembers must physically bind to at least one other protein in the complex. If we use the MBME complex estimates and combine the Ito et al. [14] and Uetz et al. [28] data, then we can make some assumptions regarding requirements for physical connectivity and explore a range of FP and FN rates for Y2H data. For a complex containing N proteins, there must be at least $N-1$ Y2H edges connecting them. Figure 24.4 demonstrates that this is largely untrue for Y2H data. For the large majority of MBME complex estimates, only a fraction of the minimum number of required edges are observed in a simple union of the Ito–Uetz Y2H data. This indicates a large number of FNs in the Y2H data.

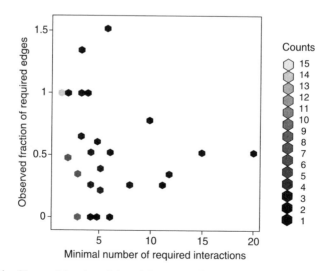

Figure 24.4   Observed fraction of the minimum number of Y2H edges necessary to connect all MBME complex members to one other complex comember.

24.7 INFLUENCE OF MEASUREMENT ERROR ON GRAPH STATISTICS

In this section, we use simulation to study the influence of stochastic and systematic measurement error on L, C, and node degree distribution for three different types of graphs: ER graphs, scale-free graphs, and overlapping cluster graphs (all simulation parameters are described subsequently). We introduce stochastic FN errors at rates of 0.05, 0.15, and 0.25, and stochastic FP rates corresponding to half of all edges being FPs. Note that the rate of FP errors corresponds to a positive predictive value (PPV) of an observed edge of 0.5. These values are guided by estimates of sensitivity for AP-MS technology [7] as well as the claim that up to half of the total number of observed interactions may be FPs [29]. For systematic errors, we simulate the sticky bait situation by selecting 0.05, 0.15, and 0.25 of all nodes to be "sticky" and then connecting them to the neighbors of their neighbors with probability 0.5. In reality, baits could also be sticky with respect to the entire hit population; we choose to work with neighbors of neighbors as just one example. We assume that the entire graph has been tested. Since methods for treatment of measurement error on tested edges will necessarily differ from those that address sampling, we leave the treatment of systematic FNs induced by errant interpretation of sampling schemes for further research. We connect results from these simulation studies back to their biological counterparts, noting areas of concern for making biological conclusions on the basis of these data. For each simulation, 100 graphs of a particular size and structure were generated and measurement error was introduced for each graph according to the rates just listed.

ER Graphs. We generate ER graphs with 50, 500, and 1000 nodes, each with approximately 5 edges per node for a total of 250, 2500, and 5000 edges, respectively. These graph sizes were chosen to reflect a range of potential networks of interest. The use of ER graphs is not necessarily intended to represent a particular type of biological network, but is instead to be used as a benchmark for comparison of the other graph types.

Scale-Free Graphs. We generate scale-free graphs with 50, 500, and 1000 nodes according to the preferential attachment model of Barabási and Albert [4] again with approximately five edges per node. Scale-free graphs have been noted in several biological settings and are included as the most likely structure for Y2H graphs.

Overlapping Cluster Graphs. Overlapping cluster graphs are intended to resemble AP-MS data since a set of complex comemberships should form a series of connected cluster graphs. We generate overlapping cluster graphs of approximately 50, 500, and 1000 nodes, using a Poisson distribution with $\lambda = 6$ to generate cluster sizes. We then randomly specifyied one node within a cluster as a member of the next cluster as well, thereby constructing a chain of clusters.

24.7.1 Path Length: L

The effects of stochastic false negatives and false positives and systematic false positives on path length are described here.

Stochastic False Negatives. As depicted in Figure 24.5, stochastic FN errors consistently increase path length for both random and scale-free graphs of all sizes under

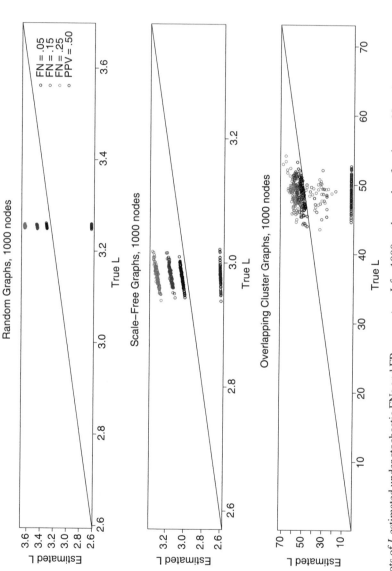

Figure 24.5 Plots of L estimated under stochastic FNs and FPs versus true L for 1000-node graphs of each type. Note that the x axes are quite different for the different graphs. Scales were set so that the x-axis range is the same as the y-axis range, with points above the 45° line indicating that the estimate was upwardly biased and points below the line indicating a downward bias.

consideration. Data are also depicted for 1000-node graphs, but the patterns appear similar for smaller graphs as well. The increase in path length is somewhat expected; since edges are randomly eliminated, this will necessarily extend the lengths of paths between pairs of nodes. Path length in the cluster graph is somewhat surprising at the first glance. For higher FN rates, average path length in fact decreases. This can be explained by a simple mistake that is often made in the calculation of L for unconnected graphs. Path lengths for pairs of nodes between which no path exists are generally excluded from the mean calculation. In this case, since the data are so highly clustered, eliminating edges tends to splinter the graph into unconnected components. Path lengths for node pairs that were once distant from each other in the original connected graph are now ignored in the overall mean calculation, and the shorter path lengths between nodes that remain connected decrease the mean. This situation shows the bias that can be introduced by computing L on unconnected graphs without accounting for unconnected pairs.

Stochastic False Positives. In general, stochastic FPs behave similarly for all three graph families by consistently decreasing L as shown in Figure 24.5. For overlapping cluster graphs, the reduction is particularly drastic. Variability in the L estimates drastically reduces; this may be an effect of the addition of a rather large number of FPs to the graph, and the consequent reduction in the variation of L due to increased saturation of the graph.

Systematic False Positives. Systematic FPs also tend to decrease estimates of L, but with more variability than observed for stochastic FPs. As illustrated in Figure 24.6, in ER graphs, systematic FPs cause consistent underestimates of L. For cluster graphs at lower stickiness rates, some values of estimated L are often quite close to the true values, but this trend decreases as stickiness rates increase. For scale-free graphs, the pattern of estimates of L under systematic FP error fall somewhere in between the range of values for random and cluster graphs.

24.7.2 Clustering Coefficient: C

Before discussing the influence of measurement error on C, we make special note that both the numerator and denominator of C are potentially subject to measurement error. In particular, C_v is the fraction of $k_v (k_v - 1)/2$ possible edges between the k_v neighbors of node v that exist, and C is the average of all C_v. For a node v, if an error is made in the observation of an edge from v to another node, this will alter both the denominator and the numerator. A FP edge will increase the denominator by k_v, a FN edge will decrease it by $k_v - 1$, and the numerator will change to include the number of edges observed given the set of edges between all pairs of neighbors observed for node v. Alternatively, errors made in the observation of edges between neighbors of v would change only the numerator. Note that an error in the observation of an edge from v to another node v' would affect both the numerator and denominator for C_v and $C_{v'}$, but would affect the numerator for $C_{v''}$ only, if both v and v' were neighbors of v''.

Stochastic False Negatives. For all three types of graphs, stochastic FNs will tend to lead to underestimates of C (Fig. 24.7). For random and scale-free graphs at lower FN rates, occasional overestimates of C may be due to a reduction of the denominator, but proportionately less reduction in the numerator. For cluster graphs in our simulation, C is always underestimated in the presence of stochastic FNs. Since C is so high in the cluster graph, if the denominator decreases as a result of FNs, the numerator tends to

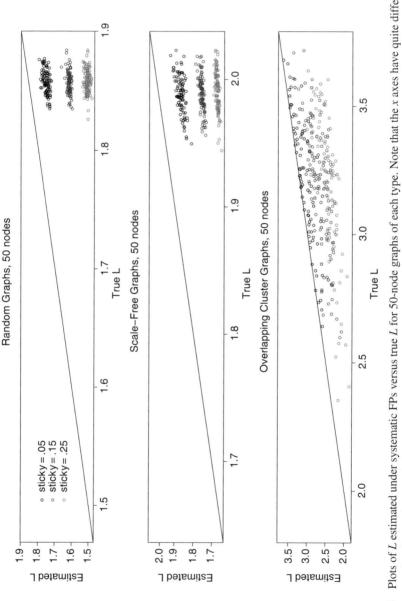

Figure 24.6 Plots of L estimated under systematic FPs versus true L for 50-node graphs of each type. Note that the x axes have quite different scales. Scales were set so that the x-axis range is the same as the y-axis range, with points above the 45° line indicating that the estimate was upwardly biased and points below the line indicating a downward bias.

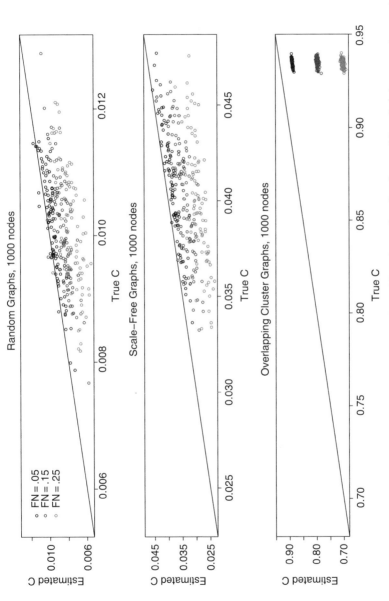

Figure 24.7 Plots of C estimated under stochastic FNs versus true C for 1000-node graphs of each type. Note that the x axes have quite different scales. Scales were set so that the x-axis range is the same as the y-axis range, with points above the 45° line indicating that the estimate was upwardly biased and points below the line indicating a downward bias.

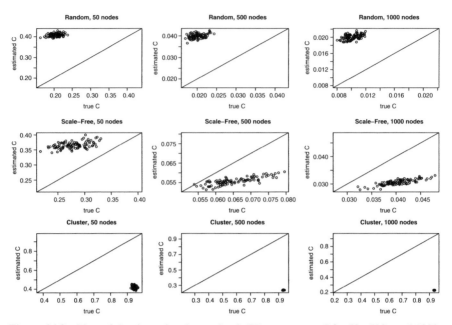

Figure 24.8   Plots of C estimated under stochastic FPs versus true C for 50-, 500-, and 1000-node graphs of each type. Note that the x axes have quite different scales. Scales were set so that the x-axis range is the same as the y-axis range, with points above the 45° line indicating that the estimate was upwardly biased, and points below the line indicating a downward bias.

decrease proportionately. The greater influence on C tends to be due to FNs between neighbors of nodes, thus decreasing the overall C.

Stochastic False Positives. Stochastic FP errors have different effects on C depending on the structure of the graph (see Fig. 24.8). For ER graphs, stochastic FPs increase C, and in general the estimates are less variable. In smaller scale-free graphs, C is overestimated and in fact closely resembles the random phenomenon. Interestingly, for larger scale-free graphs, stochastic FPs cause underestimates of C. The overestimates in the smaller graph may be due to general saturation of the graph edges. Stochastic FPs in overlapping cluster graphs dramatically decrease estimates of C. In general, nodes now have more neighbors, but the number of edges between their neighbors does not increase at the same rate conferred by their original clustering coefficient.

Systematic False Positives. For both random and scale-free graphs, systematic FPs cause a drastic increase in C. The plots in Figure 24.9 demonstrate the results for 1000-node graphs, but the plots look largely the same for 50-and 500-node graphs. Our definition of stickiness connects sticky nodes to neighbors of their neighbors with probability 0.5, and hence we expect such an increase in C. The rate of stickiness works oppositely for overlapping cluster graphs, however, since the true clustering coefficients are so high to begin. If the denominator of C_v is increased because of a FP observation, only half of the neighbors of the neighbors of v will be connected to v, and this rate is much lower than the true C_v. Thus, even very high FP probabilities (e.g., 0.5) for a very low proportion of sticky baits (e.g., 0.05) will in fact decrease C for overlapping cluster graphs.

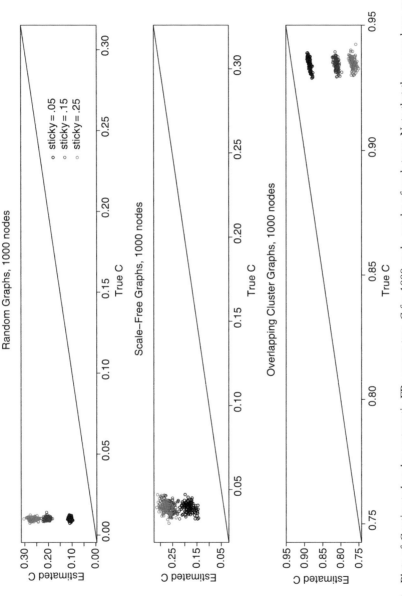

Figure 24.9 Plots of C estimated under systematic FPs versus true C for 1000-node graphs of each type. Note that the x axes have quite different scales. Scales were set so that the x-axis range is the same as the y-axis range, with points above the 45° line indicating that the estimate was upwardly biased and points below the line indicating a downward bias.

24.7.3   Node Degree Distribution

Node degree distribution is generally used to diagnose scale-free characteristics of graphs. Common practice is to plot the frequency distribution of degree versus degree on a log–log scale and then assess the fit of a straight line. Li et al. [18] note that log–log plots on the frequency scale can lead to mistaken diagnoses of ER graphs as scale-free and suggest the use of the ccdf instead. Noting their results, we will examine the effect of systematic and stochastic errors on R^2 for the fit of a straight line to the log–log plot of the cdf.

Stochastic False Negatives. Neither ER graphs nor cluster graphs are expected to have scale-free characteristics, and both are affected similarly by systematic FN observations. In particular, FN observations will tend to cause underestimates of R^2. This observation is the same in scale-free graphs (Fig. 24.10).

Stochastic False Positives. Stochastic FP observations again behave similarly for random and cluster graphs, this time increasing estimates of R^2. In contrast, although the bias is not tremendously large, estimates of R^2 are low for scale-free graphs (Fig. 24.10).

Systematic False Positives. Systematic FPs have a drastic effect on R^2 for ER graphs. Scale-free graphs are known to have a small number of "hub" nodes, and even very small numbers of sticky nodes result in an adequate number of hubs to make a ER graph look scale-free (see Fig. 24.11). On the contrary, R^2 estimates actually decrease even for very small sticky rates in scale-free graphs, possibly owing to the creation of too many hubs. In the cluster graph, stickiness tends to have a variable effect with estimates that are fairly consistent with the true R^2 values.

24.8   BIOLOGICAL IMPLICATIONS

24.8.1   Experimental Data

Table 24.1 reports L, C, and R^2 statistics for the bait-induced subgraphs for the TAP, HMS-PCI, Ito, and Uetz data. The L statistic was computed on the largest connected components of these graphs, and C and R^2 were connected using the entire bait-induced subgraph. All three statistics were computed on the underlying undirected graphs for these data. Since we know neither the FP and FN probabilities for these data, nor the combination of stochastic and systematic mechanisms by which they arise, we are unable to say how different these statistics are from their

Table 24.1   Table of L, C, and R^2 Values for Bait-Induced Subgraphs (BI) and Their Largest Connected Components (LCCs)

	Number of Nodes BI/LCC	Number of Edges BI/LCC	L	C	R^2
TAP	358/302	725/685	5.26	0.433	0.778
HMS-PCI	379/362	665/656	4.53	0.161	0.913
Ito	1096/994	1602/1545	4.83	0.0586	0.984
Uetz	228/83	190/87	7.79	0.0953	0.951

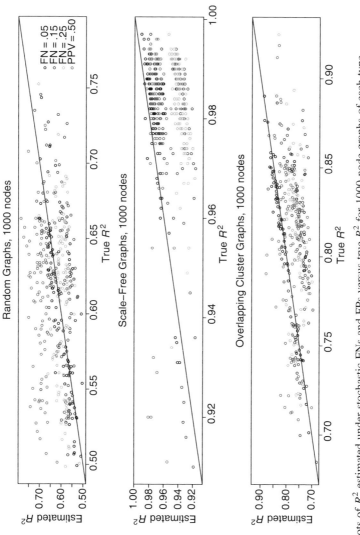

Figure 24.10 Plots of R^2 estimated under stochastic FNs and FPs versus true R^2 for 1000-node graphs of each type.

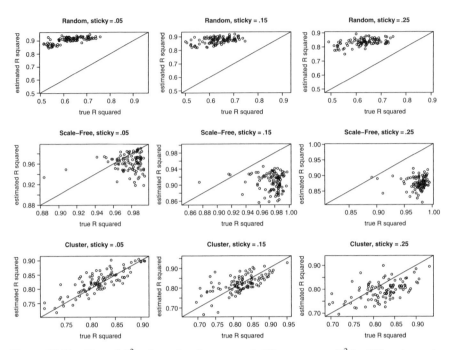

Figure 24.11   Plots of R^2 estimated under systematic FPs versus true R^2 for 500-node graphs of each type.

underlying true values. Here, we simply make a few observations on how these statistics reflect our simulation results.

By nature of the complex comembership relationships that they measure, we know that the TAP and HMS-PCI data have an overlapping cluster graph structure. Note that the values of C for these two graphs are commensurate with those observed in the stochastic FP simulations. Presumably the effects of stochastic FNs are not as prevalent since we are looking at the underlying graph; not the directed graph, and perhaps the effects of systematic FPs are ameliorated by treating only the bait-induced subgraph. In any event, the true values of C are likely higher than those based strictly on the observed data.

The overall topology of the Ito and Uetz Y2H data is not a direct result of the relationships considered; hence special attention should be paid to the implications of the calculated statistics. If the Y2H data are not subject to systematic FPs, then the reported values of C are likely lower than the true values. If, on the other hand, sticky baits are prevalent in these data, then the true values of C could well be lower. Classification of these Y2H graphs as small-world depends on the C statistic; hence the acknowledgment that the true value could vary in either direction is important to consider.

All four data graphs show fairly high values of R^2, suggesting scale-freeness for all four graphs. While these are consistent with values for true scale-free and overlapping cluster graphs, they are also consistent with random graphs that include even a very low percentage of sticky baits. The structure of the true TAP and HMS-PCI graphs is not so much in question, but this is an important consideration for the Y2H graphs.

24.8.2 Simulation Data

Our simulation studies suggest that stochastic FN observations tend to have similar implications for all three families of graphs under consideration. In particular, they tend to increase values of L, and decrease values of C and R^2 for the fit of a straight line to a log–log plot of the ccdf of node degree. Stochastic and systematic FPs, on the other hand, consistently affect L, but have different and often opposite effects on C and R^2 for ER, scale-free, and cluster graphs.

As discussed in Section 24.1, both small-world and scale-free graphs are thought to be important in biology since they are largely robust to random perturbations. For the three graph families discussed in this chapter, sources of error, both stochastic and systematic, tend to make ER and cluster graphs look more scale-free than they ought, and scale-free graphs less so. Graphs of protein–protein interactions have often been described as scale-free, and these simulations shed light on the potential involvement of measurement error in those claims.

For ER graphs, FP observations, in particular systematic FPs, tend to make ER graphs look both scale-free and small-world. False positives decrease L, increase C, and greatly increase R^2. It is of some interest to note that these effects occur even at very small rates of FP nodes. Many biological networks have been characterized as both scale-free and small-world and have not accounted for potential sources of measurement error. Our studies suggest the possibility that these networks could in reality behave like random networks, with scale-free characteristics merely as artifacts of measurement error.

Scale-free graphs are a subtype of small-world networks according to Amaral et al. [1], and therefore they should be characterized by small L, high C, and high R^2. For scale-free graphs, FN observations appear to be a larger problem than FPs since FNs increase L, decrease C, and decrease R^2.

Overlapping cluster graphs tend to be more robust to sources of error than the other two graphs, possibly because their highly structured nature dominates random noise. The estimates of L, C, and R^2 certainly do change in the presence of error, but not necessarily enough to infer scale-free or small-world behavior inappropriately. While this is a positive feature for overlapping cluster graphs, it does not necessarily provide much new information about the graph. If a particular technology is employed to probe coaffiliation relationships such as complex comembership, then the resultant graph would necessarily be a cluster graph and an investigator should already recognize this.

24.9 CONCLUSIONS

Global graph statistics have been used in several settings to characterize network topologies and make biological conclusions about the nature of these data. The simple simulation studies in this chapter illustrate the large impact of measurement error on graph statistics, and the resultant potential for misinterpretations of these data. Further rigorous research is warranted in several directions. Of perhaps primary importance is the identification of the underlying quantities that are important and the development of reliable estimates of those quantities. It remains unclear whether L relates to any important biological concept, but if it does, then better estimation procedures are badly needed. Other directions include the joint modeling of multiple types of error and development of model-based graph statistics to augment those that have already been observationally associated with different graph types. We also believe that careful characterization of sampling schemes and appropriate estimators of the relevant quantities that account for the nature of the data (e.g., using only tested edges) are required. While graphs have been successfully applied in many biological settings, more work is warranted to increase the level of statistical sophistication for drawing inference using graphs.

REFERENCES

1. Amaral, L. A. N., Scala, A., Barthelemy, M., and Stanley, H. E., Classes of small-world network, *Proc. Natl. Acad. Sci. USA* **97**(21), 11149–11152 (2000).

2. Bader, G. D. and Hogue, C. W. V., Analyzing yeast protein-protein data obtained from different sources, *Naure Biotechnol.* **20**, 991–997 (2002).

3. Balasubramanian, R., LaFramboise, T., Scholtens, D., and Gentleman, R., A graph-theoretic approach to testing associations between disparate sources of functional genomics data, *Bioinformatics* **20**, 3353–3362 (2004).

4. Barabási, A. L. and Albert, R., Emergence of scaling in random networks, *Science* **286**, 509–512 (1999).

5. Barabási, A. L. and Oltvai, Z. N., Network biology: Understanding the cell's functional organization, *Nature Rev. Genet.* **5**, 101–114 (2004).

6. Chen, J., Hsu, W., Lee, M. L., and Ng, S.-K., Increasing confidence of protein interactomes using network topological metrics, *Bioinformatics* **22**, 1998–2004 (2006).

7. Gavin, A. C. et al., Functional organization of the yeast proteome by sytematic analysis of protein complexes, *Nature* **415**, 141–147 (2002).

8. Ge, H., Liu, Z., Church, G., and Vidal, M., Correlation between transcriptome and interactome mapping data from *Saccharomyces cerevisiae*, *Nature Genet.* **29**, 482–486 (2001).

9. Gentleman, R., Scholtens, D., Ding, B., Carey, V. J., and Huber, W., Case studies using graphs on biological data, in *Bioinformatics and Computational Biology Solutions Using R and Bioconductor*, Gentleman, R., Huber, W., Carey, V., Irizarry, R., and Dudoit, S. eds., Springer, 2005.

10. Hakes, L., Robertson, D. L., and Oliver, S. G., Effect of dataset selection on the topological interpretation of protein interaction networks, *BMC Genomics* **6**, 31 (2005).

11. Han, J.-D. J., Dupuy, D., Bertin, N., Cusick, M. E., and Vidal, M., Effect of sampling on topology predictions of protein-protein interaction networks, *Nature Biotechnol.* **23**(7), 839–844 (2005).

12. Ho, Y. et al., Systematic identification of protein complexes in *Saccharomyces cerevisiae* by mass spectrometry, *Nature* **415**, 180–183 (2002).

13. Huber, W., Gentleman, R., and Carey, V. J., Graphs, in *Bioinformatics and Computational Biology Solutions Using R and Bioconductor*, Gentleman, R., Huber, W., Carey, V., Irizarry, R., and Dudoit, S., eds., Springer, 2005.

14. Ito, T., Chiba, T., Ozawal, R., Yoshida, M., Hattori, M., and Sakaki, Y., A comprehensive two-hybrid analysis to explore the yeast protein interactome, *Proc. Natl. Acad. Sci. USA* **98**, 4569–4574 (2001).

15. Jansen, R., Yu, H., Greenbaum, D., Kluger, Y., Krogan, N. J., Chung, S., Emili, A., Snyder, M., Greenblatt, J. F., and Gerstein, M., A Bayesian networks approach for predicting protein-protein interactions from genomic data, *Science* **302**, 449–453 (2003).

16. Jeong, H., Mason, S. P., Barabási, A. L., and Oltvai, Z. V., Lethality and centrality in protein networks, *Nature* **411**(6833), 41–42 (2001).

17. Kelley, R. and Ideker, T., Systematic interpretation of genetic interactions using protein network, *Nature Biotechnol.* **23**(5), 561–566 (2005).

18. Li, L., Alderson, D., Doyle, J. C., and Willinger, W., Towards a theory of scale-free graphs: Defintion, properties, and implications, *Internet Math.* **2**, 4 (2006).

19. Qi, Y., Ye, P., and Bader, J., Genetic interaction motif finding by expectation maximization—a novel statistical model for inferring gene modules from synthetic lethality, *BMC Bioinformatics* **6**, 288 (2005).

20. Ravasz, E., Somera, A. L., Mongru, D. A., Oltvai, Z. N., and Barabási, A.-L., Hierarchical organization of modularity in methabolic network, *Science* **297**, 1551–1555 (2002).

21. Sholtens, D., and Gentleman, R., Making sense of high-throughput protein-protein interaction data, *Statist. Appl. Genet. Molec. Biol.* **3**(1), Art. 39 (2004).

22. Sholtens, D., Vidal, M., and Gentleman, R., Local modeling of global interactome networks, *Bioinformatics* **21**, 3548–3557 (2005).

23. Steffen, M., Petti, A., Aach, J., D'haeseleer, P., and Church, G., Automated modelling of signal transduction network, *BMC Bioinformatics* **3**, 34 (2002).

24. Stumpf, M. P. H., Wiuf, C., and May, R. M., Subnets of scale free network are not scale-free: Sampling properties of network, *Proc. Natl. Acad. Sci. USA* **102**, 4221–4224 (2005).

25. Suthram, S., Shlomi, T., Ruppin, E., Sharan, R., and Ideker, T., A direct comparison of protein interaction confidence assignment schemes, *BMC Bioinformatics* **7**, 360 (2006).

26. Tong, A. H. et al., Systematic genetic analysis with ordered arrays of yeast deletion mutants, *Science* **294**, 2364–2368 (2001).

27. Tong, A. H. et al., Global mapping of the yeast genetic interaction network, *Science* **303**, 808–813 (2004).

28. Uetz, P., Giot, L., Cagney, G., Mansfield, T. A., Judson, R. S., Knight, J. R., Lockshon, D., Narayan, V., Srinivasan, M., and Pochart, P., et al., A comprehensive analysis of protein-protein interactions in *Saccharomyces cerevisiae*, *Nature* **403**, 623–627 (2000).

29. von Mering, C., Krause, R., Snel, B., Cornell, M., Oliver, S. G., Fields, S., and Bork, P., Comparative assessment of large-scale data sets of protein-protein interactions, *Nature* **417**, 399–403 (2002).

30. von Mering, C., Jensen, L. J., Snel, B., Hooper, S. D., Kruup, M., Foglierini, M., Jouffrey, N., Huynen, M. A., and Bork, P., STRING: Known and predicted protein-protein associations, integrated and transferred across organisms, *Nucleic Acids Res.* **33**, D433–D437 (2005).

31. Watts, D. J. and Strogatz, S. H., Collective dynamics of "small-world" networks, *Nature* **393**, 440–442 (1998).

32. Wong, S. L. et al., Combining biological network to predict genetic interactions, *Natl. Acad. Sci. USA* **101**(44), 15682–15687 (2004).

33. Ye, P., Peyser, B. D., Spencer, F. A., and Bader, J. S., Commensurate distances and similar motifs in genetic congruence and protein interaction networks in yeast, *BMC Bioinformatics* **6**, 270 (2005).

CHAPTER 25

Prediction of RNA Splicing Signals

Mark R. Segal

Department of Epidemiology and Biostatistics, University of California, San Francisco, California

25.1   INTRODUCTION

An organism extracts information from its genome via the recognition and processing of signals contained in the constituent primary nucleotide sequence. When these signals pertain to a common function, they exhibit some degree of similarity. The search for, and identification of, such characteristic sequence *motifs* constitutes one of the foremost, yet most challenging, problems in computational biology. For example, the tasks of detecting transcription factor binding sites (TFBSs) or RNA splice sites (ss), are of this flavor. The difficulties derive from the fact that these motifs are short [<25 nucleotides (nt)] and (despite the abovementioned similarity) variable. Further, in some settings, they must be isolated from lengthy ($\sim 10^{3-6}$-nt) background sequences. That these difficulties can be (partially) overcome reflects the ability of more recently available, large-sequence databases to furnish compensatory information, and the emergence of sophisticated and customized algorithms to effectively analyze these databases.

There are two broad analysis strategies that have been employed for motif finding. The first searches for novel motifs among a set of sequences that share common biologic functionality. Such methods typically build from assumptions about both signal (the motif) and nonsignal (background) sequence. They have been widely used in eliciting TFBS. The second uses the information content of experimentally verified motifs, often coupled with judiciously selected instances of negative or decoy sequences, to search for and/or characterize putative signals. This approach has been extensively used in the context of splice site recognition. It is to both such methods, and such applications, that this chapter is dedicated.

The chapter is organized as follows. A brief overview of the background biology pertaining to splicing rounds out this introductory section. Section 25.2 briefly surveys more recently

Statistical Advances in the Biomedical Sciences, edited by Atanu Biswas, Sujay Datta,
Jason P. Fine, and Mark R. Segal

443

available analytic approaches for identifying splice sites. These are distinguished by the manner in which they pursue modeling the joint distribution of a splice site sequence. The methods are reunited in using likelihood ratios to gauge performance and select models. These likelihood ratios result from applying the modeling procedure *separately* to real and decoy sequence databases. It is the inherent *comparative* nature of the splice site identification problem, coupled with the existence of these databases, that motivates our attacking it as a classification problem. Accordingly, Section 25.3 outlines some contemporary classification techniques—boosting, support vector machines, and random forests—that we subsequently apply. Section 25.4 provides results from these analyses. These serve to suggest refinements to some of the classification procedures. Comparisons and evaluations are made in terms of predictive accuracy, computational considerations, and interpretive yield. Section 25.5 contains some concluding discussion and possibilities for future work.

25.1.1 Biologic Overview of Splicing

Usually, human genes are transcribed as long precursors, with alternating long (non-coding) introns and short ($\sim$50–250 nt) internal exons. Precision on the part of the RNA splicing machinery is required in both excising introns from these primary transcripts, and ligating flanking exons, in order to produce proper messenger RNAs for subsequent protein synthesis. Approximately 15% of the point mutations that result in human genetic disease have been attributed to errors in this splicing process [37]. Further, more than 50% of human genes undergo alternative splicing [38], which confers a major source of protein diversity. These considerations underscore the importance of splicing regulation, which remains poorly understood.

A protein-coding gene in humans contains, on average, 9 exons of length 145 nt. Introns are, on the average, an order of magnitude longer and a typical gene spans 27 kilobases ($1 \text{kb} = 10^3$ nt) [38]. There are three short and degenerate splice site sequences at, or near, the intron–exon boundary: (1) the donor or 5′ splice site (5′ss) marking the exon–intron junction at the intron's 5′ end, (2) the acceptor or 3′ splice site (3′ss) marking the junction between the intron and the downstream exon, and (3) the branch site of lariat formation located 20–50 nt upstream of the 3′ss. The consensus sequence motif for the 3′ss and 5′ss features fully conserved (essential) dinucleotides at the splice junction, coupled with base preferences at flanking positions. During or shortly following transcription, components of the nuclear splicing machinery bind to the ss. This triggers formation of a macromolecular complex, the spliceosome, which consists of five small nuclear ribonucleoproteins and $\sim$145 additional proteins [53]. Through a dynamic and complex series of interactions the spliceosome recognizes boundaries and catalyzes the precise excision and ligation steps [14].

This precision, when considered in the context of the complex machinery and limited conservation of 3′ and 5′ splice sites, is surprising. If we measure information content in relative (to a uniform background) entropy "bits" (i.e., the number of binary 0/1s needed to code the signal), which here is 2-H for H given in (25.2), then the 3′ss and 5′ss each contain $\sim$8–9 bits. Thus, we expect a decoy splice site every 2^8–2^9 ($\sim$200–500) bases, whereas actual 5′ss are on average 3000 bases apart. Hence, either the characterization of ss via bits is insufficient and/or other factors are involved in splicing. The calculation of bits does not capture the strong between-position dependences known to exist in splice signals. Current attempts at more sophisticated splice signal identification, described next, seek to exploit these dependences.

One of the primary motivations for improved splice signal identification is to improve gene-finding algorithms that employ resultant exon recognition strategies. We do not address that aspect of the problem here, beyond some brief remarks in Section 25.5.

25.2 EXISTING APPROACHES TO SPLICE SITE IDENTIFICATION

There has been considerable activity in advancing methods and algorithms for splice site identification. Representative of this work are five papers published since 2000: [15,51,17,18,52]. There are some basic commonalities shared by these approaches. As mentioned above, the starting point is to employ between-position dependences to provide refined splice signal identification. Next comes recognition that the full joint probability model for a splice site sequence is inestimable because of the vast number of parameters. The methods then diverge in using differing approximation strategies that hopefully capture important dependences. However, they are reunited by common use of likelihood ratio thresholding as a means for evaluating and selecting models. These likelihood ratios are obtained by applying the model to real and decoy sequence databases.

We establish some notation that pertains to the shared features. Let X be a sequence of n random variables: $X = \{X_1, X_2, \ldots, X_n\}$ corresponding to n consecutive DNA sequence positions. Thus, each X_i takes values from the four letter nucleotide (base) alphabet $\{A, C, G, T\}$. An observed sequence is designated by $x = \{x_1, x_2, \ldots, x_n\}$. Let $p(X = x)$ denote the joint probability mass function. The state space for X has 4^n elements. In attempting to identify the (acceptor) $3'$ss, we are dealing with $n = 21$. So, it is evident here, and even for shorter motif recognition problems, that (extensive) restrictions on the allowed probability models will need to be imposed. After framing a set of restrictions that define a family of estimable probability models, the methods proceed by applying such to both real splice signal and decoy sequence data. Then, in order to arbitrate whether a given sequence x is real or decoy, appeal is made to the likelihood ratio:

$$\text{LR}(X = x) = \frac{p(X = x \mid \text{real signal model})}{p(X = x \mid \text{decoy signal model})}. \tag{25.1}$$

By thresholding LR at a series of cutpoints a receiver operating characteristics (ROC) curve characterizing the classification performance of the model is obtained. These ROC curves are used to compare performance within model families and between approaches.

25.2.1 Maximum-Entropy Models

Yeo and Burge [51] develop a framework for modeling sequence motifs based on maximum entropy. Their central idea is to approximate short sequence motif distributions with the maximal entropy distribution (MED) constrained to satisfy select marginal nucleotide frequencies, as estimated from available data. For example, so-called first-order constraints are the empirical frequencies of each nucleotide at each position of the sequence. Imposing solely first-order constraints does not capture any between-position dependences and reduces to the weight matrix model (WMM [13]). By virtue of not allowing for higher-order dependences, including those between nonadjacent sequence positions, the simplistic WMM is generally inadequate. The various more recently devised methods seek to overcome this deficiency, but adopt differing strategies to do so.

The strategy underscoring the MED approach has a compelling, information-theoretic basis; if the chosen constraints are "correct," complete, and well estimated, then the resultant MED is optimal. However, this is a big "if" as we describe below, after sketching the construction of an MED model.

The (related) principles of maximum entropy and minimum relative (or cross-) entropy date back to around the mid-1950s and have been widely used since. Briefly, the principle of maximum entropy asserts that of all candidate distributions p satisfying a set of constraints, the one that best approximates the true distribution is that with the largest Shannon entropy H, as given by

$$H(p) = -\sum p(x) \log_2 (p(x)) \tag{25.2}$$

where the sum is over all possible sequences x. In appealing to this principle, it is clear that specification of constraints is critical. While it is possible to avoid incompatible constraints by using only marginal frequencies derived from the empirical distribution, there are few guidelines or search strategies for constraint selection. Rather, Yeo and Burge rely on exhaustive evaluation where possible. The flavor of the constraints entertained, and how they capture between-position dependences, is outlined next.

Two classes of constraints—complete and specific—are employed. *Complete* constraints correspond to sets of lower-order marginal distributions of the full distribution p and can be used to capture general between-position dependences. To establish (the slightly abused) notation we illustrate for the case $n = 3$; $X = (X_1, X_2, X_3)$. There the set of all lower-order marginal distributions, obtained via summing over all omitted indices, is

$$S_X = \{p(X_1), p(X_2), p(X_3), p(X_1, X_2), p(X_2, X_3), p(X_1, X_3)\}. \tag{25.3}$$

Let $S_s^m \subseteq S_X$ be lower-order marginals of *order m* and *skip s*. In (25.3), the first three elements are first-order ($m = 1$), the last three are second-order ($m = 2$), and only the last element has nonzero skip ($s = 1$). The term S_0^1 (i.e., all first-order marginals) is included in S_s^m when $m > 1$. So, for the $n = 3$ example, we have $S_0^1 = \{p(X_1), p(X_2), p(X_3)\}$; $S_0^2 = \{S_0^1, p(X_1, X_2), p(X_2, X_3)\}$; $S_1^2 = \{S_0^1, p(X_1, X_3)\}$.

Using standard (multinomial likelihood-based) estimates, the first-order constraints ($\hat{S}_0^1$) are just the empirical frequencies of each nucleotide (A, C, G, T) at each sequence position. The maximum entropy distribution consistent with these constraints is the weight matrix model (WMM [45,13]), in which there are no between-position dependences. Alternatively, if second-order zero-skip (i.e., nearest-neighbor) constraints ($\hat{S}_0^2$) are imposed, the maximum entropy distribution is an inhomogeneous first-order Markov model, also termed a weight array model (WAM [54]). Thus, prescribing differing constraints equates to specifying differing models. Indeed, in view of the finite sample space, all possible models can potentially be so constructed.

Specific constraints are just observed frequencies for a particular member of a set of complete constraints. So, continuing with $n = 3$, there are 4 specific constraints (corresponding to each nucleotide) for each member of S_0^1 and 16 specific constraints for each second-order marginal.

Framing constraints in this fashion readily allows specification of allowed orders of between-position dependence while not imposing that such dependence pertains to adjacent positions. It is this desire to permit nonadjacent dependence that motivates position permutation as described in Section 25.2.2. Central to the approach of Yeo and Burge [51] is an iterative scaling technique [9] that provides a computationally tractable algorithm that converges to the MED satisfying a given set of constraints. Armed with this means for estimating the MED subject to constraints, a maximum entropy model (MEM) is constructed by *separately* fitting MEDs to aligned sequences corresponding to real and decoy splice signals. The classification performance of the MEM is assessed by thresholding the LR (25.1), constructed from the ratio of the respective MEDs at a series of cutpoints, thereby generating an ROC curve. Yeo and Burge [51] display a multitude of such ROC curves corresponding to differing constraint specifications. We make comparisons against the best of these in Section 25.4.

Despite the existence of results proving convergence [34] of the iterative scaling algorithm, and the optimality of constrained MED approximations and likelihood ratio thresholding, there are issues surrounding the MEM approach that invite entertaining alternate methods. Foremost are the previously mentioned concerns pertaining to constraint selection. Beyond the need for exhaustive examination thereof (absent a search strategy) and the associated computational limits, there are more basic computational barriers that arise when dealing with moderately sized motifs. These arise when tackling the 3'ss; see Section 25.3 Questions relating to the separate modeling of real and decoy sequences are discussed in Section 25.3.

25.2.2 Permuted Variable-Length Markov Models

The motivation for allowing skip constraints in pursuing MED fitting derives from the recognition that there are frequently strong nonlocal dependences within short motifs, in addition to local (nearest-neighbor) dependences. This nonlocal dependence reflects the fact that various interactions between DNA, RNA, and proteins are determined by three-dimensional conformations and so involve nucleotides that are not adjacent in the primary sequence. Zhao et al. [52] seek to capture these more distant dependences by permuting the primary signal sequence so that strongly dependent positions are moved together and can thereby be fitted using low(er)-order Markov models.

This balancing of the need to contain the order of (Markov) models and thereby avoid the (exponential) proliferation of parameters, while simultaneously retaining nonlocal dependences, underscores the various methods brought to bear on splice signal recognition. Part of the purpose of this chapter is to contrast these approaches that explicitly try to model distant dependence—via skip constraints, permutation, or Bayesian networks (Section 25.2.3)—with flexible, yet generic, classifiers that attempt no such modeling.

Permutation of signal sequence is undertaken in the context of *variable-length* Markov models (VLMM). Allowing differing model orders (memory) for different sequence positions also serves to contain the number of parameters. Construction of a permuted VLMM (PVLMM) proceeds as follows. The joint probability mass function can be factored into a product of conditional probabilities

$$p(X = x) = p(X_1 = x_1) \prod_{j=2}^{n} p(X_j = x_j | X_1^{(j-1)} = x_1^{(j-1)}) \tag{25.4}$$

where, for $i < j$, x_i^j denotes the (reverse) sequence $(x_j, x_{j-1}, \ldots, x_i)$. In this formulation the current sequence position depends on all preceding positions. However, if lower orders of dependence are *selectively* sufficient, then the abovementioned parameter savings can be attained by correspondingly lower-order modeling. Zhao et al. [52] represent this by their use of the *context function* c_j, which maps the sequence preceding position j to a (possibly) shorter string. Thus, (25.4) can be reformulated as a VLMM:

$$p(X = x) = p(X_1 = x_1) \prod_{j=2}^{n} p(X_j = x_j | c_j(X_1^{(j-1)}) = c_j(x_1^{(j-1)})). \tag{25.5}$$

For example, when $c_j(x_1^{(j-1)}) = x_{(j-k)\vee 0}^{(j-1)}$, for each position j, the VLMM corresponds to a kth-order Markov model. The VLMM attempts to choose the appropriate context at a given position. This is effected by representing the context functions as decision trees, and determining which sequence positions, and which amino acid(s) at those positions, should constitute the splits. Imposing constraints on the allowed depth of these trees or, equivalently, the order of the Markov model, helps with regard computational feasibility. Comparisons between competing models, resulting from different context functions, take recourse to penalized likelihood. The probability model (25.5) provides the likelihood, with penalties corresponding to the Akaike information criterion [1] and Bayesian information criterion [42] being entertained.

Permutation is overlaid by applying a permutation π to positions $\{1, 2, \ldots, n\}$ such that

$$p(X = x) = p(X_{\pi_{(1)}} = x_{\pi_{(1)}}) \prod_{j=2}^{n} p(X_{\pi_{(j)}} = x_{\pi_{(j)}} \mid c_j(X_{\pi_{(1)}}^{\pi_{(j-1)}}) = c_j(x_{\pi_{(1)}}^{\pi_{(j-1)}})). \qquad (25.6)$$

The objective behind introducing such permutation is to bring important nonadjacent positions together, while at the same time preserving consequential local dependence. However, it should be noted that these goals may be irreconcilable since a given position may simultaneously have local and distant dependences that cannot be captured by a one-to-one permutation mapping. Furthermore, since the number of permutations grows as $n!$ with sequence length n, finding the optimal permutation, which could be tackled by enumeration for the (donor) 5' splice site ($n = 7$ effective, i.e., not conserved positions), becomes challenging for the (acceptor) 3' splice site ($n \approx 21$ effective positions). Zhao et al. [52] propose tackling this difficulty via simulated annealing to obtain a 'near-optimal' solution by trying different starting permutations. However, aside from problems associated with sampling from the vast permutation space, runtimes become prohibitively slow, even for low-order models.

The PVLMM approach, like the MEM approach, performs separate modeling for the real and decoy splice signal data. Having determined optimal permutation and context functions for each dataset, the same thresholded likelihood ratio approach [cf. Eq. (25.1)] is used to construct an ROC curve. Additionally, biological interpretation based on components of the optimal permutation for the real splice signal data is attempted. These aspects of the PLVMM are discussed further in Section 25.4.

25.2.3 Bayesian Network Approaches

An alternate approach to simplifying the joint probability mass function is via Bayesian networks. A Bayesian network [41] can be viewed as a family of multinomial probability distributions conforming to a set of conditional independence (CI) restrictions, which, in turn, can be encoded with a directed acyclic graph (DAG). A DAG, G, is defined by the pair (V, E) where V is a set of vertices and E is a set of directed edges joining (select) vertices. Generally, vertices correspond to random variables; in applications to motif identification each vertex corresponds to a sequence position X_i. Fundamental to the DAG representation is that the absence of an edge between two vertices i and j corresponds to a CI restriction between X_i and X_j.

For a given DAG, say, G, applying the CI restrictions using the chain rule of probability gives the (unique) factorization of the probability mass function [cf. (25.5)]

$$p(X = x) = \prod_{j=1}^{n} p(X_j = x_j \mid X_{\text{pa}(j)} = x_{\text{pa}(j)}), \qquad (25.7)$$

where pa(j) are the "parent" vertices of j in G, that is the set of vertices with directed edges pointing to j. Once a DAG has been determined, subsequent estimation of model parameters is relatively straightforward. The challenge is, given a training dataset, to estimate G—this is referred to as "learning the structure of the Bayesian network". Various approaches to such estimation have been devised. Here we very briefly describe some that have been used for splice site prediction.

Cai et al. [15] generalize first-order Markov models by allowing position i to depend on any other (single) position j, rather than restricting to position $i-1$. This class of models represents a very simple Bayesian network referred to as a "tree network" since the associated graph can be depicted as a tree. Thus, $|pa(j)| = 1 \forall j$. The essential ingredients of splice site prediction using tree networks are as follows. Starting from a database of aligned splice signal sequences a weighted, undirected graph G is constructed. The weight for the edge between vertices i and j is the mutual information, M_{ij}, between the corresponding pair of sequence positions:

$$M_{ij} = \sum_u \sum_v p(X_i = u, X_j = v) \log p(X_i = u, X_j = v)/p(X_i = u)p(X_j = v) \quad (25.8)$$

where $u, v \in \{A, C, T, G\}$. From this graph a maximal spanning tree is computed. This is the acyclic graph containing all vertices that maximizes the sum of edge weights. It can be readily computed using established algorithms [20]. The tree is then oriented by specification of a root node that yields a factorization along the lines of (25.7). Cai et al. compute attendant conditional probabilities using (multinomial) maximum likelihood, which equates to using empiric frequencies. This enables computation of the right-hand-side (RHS) numerator in (25.1). Repeating this procedure for the decoy signal database provides the denominator of (25.1) with subsequent generation of an ROC curve as described previously.

The ROC results reported by Cai et al. [15] show that tree networks, while superior to the WMM (independence model), did not provide improvement over the simpler first-order Markov model. Chen et al. [18] use this to motivate consideration of more elaborate Bayesian networks. The initial steps are somewhat similar to those of Cai et al. Instead of using M_{ij} (25.8) to create a graph, they use the (approximately equivalent for large sample sizes) Pearson χ^2 statistic X_{ij}^2 for testing independence between sequence positions i and j, as computed from the associated 4×4 contingency table. Then, instead of including all (weighted) edges between vertices, only select edges are used. These are chosen by thresholding the X_{ij}^2 values corresponding to a prescribed (type I) significance level α and appealing to a referent χ_9^2 distribution. As an aside, it is well known that the χ^2 approximation can be poor for large values of the statistic as arise here [3]. However, if α is regarded solely as a tuning parameter this is immaterial.

The resulting undirected *dependence* graph is then converted into a DAG in the following ad hoc fashion. For every sequence position i, calculate $S_i = \sum_{j \in \mathcal{N}(i)} X_{ij}^2$, where $\mathcal{N}(i)$ is the set of *neighbors* of position i as defined by there being an edge connecting them to the vertex i in the dependence graph. The position with the largest S value, say, S_k, is assigned as root (zeroth layer) of the Bayesian network. The first layer contains all positions in $\mathcal{N}(k)$. Neighboring positions of each position in the first layer then constitute the second layer and so on. Note that in this network construction the same position (variable) can appear more than once as nominally distinct vertices. The process is constrained, again to combat parameter proliferation, by restricting $\tau = |pa(\cdot)| \leq 3$ and treating $\tau \in \{1, 2, 3\}$ as another tuning parameter. With the network so established, the same procedure as used by all previously described approaches to conditional probability estimation (empiric frequencies) and ROC construction [(LR thresholding per (25.1)] is applied.

Perhaps the most principled approach taken to DAG estimation in this setting is that employed by Castelo and Guigó [17], in that no a priori constraints (tree-network [15], converted constrained dependence graph [18]) on the DAG structure are imposed. Rather, armed with a scoring metric and search procedure, the DAG is learned automatically. Of course, assumptions are now embodied in the scoring metric, and searches necessarily require heuristics as searching the space of all DAGs is NP-hard. However, some compelling choices for these ingredients are available (Refs [31] and [16], respectively) and these are employed in Castel and Guigó [17]. Adopting a Bayesian framework, the score for a given DAG is the marginal likelihood using an uninformative Dirichlet prior that satisfies *equivalence* requirements (differing DAGs encoding identical conditional independence assertions have the same score). While conditional probabilities are now estimated as posterior means, the same LR thresholding is used to generate ROC curves.

25.3 SPLICE SITE RECOGNITION VIA CONTEMPORARY CLASSIFIERS

No matter how sophisticated the approach taken to modeling splice signal structure or composition, the task of evaluating the performance of such models is inherently and inescapably one of classification and/or discrimination. That is, in order to measure the success of any model in identifying real splice sites it is essential to calibrate by its performance when applied to decoy sites. The attendant summaries, for example, sensitivity and specificity, are the targets of classification analysis. Other facets of the calibration, beyond sensitivity and specificity, also deserve attention. These include computational and interpretational considerations.

As has been emphasized in Section 25.2, existing methods tackle the problem by developing separate models for real and decoy splice signal sequences and then thresholding resultant likelihood ratios. In advocating such an approach, widespread appeal is made to the Neyman–Pearson lemma and accompanying optimality results. However, this casts the problem as one of hypothesis testing about underlying model parameters, as opposed to directly targeting the abovementioned predictive accuracy measures.

There has been little by way of application of conventional classification approaches. An exception is provided by Zhang et al. [55], who obtain good results using support vector machines. This is despite the more recent emergence of a variety of flexible and powerful classification techniques. While some conventional classifiers can be obtained from up front, separately constructed, within-class models, these are either simplistic (e.g., nearest neighbors) or rely on hidden between-class attributes [e.g., pooled (over classes) within-class covariance matrix in Fisher's linear discriminant; shrinkage parameter determination in nearest shrunken centroids [47].

To showcase that classification based on separate class modeling can be suboptimal compared to procedures that engage both classes (and all data) simultaneously, consider the following illustrative problem. It is desired to discriminate between two disease subtypes based on a gene expression microarray study. Typical in this setting is a wealth of covariates and/or features (genes) and a paucity of samples and/or cases (arrays) [22,43]. By pursuing separate modeling of the subtypes, we forfeit the opportunity to detect gene sets that are (jointly) discriminatory. Rather, under this scheme, selected genes or models may afford good characterization of the parent disease but be useless for identifying subtypes. Further detail is provided in the Discussion.

Additionally, the theoretic ability to develop highly refined, within-class models is often compromised in practice by computational considerations. We have seen examples of simplifications needed to estimate DAGs in Section 25.2.3. For MEM analyses of the 21 nt $3'$ss, the sequences are partitioned into nine shorter, overlapping fragments in order to avoid the prohibitive task of storing and iterating over 4^{21} sequences. As a consequence, long-range (≥ 7) between-position dependences are lost. No results are presented for PVLMM analyses of the $3'$ss, nor could we obtain such using the software provided.

With this motivation we apply several contemporary classifiers—support vector machines (SVMs), boosting, and random forests—to the splice signal recognition. Background on random forests is provided in Section 25.3.1, where some issues pertinent to splice signal recognition are discussed. Rather than recapitulate the underpinnings of SVMs and boosting we cite some key references and provide comments on these techniques Sections 25.3.2 and 25.3.3, respectively. Regardless of the attained prediction performance of these methods, they have some appealing attributes relative to the custom splice site modeling approaches detailed in Section 25.2. These include readily available, stable, and efficient algorithms that are frequently bestowed with a variety of bells and whistles: covariate importance measures, built-in prediction error estimation, missing data-handling capabilities, and, occasionally, provision of diagnostics. This is presumably counterbalanced by the custom approaches conferring enhanced interpretational insight. We evaluate this aspect in Section 25.4.3.

25.3.1 Random Forests

We devote additional attention to detailing some of the particulars surrounding random forests since their classification performance, when applied to the $3'$ splice signal data, was somewhat anomalous, with respect to both the other classifiers considered and their reputation as highly accurate classifiers. An explanation of, and attendant remedy for, this behavior is advanced.

In a series of recent papers, Breiman has demonstrated that consequential gains in classification or prediction accuracy can be achieved by using ensembles of trees, where each tree in the ensemble is grown in accordance with the realization of a random vector. Final predictions are obtained by aggregating (voting) over the ensemble, typically using equal weights. Bagging [5] represents an early example whereby each tree is constructed from a bootstrap [24] sample drawn with replacement from the training data. The simple mechanism whereby bagging reduces prediction error for unstable predictors, such as trees, is well understood in terms of variance reduction resulting from averaging [6,29]. Such variance gains can be enhanced by reducing the correlation between the quantities being averaged. It is this principle that motivates random forests.

Random forests seek to effect such correlation reduction by a further injection of randomness. Instead of determining the optimal split of a given node of a (constituent) tree by evaluating all allowable splits on all covariates, as is done with single tree methods or bagging, a subset of the covariates drawn at random, is employed. Breiman [7,8] argues, based on a comprehensive empirical evaluation employing numerous benchmark datasets excerpted from the UCI repository, that (1) random forests enjoy exceptional prediction accuracy, and (2) this accuracy is attained for a wide range of settings of the single tuning parameter employed. After describing the essentials of random forest construction, we indicate how the $3'$ splice signal data differ from almost all the classification benchmark datasets in the repository and, indeed, why the repository is unduly narrow in scope. A related and simple refinement to the random forest algorithm is proposed.

A random forest is a collection of tree predictors $h(\mathbf{x};\theta_k)$, $k = 1, \ldots, K$ where x represents the observed input (covariate) vector of length n (here sequence positions) with the associated random vector $\mathbf{X}$ and the θ_k are independent and identically distributed (i.i.d.) random vectors.

For expository purposes we illustrate the formulation for the regression setting for which we have a numerical outcome, Y. However, identical considerations pertain to classification (categorical outcome) problems. The observed (training) data are assumed to be independently drawn from the joint distribution of $(\mathbf{X}, Y)$ and comprise N $(n + 1)$-tuples $(\mathbf{x}_1, y_1), \ldots, (\mathbf{x}_N, y_N)$.

For regression, the random forest prediction is the unweighted average over the collection: $\bar{h}(\mathbf{x}) = (1/K) \sum_{k=1}^{K} h(\mathbf{x}; \theta_k)$.

As $k \to \infty$, the law of large numbers ensures

$$E_{\mathbf{X}, Y}(Y - \bar{h}(\mathbf{X}))^2 \to E_{\mathbf{X}, Y}(Y - E_\theta h(\mathbf{X}; \theta))^2. \tag{25.9}$$

The quantity on the right is the prediction (or generalization) error for the random forest, designated PE_f^*. The convergence in (25.9) implies that random forests do not overfit.

Now define the average prediction error for an individual tree $h(\mathbf{X}; \theta)$ as

$$PE_t^* = E_\theta E_{\mathbf{X}, Y}(Y - h(\mathbf{X}; \theta))^2. \tag{25.10}$$

Assume that for all θ the tree is unbiased, namely, $EY = E_\mathbf{X} h(\mathbf{X}; \theta)$. Then

$$PE_f^* \leq \bar{\rho} PE_t^*, \tag{25.11}$$

where $\bar{\rho}$ is the weighted correlation between residuals $Y - h(\mathbf{X}; \theta)$ and $Y - h(\mathbf{X}; \theta')$ for independent θ, θ'.

The inequality (25.11) pinpoints what is required for accurate random forest regression: (1) low correlation between residuals of differing tree members of the forest and (2) low prediction error for the individual trees. Further, the random forest will, in expectation, decrease the individual tree error PE_t^*, by the factor $\bar{\rho}$. Accordingly, the randomization injected strives for low correlation.

The strategy employed to achieve these ends is as follows:

1. To keep individual error low, grow trees to maximum depth.
2. To keep residual correlation low randomize via

   a. Grow each tree on a bootstrap sample from the training data.
   b. Specify $m \ll n$ (the number of covariates/sequence positions). At each node of every tree select m covariates and pick the best split of that node based on these covariates.

Now consider Figures 25.1 and 25.2, which display two very distinct prediction error (standardized binomial deviance) profiles. Both profiles are obtained from fitting classification trees [4], with prediction error (PE) being estimated via cross-validation. The data used in Figure 25.1 are the $3'ss$ data as further described and analyzed in Section 25.4 The minimum PE is attained at about 100 splits. The minimum occurs in a plateau region, after which there is an appreciable rise in error. This increase is such that the PE at the maximal number of splits is "significantly" greater than the minimum PE; the vertical segments (contained within each circular plotting symbol) represent ± 1 standard error. Such profiles where, as a function of increasing model size and complexity (here number of splits), PE initially decreases, plateaus, and then increases are common. Indeed, prototypic depictions

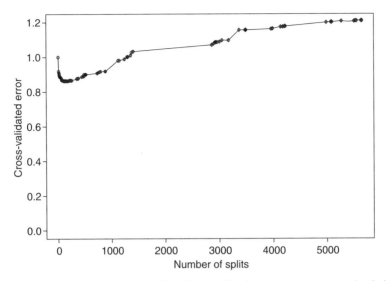

Figure 25.1 Cross-validated error profiles for classification trees grown to maximal size on the 3′ss training data.

of the relationship between PE and model complexity have this form [4 p. 87; 29, p. 38]. The presence of noise and/or redundant covariates are factors that can contribute to such profiles.

Figure 25.2 differs in that the PE at the maximal number of splits is the global minimum. That is to say, no matter how large a tree-structured predictor we fit, we don't overfit the data. This behavior is arguably unusual. The (letter recognition) data used to generate Figure 25.2 were obtained from the UCI Repository of Machine Learning Databases as converted to R

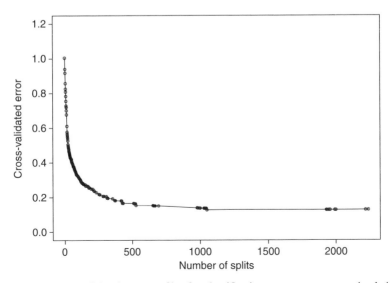

Figure 25.2 Cross-validated error profiles for classification trees grown to maximal size on the letter recognition data from mlbench.

[33], and available from the mlbench package at (http://cran.r-project.org/src/contrib/PACKAGES.html#mlbench). What is remarkable, and seemingly not appreciated, is that almost every dataset in the mlbench package exhibits this same behavior.

Our central concern, then, is that the strategy prescribed in item 1 above controls bias but not variance; such maximal trees may be highly unstable, and this instability will be reflected in inflated prediction errors. While variance control is achieved via averaging over the ensemble, there are situations, including as we show in Section 25.4 the $3'$ splice signal data, where this is not sufficient to counteract the maximal tree effect. That this behavior was not observed in empirical evaluations of random forests using the UCI repository is potentially attributable to the abovementioned property of the repository constituents.

For the R package randomForest (as with its standalone precursor), the size of the individual trees constituting the forest is controlled by a tuning parameter, nodesize. This specifies the number of cases in a node below which the tree will not split, and so determines maximal tree size. For classification forests, the default is nodesize = 1, asserted to always give good results. For data of the dimension of the $3'$ splice signal data, the maximally sized trees so created are very large indeed (>5000 splits). Also, for data of such dimensions, effecting control over tree size by varying this parameter is akward. Preferably, top-down rather than bottom-up control could be exercised. Accordingly, we introduce a new tuning parameter, anticipated to be helpful in these large sample size settings, and/or in situations where deep trees overfit. The new parameter simply controls the number of splits allowed.

25.3.2 Support Vector Machines

Detailed descriptions of support vector machines (SVMs) can be found in the literature [21,29]. A key component of SVM methodology is basis expansion, effected by transforming input vectors x, x', here (indicator representations of) real or decoy $3'$ss sequences, into a high-dimensional feature space via use of a prescribed kernel K, namely, $K(x, x') = \langle h(x), h(x') \rangle$. There are some standard choices for the kernel including *polynomial* ($K(x, x') = (1 + \langle x,x' \rangle)^d$) or *radial basis function* (RBF, $K(x, x') = \exp(-\|x-x'\|^2/\gamma)$). Since RBF kernels have been recommended as a good default [32], we adopt this choice in applying SVMs to predicting $3'$ss. It is important to note that we do not attempt to optimize kernel choice, nor do we pursue optimization of kernel parameters.

While SVMs are typically formulated as maximal margin classifiers, it is straightforward to recast them as optimizing a penalized (hinge) loss function [29].

$$\min_{\beta_0, \beta} \sum_{i=1}^{N} [1 - y_i f(x_i)]_+ + \lambda \|\beta\|^2 \tag{25.12}$$

where $y_i \in \{-1,1\}$ indicates whether the ith ($i = 1, \ldots, N$) sequence x_i is decoy or real; $f(x) = h(x)^T \beta + \beta_0$ and the subscript "+" denotes positive part. This focuses attention on the important tuning (penalty) parameter λ, for which we did undertake a limited optimization. More comprehensive evaluations await improvements of the R package SvmPath, which efficiently provides SVM solutions (for two-class problems as here) for all possible values of λ [30] but, unfortunately, cannot handle the large sample sizes of our $3'$ss data.

25.3.3 Boosting

Boosting has enjoyed considerable recent success as an effective off-the-shelf classifier. While boosting was originally presented as a procedure that combines outputs from many so-called weak classifiers (learners) to produce an ensemble, it is fundamentally distinct from bagging and random forests [29]. Insights into the basis for the success of boosting, including its resistance to overfitting, have been provided by viewing the method as additive modeling [28] and stagewise functional gradient descent [27,10].

We employ the R package gbm to classify real versus decoy $3'$ss using (shallow) classification trees as the weak learner and binomial log likelihood as the loss function. Parameters that need to be specified include number of trees (iterations), tree depth, and learning rate. As was the case for SVMs, we did not perform a comprehensive optimization for these inputs.

25.4 RESULTS

25.4.1 Data Generation

As indicated, our focus in applying contemporary classifiers is on the $3'$ (acceptor) splice site. This is motivated by the difficulties encountered by some of the modeling strategies outlined in Section 25.2 in handling the longer motifs, in contrast to the shorter $5'$ss. We used transcript datasets as constructed by Yeo and Burge [51]. Since use of computationally predicted genes could create circularities, human cDNAs were used as the starting point. Nonredundant transcripts that could be definitively aligned across the entire coding region and were not subject to alternative splicing were selected. This yielded some 12,700 introns and hence the same number of $3'$ss and $5'$ss. Real sequences were then excerpted as the sequences at positions $\{-20$ to $+3\}$ of the $3'$ss, with the consensus AG dinucleotide at positions $\{-2, -1\}$. These were then partitioned into training ($N = 8465$) and test ($N = 4233$) datasets. Decoy $3'$ss were excerpted as sequences in the exons and introns of these genes matching a minimal consensus, $Z_{18}AGZ_3$, where Z is any nucleotide. There were $N = 180,957$ training and $N = 90,494$ test decoys.

25.4.2 Predictive Performance

The essence of our results for predicting real $3'$ss is conveyed by Figure 25.3. Depicted are ROC curves corresponding to four methods: the (modified) MEM approach using the best constraint as obtained (on the same data) by [51]; random forests; boosting and support vector machines. In each case, modeling was performed using the training data and the ROC curve derived by applying selected models to the test data. The fact that only two curves are readily discerned reflects the near-identical performance of all methods except random forests. Some comments are in order.

At least as compared with SVMs and boosting, the approximation employed by MEM in order to handle sequences of length $n = 21$ by partitioning into nine shorter, overlapping sequences (see Section 25.3), does not incur any loss of predictive accuracy. On the other hand, not only could putatively superior performance be attained by SVMs and boosting since they were only cursorily optimized but also the comparison is against a highly selected and optimized MEM. That said, the performance of all methods is good. We comment on implications regarding gene finding in Section 25.5. We were unsuccessful in attempts to fit PVLMMs (using accompanying software) to the $3'$ss, presumably because of their length.

Next we turn to random forests in light of issues raised in 25.3.1 regarding growing individual trees to maximal depth. Figure 25.4 showcases prediction gains that can be realized by constraining the number of splits to 100. As noted in Section 25.3.1, it may be possible

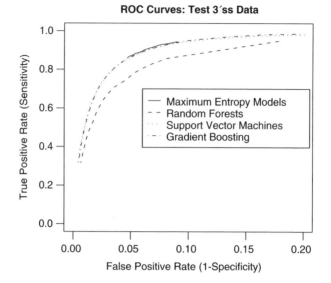

Figure 25.3 ROC curves obtained from four differing approaches applied to the 3′ss test data.

to effect similar improvement by manipulating the existing tuning parameter nodesize. However, in light of the sample sizes listed in Section , this is clearly a less direct strategy. To the extent that there is still a gap in predictive performance between random forests and SVMs/boosting/MEM it is possible to speculatively ascribe this to between-position

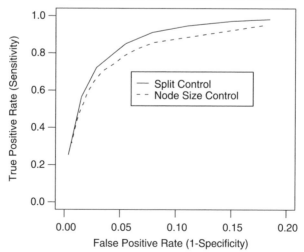

Figure 25.4 ROC curves contrasting predictive performance of random forests when individual tree size is controlled by number of splits *vs* size of terminal nodes.

dependence. Recall from (25.11) that random forests strive for low correlation between residuals from trees constituting the forest. This may be difficult to attain in settings where the covariates (sequence positions) are highly dependent. And, as described in Section 25.2, it was precisely the goal of exploiting such dependences that motivated the crop of existing techniques. Similar instances of random forests failing to realize gains in managing sequence-based predictors have been noted elsewhere [44].

By way of summary, it is apparent that off-the-shelf classifiers, employed with little optimization, are competitive with the most highly optimized, custom MEM approach. A nonoptimized MEM was, in turn, competitive with the most principled DAG approach [17]. We next address whether the class-specific approaches, by virtue of their striving for models of a (real) splice signal, confer any interpretative advantage.

25.4.3 Interpretational Yield

Several of the approaches outlined in Section 25.2 proclaim that the modeling procedure employed furnishes biological insight into the splicing process. We evaluate these claims in light of the fact that these approaches do not bestow improved predictive performance when compared with off-the-shelf classifiers.

For MEMs, it is asserted that the performance of a model informs about the set of constraints that was used. While this is collectively the case, their remain issues in interpreting and ranking the individual constraints. Conversely, the results presented for both 5'ss and 3'ss [51] indicate near-identical performance for models with differing complete constraint sets, which complicates attempts at ascribing importance to particular constraints. Further, in order to proffer biological interpretation (e.g., which positions interact with which components of the spliceosome), it is necessary to invoke specific, rather than complete, constraints. Some rankings are provided for the 5'ss; however, once again differing constraints are ranked highly by similarly fitting models. No rankings of specific constraints are given for the 3'ss. Indeed, it is not immediately obvious that the convergence results for the iterative scaling algorithm that obtains the MED while ensuring that constraints are satisfied [9] extend to the 3'ss case for specific constraints that straddle differing factors of the (overlapping) partitioned likelihood.

Similarly, for PVLMMs, only results for 5'ss are presented [52]. Biological interpretation is based on the selected permutation. Focus is placed on a subset of 5'ss sequence positions that are permuted into adjacent positions. For this subset, a series of sequence logo displays—bargraphs of information "bits" $(2-H)$ at each sequence position—for the real 5'ss are presented conditioning on complete conservation at select (permuted) adjacent positions. On the basis of differences between these plots, an attempt at interpreting spliceosome component action is made. However, there are a number of limitations surrounding this effort: (1) PVLMM barely improves on VLMM, so overinvesting in the selected permutation $\hat{\pi}$ is questionable; (2) only a subset of $\hat{\pi}$ adjacencies are used, followed by conditioning on conservation (consensus) at a further subset—the extent of this conditioning, that is, the proportion of sequences exhibiting the prescribed consensus, is unclear; and (3) perhaps most importantly, the entire approach pertains only to real 5'ss sequences. What is essential for interpretative purposes are characterizations that discriminate between real and decoy ss. That only VLMM was used for decoys not only underscores this point, but also impacts determination of $\hat{\pi}$.

The contemporary classifiers employed admit varying degrees of interpretability. SVM outputs do not include measures of variable importance. While postprocessing to attain such has been proposed, [35], the approach is problematic in the context of splice signal prediction since it relies on covariate (sequence position) independence. Variable importances for

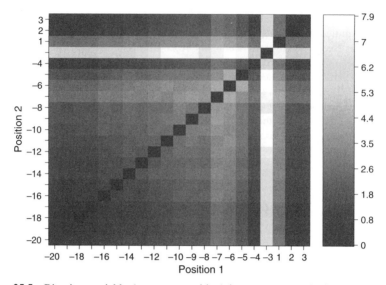

Figure 25.5 Bivariate variable (sequence position) importances as obtained from random forests.

boosting (using trees as base learners), as provided by the gbm package, build on related constructs for individual classification trees [27]. In particular, empirical improvements corresponding to splits performed on each respective covariate are summed within each tree, then averaged over all trees generated by the boosting algorithm. However, by not accommodating *surrogate splits*, this measure is vulnerable to masking [4]. More importantly, these importances are univariate, and don't capture joint covariate effects.

Random forests do provide variable importance summaries. Briefly, for each tree in the collection, prediction accuracy is computed for those (so-called out-of-bag) cases not included in the bootstrap sample used to construct that tree. A parallel computation is performed using permuted covariate values. The variable importance measure is then obtained as the difference between the two accuracies averaged over all trees, and normalized by the standard error. As with boosting, the resultant importances are univariate. However, by effecting pairwise permutation of covariate values it is possible to obtain bivariate importances, albeit at the cost of increased computation [12]. A heatmap depiction of such bivariate importances for the 21 (non-conserved) positions (omitting the AG dinucleotide at $\{-2, -1\}$) constituting the 3′ss is presented in Figure 25.5. The most striking feature is the long-range extent of bivariate importances involving the -3 position. Burge [13] makes a similar observation, based on straightforward thresholding of Pearson χ^2 tests of independence. Interestingly, he provides additional insight not in terms of interactions with the spliceosome, but rather compositional biases.

25.4.4 Computational Considerations

As indicated in Section 25.3, one of the compelling attributes of the classification procedures employed is the ready availability of stable algorithms, often replete with additional features, such as out-of-bag prediction error estimation. That is not to say that improvements aren't required; some instances have already been identified such as streamlining SVM penalty parameter optimization by extension of SvmPath, and refining variable importance measures for gbm. The large sample sizes involved here do not pose computational difficulties for boosting or random forests. However, since many SVM algorithms involve $O(N^3)$ operations, alternative approaches are desirable. Several such possibilities, including interior point methods with (approximating) low-rank kernels that require $O(Nk^2)$ operations (where k is the rank of the kernel), have emerged [23]. That said, runtimes using the R function svm were not prohibitive, even for the large sample sizes discussed here.

Conversely, as has been indicated, many of the existing approaches that seek to capture the dependency relationships of real and decoy sequences struggle as the length of the sequence increases. So, for example, we were unable to obtain results using the companion PVLMM software for the 3'ss of effective length $n \sim 21$. Similarly, the companion MEM software does not accommodate $n > 8$.

25.5 CONCLUDING REMARKS

Despite the problem of formulating precise, predictive models of RNA splicing being branded as one of the top 10 bioinformatics challenges, it is unclear whether further investment along the lines examined here—both dependence modeling and classifier application—is warranted. There are several reasons for adopting this stance. First, the predictive performance of the best approaches is already good. Additional, incremental gains will likely have a minor impact on downstream gene finding. Further, as has been illustrated here and argued elsewhere [19], the biological interpretative yield flowing from these approaches is limited. Moreover, and more importantly, the information content of the classical splice site signals has been shown to be generally modest for higher eukaryotes, and about half what is required for humans [39]. Accordingly, other factors including exonic splicing enhancers [26] and silencers [25] are known to contribute to the splicing process and ensure its high fidelity. There is also compelling evidence that splice site/exon/intron identification is such that splice sites are recognized in pairs [55], and this imparts additional information and constraints. Technological advances such as the advent of splicing specific microarrays have fueled considerable activity relating to identification of factors impacting splicing and alternative splicing [11,36,40,50], and there are numerous associated data analytic challenges. Thus, the originally stated problem of devising accurate predictive models warrants further work from this broader perspective. Additionally, there are many opportunities for further methodologic developments surrounding the techniques described herein, several having already been identified.

The suite of off-the-shelf contemporary classifiers now available afford appreciably improved predictive performance over their simpler forerunners. However, as indicated, these gains have not always been accompanied by retention of interpretability. Some more recently devised approaches that warrant further development and application with respect to gauging variable importance include the formal influence curve approachs [49] and the 1-norm SVM [56]. The latter replaces the L_2 penalty $\lambda \|\beta\|^2$ in (25.12) with an L_1 penalty $\lambda \|\beta\|$. As shown for the lasso method [46], the geometry of the L_1 penalty has the effect of inducing variable selection as λ increases. In the context of splice signal recognition, it would be of interest to determine the predictive accuracies of successively selected sequence positions, and to contrast this with the performance of existing, dependence-based modeling approaches.

Promising developments with respect to generative approaches (see below) are variable-order Bayesian networks [2]. These constitute generalizations of both (permuted) variable-length Markov models and Bayesian networks as described in Sections 25.2.2 and 25.2.3, respectively, and are accordingly correspondingly flexible. A Web-based implementation is available, and further investigation as to predictive performance for splice signal recognition is warranted.

We conclude with some comments surrounding distinctions between what we have branded separate and comparative modeling approaches, the former developing (unrelated) parametric probability models for both real and decoy sequences as a prelude to classification, and the latter pursuing classification directly and (here) nonparametrically. Ulintz et al. [48] term these approaches *generative* and *discriminative*, respectively. They opine that generative methods work well when the models are accurately specified and can then be effective even with small training set sizes. However, they further note that if data diverge from modeled distributions, classification suffers. Consequently, nonparametric discriminative methods are viewed as safer and more readily generalizable. Further, for the data of the dimensions in our splice sequence problem, these approaches will be competitive even if generative models are correctly specified.

ACKNOWLEDGMENT

The author wishes to thank Gene Yeo for providing data, Alexandre Bureau for providing software, Yuanyuan Xiao for testing software, and Ru-Fang Yeh for helpful discussions.

REFERENCES

1. Akaike, H., A new look at the statistical model identification, *IEEE Trans. Automatic Control* **19**, 716–723 (1974).

2. Ben-Gal, I., Shani, A., Gohr, A., Grau, J., Arviv, S., Shmilovici, A., Posch, S., and Grosse, I., Identification of transcription factor binding sites with variable-order Bayesian networks, *Bioinformatics* **21**, 2657–2666 (2005).

3. Bickel, P. J., Cosman, P. C., Olshen, R. A., Spector, P. C., Rodrigo, A. G., and Mullins J. I., Covariability of V3 loop amino acids, *AIDS Res. Hum. Retroviruses* **12**, 1401–1411 (1996).

4. Breiman, L., Friedman, J. H., Olshen, R. A., and Stone, C. J., *Classification and Regression Trees*, Wadsworth, Belmont, CA, 1984.

5. Breiman, L., Bagging predictors, *Machine Learn.* **24**, 123–140 (1996).

6. Breiman, L., Arcing classifiers, *Ann. Statist.* **26**; 801–849 (1998).

7. Breiman, L., Statistical modeling: *The two cultures, Statist. Sci.* **16**, 199–215 (2001).

8. Breiman, L., Random forests, *Machine Learn.* **45**, 5–32 (2001).

9. Brown, D., A note on approximations to discrete probability distributions, *Inform. Control* **2**, 386–392 (1959).

10. Bühlmann, P. and Yu, B., Boosting with the L_2 loss: Regression and classification, *J. Am. Statist. Assoc.* **98**, 324–339 (2003).

11. Burckin, T. A., Nagel, R., Mandel-Gutfreund, Y., Shiue, L., Clark, T. A., Chong, J.-L., Chang, T.-H., Squazzo, S. L., Hartzog, G. A., and Ares, M., Exploring functional

relationships between components of the transcription, splicing and mRNA export machinery by gene expression phenotype analysis, *Nature Struct. Molec. Biol.* **12**, 175–182 (2005).

12. Bureau, A., SNPs, random forests and asthma susceptibility, *Joint Statistical Meetings*, San Francisco, CA, 2003.

13. Burge, C. B., Modeling dependences in pre-mRNA splicing signals, in *Computational Methods in Molecular Biology*, Salzberg, S. L., Searls, D. B., and Kasif, S., eds. Elsevier Science, Amsterdam, 1998, pp. 129–164.

14. Burge, C. B., Tuschl, T. H., and Sharp, P. A., Splicing of precursors to mRNAs by the spliceosomes, in *The RNA World*, Gesteland, R. F., Cech, T., and Atkins, J. F., eds., Cold Spring Harbor Laboratory Press, Plainview, NY, 1999.

15. Cai, D., Delcher, A., Kao, B., and Kasif, S., Modeling splice sites with Bayes networks, *Bioinformatics* **16**, 152–158 (2000).

16. Castelo, R. and Kocka T., On inclusion-driven learning of Bayesian networks, *J. Machine Learn. Res.* **4**, 527–574 (2003).

17. Castelo, R. and Guigó, R., Splice site identification by idlBNs, *Bioinformatics* **20**, i69–i76 (2004).

18. Chen, T.-M., Lu, C.-C., and Li, W.-H., Prediction of splice sites with dependency graphs and their expanded Bayesian networks, *Bioinformatics* **21**, 471–482 (2005).

19. Claverie, J. M., From bioinformatics to computational biology, *Genome Res.* **10**, 1277–1279 (2000).

20. Cormen, T. H., Leiserson, C. E., and Rivest, R. L., *Intoduction to Algorithms*, MIT Press, Cambridge, MA, 1990.

21. Cristianini, N. and Shawe-Taylor, J., *An Introduction to Support Vector Machines*, Cambridge Univ. Press, Cambridge, UK, 2000.

22. Dudoit, S., Fridlyand, J., and Speed, T. P., Comparison of discrimination methods for the classification of tumors using gene expression data, *J. Am. Statist. Assoc.* **97**, 77–87 (2002).

23. Fine, S. and Scheinberg, K., Efficient SVM training using low-rank kernel representations, *J. Machine Learn. Res.* **2**, 243–264 (2001).

24. Efron, B. and Tibshirani, R. J., *An Introduction to the Bootstrap*, Chapman & Hall, New York, 1993.

25. Fairbrother, W. G. and Chasin, L. A., Human genomic sequences that inhibit splicing, *Molec. Cell. Biol.* **20**, 6816–6825 (2000).

26. Fairbrother, W. G., Yeh, R. F., Sharp, P. A., and Burge, C. B., Predictive identification of exonic splicing enhancers in human genes, *Science* **297**, 1007–1013 (2002).

27. Friedman J. H., Greedy function approximation: A gradient boosting machine, *Ann. Statist.* **29**, 1189–1232 (2001).

28. Friedman, J. H., Hastie, T. J., and Tibshirani, R. J., Additive logistic regression: A statistical view of boosting, *Ann. Statist.* **28**, 337–407 (2000).

29. Hastie, T. J., Tibshirani, R. J., and Friedman, J. H., *The Elements of Statistical Learning*, Springer, New York, 2001.

30. Hastie, T. J., Rosset, S., Tibshirani, R. J., and Zhu, J., The entire regularization path for the support vector machine, *J. Machine Learn. Res.* **5**, 1391–1415 (2004).

31. Heckerman, D., Geiger, D., and Chickering, D. M., Learning Bayesian networks: The combination of knowledge and statistical data, *Machine Learn.* **20**, 194–243 (1995).

32. Hsu, C. W., Chang, C. C., and Lin, C. J., *A Practical Guide to Support Vector Classification*. Technical Report, Dept. Computer Science and Information Engineering, National Taiwan Univ., 2003.

33. Ihaka, R. and Gentleman R., R: A language for data analysis and graphics, *J. Comput. Graph. Statist.* **5**, 299–314 (1996).

34. Ireland, C. and Kullback, S., Contingency tables with given marginals, *Biometrika* **55**, 179–188 (1968).

35. Jiang, T. and Owen, A. B., *Quasi-Regression for Visualization and Interpretation of Black Box Functions*. Technical Report, Statistics Dept. Stanford Univ., 2002.

36. Johnson, J. M., Castle, J., Garrett-Engele, P., Kan, Z., Loerch, P. M., Armour, C. D., Santos, R., Schadt, E. E., Stoughton, R., and Shoemaker, D. D., Genome-wide survey of human alternative pre-mRNA splicing with exon junction microarrays, *Science* **302**, 2141–2144 (2003).

37. Krawczak, M., Reiss, J., and Cooper, D. N., The mutational spectrum of single base-pair substitutions in mRNA splice junctions of human genes: Causes and consequences, *Hum. Genet.* **90**, 41–54 (1992).

38. Lander, E. S., Linton, L. M., Birren, B., Nusbaum, C., and Zody, M. C., Initial sequencing and analysis of the human genome, *Nature* **409**, 860–921 (2001).

39. Lim, L. P. and Burge, C. B., A computational analysis of sequence features involved in recognition of short introns, *Proc. Nat. Acad. Sci. USA* **98**, 11193–11198 (2001).

40. Pan, Q., Shai, O., Misquitta, C., Zhang, W., Saltzman, A. L., Mohammad, N., Babak, T., Siu, H., Hughes, T. R., Morris, Q. D., Frey, B. J., and Blencowe, B. J., Revealing global regulatory features of mammalian alternative splicing using a quantitative microarray platform, *Molec. Cell* **16**, 929–941 (2004).

41. Pearl, J., *Probabilistic Reasoning in Intelligent Systems*, Morgan Kauffmann, San Mateo, CA, 1988

42. Schwartz G., Estimating the dimension of a model, *Ann. Statist.* **6**, 461–464 (1978).

43. Segal, M. R., Dahlquist, K. D., and Conklin, B. R., Regression approaches for microarray data analysis, *J. Comput. Biol.* **10**, 961–980 (2003).

44. Segal, M. R., Barbour, J. D., and Grant, R. M., Relating HIV-1 sequence variation to replication capacity via trees and forests, *Statist. Appl. Genet. Molec. Biol.* **3**, Art. 2 (2004).

45. Staden, R., Computer methods to locate signals in nucleic acid sequences, *Nucleic Acids Res.* **12**, 505–519 (1984).

46. Tibshirani, R., Regression shrinkage and selection via the lasso, *J. Roy. Statist. Soc. Ser. B* **58**, 267–288 (1996).

47. Tibshirani, R. J., Hastie, T. J., Narasimhan, B., and Chu, G., Diagnosis of multiple cancer types by shrunken centroids of gene expression, *Proc. Natl. Acad. Sci. USA* **99**, 6567–6572 (2002).

48. Ulintz, P. J., Zhu, J., Qin, Z. S., and Andrews, P. C., Improved classification of mass spectrometry database search results using newer machine learning approaches, *Molec. Cell. Proteom.* **5**, 497–509 (2006).

49. van der Laan, M. J., Statistical inference for variable importance, *Int. J. Biostatist.* **2**, Art. 2 (2006).

50. Xiao, Y., Yang, Y. H., Burckin, T. A., Shiue, L., Hartzog, G. A., and Segal, M. R., Analysis of a splice array experiment elucidates roles of chromatin elongation factor Spt4-5 in splicing, *PLoS Comput. Biol.* **1**, e39 (2005).

51. Yeo, G. and Burge, C., Maximum entropy modeling of short sequence motifs with applications to RNA splicing signals, *J. Comput. Biol.* **11**, 377–394 (2004).

52. Zhao, X., Huang, H., and Speed, T. P., Finding short DNA motifs using permuted Markov models, *J. Comput. Biol.* **12**, 894–906 (2005).

53. Zhao, Z., Licklider, L. J., Gygi, S. P., and Reed, R., Comprehensive proteomic analysis of the human spliceosome, *Nature* **419**, 182–185 (2002).

54. Zhang, M. Q. and Marr, T. G., A weight array method for splicing signal analysis, *Comput. Appl. Biosci.* **9**, 499–509 (1993).

55. Zhang, X. H.-F., Heller, K. A., Hefter, I., Leslie, C. S., and Chasin, L. A., Sequence information for the splicing of human pre-mRNA identified by support vector machine classification, *Genome Res.* **13**, 2637–2650 (2003).

56. Zhu, J., Rosset, S., Hastie, T. J., and Tibshirani, R. J., 1-norm support vector machines, *Neural Inform. Process. Syst.* **16** (2004).

CHAPTER 26

Statistical Methods for Biomarker Discovery Using Mass Spectrometry

Bradley M. Broom and Kim-Anh Do
University of Texas M. D. Anderson Cancer Center, Houston, Texas

26.1 INTRODUCTION

A mass spectrometer is a precise tool for obtaining an accurate and detailed signature of a sample's constituent components. Mass spectrometers are used in industry and academia for both routine and research purposes, for both biological and non-biological analyses, although we concern ourselves only with the former in this chapter. Major applications of mass spectrometry in the biological sciences include the analysis of proteins, peptides, and oligonucleotides; and drug discovery, pharmacokinetics, and drug metabolism. Our focus in this chapter is on the use of mass spectrometry in proteomics, where it is now the method of choice for the analysis of complex protein samples [2]. Proteomics applications of mass spectrometry include measurement of protein abundance, detection of trace amounts of specific proteins, identification of an unknown sample, and determination of an unknown protein's amino acid sequence and folding pattern.

Mass spectrometer results are presented in the form of a molecular mass spectrum. Figure 26.1 shows a mass spectrum obtained from a study of serum proteomic features for ovarian cancer [9]. The mass spectrometer separates ionized sample molecules according to their mass (m): charge (z) ratios (m/z), so this is actually an m/z spectrum in which the horizontal axis (abscissa) represents the m/z ratio and the vertical axis (ordinate), the relative abundance. From the location and height of peaks in this spectrum, the number and relative abundance of components in the sample can be determined, as can the molecular mass of each component.

Statistical Advances in the Biomedical Sciences, edited by Atanu Biswas, Sujay Datta, Jason P. Fine, and Mark R. Segal

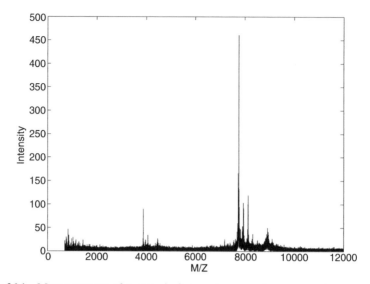

Figure 26.1 Mass spectrum of proteomic features in serum from an ovarian cancer study. The horizontal axis is m/z ratio and the vertical axis is relative abundance.

By convention, molecular masses are expressed in unified atomic mass units (uamu), although in proteomic applications daltons (Da), named in honor of John Dalton (1766–1844), are often used. An atomic mass unit is defined as exactly one-twelfth the mass of one atom of carbon-12, or approximately $931.49 \text{ MeV}/c^2$.

As diagrammed in the cartoon shown in Figure 26.2, mass spectrometers consist of three fundamental parts: an ionization source, an analyzer, and a detector. A variety of technologies can be used to implement each of these fundamental components, giving rise to a wide variety of mass spectrometers, each with its unique properties, advantages, and disadvantages. We will discuss those technologies of greatest relevance to biological studies later in the chapter. The

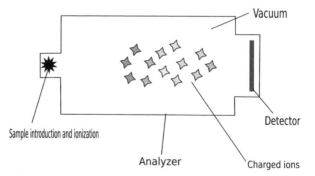

Figure 26.2 Cartoon showing the three fundamental components of a mass spectrometer: the ionizer, the analyzer, and the detector. Sample ions generated by the ionizer are directed through the analyzer, where they are separated according to their m/z ratios, and into the detector.

core of the mass spectrometer (the analyzer and the detector) is maintained under high vacuum (10^{-7}–10^{-8} torr) so that ions can be moved throughout the machine, with little chance of that movement being prevented or disrupted by other molecules. Some ionization sources also operate under high vacuum, while others operate at atmospheric pressure.

Although simple samples can be analyzed directly, the vast range of abundances in complex samples usually requires the samples to be fractionated into a series of components, each of which is analyzed individually. For example, the abundance of identified protein species in blood spans more than 10 orders of magnitude [24], with the few most abundant proteins, such as albumin, transferrin, and immunoglobin, accounting for up to 80% of the serum's total protein content. To obtain informative spectra for trace proteins in a sample containing such a wide range of protein abundances, a very stringent selection process must first be applied to significantly reduce the sample complexity. Depending on the mass spectrometer's ionization source, the fractionation device or process can be coupled directly to the mass spectrometer.

After fractionation (if required), the sample under investigation is placed into the instrument's ionization source, where molecules from the sample are ionized. Depending on the specific analysis to be conducted, the samples may be either negatively or positively ionized. For example, since proteins and peptides readily accept a proton, positive ionization is appropriate, whereas negative ionization is more suitable for the analysis of saccharides and oligonucleotides, which readily lose a proton.

A voltage differential accelerates the ions from the ionizer, through the analyzer, and into the detector. As the ionized molecules traverse the analyzer, they are separated according to their m/z ratios. In tandem mass spectrometers, more than one analyzer, not necessarily all of the same kind, are connected in tandem. Although it is possible to use other types of mass spectrometers, tandem mass spectrometers are frequently used to conduct fragmentation experiments for studying sequence and structural properties.

The detector measures the ion current and determines the abundance of the separated ions, which is recorded for subsequent analysis. Common types of detectors include photomultipliers, electron multipliers, and microchannel plate detectors. The specific type of detector used depends on the type of analyzer.

The following section describes sample ionization in more detail, with an emphasis on matrix-assisted laser desorption ionization (MALDI). Section 26.1.2 describes mass analysis, focusing specifically on time-of-flight (TOF) mass analyzers. Section 26.2 summarizes the role of mass spectrometry in biomarker discovery. Sections 26.3 and 26.4 describe the preprocessing and analysis of mass spectrometry data. Section 26.5 discusses potential statistical developments, and Section 26.6 concludes the chapter.

26.1.1 Sample Ionization

A large variety of ionization methods are used in other fields of mass spectrometry, including atmospheric-pressure chemical ionization (APCI), chemical ionization (CI), electron impact (EI), fast-atom bombardment (FAB), field desorption/field ionization (FD/FI), inductively coupled plasma (ICP), and thermospray ionization (TSP). These are not, or no longer, widely used in proteomics. For instance, ICP operates at 7000 K, a temperature at which complex biomolecules quickly decompose into their atomic components, rendering it unsuitable for most proteomics applications. We will not explore these methods further in this chapter.

For the majority of biochemical analyses, two relatively new techniques are the ionization methods of choice: electrospray ionization (ESI) and MALDI. In this chapter, we focus exclusively on MALDI. Although proteomic studies were performed before MALDI and ESI became available, using FAB, for instance, these studies were slow and expensive, and it was the development of MALDI and ESI that enabled proteomics to achieve its current popularity. Since fragmentation of large biomolecules rarely occurs in either method, both ESI and MALDI are classed as "soft" ionization methods.

In MALDI [18], a sample is dissolved in an appropriate volatile solvent, mixed with a large excess of a supporting "matrix" compound, applied to a sample plate, and dried. The sample plate is placed into the vacuum region of the mass spectrometer, where it is bombarded with pulsed laser light. The matrix compound is carefully matched to the laser's frequency, ensuring that absorption of the laser's energy by the matrix will be highly efficient. A common combination consists of sinapinic acid for the matrix compound in conjunction with a 337-nm nitrogen laser. The matrix absorbs energy from the laser, vaporizing the mixture and creating a plume of matrix and sample molecules that then form ions suitable for analysis. The highly absorbing matrix absorbs the laser energy very efficiently, protecting the analyte molecules from direct exposure to the laser energy, and hence reducing decomposition of the sample.

Surface-enhanced laser desorption/ionization (SELDI) [20] is a variant of MALDI that employs chips spotted with protein capture baits such as small molecules, antibodies, DNA, or enzymes. The analyte is applied to the chip, where proteins with affinities to the capture molecules will bind strongly to the surface. Any impurities or loosely bound proteins can then be washed away. The chip is placed into the ionization source, and the bound analytes released and ionized directly from the chip by laser desorption. Since the processes typically used to reduce sample complexity, such as affinity purification and concentration, are applied directly to the chip surface, total processing times and variability are reduced, making SELDI an attractive method for proteomics research.

The MALDI method deals well with thermolabile, nonvolatile organic compounds, especially those of high mass. It is used successfully in biochemical areas for the analysis of proteins, peptides, glycoproteins, oligosaccharides, and oligonucleotides. It is relatively straightforward to use and reasonably tolerant to buffers and other additives. The mass accuracy depends on the type and performance of the analyzer of the mass spectrometer, but most modern instruments should be capable of measuring masses to within 100 ppm, at least up to approximately 70 kDa.

As MALDI generates predominantly singly charged ions, the spectra obtained are relatively easy to interpret. For large biomolecules, the singly charged ions generated by MALDI require an analyzer that can accept the large m/z ions produced.

26.1.2 Mass Analysis

The main function of the mass analyzer is to separate, or resolve, the ions formed in the ionization source of the mass spectrometer according to their mass: charge (m/z) ratios. There are several types of mass analyzers currently available, including magnetic sector analyzers, quadrupole analyzers, quadrupole ion traps, Fourier transform analyzers, and TOF analyzers. Each type has its distinctive features and performance, including the m/z range that can be covered, the mass accuracy, and the achievable resolution. In this chapter, we consider only TOF analyzers in detail.

The TOF analyzer separates ions by simultaneously applying a fixed potential to each ion such that similarly charged ions acquire the same kinetic energy (approximately 20 keV). Consequently, the acquired velocity depends only on the ion's mass; lighter ions are accelerated more rapidly, and hence travel faster. The ion's m/z ratio can therefore be determined by measuring the time it takes for the accelerated ions to reach a detector at the end of a field free region known as the "flight" (or "drift") tube (typically 1–2 m long).

Although conceptually simple, high-resolution TOF analyzers require sophisticated design. Ideally, the ions in a TOF analyzer would start their travel through the analyzer at the same time, at the same distance from the detector, and with the same energy. In practice these conditions are not met exactly. If not corrected, these effects would cause ions of the same m/z ratio to arrive over a finite time interval, limiting resolution.

For pulsed ionization sources, such as laser ionization, time-of-flight secondary-ion mass spectrometry (TOF-SIMS), laser desorption and laser ablation, the ions can be created in a static electric extraction field within a sufficiently short time interval (a few nanoseconds). However, if the ions are extracted immediately, they will traverse the still dense plume of ions and neutral molecules desorbed by the laser. Collisions with this plume will increase both the spread of ion energies and the amount of fragmentation. In delayed ion extraction, both of these effects are mitigated by allowing the plume to expand before the ions are extracted by the application of a high-voltage pulse with a fast risetime (less than, say, 10 ns). For continuous ionization processes, such as EI, however, the ions must be collected over a time interval (e.g., several microseconds). After sufficient ions have been collected, the ionization process can be stopped and a high-voltage pulse applied to inject the collected ions into the TOF analyzer. Alternatively, a continuous stream of slow-moving ions can be converted into pulses of rapidly moving ions by applying the accelerating voltage pulse in a direction orthogonal to the stream.

The length of the flight path is not identical for all ions, since ion creation takes place in a finite volume. Wiley and McLaren [47] showed that by carefully shaping the potential distribution of the ion extraction field, ions with further to travel could be given slightly more kinetic energy so that an ion's arrival time at the detector is independent of its starting position. This is known as *spatial focusing*.

Ions can also have different initial energies, due to space charge effects and processes such as fragmentation, desorption, and ablation. A *reflectron TOF spectrometer* compensates for this initial energy distribution by reflecting the ions from a repelling electric field called an *ion reflector* or *reflectron*. Faster ions travel further into the repelling electric field, and hence take longer to reach the detector. A single-stage reflector produces an homogeneous electrostatic field that can provide only a first-order correction of the initial energy spread. A multiple-stage reflector with a carefully designed quadratic field can correct for all aberrations caused by the initial energy distribution such that all the ions (of the same m/z ratio) arrive at the detector at the same time. This is known as *energy focusing*.

Time-of-Flight analyzers place strong demands on the detector and data acquisition systems. After each laser pulse, ions from the complete mass distribution will arrive at the detector within a very short time interval. To obtain a detailed mass spectrum, the detector and data acquisition system must have very fast response and recovery times, and very high data throughput rates. Most TOF spectrometers employ multichannel plate (MCP) detectors, which can detect many ions at the same time. To reduce statistical uncertainties, TOF spectra are obtained by averaging the results of hundreds to thousands of individual laser pulses. The lasers are typically pulsed between 10 and 100 times per second.

Time-of-Flight analyzers are ideal for pulsed or spatially confined ionization sources, do not require ion beam scanning, and can provide a complete mass spectrum for each ionization event. They offer mass resolutions greater than 10,000, a mass range up to 500 kDa, high ion transmission (greater than 10%), and large acceptance volumes, and can obtain spectra for extremely small sample amounts; TOF analyzers also are relatively inexpensive.

26.2   BIOMARKER DISCOVERY

High-resolution mass spectrometry of the proteome is a promising approach to the identification of novel biomarkers. In this approach, mass spectrums are obtained from many cases and controls, and then compared to identify peaks that differ significantly between the two groups. Any peak so identified is then isolated and characterized (probably using mass spectrometry methods).

Researchers have recently begun to explore the potential diagnostic utility of this approach, and, specifically, to determine whether such peaks can serve as biomarkers of the early stages of diseases such as cancer [1,28,21,22,35,46,49]. These explorations have focused on spectra obtainable from readily available biological fluids such as blood, urine, or saliva.

Although the approach itself is conceptually simple, the biomarkers to be found (if any) are expected to be present in extremely small amounts relative to the components of the normal proteome. Consequently, biomarker discovery studies are challenging statistically. Specific challenges of such studies include precisely aligning the multiple spectra obtained, and separating true differences between the cases and controls from normal sample variability and from minute variations in sample handling. In the following section, we survey a number of statistical methods for addressing these challenges.

Mass spectrometry data from several biomarker discovery studies is publicly available. The best known is probably the Clinical Proteomics program, jointly run by the NCI and FDA [21]. The databank consisting of various SELDI and Qstar datasets is currently located at http://home.ccr.cancer.gov/ncifdaproteomics/. However, researchers should be cautioned about the quality of these data [4]. Both SELDI data and MATLAB scripts for processing and analysis are also made available at M. D. Anderson Cancer Center; see http://bioinformatics.mdanderson.org.

26.3   STATISTICAL METHODS FOR PREPROCESSING

Many mass spectrometry instruments have associated software that will perform peak detection and quantification automatically, but these may not address all the necessary preprocessing steps. Processing issues involve a partial list of important steps, including

- Spectral calibration
- Correcting for matrix noise
- Spectral denoising
- Baseline estimation and subtraction
- Peak detection and quantification
- Normalization
- Identifying harmonics or common patterns and modifications

A common problem in practice is that the same peak may drift slightly over time due to changes in the instrument. For all analyses, it is essential to calibrate the instrument's m/z scale using a standard sample of a type similar to that of the sample being analyzed (e.g., a protein calibrant for a protein sample). Possible factors affecting the sampling conditions of the mass spectrometer, include, but are not necessarily limited to

- Variations in ionization efficiency
- Possible clogging or erosion of internal apertures, resulting in altered interface transport efficiency
- Different matrix or matrix concentration in the samples, resulting in possible matrix suppression
- Environmental temperature or humidity fluctuations

To address this problem, one can run a "calibration sample" consisting of only a small number of proteins with a priori known identities. Since the masses of the peptides are known, the flight times are empirically observed, and one can thus fit a quadratic model

$$m/z = at^2 + bt + c.$$

The model parameters are the least-square solution to this equation and are assumed to hold for several samples. As these parameters can change over time, investigators should perform visual checks that some of the larger peaks align across samples [23].

A problem unique to MALDI spectra is matrix noise, resulting from the breaking free of other material besides the peptide of interest when a sample is laser-blasted. This occurs at the very low m/z end of the spectrum, producing an unstable effect [29]. Although this problem can be addressed by imposing a threshold m/z cutoff value, excluding values below some chosen m/z cutoff, such a threshold can be affected by other machine settings such as the laser intensity.

The spectra can be modeled mathematically as

$$Y_i(t) = k_i S_i(t) + B_i(t) + \epsilon_{it},$$

where Y_i (j) is the intensity of spectrum i at time index t, k_i is a normalization factor, S_i is the protein signal of interest (a set of peaks), B_i is a smooth underlying baseline, and $\epsilon \sim N(0, \sigma^2(t))$ is some high-frequency noise. One particular problem is simply that peaks can have different shapes in different parts of the m/z range; higher m/z peaks are broader. Some factors that can contribute to this broadening are uncertainty in the initial velocity of the peptide, isotopic spread, and the nonlinearity of the clock tick to m/z mapping.

There are a number of lowpass denoising filters for spectra (e.g., Savitzky–Golay [39], least squares [16], DISPO [50]). Bioinformatics researchers at M. D. Anderson Cancer Center prefer the wavelet-based denoising filter, which adapts naturally to the multiscale nature of the data. The overall denoising algorithm involves mapping spectra to the wavelet domain, applying hard thresholding of the wavelet coefficients, and subsequently performing an inverse mapping back to the spectral domain [10].

Following the spectra smoothing process is baseline estimation. Often, simple ad hoc methods suffice, such as fitting a local maximum to avoid negative-valued intensities from subtraction, adapting the moving window to increasing m/z values.

Spectra also need to be normalized before comparing peak intensities across spectra. One common method is to use the summed intensities for the entire spectrum, after denoising and baseline subtraction [37,30]

Finally, if the analysis method to be used (see text below) operates on specific features, such as peaks, within the spectra, these features must be identified and their (relative) magnitude determined. Peaks, for example, can be identified using a maximum finder. Feature detection can be facilitated by considering a set of spectra rather than a single spectrum. Assuming that the spectra have been roughly aligned, one can perform peak detection on the average spectrum within a group [33].

We will now review the preprocessing of Wu's ovarian cancer dataset [48] by Tibshirani et al. [44]. The dataset consists of MALDI-MS spectra collected from serum samples of 89 subjects (42 noncancer controls and 47 cancer cases) measured at 91,360 sites with a range of 800–3500 Da with 0.019-Da intervals. For the first step of peak extraction, a simple peak-finding method was used, looking for m/z values (sites) whose intensity is higher than that of the surrounding 100 sites and also higher than the estimated average background at that particular site. A *supersmoother* with a span of 0.002 was applied to the raw spectra, and the estimated peak widths in the smoothed spectra were approximately 0.5% of the corresponding m/z value. The m/z values were then log-transformed, rendering the peak widths approximately constant across the m/z range. The peak extraction process resulted in 14,067 peaks from the individual spectra. Next, to perform peak alignment, complete linkage hierarchical clustering was applied to the set of extracted peaks. The centroid or mean position of each cluster is viewed as the representative position for the peak. For this particular dataset, pruning the dendogram at the height of log(0.005) produced 192 clusters, where the distance between any two peaks within a cluster is at most log(0.005). The next step is the search for common peaks in individual spectra and any simple ad hoc method suffices to produce a summary of spectrum peak heights y_{ij} for i m/z values and j observations.

For a second example of preprocessing, we will consider a motivating dataset from the First Annual Proteomics Data Mining Conference. This dataset consists of MALDI-MS spectra of serum for 24 individuals with lung cancer and 17 normal individuals (without cancer). For each subject (sample), the raw data contained recordings of 20 fractions. Each such spectrum had readings for 60,831 m/z values. Thus there were 41 subjects, each with 20 fractions, consisting of 60,831 observations for each fraction.

The research challenge was group comparison experiments with data from samples under two biologic conditions, in this case lung cancer and normal samples, and the ultimate goal was to identify protein biomarkers that distinguish between samples from the different conditions. Assuming the concept that the upregulation or downregulation of certain proteins is the consequence of a transformed cancerous cell and its clonal expansion, an early detection research project may focus on the identification of such early molecular signs of lung cancer via the assessment of protein profiles from specific biological specimens. Researchers can thus analyze the collected protein profiles and identify signature *fingerprints* for the classification between lung cancer and normal states. Thus researchers can ultimately study the biological significance of those specific proteins or peptides associated with the identified signature profiles. Such advances can eventually lead to a clinical detection tool. Different research groups attempt to develop techniques to classify or cluster the same dataset.

Baggerly et al. [3] preprocessed this dataset as follows:

1. *Baseline subtraction*—a baseline, computed using a windowed local minimum technique, is subtracted from the data. This baseline correction has to be performed separately for each fraction in each sample. Thus, this is a crucial step in the preprocessing, as the fractions cannot otherwise be combined meaningfully.

2. *Sinusoidal noise removal*—using a Fourier transform, a periodic noise most likely associated with electrical activity is removed.

3. *Current normalization*—this was effected by dividing by the total current over all the readings.

4. *Defractionation*—the normalized fractions were combined to generate one spectrum per patient.

5. *Windowed dimension reduction*—taking the maximum intensity in each window (of 200 readings) and taking windows in which at least 8 of the samples contained a peak, the dimensionality was reduced from 60,381 to 506, to generate a 506-peak dataset.

The same preprocessing steps were used by Muller et al. [34] in their later analysis of the same dataset.

26.4 STATISTICAL METHODS FOR MULTIPLE TESTING, CLASSIFICATION, AND APPLICATIONS

In this section we review current methodologies, including some complex Bayesian and functional methods for analyzing mass spectrometry data. These methods can be divided into two broad categories: (1) methods that first identify features, such as peaks, in the mass spectrometry data and then analyze the variation in the values of those features, and (2) methods that directly analyze the entire spectrum as functional data.

26.4.1 Multiple Testing and Identification of Differentially Expressed Peaks

Most applications are motivated by the goal to select m/z values that are significantly associated with some phenotypic trait; thus, using a multiple testing procedure, we would identify proteins that are truly different in mean intensities between two groups or more. The overall testing procedure involves creating reasonable test statitics relative to some parameter of interest (e.g., quantiles or means) and determining how to derive joint inference for these that uses the known dependence structure in the data. The investigator makes specific choices on the parameters on interest, null hypotheses, test statistic, and specific type I error rate. The different types of type I error rates allow investigators different controls such as

- Controlling the probability of making at least one false-positive decision is known as the *familywise error rate* (FWER). This is a stringent method and produces error rates that are too conservative to be useful in high-throughput data analyses.
- Controlling the probability of making more than a predefined number of false-positive decisions is known as the generalized familywise error rate.
- Controlling the expected proportion of false positives amongst all positives is called false discovery rate (FDR). This method does not provide a probabilistic bound that the proportion of false positives is smaller than a predefined threshold, for instance, 0.05.
- Controlling the proportion of false positives among all positives at a user-defined value q is called the *tail probability* of the proportion of false positives, TPPFP(q). This method is favorable to the FDR method, particularly in the setting where test statistics are highly dependent, thus inducing high variability in the expected number of false positives.

Benjamini and Hochberg [5] first introduced a Bonferroni-like method to control the FDR. Storey [43] subsequently argued that the FDR is interesting only when positive decisions have occurred, thus introducing a procedure to control the conditional FDR or the *positive FDR*. Extensions of this work are described in Storey and Tibshirani [42], including methods to estimate the positive FDR and simulation studies conducted under various stages of coregulation. Other authors [25,38] provide comparison studies of selected methods where procedures controlling FDR exhibit higher power than do those controlling FWER. Bayesian FDRs have also been discussed in Genovese and Wasserman [12,13] and Do et al. [11], among others. Scheid and Spang [40] introduced a binwise FDR method to separate two overlapping score distributions by estimating their proportions after binning and evaluating its performance in a simulation study. Their estimator utilized prior probabilities for gene classes. The binwise FDR results in a discretized estimation of mixture probabilities conditional on the chosen binning. Many authors have developed TPPFP multiple testing procedures; for example, Lehmann and Romano [26] discussed marginal stepdown procedures, Genovese and Wasserman [12,13] described the inversion method for independent test statistics and its conservative version for general dependence structures under a Bayesian framework, and van der Laan et al. [45] proposed a new resampling-based multiple testing procedure to asymptotically control the TPPFP(q) by fitting an empirical Bayes two-component mixture model to the data to obtain an upper bound for marginal posterior probabilities of the null being true, conditional on the data. Birkner et al. [7] applied this method to SELDI-TOF mass spectrometry proteomic data. User-friendly tools for differrential expression analysis in R can be found in OOMPA, at the Website bioinformatics.mdanderson.org/Software. Methods implemented in the latest release include two-sample t test, fixed-effect linear models with ANOVA, beta-uniform mixture model controlling for FDR, Wilcoxon rank-sum test with empirical Bayes, significance analysis of microarrays (SAM), and Dudoit's adjustment of p values to control the FWER.

26.4.2   A Peak Probability Contrast (PPC) Procedure for Sample Classification

For pattern classification from protein spectra, Tibshirani et al. [44] proposed an algorithm in which the optimal discriminating split point for the height of each peak depends on the quantiles of all measurements at a peak position. After defining $p_{ik}(\alpha)$ to be the proportion of spectra in group k (control or cancer case) with a peak at site i larger than the α quantile of all the peaks at this site, the critical step is to choose $\hat{\alpha}\,(i)$ that maximizes $\Delta_i = |p_{i,\text{case}} - p_{i,\text{control}}|$. The resulting class probabilities are $\hat{p}_{ik}$ at the maximized value. To perform class prediction for a new spectrum, a binary profile can be created for each patient using the optimal discriminating split points, and compared to each probability centroid vector corresponding to the case and control groups, based on some kind of distance measure. To assess the sensitivity and specificity of this discrimination procedure, cross-validation can be carried out coupled with this whole process. Further, the significance of each peak can also be assessed by an adaptation of the FDR concept to this method.

26.4.3   A Semiparametric Model for Protein Mass Spectroscopy

Müller et al. [34] developed a mixture-of-Beta model for protein mass charge spectra.

   Their model includes a hierarchical prior with indicator parameters related to differential expression of proteins. The mass spectrometry data may be viewed as frequencies of detector events reported on a grid m_i, $i = 1, \ldots, M$ of mass/charge values. Typically M is large, say, 50,000. Let y_{ti} denote the observed count for sample t at grid value m_i. The main assumption is that the recorded counts (*empirical spectrum*) arise partially from the proteins of interest (*protein spectrum*) and noise (*baseline*). The baseline corresponds to counter events from protein fragments, matrix dispersed by the laser impact, detector noise, and other artifacts of the experiment unrelated to the proteins of interest. Let f_t denote the latent protein spectrum for sample t, with B_t for the baseline, and $p_t = (1 - p_{0t}) f_t + p_{0t} B_t$, for the *mean spectrum* for sample t. Further, let Beta(x; m, s) denote a Beta distribution on the random variable x with mean m and standard deviation s, using a nonstandard parameterization of the Beta model. Thus f_t can be represented as a mixture of Beta kernels. Let $x = x_t$ index the biologic condition of sample t, with $x_t \in \{0,1\}$ for a two-group comparison, and assume

$$f_t(m) = \sum_{g=1}^{G} w_{xg} \text{ Beta}(m; \ \epsilon_g, \alpha_g). \tag{26.1}$$

The weights are indexed by x; that is, equal weights are assumed for all samples under the same biologic condition. The size of the mixture G defines the number of proteins. It is part of the unknown parameter vector, making (26.1) a random size mixture of Beta kernels. A similar mixture of Beta kernels parameterizes the baseline functions B_t. A hierarchical prior can be defined with a positive prior probability for ties of weights w_{xg} across biologic conditions. An indicator $\lambda_g = I(w_{0g} = w_{1g})$ is introduced to allow easy posterior inference about proteins with differential expression across the two biologic conditions. Posterior inference can be implemented by setting up Markov chain Monte Carlo (MCMC) posterior simulation. A complication arises from the random size of the mixture (26.1) because changing the mixture size G induces changes in the dimension of the parameter vector. Reversible jump MCMC (RJMCMC) simulation methods can be used to define the posterior simulation across the variable-dimension parameter space, facilitated by split/merge and birth/death moves. The birth/death moves make use of a reference solution obtained by some ad hoc method up front, before initiating the posterior simulation. Figure 26.3 shows some aspects of the posterior inference.

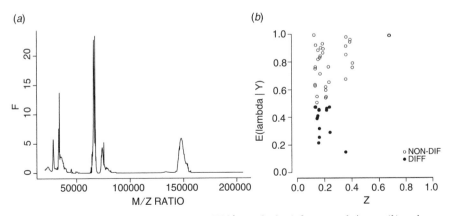

Figure 26.3 (*a*) Imputed mean spectrum $E(f_t \,|\, x_t = 0$, data) for normal tissue; (*b*) and posterior probability of nondifferential expression for each protein, $E(\lambda_g \,|\, y)$ versus kernel mean mass ϵ_g. Peaks with marginal posterior probability of differential expression beyond 0.5 are marked as solid dots.

26.4.4   Smoothed Principal-Component Analysis (PCA) for Proteomic Spectra

Methodological aspects of functional data analysis (FDA) for a family of observed curves have been considered by a number of authors [36,15,17,27].

Concurrently, Dean Billheimer (Vanderbilt University) [6] and Kim-Anh Do have considered methods for analyzing mass spectrometry utilizing the concept of smoothed principal components, exploiting the nature of proteomics spectra as functional data. The mass spectrometry process may be viewed as a process generating data through time or continuous m/z values. The motivation for FDA is to treat the entire measured function not as a single observation, but rather as a closely spaced sequence of measurements. The advantage of this approach is that we may incorporate continuity and smoothness constraints if known for the data-generating process. With FDA, one needs to perform two main steps before data analysis:

- *Representing the data via basis functions*—this plays the key role in defining smoothness and continuity conditions of the resulting data.

- *Data registration or feature alignment*—this transforms the functional argument axis to align key features of the response. The transformation may entail a linear shift to correct for an arbitrary starting time, or a nonlinear transformation for more complicated cases. A MALDI-TOF spectrum consists of measurements of a continuous intensity process at discretely sampled timepoints, subsequently converted to an equivalent m/z axis through an analytically derived transformation. The m/z axis is more precisely calibrated using peaks with known mass values. We adapt and implement the methods of Ramsay and Silverman [36].

First, a dimension reduction step can be performed via a smoothed PCA. Once dimension has been appropriately reduced, there are at least two ways of estimating population densities and formalizing the discrimination. One is nonparametric and based, for example, on kernel methods; another is parametric and founded, for example, on fitting a Gaussian density. The latter approach is generally preferable for real-time classification of future proteomics profiles, since it summarizes empirical training sample information using a relatively small number of parameters. The overall approach can be summarized as follows:

1. Start with the $G \times N$ data matrix X. Here G is the number of peaks of proteins and N is the total number of normal and tumor samples, and x_{gt} denotes the observed response in sample t for peak g, that is x_{gt} is the weight of identified peak g. Subtract the column means, so that $\sum_g x_{gt} = 0$ in each column.

2. Fourier transform each sample of the data, choosing m basis functions. The transformed data matrix is still denoted as X, now with dimension $(m \times N)$.

3. Let $S = (I + \alpha Q)^{-0.5}$, where α is a given smoothing parameter and Q is the derivative operator. Operate by S on the transformed data.

4. For each sample (column) t, perform PCA on the smooth Fourier transform to obtain eigenfunctions $\{v_j\}$ and eigenvalues $\{e_j\}$, leaving one sample (x_s, say) out from the data.

5. Estimate C_{-s} by eigenvalues and eigenfunctions obtained from step 4: $C_{-s} = \sum_j e_j \cdot v_j \cdot v_j^T$.

6. Applying the general idea of quadratic discriminant analysis, define a discriminant score for sample s as follows. Let μ_k ($k = 1, 2$) denote the smoothed mean curves under the

two biologic conditions of interest, and let $d_k = (X_s - \mu_k)^T C_{-s}^{-1} (X_s - \mu_k) + \ln|C_{-s}| - 2\ln\pi_k$, where π_k is a prior probability (as default, use 0.5).

7. Sample s is assigned to the class $k = 1$ if $d_1 < d_2$; otherwise it is assigned to class 2.
8. Calculate the sensitivity and specificity fractions.

The sensitivity of the combined method can be assessed with respect to (1) the choice of (1) basis functions or (2) discriminant function: linear, quadratic (with a full or shrunken covariance matrix), or a kernel-based nonparametric method. One can identify intervals on the functional curve that exhibit extreme variation and use proteins corresponding to these intervals in a subsequent peak selection procedure to identify important peaks.

26.4.5 Wavelet-Based Functional Mixed Model and Application

The methodologies described above consist of two steps: peak detection followed by differential peak comparison. Although conceptually simple, this approach may miss important low-intensity differences between the spectra if the peak detection method is not sufficiently sensitive. In addition, proteomic spectra are notoriously sensitive to sample preparation and handling and the precise environmental conditions under which the spectra are obtained. Even when these factors are not confounded with the outcomes of interest (by good experimental design), modeling their effects is still important.

Morris et al. [31,32] address both of these issues in a single framework by applying Bayesian functional modeling to the spiky proteomic data, represented by wavelet-based functional mixed models. Their method yields posterior estimates for both the overall mean spectrum and the fixed-effect functions, each of which can be used to identify differentially expressed peaks associated with that effect, while adjusting for the potentially nonlinear effects of the other factors.

The application dataset to which Morris et al. applied their method was a SELDI-TOF experiment, conducted at M. D. Anderson Cancer Center, to study proteins in the serum of mice implanted with cancer tumors. Each mouse was implanted with a tumor from one of two cancer cell lines: A375P, a human melanoma cancer cell line with low metastatic potential; or PC3MM2, a highly metastatic human prostate cancer cell line. Each tumor was implanted into one of two organs (brain or lung). Later, serum samples were extracted from the mice and placed on SELDI chips. The primary goals of this study were to assess whether differential protein expression was more tightly coupled to the host organ site or to the donor cell line type, and to identify any protein peaks differentially expressed by organ site, by cell line, and/or by their interaction.

For each mouse, two spectra were obtained, one at low laser intensity and one at high intensity, measuring low- and high-molecular-weight proteins, respectively. Instead of independently modeling the spectra obtained at the different laser intensities, Morris et al. combined information across the two laser intensities and modeled an additional fixed-effect function to account for the change in laser intensity.

Since the experiment involved 16 nude mice with two spectra per mouse, the dataset is represented by 32 spectra, $Y_i(t)$, $i = 1, \ldots, 32$.

The functional mixed model to fit these spectra is given by

$$Y_i(t) = \sum_{j=0}^{4} X_{ij}\beta_j(t) + \sum_{k=1}^{16} Z_{ik}U_k(t) + E_i(t). \qquad (26.2)$$

In the first term, β_j, $j = 0, \ldots, 4$ represent the overall mean spectrum and the four fixed-effect functions and X_{ij} represent covariates that model the correspondence between the samples i and the fixed-effect functions j. Specifically

- $X_{i0} = 1$ and corresponds to the overall mean spectrum $\beta_0(t)$.
- X_{i1} corresponds to the cell line main effect function $\beta_1(t)$ and equals 1 or -1 corresponding to whether the mouse was injected with the A375P or PC3MM2, respectively.
- X_{i2} corresponds to the organ main effect function $\beta_2(t)$.
- $X_{i3} = X_{i1} * X_{i2}$ corresponds to the organ–cell line interaction function $\beta_3(t)$.
- $\beta_4(t)$ models the laser intensity effect, with corresponding covariate $X_{i4} = 1$ or -1 if the spectrum came from low- or high-intensity scans, respectively.

The second term consists of random-effect functions $U_k(t)$ for each mouse, to model the correlation between repeatedly measured spectra. Thus $Z_{ik} = 1$ if and only if spectrum i came from mouse k. The third term models residual errors.

Before fitting this model, the covariance matrices associated with U and E must be constrained. Morris et al. use wavelet shrinkage regularization because it allows more flexible structures that are better suited to spiky proteomic spectra than do earlier models based on smoothing splines. Briefly, a wavelet series approximation consists of a sum of wavelet basis functions over a series of scales J, where larger J represent exponentially coarser levels of detail. For fitting this model, Morris et al. used the Daubechies wavelet with vanishing fourth moments and performed the discrete wavelet transform down to $J = 11$ levels. A modified empirical Bayes procedure with appropriate constraints was used to estimate the regularization parameters that determine the relative tradeoff between variance and bias in the nonparametric estimation. The MCMC procedure was implemented in MATLAB with 1000 burn-in iterations, followed by 20,000 iterations, retaining every tenth iteration for final analyses.

Figure 26.4 contains the posterior means and 95% posterior credible bands for the organ and cell line main effect functions, the interaction function, and the laser intensity effect function. The organ main effect function $\beta_1(x)$ can be interpreted as the difference between the mean spectra for lung and brain-injected animals at m/z value x, after adjusting for the functional effects of cell line, cell line–organ interaction, and laser intensity. The spiky nature of these fixed-effect functions indicate that differences in spectra between treatment groups are localized, and highlights the importance of using adaptive regularization methods with these data. To identify locations within the curves at which there is strong evidence of significant effects, one may employ plots of pointwise posterior probabilities $\Pr(\beta_j(x) > 0|\mathbf{Y})$.

Peak detection can be performed by using the posterior mean estimate of the overall mean spectrum $\beta_0(t)$. Morris et al. first applied the first difference operator ∇ to the regularized estimate of the mean spectrum $\Gamma_0(t) = \nabla\beta_0(t) = \beta_0(t + 1) - \beta_0(t)$. A location t was considered to be a *peak* if its first difference and the first difference immediately preceding it were positive ($\Gamma_0(t - 1) > 0$ and $\Gamma_0(t) > 0$) and the first differences for the two locations immediately following it were negative ($\Gamma_0(t + 1) < 0$ and $\Gamma_0(t + 2) < 0$). This condition ensured that this location was a local maximum, and the left and right slopes of the peak were monotone for at least two adjacent points.

This procedure yields a small subset of peaks from the total number of observations within the spectrum. The resulting peaks can be tentatively identified by searching for their estimated m/z values in TagIdent, a database (available at http://us.expasy.org/tools/tagident.html) containing the molecular masses and pH for proteins observed in a variety of species. It is

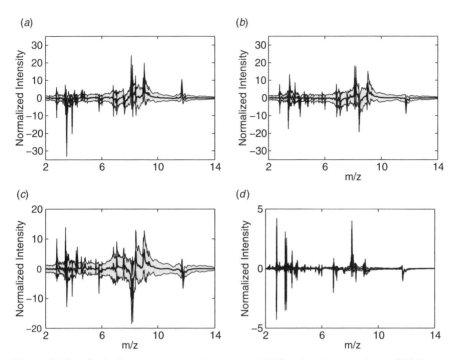

Figure 26.4 *Fixed-effect curves*: posterior mean and 95% pointwise posterior credible bands for (*a*) organ main effect function, $\beta_1(x)$; (*b*) cell line main effect function, $\beta_2(x)$; (*c*) organ–cell line interaction function, $\beta_3(x)$; (*d*) laser intensity effect function, $\beta_4(x)$.

possible to search for proteins emanating from either the source (human) or the host (mouse) whose molecular masses are within the estimated mass accuracy (0.3%) of the SELDI instrument from the peak concerned. If required, each specific peak can be definitively identified by performing an additional MS/MS experiment.

In summary, the Bayesian wavelet-based functional mixed model may be used to model proteomic spectra since it results in adaptive regularization of the fixed-effect functions, avoids attenuation of the effects at the peaks, and is reasonably flexible in modeling the between-curve covariance structures, accommodating autocovariance structures induced by peaks and heteroscedasticity allowing different between-spectrum variances for different peaks.

26.4.6 A Nonparametric Bayesian Model Based on Kernel Functions

Clyde et al. [8] describe nonparametric statistical models for spectra that permit simultaneous filtering of noise and removal of baseline trends in conjunction with peak identification, quantification, and, ultimately, classification.

The model is described for a single spectrum, which may be either a raw spectrum or the average of spectra from several laser shots or individuals. The raw data consist of a time series of intensities of ions striking the detector at recorded time intervals (each clock tick is 4 ns). Clyde et al. [8] develop the model for intensity as a TOF function rather than with m/z, denoting the observed intensity measurement at observed TOF $t \in T \equiv [t_0, t_n]$ as Y_t, with expected intensity $E[Y_t]$ given by the function $f(t)$. The mean intensity is expressed through a linear combination of kernel functions

$$f(t) = b(t) + \sum_{j=1}^{J} k(t; \tau_j, \omega_j)\gamma_j, \qquad (26.3)$$

where $k(t;\tau;\omega)$ is proportional to a nonnegative density, such as a Gaussian or Cauchy kernel, and $b(t)$ represents the systematic background process; J represents the number of kernels or peaks or proteins in the spectrum. The kernel basis functions can incorporate variations in peak shapes, including spikiness in the time domain. We may interpret τ_j as the expected TOF for protein j; γ_j (the area under the curve) as a measure of the concentration or abundance of protein j; the parameter ω_j controls the width of peaks. A specific mass spectrometer is characterized by its resolution, which directly affects one's ability in determining whether a given peak in the spectrum corresponds to a single protein or more. For a symmetric single ion peak with expected TOF τ_j, the resolution ρ_j at peak j is defined as

$$\rho_j = \frac{\tau_j}{\Delta\tau_j} \qquad (26.4)$$

where $\Delta\tau_j$ is the full peak width at 50% of the maximum height [41]. Available prior information about resolution can be translated into prior knowledge about ω_j, which will aid in resolving the number of proteins in a peak.

Prior functions need to be proposed for subsequent posterior inference about the unknown function f. For example, one may choose any positive number $\nu^+ > 0$ and let $J \sim \text{Poisson}(\nu^+)$ with mean ν^+. Conditionally on J, let $(\gamma_j, \tau_j, \omega_j)$ be i.i.d. following $\pi(d\gamma, d\tau, d\omega)$, where π is a probability distribution on $\mathbb{R}^+ \times \mathbb{T} \times \mathbb{R}^+$. Thus $f(t)$ can be represented as a convolution of kernels

$$f(t) = b(t) + \int_{\mathbb{T} \times \mathbb{R}^+} k(t; \tau, \omega)\Gamma(d\tau, d\omega), \qquad (26.5)$$

where

$$\Gamma(d\tau, d\omega) = \sum_{j=1}^{J} \gamma_j \delta_{\tau_j}(d\tau)\delta_{\omega_j}(d\omega)$$

is a discrete random Borel measure on $\mathbb{T} \times \mathbb{R}^+$ with a random number J jumps of random heights γ_j at the random points (τ_j, ω_j).

The gamma random field can be used to construct a prior for $f(t)$, ensuring that prior beliefs are specified coherently across all possible partitions of time of flight. The joint distribution for τ_j and ω_j may be specified by (1) taking τ_j uniform over $\mathbb{T}$ when there is no prior information on the distribution of the mass/charge of expected proteins and (2) utilizing existing databases of proteins and associated masses to construct a more informative prior for τ_j.

A hierarchical representation of the priors is as follows:

$$f(t) = \beta\left\{\beta_0 + k_b(t, \tau_0, \omega_0)\gamma_0 + \sum_{j=1}^{J} k(t, \tau_j, \omega_j)\gamma_j\right\},$$

where

$$
\begin{aligned}
\gamma_j \mid J, \epsilon \ \text{i.i.d truncated Gamma}(0, 1; \epsilon) && \text{for} \quad j = 0, \ldots, J, \\
\tau_j \mid J \ \text{i.i.d uniform}(\mathbb{T}) && \text{for} \quad j = 1, \ldots, J, \\
\rho_j \mid J \ \text{i.i.d lognormal}(\varrho, 0.05) && \text{for} \quad j = 1, \ldots, J, \\
\omega_j = g_k(\tau_j, \rho_j) && \text{for} \quad j = 1, \ldots, J, \\
J \mid \alpha, \epsilon \sim \text{Poisson}(\nu_\epsilon^+(\alpha, 1)). &&
\end{aligned}
$$

The next step is to specify an appropriate likelihood in order to conduct posterior inference about parameters in the model. Exploratory analysis can aid in subsequent modeling of the intensities, for example, a gamma model for intensities Y_t. The posterior distribution of all unknowns is proportional to the likelihood of the data based on the gamma model hierarchy. Since marginal posterior distributions for most quantities of interest are not available analytically, investigators can resort to simulation of the posterior distribution via a RJMCMC algorithm [14], implemented by birth/death or split/merge moves. Efficient computation is possible since the specification of the likelihood described above can avoid the inversion of large matrices that often arise in Gaussian approaches and memory requirements can be economized by computing the kernels only as needed.

In summary, the kernel model of Clyde et al. [8] can capture the asymmetry features of MS peaks and can differentiate between single-protein peaks versus multiple-protein peaks. The tradeoff for this flexibility is that investigators will need to specify (1) prior information on resolution, particularly for peaks that are wider than expected, suggesting that multiple kernels or proteins will be required for a good model fit; and (2) the minimum peak size to prevent overfitting.

26.5 POTENTIAL STATISTICAL DEVELOPMENTS

A future goal in statistical developments at M. D. Anderson Cancer Center is to develop inference for protein mass/charge spectra that includes protein identity as part of the model. Models can be developed for protein mass/charge spectra that replace stepwise inference by joint inference in a single model. In particular, this includes baseline subtraction, denoising, peak detection, comparison across biologic conditions, and identification of peaks as specific proteins. The latter is a critical bottleneck of current methods. The problem is common to many applications of mixture models; one aspect is known as the *mode-matching problem*. The problem arises whenever it is desired to make inference about specific terms in a mixture model. The problem is easiest described in the context of summarizing MCMC output. Assume that we want to report inference on unique terms in the mixture (26.1) and the corresponding locations ϵ_g. Trying to report ergodic averages of ϵ_g across MCMC iterations, it is unclear how different terms should be matched across iterations. When should two imputed values ϵ_g be counted as arising from the same protein, and when should they be considered two separate proteins [19]? The problem can be avoided by explicitly including protein identity in the model, by defining a prior on ϵ_g as a distribution on the list of, for example, human proteins (e.g., see http://us. expasy.org). The approach cannot deterministically resolve questions of allocating a specific peak to two proteins with very similar masses. However, it will attach appropriate probabilities to possible ways of allocation. Typically, known functions and descriptions allow us to set informative weights; otherwise, uniform prior weights may be used.

Current work by Kim-Anh Do and Peter Mueller can be described in the context of building the model for two-group comparison, the primary goal of the experiment is to identify proteins with differential abundance under $x = 0, 1$. Recall the notation from Section 26.4.3; the observed spectrum for sample t at mass grid m_i is denoted y_{ti}; the mean spectrum, without sampling noise, is

$$p_t = (1 - p_{0t})f_t + p_{0t}B_t,$$

consisting of a sum of protein peaks f_t and a baseline B_t. The protein peaks f_t are represented as a mixture of Beta kernels (26.1); indicator parameters $\lambda_g = I(w_{g0} = w_{g1})$ record whether the gth protein is equally abundant under both biologic conditions. Assuming a multinomial sampling model, the observed counts y_{ti} are considered as a histogram of samples generated from the unknown mean spectrum p_t. Alternatively, one could consider a regression likelihood by assuming that the observed y_{ti} arise as the true p_t plus a random residual noise. The latter is implicitly assumed in methods that define some variation of smoothing. The former more closely corresponds to an idealized description of the experimental setup. Conditional on λ_g, a Dirichlet prior can be defined for the weights w_{gx}. Let $\Gamma_1 = \{g : \lambda_g = 1\}$ denote the set of nondifferentially expressed proteins, and let $W_1 = \Sigma_{\Gamma_1} w_{g0}$ and $W_0 = 1 - W_1$. Independent Dirichlet priors can be assumed for $(w_{g0}/W_1; g \in \Gamma_1)$ and $(w_{gx}/W_0; g \notin \Gamma_1)$, $x = 0$ and 1; that is, for the subsets of weights for nondifferentially and differentially expressed proteins, rescaled to sum to one. Posterior inference can be carried out by posterior MCMC simulation. Changing G, and changing λ_g the dimension of the parameter vector, requiring RJMCMC. Some moves, including changing a single w_{gx} and changing ϵ_g in the generalization with multiple charges, require deterministic adjustment of other parameters when proposing a change in one parameter, but involve no dimension change. For example, when a change in w_{gx} is proposed, the remaining $w_{g'x}$ need to be rescaled to maintain the unit sums. MCMC transition probabilities can be developed that include a deterministic transformation. The nature of the transition probability is like RJMCMC, but without the dimension change and auxiliary variables. The transition probabilities are straightforward, but to our knowledge current literature does not include any discussion of such moves. Hierarchical extensions that explicitly allow for sample-to-sample variations can also be considered. The Dirichlet prior on the weights gives rise to awkward implications in the implementation of the MCMC. These could be avoided by dropping the conditioning on the total number of events, that is, by replacing the multinomial sampling model with a Poisson sampling model. The weights would then be replaced by Poisson means.

Another future goal is to develop a centered model: a simplified, computationally efficient version of the model in Section 26.4.3. Recall the notation y_{ti} denoting the recorded count in sample t for mass m_i, and $f_t(m)$ and $B_t(m)$ denoting the cleaned protein spectrum and the baseline for sample t, evaluated at mass/charge m. Let $\bar{y}_{xi}$ denote the average spectrum under condition x, and let $\bar{y}_i = \frac{1}{2}(\bar{y}_{0i} + \bar{y}_{1i})$ denote the average across conditions. Now consider the transformed data $y_t^* = y_t - \bar{y}$, where $\lambda_g = I(w_{0g} = w_{1g})$ is an indicator for nondifferential expression of protein g. Assuming a representation of f_t as in (26.1), in the centered data, terms corresponding to peaks arising from nondifferentially expressed proteins cancel out, and we are left with modeling only differentially expressed genes. A model similar to (26.1) can be specified

$$f_t^*(m) = \sum_{g=1}^{G^*} w_{xg}^* \, \mathrm{Beta}(m; \epsilon_g, \alpha_g),$$

where G^* is the number of proteins with differential abundance. The terms in f_t^* now correspond only to proteins that are differentially expressed. Further, for the transformed data it is more convenient to use a regression likelihood, with mean function f_t^*. The weights for differentially expressed proteins g under the two conditions $x = 0, 1$ show up with opposite signs, $w_{0g}^* = -w_{1g}^*$.

26.6 CONCLUDING REMARKS

In this chapter, we have seen that, relative to the components of the normal proteome, potential proteomic biomarkers are expected to be present in extremely small amounts. Consequently, their identification—specifically, separating true differences between cases and controls from normal sample variability and from minute variations in sample handling—is a significant statistical challenge. In this chapter, we reviewed the statistical preprocessing and analysis of proteomic datasets, with an emphasis on the analytic rigor required. We also reviewed several more recent statistical methods for the analysis of proteomic data, including Bayesian functional data approaches.

In the near future, molecular profiling will likely become a routine adjunct to the pathological reporting of most human diseases such as cancer. The challenge of translating proteomic pathway profiling to the bedside is to develop technology that can efficiently use small volumes of tumor tissue, routinely obtained at biopsy, to assess multiple cell signaling pathways simultaneously. Future research in mass spectrometry, driven by the more recent National Institute of Health initiative in clinical proteomics, will focus on multidisciplinary consortia to develop improved methods in standardization, reproducibility, and comparability within and among research institutions. In particular for mass spectrometry, new methods and platforms such as liquid chromatography mass spectrometry and protein arrays will be further investigated.

For the full potential for these new methods and platforms to be fully realized, new statistical methods will need to be developed. Additional intricate and difficult issues to study will also arise from the complex interactions within the human proteome, including multiple posttranslational protein modifications.

ACKNOWLEDGMENT

Kim-Anh Do was partially supported by NIH University of Texas SPORE in Prostate Cancer grant CA90270 and University of Texas SPORE in Breast Cancer grant CA116199.

REFERENCES

1. Adam, B. L., Qu, Y., Davis, J. W., Ward, M. D., Clements, M. A., Cazares, L. H., Semmes, O. J., Schellhammer, P. F., Yasui, Y., Feng, Z., and Wright, G. L., Jr., Serum protein fingerprinting coupled with a pattern-matching algorithm distinguishes prostate cancer from benign prostate hyperplasia and healthy men, *Cancer Res.* **62**, 3609–3614 (2002).

2. Aebersold, R. and Mann, M., Mass spectrometry-based proteomics, *Nature* **422**, 198–207 (2003).

3. Baggerly, K. A., Morris, J. S., Wang, J., Gold, D., Xiao, L. C., and Coombes, K. R., A comprehensive approach to analysis of MALDI-TOF proteomics spectra from serum samples, *Proteomics* **3**, 1667–1672 (2003).

4. Baggerly, K. A., Morris, J. S., and Coombes, K. R., Reproducibility of SELDI-TOF protein patterns in serum: Comparing datasets from different experiments, *Bioinformatics* **20**(5), 777–785 (2004).

5. Benjamini, Y. and Hochberg, Y., Controlling the false discovery rate: A practical and powerful approach to multiple testing, *J. Roy. Statist. Soc. B* **57**, 289–300 (1995).

6. Billheimer, D., A functional data approach to MALID-TOF MS protein analysis (unpublished manuscript).

7. Birkner, M. D., Hubbard, A. E., van der Laan, M. J., Skibola, C. F., Hegedus, C. M., and Smith, M. T., Issues of processing and multiple testing of SELDI-TOF MS proteomic data, *Statist. Appl. Genet. Molec. Biol.* **5**(1) (2006).

8. Clyde, M., House, L., and Wolpert, R., Nonparametric models for proteomic peak identification and quantification, in Do, K.-A., Müller, P., and Vannucci, M., eds., *Bayesian Inference for Gene Expression and Proteomics*, Cambridge Univ. Press, 2006, pp. 293–308.

9. Conrads, T. P., Fusaro, V. A., Ross, S., Johann, D., Rajapakse, V., Hitt, B. A., Steinberg, S. M., Kohn, E. C., Fishman, D. A., Whitely, G., Barrett, J. C., Liotta, L. A., Petricoin III, E. F., and Veenstra, T. D., High-resolution serum proteomic features for ovarian cancer detection, *Endocr. Related Cancer* **11**, 163–178 (2004).

10. Coombes, K. R., Tsavachidis, S., Morris, J. S., Baggerly, K. A., Hung, M.-C., and Kuerer, H. M., Improved peak detection and quantification of mass spectrometry data acquired from surface-enhanced laser desorption and ionization by denoising spectra with the undecimated discrete wavelet transform, *Proteomics* **5**(16), 4107–4117 (2005).

11. Do, K.-A., Müller, P., and Tang, F., A bayesian mixture model for differential gene expression, *J. Roy. Statist. Soc. C* **54**(3), 627–644 (2005).

12. Genovese, C. and Wasserman, L., A stochastic process approach to false discovery control, *Ann. Statist.* **32**(3), 1035–1061 (2004).

13. Genovese, C. R. and Wasserman, L., Exceedance control of the false discovery proportion, *J. Am. Statist. Assoc.* **101**(476), 1408–1417 (2006).

14. Green, P. J., Reversible jump Markov chain Monte Carlo computation and Bayesian model determination, *Biometrika* **82**, 711–732 (1995).

15. Hall, P., Poskitt, D. S., and Presnell, B., A functional data-analytic approach to signal discrimination, *Technometrics* **43**, 1–9 (2001).

16. Hamming, H., *Digital Filters*, 2nd ed., Prentice-Hall, Englewood Cliffs, NJ, 1983.

17. Hastie, T., Buja, A., and Tibshirani, R., Penalized discriminant analysis, *Ann. Statist.* **23**, 73–102 (1995).

18. Hillenkamp, F., Karas, M., Beavis, R. C., and Chait, B. T., Matrix-assisted laser desorption/ionization mass spectrometry of biopolymers, *Anal. Chem.* **63**, 1193A–1203A (1991).

19. Holmes, C. C., Jasra, A., and Stephens, D. A., Markov chain monte carlo methods and the label switching problem in bayesian mixture modeling, *Statist. Sci.* **20**, 50–67 (2005).

20. Hutchens, T. W. and Yip, T.-T., New desorption strategies for the mass spectrometric analysis of macromolecules, *Rapid Commun. Mass Spectrom.* **7**(7), 576–580 (1993).

21. Petricoin III, E. F., Ardekani, A. M., Hitt, B. A., Levine, P. J., Fusaro, V. A., Steinberg, S. M., Mills, G. B., Simone, C., Fishman, D. A., Kohn, E. C., and Liotta, L. A., Use of proteomic patterns in serum to identify ovarian cancer, *The Lancet* **359**, 572–577 (2002).

22. Petricoin III, E. F., Ornstein, D. K., Paweletz, C. P., Ardekani, A., Hackett, P. S., Hitt, B. A., Velassco, A., Trucco, C., Wiegand, L., Wood, K., Simone, C. B., Levine, P. J., Linehan, M., Emmert-Buck, M. R., Steinberg, S. M., Kohn, E. C., and Liotta, L. A., Serum proteomic patterns for detection of prostate cancer, *J. Natl. Cancer Inst.* **94**(20), 1576–1578 (2002).

23. Jeffries, N., Algorithms for alignment of mass spectrometry proteomic data, *Bioinformatics* **21**, 3066–3073 (2005).

24. Katz, J. E., Mallick, P., and Agus, D. B., A perspective on protein profiling of blood, *BJU Int.* **96**, 477–482 (2005).

25. Keselman, H. J., Cribbie, R., and Holland, B., Controlling the rate of type I error over a large set of statistical tests, *Br. J. Math. Statist. Psychol.* **55**, 27–39 (2002).

26. Lehmann, E. L. and Romano, J. P., Generalizations of the familywise error rate, *Ann. Statist.* **33**(3), 1138–1154 (2005).

27. Leurgans, S., Moyeed, R., and Silverman, B., Canonical correlation analysis when the data are curves, *J. Roy. Statist. Soc. B* **55**, 725–740 (1993).

28. Li, J., Zhang, Z., Rosenzweig, J., Wang, Y. Y., and Chan, D. W., Proteomics and bioinformatics approaches for identification of serum biomarkers to detect breast cancer, *Clin. Chem.* **48**(8), 1296–1304 (2002).

29. Malyarenko, D. I., Cooke, W. E., Adam, B.-L., Malik, G., Chen, H., Tracy, E. R., Trosset, M. W., Sasinowski, M., Semmes, O. J., and Manos, D. M., Enhancement of sensitivity and resolution of surface-enhanced laser desorption/ionization time-of-flight mass spectrometric records for serum peptides using time-series analysis techniques, *Clin. Chem.* **51**, 65–74 (2005).

30. McShane, L. M., Altman, D. G., Sauerbrei, W., Taube, S. E., Gion, M., and Clark, G. M., Reporting recommendations for tumor marker prognostic studies (REMARK), *J. Natl. Cancer Inst.* **97**, 1180–1184 (2005).

31. Morris, J. S., Brown, P. J., Baggerly, K. A., and Coombes, K. R., Analysis of mass spectrometry data using Bayesian wavelet-based functional mixed models, in Do, K.-A., Müller, P., and Vannucci, M., eds., *Bayesian Inference for Gene Expression and Proteomics*, Cambridge University Press, 2006, pp. 269–292.

32. Morris, J. S., Brown, P. J., Herrick, R. C., Baggerly, K. A., and Coombes, K. R., *Bayesian Analysis of Mass Spectrometry Proteomics Using Wavelet Based Functional Mixed Models*, UT MD Anderson Cancer Depatment of Biostatistics and Applied Mathematics Working Paper Series, Working Paper 22, (http://www/bepress.com/mdandersonbiostat/paper22), March 2006.

33. Morris, J. S., Coombes, K. R., Koomen, J., Baggerly, K. A., and Kobayashi, R., Feature extraction and quantification for mass spectrometry data in biomedical applications using the mean spectrum, *Bioinformatics* **21**(9), 1764–1775 (2005).

34. Müller, P., Do, K.-A., Bandyopadhyay, R., and Baggerly, K., *A Bayesian Mixture Model for Protein Biomarker Discovery*, Technical Report, M. D. Anderson Cancer Center, 2006.

35. Rai, A. J., Zhang, Z., Rosenzweig, J., Shih, I. M., Pham, T., Fung, E. T., Sokoll, L. J., and Chan, D. W., Proteomic approaches to tumor marker discovery: Identification of biomarkers for ovarian cancer, *Arch. Pathol. Lab. Med.* **126**, 1518–1526 (2002).

36. Ramsay, J. O. and Silverman, B. W., *Functional Data Analysis*, Springer-Verlag, New York, 1997.

37. Ransohoff, D. F., Bias as a threat to the validity of cancer molecular-marker research, *Nature Rev. Cancer* **5**(2), 142–149 (2005).

38. Reiner, A., Yekutieli, D., and Benjamini, Y., Identifying differentially expressed genes using false discovery rate controlling procedures, *Bioinformatics* **19**, 368–375 (2003).

39. Savitzky, A. and Golay, M. J. E., Smoothing and differentiation of data by simplified least squares procedures, *Anal. Chem.* **36**, 1627–1639 (1964).

40. Scheid, S. and Spang, R., A false discovery rate approach to separate the score distributions of induced and non-induced genes, in *Proc. 3rd Int. Workshop on Distributed Statistical Computing*, 2003.

41. Siuzdak, G., *The Expanding Role of Mass Spectrometry in Biotechnology*, MCC Press, 2003.

42. Storey, J. D. and Tibshirani, R., *Estimating False Discovery Rates under Dependence, with Applications to DNA Microarrays*, Technical Report, Stanford Univ., 2001.

43. Storey, J. D., The positive false discovery rate: A bayesian interpretation and the q-value, *Ann. Statist.* **31**(6), 2013–2035 (2003).

44. Tibshirani, R., Hastie, T., Narasimhan, B., Soltys, S., Shi, G., Kong, A., and Le, Q.-T., Sample classificationfrom protein mass spectrometry, by "peak probability contrasts," *Bioinformatics* **20**, 3034–3044 (2004).

45. van der Laan, M. J., Birkner, M. D., and Hubbard, A. E., Resampling based multiple testing procedure controlling tail probability of the proportion of false positives, *Statist. Appl. Genet. Molec. Biol.* **4**(1) (2005).

46. Vlahou, A., Schellhammer, P. F., Mendrinos, S., Patel, K., Kondylis, F. I., Gong, L., Nasim, S., and Wright, G. L., Jr., Development of a novel proteomic approach for the detection of transitional cell carcinoma of the bladder in urine, *Am. J. Pathol.* **154**, 1491–1502 (2001).

47. Wiley, W. C. and McLaren, I. H., Time of flight mass spectrometer with improved resolution, *Rev. Sci. Instrum.* **26**, 1150–1157 (1955).

48. Wu, B., Abbott, T., Fishman, D., McMurray, W., More, G., Stone, K., Ward, D., Williams, K., and Zhao, H., Comparison of statistical methods for classification of ovarian cancer using mass spectrometry data, *Bioinformatics* **19**(13), 1636–1643 (2003).

49. Zhang, Z., Jr., Bast, R. C., Yu, Y., Li, J., Sokoll, L. J., Rai, A. J., Rosenzweig, J. M., Cameron, B., Wang, Y. Y., Meng, X.-Y., Berchuck, A., van Haaften-Day, C., Hacker, N. F., de Bruijn, H. W. A., van der Zee, A. G. J., Jacobs, I. J., Fung, E. T., and Chan, D. W., Three biomarkers identified from serum proteomic analysis for the detection of early stage ovarian cancer, *Cancer Res.* **64**, 5882–5890 (2004).

50. Ziegler, H., Properties of digital smoothing polynomial (DISPO) filters, *Appl. Spectrosc.* **35**(1), 88–92 (1981).

CHAPTER 27

Genetic Mapping of Quantitative Traits: Model-Free Sib-Pair Linkage Approaches

Saurabh Ghosh and Partha P. Majumder
Human Genetics Unit, Indian Statistical Institute, Kolkata, India

27.1 INTRODUCTION

Most common genetic disorders such as cardiovascular disease, type 2 diabetes, and asthma are complex in nature. Unlike simple Mendelian disorders (e.g., cystic fibrosis, which can be explained by mutations in a single gene), complex disorders are controlled by multiple loci, each with minor gene effects and possibly interacting epistatically. Thus, no single locus is either necessary or sufficient to explain the pathogenesis of a complex disorder. Hence, compared to Mendelian disorders, it is statistically more challenging to develop methods to identify loci that control complex disorders.

With the availability of densely spaced polymorphic DNA markers, there has been an active interest in developing statistical methods for linkage mapping of complex traits using data on nuclear families or larger pedigrees. Linkage analyses identify regions on the genome that exhibit increased sharing of alleles among relatives with similar phenotypes. Most complex disorders have binary endpoints defined by the affectation status of an individual. Thus, the phenotypic variation between individuals is minimal. For example, in a genetic study on type 2 diabetes, all individuals diagnosed with the disease were indistinguishable in the analyses. However, these clinical endpoint traits are typically determined on the basis of values of a set of heritable quantitative characters. For example, increased body mass index (BMI), fasting blood sugar levels, and insulin levels are precursors of type 2 diabetes. These precursor variables are continuous in nature and contain more information on interindividual variablity than binary clinical endpoints. Thus, it has been argued that analyzing quantitative precursors,

Statistical Advances in the Biomedical Sciences, edited by Atanu Biswas, Sujay Datta,
Jason P. Fine, and Mark R. Segal
Copyright © 2008 John Wiley & Sons, Inc.

including mapping genes controlling them, may be statistically a more powerful strategy for deciphering the architecture of a complex disease.

Binary traits can be completely modeled using allele frequencies and genotypic penetrances, defined as the conditional probabilities of affectation given the genotypes. However, quantitative traits require a stronger layer of modeling: the probability distribution of the trait values. There are currently two popular classes of statistical methods for mapping quantitative trait loci (QTL): (1) Haseman–Elston regression [12] and its extensions implemented in the statistical package SAGE, and (2) likelihood-based methods such as variance components [3,1] implemented in the statistical package SOLAR and score tests [21,22]. The first class of methods is essentially distribution-free; that is, it does not require specification of the probability distribution of trait values, but can use only small families. The second class of methods can use data on large pedigrees but is parametric in nature; that is, it requires strong statistical assumptions, such as multivariate normality of trait values for different members of a pedigree. The statistical tradeoff between the two classes of methods is that while the parametric methods are statistically more powerful when the distributional assumptions are valid, the distribution-free methods are more robust to violations in the underlying assumptions. However, it is often very difficult to verify the distributional assumptions, particularly for high-dimensional data. Hence, model-free methods have gained popularity given that the data demand is less and the probability of a false-positive inference is low.

The focus of this chapter is restricted to analyses of sib-pair data. We discuss the basic framework of the distribution-free linkage approaches motivated by the classical Haseman–Elston regression [12] for independent sib-pairs and show that even the Haseman–Elston approach and its extensions (e.g., the incorporation of relative pairs other than sibs [2], simultaneous analyses of data on different types of relative pairs [14], multipoint interval mapping [13], squared sums instead of squared differences [6], a weighted combination of squared sums and squared differences [24], population mean-corrected cross-products based on larger sibships [7], and a reverse regression strategy [20]) make certain model assumptions, which can be relaxed to improve on linkage inferences. Although these methods do not make any prior assumptions on the probability distribution of the quantitative trait values, they are not nonparametric in the strict statistical sense because they assume a linear relationship between a suitable function of the quantitative traits of a relative pair and the estimated identity-by-descent (i.b.d.) score at a marker locus. However, the true nature of relationship between these variables is governed by various biological parameters such as the recombination distance between the QTL and the marker locus, interference, and dominance at the QTL. Hence, it is of interest to explore for appropriate nonparametric alternatives that are robust with respect to these parameters. In particular, we propose a modified version of the nonparametric regression approach suggested by us earlier [11] and compare its performance, using Monte Carlo simulations, with some of the methods belonging to the Haseman–Elston class of regression. We also discuss one application [10] of the proposed nonparametric regression to real data.

27.2   THE BASIC QTL FRAMEWORK FOR SIB-PAIRS

The smallest family structure appropriate for linkage analyses is a nuclear family comprising a pair of parents and two offspring. Linkage studies based on such family structures are called *independent sib-pair analyses*. The basic QTL paradigm for sib-pairs is that siblings inheriting similar QTL alleles from their parents will have similar quantitative trait values. Since data are available at marker loci and not at the QTL, the paradigm extends to any marker locus that is linked to the QTL, that is, physically close to the QTL. On the other hand, marker loci unlinked to the QTL will not exhibit such property. This forms the basis of the linkage tests. For parametric sib-pair linkage methods, the allelic similarity is modeled via the genotypic probabilities of the parents and the transmission probabilities of the parental alleles to the two sibs. For nonparametric sib-pair methods, the allelic similarity is measured by a parameter termed the *i.b.d. score*, defined as the proportion of parental alleles shared by the two siblings.

27.3 THE HASEMAN–ELSTON REGRESSION FRAMEWORK

Assume that a quantitative trait Y is controlled by an autosomal biallelic locus with alleles A_1 and a_1. No assumption is made as to the nature of the probability distribution of the trait values. The underlying population is assumed to be in Hardy–Weinberg equilibrium with respect to the trait locus. Suppose that $\{(y_{j1}, y_{j2}) : j = 1, 2, \ldots, n\}$ are the quantitative trait values of n sib-pairs. It is assumed that the conditional expectations of Y given the genotypes A_1A_1, A_1a_1 and a_1a_1 are α, β, and $-\alpha$, respectively, the variance of Y conditioned on any genotype is equal, σ^2, and the correlation coefficient between the trait values of any sib-pair is ρ. The parameter β is known as the *dominance parameter*. If β is equal to 0 (i.e., the expectation of the quantitative trait values conditioned on the heterozygous genotype A_1a_1 is equal to the mean of the expectations conditioned on the two homozygous genotypes A_1A_1 and a_1a_1), we say that there is no dominance at the QTL. Define $y_j = (y_{j1} - y_{j2})^2$, that is, the squared difference in trait value for the jth sib-pair. We note here, that y_j is not affected by location shifts of Y, and hence the structure of the conditional expectations of Y given the different QTL genotypes can be assumed without loss of generality.

Suppose that we have genotype data on M ordered marker loci located on the same chromosome. Let π_{mj} denote the proportion of alleles shared i.b.d. at the mth marker locus for the jth sib-pair, $m = 1, 2, \ldots, M; j = 1, 2, \ldots, n$. If $f_{mj}^{(r)}$ denotes the probability that the jth sib-pair shares r alleles i.b.d. at the mth marker locus, $r = 1, 2$, then the estimator of π_{mj} is given by $\hat{\pi}_{mj} = f_{mj}^{(2)} + \frac{1}{2}f_{mj}^{(1)}$. Haseman and Elston [12] have explicitly calculated $f_{mjil}^{(r)}$ for different parental mating types and in the case of missing parental information, they have suggested an algorithm considering phenosets [5]. Haseman and Elston [12] showed that, if $\beta = 0$ (i.e., if there is no dominance at the trait locus), the conditional expectation of y_j given $\hat{\pi}_{mj}$ (i.e., the expected square difference in sib-pair quantitative trait values conditioned on the estimated i.b.d. score at a marker locus) is $\beta_0 + \beta_1\hat{\pi}_{mj}$, where $\beta_1 = c_0 \times (1 - 2\theta)^2$; c_0 is a negative constant and θ is the recombination fraction between the QTL and the marker locus. Hence, $\beta_1 = 0$ implies $\theta = 0.5$ (i.e., the marker is not linked to the QTL) and $\beta_1 < 0$ implies $\theta < 0.5$ (i.e., the marker is linked to the QTL). Thus, the test for linkage between the QTL and a marker locus can be performed on the basis of a linear regression of y_j on $\hat{\pi}_{mj}$. With increase in doinance at the QTL, the overlap between the distributions of trait values for the major homozygous genotype and the heterozygous genotype increases and leads to departures from the abovementioned linear regression of y_j on $\hat{\pi}_{mj}$. Thus, the power of this test decreases with increase in dominance at the QTL.

27.4 NONPARAMETRIC ALTERNATIVES

Because the Haseman–Elston method, in spite of being distribution-free, is adversely affected by the presence of dominance at the QTL, we [11] proposed a two-stage nonparametric approach. In the first stage, a statistic based on Spearman's rank correlation between y_j and $\hat{\pi}_{mj}$ [17] was used on coarsely spaced markers to identify regions showing significant correlation between the variables; in the second stage, a nonparametric regression based on kernel smoothing [19] was used to empirically estimate the nature of functional relationship between the two variables using data on densely spaced markers in regions identified in the first stage. The marker interval yielding the lowest residual sum of squares in the nonparametric regression was identified in the second stage, but no statistical test was performed. However, we found that since the computation of the rank correlation involves summarization of the squared sib-pair trait differences into ranks, the test is not very powerful and leads to a high rate of false negatives. This prompted us to explore whether we could develop a method based entirely on the nonparametric regression step. Moreover, with the availability of densely spaced polymorphic markers such as single-nucleotide polymorphisms (SNPs), it is more meaningful to use multipoint i.b.d. scores (i.e., use simulataneous information on all the available markers) instead of two-point scores as used in our original method.

27.5 THE MODIFIED NONPARAMETRIC REGRESSION

Following Ghosh and Majumder [11], we assume a nonparametric regression model of y_j and $\hat{\pi}_{jp}$ at any arbitrary point p on the genome

$$y_j = \psi(\hat{\pi}_{jp}) + e_j; \quad j = 1, 2, \ldots, n,$$

where ψ is a real-valued function of $\hat{\pi}_{jp}$ and the e_j terms are homoskedastic random errors with mean 0.

The estimated i.b.d. score at any arbitrary point p of the genome $\hat{\pi}_{jp}$ is a multipoint score based on genotypic information simultaneously on all available markers on that chromosome. We use the linear regression method of Fulker et al. [9] to estimate the multipoint i.b.d. scores.

The functional form of ψ is estimated using a kernel-smoothing technique [19]. The kernel function used is

$$\kappa(t) = \begin{cases} \frac{3}{4}(1 - t^2), & \text{if } |t| < 1; \\ 0 & \text{otherwise} \end{cases} \tag{27.1}$$

The estimator of y_j is given by:

$$\hat{y}_j = \hat{\psi}(\hat{\pi}_{jp})$$

$$= \frac{\sum_{i=1}^{n} \kappa\left(\dfrac{\hat{\pi}_{jp} - \hat{\pi}_{ip}}{h}\right) y_i}{\sum_{i=1}^{n} \kappa\left(\dfrac{\hat{\pi}_{jp} - \hat{\pi}_{ip}}{h}\right)},$$

where h is the "optimal" window length in the kernel smoothing procedure obtained by minimizing the residual sum of squares in the preceding regression. Since nonparametric regression tends to overfit data, we use a "leave one out" technique for computing the residual sum of squares [i.e., leave out y_j when predicting $\psi(\hat{\pi}_{jp})$].

27.5.1 Evaluation of Significance Levels

Because of the complex asymptotic behavior of the nonparametric regression estimator based on kernel smoothing, it is difficult to assess the performance of the proposed estimator in detecting linkage. For this purpose, we develop a diagnostic measure using an analog of R^2, the square of the correlation coefficient between the response variable and the explanatory variable. For linear regression, we know that

$$R^2 = \frac{\text{explained sum of squares by regression}}{\text{total variation in } y}$$

$$= 1 - \frac{\text{residual sum of squares by regression}}{\text{total variation in } y}$$

Our proposed diagnostic measure is:

$$\Delta = 1 - \frac{\sum_{j=1}^{n}\left\{y_j - \widehat{\psi}(\widehat{\pi}_j)\right\}^2}{\sum_{j=1}^{n}(y_j - \bar{y})^2}$$

We note that, like R^2, Δ is also a location- and scale-invariant measure. Under no linkage (the null hypothesis), Δ is expected to be close to 0. However, it is not easy to obtain either the exact or even the asymptotic sampling distribution of Δ under the null. While it is clear that the upper bound for Δ is 1, a sharp lower bound is not very obvious. However, a crude lower bound can be obtained as

$$1 - \frac{\sum_{j=1}^{n}\max\left\{(y_j - y_m)^2, (y_j - y_M)^2\right\}}{\sum_{j=1}^{n}(y_j - \bar{y})^2},$$

where y_m and y_M are respectively the minimum and the maximum squared sib-pair trait values observed in the data. One has to use Monte Carlo resampling techniques to obtain empirical thresholds under the null hypothesis of no linkage. We generate estimated i.b.d. scores using a multinomial random-number generator with cell probabilities equal to the marginals of estimated i.b.d. proportions at the marker locus nearest to the point p of interest, but preserve the quantitative trait values observed in the original dataset. We note that the marginal probabilities of estimated i.b.d. scores are provided in Table V of Haseman and Elston [12] for biallelic markers and can be easily generalized for polymorphic markers. The statistic Δ is computed for each resampled set, and the proportion of replications in which Δ exceeds the observed value of Δ, computed from the dataset, is an empirical estimator of the p value of the test.

Since the proposed Δ statistic does not consider the direction of the relationship between squared sib-pair trait differences and estimated i.b.d. scores, there may be concern that the rate of false positives is inflated owing to a random positive relationship between the variables under the null hypothesis of no linkage. One way to circumvent this problem is to ensure that the product moment (or rank) correlation between the variables is negative at the significant linkage regions.

27.6 COMPARISON WITH LINEAR REGRESSION METHODS

Elston et al. [7] suggested that instead of squared sib-pair differences, it may be statistically more powerful to use the population-mean-corrected cross-products of sib-pair trait values as the response variable in the linear regression setup. They showed that, as in the classical Haseman–Elston regression [12], the expected population-mean-corrected cross-product of sib-pair trait values conditional on the estimated i.b.d. score at a marker locus is a linear function of the estimated i.b.d. score with the slope parameter equal to 0 if and only if the QTL and the marker locus are unlinked. We note that the population-mean-corrected cross-product is also unaffected by location shifts in trait values. We compare, using simulated data, the powers of the two proposed nonparametric methods (based on ranks and kernel smoothing) with the two linear regression methods discussed in Elston et al. [7]: one using the squared differences of sib-pair trait values and the other using the population mean corrected cross-products of sib-pair trait values.

We consider three probability distributions for the quantitative trait values: normal, location-shifted χ^2, and location-shifted Poisson. Under the assumption that the QTL is biallelic, the distribution of trait values is a mixture of three components. When each component distribution is symmetric, a normal distribution is a good fit to the data. When the distributions of the components are skewed, a χ^2 distribution may be appropriate. For data involving counts (e.g., number of symptoms associated with a clinical manifestation), a Poisson distribution may be a good fit.

We generate data on a quantitative trait for 200 sib-pairs and marker genotypes at 10 ordered markers on a chromosome. We assume that the recombination distance between two consecutive marker loci is 0.05 and the quantitative trait locus is in between the third and fourth markers at a recombination distance 0.02 from the third marker. We also assume that the each of the odd-numbered markers has four equifrequent alleles, while each of the even numbered markers has three equifrequent alleles. The steps are as follows:

1. We generate the trait i.b.d. scores of the sib-pairs using a trinomial random-number generator with cell probabilities $(\frac{1}{4}, \frac{1}{2}, \frac{1}{4})$.

2. We generate the trait genotypes of the sib-pairs using a 9-variate random-number generator with cell probabilities given by the conditional trait genotype distribution of sib-pairs given their trait i.b.d. score as provided in Table I of Haseman and Elston [12].

3. We generate the marker i.b.d. scores of the sib-pairs for the two marker loci flanking the trait locus using the conditional distribution of marker i.b.d. score given trait i.b.d. score as provided in Table IV of Haseman and Elston [12].

4. We sequentially generate the i.b.d. scores of the sib-pairs for each nonflanking marker conditioned on the i.b.d. score at the last marker generated using Table IV of Haseman and Elston [12].

5. We generate the estimated i.b.d. score of each sib-pair at each marker using the conditional distribution of the estimated marker i.b.d. score given the marker i.b.d. score using Table V of Haseman and Elston [12].

6. We generate the quantitative trait values of the sib-pairs from (a) a bivariate normal distribution, (b) a bivariate distribution with location-shifted χ^2 marginals, and (c) a bivariate distribution with location-shifted Poisson marginals such that the mean vector has components α β, or $-\alpha$, according to whether the trait genotype is A_1A_1, A_1a_1, or a_1a_1, respectively, and the dispersion matrix is given by $\sigma^2 \{(1 - \rho)I_2 + \rho 11'\}$.

We use the linear regression method of Fulker et al. [9] to estimate the multipoint i.b.d. scores at arbitrary points on the genome based on information on all the marker loci generated. The 5% thresholds for the tests are obtained from a standard normal distribution for the rank-based method and the two linear regression procedures of Elston et al. [7] and via 1000 resampled datasets based on Δ for the non-parametric regression method. We perform 10,000 replications to obtain empirical powers for our test procedures. In all our simulations, we used fixed parameter values $\alpha = 3$, $\sigma^2 = 1$, and $\rho = 0.5$.

27.7 SIGNIFICANCE LEVELS AND EMPIRICAL POWER

Since the significance levels of our proposed test procedure are determined emprically based on resampled datasets, it is of interest to evaluate the rate of false positives. For this purpose, we generate 1000 replications of estimated i.b.d. scores at a hypothetical marker locus with four equifrequent alleles using the probability distribution provided in Table V of Haseman and Elston [12]. Using the empirical 5% thresholds (provided in Tables 27.1–27.3 in parentheses) based on the ordered Δ values from the 1000 resampled datasets, we find that for all our simulation parameter values and probability distributions, the rates of false positives (i.e., obtaining a significantly positive Δ value under the null hypothesis of no linkage) are between 0.045 and 0.054 (details omitted for brevity). Thus, the resampling strategy maintains an appropriate rate of false positives.

We evaluate the empirical powers of our proposed test procedures for a point in between the third marker and the trait locus, at a recombination distance 0.01 from the trait locus, for different frequencies (p) of A_1, and different levels of dominance (β) such that the heritability of the

Table 27.1 Empirical Powers of Different Test Procedures for a Normally Distributed Trait and Simulation Parameter Values of $\alpha = 3$, $\sigma^2 = 1$, $\rho = 0.7$, $\theta = 0.01$; and Varying Trait Allele Frequency (p) and Degree of Dominance at Trait Locus (β)[a]

		Empirical Power			
p	β	R	E1	E2	KS
0.5		(−1.645)	(−1.645)	(1.645)	(0.118)
	0	0.702	0.899	0.912	0.878
	1	0.681	0.825	0.838	0.852
	2	0.635	0.685	0.702	0.769
0.7		(−1.645)	(−1.645)	(1.645)	(0.121)
	0	0.666	0.817	0.833	0.804
	1	0.615	0.757	0.768	0.779
	2	0.579	0.658	0.670	0.728
0.9		(−1.645)	(−1.645)	(1.645)	(0.125)
	0	0.613	0.754	0.770	0.749
	1	0.572	0.705	0.719	0.733
	2	0.538	0.627	0.634	0.703

[a]Symbols key: R—rank correlation; E1—Elston et al. [7] with squared sib-pair trait difference; E2—Elston et al. [7] with population mean-corrected cross-product of sib-pair trait values; KS—nonparametric regression based on kernel smoothing. Figures in parentheses indicate 5% thresholds for the test procedures.

Table 27.2   Empirical Powers of Different Test Procedures for a χ^2 Distributed Trait and Simulation Parameter Values of $\alpha = 3$, $\sigma^2 = 1$, $\rho = 0.7$, $\theta = 0.01$; and Varying Trait Allele Frequency (p) and Degree of Dominance at Trait Locus $(\beta)^a$

		Empirical Power			
p	β	R	E1	E2	KS
0.5		(−1.645)	(−1.645)	(1.645)	(0.124)
	0	0.692	0.890	0.902	0.876
	1	0.662	0.817	0.823	0.846
	2	0.600	0.671	0.677	0.765
0.7		(−1.645)	(−1.645)	(1.645)	(0.129)
	0	0.664	0.805	0.819	0.801
	1	0.628	0.738	0.744	0.772
	2	0.574	0.650	0.648	0.724
0.9		(−1.645)	(−1.645)	(1.645)	(0.135)
	0	0.607	0.743	0.747	0.745
	1	0.583	0.694	0.695	0.730
	2	0.541	0.616	0.611	0.698

^{a}Symbols key: R—rank correlation; E1—Elston et al. [7] with squared sib-pair trait difference; E2— Elston et al. [7] with population mean-corrected cross-product of sib-pair trait values; KS—nonparametric regression based on kernel smoothing. Figures in parentheses indicate 5% thresholds for the test procedures.

Table 27.3   Empirical Powers of Different Test Procedures for a Poisson Distributed Trait and Simulation Parameter Values of $\alpha = 3$, $\sigma^2 = 1$, $\rho = 0.7$, $\theta = 0.01$; and Varying Trait Allele Frequency (p) and Degree of Dominance at Trait Locus $(\beta)^a$

		Empirical Power			
p	β	R	E1	E2	KS
0.5		(−1.645)	(−1.645)	(1.645)	(0.136)
	0	0.684	0.872	0.881	0.873
	1	0.659	0.805	0.802	0.841
	2	0.593	0.652	0.645	0.760
0.7		(−1.645)	(−1.645)	(1.645)	(0.141)
	0	0.658	0.788	0.786	0.792
	1	0.615	0.720	0.703	0.768
	2	0.572	0.629	0.613	0.715
0.9		(−1.645)	(−1.645)	(1.645)	(0.149)
	0	0.604	0.727	0.724	0.744
	1	0.575	0.673	0.659	0.720
	2	0.527	0.600	0.582	0.691

^{a}Symbols key: R—rank correlation; E1—Elston et al. [7] with squared sib-pair trait difference; E2— Elston et al. [7] with population mean-corrected cross-product of sib-pair trait values; KS—nonparametric regression based on kernel smoothing. Figures in parentheses indicate 5% thresholds for the test procedures.

trait varies between 33% and 82%. The results are provided in Table 27.1 (normal distribution), Table 27.2 (χ^2 distribution), and Table 27.3 (Poisson distribution). The four analyses are referred to as follows: the rank correlation (R), the nonparametric regression based on kernel smoothing (KS), Elston [7] with squared sib-pair trait difference (E1), and Elston [7] with mean-corrected cross-product of sib-pair trait values (E2). We note here that the empirical powers for the KS procedure are computed with the additional restriction that the product–moment correlations are negative. This, as mentioned in a previous section, is done to avoid false positives arising from random positive correlation between the underlying variables under the null hypothesis of no linkage. From Tables 27.1 and 27.2, we find that for normal and χ^2 trait values E2 and E1 yield more power than R and KS when there is no dominance at the trait locus ($\beta = 0$). However, the power of KS is only marginally less than the two linear regression procedures of Elston [7]. On the other hand, as dominance increases (i.e., $\beta = 1$), KS outperforms both E1 and E2, although R is still less powerful than E1 and E2. When dominance is high (i.e., $\beta = 2$), KS has a much higher power than do E1 and E2. This is expected because there is deviation in the underlying linear relationship in the regression strategies of Elston [7] as dominance increases at the trait locus. We also find that these inferences are valid for different levels of heterozygosity at the trait locus. However, E1 and E2 are more adversely affected by decrease in heterozygosity than are R and KS. We note here that E2 is more powerful than E1, except for high dominance ($\beta = 2$) and low heterozygosity at the trait locus. From Table 27.3, we find that for Poisson trait values, KS is more powerful than E1 and E2, even in the absence of dominance, except when heterozygosity is very high ($p = 0.5$). The difference in power increases with increase in dominance. The rank correlation R is uniformly less powerful than both E1 and E2 for all levels of dominance, although the difference decreases as dominance increases. We also find that E1 is more powerful than E2 except for no dominance and high heterozygosity at the trait locus. We note here that compared to normal trait values, χ^2 trait values yield less powers of all the procedures. Similarly for Poisson trait values, the powers of all the procedures are lower than those for χ^2 trait values. However, the decrease is much more pronounced for E1 and E2, while it is marginal for R and KS.

27.8 AN APPLICATION TO REAL DATA

The Collaborative Study on the Genetics of Alcoholism (COGA) is a multicenter research program established to detect and map susceptibility genes for alcohol dependence and related phenotypes. Oscillations of brain activity play an important role in the functional organization of neuronal activity that unlies sensory and cognitive processing. Electroencephalogram (EEG) waves reflect the mean excitation of the pool of neurons and are considered to be an ideal endophenotype in the genetic study of alcoholism. Ghosh et al. [10] used the nonparametric regression method to perform a genomewide linkage scan on Beta 2 EEG Waves (frequency range 16–20 Hz) using genotype dataon 405 microsatellite markers with average heterozygosity 0.74 and average intermarker distance 10.9 cm. The analysis was based on 99 independent sib-pairs. The nonparametric regression was performed at every 1 cm on the genome, resulting in a large number of tests over the genome. Hence, 100,000 Monte Carlo resampled sets (as described in Section 27.5.1) were generated to obtain the empirical p values. Significant linkage evidence ($p < 0.00001$) was obtained on chromosomes 1, 4, 5, and 15 very close to potential candidate gene clusters such as *GABRA*, *GABRB*, *ADH*, and *CHRNA7*. We note here that a variance components analysis using the SOLAR software [16] on entire pedigrees of the COGA dataset has also provided a significant linkage finding near the *GABRA* receptor gene on chromosome 4. The regions harboring these genes are being followed up with association and functional studies to assess the roles of these genes for regulating Beta 2 EEG waves.

27.9   CONCLUDING REMARKS

The regression procedure of the classical Haseman–Elston method [12] and its extensions such as Elston et al. [7] assume a linear functional relationship between squared sib-pair trait differences or population-mean-corrected cross-product of sib-pair trait values and the estimated i.b.d. score at a marker locus. While this assumption holds good when the dominance at the trait locus is low, there is increasing deviation from a linear relationship with increase in dominance at the trait locus. Thus, the proposed nonparametric regression, which does not assume any such functional relationship, outperforms the linear regression procedures as dominance increases. While the rank correlation procedure is also completely model-free, it suffers from the inherent limitation of excessive summarization. The observed trait values are not directly utilized in the test statistic and contribute indirectly via ranks. This leads to loss of information and hence reduction in power. However, despite this limitation, it still performs better than do the linear regression procedures in the presence of high dominance.

Elston et al. [7] observed that a linear regression with population-mean-corrected cross-products of sib-pair trait values as the response variable is sometimes more powerful than the squared sib-pair trait differences as used in the traditional Haseman–Elston [12] approach. Although, for brevity, we are not presenting the results, we performed some simulations and found that for both the rank correlation and kernel smoothing procedures, the squared sib-pair trait values yielded more power than did the population-mean-corrected cross-product for all our simulation parameters.

Current methods use LOD scores as a diagnostic to evaluate the significance of linkage peaks. Since our proposed rank correlation and kernel smoothing methods are nonparametric, a direct comparison with likelihood-based LOD scores is not possible. However, if we consider the p values of our linkage peaks, we can theoretically obtain the LOD scores that would yield these p values. For example, a p value < 0.0001 can be attained for a LOD score greater than 3.29, while a p value < 0.001 can be attained for a LOD score greater than 2.35.

Studies have shown that larger sibships are more informative on linkage than are independent sib-pairs. Most model-free sibship methods, such as those of Elston et al. [7], divide each sibship into all possible sib-pairs and perform a weighted regression. The weights must be estimated on the basis of the correlation of squared differences (or mean-corrected cross-products) of sib-pair trait values between different sib-pairs. Ghosh and Reich [25] developed a method to integrate sibship data into a so-called "contrast function" which circumvents the problem of assigning a priori weights to different sib-pairs. Their linkage method was based on a linear regression of the squared contrast function of a sibship on a quadratic function of the matrix of i.b.d. scores at a marker locus for the different sib-pairs within the sibship and can be viewed as a direct extension of the sib-pair based classical Haseman–Elston regression [12] to larger sibships. However, since the method is based on a linear regression approach, it is susceptible to the presence of dominance at the QTL. Thus, it is of interest to compare the performance of a nonparametric regression of the squared contrast function on the matrix of marker i.b.d. scores based on kernel smoothing with the linear regression approach. Moreover, analytic and simulation studies have shown that squared differences do not carry sufficient linkage information [23], and inclusion of other transformations of sib-pair trait values [24,8,22] results in increase of statistical power to detect linkage. In particular, Ghosh and Reich [25] showed that the simultaneous use of the mean quantitative trait value of a sibship and the contrast function was a more powerful strategy compared to only the contrast function. We are currently exploring a nonparametric regression-based test incorporating both the mean and the contrast functions.

There is increasing evidence [4,18,15] that sib-pairs ascertained through extreme phenotypic values carry more linkage information compared to unselected samples. However, the availability of such data is rare, and it may be more practical to ascertain sib-pairs through one proband having an extreme phenotypic value. Likelihood-based methods that require explicit specification of the probability distribution of the underlying quantitative trait may provide misleading linkage inferences since the likelihood needs to be computed using an appropriate truncated distribution under ascertainment. On the other hand, the model-free methods are more robust with respect to any bias due to selected sampling. In particular, since we obtain empirical p values using resampled datasets, our method is not likely to be affected by ascertainment bias. However, extensive simulations need to be carried out to validate this hypothesis.

We emphasize that many currently used distribution based and distribution free linkage methods are theoretically valid under conditions like normality of trait values and/or the absence of dominance at the trait locus. Quantitative traits such as age of disease onset and number of symptoms associated with disease diagnosis are likely to have skewed and kurtotic distributions. Deviation from normality induces skewness and/or kurtosis in the quantitative trait distribution conditioned on the genotypes, thereby lowering the powers of these test procedures. A major advantage of our nonparametric regression method is that it does not involve modeling of trait parameters or assumption a specific functional relationship between quantitative trait values and marker i.b.d. scores, and hence is more robust to violations in model assumptions.

ACKNOWLEDGMENT

This work has been supported by NIH Grant R01-TW-6604 from the Fogarty International Center.

REFERENCES

1. Almasy, I. and Blangero, J., Multipoint quantitative-trait linkage analysis in general pedigrees, *Am. J. Hum. Genet.* **62**, 1198–1211 (1998).

2. Amos, C. I. and Elston, R. C., Robust methods for the detection of genetic linkage for quantitative data from pedigrees, *Genet. Epidemiol.* **6**, 349–360 (1989).

3. Amos, C. I., Robust variance-components approach for assessing genetic linkage in pedigrees, *Am. J. Hum. Genet.* **54**, 535–543 (1994).

4. Carey, G. and Williamson, J., Linkage analysis of quantitative traits: Increased power by using selected samples, *Am. J. Hum. Genet.* **49**, 786–796 (1991).

5. Cotterman, C. W., Factor union phenotype system, in *Computer Applications in Genetics*, Univ. Hawaii Press, Honululu, 1969, pp. 1–19.

6. Drigalenko, E., How sib-pairs reveal linkage, *Am. J. Hum. Genet.* **63**, 1242–1245 (1998).

7. Elston, R. C., Buxbaum, S., Jacobs, K. B., and Olson, J. M., Haseman and Elston revisited, *Genet. Epidemiol.* **19**, 1–17 (2000).

8. Forrest, W., Weighting improves the "new Haseman Elston" method, *Hum. Hered.* **52**, 47–54 (2001).

9. Fulker, D. W., Cherny, S. S., and Cardon, L. R., Multipoint interval mapping of quantitative trait loci, using sib-pairs, *Am. J. Hum. Genet.* **56**, 1224–1233 (1995).

10. Ghosh, S., Begleiter, H., Porjesz, B., Chorlian, D. B., Edenberg, H. J., Foroud, T., Goate, A., and Reich, T. Linkage mapping of Beta 2 EEG Waves via non-parametric regression, *Am. J. Med. Genet.* **118B**, 66–71 (2003).

11. Ghosh, S. and Majumder, P. P., A two-stage variable stringency semiparametric method for mapping quantitative trait loci with the use of genome-wide scan data on sib pairs, *Am. J. Hum. Genet.* **66**, 1046–1061 (2000).

12. Haseman, J. K. and Elston, R. C., The investigation of linkage between a quantitative trait and a marker locus, *Behav. Genet.* **2**, 3–19 (1972).

13. Olson, J. M., Robust multipoint linkage analysis: An extension of the Haseman-Elston method, *Genet. Epidemiol.* **12**, 177–193 (1995).

14. Olson, J. M. and Wijsman, E. M., Linkage between quantitative trait and marker loci: Methods using all relative pairs, *Genet. Epidemiol.* **10**, 87–102 (1993).

15. Peng, J. and Siegmund, D., Mapping quantitative traits with random and with ascertained sibships, *Proc. Natl. Acad. Sci. USA* **101**, 7845–7850 (2004).

16. Porjesz, B., Begleiter, H., Wang, K., Almasy, L., Chorlian, D., Stimus, A. T., Kuperman, S., O'Connor, S. J., Rohrbaugh, J., Bauer, L. O., Edenberg, H. J., Goate, A., Rice, J. P., and Reich, T., Linkage and linkage disequilibrium mapping of ERP and EEG phenotypes, *Biol. Psychol.* **61**(1–2), 229–248 (2002).

17. Randles, R. H. and Wolfe, D. A., *Introduction to the Theory of Nonparametric Statistics*, Wiley, New York, 1979.

18. Risch, N. and Zhang, H., Extreme discordant sib pairs for mapping quantitative trait loci in humans, *Science* **268**(5217), 1584–1589 (1995).

19. Silverman, B. W., *Density Estimation for Statistics and Data Analysis*, Chapman & Hall, London, 1986.

20. Sham, P. C., Purcell, S., Cherny, S. S., and Abecasis, G. R., Powerful regression-based quantitative-trait linkage analysis of general pedigrees, *Am. J. Hum. Genet.* **71**, 238–253 (2002).

21. Tang, H. K. and Siegmund, D., Mapping quantitative trait loci in oligogenic models, *Biostatistics* **2**, 147–162 (2001).

22. Wang, K. and Huang, J., A score-statistic approach for the mapping of quantitative-trait loci with sibships of arbitrary size, *Am. J. Hum. Genet.* **70**, 412–424 (2002).

23. Wright, F. A., The phenotypic difference discards sib-pair QTL linkage information, *Am. J. Hum. Genet.* **60**, 740–742 (1997).

24. Xu, X., Weiss, S., Xu, X., and Wei, I. J., A unified Haseman-Elston method for testing linkage with quantitative traits, *Am. J. Hum. Genet.* **67**, 1025–1028 (2000).

25. Ghosh, S. and Reich, T., Integrating sibship data for mapping quantitative trait loci, *Am. Hum. Genet.* **66**, 169–182 (2002).

PART V

Miscellaneous Topics

C H A P T E R 28

Robustness Issues in Biomedical Studies

Ayanendranath Basu

Indian Statistical Institute, Kolkata, India

28.1 INTRODUCTION: THE NEED FOR ROBUST PROCEDURES

The science of statistics deals primarily with the collection, analysis, and interpretation of data. However, the inferential part of statistical theory depends not only on the observed data but also on the assumed scenario under which inference is performed. In classical parametric theory the models are generally formulated under an exact set of conditions. The optimal statistical procedures in this context are derived under this set of conditions. When some of these conditions fail to hold, their impact may be quite drastic on the abovementioned optimal procedures. In this sense classical parametric inference can be somewhat limited in scope in spite of its technical richness. Even slight deviations from the assumed conditions can seriously damage the inference procedure. For real-life data all parametric models are only approximations to reality, and minor deviations from the parametric assumptions are never entirely unexpected. Thus modifications of the procedures that are optimal or near-optimal at the model should also be considered that retain reasonable inferential properties even when the true distribution is close to but not exactly in the model. This is the main motivation for generating "robust" procedures. The word "robustness" may have similar but slightly different meanings to different individuals; in an intuitive sense we will take this word to mean insensitivity of the procedure to minor deviations from the assumed set of conditions. According to Hampel et al. [31], "Robust statistics, as a collection of related theories, is the statistics of approximate parametric models."

One branch of statistics that deals with the issue of minimizing parametric assumptions and performing statistical inference in a more general setup is *nonparametric statistics*. As nonparametric models are usually based on a minimal set of assumptions, the chances of model violations are low. Thus the robustness issue is automatically resolved, at least in part.

Statistical Advances in the Biomedical Sciences, edited by Atanu Biswas, Sujay Datta, Jason P. Fine, and Mark R. Segal

However, the use of nonparametric statistics also does not satisfactorily resolve the robustness issue. Nonparametric models may have more general applicability than parametric models but cannot match the efficiency of the latter when the parametric model correctly describes the data-generating distribution. Parametric methods are simple to use and allow the imposition of nice, smooth, and mathematically tractable structures on the model under which the data are analyzed; the vast and rich literature on parametric models, including fundamental and powerful procedures such as sufficiency reduction and maximum-likelihood estimation, make parametric estimation a very attractive tool for the statistician. Robust methods are in fact closer in spirit to classical parametric methods than nonparametric ones. However, nonparametric methods are often included in the robustness literature, and we will provide appropriate references for the same at several places in this chapter.

Ironically, the properties of parametric inference, which are useful in demonstrating the efficiency of the methods under model conditions, are often themselves responsible for their lack of robustness when the model conditions fail. Sufficiency, for example, is a highly nonrobust concept. The maximum-likelihood estimator (MLE) is notoriously nonrobust for many parametric models. The ideas of efficiency and robustness are often viewed as conflicting concepts. Many common robust procedures essentially represent attempts to robustify the method based on maximum likelihood.

The basic ideas of robust statistics have been in use for a long time (see, e.g., Ref. 68), but the mathematical framework was developed only between the late 1950s and the early 1970s. The pioneering works of Hodges and Lehmann [32] and Tukey [72] helped demonstrate the nonrobustness of several classical procedures. The fundamental work done by Huber [36–39] and Hampel [27–30] laid the theoretical foundation of "robust statistics." The Princeton Robustness Study of Andrews et al. [2] also deserves special mention, especially from the computational point of view.

The rest of the chapter is organized as follows. In Section 28.2 we discuss some of the basic robustness tools such as the M-estimators and influence function, and also provide some references for alternative approaches to robustness. In Section 28.3 we provide a survey of the more recent literature dealing with the use of robust methods in biomedical applications. In Sections 28.4–28.6 we choose three important topics in biomedical research and discuss a selected robust method appropriate for handling each such case. Concluding remarks are given in Section 28.7.

28.2   STANDARD TOOLS FOR ROBUSTNESS

In this chapter our discussion will remain mostly intuitive and expository. In this section we will focus on some of the basic robustness tools such as M-estimators, influence function, and breakdown point. We will discuss their interpretations and in which situations they are relevant. We will generally refrain from making specific distinctions between terms such as *qualitative robustness*, *quantitative robustness*, *B-robustness*, and "the minimax approach to robustness," although they are closely related to the tools that we will discuss in this section.

Let F represent the unknown, underlying true distribution; we will denote our parameter of interest by $T(F)$, where $T(\cdot)$ is a function from $\mathcal{F}$ to the real line $\mathbb{R}$ (or the p-dimensional real plane if $T(F)$ is a vector of order p), and $\mathcal{F}$ is a convex set of distribution functions containing all plausible models and all empirical distribution functions. We will refer to T as a *statistical functional*. See Fernholz [20] for a nice discussion of statistical functionals; also see Serfling [64].

28.2.1 M-Estimators

Estimators $T_n (X_1, \ldots, X_n)$ of the parameter $\theta = T(F)$ obtained as minimizers of objective functions of the form

$$\sum_{i=1}^{n} \rho(X_i, \theta) \qquad (28.1)$$

over θ, or as the solutions of the corresponding estimating equations

$$\sum_{i=1}^{n} \psi(X_i, \theta) = 0, \qquad (28.2)$$

are called *M-estimators* (maximum-likelihood-type estimators) of θ where $X_1, \ldots, X_n$ represent a random sample from the true distribution F and $\{F_\theta : \theta \in \Theta\}$ represents the parametric form modeling it. The particular choice $\rho(x, \theta) = -\log f_\theta(x)$, or $\psi(x, \theta) = (\partial/\partial\theta) \log f_\theta(x)$, generates the negative of the log likelihood and the maximum-likelihood score equation, respectively; the relevant estimator in this context is the MLE. It is intuitively obvious that if one has control over the functions ρ or ψ, one can choose them in a way to limit the impact of large residuals.

Occasionally, the expression in Equation (28.1) may not have a minimum, but in such cases there would usually be an equivalent formulation of $\rho(\cdot, \cdot)$ where one will not encounter this problem. Equation (28.2), on the other hand, may have more than one solution depending on the local minima of the minimization problem so that the correct solution has to be chosen with care. See Huber [42] for more discussion on these issues.

28.2.2 Influence Function

Huber [36,40] considered and demonstrated the existence of M-estimators of location and regression with minimax asymptotic variance over a specified neighborhood of a given distribution. This generated the minimax approach to robustness. Hampel [27,30] assessed the robustness of an estimator by examining the behavior of the first Gateaux derivative of the corresponding functional at the underlying model distribution. This has led to the infinitesimal approach to robustness and the concept of the influence function.

We say that a functional T is Gateaux-differentiable [41,60] at the distribution F (which is in the domain of the functional T) if there exists a real function b such that for all G in the domain of T the relation

$$\frac{\partial}{\partial\alpha}[T((1 - \alpha)F + \alpha G)]_{\alpha=0} = \int b(x)\, dG(x) \qquad (28.3)$$

holds. The left-hand side (LHS) of this equation is the directional derivative of T at F in the direction of G. By putting $G = F$ in (28.3), it can be easily seen that

$$\int b(x)\, dF(x) = 0,$$

so that $dG(x)$ may be replaced with $d(G - F)(x)$ in Equation (28.3). This function $b(x)$ is independent of G, and is called the *influence function* of the functional T at the distribution F. However, Equation (28.3) only defines the influence function implicitly. An explicit formulation of the influence function can be obtained if, in particular, one chooses $G = \Delta_y$ (the probability measure that puts all its mass on point y). Denoting the value of the influence function of the functional T for the distribution F at the point y by $IF(y, T, F)$, it can be explicitly defined as

$$IF(y, T, F) = \frac{\partial}{\partial \alpha} \left[T((1 - \alpha)F + \alpha\Delta_y) \right]_{\alpha=0}. \tag{28.4}$$

The influence function represents the asymptotic effect of an infinitesimal contamination at point y on the estimate; it is a measure of the effect of the contamination on the asymptotic bias. Small values of the influence function $IF(y, T, F)$ are preferable, since that would mean that such contaminations do not seriously affect the estimate. Boundedness of the influence function is often cited as a desirable property for robustness. The influence function is also closely related to the asymptotic variance of the estimator. See Reeds [60], Fernholz [20], and Hampel et al. [31] for more discussions on this issue.

In actual practice, the existence of the influence function requires a condition weaker than Gateaux differentiability, as has been demonstrated by Huber [41]; this makes its range of applicability very large, as it can be calculated in most realistic conditions without having to worry about complicated regularity conditions. And despite the technical nature of its definition, the actual evaluation of the influence function via Equation (28.4) is often a routine matter.

28.2.3 Breakdown Point

While the influence function quantifies the impact of an infinitesimal contamination on a functional, the breakdown point measures the proportion of data contamination that the functional can withstand before being arbitrarily perturbed. When a sample of fixed size n is under consideration, the *finite sample breakdown point* ϵ_n^* of an estimator is the smallest fraction of these n observations that, when replaced by arbitrary numbers, can completely destroy the estimator and render it meaningless. Thus for the sample mean the finite sample breakdown point is $1/n$, while for the α-trimmed mean (see Section 28.2.4) the finite sample breakdown point is $\epsilon_n^* = ([\alpha n] + 1)/n$, where $[\cdot]$ represents the box function.

As a measure of global stability, it may be more meaningful and sometimes easier to determine the asymptotic breakdown point of the functional. Given a functional $T(\cdot)$, its asymptotic breakdown point quantifies the separation between G and F before $T(G)$ is arbitrarily far from $T(F)$. Normally the breakdown point does not depend on F; in this sense it is essentially a global property of T. The asymptotic breakdown point is usually the limit of the finite sample breakdown point ϵ_n^*. Thus the asymptotic breakdown point of the median is $\frac{1}{2}$, while the asymptotic breakdown point of the α-trimmed mean is α. Since the asymptotic breakdown point represents the maximum permitted percentage of the perturbation minority such that it has only limited impact on the functional, the upper bound of the asymptotic breakdown point is $\frac{1}{2}$.

28.2.4 Basic Miscellaneous Procedures

Given the data $X_1, \ldots, X_n$, the sample mean is in general the most common estimator of the population mean in statistical literature. Often it is the minimum varianced unbiased estimator, although there are models where the sample mean is not sufficient for the population mean parameter. The sample mean can be used for estimating other parameters as well, such as the scale parameter of the exponential model. However, since the estimator needs only one bad value to be arbitrarily perturbed, from the robustness viewpoint the sample mean is generally an unsatisfactory estimator.

The sample median, on the other hand, enjoys an inherent robustness property. If the order statistics of the data are represented by $X_{(1)}, \ldots, X_{(n)}$, the sample median is $X_{(m+1)}$ when $n = 2m + 1$ is odd, and equals $(X_{(m)} + X_{(m+1)})/2$ when $n = 2m$ is even. The finite sample breakdown point of the sample median is approximately $\frac{1}{2}$; the median functional at the distribution function F, defined by

$$T_{\text{median}}(F) = F^{-1}\left(\tfrac{1}{2}\right),$$

has asymptotic breakdown point equal to $\frac{1}{2}$. For the mean functional $T_{\text{mean}}(F)$, the influence function is given by

$$IF(y, T_{\text{mean}}, F) = y - T_{\text{mean}}(F)$$

and hence is an unbounded function of the argument y unless the distribution of F has bounded support. On the other hand, under the assumption that F has a density f that is continuous and positive at $T_{\text{median}}(F)$, it is quite easy to show that the influence function of the median functional is bounded (so is, in fact, the influence function of any of the other quantiles under similar assumptions).

Other common and elementary procedures include the trimmed mean and the winsorized mean. The $100\alpha\%$ trimmed mean is the mean of the remaining observations after having deleted the largest $100\alpha/2\%$ observations as well as the smallest $100\alpha/2\%$ observations. In a sample of size n, this amounts to discarding the smallest and the largest $[n\alpha/2]$ observations. One can consider one-sided trimmed means also, where only the largest (or smallest) observations constitute potential outliers. The $100\alpha\%$ winsorized mean is similar in spirit to the trimmed mean, except that in the winsorized case the values in the lower tail that would have been discarded by the trimmed mean are replaced by the smallest value that is retained, while the values in the upper tail that would have been discarded by the trimmed mean are replaced by the largest value that is retained. Thus, given n observations, any observation larger than $X_{(m)}$ is replaced by $X_{(m)}$, where $m = n - [n\alpha/2]$. Similarly in the lower tail.

28.2.5 Alternative Approaches

There are some interesting and useful alternative approaches to the mainstream robust techniques represented by the M-estimators and their variants. We will briefly touch on these in this subsection, but applications of some of these methods will be illustrated in detail for some specific cases in the later sections.

The first such approach is the one based on adaptive procedures. Such procedures adjust to the current knowledge about the unknown distribution and its parameters. See Hogg [33] and Welsch [77] for some representative—if old—descriptions of adaptive procedures.

A second approach is that based on minimum-distance procedures where one chooses an estimator based on the minimization of a "distance" between the members of a parametric family and a nonparametric estimate of the population distribution. This approach itself branches into two related but substantially different approaches; one is based on the minimization of a distance between the model distribution function and the empirical distribution function (see, e.g., Ref. 53), while the other is a density-based approach that minimizes a measure of discrepancy between the model density and a nonparametric estimate of the true density (see, e.g., Refs. 6, 49, and 66).

Another approach is to modify Bayesian techniques in order to impart a certain amount of robustness in these methods (see, e.g., Refs. 12 and 43).

28.3 THE ROBUSTNESS QUESTION IN BIOMEDICAL STUDIES

Robustness is an active research area and has traveled a long way since its infancy in the early 1960s. Scores of subspecializations have now cropped up within the main theme of robustness. Clearly it is not possible to present a comprehensive description of all these within the short span of the present chapter. In this section we will present a brief literature survey of the more recent robustness literature specifically in the context of biomedical studies.

Application of robust methods in the biomedical literature is extensive and varied. Since biomedical data are hardly, if ever, available from controlled laboratory settings, they are almost always subject to noise. Robust methods try to model the bulk of the data after discounting for the contamination, which is assumed to be a small and insignificant portion of the data. Discounting the aberrant portion may be done in many innovative ways and depends on the field of application.

Most robust techniques guard against the influence of outliers. Outliers that are incorporated into a multivariate calibration model can significantly reduce the performance of the model. However, in the case of multiple outliers, the standard methods of outlier detection may sometimes fail. Pell [54] examined the use of robust principal-component regression and iteratively reweighted partial least squares for multiple outlier detection.

Nonparametric methods that do not give undue weight to observations inconsistent with a parametric model have also received their share of attention. An interesting application of nonparametric analysis of recurrent events in the presence of a terminal events such as death has been developed by Ghosh and Lin [24]. They considered the marginal mean of the cumulative number of recurrent events over time and presented a simple nonparametric estimator with some optimum asymptotic properties. They also developed nonparametric statistics for comparing two mean frequency functions and for combining data on recurrent events and death. The asymptotic null distributions of all the statistics have also been developed.

A common scenario in biomedical studies is multistate events data in which a single subject is at risk for multiple events. Subjects are followed over a period of time and may experience events of multiple types. Survival analysis, repeated-events data, competing risks data, and the illness–death model are all examples of multistate events data that can be represented as a stochastic process $X(t)$ with its value at time t denoting a subject's state. The analysis of multistate events data is complicated by multiple events within a subject. Glidden [25] considered nonparametric estimation of the vector of probabilities of state membership at time t and developed robust confidence bands for these curves, taking into account possibly non-Markov transitions.

Another novel application of a nonparametric method is the estimation of gap-time survival functions for ordered multivariate failure data. Times between sequentially ordered events or gap times are of much interest in biomedical studies. Gap times are usually right-censored, and within-subject failure times are not independent. For example, in a cancer study an individual's times from incidence to remission and from remission to recurrence cannot be assumed to be independent. Hence analyses of second and subsequent gap times include induced dependent censoring and nonidentifiability of marginal distributions. Schaubel and

Cai [63] constructed one-sample estimators of conditional gap-time-specific survival functions through a nonparametric approach and proposed methods for confidence bands. Their estimators are uniformly consistent, and standardized estimators converge weakly to a zero mean Gaussian process whose covariance function can be consistently estimated.

Sometimes very simple robust estimators can be used to derive attractive results in a complicated case. Kafadar and Prorok [44] developed a technique to estimate the difference in location parameters of two survival curves that is robust to the assumptions of duration of preclinical disease. This is useful in evaluating the effect of a cancer screening program. In this evaluation the important measures are the average lead time (time by which diagnosis is advanced by detection before onset of clinical trials) and average benefit time (time by which survival is extended). Randomized clinical trials allow the separation of these effects so that lead time and benefit time can be estimated for the population offering screening. Here treatment means the invitation to participate in a screening program. In screening trials a population of ostensibly healthy individuals is recruited from which cases evolve as the trial proceeds. The estimation of average lead time and average benefit time is conducted on cases that arise only after the trial has started. Because of length-biased sampling, the cases in the two groups may not be comparable. The identification of comparable cases in estimating the mean difference between survival curves adds a component of substantial variability, to the estimators for average lead time and average benefit time. To reduce the variability, one of the proposed strategies is to use $100\alpha\%$-trimmed means.

Robust methods are also useful in comparing case–cohort estimators of survival data. A case–cohort sample of adoptees was collected to investigate genetic and environmental influence on premature death. Petersen et al. [55] compared six regression coefficient estimators and two different estimators of their variances, of which the robust variance estimator showed a better overall performance.

In many applications the precise form of the model underlying the data may not be known, but several plausible choices designated by a family Ψ may be available. Often optimal tests for each member of the model exist, but these tests may have poor power under another model. Several approaches have been developed to obtain a single test with good power properties over the range of the models. Tarone [71], Fleming and Harrington [22], and Friedlin et al. [23] investigate this issue from different angles.

To avoid making any assumptions about the distribution of baseline and posttest responses, semiparametric estimation of treatment effects has been proposed by Leon et al. [48]. Rank-based procedures for testing noninferiority and equivalence hypotheses for continuous data arising from multicenter clinical trials have been developed for mixed models [59]. Silvapulle [65] developed a robust test to evaluate a treatment in comparison to a placebo in two or more groups of patients. In the presence of qualitative interaction or crossover interaction between patient groups and treatments, Silvapulle developed a test using M-estimators that is power-robust against long-tailed error distributions.

Some of the methods to analyze longitudinal data in medical studies has also been subject to heavy-tailed distributions. A model fit by generalized estimating equations has been used extensively for this purpose. Since the generalized estimating equation (GEE) tries to minimize a quadratic form of residuals, the method is not robust. To rectify this deficiency, Hu and Lachin [35] introduced a family of truncated robust estimating equations. Like GEE, their equations also assess the covariate effects in the generalized linear model in the complete population of observations and both are approximately unbiased. GEE, as expected, is more efficient with normal data, but efficiency decreases rapidly when data become contaminated or are heavy-tailed. Also, GEE may be sensitive to the working correlation specification; different working correlation structures may lead to different conclusions about the effect of the treatment. In such cases robust estimating equations perform much better, and in the applications considered by Hu and Lachin, it was observed that robust estimating equations consistently conclude that treatment effect is highly significant.

Robust procedures have also been used in the analysis of mean and covariance structures. Several robust methods have been proposed by Yuan and Bentler [81] for model fitting and testing. These include direct estimation of structured parameters using M-estimation and a two-stage procedure based on M- and S-estimators of population covariances. Two distribution-free test statistics have also been proposed to judge the adequacy of a hypothesized model.

Robust estimates of location and scale have been used in determining a prediction interval by Horn et al. [34]. They compared four reference intervals, and the robust estimator showed the best performance for small sample sizes. Simple robust procedures have been proposed by Emerson et al. [19] for meta-analyses. Meta-analyses often use a random-effect model to incorporate unexplained heterogeneity of study results. When combining risk differences in sets of 2×2 tables, it was found that a 20% trimmed weighted version of the DerSimonian–Laird [17] procedure is attractive as it offers resistance against the impact of highly anomalous results.

In the following three sections we discuss selected robust methods in the areas of logistic regression, censored survival data, and adaptive estimation in clinical trials.

28.4   ROBUST ESTIMATION IN THE LOGISTIC REGRESSION MODEL

Case–control or retrospective studies are commonly used to investigate effects of covariates on target disease outcomes. If the disease outcome is binary, logistic regression is a natural choice as an analytical tool. Maximum-likelihood methods are generally applied to obtain estimates of the logistic regression parameters. However, since atypical observations may have dramatic impacts on the maximum-likelihood fits, several robust alternative methods of estimation have been proposed in the literature. Here we provide a brief review of some robust methods in logistic regression followed by a detailed description of a selected method.

Besag [8] has discussed some resistant alternatives to the MLE for the generalized linear model. Another approach was taken by Pregibon [56], who defined a robust estimator as the minimizer of a certain loss function of the sum of deviances of the observations. In an important paper Copas [15] contrasts two forms of robust estimates for logistic regression parameters and concluded that a misclassification maximum-likelihood estimate is preferable over the robust estimates due to Pregibon [56], as the latter is inconsistent at the logistic model. In another paper Kunsch et al. [47] proposed the downweighting of aberrant observations through elliptical contours. Carroll and Pederson [13] developed robust estimates of logistic regression parameters that belong to the Mallows class. Bianco and Yohai [9] also proposed a corrected version of Pregibon's estimator that they showed to be consistent and asymptotically normal.

Prentice and Pyke [57] showed that one can get the MLE of the logistic slope parameter with case–control sampling from a standard prospective logistic regression program and that the resulting standard errors are asymptotically correct. Wang and Carroll [74] extended this idea to produce robust estimators of the case–control parameters via prospective methodology. They focus specifically on estimates that downweight observations on the basis of one of the following three factors: (1) leverage, (2) extreme fitted values, and (3) likelihood of misclassification. These estimators are consistent and asymptotically normally distributed under the case–control sampling scheme, and the prospective formulas for asymptotic covariance estimates may be used without modification in case–control studies.

However, none of the downweighting techniques, either by using elliptical contours or in terms of extreme predicted probabilities, are highly effective for case–control studies. This is mainly because the marginal distribution of the covariates is a mixture of two distinct distributions: one for the cases and the other for the controls. In the extreme case where the cases indicate rare diseases and their number is very small compared to the controls, and if the cases and the controls have sufficiently separated centers, the entire group of cases may be considered as outliers and receive little or no weight.

Estimation of logistic regression parameters in the presence of outliers using minimum distance methods provides an attractive alternative. Bondell [11] considered a robust estimation procedure based on the minimization of the weighted Cramer–von Mises procedure. The minimum distance idea can also be used to construct a natural class of goodness-of-fit statistics for testing the validity of the logistic regression model.

In the following we discuss in detail another estimator that is also closely linked to the minimization of a distance (although in a different sense) and can be very useful in some specific logistic regression problems. The approach is based on a study by Markatou et al. [51], and is useful when the covariate X is discrete with multiple Y observations at each level combination of X. The approach constructs weighted maximum-likelihood score equations and studies the corresponding M-estimators for discrete distributions. The method borrows heavily from Simpson [66] and Lindsay [49], as well as the minimum Hellinger distance and related ideas of Beran [5–7]. Other related approaches in weighted likelihood estimation include the work of Green [26], who extensively discussed the theory and use of iteratively reweighted least squares for maximum likelihood, and suggested the replacement of the usual maximum-likelihood score equations with weighted score equations. Field and Smith [21] also suggested another weighted likelihood estimation method. The concepts of minimum distance and weighed likelihood have also been used to develop robust tests of hypotheses. Simpson [67] proposed the Hellinger deviance test, a robust alternative to the likelihood ratio test. Agostinelli and Markatou [1] proposed robust tests of hypothesis based on the weighted likelihood ideas.

To explain the proposed weighted likelihood method in a general setting, let $X_1, \ldots, X_n$ be a random sample from a discrete distribution with probability mass function $f_\theta(x)$; without loss of generality, let the sample space be $\chi = \{0, 1, \ldots, \}$ or a subset of it. Let $u_\theta(x) = \nabla_\theta \log f_\theta(x)$ be the maximum-likelihood score function, ∇_θ representing the gradient with respect to θ. Under standard regularity conditions, the weighted MLE of θ will be obtained as a solution of the estimating equation

$$\sum_{i=1}^n w(X_i, F_\theta, F_n) u_\theta(X_i) = 0, \tag{28.5}$$

where F_θ is the model distribution function and F_n is the empirical distribution function. The weight function $w(X, \cdot, \cdot)$ is selected in such a way that it has a value close to 1 if there is no evidence of model violation at X from the empirical distribution function. It has a value close to zero or exactly zero at X if the empirical cumulative distribution function indicates lack of fit at X. Thus the weight function downweights observations that are inconsistent with the model. If the assumed model is correct, the weight assigned to each observation should be asymptotically equal to 1, which is necessary for the estimator to be asymptotically efficient at the model.

In determining which observations to downweight, the abovementioned method uses a probabilistic approach instead of a geometric interpretation; in the latter a point is labeled as an outlier if it is geographically well separated from the bulk of the data. In the probabilistic interpretation an observation is an outlier if it is very unlikely to occur if the fitted model were true. Davies and Gather [16] have also defined outliers in terms of their position relative to the model that most of the observations follow.

For any value x in the sample space χ, we define the Pearson residual as

$$\delta(x) = \frac{d(x)}{f_\theta(x)} - 1,$$

where $d(x)$ is the relative frequency of the value x in the sample and $f_\theta(x)$ is the corresponding value under the model when the parameter value is θ. If the model is correctly specified, then the residuals $\delta(x)$ converge to 0 almost surely. Wherever convenient, we will simply write δ for $\delta(x)$ unless there is a scope for confusion. Probabilistic outliers will manifest themselves through large positive values of δ at the corresponding x, since here the observed $d(x)$ will be much larger than the expected model probability $f_\theta(x)$.

Minimum-disparity estimators of the parameter θ are obtained by minimizing a disparity—a density-based divergence between the empirical density $d(\cdot)$ and the model density $f_\theta(x)$. Such a measure is defined by

$$\rho_G(d, f_\theta) = \sum_x G(\delta(x)) f_\theta(x),$$

where G is a real-valued, convex, thrice-differentiable function on $[-1, \infty)$ with $G(0) = 0$. Under differentiability and appropriate regularity conditions, the minimum-disparity estimating equation has the form

$$\sum_x A(\delta(x)) \nabla_\theta f_\theta(x) = 0, \tag{28.6}$$

where $A(\delta) = (1 + \delta)G'(\delta) - G(\delta)$, Where G' is the derivative of G. Without changing the estimating properties of the disparity, it may be recentered and scaled so that the function $A(\delta)$ satisfies $A(0) = 0$ and $A'(0) = 1$. Thus when the assumed model is correct, $A(\delta)$ converges to 0 almost surely. The function $A(\delta)$ is called the *residual adjustment function* (RAF) of the disparity.

The estimating equation (28.6) can be rewritten as

$$\sum_x \frac{A(\delta(x)) + 1}{\delta(x) + 1} u_\theta(x) \, d(x) = 0.$$

Rewriting the sum on the left-hand side (LHS) as the sum over the sample index i rather than over the sample space, this estimating equation turns out to be exactly in the form of (28.5) with $w(x, F_\theta, F_n) = (A(\delta(x)) + 1)/(\delta(x) + 1)$. Clearly the weights converge to 1 under the model for all x since $A(\delta)$ and δ both converge to 0 for all x; thus the estimating equation resembles the likelihood score equation when the sample size is large. On the other hand, if there are disparities for which the residual adjustment function has a strong downweighting effect on the large positive δ outliers—such as the Hellinger distance or the negative exponential disparity [49]—outlying observations will have a substantially reduced weight in Equation (28.5).

Under the model, the weighted likelihood estimators obtained as the solution of Equation (28.5) where the weights are constructed in the manner described above are asymptotically fully efficient. The estimators also have the same influence function as that of the MLE under the model. However, this implies that the influence function of the weighted likelihood estimator can be potentially unbounded. In this case the influence function turns out to be a misleading indicator of the robustness properties of the weighted likelihood estimators. The appropriate analysis using higher-order terms (viewing the influence function approach as the first-order analysis) correctly captures the robustness properties of these estimators [49].

Now consider the logistic regression scenario. Suppose that there are m different covariate patterns and multiple observations at each covariate pattern x_i. Thus there are n_i binary observations corresponding to x_i, and let Y_i be the number of observations among them that are equal to 1. Hence Y_i has a binomial distribution with parameters (n_i, p_i). Let $d_i = y_i/n_i$, where y_i represents the observed value of Y_i, and let

$$p_i = \frac{\exp(x_i^T \beta)}{1 + \exp(x_i^T \beta)}.$$

We will define the Pearson residual as

$$\delta_1(x_i) = \frac{d_i}{p_i} - 1. \tag{28.7}$$

The estimation procedure should downweight the ith case if the corresponding Pearson residual $\delta_1(x_i)$ is a large positive value. However, any observation corresponding to a given covariate pattern x_i generates a two-cell Bernoulli distribution, and negative Pearson residuals for the "1-cells" (successes), actually indicate positive residuals for the "0-cells" (failures). Thus it is also necessary to define

$$\delta_0(x_i) = \frac{1 - d_i}{1 - p_i} - 1$$

as the residual for the "0-cell." The estimators are then obtained by minimizing a weighted sum of distances. For a given function G with the requisite properties, we minimize, with respect to β, the objective function

$$\sum_{i=1}^{m} n_i \left[\frac{\exp(x_i^T \beta)}{1 + \exp(x_i^T \beta)} G(\delta_1(x_i)) + \frac{1}{1 + \exp(x_i^T \beta)} G(\delta_0(x_i)) \right].$$

By equating the derivative to 0, one gets the corresponding estimating equation as

$$\sum_{i=1}^{m} n_i [A(\delta_1(x_i)) - A(\delta_0(x_i))] \nabla_\beta \left\{ \frac{\exp(x_i^T \beta)}{1 + \exp(x_i^T \beta)} \right\} = 0, \tag{28.8}$$

which differs from the corresponding maximum-likelihood score equation

$$\sum_{i=1}^{m} n_i [\delta_1(x_i) - \delta_0(x_i)] \nabla_\beta \left\{ \frac{\exp(x_i^T \beta)}{1 + \exp(x_i^T \beta)} \right\} = 0$$

only through the form of the RAF $A(\delta)$. Equation (28.8) can be rewritten as

$$\sum_{i=1}^{m} [w_1(x_i) y_i (1 - p_i) - w_0(x_i)(n_i - y_i) p_i] x_i = 0, \tag{28.9}$$

where

$$w_1(x_i) = [A(\delta_1(x_i)) + 1]/[\delta_1(x_i) + 1]$$
$$w_0(x_i) = [A(\delta_0(x_i)) + 1]/[\delta_0(x_i) + 1]$$

and $\nabla_\beta p_i = p_i(1 - p_i)x_i$. Equation (28.9) can be solved iteratively as a weighted likelihood procedure. At every stage, the current values of β may be used to construct new weights w_1, w_0, and the equation can be solved to obtain new estimates of the parameters treating the weights as fixed constants. The process is repeated until convergence. The method generates fully efficient estimators of the parameter vector β; additionally, the estimators are endowed with strong robustness properties.

In the following we present an example to illustrate the performance of the method described above in real situations. The data analysis presented here has been reproduced from Markatou et al. [51] with the kind permission of Elsevier. The example involves data that were generated from a toxicological experiment and are presented in O'Hara Hines and Carter [52, p. 13]. Six different concentrations of the toxicant potassium cyanate (KSCN) were applied to 48 vials of trout fish eggs. Each vial contained 61–179 eggs. For half of the vials, the eggs were allowed to water-harden for several hours before the toxicant was applied. For the others, the toxicant was applied just after fertilization. The number of dead eggs in each vial was counted after 19 days of the start of the experiment.

The proportion of dead eggs in each vial is treated as the response, and a logistic regression model is fitted to the data with covariates for water hardening (0 if the toxicant was applied before and 1 if it was applied after water hardening), and for a linear and quadratic term in log concentration of the toxicant. The quadratic term in log concentration is used to describe a sharp increase in mortality caused by the two highest concentrations. Weighted likelihood estimation is used to fit the model with weight functions based on the RAFs of the negative exponential disparity and the Hellinger distance, respectively. Table 28.1 gives the weights of those cases that did not receive a weight of nearly or exactly 1. We have two columns of weights; column 1 corresponds to the weights of the response cells and column 0, to those of the nonresponse cells, for any given x_i. The parameters α, β_1, β_2, and β_3 are the intercept,

Table 28.1   Weights for KSCN Example

Case	Negative Experimental Weights		Hellinger Weights	
	Column 1	Column 0	Column 1	Column 0
12	0.8905032	1.0000000	0.8552530	0.9965012
13	—	—	0.0000000	0.9977647
14	—	—	0.3279548	0.9982219
28	—	—	0.4098746	0.9993424
32	—	—	0.0000000	0.9993965
34	0.8940925	1.0000000	0.8541044	0.9973958
35	—	—	0.2994592	0.9996005
36	—	—	0.7742827	0.9997611
37	0.8532178	1.0000000	0.8281611	0.9836277
38	0.6217503	1.0000000	0.7023431	0.8955089
39	0.7126061	1.0000000	0.7485037	0.9477368
40	—	—	0.0000000	0.9973932
41	—	—	0.8588582	0.9943002
42	—	—	0.3038337	0.9886016
43	—	—	0.0000000	0.9866767
44	—	—	0.0000000	0.9828289

and the slope parameters associated with water hardening, log concentration and squared log concentration, respectively.

The negative exponential RAF downweights observations 12, 34, 37, 38, and 39. After inspection of the data, it is seen that observation 12 with weight 0.8905 and observation 34 with weight 0.8940 have, respectively, the highest numbers of dead eggs at concentration level 360, after and before water hardening. Observations 37, 38, and 39 at concentration level 720, prior to water hardening, are also downweighted as having high mortality. Notice that observations 38 and 39 received lowest weights. Examination of these observations showed that the mortality was high compared to all four replicates at the next higher concentration level at the same water hardness level. O'Hara Hines and Carter [52] have considered observations 38, 39, and 26 as possible outliers. The weighted likelihood method gives observation 26 a weight of nearly 1, indicating that it is consistent with the fitted model. An analog of Cook's statistic also identified observations 38 and 39 as potential outliers.

When the Hellinger RAF is used for the construction of the weights, observations 13, 32, 40, 43, and 44 received a weight of 0. Examination of those observations reveals that observation 32 has a 0 response, while observations 40, 43, and 44 have the lowest mortality at concentration levels 720 and 1440, respectively, at the same water-hardening level. For similar reasons observation 42 receives a weight of 0.3038, while observation 41 receives a weight of 0.8588. Observation 13, having the lowest number of dead eggs at concentration level 720 after water hardening is applied, receives a weight of 0, suggesting its incompatibility with the fitted model.

28.5 ROBUST ESTIMATION FOR CENSORED SURVIVAL DATA

In biomedical and industrial settings the statistician routinely encounters survival data. The problem with such data is that they are seldom fully observed and their analysis is complicated by various censoring mechanisms that come into play. Here we will describe the method proposed by Basu et al. [3], one of the more recent methods of estimating the model parameters robustly under censored survival data.

The method is an adaptation of the approach of Basu et al. [4], which considered robust estimation of the parameters when independent and identically distributed (i.i.d.) observations are available from the true distribution, which is modeled by a parametric family. Basu et al. [4] minimize a family of density-based divergences; the divergences are indexed by a single tuning parameter α, and represent data-based measures of discrepancy between the true density and the assumed model density. A major advantage of the method is that it does not require additional accessories such as kernel density estimation or other forms of nonparametric smoothing to produce nonparametric density estimates of the true underlying density function. The empirical distribution function itself is sufficient for the purpose of constructing the divergence in the case of i.i.d. data. For the right-censoring scenario, one can replace the empirical distribution function with the corresponding estimate of the cumulative distribution function based on the Kaplan–Meier estimate [45] of the survival curve. Thus in this situation one can also construct the data-based estimate of the divergence measure without having to take recourse to nonparametric smoothing techniques.

First we discuss in brief the method proposed by Basu et al. [4] for i.i.d. data. Let the parametric model $\{F_\theta\}$ be indexed by an unknown p-dimensional parameter $\theta \in \Theta \subset \mathbb{R}^p$. We will assume that F_θ has a density f_θ with respect to the dominating measure; let $\mathcal{G}$ represent the class of all distributions having densities with respect to this dominating measure. Basu et al. [4] defined the density power divergence $d_\alpha(g, f)$ between two density functions g and f as

$$d_\alpha(g,f) = \int \left\{ f^{1+\alpha}(x) - \left(1 + \frac{1}{\alpha}\right) g(x) f^\alpha(x) + \frac{1}{\alpha} g^{1+\alpha}(x) \right\} dx \quad \text{for} \quad \alpha > 0. \qquad (28.10)$$

In this form the divergence is not directly defined if α equals 0, but can be defined in the limiting sense as

$$d_0(g,f) = \lim_{\alpha \to 0} d_\alpha(g,f) = \int [g(x) \log(g(x)/f(x)) + (f(x) - g(x))] dx \qquad (28.11)$$

Since $\int (f(x) - g(x)) \, dx = 0$, the RHS of Equation (28.11) represents a version of the Kullback–Leibler divergence [46]. On the other hand, $\alpha = 1$ leads to the squared L_2 distance $\int \{(g(x) - f(x))\}^2 dx$. The density power divergence is a nonnegative measure for all $\alpha \geq 0$.

Under the parametric model $\{F_\theta\}$, the minimum-density power divergence estimator of θ corresponding to the tuning parameter α at the target density g is obtained by minimizing $d_\alpha(g, f_\theta)$ over $\theta \in \Theta$. However, when f is replaced by f_θ on the RHS of Equation (28.10), the third term of the integral is independent of the parameter, and the minimization of the divergence with respect to θ is equivalent to the minimization of

$$\int f_\theta^{1+\alpha}(x) dx - \left(1 + \frac{1}{\alpha}\right) \int f_\theta^\alpha(x) dG(x). \qquad (28.12)$$

Thus, given a random sample $X_1, \ldots, X_n$ from the true unknown distribution G, one can minimize a data-based estimate of the expression in (28.12) given by

$$\int f_\theta^{1+\alpha}(x) dx - \left(1 + \frac{1}{\alpha}\right) \int f_\theta^\alpha(x) dG_n(x) = \int f_\theta^{1+\alpha}(x) dx$$
$$- \left(1 + \frac{1}{\alpha}\right) n^{-1} \sum f_\theta^\alpha(X_i), \qquad (28.13)$$

where G_n is the empirical distribution function. Under differentiability conditions on the model and other regularity conditions, the minimum-density power divergence estimating equation has the form

$$\int u_\theta(x) f_\theta^{1+\alpha}(x) \, dx - n^{-1} \sum_{i=1}^n u_\theta(X_i) f_\theta^\alpha(X_i) = 0, \qquad (28.14)$$

where $u_\theta(\cdot) = (\partial/\partial\theta) \log f_\theta(\cdot)$ is the likelihood score function. Thus if $\{F_\theta\}$ is a location model, the estimating equation becomes

$$\sum_{i=1}^n u_\theta(X_i) f_\theta^\alpha(X_i) = 0,$$

and observations that are less likely under the model are subjected to greater downweighting through the presence of the $f_\theta^\alpha(\cdot)$ term. Notice also that the minimum-density power divergence estimating equation is unbiased under the model. One gets the maximum-likelihood score equation for $\alpha = 0$. Larger values of α lead to greater robustness, but efficiency decreases with α also. We will refer to these estimators as minimum-divergence estimators (MDEs).

In case of the right-censoring problem, each random observation X_i from the target distribution G is associated with a random observation C_i from the censoring distribution H, and one observes only $Y_i = \min(X_i, C_i)$ together with δ_i (which is the indicator of the event $X_i < C_i$) but not the X_i. It is assumed that the variable of interest X and the censoring variable C are independent. It is also necessary to assume that the distribution G and the censoring distribution H have no common points of discontinuity. Under this, and some other appropriate regularity conditions, the Kaplan–Meier estimate [45] of the survival function $\hat{S}_n(x)$ converges almost surely to the true survival function $S(x)$. Thus in the right-censoring problem one can replace G_n in (28.13) with $\hat{G}_n = 1 - \hat{S}_n$, which provides a consistent estimator of the true distribution function in this context.

The best-fitting parameter is then the minimizer of

$$D(\theta) = \int f_\theta^{1+\alpha}(x)dx - \left(1 + \frac{1}{\alpha}\right) \int f_\theta^\alpha(x)d\hat{G}_n(x) = \int V_\theta(x)d\hat{G}_n(x)$$

where $V_\theta(x) = \int f_\theta^{1+\alpha}(x)dx - (1 + \frac{1}{\alpha})f_\theta^\alpha(x)$ and $\hat{G}_n$ is the Kaplan–Meier estimate of the cumulative distribution function G. The minimum density power divergence estimators of the parameter θ are obtained as the solution of the equation

$$\int \psi_\theta(x)d\hat{G}_n(x) = 0,$$

where the elements of $\psi_\theta(\cdot)$ represent the partial derivatives of $V_\theta(x)$ with respect to the components of θ. Thus, to determine the asymptotic distribution of the minimum-density power divergence estimator in the case of censored survival data, one needs a law of large numbers and central-limit theorem-type results for general functionals $\int \phi(x)d\hat{G}_n(x)$ of the Kaplan–Meier estimator. Fortunately, a series of papers in the 1990s by W. Stute and J. L. Wang [69,70,75] provided just this theoretical structure, including strong consistency and asymptotic normality results. When aided by these results, it is a fairly routine task to show that the proposed estimator of Basu et al. [3] has an asymptotic normal distribution with a limiting covariance matrix depending, among other things, on the true distribution G and the censoring distribution H. See Basu et al. [3] for details on the theoretical derivations.

Next we provide an example illustrating the application of this method on some real-life data presented in Efron [18]. Analysis of these data has been reproduced here from Basu et al. [3] under the kind permission of the Institute of Statistical Mathematics. The data relate to a study comparing radiation therapy alone (arm A) and radiation therapy and chemotherapy (arm B) for the treatment of head and neck cancer. There were 51 patients assigned to arm A of the study, 9 of which were lost to follow-up and, therefore, censored; 45 patients were assigned to arm B of the study of which 14 were lost to follow-up. The censoring levels are quite high in these datasets, be equal to approximately 20% and 30%, respectively. Efron [18] makes various analyses of these data, which show radiation and chemotherapy B to be more effective in terms of survival times. In this example the fit of the Weibull model to these data is studied.

The MLEs and the MDEs of the two Weibull parameters (where a represents the scale parameter and b represents the shape parameter) are given for various values of the tuning parameter α in Tables 28.2 and 28.3. There are very significant changes in both parameter

Table 28.2   Analysis of Efron Data Assuming Weibull Model: Aim A

	α	Scale $\hat{a}$	Shape $\hat{b}$
MLE	0	399.24	0.91
MDE	0.001	418.18	0.98
	0.01	417.72	0.98
	0.1	412.72	0.99
	0.2	402.51	1.00
	0.25	395.31	1.02
	0.5	321.90	1.16
	0.75	252.85	1.44
	1.0	249.47	1.47

Table 28.3   Analysis of Efron Data Assuming Weibull Model: Arm B

	α	Scale $\hat{a}$	Shape $\hat{b}$
MLE	0	925.45	0.76
MDE	0.001	789.23	0.91
	0.01	790.07	0.91
	0.1	791.81	0.90
	0.2	789.26	0.90
	0.25	785.13	0.90
	0.5	726.72	0.93
	0.75	551.53	1.03
	1.0	343.07	1.31

estimates, with α including a change from $\hat{b} < 1$ (MLE and small α MDE) to $\hat{b} > 1$ (larger α MDE). Figures 28.1 and 28.2 illustrate the results for arms A and B, respectively. A kernel density estimate, formed by kernel smoothing the Kaplan–Meier estimator (e.g., see Ref. 73, Sec. 6.2.3) is shown in each figure; the bandwidth is subjectively chosen not to over-smooth the data. It is clearly shown by the kernel density estimates that in each case there is a main body of data to the left, together with some much more long-lived individuals to the right. The MLE Weibull fits are monotone decreasing because $\hat{b} < 1$, providing an unacceptable compromise between accommodating the main body and the long tail of the data, and conse-quently failing to capture either. The robust Weibull fits, with $\hat{b} > 1$, provide a wholly better fit to the main body of the data at the expense of essentially ignoring the long tail. As such, this is entirely successful in terms of robust fitting.

It may perhaps be argued that the contamination is in fact of interest and should also be modeled. The results of fitting two-component Weibull mixtures to the data by maximum likelihood are, therefore, also shown on Figures 28.1 and 28.2. (The corresponding parameter estimates, in an obvious notation, are given in Table 28.4.) In Figure 28.1, the Weibull mixture confirms the robust Weibull fit as being appropriate for the main body of data

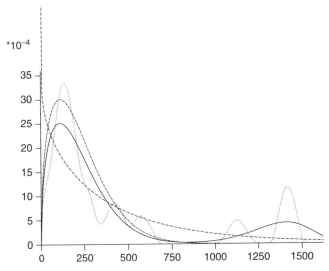

Figure 28.1 Kernel density estimate (dotted line), MLE Weibull fit (dashed line), MDE $\alpha = 1$ Weibull fit (dotted–dashed line), and MLE two-component mixture Weibull fit (solid line) for arm A of the Efron [18] data.

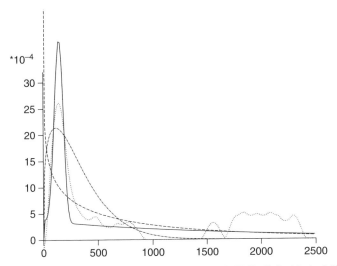

Figure 28.2 Kernel density estimate (dotted line), MLE Weibull fit (dashed line), MDE $\alpha = 1$ Weibull fit (dotted–dashed line), and MLE two-component mixture Weibull fit (solid line) for arm B of the Efron [18] data.

and adds a small second component to cover the tail. In Figure 28.2, the Weibull mixture takes a rather different form, that of a narrow peak to the left and a long flat tail to the right. Further alternative parametric models with heavier tails might be a betterway to model in this case.

Table 28.4   Maximum-Likelihood Parameter Estimates for Efron Data Assuming Two-Component Weibull Mixture Model

Component	$\hat{p}$		Arm A 0.82	Arb B 0.35
First	Scale $\hat{a}$		241.53	156.00
	Shape $\hat{b}$		1.47	4.08
Second	Scale $\hat{a}$		1428.11	1800.00
	Shape $\hat{b}$		9.17	0.90

28.6   ADAPTIVE ROBUST METHODS IN CLINICAL TRIALS

Here we provide an example of a situation—within the clinical trials framework—where the aim is not to theoretically develop an advanced robust estimator specific to the situation but to demonstrate that the use of very elementary robust estimators can lead to substantial gains from the ethical angle. In clinical trials, where the interest of the experimenter is in comparing two drugs, A and B, it is important from the ethical standpoint that as few individuals be allocated to the poorer treatment as possible during the course of decisionmaking. Since the statistical determination of the better treatment is not available prior to the experiment (after all, that is what the clinical trial is trying to determine), to conform to the abovementioned ethical view, it is imperative that one follow an adaptive design that specifies a higher allocation probability for the treatment currently considered superior when allocating the next subject. In the following we will describe such an adaptive design following Biswas and Basu [10]. Notice that such a goal cannot be achieved with a fixed-sample-size trial, which predetermines the number of subjects allocated to each treatment.

There are several examples of adaptive allocation designs based on a dichotomous response, but the work of Biswas and Basu is among the rare few that consider an adaptive allocation design based on the actual measurements of a continuous response. Here we discuss the method in brief. Let the responses under treatments A and B possess absolutely continuous distribution functions F_A and F_B, with respect to the Lebesgue measure. Our interest is in the mean parameters μ_A and μ_B, and we wish to choose between the following hypotheses:

$$H_1 : \mu_A = \mu_B, \quad H_2 : \mu_A \geq \mu_B, \quad H_3 : \mu_A \leq \mu_B.$$

We consider the treatment with the larger mean to be superior.

The design proposed by Biswas and Basu deterministically assigns the first subject to treatment A and the second subject to treatment B. However, starting from the third, the incoming subjects are allocated to one of the treatments based on an adaptive design described below which utilizes the entire information on the continuous responses for the previously allocated subjects up to that point. Let W_i be the response for the ith subject, and let δ_i be the indicator that takes the value 1 if the ith subject is allocated to treatment A and is 0 otherwise. The number of subjects N_{A_j} and N_{B_j} treated by treatments A and B, respectively, after the entry of the jth subject can be expressed as

$$N_{A_j} = \sum_{i=1}^{j} \delta_i, \quad N_{B_j} = \sum_{i=1}^{j} (1 - \delta_i) = j - N_{A_j}.$$

The rule for allocating the $(j + 1)$th patient after having observed the responses $W_1, \ldots, W_j$ and the corresponding indicators $\delta_1, \ldots, \delta_j$ are as follows:

1. Choose a continuous cumulative distribution function $G(\cdot)$ that is symmetric about 0, i.e. $G(0) = \frac{1}{2}$, and $G(-x) = 1 - G(x)$ for all real x. The $N(0,1)$ distribution function $\Phi(\cdot)$ is the most prominent example.

2. Determine the observed sample means for the two populations at this stage as

$$\hat{\mu}_{A_j} = \frac{\sum_{i=1}^{j} \delta_i W_i}{\sum_{i=1}^{j} \delta_i}, \quad \hat{\mu}_{B_j} = \frac{\sum_{i=1}^{j} (1 - \delta_i) W_i}{\sum_{i=1}^{j} (1 - \delta_i)}.$$

3. Allocate the $(j + 1)$th patient to treatment A with probability $G((\hat{\mu}_{A_j} - \hat{\mu}_{B_j})/c)$ and treatment B with probability $1 - G((\hat{\mu}_{A_j} - \hat{\mu}_{B_j})/c)$. The parameter c is a tuning parameter. The allocation should favor the treatment that has led to larger average responses in the past. This scheme is able to achieve that.

4. One could also choose G as a function of j. In that case the decision rule can be framed in such a way that the same magnitude of difference between the means is treated as more significant for larger j and the allocation probability to the currently better treatment can be suitably magnified.

On the basis of a targeted n observations, one can choose the hypothesis H_2 or H_3 to be the plausible one if the the observed difference of means (or a scaled version of it) is greater or smaller, respectively, than a prespecified cutoff. Otherwise one chooses the hypothesis H_1. Clearly, one can also use an early stopping criterion that will allow termination of the experiment prior to observing n subjects if the accumulated information at the point of termination is considered to provide sufficient evidence in support of one of the hypotheses.

The ACTG 076 trial conducted by the AIDS Clinical Trial Group (see, e.g., Ref. 14) provides a case in question where the use of an adaptive design could have provided a substantial improvement over the conventional design from the ethical standpoint. The aim was to determine whether the drug zidovudine (AZT) could reduce vertical HIV transmission from the infected mother to the child, and out of 476 pregnant women who had enrolled for the study, an equal number (238) of women and their infants were allocated to both the AZT group and the placebo group. Yao and Wei [78] showed that an appropriate adaptive design based on the "randomized play the winner rule" could have done the allocation in the ratio 300–176 and about 11 newborns could have been saved during the course of decisionmaking without significant loss in efficiency. This provides a quite compelling argument in favor of the use of adaptive designs in clinical trials. Although we will be dealing with continuous rather than dichotomous responses, it is obvious that such ethical gains are likely in this case as well.

In the discussion above, the adaptive designed proposed by Biswas and Basu has been presented entirely in terms of the sample means $\hat{\mu}_{A_j}$ and $\hat{\mu}_{B_j}$ of the two treatments. However since in this approach one makes a direct comparison of the sample means on the basis of the observations of the continuous responses, the robustness question deserves consideration; it is especially important in view of the fact that the sample mean can be a notoriously nonrobust estimator of the population mean. A few outlying observations may be sufficient to alter the hierarchy between the two means and hence lead to larger allocations to the poorer treatment than are ethically desirable. A few extreme outliers can, in fact, move the allocation proportion in the wrong direction in comparison with the conventional fixed sample size design, unlike the ACTG076 trial described above. In case of adaptive designs involving dichotomous responses, it is not possible to identify an observation as an outlier by looking at the corresponding indicator function. But in case of the continuous responses, where the actual magnitudes of the responses have been observed, the robustness question is a very pertinent one. Biswas and Basu have used two simple estimators of the mean parameter under two different models: the M-estimator based on Huber's ψ function (see, e.g., Ref. 31) for the normal model, and the weighted likelihood estimator of Field and Smith [21], and demonstrated through their simulations that small proportions of data contamination can lead to substantially larger allocations to the poorer treatment through this adaptive scheme. Adaptive designs—in the context of clinical trials—is a double-edged sword; a few bad observations can completely ruin its entire purpose. Adaptive designs should always be used in conjunction with robust procedures, which is the point Biswas and Basu try to establish.

Here we present an example of the application of this continuous adaptive design (CAD) on a dataset that is a part of the fluoxetine trial, a famous real-life adaptive trial. This example is reproduced from Biswas and Basu [10] under the kind permission of Sankhya. A particular permutation of the first 20 observations from each of the two treatments, A and B, generates the following observations:

A: 4, 2, -20, 0, -21, -3, -16, -9, 3, 0, -6, -7, -3, -3, -4, -16, -6, -11, -3, -16

B: -1, -1, -12, -2, -11, -17, -5, -12, -10, -21, -7, -8, -20, -4, 2, -14, -1, -8, -16, -15.

For each treatment a response from the top of the stack of that treatment is chosen whenever an observation from that treatment is required. For illustration, we consider the normal model with equal variances and estimate the means using the M-estimators with Huber's ψ function for several different values of the tuning parameter b (see, e.g., Ref. 10, p. 31). The results (number of allocations to treatment A) are presented in Table 28.5 along with the allocation obtained by using the sample mean. The experiment was terminated when 20 individuals had entered the study. Notice that the robust methods generally allocated more observations to treatment A, the treatment with the larger mean, compared to the method based on sample mean, probably because of the latter's inability to deal with the three very small values, -20, -21, and -16, early on in the chain of treatment A values.

Table 28.5   Number of Allocations to Treatment A Under Different Estimates

Scaling Constant c	Estimates			
	M-Estimates with Tuning Parameter b			
	$b=1.25$	$b=1.50$	$b=2.00$	Sample Mean
2.5	12	10	10	10
5.0	11	11	11	10
7.5	11	11	10	10
10.0	11	10	10	9

28.7 CONCLUDING REMARKS

In this chapter we have tried to give a flavor of the importance of the robustness question in biomedical studies. This is too large a research area to be satisfactorily covered in the short span of this chapter, and many important areas have been left untouched or mentioned only superficially. In particular, we have not discussed the linear model in any great detail (except in the case of logistic regression); the rich literature in this area includes the works of Yohai and his colleagues [79,80] and Rousseeuw and his colleagues [61,62]. Two other useful resources that cover different angles of the robustness question in the biomedical setup are Rao and Chakraborty [58] and Maddala and Rao [50]. The references cited therein give substantial material for future reading pertinent to the theme of this chapter.

ACKNOWLEDGMENT

Some of the examples reproduced here are from articles copyrighted to different journals or organizations. They have been reproduced with the permission of the appropriate authorities. A list is provided below, and their assistance in permitting these reproductions is gratefully acknowledged. The example and tables in Section 28.4 have been reproduced from *Journal of Statistical Planning and Inference*, Vol. 57, Markatou, M., Basu, A. and Lindsay, B. G. (authors), Weighted likelihood estimating equations: The discrete case with applications to logistic regression, pp. 215–232 (1997), with kind permission from Elsevier.

The example, tables, and figures in Section 28.5 have been reproduced from *The Annals of the Institute of Statistical Mathematics*, Vol. 58, Basu, S., Basu, A. and Jones, M. C. (authors), Robust and efficient estimation for censored survival data, pp. 341–355 (2006), with kind permission from Institute of Statistical Mathematics.

The example and tables in Section 28.6 have been reproduced from Sankhya B, Vol. 63, Biswas, A. and Basu, A. (authors), Robust adaptive designs in clinical trials for continuous responses, pp. 27–42, 2001, with kind permission from Sankhya.

REFERENCES

1. Agostinelli, C. and Markatou, M., Tests of hypothesis based on the weighted likelihood methodology, *Statistica Sinica* **11**, 499–514 (2001).

2. Andrews, D. F., Bickel, P. J., Hampel, F. R., Huber, P. J., Rogers, W. H., and Tukey, J. W., *Robust Estimates of Location: Survey and Advances* Princeton Univ. Press, 1972.

3. Basu, S., Basu, A., and Jones, M. C., Robust and efficient parametric estimation for censored survival data. *Ann. Inst. Statist Math* **58**, 341–355 (2006).

4. Basu, A., Harris, I. R., Hjort, N. L., and Jones, M. C., Robust and efficient estimation by minimising a density power divergence, *Biometrika* **85**, 549–559 (1998).

5. Beran, R. J., Robust location estimates, *Ann. Statist.* **5**, 431–444 (1977).

6. Beran, R. J., Minimum Hellinger distance estimation for parametric models, *Ann. Statist* **5**, 444–463 (1977).

7. Beran, R. J., Robust estimation in models for independent nonidentically distributed data, *Ann. Statist.* **10**, 415–428 (1982).

8. Besag, J., On resistant techniques and statistical analysis, *Biometrika* **68**, 463–469 (1981).

9. Bianco, A. M. and Yohai, V. J., Robust estimation in the logistic regression model, in *Robust Statistics, Data Analysis, and Computer Intensive Methods*, Rieder, H., ed., Lecture Notes in Statistics, Springer-Verlag, New York, 1996, Vol. 109, pp. 17–34.

10. Biswas, A. and Basu, A., Robust adaptive designs in clinical trials for continuous responses, *Sankhya B* **63**, 27–42 (2001).

11. Bondell, H. D., Minimum distance estimation for the logistic regression model, *Biometrika* **92**, 724–731 (2005).

12. Box, G. E. P., Leonard, T., and Wu, C.F., *Scientific Inference, Data Analysis and Robustness*, Academic Press, New York, 1983.

13. Carroll, R. J. and Pederson, S., On robustness in the logistic regression model, *J. Roy. Statist. Soc. B* **55**, 693–706 (1993).

14. Connor, E. M., Sperling, R. S., Gelber, R., Kiselev, P., Scott, G., O'Sullivan, M. J., Van Dyke, R., Bey, M., Shearer, W., Jacobson, R. L., Jiminez, E., O'Neill, E., Bazin, B., Delfraissy, J., Culname, M., Coombs, R., Elkins, M., Moye, J., Stratton, P., and Balsey, J., Reduction of maternal-infant transmission of human immunodeficiency virus type 1 with zidovudine treatment, *New Eng. J. Med.* **331**, 1173–1180 (1994) (report written for the Pediatric AIDS Clinical Trial Group Protocol 076 Study Group).

15. Copas, J. B., Binary regression models with contaminated data, *J. Roy. Statist. Soc. B* **50**, 225–265 (1988).

16. Davies, L. and Gather, U., The identification of multiple outliers, *J. Am. Statist. Assoc.* **88**, 782–792 (1993).

17. DerSimonian, R. and Laird, N., Meta-analysis in clinical trials, *Controlled Clinical Trials*, **7**(3), 177–188 (1986).

18. Efron, B., Logistic regression, survival and the Kaplan-Meier curve, *J. Am. Statist. Assoc.* **83**, 414–425 (1988).

19. Emerson, J. D., Hoaglin, D. C., and Mosteller, F., Simple robust procedures for combining risk differences in sets of 2×2 tables, *Statist. Med.* **15**, 1465–1488 (1996).

20. Fernholz, L. T., *Von Mises Calculus for Statistical Functionals*, Lecture Notes in Statistics, Vol. 19, Springer, New York, 1983.

21. Field, C. A. and Smith, B., Robust estimation—a weighted maximum likelihood approach, *Int. Statist. Rev.* **62**, 405–424 (1995).

22. Fleming, T. R. and Harrington, D. H. *Counting Processes and Survival Analysis*, Wiley, New York, 1991.

23. Friedlin, H., Podgor, M. J., and Gastwirth, J. L., Efficiency robust tests for survival or ordered categorical data, *Biometrics* **55**, 883–886 (1999).

24. Ghosh, D. and Lin, D. Y., Nonparametric analysis of recurrent events and death, *Biometrics* **58**, 554–562 (2000).

25. Glidden, D. V., Robust inference for event probabilities with non-Markov event data, *Biometrics* **58**, 361–368 (2002).

26. Green, P. J., Iteratively reweighted least squares for maximum likelihood estimation, and some robust and resistant alternatives, *J. Roy. Statist. Soc. B* **46**, 149–192 (1984).

27. Hampel, F. R., *Contributions to the Theory of Robust Estimation*, Ph.D. dissertation, Univ. California, Berkeley, 1968.

28. Hampel, F. R., A general qualitative definition of robustness, *Ann. Math. Statist.* **42**, 1887–1896 (1971).

29. Hampel, F. R., Robust estimation: A condensed partial survey, *Z. Wahrsch. verw. Geb.* **27**, 87–104 (1973).

30. Hampel, F. R., The influence curve and its role in robust estimation, *J. Am. Statist. Assoc.* **69**, 383–393 (1974).

31. Hampel, F. R., Ronchetti, E. M., Rousseeuw, P. J., and Stahel, W. A., *Robust Statistics: The Approach Based on Influence Functions*, Wiley, New York, 1986.

32. Hodges, J. L., Jr. and Lehmann, E. L., The efficiency of some nonparametric competitors of the *t*-test, *Ann. Math. Statist.* **27**, 324–335 (1956).

33. Hogg, R. V., Adaptive robust procedures: A partial review and some suggestions for future applications and theory, *J. Am. Statist. Assoc.* **69**, 909–927 (1974).

34. Horn, P. S., Pesce, A. J., and Copeland, B. E., A robust approach to reference interval estimation and evaluation, *Clin. Chem.* **44**, 622–631 (1998).

35. Hu, M. and Lachin, J. M., Application of robust estimating equations to the analysis of quantitative longitudinal data, *Statist. Med.* **20**, 3411–3428 (2001).

36. Huber, P. J., Robust estimation of a location parameter, *Ann. Math. Statist.* **35**, 73–101. (1964).

37. Huber, P. J., A robust version of the probability ratio test, *Ann. Math. Statist.* **36**, 1753–1758 (1965).

38. Huber, P. J., The behavior of maximum likelihood estimates under nonstandard conditions, in *Proc. 5th Berkeley Symp. Mathematical Statistics and Probability*, Univ. California Press, Berkeley, 1967, Vol. 1, pp. 221–233.

39. Huber, P. J., Robust statistics: A review, *Ann. Math. Statist.* **43**, 1041–1067 (1972).

40. Huber, P. J., Robust regression: Asymptotics, conjectures and Monte Carlo, *Ann. Statist.* **1**, 799–821 (1973).

41. Huber, P. J., *Robust Statistical Procedures*, SIAM, Philadelphia, 1977.

42. Huber, P. J., *Robust Statistics*, Wiley, New York, 1981.

43. Kadane, J. B., ed., *Robustness of Bayesian Analysis*, Elsevier/North-Holland, Amsterdam, 1984.

44. Kafadar, K. and Prorok, P. C., Estimating the difference in location parameters of two survival curves, with application to cancer screening, *J. Statist. Inform. Plan.* **57**, 165–179 (1997).

45. Kaplan, E. L. and Meier, P., Nonparametric estimation from incomplete observations, *J. Am. Statist. Assoc.* **53**, 457–481 (1958).

46. Kullback, S. and Leibler, R. A., On information and sufficiency, *Ann. Math. Statist* **22**, 79–86 (1951).

47. Kunsch, H. R., Stefanski, L. A., and Carroll, R. J., Conditionally unbiased bounded influence estimation in general regression models with applications to generalized linear models, *J. Am. Statist. Assoc.* **84**, 460–466 (1989).

48. Leon, S., Tsiatis, A. A., and Davidian, M., Semiparametric estimation of treatment effect in a pretest-posttest study, *Biometrics* **59**, 1046–1055 (2003).

49. Lindsay, B. G., Efficiency versus robustness: The case for minimum Hellinger distance and related methods, *Ann. Statist.* **22**, 1081–1114 (1994).

50. Maddala, G. S. and Rao, C. R., eds. *Handbook of Statistics, Robust Inference* Vol. 15, Elsevier, Amsterdam, 1997.

51. Markatou, M., Basu, A., and Lindsay, B. G., Weighted likelihood estimating equations: The discrete case with applications to logistic regression, *J. Statist. Plan. Inform.* **57**, 215–232 (1997).

52. O'Hara Hines, R. J. and Carter, E. M., Improved added variable and partial residual plots for the detection of influential observations in generalized linear models, *Appl. Stat.* **42**, 3–20 (1993).

53. Parr, W. C. and Schucany, W. R., Minimum distance and robust estimation, *J. Am. Statist. Assoc.* **75**, 616–624 (1980).

54. Pell, R. J., Multiple outlier detection for multivariate calibration using robust statistical techniques, *Chem. Int. Lab. Syst.* **52**, 87–104 (2000).

55. Petersen, L., Sorensen, T. I. A., and Andersen, P. K., Comparison of case-cohort estimators based on data on premature death of adult adoptees, *Statist. Med.* **22**, 3795–3803 (2003).

56. Pregibon, D., Resistant fits for some commonly used logistic models with medical applications, *Biometrics* **38**, 485–498 (1982).

57. Prentice, R. L. and Pyke, R., Logistic disease incidence models and case control studies, *Biometrika* **66**, 403–412 (1979).

58. Rao, C. R. and Chakraborty, R., eds., *Handbook of Statistics*, Applications in Biology and Medicine, Vol. 8, Elsevier, Amsterdam, 1991.

59. Rashid, M. M., Rank-based tests for non-inferiority and equivalence hypotheses in multi-centre clinical trials using mixed models, *Statist. Med.* **22**, 291–311 (2003).

60. Reeds, J. A., *On the Definition of von Mises Functionals*, Ph.D. thesis, Harvard Univ., Cambridge, MA, 1976.

61. Rousseeuw, P. J., Least median of squares regression, *J. Am. Statist. Assoc.* **79**, 871–880 (1984).

62. Rousseeuw, P. J. and Leroy, A. M., *Robust Regression and Outlier Detection*. Wiley, New York, 1987.

63. Schaubel, D. E. and Cai, J., Non-parametric estimation of gap time survival functions for ordered multivariate failure time data, *Statist. Med.* **23**, 1885–1900 (2004).

64. Serfling, R. J., *Approximation Theorems of Mathematical Statistics*, Wiley, New York, 1980.

65. Silvapulle, M. J., Tests against qualitative interaction: Exact critical values and robust tests, *Biometrics* **57**, 1157–1165 (2001).

66. Simpson, D. G., Minimum Hellinger distance estimation for the analysis of count data, *J. Am. Statist. Assoc.* **82**, 802–807 (1987).

67. Simpson, D. G., Hellinger deviance tests: Efficiency, breakdown points and examples, *J. Am. Statist. Assoc.* **84**, 107–113 (1989).

68. Stigler, S. M., Simon Newcomb, Percy Daniell and the history of robust estimation 1885–1920, *J. Am. Statist. Assoc.* **68**, 872–879 (1973).

69. Stute, W., The central limit theorem under random censorship, *Ann. Statist.* **21**, 422–439 (1995).

70. Stute, W. and Wang, J. L., A strong law under random censorship, *Ann. Statist.* **21**, 1591–1607 (1993).

71. Tarone, R. E., On the distribution of the maximum of the log-rank statistics and the modified Wilcoxon statistic, *Biometrics* **37**, 79–85 (1981).

72. Tukey, J. W., A survey of sampling in contaminated distributions, in *Contributions to Probability and Statistics*, Olkin, I., ed., Stanford Univ. Press, 1960.

73. Wand, M. P. and Jones, M. C., *Kernel Smoothing*, Chapman & Hall, London, 1995.

74. Wang, C. Y. and Carroll, R. J., On robust estimation in case-control studies, *Biometrika* **80**, 237–241 (1993).

75. Wang, J. L., M-estimators for censored data: Strong consistency, *Scand. J. Statist.* **22**, 197–206 (1995).

76. Wang, J. L., Asymptotic properties of M-estimators based on estimating equations and censored data, *Scand. J. Statist.* **26**, 297–318 (1999).

77. Welsh, A. H., *Some Problems in Adaptive Estimation*, Ph.D. dissertation, Australian National Univ., Canberra, 1984.

78. Yao, Q. and Wei, L. J., Play the winner for phase II/III clinical trials, *Statist. Med.* **15**, 2413–2423 (1996).

79. Yohai, V. J., High breakdown point and high efficiency robust estimates for regression, *Ann. Statist.* **15**, 642–656 (1987).

80. Yohai, V. J. and Maronna, R. A., Asymptotic behavior of M-estimators for the linear model, *Ann Statist.* **17**, 258–268 (1979).

81. Yuan, K. H. and Bentler, P. M. Robust mean and covariance structure analysis, *Br. J. Math. Statist. Psychol.* **51**, 63–88 (1998).

CHAPTER 29

Recent Advances in the Analysis of Episodic Hormone Data

Timothy D. Johnson

Department of Biostatistics, University of Michigan, Ann Arbor, Michigan

Yuedong Wang

Department of Statistics and Applied Probability, University of California, Santa Barbara, California

29.1 INTRODUCTION

Many questions that are important to investigators make the analysis of hormonal time series data challenging. Typical questions are

1. How many pulses occur per unit time in healthy individuals? In diseased individuals?
2. How much hormone is released during each episode? During the data collection period?
3. How rapidly is the hormone eliminated from the bloodstream?
4. Do pulses occur uniformly throughout the day, or does the event rate follow some non-homogeneous process?

In order to answer these questions, the first thing one must usually do is distinguish between a pulse and noise. This is perhaps the most challenging issue facing an analysis.

Hormone secretion can be broadly classified into two categories: episodic (or pulsatile) and rhythmic. *Episodic* secretion is characterized by the release of large masses of hormone in a short period of time. These secretion events can have either regular or irregular periods. *Rhythmic* hormone secretion is characterized by a slowly varying ebb and flow of hormone release much like the ocean tide. The period of secretion may be diurnal or it may occur

Statistical Advances in the Biomedical Sciences, edited by Atanu Biswas, Sujay Datta, Jason P. Fine, and Mark R. Segal

527

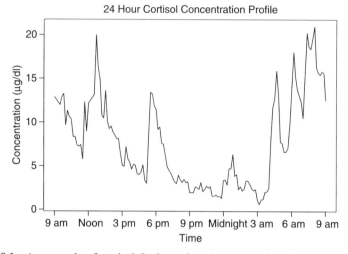

Figure 29.1 An example of cortisol (hydrocortisone) concentrations from a female subject suffering from depression over a 24-h time period. Note the pulsatile nature of the concentration. Plasma concentration levels of cortisol were obtained at 10-min intervals.

over shorter or longer periods of time. Here we concentrate on two new statistical methods developed for the analyses of hormones that display both components of episodic and (basal) rhythmic secretion, such as that displayed in Figure 29.1.

One difficulty associated with the analyses of hormone secretion is that it is not a directly observable phenomenon. Rather, plasma concentrations of hormone are observed. A common assumption is that plasma concentration is the convolution of the secretion of hormone into the circulatory system and the elimination of hormone from this system. Further, in most experiments, hormone elimination is not directly observable and a functional form is assumed, usually exponential. If both secretion and elimination were independently observable, then the direct convolution of these two observed phenomena would be possible. Since they are not, we have an inverse problem. Inverse problems typically arise when one only has indirect observations of a quantity of interest. Furthermore, inverse problems are most often nonlocal; dependence of the functional value at a particular point depend on physical conditions distant to that point [15]. In the present case we wish to glean information about the secretion and elimination functions given the concentration. This is known as *deconvolution* and is a well-known ill-posed problem [37].

This ill-posed problem can be handled by two methods. The first is to simply analyze the observed concentration profile and assume that patterns found in the concentration can be used as a surrogate for secretion patterns. This is most likely a fair assumption; however, only a limited amount of information about the secretion can be gleaned. The second method assumes a model for the secretion of hormone into the circulatory system and a model for the elimination of hormone from the system. However, even in this case, the problem of deconvolution is ill-posed and methods to deal with it must be used, such as regularization.

Many methods have been developed to analyze pulsatile hormone data. A primary focus of each is on the identification of the number and location of pulses. These methods have been categorized into criterion-based methods and model-based methods [24]. From a historical standpoint, criterion-based methods were the first methods developed. In general, criterion-based methods use test statistics to identify "pulses" in the observed concentration profile. Typically, criterion-based methods use the coefficient of variation (CV) from the assay as the true CV, which ignores other sources of variation. Pulses are identified as locations in the data where the relative variability (relative to the assay CV) is higher than some threshold. Hence estimated quantities related to variation are biased, such as the number of pulses. This leads to overidentification of the true number of pulses [24]. Mauger and Brown identify seven criterion-based methods: the Goodman–Karsch rule [8], ULTRA [34], PULSAR [25], the cycle detector [3], CLUSTER [35], DETECT [29], and Pulsefit [27]. Technically, Pulsefit has an underlying model associated with it. However, since it requires the complete decay of a pulse to baseline before another event occurs, which is an oversimplification, Mauger and Brown include it in their list of criterion-based methods.

One of the earliest models developed is due to O'Sullivan and O'Sullivan [30], who model the hormone concentration in blood as the convolution of a pulse function and a point process plus an error term. However, their estimation procedure is rather ad hoc and they assume a zero basal concentration. Diggle and Zeger [5] entertain a non-Gaussian autoregressive model. Their model is a mixture of zero mean Gaussian noise and gamma-shaped pulses. The mixing probability is allowed to depend on past observations, thus incorporating feedback into their model. Kushler and Brown [21] develop a model that assumes no functional form for hormone secretion; instead, they assume instantaneous hormone secretion. Veldhuis and Johnson [36] develop a model where hormone concentration is the convolution of exponential decay and Gaussian-shaped secretion events. Their model is conditional on the number of pulses. Thus, they rely on a criterion-based method to preselect pulses. Their model assumes constant basal concentration. Komaki [20] develops a state space representation of a non-Gaussian time-series model. His model assumes a zero mean basal concentration. Keenan et al. [18] develop a stochastic differential equation model, the solution of which is a stochastic version of the Veldhuis–Johnson deconvolution model. However, they entertain a generalized gamma family of densities for the shape of the pulsatile events as opposed to Gaussian-shaped events. They also assume a constant basal concentration and estimate parameters conditional on pulse number and location. Thus, their method also relies on criterion-based methods to identify pulse number and location. Guo et al. [10] also entertain a state space model; more precisely, a multiprocess dynamic linear model. Their model is the first one to include a changing basal concentration. They model it using the smoothing–spline approach to function estimation. Johnson [12] is the first attempt at a fully Bayesian approach to this problem. He assumes the deconvolution model of Veldhuis and Johnson [36] with constant basal concentration. However, the number of events and their locations are assumed to be random variables and are jointly estimated along with all other model parameters. Further, his method appears to be the first model that allows random variation of individual pulse parameters to account for biological variation between pulses within a subject that may be of scientific interest. Johnson [12] models these pulse parameters in a hierarchical fashion. The posterior number of pulses, or secretion events, is taken to be the mode of the marginal posterior distribution of this number. All further analyses are conditional on this number of pulses.

The two most recent methods, both of which incorporate a changing basal secretion function, are those due to Yang et al. [42] and Johnson [13]. Yang et al. [42] fit a nonlinear mixed-effect partial spline model to pulsatile hormone data. Their model relies on some external method to identify an initial number of pulses along with their locations. The basal concentration is assumed to be a nonconstant function and is considered a nuisance. Nevertheless, it must be accounted for in the analysis of hormones that suggest a nonconstant basal concentration. They suggest the use of the Bayesian information criterion (BIC) as a model selection procedure for final selection of the number of pulses. The most recent approach is that of

Johnson [13]. His approach generalizes the early fully bayesian model to include a varying basal concentration. Further, he does not condition on the mode of the number of events, as he did earlier [12]. Rather, he takes the more germane Bayesian approach of model averaging (over the number and location of pulse functions) to obtain a better overall fit to the data and to account for variation across models.

The remainder of this chapter outlines and compares these last two models. Both models are based on a general biophysical model outlined in Section 29.2 Section 29.3 outlines the Bayesian modeling approach and is applied to a real dataset. The nonlinear mixed-effect partial spline model is introduced in Section 29.4 and is applied to the same dataset. We conclude with a comparison of the two methods, including each method's strengths and weaknesses, in Section 29.5 as well as a brief discussion on directions for future research.

29.2 A GENERAL BIOPHYSICAL MODEL

In this section we present two model-based methods for the analysis of episodic release of hormone. One of the methods is Bayesian, while the other takes a frequentist stance. Both methods are more or less based on the same model. Only the statistical paradigm used in estimation differs. The biophysical model presented follows that in Yang et al. [42].

Both *in vivo* and *in vitro* studies of the endocrine system suggest the existence of two physiologically distinct modes of hormonal secretion: *basal* secretion, which is characterized by a slowly varying (perhaps constant) release or granule leakage of hormone, and *episodic* or *pulsatile* secretion, which is characterized by a large amount of hormone released into the circulatory system in a relatively short time. This release is typically caused by some type of signaling mechanism, such as another hormone, that triggers the release of the hormone in question [2].

To this end, let $s(t)$ denote the secretion rate. Then s can be written as

$$s(t) = b(t) + \Psi(t),$$

where $b(t)$ is the basal component and $\Psi(t)$ is the pulsatile component of the hormone secretion rate. The pulsatile component will take a parametric form and is made up of piecewise additive pulse functions, $\Psi_k(t; \tau_k, \gamma_k)$, where τ_k is the location, or occurrence time, of the pulse and γ_k is a vector of pulseshape parameters:

$$\Psi(t) = \sum_{k=1}^{K} \Psi_k(t; \tau_k, \gamma_k). \tag{29.1}$$

Typically the number of pulses K is unknown and is to be estimated along with the τ_k and γ_k.

There are several mechanisms for the clearance of hormone from the circulatory system, including cellular binding of hormone to target cells, enzymatic cleavage, and glomerular filtration. Typically these mechanisms cannot be modeled individually, and a single elimination function is used to model the overall clearance rate. This elimination or "decay function" E, which describes the removal of hormone from the circulatory system, is typically chosen to be exponential or biexponential unless experimentation suggests some other form.

Let $c(t)$ denote the true concentration level at time t. A standard model for the true concentration level at time t is [18]

$$
\begin{aligned}
c(t) &= c(0)E(t) + g(t) + (s * E)(t) \\
&= c(0)E(t) + g(t) + (b * E)(t) + (\Psi * E)(t),
\end{aligned}
\tag{29.2}
$$

where the asterisk ($*$) represents the convolution operator defined by $(u * v)(t)\,0 = \int_0^t v(s)v(t - s)ds$, $c(0)$ is the initial concentration at time zero (the beginning of the experiment), and g represents microscopic biological variation.

In many experiments the main interest lies in the number of episodic secretion events and the amount of hormone secreted during each event. Thus we treat each of the first three summands on the right-hand side (RHS) of (29.2) as a nuisance and write them as a single function

$$
f(t) = c(0)E(t) + g(t) + (b * E)(t).
\tag{29.3}
$$

Function f is unobservable, and we approximate it using a spline representation. Details of how this is accomplished are given in Sections 29.3 and 29.4.

Let Y_t denote the observed concentration at time t (neither method described below requires that the observed times series be equally spaced in time). Let ε_t denote the error at time t. Let n denote the number of observations in the time series. Then our model, for $j = 1, \ldots, n$, is

$$
\begin{aligned}
Y_j &= c(t_j) + \varepsilon_j = f(t_j) + (\Psi * E)(t_j) + \varepsilon_j \\
&= f(t_j) + \sum_{k=1}^{K} \int_0^{t_j} \Psi_k(v;\, \tau_k, \gamma_k)E(t_j - v)dv + \varepsilon_j.
\end{aligned}
\tag{29.4}
$$

It is common to assume that the $\varepsilon_t \overset{\text{i.i.d.}}{\sim} N(0, \sigma)$. However, in this work we allow for more general error structures. In particular, let $\varepsilon = (\varepsilon_1, \ldots, \varepsilon_n)^T$. Then we assume $\varepsilon \sim N_n (0, \sigma^2\Lambda)$.

Model (29.4) is the starting point for both the Bayesian and frequentist methods. There are several challenging issues in fitting this model to data, including the following:

1. The number of episodic events K is unknown, as is the location of these events.

2. Deconvolution is an ill-posed problem.

3. The function $f(t)$ is latent and must be approximated.

29.3 BAYESIAN DECONVOLUTION MODEL (BDM)

One of the biggest challenges to overcome when modeling (29.4) is determination of the number of episodic events and their locations. From the frequentist perspective, deconvolution of episodic hormone profiles is a two-stage process: (1) the number and approximate locations of episodes is determined; and (2) conditional on the number of episodes K, parameters in model (29.4) are estimated. Thus an error in the first stage carries over to the second. From the Bayesian perspective, this is not a problem. The quantity K can be assumed to be an unknown random variable and estimated from the data in a fully Bayesian model simultaneously with all other parameters. However, as K varies, so do the number of parameters γ_k

and locations τ_k. Thus, by allowing K to change, we have a variable-dimension parameter space. Standard MCMC techniques, such as Gibbs sampling [10] and the Metropolis–Hastings algorithm [26,11], require fixed parameter spaces. There have been, however, more recent advances in MCMC simulation that allow for variable-dimension parameter spaces. Two of these are the reversible jump Markov chain Monte Carlo (RJMCMC) [9] and the birth–death Markov chain Monte Carlo (BDMCMC) [32].

We begin by making a slight modification to the deconvolution model introduced in Equation (29.4). The modification is motivated by the fact that the observed concentration must be nonnegative. Hence, the error term ε_j is strictly nonnegative and the normality of the error may be called into question. Further, any symmetric error structure may be inappropriate [31]. Here we allow the error term to depend on the mean by taking a log transformation of the data:

$$\ln(Y_j + 1) = \ln\left(f(t_j) + \sum_{k=1}^{K} \int_0^{t_j} \Psi_k(v; \tau_k, \gamma_k) E(t_j - v) dv \right) + \varepsilon_j. \qquad (29.5)$$

Note that we have added 1 to the observed concentration prior to taking the natural logarithm. This is done to help with model fit [13]. Further, K will no longer have the interpretation of the number of episodic secretion events. Rather, we think of it as the number of component functions that, when summed or superimposed, make up the pulse function $\Psi(t)$. This interpretation allows for an episodic secretion event to be made up of the superposition of two or more component functions $\Psi_k(t; \tau_k, \gamma_k)$, thereby allowing more flexibility in model fitting (analogous to the way a mixture distribution of normal densities can be used to fit arbitrarily shaped densities, including multimodal densities).

We assume that hormone elimination is exponential: $E(t) = \exp(-\delta t)$. Further, we assume that the component functions are Gaussian-shaped: $\Psi_k(t; \tau_k, \gamma_k) = \alpha_k \exp[-0.5(t - \tau_k)^2/v_k^2]/\sqrt{2\pi v_k^2}$, hence the parameter set $\gamma_k = \{\alpha_k, v_k^2\}$. Each α_k is the amount of hormone secreted that is attributable to the kth component function and $\sum_{k=1}^{K} \alpha_k$ is thus the total amount of hormone released from all functions, hence from all episodic events. Further we assume that the errors $\varepsilon_j \sim N(0, \sigma)$ independently of one another. Johnson [13] considers other forms for the error structure, allowing correlated errors, but found that this error structure gave the best overall fit to the data.

Thus the likelihood formulation of our model is

$$\left[\ln(Y_j + 1) \mid \Theta\right] \sim N[\ln[c(t_j)], \sigma], \qquad (29.6)$$

where Θ is the collection of all model parameters.

Johnson [12] considers a model similar to that in Equation (29.5). However, he considers only a constant function $f(t_j) \equiv c$. Here, we consider a more general approximation of $f(t_j)$ by using a cubic B-spline approximation [4]. In particular, let P denote the number of interior knots at locations $\{\xi_i\}_{i=1}^{P}$. We consider both P and $\{\xi_i\}$ to be unknown quantities to be estimated. Further, conditional on P, let $\{\beta_i\}_{i=1}^{(P+4)}$ denote the set of B-spline coefficients and $X(\{\xi_i\})$ be a design matrix whose rows are the basis functions for the B-spline representation of the function f. Note that X depends on both the number and locations of the knots. For notational clarity, we henceforth drop the dependence of X on the number and locations of the knots. The dimension of X is $N \times (P + 4)$ where N is the number of observations. Note that eight additional knots (in addition to the interior knots) must be specified for a cubic B-spline. Four of the knots are placed at 0, the beginning of the data, and the other four are placed at 24, the end of the data. Then f can be approximated at the observation times t_j by $f(t_j) \approx \langle X_j^T, \beta_P \rangle$, where X_j^T, is the jth row of X and $\beta_P = (\beta_1, \ldots, \beta_{P+4})^T$.

We are now in a position to specify priors on all model parameters Θ. We present priors based on the factorization of the full joint prior distribution given at the end of this section.

We begin with parameters of the B-spline approximation to f. The number of knots P is assigned a negative binomial (Negbin) prior with mean 3 and variance 6: $P \sim \text{Negbin}(3, 1)$. The number of knots can be regarded as a smoothing parameter in a B-spline representation [4] with fewer knots resulting in smoother approximations to f. Since we believe that f should be a rather smooth function, we choose to assign P a negative binomial prior with a small mean. Conditional on P, the a priori locations $\{\xi_i\}$ of the knots are assumed independent and uniformly distributed over the support of the sample space ([0,24]). The conditional joint prior density is thus $p(\xi_1, \ldots, \xi_P | P) = 24^{-P}$.

The last set of parameters that require a prior distribution, for the specification of f, is $\{\beta_i\}$. Each summand that contributes to the function f in Equation (29.3) is necessarily nonnegative as each summand represents a component of the overall concentration. Therefore, it is natural to approximate the nonnegative function f, with a nonnegative approximation. Further, the basis functions of the B-spline representation of any function are nonnegative. Hence a sufficient condition for a B-spline function to be nonnegative is that each B-spline coefficient β_i, $i = 1, \ldots, P$, is nonnegative [4]. We handle this constraint by modeling the natural log of the B-spline coefficients hierarchically: $\ln(\beta_i) | P, \beta, \psi \sim N(\beta, \psi)$, with $\beta \sim N(3, 1)$ and $\psi^2 \sim IG(2.1, 2)$. (Our simulation studies show that results are rather insensitive to the choice of these priors. However, the acceptance rate in the reversible jump MCMC step is quite sensitive. These priors were chosen to give reasonable acceptance rates.) Furthermore, the offset of 1 added to the observed concentration must be accounted for in the estimation of the function f; otherwise the estimate of f can include negative values—an impossibility. We do so by adding 1 to the estimate of the function f.

The priors on the parameters of the component functions $\Psi_k(t, \tau_k, \gamma_k)$ are specified hierarchically. We begin with the number of components K, which we give a negative binomial prior with mean 10 and variance 20: $K \sim \text{Negbin}(10, 1)$. This prior reflects our belief that the number of secretion events should be about 10 in a 24-h period. However, it is also sufficiently variable to allow fewer events and allow for the possibility that secretion events may be made up of several component functions. Given K, the locations of the component functions, $\{\tau_k\}$ are assumed to be independently distributed as uniform random variables on the support of the sample space. Their joint prior density, conditional on K, is $p(\tau_1, \ldots, \tau_K | K) = 24^{-K}$.

The other parameters necessary for specification of the K component functions are the α_k and v_k^2, $k = 1, \ldots, K$. Since the $v_k^2 > 0$, we place a normal distribution on the log of the v_k^2:

$$\ln(v_k^2) | K, v, \zeta \sim N(\ln(v), \zeta), \qquad k = 1, \ldots, K,$$

$$\ln(v) \sim N(-1, 1), \quad \zeta^2 \sim IG(5, 2).$$

The mass secreted from each component k, namely, α_k, must also be positive. We also specify this prior hierarchically:

$$\ln(\alpha_k) | K, \alpha, \vartheta \sim N(\ln(\alpha), \vartheta), \qquad k = 1, \ldots, K,$$

$$\ln(\alpha) \sim N(3, 1), \quad \vartheta^2 \sim IG(5, 2).$$

These prior and hyperprior distributions are based partly on mathematical convenience and on satisfying the positivity constraints. The parameter values of these distributions were derived according to what we believe to be biologically relevant values. Simulation studies (unpublished) show that the results are only modestly sensitive to the variances of these distributions. However, uninformative priors are not sensible here, because of the ill-posed nature of the problem. The last two parameters that require prior distributions are the decay rate δ and the model error variance σ^2. We choose to model the decay function in terms of the hormone half-life instead of the decay rate δ. The two are related by $t_{1/2} = \ln(2)/\delta$. During MCMC simulation of the posterior, we found that modeling the removal of hormone from the system in terms of the half-life resulted in better simulation performance. When the decay rate δ was used, it often happened that both δ and the α_k would simultaneously escape to very large numbers—numbers that are biologically impossible—never to return. The most likely reason for this phenomenon is the ill-posed nature of the problem. A large α_k can be counteracted by a large decay rate in the convolution model with very little change in the concentration. This does not happen when we use the half-life because as $\delta \to \infty$, $t_{1/2} \to 0$. Since the half-life is a strictly positive number, we assign it a lognormal prior: $\ln(t_{1/2}) \sim N(-1, 1)$. This prior has 90% of it's mass between 0.07 and 1.9, while only 1.6×10^{-4} of the mass is less than 0.01, thus controlling the half-life from becoming too small (for cortisol, half-lives are typically in the range of 30 min–1 h). The model error variance is given a vague, proper prior: $\sigma^2 \sim IG(0.001, 0.001)$.

Given these prior distributions, we can write the full joint prior distribution as the product of the marginal and conditional prior factors given above:

$$
p\left[\{\beta_i\}, \beta, \psi^2, \{\xi_i\}, P, \{\tau_j\}, \{v_j^2\}, v, \zeta^2, \{\alpha_j\}, \alpha, \vartheta^2, K, t_{1/2}, \sigma^2\right]
$$

$$
= \prod_{i=1}^{P} \left[p(\{\beta_i\}|P, \beta, \psi^2) p(\{\xi_i\} \mid P) \right]
$$

$$
\times \prod_{j=1}^{K} \left[p(\tau_j \mid K) p(\alpha_j \mid K, \alpha, \vartheta^2) p(v_j^2 \mid K, v, \zeta^2) \right]
$$

$$
\times p(P) p(\beta) p(\psi^2) p(K) p(v) p(\zeta^2) p(\alpha) p(\vartheta^2) p(t_{1/2}) p(\sigma^2).
$$

29.3.1   Posterior Processing

All our analyses are based on marginalizing over the number of component functions; that is, we average over all models indexed by the number of component functions, K. To effect a one-to-one correspondence between the number of component functions and the number of secretion events, a conditional analysis may be more appropriate. In this case, we suggest that the conditional prior on the location of the component functions, that is, $p(\tau_1, \ldots, \tau_K \mid K)$, be changed. In this situation we might expect that the secretion events are somewhat spread out over time, and the prior should reflect this belief. Therefore, a suitable prior might be every third-order statistic from $3K + 2$ uniform random variables over the range of the data. This is the prior chosen by Johnson [12].

We approach the problem from a different angle and believe that it is reasonable to assume that several component functions could possibly make up a secretion event, giving the model more flexibility in fitting the data. However, this approach causes another problem: determination of the number and locations of secretion events, which in many studies is of primary interest. One solution to this problem is to use the estimated marginal posterior distribution of the locations of the component functions. The estimated density of this distribution is

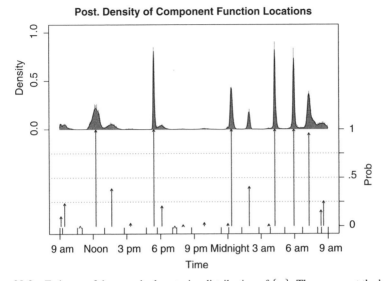

Figure 29.2   Estimate of the marginal posterior distribution of $\{\tau_j\}$. The arrows at the bottom denote candidate locations for secretion events as obtained by a model-based clustering algorithm. The probabilities of an event for the 23 candidate locations (intervals) are given in Table 29.1 and are shown above by the height of the arrows. The "rug" at the bottom of the figure demarcates the endpoints of the intervals on which probabilities are calculated.

given in Figure 29.2. The density is multimodal, and it is reasonable to assume that each mode i represents the most likely position of secretion event i [i.e., its maximum a posteriori (MAP) estimate]. The arrows at the bottom of the figure show where all the modes occur (some of the modes are too small to be detected in the graph). The modes are located by first smoothing the histogram using a mixture of Dirichlet process (MDP) priors to estimate the density [7,1,6,28] with a smooth, continuous curve. Details can be found in Johnson [13]. The MDP density estimate is shown in Figure 29.2 as the solid black line. The modes of the density estimate can be obtained by analytical differentiation of the MDP density estimate on a fine grid and a numerical search for those times where the first derivative is zero and the second derivative is negative (see Johnson [13] for a justification of this approach).

One nice feature of the Bayesian approach, and this method of determining locations of "candidate" secretion events, is that we can estimate the probability that a secretion event SE_i occurs in a neighborhood surrounding its MAP estimate M_i. Define the probability of an event occurring in the neighborhood around the MAP estimate of secretion event i as the ratio of the number of iterations that contain at least one component function in the neighborhood to the total number of iterations. Note that we do not take the ratio of the total number of component functions within the neighborhood because more than one component function might make up a secretion event. In fact, a closer examination revealed that two or more component functions are used to make up the secretion event whose MAP estimate is at 12:12 p.m. roughly 50.6% of the time, where the neighborhood is defined as the interval whose endpoints are the two surrounding minima. All probabilities given in Figure 29.2 (where the heights of the arrows indicate probabilities) and Table 29.1 are defined on intervals that have the surrounding local minima, $M_i \in (m_{i-1}, m_i)$, as their endpoints.

Table 29.1 Posterior Probabilities of Event Candidates (Cand.)[a]

Cand.	Time	Pr	Mass	Cand.	Time	Pr	Mass
1	9:06a	0.103	1.123	11	9:54p	0.037	0.124
2	9:26a	0.237	2.362	12	12:00a	0.027	0.079
3	10:50a	0.004	0.023	13	12:20a	0.999	7.250
4	12:16p	0.996	22.424	14	1:56a	0.410	1.858
5	1:41p	0.391	3.998	15	3:41a	0.020	0.046
6	3:20p	0.032	0.184	16	4:15a	0.999	22.323
7	5:25p	0.999	15.682	17	5:57a	0.999	21.191
8	6:07p	0.210	2.555	18	7:17a	0.959	25.561
9	7:15p	0.001	0.003	19	8:21a	0.163	1.857
10	8:00p	0.015	0.060	20	8:35a	0.257	3.153

[a]Times correspond to location modes, and probabilities correspond to the intervals (m_{i-1}, m_i), $i = 1, \ldots,$ 20, shown in Figure 29.2.

29.3.2 An Example

The cortisol time series shown in Figure 29.1 is fitted with the Bayesian deconvolution model. We run the MCMC sampler for 525,000 iterations with a burn-in of 25,000. The final Markov chain is thinned by saving every 250th iteration. Thus the final posterior is estimated with 2000 samples. The sampler is coded in the C programming language. The simulation takes approximately 28 min on a PowerPC G5, 2.7-GHz computer. The estimated posterior predictive mean and the spline approximation to $f(t)$ are displayed along with the data in Figure 29.3, from

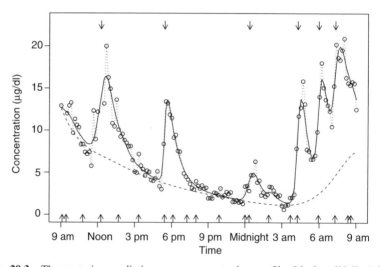

Figure 29.3 The posterior predictive mean concentration profile (black solid line) is displayed on top of the data (open circles and dotted line). The dashed line is the estimated posterior mean of the nuisance function: $E(f(t) \mid Y)$. The arrows at the bottom indicate the MAP estimates of all candidate secretion events, those at the top indicate the MAP estimates for all events whose probability of occurring between two local minima is greater than 50%.

which it appears that the model fits the data well. Bayesian deleted residuals (not shown) verify this conclusion. Also, the Bayesian χ^2 goodness-of-fit statistic further verifies that there is no overall lack of fit or evidence of overfitting [14]. The arrows at the top of the figure indicate the most probable locations of the major secretion events (a *major secretion event location* is defined as a location that is detected in our modeling procedure over 50% of the time). Our method detected six major secretion events in this dataset.

The marginal estimated posterior density of the locations τ_k of the component functions $\Psi_k(t; \tau_k, \gamma_k)$ is shown in Figure 29.2. We note that this density estimate is marginal over all other parameters, including K, the number of component functions. The arrows under the density indicate the MAP estimate of each candidate secretion event, based on our posterior processing described in Section 29.3.1. The heights of the arrows indicate the probabilities that an event has occurred in the intervals whose endpoints are demarcated by the "rug" at the bottom of the figure.

The posterior probability of each of the 20 candidate secretion events, their MAP estimates, and the mean mass of hormone secreted are tabulated in Table 29.1. It is interesting to compare the times given in Table 29.1 for all 20 candidate secretion locations and the concentration pattern in Figures 29.3. Note that the times appear prior to the peaks in concentration as a result of the convolution. Further, one can see that each candidate event appears to be associated with a peak, spike, or bulge in the concentration. For example, consider the sixth candidate that occurs at 3:20 p.m. It is detecting a small spike in concentration near that same time, as can be clearly seen in Figure 29.1. However, it appears in only 3% of the posterior samples; the rest of the time the model considers it noise. Another example is candidate 14, which has a posterior probability of 0.41. It is picking up the bulge in the data that is evident at about 2:00 a.m. The other 59% of the time it is considered noise. However, because of model averaging, the overall fit to the data around 2:00 a.m. is better than it would be if we had conditioned on the number of secretion events detected with a threshold of 0.5.

29.4 NONLINEAR MIXED-EFFECTS PARTIAL-SPLINES MODELS

Again, we assume the model (29.4) and let $\Psi_k(v; \tau_k, \gamma_k) = a_k \Psi (v; \tau_k, \gamma_k)$, where a_k is referred to as *amplitude*. Define a pulseshape function as $p(t; \tau, \gamma) = \int_0^t \Psi (v; \tau, \gamma)E(t - v)dv$ where τ is a pulse location and γ are pulseshape parameters. We model f nonparametrically using a polynomial spline

$$W_m = \left\{ f : f, f', \dots, f^{(m-1)} \text{ absolutely continuous, } \int_0^1 (f^{(m)})^2 dt < \infty \right\}. \quad (29.7)$$

A common choice of the order m is $m = 2$, which corresponds to a cubic spline.

All frequentist approaches except Yang et al. [42] treat the amplitudes as deterministic parameters and ignore variations in pulseshape parameters. More recent studies indicate that the amplitudes and pulseshape parameters vary during the day [18,19,17], and it is of scientific interest to model the variation between pulses.

Let $\boldsymbol{a} = (a_1, \dots, a_K)^T$, $\gamma = (\gamma_1^T, \dots, \gamma_K^T)^T$, and $\phi = (\boldsymbol{a}^T, \gamma^T)^T$. To model variation between pulses, we assume the following linear mixed model

$$\phi = A\beta + \boldsymbol{B}\boldsymbol{b}, \quad \boldsymbol{b} \sim N(0, \sigma^2 \boldsymbol{D}), \quad (29.8)$$

where β and $\boldsymbol{b}$ are fixed and random effects and A and $\boldsymbol{B}$ are design matrices for the fixed and random effects, respectively. Note that the form of (29.8) is general, which may be used to construct various second-stage models.

Let $\boldsymbol{Y} = (Y_1, \ldots, Y_n)^T$, $\boldsymbol{f} = (f(t_1), \ldots, f(t_n))^T$, $\varepsilon = (\varepsilon_1, \ldots, \varepsilon_n)^T$, and $\eta = (\sum_{k=1}^{K} a_k p(t_1; \tau_k, \xi_k), \ldots, \sum_{k=1}^{K} a_k p(t_n; \tau_k, \xi_k))^T$. A NMPSM is the combination of the first-stage model (29.4) and the second stage model (29.8)

$$
\begin{aligned}
y &= f + \eta + \varepsilon, \quad \varepsilon \sim N(0, \sigma^2 \Lambda), \\
\phi &= A\beta + Bb, \quad b \sim N(0, \sigma^2 D).
\end{aligned}
\tag{29.9}
$$

Here, Λ and D are assumed to depend on an unknown parameter vector θ.

The pulse detection analysis involves estimation of the following parameters: the number of pulses K, pulse locations $\tau = (\tau_1, \ldots, \tau_K)^T$, β, f, θ, σ^2, and $\boldsymbol{b}$. Since the total number of parameters depends on the unknown parameter K, it is difficult to estimate all the parameters simultaneously in a frequentist framework. Yang et al. [42] propose the following algorithm:

1. *Initialize*—identify potential pulse locations and provide initial values. Denote the total number of potential pulses as $K_{\max}$. Specify a lower bound for the number of pulses $K_{\min}$.
2. *Pulse detection*:
   a. For $K = K_{\max} \ K_{\max} - 1, \ldots, K_{\min}$, repeat
      i. Fit the model (29.9) and compute t statistics t_k, $k = 1, \ldots, K$.
      ii. Delete the location with the smallest $|t_k|$.
   b. Select the final model using one of the AIC, BIC, RIC, and GCV criteria.
3. *Parameter estimation*—fit the final model.

At step 1, we focus on finding all possible pulses and are less concerned with false identifications. Quantity $K_{\min}$ may be taken as zero. When K and τ are fixed in steps 2 and 3, we fit the corresponding NMPSM to estimate parameters β, f, θ, σ^2, and $\boldsymbol{b}$. The estimation procedure iterates between (a) fixing σ^2 and θ as their current estimates, estimate β, f, and $\boldsymbol{b}$ by minimizing a double-penalized log likelihood; and (b) fixing β, f, and $\boldsymbol{b}$ as their current estimates, estimate θ and σ^2 by maximizing an approximate profile likelihood. Details can be found in Yang et al. [42] and Ke and Wang [16].

We now provide more details about step 2 in the algorithm above. The t statistic is defined as $t_k = \hat{\alpha}_k / \sqrt{\hat{\text{var}}(\hat{\alpha}_k)}$, $k = 1, \ldots, K$, where $\hat{\text{var}}(\hat{\alpha}_k)$ is the approximate variance of $\hat{\alpha}_k$ after linearization (Theorem 1 in Ref. 38). Step 2a creates a nested sequence of pulse locations with their corresponding models denoted as $\mathcal{M}_{K_{\max}}, \ldots, \mathcal{M}_{K_{\min}}$. We define the total degrees of freedom for $\mathcal{M}_K$ as

$$
\text{df}_K \equiv \text{tr}\tilde{H}(\hat{\lambda}) + \text{IDF} \times \text{df}_P(K),
\tag{29.10}
$$

where $\tilde{H}(\hat{\lambda})$ is the smoother matrix for the nonparametric function f, $\hat{\lambda}$ is an estimate of the smoothing parameter λ by the GCV or the GML method [37,39], $\text{df}_P(K)$ is the number of parameters associated with pulses, and IDF (inflated degrees of freedom) accounts for the extra cost involved in selecting pulse locations [23]. A good choice of IDF is around 1.2.

We estimate K as the minimizer of $\hat{K}$ of the following criterion:

$$rss(K) + a\sigma^2 \mathrm{df}_K, \tag{29.11}$$

where $rss(K)$ is the residual sum of squares of model $\mathcal{M}_K$ and a is a constant that balances the tradeoff between goodness-of-fit and model complexity; $a = 2$, $a = \log n$, and $a = 2\log \mathrm{df}_{K_{max}}$ correspond to the AIC, BIC, and RIC criteria, respectively. We estimate σ^2 based on the biggest model $\mathcal{M}_{K_{max}}$. We may also use the GCV criterion

$$\frac{rss(K)}{(1 - \mathrm{df}_K/n)^2}.$$

We estimate pulse locations $\hat{\tau}$ as those pulse locations in the final model $\mathcal{M}_K$. Simulations show that all four model selection procedures work very well [42]. Models BIC and RIC perform slightly better. The whole procedure is quite stable, and there is no sign of overfitting. We used a fixed IDF to correct bias incurred by the adaptive pulse selection. One potential future research topic is to estimate the total degrees of freedom df_K in (29.10) by a data-driven procedure such as the generalized degrees of freedom [43], or to replace the penalty to model complexity [second term in (29.11)] by the covariance penalty [33].

We have developed an R package called PULSE for pulse detection based on NMPSMs. The package is available at http://www.pstat.ucsb.edu/faculty/yuedong/software.html. It consists of three main functions, `pulini`, `puldet`, and `pulest`, for the three steps in the algorithm presented above and other pulse detection and utility functions. Information about the utility functions `pul.control`, `summary.puldet`, and `summary.pulest` can be found in Yang et al. [41].

So far we have left the form of the pulseshape function p unspecified. Several prototype pulseshape functions are used in the literature. One simple and useful pulseshape function is the following double-exponential pulse function

$$p(t; \tau, \gamma) = \begin{cases} \exp\{\gamma_1(t - \tau)\}, & t < \tau, \\ \exp\{-\gamma_2(t - \tau)\}, & t \geq \tau, \end{cases} \tag{29.12}$$

where τ is the pulse location, γ_1 is the infusion rate, and γ_2 is the decay rate. The double-exponential pulseshape function is specified by `type = c("dblexp")` in our R functions. We have various options for each of the three parameters: infusion rate, decay rates, and amplitudes. Because of fast infusion relative to the sampling rate, the infusion rates are usually difficult to estimate since there are very few observations providing information about them. Furthermore, the parameters of interest are usually the decay rates and amplitudes. To improve numerical stability, we usually assume a common parameter for all infusion rates. The decay rates and amplitudes can be specified as common, fixed, random, or mixed, depending on the purpose of a fit. For example, to obtain initial values for fixed parameters, we may specify both of them as common. To detect pulse locations among the initial locations, we may specify amplitudes as random. For the final fit, we may use the most general model by specifying both of them as mixed.

We now use the same cortisol data to illustrate R functions in the PULSE package. We first load the ASSIST library (available at http://cran.r-project.org) and PULSE functions into R:

```
> library(assist)
> source("ssrfuns.R")
> source("puldet.R")
> source("pulest.R")
```

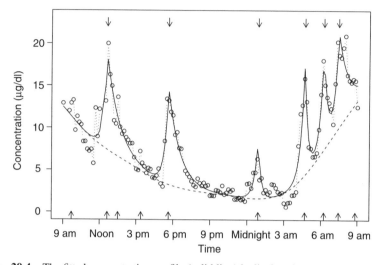

Figure 29.4   The fitted concentration profile (solid line) is displayed on top of the data (open circles and dotted line). The estimate of the baseline function f is displayed as the dashed line. Initial and final pulse locations are marked below and above, respectively.

```
> source("pulini.R")
> source("baseini.R")
```

The data frame cort consists of two variables, time and conc. Our first step is to detect potential pulse locations using pulini. Many existing pulse detection and change point detection methods may be used. When pulse locations are peaks of the double-exponential function, the mean function has change points in the first derivative at these positions. We use one method called "pcp" in Yang [40] to detect change points in the first derivative:

```
> pl <- pulini (time, conc, data=cort, method="pcp", alpha=0.6)
```

This method identifies 10 potential pulse locations, which are marked at the bottom of Figure 29.4. One may also use the CLUSTER method in Veldhuis and Johnson [35] by specifying method="CLUSTER".

Before fitting a NMPSM, we need initial values for the baseline function and parameters that can be derived by baseini and pulest, respectively:

```
> bl <- baseini(time, conc, data=cort, puloc=pl,
                method="shift", smooth="spline")
> fix.ini < - pulest(time, conc, data=cort, baseline=0, puloc=pl,
    start=list(fixed=c(1,1,1), Inif=bl), type=c("dblexp"),
    params=list(infrate="common", decrate="common",
    amplitude="common"), control=list(pul=list(TOLr=0.005,
    IDF=1, trace=F)))$coef$fixed
```

Alternative options for method and smooth in the baseini function are method = "select" and smooth = "loess". We only need initial values for fixed parameters. Therefore, we fit a simple fixed-effect model with common parameters for the infusion rate, decay rate, and amplitude.

Now we are ready for the second step: pulse detection. We use cubic spline to model the baseline function. Note that we set puloc = pl, the initial locations selected by the change point method. It is a good practice to check these locations before calling puldet to find obvious omissions and/or false locations. Among the initial locations, we need to detect those that have amplitudes significantly different from zero. For this purpose, we specify amplitudes as random, which is equivalent to shrinking them toward zero:

```
> det <- puldet(time, conc, data=cort, type=c("dblexp"),
    baseline=list(nb=~ time, rk=cubic(time)),
    start=list(puloc=pl, kmin=5, fixed=fix.ini[ 1:2] , Inif=bl),
    params=list(infrate="common", decrate="mixed",
    amplitude="random"))
> det
...
```

Fitting Table:

	BIC	RIC	AIC	GCV	DROP	DF
10	480.2258	598.1166	317.0249	361.0692	10	34.23649
9	460.2917	569.9182	308.5313	344.0252	4	31.83649
8	443.9538	545.3161	303.6339	333.8710	1	29.43649
7	431.7511	524.8492	302.8717	330.6114	3	27.03649
6	414.8130	499.6468	297.3740	320.3573	6	24.63649
5	423.5102	500.0799	317.5118	346.8865	9	22.23649

```
Initial location(s): 0.691 3.626 4.489 6.388 8.633 15.885
                     19.683 21.237 22.446 23.827

Location(s) selected:
with AIC,  6 pulse(s): 3.626 8.633 15.885 19.683 21.237 22.446
with GCV,  6 pulse(s): 3.626 8.633 15.885 19.683 21.237 22.446
with BIC,  6 pulse(s): 3.626 8.633 15.885 19.683 21.237 22.446
with RIC,  6 pulse(s): 3.626 8.633 15.885 19.683 21.237 22.446
```

All criteria select the same six locations. The final step is to fit the final model and derive estimates of the parameters. Since the estimation does not involve a selection process, we set IDF = 1:

```
> fit <- pulest(time, conc, data=cort, type=c("dblexp"),
    puloc=det$ detloc$ BIC$ puloc, start=list(fixed=fix.ini,
    Inif=bl), params=list(infrate="common", decrate="mixed",
    amplitude="mixed"), control=list(pul=list(IDF=1)))

> summary(fit)
...
```

```
Estimation at the selected model:

Parametric component
```

```
Fixed effects:
     Value Std.Error    t-value         p-value
l 1.1545492 0.1322566  8.729613 1.294610e-14
r 0.7369978 0.2435994  3.025450 3.009351e-03
a 2.3750959 0.1964722 12.088710 8.060218e-23
Random effects:
     A1       A2      A3      A4     A5      A6
  0.0412   0.0540 -0.4386 0.2153 0.0822 0.0457
     R1       R2      R3      R4     R5      R6
 -0.7072 -0.6738  0.7370 0.7228 0.5071 -0.5858

Non-parametric
estimate of smoothing parameter: 7.14286e-06
Degrees of Freedom of the baseline (df(base)): 7.836487

Residual standard deviation: 1.328 on 126.164 degrees
                                      of freedom
```

29.5 CONCLUDING REMARKS

The statistical analyses and models used to analyze pulsatile hormone time series have become more sophisticated over the years. In this chapter we have presented two new methods for analyzing a general, realistic biophysical model. For rather clean datasets, such as the one presented, the two approaches often lead to similar results. Both methods give reasonable results for the dataset analyzed, and there are only minor differences in the fit that can be seen in Figures 29.3 and 29.4. These differences may be attributable to the differences in the pulseshape function, the fact that a log transformation of the concentration was applied prior to model fitting in the BDM method, and differences in the baseline function approximation. The two most apparent differences are in the shape and height of the fitted concentration pulses. The NMPSM has sharper peaks that are higher than the corresponding peaks fitted with the BDM. The sharp peaks are due to the double-exponential waveform (29.12) assumed in the NMPSM. This form may also account for the relatively higher peaks. The difference in peak height may also be partially due to the log transformation used in the BDM analysis. Another minor difference occurs in the shape around 2 a.m. Around this time there is a shoulder in the BDM fitted concentration profile that is absent in the NMPSM fitted profile. This shoulder is due to the marginal approach taken in the BDM. The advantage of marginalizing is that it results in a better overall fit to the data than taking a conditional approach (within the Bayesian framework).

One advantage of the Bayesian approach over the frequentist approach is that it is more coherent. All parameters are estimated simultaneously (in the full posterior distribution), while the frequentist approach divides estimation into two stages: (1) pulse detection (estimation of K and τ) using a model selection procedure and (2) estimation of the parameters $\beta, f, \theta \sigma^2$, and b conditional on detected pulses. Therefore, the uncertainty in the first-stage analysis is ignored. A second advantage is that it provides posterior probabilities, not only a number of pulses together with their locations. Further, the Bayesian approach does not rely on initial pulse detection by some ad hoc model selection procedure.

A disadvantage of the Bayesian approach is that is it rather difficult to elicit prior information and to implement the advanced MCMC simulation (code is available from the first author). Further, because of the ill-posed nature of the deconvolution problem and its associated instability, uninformative prior specification and an objective Bayesian approach may not be feasible. Finally, the Bayesian approach is computationally intensive. The MCMC simulation of the cortisol dataset takes about 28 minutes on a Dual 2.7-GHz PowerMac G5. Further the MDP postprocessing takes an additional 15 min. The NMPSM method takes approximately 3 min on a PC with Dual Xeon 2.4-GHz processors.

Although great strides have been made in the modeling and analysis of pulsatile hormone data, there remains much room for future research. We have identified several areas where future research is welcome. First, many model-based approaches have appeared since 1992 or so. However, there has been no systematic evaluation and comparison of these models. It is time to take a step back and compare the relative merits of these different approaches.

Looking forward, as statistical methods and procedures have advanced, more realistic models have been entertained. As these methods continue to advance, even more realistic models will be plausible. These models may include feedback mechanisms, both autofeedback and feedback from other hormones and the central nervous system.

Pulse detection is only the initial step toward answering scientific and clinical questions raised at the beginning of this chapter. All current methods are based on a two-stage analysis. At the first stage, one models each subject separately to derive summary statistics such as frequencies, locations, and masses of pulses, half-lives, pulseshape parameters, and basal secretion rates. At the second stage, one analyzes each summary variable as a function of covariates. We are currently working on building integrated Bayesian/frequentist models for all subjects to provide coherent inference for the population. Pooling data from all subjects, we expect the integrated approach to be more efficient and robust. It will allow us to investigate covariate effects and variation between subjects as well as variation between pulses within a subject. See Liu and Wang [22] for an alternative approach that does not require pulse detection.

ACKNOWLEDGMENT

The authors would like to thank Dr. Elizabeth A. Young of the Department of Psychiatry and Mental Health Research Institute, University of Michigan, for the use of the cortisol hormone data. We are also grateful to the referees for their helpful suggestions. The work of Timothy D. Johnson was funded by the NIH grant P60 DK20572. The wrok of Yuedong Wang was supported by NIH grant R01 GM58533.

REFERENCES

1. Antoniak, C. E., Mixtures of Dirichlet processes with applications to Bayesian nonparametric problems, *Ann. Statist.* **2**, 1152–1174 (1974).

2. Berne, R. M. and Levy, M. N., eds., *Physiology*, 3rd ed., Mosby Year Book, St. Louis, MO, 1993.

3. Clifton, D. K. and Steiner, R. A., Cycle detection: A technique for estimating the frequency and amplitude of episodic fluctuations in blood hormone and substrate concentrations, *Endocrinology* **112**, 1057–1064 (1983).

4. de Boor, C., *A Practical Guide to Splines*, Springer-Verlag, New York, 1978.

5. Diggle, P. J. and Zeger, S. L., A non-Gaussian model for time series with pulses, *J. Am. Statist. Assoc.* **84** 354–359 (1989).

6. Escobar, M. D. and West, M., Bayesian density estimation and inference using mixtures, *J. Am. Statist. Assoc.* **90**, 577–588 (1995).

7. Ferguson, T. S., A Bayesian analysis of some nonparametric problems, *Ann. Statist.* **1**, 209–230 (1973).

8. Goodman, R. L. and Karsch, F. J., Pulsatile secretion of luteinizing hormone: Differential suppression by ovarian steroids, *Endocrinology* **107**, 1286–1289 (1980).

9. Green, P. J., Reversible jump Markov chain Monte Carlo computation and Bayesian model determination, *Biometrika* **82**, 711–732 (1995).

10. Guo, W., Wang, Y., and Brown, M. B., A signal extraction approach to modeling hormone time series with pulses and a changing baseline, *J. Am. Statist. Assoc.* **94**, 746–756 (1999).

11. Hastings, W. K., Monte Carlo sampling methods using Markov chains and their applications, *Biometrika* **57**, 97–109 (1970).

12. Johnson, T. D., Bayesian deconvolution analysis of hormone concentration profiles, *Biometrics* **59**, 650–660 (2003).

13. Johnson, T. D., Detecting pulsatile hormone secretion events: A Bayesian approach, Univ. Michigan Dept. Biostatistics Working paper series, Working Paper 56 (http://www.bepress.com/umichbiostat/paper56), 2006.

14. Johnson, V. E., A Bayesian chi-squared test for goodness of fit, *Ann. Statist.* **32**, 2361–2384 (2004).

15. Kaipio, J. and Somersalo, E., *Statistical and Computational Inverse Problems*, Springer, New York, 2005.

16. Ke, C. and Wang, Y., Semi-parametric nonlinear mixed effects models and their applications (with discussion), *J. Am. Statist. Assoc.* **96**, 1272–1298 (2001).

17. Keenan, D. M. and Veldhuis, J. D., Cortisol feedback state governs adrenocorticotropin secretory-burst shapes, frequency, and mass in a dual-waveform construct: Time of day-dependent regulation, *Am. J. Physiol.* **285**, R950–R961 (2003).

18. Keenan, D. M., Veldhuis, J. D., and Yang, R., Joint recovery of pulsatile and basal hormone secretion by stochastic nonlinear random-effects analysis, *Am. J. Physiol.* **275**, R1939–R1949 (1998).

19. Keenan, D. M. et al., Physiological control of pituitary hormone secretory-burst mass, frequency, and waveform: A statistical formulation and analysis, *Am. J. Physiol.* **285**, R664–R673 (2003).

20. Komaki, F., State-space modelling of time series sampled from continuous processes with pulses, *Biometrika* **80**, 417–429 (1993).

21. Kushler, R. H. and Brown, M. B., A model for the identification of hormone pulses, *Statist. Med.* **10**, 329–340 (1991).

22. Liu, A. and Wang, Y., Modeling of hormone secretion-generating mechanisms with splines: A pseudo-likelihood approach, *Biometrics* (in press).

23. Luo, Z. and Wahba, G., Hybrid adaptive splines, *J. Am. Statist. Assoc.* **92**, 107–116 (1997).

24. Mauger, D. T. and Brown, M. B., A comparison of methods that characterize pulses in a time series, *Statist. Med.* **14**, 311–325 (1995).

25. Merriam, G. R. and Wachter, K. W., Algorithms for the study of episodic hormone secretion, *Am. J. Physiol.* **243**, 310–318 (1982).

26. Metropolis, N. et al., Equations of state calculations by fast computing machines, *J. Chem. Phys.* **21**, 1087–1091 (1953).

27. Munson, P. J. and Rodbard, D., Pulse detection in hormone data: Simplified efficient algorithm, in *Proc. Statistical Computing Section, American Statistical Association*, 1989, pp. 295–300.

28. Neal, R. M., Markov chain sampling methods for Dirichlet process mixture models, *J. Comput. Graph. Statist.* **9**, 249–265 (2000).

29. Oerter, K. E., Guardabasso, V., and Rodbard, D., Detection and characterization of peaks and estimation of instantaneous secretory rate for episodic pulsatile hormone secretion, *Comput. Biomed. Res.* **19**, 170–191 (1986).

30. O'Sullivan, F. O. and O'Sullivan, J., Deconvolution of episodic hormone data: An analysis of the role of season on the onset of puberty in cows, *Biometrics* **44**, 339–353 (1988).

31. Rodbard, D., Rayford, P. L., and Ross, G. T., Statistical quality control. in *Statistics in Endocrinology*, McArthur, J. W. and Colton, T., eds., MIT Press, Cambridge, MA, 1970, pp. 411–429.

32. Stephens, M., Bayesian analysis of mixture models with an unknown number of components—an alternative to reversible jump methods, *Ann. Statist.* **28**, 40–74 (2000).

33. Tibshirani, R. and Knight, K., The covariance inflation criterion for adaptive model selection, *J. Roy. Statist. Soc. Ser. B, Statist. Methodol.* **61**, 529–546 (1999).

34. Van Cauter, E. L. et al., Quantitative analysis of spontaneous variation in plasma prolactin in normal man, *Am. J. Physiol.* **241**, 355–363 (1981).

35. Veldhuis, J. D. and Johnson, M. L., Cluster analysis: A simple, versatile and robust algorithm for endocrine pulse detection, *Am. J. Physiol.* **250**, 486–493 (1986).

36. Veldhuis, J. D. and Johnson, M. L., Deconvolution analysis of hormone data, in *Methods in Enzymology*, Brand, L. and Johnson, M. L., eds., Academic Press, San Diego, CA, 1992, pp. 539–575.

37. Wahba, G., *Spline Models for Observational Data*, Society for Industrial and Applied Mathematics, Philadelphia, 1990.

38. Wang, Y., Mixed-effects smoothing spline ANOVA, *J. Roy. Statist. Soc. Ser. B* **60**, 159–174 (1998).

39. Wang, Y. and Ke, C., *ASSIST: A Suite of s Functions Implementing Spline Smoothing Techniques. Manual for the ASSIST Package* (available at http://www.pstat.ucsb.edu/faculty/yuedong/software), 2002.

40. Yang, Y., *Detecting Change Points and Hormone Pulses Using Partial Spline Models*, Ph.D. thesis, Dept. Statistics and Applied Probability, Univ. California—Santa Barbara, 2002.

41. Yang, Y., Liu, A., and Wang, Y., *PULSE: A Suite of R Functions for Detecting Pulsatile Hormone Secretions. Manual for the PULSE Package* (available at http://www.pstat.ucsb.edu/faculty/yuedong/software), 2004.

42. Yang, Y., Liu, A., and Wang, Y., Detecting pulsatile hormone secretions using nonlinear mixed effects partial spline models, *Biometrics* **62**, 230–238 (2005).

43. Ye, J. M., On measuring and correcting the effects of data mining and model selection, *J. Am. Statist. Assoc.* **93**, 120–131 (1998).

CHAPTER 30

Models for Carcinogenesis

Anup Dewanji

Indian Statistical Institute, Kolkata, India

30.1 INTRODUCTION

In the past few decades (as of 2007) a vast biostatistical literature has appeared on exposure–response or dose–response analyses for experimental (or, toxicological) and epidemiological studies in which the endpoint of interest is cancer. Along with the development of sophisticated statistical tools for the analyses of such data, there was also significant development in the area of modeling the cancer process, or carcinogenesis [53,39]. The ultimate interest lies, of course, in modeling the dose–response relationship between a disease outcome (i.e., cancer) of interest and exposure level of a primary agent under study, so that a cancer risk assessment can be carried out [38].

Ideally, risk assessments should be based on epidemiological studies involving human subjects, since the risk on their health and well-being is of primary interest and also the estimates of risk for levels of exposure, that are close to those of human living conditions, can be directly obtained. However, the exposure levels and disease outcomes are often measured with less precision in epidemiological studies than in experimental studies, leading to possible bias in risk estimates. Also, exposure to multiple agents (confounders) makes it difficult to investigate the effect of a single primary agent. When appropriate epidemiological studies are not available, one has to rely on experimental data in which the levels of exposure are typically higher than those in the general population. Although the problem of precision in the measurement of exposure and disease outcome and that of potential confounders can be largely controlled in experimental studies with laboratory animals, the twin problems of low dose and interspecies extrapolation become difficult issues. These two are among the most contentious scientific issues of the day.

An appropriate modeling of the relationship between the different exposure levels and the disease under study may be able to address the issues in a meaningful way, but there always remains the question of validity of a particular model being used. While purely standard

Statistical Advances in the Biomedical Sciences, edited by Atanu Biswas, Sujay Datta, Jason P. Fine, and Mark R. Segal

statistical models have been used to address them [23,29], the consensus now appears to be moving toward the use of models with strong biological underpinnings, referred to as *biologically based models*. For example, physiologically based pharmacokinetic (PBPK) models [1,54] have been developed by consideration of the uptake, distribution, and disposal of agents of interest by the subjects (human or animal). Interspecies differences in response to exposure to environmental agents can often be explained, at least partially, in terms of differences in uptake and distribution of the agents. Thus, the PBPK models have advanced broadly our understanding of differential species toxicology, and these models are important tools in risk assessment [55].

Cancer models for dose–response studies span a hierarchy that reflects the ability to incorporate different kinds of information regarding carcinogenesis. Early attempts to model carcinogenicity studies using simple statistical models indicate the modest amount of data that were available; these models do not incorporate any information about the mechanism of action. In most cases, there were no supplementary studies that could be used to identify a mechanism of action. These models were fitted to available data in order to describe the apparent dose–response relationship without reference to any specific mechanism of action. The next are biologically based models that are developed with parameters associated with specific biological mechanisms of action based on information from laboratory studies. Such models generally require extensive data, which also helps determine how well they conform with the model characteristics. A step further is to include the pharmacokinetic information in developing the models for carcinogenic risk assessment. In this case, the model and the dose–response shape are determined not only by the characteristics of the carcinogenic mechanism taking place at the target tissue but also by the pharmacokinetic processes involving absorption and distribution of the exposure dose, which are seldom accounted for by models of carcinogenesis. This is an important and essential step, as discussed before.

When discussing exposure– or dose–response relationship, it is important to define clearly what response one is referring to. It is also important to have a clear idea of the various commonly used measures of disease frequency. The two fundamental measures used in epidemiology and toxicology are the incidence (or hazard) rate and the probability of disease. The incidence or hazard rate measures the rate (per person per unit of time) at which new cases of a disease appear in the population under study. Because the incidence rates for many chronic diseases, including cancer, vary strongly with age, a commonly used measure of frequency in epidemiologic studies is the age-specific incidence rate, usually reported in 5-year age categories. For example, the age-specific incidence rate per year in the 5-year age group of 35–39 may be estimated as the ratio of the number of new cases of cancer occurring in that age group in a single year and the number of individuals in that age group who are cancer-free at the beginning of the year. This rate may depend on the single year of interest reflecting the effect of a particular cohort. Strictly speaking, the denominator of the ratio should be the person-years at risk during the year to account for individuals who do not contribute fully during the year. Mathematically, incidence rate is an instantaneous concept defined as

$$h(t) = \lim_{\Delta \to 0} P[t \leq T < t + \Delta | T \geq t],$$

where T denotes the time of response for a subject. The dependence of $h(t)$ on the exposure level and other history is suppressed here to simplify notation. The other commonly used measure is the probability that an individual will develop cancer in a specified period of time. For risk assessment, interest is most often focused on the lifetime probability, often

549

called *lifetime risk of developing cancer*. Although the lifetime can be arbitrarily defined, let us denote this by L. The relationship between the lifetime risk R and the incidence rate can be expressed by the equation

$$R = 1 - \exp\left[-\int_0^L h(s)ds\right].$$

In general, the probability $P(t)$ of developing the disease of interest by age t, can be expressed as

$$P(t) = 1 - \exp\left[-\int_0^t h(s)ds\right].$$

When the incidence rate is small, as is true for most chronic diseases, $P(t)$ can be approximated by $\int_0^t h(s)ds$, the cumulative incidence function. Usually, in epidemiological studies, there is direct information on $h(t)$ and, in toxicological studies, there is direct information on $P(t)$ or R. It is, therefore, convenient to model the cancer process in terms of $h(t)$ or $P(t)$, as a function of the exposure history. Simple statistical models exist for the lifetime risk R as a function of the exposure level, say, d, assumed constant throughout the lifetime.

In the following sections, we describe the different classes of cancer models, depending on the ability to incorporate information, and also indicate the commonly used statistical methods to analyze data using these models. In Section 30.2, we describe some standard simple statistical models for cancer. Section 30.3 slowly builds on with the description of multistage models, followed by a description of two-stage models incorporating cell kinetics in Section 30.4. The models of these two sections can be called *biologically based models*, although the one of Section 30.4 is biologically more realistic. Section 30.5 briefly discusses the PBPK models and Section 30.6 describes the different statistical methods for analysis. Section 30.7 ends with some discussion.

30.2 STATISTICAL MODELS

Quantal response data relates to the final incidence of cancer (binary) for each individual at each dose level requiring models for the lifetime risk R. Early models for this were *tolerance distribution models*, which are based on the assumption that each individual in a population has a tolerance for the carcinogenic agent. If the dose of the agent exceeds an individual's tolerance, a tumor will develop; otherwise, no tumor will develop. Tolerances are assumed to vary from individual to individual across the population following a standard probability distribution, such as the normal distribution. Normal tolerance distribution, known as the *probit* model, assumes that the tolerances have a normal distribution with mean μ and standard deviation σ. Therefore, the probability that an individual at exposure level d developes cancer (a tumor), $R(d)$, is the probability that the individual's tolerance is lower than d, which is given by

$$R(d) = \int_{-\infty}^{(d-\mu)/\sigma} \frac{1}{\sqrt{2\pi}} \exp\left[\frac{-z^2}{2}\right] dz = \Phi\left(\frac{d-\mu}{\sigma}\right),$$

where $\Phi(\cdot)$ denotes the standard normal cumulative distribution function. The *log-probit* model assumes lognormal distribution for the tolerances (i.e., the logarithms of tolerances have a normal distribution with mean μ and standard deviation σ). In this case, the lifetime risk at exposure level d is given by

$$R(d) = \int_{-\infty}^{(\log d - \mu)/\sigma} \frac{1}{\sqrt{2\pi}} \exp\left[\frac{-z^2}{2}\right] dz = \Phi\left(\frac{\log d - \mu}{\sigma}\right).$$

Another common model is the *logit* model in which the tolerances are assumed to follow a logistic distribution leading to the model for lifetime risk as

$$R(d) = \frac{\exp[\alpha + \beta d]}{1 + \exp[\alpha + \beta d]}.$$

There are some *mechanistic* models, so called because they would postulate a hypothetical mechanism of action not validated by laboratory information. The *one-hit* model is the simplest such model postulating that cancer is the result of a single event in a single cell. The model has only one parameter and describes a dose–response relationship with a fixed shape that is virtually linear in the low and middle dose ranges. The lifetime risk is given by $R(d) = 1 - \exp[-qd]$, $q > 0$. Note that this can be derived by assuming an exponential distribution for the tolerances. This model is typically fitted to a single dose point, usually the lowest dose with an increased incidence of cancer. The one-hit model should not be used to fit a dataset with more than one doses if the responses do not follow the model's fixed linear shape [56]. A general expression for the one-hit model is given by

$$R(d) = 1 - \exp[-(q_0 + q_1 d)], q_0, q_1 > 0,$$

which can also be obtained from tolerance distribution.

The *multihit* model postulates that cancer is the result of a fixed number of identical events (or "hits") in a tissue [51]. The shape of the model is governed by the number of hits assumed necessary for the induction of cancer. The more hits required, the lower the probability of cancer at low doses, but the faster that probability rises at higher doses. If k hits are required, the probability of developing cancer is

$$R(d) = \int_0^{qd} z^{k-1} \exp(-z)/(k-1)! dz,$$

with the general expression given by

$$R(d) = \int_0^{q_0 + q_1 d} z^{k-1} \exp(-z)/(k-1)! dz.$$

The *Weibull* model [48], given by $R(d) = 1 - \exp[-qd^k]$, $q, k > 0$, exhibits a dose–response relationship that is either sublinear (shape parameter $k < 1$), linear ($k = 1$), or supralinear ($k > 1$). A quantal response model currently in wide use by regulatory agencies, the *linearized multistage* (LMS) model, has been derived from the *multistage* model (see the next section) after making a number of approximations. The lifetime risk at exposure d is given by $R(d) = 1 - \exp[-(q_0 + q_1 d + \cdots + q_k d^k)]$. The approximations for this expression are, however, unlikely to hold with experimental data when the probability of cancer is kept high. Statistically, the polynomial function of dose permits the description of nonlinear dose–response relationships. A special case of the LMS model is the one-hit model in which $k = 1$.

In some studies, information on the ages at which the tumor responses (e.g., time to tumor onset, time to death with or without tumor) were observed may be available. Incorporation of this information has been shown to improve the analysis by removing certain bias due to premature deaths of study individuals without developing tumor. This can be achieved by the use of *time-to-event* models [23] usually through the modeling of hazard rates. Probably the most widely used in the statistical literature is the Weibull model [25]. A time-to-event version of the LMS model is also widely used for analyses. Some important examples of *time-to-event* models are described in the following:

1. *Multievent Models.* Suppose that a tumor arises as a result of the occurrence of a number (say, k) of biological events so that the time to tumor T can be written as $T = \max\{T_1, \ldots, T_k\}$, where the T_i values are independent random variables representing the times at which the k independent biological events occur. Then, denoting the cumulative hazard function of T_i at dose d by $\Lambda i(t, d)$, we have the probability of tumor by time t at dose d as

$$P(t, d) = \prod_{i=1}^{k} [1 - \exp\{-\Lambda_i(t, d)\}]$$

$$\approx \prod_{i=1}^{k} \Lambda_i(t, d), \quad \text{for small } d \text{ and } t$$

$$= \left\{ \prod_{i=1}^{k} \psi(d) \right\} t^k,$$

where $\lambda_i(t, d) = \psi(d)$ is the hazard rate of T_i (assumed constant) at dose d, for $i = 1, \ldots, k$. This leads to a *Weibull* distribution for T with hazard rate at dose d given by $k \left[\prod_{i=1}^{k} \psi(d) \right] t^{k-1}$.

2. *Relative and Additive Risk Models.* We introduced the concept of incidence or hazard rate $h(t)$ in Section 30.1 as being the appropriate statistical concept that captures the epidemiological idea of an incidence and can relate to an individual's risk of developing cancer, given his/her history. It, therefore, makes sense to model individual's risk in terms of the hazard rate. It is necessary to express the possible dependence of $h(t)$ on the exposure level d and other history, referred to collectively as the *vector of covariates* for an individual at time t and denoted by $z(t)$. Hence, $h(t)$ may be denoted by $h(t, z(t))$; if $z(t)$ includes only the exposure level d, it may be denoted by $h(t, d)$. Let $z_0(t)$ represent some standard (known) covariate history; for example, when $z(t)$ includes only exposure level, $z_0(t)$ could be just $\{d = 0\}$. Relative risk models attempt to describe risks in populations by focusing on the ratio of $h(t, z(t))$ and $h(t, z_0(t))$, termed the *relative risk function* [of time t as well as the covariate history $z(t)$] and denoted by RR$(t, z(t))$; that is, RR$(t, z(t))$ is the relative risk of an individual with covariate history $z(t)$ over an individual with covariate history $z_0(t)$ at time t. Commonly, the various forms of RR$(t, z(t))$ assume it to be independent of time t, leading to the well-known class of proportional hazards models of Cox [9]. A general class of models will consider dependence on time t through $z(t)$ only. The most commonly used forms of RR$(t, z(t))$ consider

multiplicative and additive functions of the covariates. The multiplicative model is given by $\mathrm{RR}(t, z(t)) = \exp[\beta_1 z_1 + \cdots + \beta_m z_m]$, and the additive model by $\mathrm{RR}(t, z(t)) = 1 + \beta_1 z_1 + \cdots + \beta_m z_m$, where $z_1, \ldots, z_m$ are the covariates of interest (including the exposure level d) and $\beta_1, \ldots, \beta_m$ are the corresponding parameters to be estimated from data. Quite often the relative risk cannot be adequately described by either a multiplicative or an additive model. For example, when the incidence rate is obtained from the TSCE model of Section 30.4, the expression for $\mathrm{RR}(t, z(t))$ is neither multiplicative nor additive; in fact, it depends on both t and $z(t)$ in a very complicated way. In contrast, the hazard rate for the *additive risk* model has the general form

$$\lambda(t, d) = \lambda_0(t) + \gamma\{z(t)\beta\},$$

where $\gamma(\cdot)$ is a suitable chosen function; for example, $\gamma(x) = x$.

3. *General Product Hazards Models*. The cumulative hazard function is of the proportional hazards form

$$\Lambda(t, d) = g(d)H(t),$$

where $g(\cdot)$ is a positive convex function of dose and $H(\cdot)$ is a positive nondecreasing function of time.

4. **Log-Linear Models.** The event time T can be written as

$$\log T = z(t)\beta + \sigma W,$$

where $\sigma > 0$ and W is an error variable. When $z(t)$ is time-independent and, for example, W follows a standard normal distribution, the corresponding event time T follows a *log-normal* distribution; when W follows an extreme-value distribution, the corresponding event time T follows a *Weibull* distribution.

Around the 1980s, the limitations of many of these models became apparent. Risk estimates spanning a wide range could be computed with different models, which give reasonably good fit to data, but there was no biological information to help in selecting one of these models over another. Several of these models were severely constrained in shape and were not adequate to empirically describe some datasets with multiple doses showing a nonlinear dose–response relationship. Multistage models, which had been undergoing mathematical development, came into use as they were able to fit the newer animal carcinogenicity studies that tested several dose groups. It is to be noted that the biologically based models, described in the next two sections, are time-to-event models; however, they can also be used for quantal response data when time-to-tumor information is not available.

30.3 MULTISTAGE MODELS

The *Armitage–Doll multistage* model was developed in the 1950s [47,3–5] with the aim of mathematically describing the basic processes leading to the development of cancer and may be said to be the first significant step toward biologically based modeling [30]. The purpose of proposing the multistage model was to appropriately take into account several factors: the rapid increase of cancer mortality and incidence rates with age and the principle

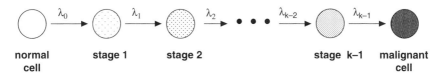

Figure 30.1 Pictorial representation of the multistage model.

that a specific number of changes are needed before the induction of a tumor, as well as some other findings resulting from cancer epidemiology, animal carcinogenicity studies, and *in vitro* studies [2]. During the last few decades, a large number studies have provided important contributions to the development of mathematical theory of the model and of computational methods; the model implications have also been investigated in detail [16,17,56,13,30]. The multistage model has been widely used for risk assessment by the USEPA and FDA, as well as by many international cancer agencies, including the International Agency for Research on Cancer (IARC) and the World Health Organisation (WHO).

This model is based on the assumption that a single normal cell may become fully malignant only after it has undergone a sequence of k (where $k \geq 1$) irreversible and heritable changes. It is assumed that the probability of each change is very low and that the tissue under study is initially formed by a large number of normal cells. The sequence of changes is generally assumed to take place, according to a specific order, spontaneously or when induced by the environmental exposure. The malignancy (the first malignant cell) arises when one single susceptible cell sustains a number (k) of critical changes to take it from a normal tissue cell to a malignant cell, which then grows (after a short lag of time, usually assumed to be negligible) into a malignant tumor (see Fig. 30.1).

Suppose that there are N normal cells susceptible to malignant transformation in the tissue of interest, and let us assume that these N cells act independently. Let $p(t)$ be the probability that a specific susceptible cell is malignant by time t. Then the overall hazard for malignancy is given by $h(t) = Np'(t)/[1 - p(t)]$. Assuming the rate of change (transition) from ith stage to $(i+1)$th stage, for $i = 0, \ldots, k-1$, with the 0th stage meaning normal and the kth stage meaning malignant, to be λ_i, independent of time t, we have the waiting-time distribution from stage i to stage $(i+1)$ as exponential with rate parameter λ_i. Let $p_i(t)$ be the probability that a cell is in stage i at time t. Then, we have $p(t) = p_k(t)$ and $h(t) = Np'_k(t)/[1 - p_k(t)]$. If we now assume that malignancy at the cell level is a rare event [i.e., $p_k(t) \approx 0$], we may approximate the hazard rate by $h(t) \approx Np'_k(t)$. In this case, Taylor series expansion [36,39] leads to the approximation

$$h(t) \approx Np'_k(t) = \frac{N\lambda_0 \cdots \lambda_{k-1}}{(k-1)!} t^{k-1} [1 - \bar{\lambda} + f(\lambda, t)],$$

where $\bar{\lambda} = \sum_{i=0}^{k-1} \lambda_i/k$ is the mean of the transition rates and $f(\lambda, t)$ involves the second- and higher-order terms of the transition rates. Retention of only the first term in this series expansion leads to the Armitage–Doll approximation:

$$h(t) \approx \frac{N\lambda_0 \cdots \lambda_{k-1}}{(k-1)!} t^{k-1}.$$

Therefore, with the two approximations—(1) $p_k(t) \approx 0$ and (2) $[\bar{\lambda} + f(\lambda, t)]$ is negligibly small compared to 1—this model predicts an age-specific incidence curve that increases with a power of age that is one less than the number of distinct stages involved in malignant transformation. Note that the age-specific incidence rate (or hazard rate) is a measure of the rate of appearance of malignant tumors in a previously tumor-free tissue. It has indeed been observed that for many human carcinomas, the age-specific incidence rate increases roughly like a power of the age, and the Armitage–Doll model, as already mentioned, was originally proposed to explain this observation. For k ranging from 2 to 6, this model is in agreement with many age-specific incidence curves derived from epidemiologic data available for many types of tumor [2]. This particular form of hazard rate is known as the *Weibull* model with corresponding survival function (probability of remaining tumor-free) at time t is given by $S(t) = \exp[-At^k]$, where $A = N\lambda_0 \cdots \lambda_{k-1}/k!$

Since the Armitage–Doll model does not allow for cell death, it is immediately clear that any susceptible cell eventually becomes malignant with probability 1. Further, since the time to malignant transformation is the sum of k exponential waiting-time distributions, it follows that $h(t)$ is a monotone increasing function even without the approximation. Since the probability $p_k(t)$ satisfies the Kolmogorov equation $p'_k(t) = \lambda_{k-1}p_{k-1}(t)$ (from the theory of pure birth process), the hazard rate can also be written as

$$h(t) = \frac{Np'_k(t)}{1 - p_k(t)} = N\lambda_{k-1}E[X_{k-1}(t)|X_k(t) = 0],$$

where $X_i(t)$ is a sequence of random variables associated with each cell such that $X_i(t) = 1$ if the cell is in stage i at time t and 0 otherwise. When $p_k(t) \approx 0$, this conditional expectation may be approximated by the corresponding unconditional expectation and $h(t) \approx Np'_k(t) = N\lambda_{k-1}E[X_{k-1}(t)]$. Expressions similar to these can be written for the hazard function of the two-stage clonal expansion model in the next section.

For the Armitage–Doll model to hold, the transition rates λ_i have to be constant and very small. An example of how poorly this approximation may do is discussed in Moolgavkar [36,39]. In addition, in animal experiments, precisely where this model has been widely used, the probability of tumor may be too large (because of usually higher levels of exposure) for $p_k(t) \approx 0$ to hold, in which case this approximate model should be avoided altogether.

In order to describe the dose–response relationship, the dose dependence of the transition rates is commonly assumed to be linear with $\lambda_i(d) = a_i + b_id$, for $i = 0, \cdots, k - 1$, where the coefficients $a_i > 0, b_i \geq 0$. As a consequence, the survival function at dose d can be written as

$$S(t,d) = \exp\left[-N(a_0 + b_0d) \cdots (a_{k-1} + b_{k-1}d)t^k/k!\right]$$

so that the probability of tumor by time t at dose d can be written as

$$P(t,d) = 1 - S(t,d) = 1 - \exp\left[-(q_0 + q_1d + \cdots + q_kd^k)t^k/k!\right],$$

where q_i are the coefficients of the kth-degree polynomial resulting from the product of the k transition rates $(a_i + b_id)$. One can write $P(t, d)$ as $1 - \exp[-g(d)t^k]$, or the hazard rate as $h(t, d) = g(d)t^{k-1}$, where $g(d)$ is a kth-degree polynomial in dose d. If the data to which the model have to be fitted refer to lifetime exposure, the factor t^k in $P(t, d)$ reduces to a constant ($t =$ lifetime), so that the lifetime risk is given by $R(d) = 1 - \exp[-g(d)] = 1 - \exp[-(q_0 + q_1d + \cdots + q_kd^k)]$, where the constant t^k is absorbed in the q_i coefficients (see the LMS model in Section 30.8).

It is worthwhile to mention that in the above model, the function $g(d) = q_0 + q_1 d + \cdots + q_k d^k$ is restricted only to those polynomials that can be factored into the product of linear terms $a_i + b_i d$ with $a_i > 0$ and $b_i \geq 0$. However, if this restriction is replaced by a weaker condition that each $q_i \geq 0$ with $q_0 > 0$, the resulting *generalized multistage model* is more flexible and can fit more datasets.

Generally of interest is the *excess risk*, or the risk above the background, defined as

$$E(d) = \frac{R(d) - R(0)}{1 - R(0)} = 1 - \exp\left[-(q_1 d + \cdots + q_k d^k)\right].$$

As d approaches zero (for low-dose extrapolation), the marginal increase in the cancer risk for a small increment of exposure can be determined by the slope (or first derivative) of $E(d)$. Thus, at low doses, the incremental risk for a small dose sufficiently close to zero is approximately $E(d) \approx q_1 d$. For more details on this linearizing at low doses, referred to as the *linearized multistage model*, see Cogliano et al. [8].

Note that the transition rates λ_i terms are so far assumed to be independent of time; in particular, when the rates depend on the exposure d of a specific agent as $\lambda_i(d) = a_i + b_i d$, the value of the exposure d is usually considered to be substantially constant over time. This assumption is generally valid in the case of carcinogenic experiments, in which the experimental animals are exposed to a constant exposure for approximately the whole duration of the study (usually, the lifetime of the species), and may also hold for some patterns of human exposure to environmental carcinogens. This is not generally valid for all human exposures. The time-dependent exposure patterns may be specified by a function $d(t)$ that can very well be approximated by a piecewise constant function leading to piecewise constant transition rates [11]. The use of the multistage model has been extended to consider such time-dependent exposure patterns, and the consequences have been investigated in detail by several authors [11,26,45].

30.4 TWO-STAGE CLONAL EXPANSION MODEL

One notable feature of the Armitage–Doll multistage model is that the incidence rate is always increasing in age. Many adult human cancers do exhibit increasing incidence rate and, hence, can be adequately described by the multistage model. However, there are some with non-increasing incidences (e.g., breast and also some childhood cancers such as retinoblastoma) that cannot be explained by this model [41,37]. This model is also not suitable for analyzing data from animal experiments for reasons stated in the previous section. The two-stage clonal expansion (TSCE) model, by introduction of cell kinetics such as cell growth and death or differentiation, provides an alternative that is biologically more plausible and, at the same time, can explain many of these exceptional incidences.

The TSCE model posits that malignant transformation of a normal susceptible cell is the result of two specific, rate-limiting, hereditary (at the level of the cell) and irreversible events. This model is best interpreted within the initiation–promotion–progression paradigm of chemical carcinogenesis. Initiation, which confers a growth advantage on the cell, is a rare event and can be regarded as the first rate-limiting step. Mathematically, the process of initiation (arrival of normal cells into the initiated compartment) can reasonably be modeled by a time-dependent Poisson process (details to follow). The TSCE model also posits that promotion consists of the clonal expansion of these initiated cells by a stochastic birth–death process. Finally, one of the initiated cells may be converted into a malignant cell, and this conversion (progression) may involve one or more mutations (see Fig. 30.2). A number of works [40,41,43,21] discuss the biology underlying the model and develop the mathematical and statistical tools required to fit the model to data. The model also conforms with observed incidences from many epidemilogic studies on different types of cancer [37]. When more biological information on the pathway to malignancy (e.g., on the intermediate lesions) is available, the model can be extended to include this information [14,15]. Slightly different versions of the model have been considered in the past by Kendall [24], Neyman and Scott [46], and more recently by Portier and Kopp-Schneider [49].

Because the TSCE model explicitly considers both genomic events and cell proliferation kinetics, it provides a flexible tool for incorporating both genotoxic and nongentoxic carcinogens in cancer risk assessment. Within the framework of the model, an environmental agent acts by affecting one or more of its parameters. An agent may affect the rate of one or both stages (initiation and progression) or of cell proliferation kinetics of normal or initiated cells. A purely genotoxic agent would be expected to increase the rates of initiation or progression or both, while a purely promoting agent might increase the cell division rate or decrease the rate of death or differentiation (known as *apoptosis*) of initiated cells (and possibly also of normal cells, but this would be expected to have less effect on cancer incidence). Because the TSCE model explicitly considers initiation and promotion, it can be used for analyses of the intermediate lesions. For the purpose of quantitative risk assessment, analysis of intermediate lesions allows the dose–response curve to be extended to lower doses than would be possible with consideration of malignant lesions alone because intermediate

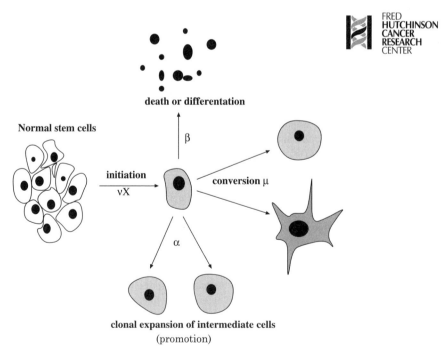

Figure 30.2   Pictorial representation of the two-stage clonal expansion model.

lesions generally develop at doses that are too low for the appearance of malignant lesions with typical experimental protocol.

Although stochastic growth of normal susceptible cells can, in theory, be accommodated, it is convenient and mathematically simpler to assume the growth of normal cells to be deterministic. This is a reasonable assumption because the number of normal cells is large and probably still under tight homeostatic control, whereas intermediate or initiated cells are assumed to undergo a stochastic process because their numbers are small compared to the number of normal cells in the tissue. Furthermore, the process of initiation likely results in the loosening of homeostatic control leading to the positive net growth of intermediate lesions with rates that are typically increased over background growth rates of normal cells. As a result, statistical fluctuations become more important in the intermediate compartment and need to be considered.

Here we summarize the basic assumptions required for the mathematical development of the model. Let $X(t)$ be the number of normal susceptible cells at time t. Then, initiated cells arise from normal cells according to a nonhomogeneous Poisson process with intensity v $(t)X(t)$, where $v(t)$ is the first mutation rate per cell at time t. Intermediate cells then either divide at rate $\alpha(t)$, die (or differentiate) at rate $\beta(t)$, or divide into one intermediate and one malignant cell at rate $\mu(t)$. Note that the rate parameters may depend on time, in particular, through the time-dependent exposure pattern, which may affect the rate parameters in α, say, piecewise constant manner. Because of the presence of cell death, however, intermediate cells or their clones may become extinct before giving rise to malignant progeny, thereby removing the corresponding initiation at the tissue level.

Here, only an outline of the mathematical steps is given. More details can be found in the review article by Moolgavkar and Luebeck [43]. Let $Y(t)$ and $Z(t)$ represent the number of intermediate and malignant cells, respectively, at time t, and let

$$\Psi(y, z; t) = \sum_{j,k} P_{j,k}(t) y^j z^k$$

be the corresponding joint probability generating function with

$$P_{j,k}(t) = \Pr[Y(t) = j, Z(t) = k \mid Y(0) = 0, Z(0) = 0].$$

Noting that the process $(Y(t), Z(t))$ is Markovian, Ψ satisfies the Kolmogorov forward differential equation

$$\Psi'(y, z; t) = \frac{\partial \Psi(y, z; t)}{\partial t} = (y - 1)v(t)X(t)\Psi(y, z; t)$$

$$\times \{\mu(t)yz + \alpha(t)y^2 + \beta(t) - [\alpha(t) + \beta(t) + \mu(t)]y\}\frac{\partial \Psi}{\partial y},$$

(30.1)

with initial condition $\Psi(y, z; 0) = 1$. Note that $\Psi(1, 0; t) = \Pr[Z(t) = 0]$ is the survival function $S(t)$ for time T of appearance of the first malignant cell, while $P(t) = 1 - S(t)$ is the probability of tumor (first malignancy) by time t for this model. The hazard (incidence) function is then given by

$$h(t) = \frac{P'(t)}{1 - P(t)} = \frac{-\Psi'(1, 0; t)}{\Psi(1, 0; t)}.$$

It follows immediately from the Kolmogorov equation (30.1) that

$$\Psi'(1,0;\,t) = -\mu(t)\frac{\partial\Psi}{\partial y}(1,0;\,t),$$

and thus

$$h(t) = \mu(t)E[Y(t)Z(t) = 0],$$

where E denotes the expectation and where we have used the relationship

$$E[Y(t)\,|\,Z(t)=0] = \frac{\partial\Psi}{\partial y}\frac{(1,0;\,t)}{\Psi(1,0;\,t)}.$$

As in the previous section, this conditional expectation, for rare tumors when $\Pr[Z(t)=0] \approx 1$, can be replaced by the corresponding unconditional expectation to get an approximate expression for the hazard rate as given by

$$h(t) \approx \mu(t)E[Y(t)] = \mu(t)\int_0^t \nu(s)X(s)\exp\left[\int_s^t (\alpha(u) - \beta(u))du\right]ds,$$

where the expression for $E[Y(t)]$ can be easily obtained from (30.1). However, this approximate form, besides being inapplicable in animal carcinogenicity studies as discussed before, lacks in flexibility in the sense that it essentially depends on two parameter combinations: (1) the net cell proliferation rate $(\alpha - \beta)$ and (2) the product $\mu \times \nu X$. From this nothing can be learned about the roles of α or β alone. This constitute a serious shortcoming. We shall see that the computation of the exact hazard can be carried out as a recursive procedure in the case of piecewise constant rate parameters. This situation of piecewise constant exposure pattern, which in effect leads to piecewise constant rate parameters, is indeed most frequently encountered among epidemiologic and experimental data.

The exact solution of the Kolmogorov equation (30.1) involves solving the characteristic equations associated with it [43], which are as given by

$$\frac{dy}{ds} = -\{\mu(s)yz + \alpha(s)y^2 + \beta(s) - [\alpha(s)$$

$$+ \beta(s) + \mu(s)]y\} = -R(y,s), \text{say},\qquad(30.2)$$

$$\frac{dz}{ds} = 0 \ (z \text{ is constant along characteristics})$$

$$\frac{dt}{ds} = 1, \quad \frac{d\Psi}{ds} = (y-1)\nu(s)X(s)\Psi.$$

The ordinary differential equation for Ψ may be solved along characteristics to yield

$$\Psi(y(t),z;\,t) = \Psi_0 \exp\left\{\int_0^t [y(s,t) - 1]\nu(s)X(s)ds\right\},$$

where $\Psi_0 = \Psi(y(0), z; 0) = 1$ is the initial value of Ψ, and the explicit dependence of y on both s and t is acknowledged. We are interested in computing $\Psi(1,0;\,t)$ for any t, and thus we need to find the values of $y(s, t)$ along the characteristic through $(y(0), 0, 0)$, where $y(0)$ is the initial value of y, and $(y(t), 0, t)$ with $y(t) = 1$. Now, along the characteristic, y satisfies the differential equation $dy/ds = -R(y, s)$, and this is just a Ricatti equation, which can be readily integrated in closed form if the parameters of the model are piecewise constant. To be precise, the Ricatti equation for y can be solved to yield a value for $y(u, t)$ for any u, with initial condition $y(t, t) = 1$. Then, the survival function is

$$S(t) = \Psi(1, 0; t) = \exp\left\{\int_0^t [y(s, t) - 1]v(s)X(s)ds\right\}, \tag{30.3}$$

and the exact hazard rate is given by

$$h(t) = -\Psi'(1, 0; t)/\Psi(1, 0; t) = -\int_0^t v(s)X(s)y_t(s, t)ds,$$

where y_t denotes the derivative of y with respect to t.

Assume that there are n intervals $[t_{i-1}, t_i]$ with $i = 1, 2, \ldots, n$, covering the time period $[t_0 = 0, t_n = t]$. Then the solution of (30.2), $y(s, t)$, can be computed recursively starting from $s = t = t_n$ using the boundary condition $y(t, t) = 1$. For $s \in [t_{i-1}, t_i]$, we have [43]

$$y(s, t) = \frac{B_i - A_i \dfrac{y(t_i, t) - B_i}{y(t_i, t) - A_i} \exp[\alpha_i(A_i - B_i)(s - t_i)]}{1 - \dfrac{y(t_i, t) - B_i}{y(t_i, t) - A_i} \exp[\alpha_i(A_i - B_i)(s - t_i)]},$$

where $A_i(B_i)$ are the lower (upper) root of the quadratic form $\alpha_i x^2 - [\alpha_i + \beta_i + \mu_i]x + \beta_i$. The constant parameters α_i, β_i, and μ_i refer, respectively, to the cell division, cell death, and second mutation rate in the time interval $[t_{i-1}, t_i]$. When $v(s)X(s)$ is also piecewise constant over time (vX_i in the ith interval, say), the time integral in (30.3) can be computed in explicit form. For $s \in [t_{i-1}, t_i]$, we can rewrite the integrand $[y(s, t) - 1]$ as

$$y(s, t) - 1 = \frac{C_i}{1 - r_i \exp[\delta_i(s - t_i)]} + (A_i - 1),$$

where $C_i = B_i - A_i$, $r_i = (y(t_i, t) - B_i)/(y(t_i, t) - A_i)$ and $\delta_i = \alpha_i(A_i - B_i)$. The survival function (30.3) can then be computed as

$$S(t) = \exp\left[-\sum_{i=1}^n H_i\right] \text{ with}$$

$$H_i = -vX_i \int_{t_{i-1}}^{t_i} [y(s, t) - 1]ds$$

$$= -vX_i \left[(B_i - 1)(t_i - t_{i-1}) + \ln\left(\frac{1 - r_i}{1 - r_i \exp[-\delta_i(t_i - t_{i-1})]}\right)\middle/\alpha_i\right].$$

It is to be noted that the hazard rate for this TSCE model, unlike that of the Armitage–Doll multistage model, approaches a finite asymptote as a function of age [43]. It is interesting to note that for constant rate parameters, the hazard rate is dependent on only three combinations of the four parameters α, β, νX and μ [21]. Thus, for instance, in the case of constant exposures, tumor incidence data alone are not sufficient to estimate all the biological parameters of the model. To see this, consider the solution of $y(s,t)$ of the Ricatti equation (30.2). Obviously, $y(s,t)$ solves an equation of the form $y' = -\alpha(A - y)(B - y)$, where A and B are the two roots of the Ricatti equation. It is easy to see that the transform $w = \alpha(y - 1)$ solves a similar equation when α is constant, that is, $w' = -[\alpha(A - 1) - w][\alpha(B - 1) - w]$. Therefore, the solution w depends only on the two modified roots $\alpha(A - 1)$ and $\alpha(B - 1)$ and the hazard rate on these two roots and $\nu X/\alpha$. Of course, other parametrizations can be given that are merely combinations of the three given here. This TSCE model has been applied to data from many epidemiologic and toxicological studies. For a comprehensive discussion of some of those, see Moolgavkar and Luebeck [44].

30.5 PHYSIOLOGICALLY BASED PHARMACOKINETIC MODELS

The pharmacokinetic processes that govern the absorption, distribution, accumulation, detoxification, excretion, as well as the chemical transformation of the exposure dose in the target tissue, may largely influence the shape of carcinogenic dose–response relationship. The incorporation of this mechanism, the model, and the associated data in carcinogenic risk assessment is considered today as an essential step. This is because the dose–response shape is determined not only by the characteristics of the carcinogenic processes taking place at the target tissue, which are expected to be described by the models of the previous sections, but also by the parameters of the pharmacokinetic processes, generally not accounted for by the models of carcinogenesis. In the case of multistage model, for example, this implies that the dose–response shape is determined not only by the number and types of the different carcinogenic stages but also by the characteristics of the pharmacokinetic mechanism. As is well known, reference to pharmacokinetic theory and data has made it possible to give a suitable and satisfactory explanation and interpretation of the shape of some specific classes of dose–response curves, as well as to appropriately extend the application of multistage and other mathematical models to the analysis [1].

Pharmacokinetics may be accounted for in cancer risk modeling by considering it within a model for carcinogenesis (therefore, by modifying the model). In practice, the exposure (or, administered) dose d of the model is substituted by a function $g(d)$ of d, up to some known or unknown parameters, describing the concentration of the substance that is estimated or modeled to be active at the target tissue. The form of this function depends on the pharmacokinetics of the involved agent in the tissue under study. In other words, pharmacokinetics is simply accounted for by making reference to the doses effectively active at the target tissue that are estimated through pharmacokinetic data and models. This class of models describing the relationship between the administered dose d and the effective dose $g(d)$ at the target tissue is called the *physiologically based pharmacokinetic* (PBPK) models since we are concerned with the effective dose in terms of the administered dose. These models are important because the effective dose at a target tissue is more closely related to incidence of cancer than is the external dose. It is to be noted that the pharmacokinetics, and hence the models, depend on the chemical agent under study and also the target tissue.

As an example, inclusion of pharmacokinetics in one-hit and multistage models has allowed satisfactory fitting of these models to observed dose–response relationships in experiments, whose downward curvature (convexity) and supralinear trend could be explained by the hypothesis that a main carcinogenic agent was an active metabolite of the administered dose, which was produced by a saturable metabolic process [19,51]. The metabolic process, whose final product is an active carcinogenic metabolite, is described by a Michaelis–Menten process that, under steady-state conditions, is assumed to lead an exposure dose d to a concentration of the active metabolite in the target tissue given by $g(d) = K_1 d/(1 + K_2 d)$, where K_1 and K_2 are constants. In order to appropriately take the metabolic processes into account through the multi-stage model, say, the exposure dose d may simply be substituted by $g(d) = d/(1 + K_2 d)$, while the constant K_1 in the numerator is absorbed by the constants of the multistage model.

It is interesting to note that this modification of the model needed by the saturation processes taking place generally at high doses (present in some animal experiments) may not be expected to cause significant nonlinearities at very low doses, if the saturation process is governed by the Michaelis–Menten law. This is because, in the low-dose region when d is far lower than $1/K_2$, the function $g(d)$ is substantially linear and, therefore, does not change the low-dose mathematical form of the original multistage or other models that are being used.

Whenever appropriate epidemiologic data is not available, cancer risk assessments need to rely on the use of animal data to predict the risk of chemical exposures to the human population. Interspecies extrapolation is a necessary element in this process and has been performed by scaling the doses used in a dose–response relationship according to body weight or surface area. Thus, a dose represented in the appropriate units, mg kg^{-1} day^{-1} (body weight scaling) or mg/m^2 (surface area scaling), is assumed to result in the same cancer incidence across species. As an improvement to the risk assessment process, the PBPK models have been used to estimate target tissue doses and facilitate interspecies extrapolation by investigating how the pharmacokinetic parameters (of the PBPK models) vary over the species.

There have been several studies utilizing PBPK models in risk assessment. The twofold role of a PBPK model has been to (1) predict a measure of animal tissue (effective) dose $g(d)$ to be used in the dose–response curve and (2) determine the human administered dose corresponding to a tissue dose at a given level of risk. Inherent in role 2 is the assumption that the PBPK model is valid at the low dose identified by the cancer model. The underlying question is whether the model, parametrized under higher-dose conditions, provides a reasonable representation of the kinetics at low doses. To answer this question, kinetic data obtained at low administered doses are required and would improve the risk assessment process by eliminating the uncertainty introduced by the high- to low-dose extrapolation of the PBPK model. The following steps outline the use of a PBPK model in cancer risk assessment:

1. The PBPK model is used to calculate the effective dose in the animal tissue.
2. A cancer model is fitted to the cancer incidence (bioassay data) versus effective dose (from PBPK model) in animals.
3. The value of the effective dose in the animal tissue at a specified risk level is determined from the cancer model. It is assumed that the effective dose has the same effect across species or scales allometrically to determine the effective dose value in human.
4. The PBPK model for human is used to determine the administered dose corresponding to the effective dose value at the specified risk level.

Variations of this basic methodology has been employed in many examples dealing with different chemical agents. The reader is referred to Luebeck et al. [33, Sec. 6.8] for a comprehensive discussion of many of these applications.

Interspecies extrapolation is also an important issue in implementing the third step of the methodology described in the previous paragraph. Many of the physiologic and metabolic parameters used in pharmacokinetic modeling are directly correlated with the body weight of the particular organism. These physiologic parameters generally vary with body weight (BW) according to a power function expressed as $y = a(BW)^b$, where y is a physiologic parameter of interest and a and b are constants [12,31]. If b equals 1, the physiologic parameter y correlates directly with body weight. If b equals two-thirds, the parameter y correlates with surface area. Travis et al. [55] assume that certain physiologic and metabolic processes scale across species with the 0.75 power of body weight. While there is a substantial body of empirical data to suggest that this assumption is at least approximately correct, it is far from universally accepted.

Another concept is the presence of a biologically variable timescale between species. Hill [22] first suggested that body size served as the regulating mechanism for an internal biological clock, making the rate of all biological events constant across species when compared with per unit physiologic time. His conclusions have been supported by many authors [31], who have shown that breath duration, heartbeat duration, longevity, pulse time, breathing rates, and blood flow rates are approximately constant across species when expressed in internal time units. This time unit has been termed *physiologic time* (t') and can be defined in terms of chronological time (t) and body weight (BW) as, for example, $t' = t/(BW)^{0.25}$. Thus, while chronologic time is the same for all species, physiologic time differs for each species. The value of this concept is that all species have approximately the same physiologic and metabolic rates when measured in the physiologic timeframe.

Interspecies extrapolation of toxic effects attempts to find a measure of administered dose (i.e., $mg\,kg^{-1}\,day^{-1}$ or $mg\,m^{-2}\,day^{-1}$) that produces the same measure of effect in all species. It is understood that any such extrapolation procedure is only approximately correct and should be used only when species-specific data are unavailable. Historically, it has been assumed that a single extrapolation procedure would work for all chemicals regardless of their action mechanisms. Travis et al. [55] demonstrated that, regardless of action mechanism, the appropriate metric was dose (mg/kg) per unit of physiological time, which, at low doses, is equivalent to $mg\,kg^{-0.75}\,day^{-1}$.

30.6   STATISTICAL METHODS

We will discuss here some of the main statistical tools that have been developed over the last few decades (as of 2007) for analyses of epidemiologic and toxicologic data. Important issues like sampling and data analysis in epidemiologic studies to ensure appropriate interpretation of results in the presence of possible confounding will not be discussed here as that will be outside the scope of this work. Any book on epidemiology (e.g., see Ref. 52) will discuss these issues in detail.

For analysis of epidemiologic data, there are two broad classes of models, namely, relative risk models and Poisson models. When information is available on each individual in an epidemiologic study, the relative risk models are widely used, whereas the Poisson models are used when data are available at a group level, generally cross-tabulated by age groups and exposures of interest. For analysis of toxicologic data also, there are two broad classes of models. The first is the class of quantal response models, which are used to investigate the relationship between exposure (or dose) d and lifetime risk $R(d)$ or lifetime probability of response (or tumor, if the experiment under consideration is a cancer bioassay). In some experiments, particularly if serial sacrifices are performed, information may be available not only on whether a particular animal has response (or, tumor) or not but also on when (age of the animal) the response is observed. In this case, this extra piece of information can be accommodated by using one of the time-to-response (or time-to-tumor) models. As mentioned before, incorporation of this extra information reduces bias induced by mortality from other causes. In the following paragraphs, we discuss briefly the statistical methods used for the four broad classes of models as mentioned above.

1. *Relative Risk Models.* Since the relative risk is defined in terms of hazard rates, and data are available on each individual in the study cohort, the standard techniques of analyzing survival data with covariates can be employed for specific parametric models. For more general models, special methods of analysis also exist, including Cox's partial likelihood method [10] for proportional hazards model with arbitrary baseline hazard. There is a vast biostatistics literature on the application of relative risk models to the analyses of various study designs encountered in epidemiology, including cohort and case–control studies, with adjustment for confounding, [6,7]. It is to be noted that the analysis of such data at the individual level need not rely on relative risk models alone. In fact, any model for the hazard rate $h(t)$ with identifiable parameters should suffice.

2. *Poisson Models.* Quite often information is available, not on individual members of a study cohort, but on subgroups that are reasonably homogeneous with respect to important characteristics, including exposure level, determining disease (cancer) incidence. A well-known example is provided by the British Doctors' study of tobacco smoking and lung cancer [18]. For the cohort of individuals in this study, information on the number of lung cancer deaths is cross-tabulated by daily level of smoking (reported in fairly narrow ranges) and 5-year age categories. For Poisson regression, the number of cases (the number of lung cancer deaths in the example of the British Doctors' study) in each cell of the cross-tabulated data is assumed to have a Poisson distribution with expectation that it is a function of the covariates of interest, including exposure level. The number of cases in different cells of the cross-tabulated data are assumed to be independent. Suppose that the data are cross-tabulated in I distinct cells, and let E_i be the expectation of the number of cases in the ith cell with the corresponding observed number O_i. Then, the usual Poisson likelihood can be written down for the ith cell, and the total likelihood is the product of all these likelihood contributions over the I cells. The expectations E_i are made functions of the covariates of interest involving some unknown parameters that are to be estimated. Generally $\log(E_i)$ is modeled as a linear function of the covariates, where the coefficients are unknown parameters. More elaborate functions can also be used in terms of the hazard function derived from the biologically based models mentioned in the previous sections. Moolgavkar et al. [42] modeled this expectation by the hazard function derived from the TSCE model to analyze the British Doctors' data.

3. *Quantal Response Models.* There is a whole class of models, the tolerance distribution models, as discussed in Section 30.2. There is also the linearized multistage (LMS) model with its basis in biological considerations. More generally, any time-to-tumor model based on biological considerations (multistage, TSCE, and/or PBPK models) leads to an expression for lifetime risk $R(d)$ by computing the probability of tumor by time t, $P(t, d)$, at a constant lifetime t. The quantal response models can be fit to quantal response data using the maximum-likelihood approach provided that the number of dose groups is larger than the number of unknown parameters. The likelihood is of the product binomial form, that is, a product of binomial likelihoods, which correspond to the binomial observations from the different dose groups. Formally, if there are

n_i observations in the ith dose group with dose level d_i having y_i number of cases, for $i = 0, 1, \ldots, k$, then the likelihood is proportional to $\prod_{i=0}^{k} [R(d_i)]^{y_i} [1 - R(d_i)]^{n_i - y_i}$. If there are individual specific covariates, in addition to the exposure or dose level, the likelihood will be a product of Bernoulli terms contributed by each individual [27].

4. *Time-to-Event Models*. In some bioassay data, information on the ages at which tumors were observed may be available. As discussed at the end of Section 30.2, this information can be incorporated in the analyses by using time-to-event or time-to-tumor models (including the biologically based ones). The fundamental concept required to fit these models to such data is that of the hazard rate, which we discussed before. A number of standard time-to-tumor models were reviewed by Kalbfleisch et al. [23]. The form of the likelihood depends on the amount of information available on tumor onset times, on whether the tumor is rapidly fatal, and on the relationship between tumor mortality and death due to other causes [28].

30.7   CONCLUDING REMARKS

In this work, we present a summary of models (not necessarily an exhaustive one) used to describe the carcinogenic process, particularly for the purpose of carcinogenic risk assessment based on data from epidemiologic and experimental studies. Starting from a more standard purely statistical models to the biologically based TSCE model, possibly with incorporation of pharmacokinetic considerations, there is a wide range of models that exists at this time. We also discuss the different common statistical methods to analyze different kinds of data using these models.

The multistage model has a long history and has been used largely in low-dose risk assessment and carcinogen regulation. It has been studied, discussed, criticized, and applied for more than 30 years, and theoretical and practical developments of the various details and related aspects have continued for so long that a large body of knowledge has accumulated on the subject. Its limitations, however, have been understood only recently relatively. For example, although this model can adequately explain most adult tumors, the risk of which usually increases with age, it fails to account for nonmonotonic risk as observed in, for example, any childhood cancer or breast cancer. Also, it is difficult to incorporate time-dependent exposure patterns, which is only too common in most epidemiologic studies. Since the development of the TSCE model incorporating cell proliferation, it is now possible to amend most of these inadequacies, including the time-dependent exposure patterns. A number of applications of biologically based models in the analysis of epidemiologic and experimental data have been reported and discussed in Luebeck et al. [34] and Moolgavkar and Luebeck [44].

Although the multistage and the TSCE models are based on biological considerations, they do not entirely capture the complex biological reality of the carcinogenic process. For example, the multistage model ignores cell proliferation, whereas the TSCE model posits only a single intermediate stage with altered cell kinetics. Furthermore, there may be more than a single pathway to malignancy in a tissue. Attempts have been made to develop more general models that accommodate some of these possibilities [49,32,35,20]. The problem with such general models is that there is little quantitative information for estimation of the model parameters. Unfortunately, quantitative risk assessment using a biologically based model requires more data than are usually available, in order to circumvent he problem of parameter identifiability. Nonetheless, use of models based on current understanding of carcinogenesis does identify data gaps that need to be filled up. The use of such models can also suggest plausible explanations for observed exposure–response relationship. For example, in 1999 Lutz and Kopp-Schneider proposed a mechanism for the J-shaped exposure–response relationship reported for some carcinogens.

ACKNOWLEDGMENT

The author is grateful to Dr. E. G. Luebeck for many helpful discussion and supplying Figures 30.1 and 30.2. The author also thanks the two anonymous referees for their helpful comments.

REFERENCES

1. Anderson, M. W., Hoel, D. G., and Kaplan, N. L., A general scheme for the incorporation of pharmacokinetics in low-dose risk estimation for chemical carcinogens: Example—vinyl chloride, *Toxicol. Appl. Pharmacol.* **55**, 154–161 (1980).

2. Armitage, P., Multistage models of carcinogenesis, *Environ. Health Perspect.* **63**, 195–201 (1985).

3. Armitage, P. and Doll, R., The age distribution of cancer and a multistage theory of carcinogenesis, *Br. J. Cancer* **8**, 1–12 (1954).

4. Armitage, P. and Doll, R., A two-stage theory of carcinogenesis in relation to the age distribution of human cancer, *Br. J. Cancer* **11**, 161–169 (1957).

5. Armitage, P. and Doll, R., Stochastic models for carcinogenesis, in *Proc. 4th Berkeley Symp. Mathematical Statistics and Probability*, Neyman, J., ed., Univ. California Press, Berkeley, 1961, Vol. IV, 19–38.

6. Breslow, N. E. and Day, N. E., *Statistical Methods in Cancer Research*, Vol. I: *The Analysis of Case-Control Studies*, IARC Scientific Publications No. 32, International Agency for Research on Cancer, Lyon, 1980.

7. Breslow, N. E. and Day, N. E., *Statistical Methods in Cancer Research*, Vol. II: *The Design and Analysis of Cohort Studies*, IARC Scientific Publications No. 82, International Agency for Research on Cancer, Lyon, 1987.

8. Cogliano, V. J., Luebeck, E. G., and Zapponi, G. A., The multistage model of carcinogenesis: A critical review of its use, in *Perspectives on Biologically Based Cancer Risk Assessment*, Cogliano, V. J., Luebeck, E. G., and Zapponi, G. A., eds., NATO Committee on the Challenges of Modern Society, Kluwer Academic/Plenum Publishers, New York, 1999, pp. 183–204.

9. Cox, D. R., Regression models and life tables (with discussion), *J. Roy. Statist. Soc. B* **34**, 187–220 (1972).

10. Cox, D. R., Partial likelihood, *Biometrika* **62**, 269–276 (1975).

11. Crump, K. S. and Howe, R. B., The multistage model with a time dependent dose pattern: Application to carcinogenic risk assessment, *Risk Anal.* **4**, 163–176 (1984).

12. Davidson, I. W., Parker, J. C., and Beliles, R. P., Biological basis for extrapolation across mammalian species, *Regul. Toxicol. Pharmacol.* **6**, 211–237 (1986).

13. Day, N. E. and Brown, C. C., Multistage models and primary prevention of cancer, *J. Natl. Cancer Inst.* **64**, 977–989 (1980).

14. Dewanji, A., Venzon, D. J., and Moolgavkar, S. H., A stochastic two-stage model for cancer risk assessment. II. The number and size of premalignant clones, *Risk Anal.* **9**, 179–187 (1989).

15. Dewanji, A., Moolgavkar, S. H., and Luebeck, E. G., Two-mutation model for carcinogenesis: Joint analysis of premalignant and malignant lesions, *Math. Biosci.* **104**, 97–109 (1991).

16. Doll, R., The age distribution of cancer: Implications for models of carcinogenesis, *J. Roy. Statist. Soc. A* **134**, 133–166 (1971).

17. Doll, R., An epidemiological perspective on the biology of cancer, *Cancer Res.* **38**, 3573–3583 (1978).

18. Doll, R. and Peto, R., Cigarette smoking and bronchial carcinoma: Dose and time relationships among regular smokers and life-long non-smokers, *J. Epidemiol. Commun. Health* **32**, 303–313 (1978).

19. Gehring, P. J., Watanabe, P. G., and Park, C. N., Risk of angiosarcoma in workers exposed to vinyl chloride as predicted from studies in rats, *Toxicol. Appl. Pharmacol.* **49**, 15–21 (1979).

20. Hazelton, W. D., Clements, M. S., and Moolgavkar, S. H., Multistage carcinogenesis and lung cancer mortality in three cohorts, *Cancer Epidemiol. Biomark. Prevent.* **14**, 1171–1181 (2005).

21. Heidenreich, W. F., Luebeck, E. G., and Moolgavkar, S. H., Some properties of the hazard function of the two-mutation clonal expansion model, *Risk Anal.* **17**, 391–399 (1997).

22. Hill, A. V., The dimensions of animals and their muscular dynamics, *Proc. Roy. Inst. Great Britain* **34**, 450–471 (1950).

23. Kalbfleisch, J. D., Krewski, D., and Van Ryzin, J., Dose-response models for time-to-response toxicity data, *Can. J. Statist.* **11**, 25–49 (1983).

24. Kendall, D. G., Birth-and-death processes, and the theory of carcinogenesis, *Biometrika* **47**, 13–21 (1960).

25. Kodell, R. L. and Nelson, C. J., An illness-death model for the study of the carcinogenic process using survival/sacrifice data, *Biometrics* **36**, 267–277 (1980).

26. Kodell, R. L., Gaylor, D. W., and Chen, J. J., Consequences of using average lifetime dose rate to predict risks from intermittent exposures to carcinogens, *Risk Anal.* **7**, 339–345 (1987).

27. Krewski, D. and Van Ryzin, J., Dose-response models for quantal response toxicity data, in *Statistics and Related Topics*, Csorgo, M., Dawson, D. A., Rao, J. N. K., and Saleh, E., eds., North-Holland, Amsterdam, 1981, pp. 201–231.

28. Krewski, D., Crump, K. S., Farmer, J., Gaylor, D. W., Howe, R., Portier, C., Salsburg, D., Sielken, R. L., and Van Ryzin, J., A comparison of statistical methods for low dose extrapolation utilizing time-to-tumor data, *Fund. Appl. Toxicol.* **3**, 140–156 (1983).

29. Krewski, D., Murdoch, D., and Dewanji, A., Statistical modeling and extrapolation of carcinogenesis data, in *Modern Statistical Methods in Chronic Disease Epidemiology*, Moolgavkar, S. H. and Prentice, R. L., eds., Wiley, New York, 1986, pp. 259–282.

30. Krewski, D., Goddard, M. J., and Zielinski, J. M., Dose-response relationships in carcinogenesis, in *Mechanisms of Carcinogenesis in Risk Identification*, Vanio, H., Magee, P. N., McGregor, D. B., and McMichael, A. J., eds., IARC, Lyon, 1992, pp. 579–599.

31. Lindstedt, S. L., Allometry: Body size constraints in animal design, in *Drinking Water and Health. Pharmacokinetics in Risk Assessment*, Vol. 8, National Academy Press, Washington, DC, 1987.

32. Little, M. P., Are two mutations sufficient to cause cancer? Some generalizations of the two-mutation model of carcinogenesis of Moolgavkar, Venzon, and Knudson, and of the multistage model of Armitage and Doll, *Biometrics* **51**, 1278–1291 (1995).

33. Luebeck, E. G., Watanabe, K., and Travis, C., Biologically based models of carcinogenesis, in *Perspectives on Biologically Based Cancer Risk Assessment*, Cogliano, V. J., Luebeck, E. G., and Zapponi, G. A., eds., NATO Committee on the Challenges of Modern Society, Kluwer Academic/Plenum Publishers, New York, 1999, pp. 205–241.

34. Luebeck, E. G., Travis, C., and Watanabe, K., Informative case studies, in *Perspectives on Biologically Based Cancer Risk Assessment*, Cogliano, V. J., Luebeck, E. G., and Zapponi, G. A., eds., NATO Committee on the Challenges of Modern Society, Kluwer Academic/Plenum Publishers, New York, 1999, pp. 275–308.

35. Lutz, W. K. and Kopp-Schneider, A., Threshold dose response for tumor induction by genotoxic carcinogens modeled via cell-cycle delay, *Toxicol. Sci.* **49**, 110–115 (1999).

36. Moolgavkar, S. H., The multistage theory of carcinogenesis and the age distribution of cancer in man, *J. Natl. Cancer Inst.* **61**, 49–52 (1978).

37. Moolgavkar, S. H., Carcinogenesis modeling: From molecular biology to epidemiology, *Annul. Rev. Public Health* **7**, 151–169 (1986).

38. Moolgavkar, S. H., ed., *Scientific Issues in Quantitative Cancer Risk Assessment*. Birkhäuser, Boston, 1990.

39. Moolgavkar, S. H., Stochastic models of carcinogenicity, in *Handbook of Statistics*, Rao, C. R. and Chakraborty, R., eds., Elsevier Science Publishers B.V., 1991, Vol. 8, pp. 373–393.

40. Moolgavkar, S. H. and Venzon, D. J., Two-event models for carcinogenesis: Incidence curves for childhood and adult tumors, *Math. Biosci.* **47**, 55–77 (1979).

41. Moolgavkar, S. H. and Knudson, A. G., Mutation and cancer: A model for human carcinogenesis, *J. Natl. Cancer Inst.* **66**, 1037–1052 (1981).

42. Moolgavkar, S. H., Dewanji, A., and Luebeck, E. G., Cigarette smoking and lung cancer: Reanalysis of the British Doctors' data, *J. Natl. Cancer Inst.* **81**, 415–420 (1989).

43. Moolgavkar, S. H. and Luebeck, E. G., Two-event model for carcinogenesis: Biological, mathematical and statistical considerations, *Risk Anal.* **10**, 323–341 (1990).

44. Moolgavkar, S. H. and Luebeck, E. G., Dose-response modeling for cancer risk assessment, in *Human and Ecological Risk Assessment: Theory and Practice*, Paustenbach, D. J., ed., Wiley, New York, 2002, pp. 151–188.

45. Murdoch, D. J. and Krewski, D., Carcinogenic risk assessment with time-dependent exposure patterns, *Risk Anal.* **8**, 521–530 (1988).

46. Neyman, J. and Scott, E., Statistical aspects of the problem of carcinogenesis, in *Proc. 5th Berkeley Symp. Mathematical Statistics and Probability*, Univ. California Press, Berkeley, 1967, pp. 745–776.

47. Nordling, C. O., A new theory of the cancer inducing mechanism, *Br. J. Cancer* **7**, 68–72 (1953).

48. Peto, R. and Lee, P. N., Weibull distributions for continuous-carcinogenesis experiments, *Biometrics* **29**, 457–470 (1973).

49. Portier, C. and Kopp-Schneider, A., A multistage model of carcinogenesis incorporating DNA damage and repair, *Risk Anal.* **11**, 535–543 (1991).

50. Rai, K. and Van Ryzin, J., A generalized multihit dose-response model for low-dose extrapolation, *Biometrics* **37**, 341–352 (1981).

51. Rai, K. and Van Ryzin, J., A dose-response model incorporating nonlinear kinetics, *Biometrics* **43**, 95–105 (1987).

52. Rothman, K. J. and Greenland, S., *Modern Epidemiology*. Lippincott-Raven, New York, 1998.

53. Tan, W. Y., *Stochastic Models for Carcinogenesis*, Marcel Dekker, New York, 1991.

54. Travis, C. C., Pharmacokinetics, in *Carcinogenic Risk Assessment*, Travis, C. C., ed., Plenum Press, New York, 1988.

55. Travis, C. C., White, R. K., and Ward, R. C., Interspecies extrapolation of pharmacokinetics, *J. Theor. Biol.* **142**, 285–304 (1990).

56. Whittemore, A. and Keller, J., Quantitative theories of carcinogenesis, *SIAM Rev.* **20**, 1–30 (1978).

Index

Statistical Advances in the Biomedical Sciences, edited by Atanu Biswas, Sujay Datta,
Jason P. Fine, and Mark R. Segal
Copyright © 2008 John Wiley & Sons, Inc.

576

Markov process (*Continued*)
  non-Gaussian, 132
  unobserved, 126
Mass spectrometry, 420, 465
Matched-pair design, 71
Maximum a posteriori (MAP), 535, 537
Maximum entropy model (MEM), 445
Maximum likelihood, 71, 111, 128, 354, 374,
    406, 502, 508
  estimator (MLE), 58, 115, 125, 128, 129,
    134, 145, 147, 150, 165, 170, 502,
    503, 508, 510, 515, 516, 518
  inference, 159
  misclassification estimate, 508
  nonparametric estimation, 171
  restricted, 111
  semiparametric estimation, 162, 164,
    170, 171
  score equation, 503, 509, 511, 515
Maximum tolerated schedule, 12
Mean integrated squared error, 103
Mean squared prediction error (MSPE), 109,
    110, 111, 118
Measurement error, 87, 90, 136, 141, 145,
    146, 147, 148, 151, 152, 153, 425
  in binary regression, 142
  in covariates, 142, 144, 147, 149, 150, 152
  in predictors, 143
  linear, 141, 142
  model, 141, 142
  nondifferential, 144, 146
  nonlinear, 142
Mechanistic model, 136, 550
  multi-hit, 550
  one-hit, 550
Median inhibitory concentration, 85, 86, 87
M-estimator, 502, 503, 505, 508, 509, 520
  based on Huber's ψ function, 520
  of location, 503
  of regression, 503
  power-robust, 507
  semiparametric, 166
Metropolis-Hastings algorithm, 89, 532, 167
Meta-analysis, 74, 319, 325, 508
Michigan ECMO trial, 46
Microarrays
  array comparative genomic hybridization
    (aCGH), 385, 400
  biomarker identification, 309

censored survival phenotypes, 351, 385
classification, 327
clustering, 328, 365
correlated gene expression, 309, 327, 341
differential expression (DE), 314, 328,
    336, 341
meta analysis, 319, 325
normalization, 344
single nucleotide polymorphism (SNP),
    385, 400
timecourse experiments, 309, 315, 377
Minimum-disparity
  estimating equation, 510
  estimator, 510
Minimum-distance procedure, 505, 509
Minimum-divergence
  estimator (MDE), 515, 516
Minimum spanning trees, 106
Minimum-variance unbiased estimator,
    58, 504
Minimum-volume confidence
  rectangle, 167
Misclassification, 141, 143, 147, 148, 149,
    151, 152, 508
  maximum-likelihood estimate, 508
Missing
  proportion, 153
  response, 152
Misspecified likelihood estimation, 166
Mixed-effect model
  Bayesian nonlinear, 84
  hierarchical nonlinear, 84, 92
  nonlinear, 92
  nonlinear partial spline, 529, 530, 537
Mixed model, 507
Mixture model, 342, 365
Mixture of Dirichlet processes (MDP), 535
Model comparison
  likelihood-based, 136
Model fitting, 508
Model free designs, 12
Model selection, 125, 529
Modified play-the-winner, 36
Monte Carlo, 99, 126, 132
  error, 133
  p value, 105, 106
  randomness, 128
  sequential (SMC), 126, 127, 128, 129, 136
  simulation, 71

BHAT and MILLER · Elements of Applied Stochastic Processes, *Third Edition*
BHATTACHARYA and WAYMIRE · Stochastic Processes with Applications
BILLINGSLEY · Convergence of Probability Measures, *Second Edition*
BILLINGSLEY · Probability and Measure, *Third Edition*
BIRKES and DODGE · Alternative Methods of Regression
BISWAS, DATTA, FINE, and SEGAL · Statistical Advances in the Biomedical Sciences: Clinical Trials, Epidemiology, Survival Analysis, and Bioinformatics
BLISCHKE AND MURTHY (editors) · Case Studies in Reliability and Maintenance
BLISCHKE AND MURTHY · Reliability: Modeling, Prediction, and Optimization
BLOOMFIELD · Fourier Analysis of Time Series: An Introduction, *Second Edition*
BOLLEN · Structural Equations with Latent Variables
BOLLEN and CURRAN · Latent Curve Models: A Structural Equation Perspective
BOROVKOV · Ergodicity and Stability of Stochastic Processes
BOULEAU · Numerical Methods for Stochastic Processes
BOX · Bayesian Inference in Statistical Analysis
BOX · R. A. Fisher, the Life of a Scientist
BOX and DRAPER · Response Surfaces, Mixtures, and Ridge Analyses, *Second Edition*
* BOX and DRAPER · Evolutionary Operation: A Statistical Method for Process Improvement
BOX and FRIENDS · Improving Almost Anything, *Revised Edition*
BOX, HUNTER, and HUNTER · Statistics for Experimenters: Design, Innovation, and Discovery, *Second Editon*
BOX and LUCEÑO · Statistical Control by Monitoring and Feedback Adjustment
BRANDIMARTE · Numerical Methods in Finance: A MATLAB-Based Introduction
BROWN and HOLLANDER · Statistics: A Biomedical Introduction
BRUNNER, DOMHOF, and LANGER · Nonparametric Analysis of Longitudinal Data in Factorial Experiments
BUCKLEW · Large Deviation Techniques in Decision, Simulation, and Estimation
CAIROLI and DALANG · Sequential Stochastic Optimization
CASTILLO, HADI, BALAKRISHNAN, and SARABIA · Extreme Value and Related Models with Applications in Engineering and Science
CHAN · Time Series: Applications to Finance
CHARALAMBIDES · Combinatorial Methods in Discrete Distributions
CHATTERJEE and HADI · Regression Analysis by Example, *Fourth Edition*
CHATTERJEE and HADI · Sensitivity Analysis in Linear Regression
CHERNICK · Bootstrap Methods: A Guide for Practitioners and Researchers, *Second Edition*
CHERNICK and FRIIS · Introductory Biostatistics for the Health Sciences
CHILÈS and DELFINER · Geostatistics: Modeling Spatial Uncertainty
CHOW and LIU · Design and Analysis of Clinical Trials: Concepts and Methodologies, *Second Edition*
CLARKE and DISNEY · Probability and Random Processes: A First Course with Applications, *Second Edition*
* COCHRAN and COX · Experimental Designs, *Second Edition*
CONGDON · Applied Bayesian Modelling
CONGDON · Bayesian Models for Categorical Data
CONGDON · Bayesian Statistical Modelling
CONOVER · Practical Nonparametric Statistics, *Third Edition*
COOK · Regression Graphics
COOK and WEISBERG · Applied Regression Including Computing and Graphics
COOK and WEISBERG · An Introduction to Regression Graphics
CORNELL · Experiments with Mixtures, Designs, Models, and the Analysis of Mixture Data, *Third Edition*
COVER and THOMAS · Elements of Information Theory
COX · A Handbook of Introductory Statistical Methods
* COX · Planning of Experiments
CRESSIE · Statistics for Spatial Data, *Revised Edition*
CSÖRGŐ and HORVÁTH · Limit Theorems in Change Point Analysis
DANIEL · Applications of Statistics to Industrial Experimentation
DANIEL · Biostatistics: A Foundation for Analysis in the Health Sciences, *Eighth Edition*

*Now available in a lower priced paperback edition in the Wiley Classics Library.

* DANIEL · Fitting Equations to Data: Computer Analysis of Multifactor Data, *Second Edition*

DASU and JOHNSON · Exploratory Data Mining and Data Cleaning

DAVID and NAGARAJA · Order Statistics, *Third Edition*

* DeGROOT, FIENBERG, and KADANE · Statistics and the Law

DEL CASTILLO · Statistical Process Adjustment for Quality Control

DeMARIS · Regression with Social Data: Modeling Continuous and Limited Response Variables

DEMIDENKO · Mixed Models: Theory and Applications

DENISON, HOLMES, MALLICK and SMITH · Bayesian Methods for Nonlinear Classification and Regression

DETTE and STUDDEN · The Theory of Canonical Moments with Applications in Statistics, Probability, and Analysis

DEY and MUKERJEE · Fractional Factorial Plans

DILLON and GOLDSTEIN · Multivariate Analysis: Methods and Applications

DODGE · Alternative Methods of Regression

* DODGE and ROMIG · Sampling Inspection Tables, *Second Edition*

* DOOB · Stochastic Processes

DOWDY, WEARDEN, and CHILKO · Statistics for Research, *Third Edition*

DRAPER and SMITH · Applied Regression Analysis, *Third Edition*

DRYDEN and MARDIA · Statistical Shape Analysis

DUDEWICZ and MISHRA · Modern Mathematical Statistics

DUNN and CLARK · Basic Statistics: A Primer for the Biomedical Sciences, *Third Edition*

DUPUIS and ELLIS · A Weak Convergence Approach to the Theory of Large Deviations

EDLER and KITSOS · Recent Advances in Quantitative Methods in Cancer and Human Health Risk Assessment

* ELANDT-JOHNSON and JOHNSON · Survival Models and Data Analysis

ENDERS · Applied Econometric Time Series

† ETHIER and KURTZ · Markov Processes: Characterization and Convergence

EVANS, HASTINGS, and PEACOCK · Statistical Distributions, *Third Edition*

FELLER · An Introduction to Probability Theory and Its Applications, Volume I, *Third Edition*, Revised; Volume II, *Second Edition*

FISHER and VAN BELLE · Biostatistics: A Methodology for the Health Sciences

FITZMAURICE, LAIRD, and WARE · Applied Longitudinal Analysis

* FLEISS · The Design and Analysis of Clinical Experiments

FLEISS · Statistical Methods for Rates and Proportions, *Third Edition*

† FLEMING and HARRINGTON · Counting Processes and Survival Analysis

FULLER · Introduction to Statistical Time Series, *Second Edition*

† FULLER · Measurement Error Models

GALLANT · Nonlinear Statistical Models

GEISSER · Modes of Parametric Statistical Inference

GELMAN and MENG · Applied Bayesian Modeling and Causal Inference from Incomplete-Data Perspectives

GEWEKE · Contemporary Bayesian Econometrics and Statistics

GHOSH, MUKHOPADHYAY, and SEN · Sequential Estimation

GIESBRECHT and GUMPERTZ · Planning, Construction, and Statistical Analysis of Comparative Experiments

GIFI · Nonlinear Multivariate Analysis

GIVENS and HOETING · Computational Statistics

GLASSERMAN and YAO · Monotone Structure in Discrete-Event Systems

GNANADESIKAN · Methods for Statistical Data Analysis of Multivariate Observations, *Second Edition*

GOLDSTEIN and LEWIS · Assessment: Problems, Development, and Statistical Issues

GREENWOOD and NIKULIN · A Guide to Chi-Squared Testing

GROSS and HARRIS · Fundamentals of Queueing Theory, *Third Edition*

* HAHN and SHAPIRO · Statistical Models in Engineering

HAHN and MEEKER · Statistical Intervals: A Guide for Practitioners

HALD · A History of Probability and Statistics and their Applications Before 1750

HALD · A History of Mathematical Statistics from 1750 to 1930

*Now available in a lower priced paperback edition in the Wiley Classics Library.
†Now available in a lower priced paperback edition in the Wiley–Interscience Paperback Series.

KHURI, MATHEW, and SINHA · Statistical Tests for Mixed Linear Models

KLEIBER and KOTZ · Statistical Size Distributions in Economics and Actuarial Sciences

KLUGMAN, PANJER, and WILLMOT · Loss Models: From Data to Decisions, *Second Edition*

KLUGMAN, PANJER, and WILLMOT · Solutions Manual to Accompany Loss Models: From Data to Decisions, *Second Edition*

KOTZ, BALAKRISHNAN, and JOHNSON · Continuous Multivariate Distributions, Volume 1, *Second Edition*

KOVALENKO, KUZNETZOV, and PEGG · Mathematical Theory of Reliability of Time-Dependent Systems with Practical Applications

KOWALSKI and TU · Modern Applied U-Statistics

KVAM and VIDAKOVIC · Nonparametric Statistics with Applications to Science and Engineering

LACHIN · Biostatistical Methods: The Assessment of Relative Risks

LAD · Operational Subjective Statistical Methods: A Mathematical, Philosophical, and Historical Introduction

LAMPERTI · Probability: A Survey of the Mathematical Theory, *Second Edition*

LANGE, RYAN, BILLARD, BRILLINGER, CONQUEST, and GREENHOUSE · Case Studies in Biometry

LARSON · Introduction to Probability Theory and Statistical Inference, *Third Edition*

LAWLESS · Statistical Models and Methods for Lifetime Data, *Second Edition*

LAWSON · Statistical Methods in Spatial Epidemiology

LE · Applied Categorical Data Analysis

LE · Applied Survival Analysis

LEE and WANG · Statistical Methods for Survival Data Analysis, *Third Edition*

LePAGE and BILLARD · Exploring the Limits of Bootstrap

LEYLAND and GOLDSTEIN (editors) · Multilevel Modelling of Health Statistics

LIAO · Statistical Group Comparison

LINDVALL · Lectures on the Coupling Method

LIN · Introductory Stochastic Analysis for Finance and Insurance

LINHART and ZUCCHINI · Model Selection

LITTLE and RUBIN · Statistical Analysis with Missing Data, *Second Edition*

LLOYD · The Statistical Analysis of Categorical Data

LOWEN and TEICH · Fractal-Based Point Processes

MAGNUS and NEUDECKER · Matrix Differential Calculus with Applications in Statistics and Econometrics, *Revised Edition*

MALLER and ZHOU · Survival Analysis with Long Term Survivors

MALLOWS · Design, Data, and Analysis by Some Friends of Cuthbert Daniel

MANN, SCHAFER, and SINGPURWALLA · Methods for Statistical Analysis of Reliability and Life Data

MANTON, WOODBURY, and TOLLEY · Statistical Applications Using Fuzzy Sets

MARCHETTE · Random Graphs for Statistical Pattern Recognition

MARDIA and JUPP · Directional Statistics

MASON, GUNST, and HESS · Statistical Design and Analysis of Experiments with Applications to Engineering and Science, *Second Edition*

McCULLOCH and SEARLE · Generalized, Linear, and Mixed Models

McFADDEN · Management of Data in Clinical Trials, *Second Edition*

* McLACHLAN · Discriminant Analysis and Statistical Pattern Recognition

McLACHLAN, DO, and AMBROISE · Analyzing Microarray Gene Expression Data

McLACHLAN and KRISHNAN · The EM Algorithm and Extensions, *Second Edition*

McLACHLAN and PEEL · Finite Mixture Models

McNEIL · Epidemiological Research Methods

MEEKER and ESCOBAR · Statistical Methods for Reliability Data

MEERSCHAERT and SCHEFFLER · Limit Distributions for Sums of Independent Random Vectors: Heavy Tails in Theory and Practice

MICKEY, DUNN, and CLARK · Applied Statistics: Analysis of Variance and Regression, *Third Edition*

* MILLER · Survival Analysis, *Second Edition*

MONTGOMERY, PECK, and VINING · Introduction to Linear Regression Analysis, *Fourth Edition*

MORGENTHALER and TUKEY · Configural Polysampling: A Route to Practical Robustness

---

*Now available in a lower priced paperback edition in the Wiley Classics Library.

*Now available in a lower priced paperback edition in the Wiley Classics Library.

†Now available in a lower priced paperback edition in the Wiley–Interscience Paperback Series.

SCHIMEK · Smoothing and Regression: Approaches, Computation, and Application
SCHOTT · Matrix Analysis for Statistics, *Second Edition*
SCHOUTENS · Levy Processes in Finance: Pricing Financial Derivatives
SCHUSS · Theory and Applications of Stochastic Differential Equations
SCOTT · Multivariate Density Estimation: Theory, Practice, and Visualization
† SEARLE · Linear Models for Unbalanced Data
† SEARLE · Matrix Algebra Useful for Statistics
† SEARLE, CASELLA, and McCULLOCH · Variance Components
SEARLE and WILLETT · Matrix Algebra for Applied Economics
SEBER and LEE · Linear Regression Analysis, *Second Edition*
† SEBER · Multivariate Observations
† SEBER and WILD · Nonlinear Regression
SENNOTT · Stochastic Dynamic Programming and the Control of Queueing Systems
* SERFLING · Approximation Theorems of Mathematical Statistics
SHAFER and VOVK · Probability and Finance: It's Only a Game!
SILVAPULLE and SEN · Constrained Statistical Inference: Inequality, Order, and Shape Restrictions
SMALL and McLEISH · Hilbert Space Methods in Probability and Statistical Inference
SRIVASTAVA · Methods of Multivariate Statistics
STAPLETON · Linear Statistical Models
STAUDTE and SHEATHER · Robust Estimation and Testing
STOYAN, KENDALL, and MECKE · Stochastic Geometry and Its Applications, *Second Edition*
STOYAN and STOYAN · Fractals, Random Shapes and Point Fields: Methods of Geometrical Statistics
STREET and BURGESS · The Construction of Optimal Stated Choice Experiments: Theory and Methods
STYAN · The Collected Papers of T. W. Anderson: 1943–1985
SUTTON, ABRAMS, JONES, SHELDON, and SONG · Methods for Meta-Analysis in Medical Research
TAKEZAWA · Introduction to Nonparametric Regression
TANAKA · Time Series Analysis: Nonstationary and Noninvertible Distribution Theory
THOMPSON · Empirical Model Building
THOMPSON · Sampling, *Second Edition*
THOMPSON · Simulation: A Modeler's Approach
THOMPSON and SEBER · Adaptive Sampling
THOMPSON, WILLIAMS, and FINDLAY · Models for Investors in Real World Markets
TIAO, BISGAARD, HILL, PEÑA, and STIGLER (editors) · Box on Quality and Discovery: with Design, Control, and Robustness
TIERNEY · LISP-STAT: An Object-Oriented Environment for Statistical Computing and Dynamic Graphics
TSAY · Analysis of Financial Time Series, *Second Edition*
UPTON and FINGLETON · Spatial Data Analysis by Example, Volume II: Categorical and Directional Data
VAN BELLE · Statistical Rules of Thumb
VAN BELLE, FISHER, HEAGERTY, and LUMLEY · Biostatistics: A Methodology for the Health Sciences, *Second Edition*
VESTRUP · The Theory of Measures and Integration
VIDAKOVIC · Statistical Modeling by Wavelets
VINOD and REAGLE · Preparing for the Worst: Incorporating Downside Risk in Stock Market Investments
WALLER and GOTWAY · Applied Spatial Statistics for Public Health Data
WEERAHANDI · Generalized Inference in Repeated Measures: Exact Methods in MANOVA and Mixed Models
WEISBERG · Applied Linear Regression, *Third Edition*
WELSH · Aspects of Statistical Inference
WESTFALL and YOUNG · Resampling-Based Multiple Testing: Examples and Methods for p-Value Adjustment
WHITTAKER · Graphical Models in Applied Multivariate Statistics

---

*Now available in a lower priced paperback edition in the Wiley Classics Library.
†Now available in a lower priced paperback edition in the Wiley–Interscience Paperback Series.

*Now available in a lower priced paperback edition in the Wiley Classics Library.